T0180317

Lecture Notes in Computer Science 13109

More information about this subseries at https://link.springer.com/bookseries/7407

Teddy Mantoro · Minho Lee ·
Media Anugerah Ayu · Kok Wai Wong ·
Achmad Nizar Hidayanto (Eds.)

Neural Information Processing

28th International Conference, ICONIP 2021
Sanur, Bali, Indonesia, December 8–12, 2021
Proceedings, Part II

 Springer

Editors
Teddy Mantoro (iD)
Sampoerna University
Jakarta, Indonesia

Media Anugerah Ayu (iD)
Sampoerna University
Jakarta, Indonesia

Achmad Nizar Hidayanto (iD)
Universitas Indonesia
Depok, Indonesia

Minho Lee (iD)
Kyungpook National University
Daegu, Korea (Republic of)

Kok Wai Wong (iD)
Murdoch University
Murdoch, WA, Australia

ISSN 0302-9743   ISSN 1611-3349   (electronic)
Lecture Notes in Computer Science
ISBN 978-3-030-92269-6   ISBN 978-3-030-92270-2   (eBook)
https://doi.org/10.1007/978-3-030-92270-2

LNCS Sublibrary: SL1 – Theoretical Computer Science and General Issues

This Springer imprint is published by the registered company Springer Nature Switzerland AG
The registered company address is: Gewerbestrasse 11, 6330 Cham, Switzerland

Preface

Welcome to the proceedings of the 28th International Conference on Neural Information Processing (ICONIP 2021) of the Asia-Pacific Neural Network Society (APNNS), held virtually from Indonesia during December 8–12, 2021.

The mission of the Asia-Pacific Neural Network Society is to promote active interactions among researchers, scientists, and industry professionals who are working in neural networks and related fields in the Asia-Pacific region. APNNS has Governing Board Members from 13 countries/regions – Australia, China, Hong Kong, India, Japan, Malaysia, New Zealand, Singapore, South Korea, Qatar, Taiwan, Thailand, and Turkey. The society's flagship annual conference is the International Conference of Neural Information Processing (ICONIP).

The ICONIP conference aims to provide a leading international forum for researchers, scientists, and industry professionals who are working in neuroscience, neural networks, deep learning, and related fields to share their new ideas, progress, and achievements. Due to the current COVID-19 pandemic, ICONIP 2021, which was planned to be held in Bali, Indonesia, was organized as a fully virtual conference.

The proceedings of ICONIP 2021 consists of a four-volume set, LNCS 13108–13111, which includes 226 papers selected from 1093 submissions, representing an acceptance rate of 20.86% and reflecting the increasingly high quality of research in neural networks and related areas in the Asia-Pacific. The conference had four main themes, i.e., "Theory and Algorithms," "Cognitive Neurosciences," "Human Centred Computing," and "Applications."

The four volumes are organized in topical sections which comprise the four main themes mentioned previously and the topics covered in three special sessions. Another topic is from a workshop on Artificial Intelligence and Cyber Security which was held in conjunction with ICONIP 2021. Thus, in total, eight different topics were accommodated at the conference. The topics were also the names of the 20-minute presentation sessions at ICONIP 2021. The eight topics in the conference were: Theory and Algorithms; Cognitive Neurosciences; Human Centred Computing; Applications; Artificial Intelligence and Cybersecurity; Advances in Deep and Shallow Machine Learning Algorithms for Biomedical Data and Imaging; Reliable, Robust, and Secure Machine Learning Algorithms; and Theory and Applications of Natural Computing Paradigms.

Our great appreciation goes to the Program Committee members and the reviewers who devoted their time and effort to our rigorous peer-review process. Their insightful reviews and timely feedback ensured the high quality of the papers accepted for publication. Finally, thank you to all the authors of papers, presenters, and participants at the conference. Your support and engagement made it all worthwhile.

December 2021

Teddy Mantoro
Minho Lee
Media A. Ayu
Kok Wai Wong
Achmad Nizar Hidayanto

Organization

Honorary Chairs

Jonathan Chan   King Mongkut's University of Technology Thonburi, Thailand

Lance Fung   Murdoch University, Australia

General Chairs

Teddy Mantoro   Sampoerna University, Indonesia

Minho Lee   Kyungpook National University, South Korea

Program Chairs

Media A. Ayu   Sampoerna University, Indonesia

Kok Wai Wong   Murdoch University, Australia

Achmad Nizar   Universitas Indonesia, Indonesia

Local Arrangements Chairs

Linawati   Universitas Udayana, Indonesia

W. G. Ariastina   Universitas Udayana, Indonesia

Finance Chairs

Kurnianingsih   Politeknik Negeri Semarang, Indonesia

Kazushi Ikeda   Nara Institute of Science and Technology, Japan

Special Sessions Chairs

Sunu Wibirama   Universitas Gadjah Mada, Indonesia

Paul Pang   Federation University Australia, Australia

Noor Akhmad Setiawan   Universitas Gadjah Mada, Indonesia

Tutorial Chairs

Suryono   Universitas Diponegoro, Indonesia

Muhammad Agni Catur Bhakti   Sampoerna University, Indonesia

Proceedings Chairs

Adi Wibowo	Universitas Diponegoro, Indonesia
Sung Bae Cho	Yonsei University, South Korea

Publicity Chairs

Dwiza Riana	Universitas Nusa Mandiri, Indonesia
M. Tanveer	Indian Institute of Technology, Indore, India

Program Committee

Abdulrazak Alhababi	Universiti Malaysia Sarawak, Malaysia
Abhijit Adhikary	Australian National University, Australia
Achmad Nizar Hidayanto	University of Indonesia, Indonesia
Adamu Abubakar Ibrahim	International Islamic University Malaysia, Malaysia
Adi Wibowo	Diponegoro University, Indonesia
Adnan Mahmood	Macquarie University, Australia
Afiyati Amaluddin	Mercu Buana University, Indonesia
Ahmed Alharbi	RMIT University, Australia
Akeem Olowolayemo	International Islamic University Malaysia, Malaysia
Akira Hirose	University of Tokyo, Japan
Aleksandra Nowak	Jagiellonian University, Poland
Ali Haidar	University of New South Wales, Australia
Ali Mehrabi	Western Sydney University, Australia
Al-Jadir	Murdoch University, Australia
Ana Flavia Reis	Federal Technological University of Paraná, Brazil
Anaissi Ali	University of Sydney, Australia
Andrew Beng Jin Teoh	Yonsei University, South Korea
Andrew Chiou	Central Queensland University, Australia
Aneesh Chivukula	University of Technology Sydney, Australia
Aneesh Krishna	Curtin University, Australia
Anna Zhu	Wuhan University of Technology, China
Anto Satriyo Nugroho	Agency for Assessment and Application of Technology, Indonesia
Anupiya Nugaliyadde	Sri Lanka Institute of Information Technology, Sri Lanka
Anwesha Law	Indian Statistical Institute, India
Aprinaldi Mantau	Kyushu Institute of Technology, Japan
Ari Wibisono	Universitas Indonesia, Indonesia
Arief Ramadhan	Bina Nusantara University, Indonesia
Arit Thammano	King Mongkut's Institute of Technology Ladkrabang, Thailand
Arpit Garg	University of Adelaide, Australia
Aryal Sunil	Deakin University, Australia
Ashkan Farhangi	University of Central Florida, USA

Atul Negi	University of Hyderabad, India
Barawi Mohamad Hardyman	Universiti Malaysia Sarawak, Malaysia
Bayu Distiawan	Universitas Indonesia, Indonesia
Bharat Richhariya	IISc Bangalore, India
Bin Pan	Nankai University, China
Bingshu Wang	Northwestern Polytechnical University, Taicang, China
Bonaventure C. Molokwu	University of Windsor, Canada
Bo-Qun Ma	Ant Financial
Bunthit Watanapa	King Mongkut's University of Technology Thonburi, Thailand
Chang-Dong Wang	Sun Yat-sen University, China
Chattrakul Sombattheera	Mahasarakham University, Thailand
Chee Siong Teh	Universiti Malaysia Sarawak, Malaysia
Chen Wei Chén	Chongqing Jiaotong University, China
Chengwei Wu	Harbin Institute of Technology, China
Chern Hong Lim	Monash University, Australia
Chih-Chieh Hung	National Chung Hsing University, Taiwan
Chiranjibi Sitaula	Deakin University, Australia
Chi-Sing Leung	City University of Hong Kong, Hong Kong
Choo Jun Tan	Wawasan Open University, Malaysia
Christoph Bergmeir	Monash University, Australia
Christophe Guyeux	University of Franche-Comté, France
Chuan Chen	Sun Yat-sen University, China
Chuanqi Tan	BIT, China
Chu-Kiong Loo	University of Malaya, Malaysia
Chun Che Fung	Murdoch University, Australia
Colin Samplawski	University of Massachusetts Amherst, USA
Congbo Ma	University of Adelaide, Australia
Cuiyun Gao	Chinese University of Hong Kong, Hong Kong
Cutifa Safitri	Universiti Teknologi Malaysia, Malaysia
Daisuke Miyamoto	University of Tokyo, Japan
Dan Popescu	Politehnica University of Bucharest
David Bong	Universiti Malaysia Sarawak, Malaysia
David Iclanzan	Sapientia Hungarian Science University of Transylvania, Romania
Debasmit Das	IIT Roorkee, India
Dengya Zhu	Curtin University, Australia
Derwin Suhartono	Bina Nusantara University, Indonesia
Devi Fitrianah	Universitas Mercu Buana, Indonesia
Deyu Zhou	Southeast University, China
Dhimas Arief Dharmawan	Universitas Indonesia, Indonesia
Dianhui Wang	La Trobe University, Australia
Dini Handayani	Taylors University, Malaysia
Dipanjyoti Paul	Indian Institute of Technology, Patna, India
Dong Chen	Wuhan University, China

Donglin Bai   Shanghai Jiao Tong University, China
Dongrui Wu   Huazhong University of Science & Technology, China
Dugang Liu   Shenzhen University, China
Dwina Kuswardani   Institut Teknologi PLN, Indonesia
Dwiza Riana   Universitas Nusa Mandiri, Indonesia
Edmund Lai   Auckland University of Technology, New Zealand
Eiji Uchino   Yamaguchi University, Japan
Emanuele Principi   Università Politecnica delle Marche, Italy
Enmei Tu   Shanghai Jiao Tong University, China
Enna Hirata   Kobe University, Japan
Eri Sato-Shimokawara   Tokyo Metropolitan University, Japan
Fajri Koto   University of Melbourne, Australia
Fan Wu   Australian National University, Australia
Farhad Ahamed   Western Sydney University, Australia
Fei Jiang   Shanghai Jiao Tong University, China
Feidiao Yang   Microsoft, USA
Feng Wan   University of Macau, Macau
Fenty Eka Muzayyana   UIN Syarif Hidayatullah Jakarta, Indonesia
  Agustin
Ferda Ernawan   Universiti Malaysia Pahang, Malaysia
Ferdous Sohel   Murdoch University, Australia
Francisco J. Moreno-Barea   Universidad de Málaga, Spain
Fuad Jamour   University of California, Riverside, USA
Fuchun Sun   Tsinghua University, China
Fumiaki Saitoh   Chiba Institute of Technology, Japan
Gang Chen   Victoria University of Wellington, New Zealand
Gang Li   Deakin University, Australia
Gang Yang   Renmin University of China
Gao Junbin   Huazhong University of Science and Technology,
                                     China
George Cabral   Universidade Federal Rural de Pernambuco, Brazil
Gerald Schaefer   Loughborough University, UK
Gouhei Tanaka   University of Tokyo, Japan
Guanghui Wen   RMIT University, Australia
Guanjin Wang   Murdoch University, Australia
Guoqiang Zhong   Ocean University of China, China
Guoqing Chao   East China Normal University, China
Sangchul Hahn   Handong Global University, South Korea
Haiqin Yang   International Digital Economy Academy, China
Hakaru Tamukoh   Kyushu Institute of Technology, Japan
Hamid Karimi   Utah State University, USA
Hangyu Deng   Waseda University, Japan
Hao Liao   Shenzhen University, China
Haris Al Qodri Maarif   International Islamic University Malaysia, Malaysia
Haruhiko Nishimura   University of Hyogo, Japan
Hayaru Shouno   University of Electro-Communications, Japan

He Chen	Nankai University, China
He Huang	Soochow University, China
Hea Choon Ngo	Universiti Teknikal Malaysia Melaka, Malaysia
Heba El-Fiqi	UNSW Canberra, Australia
Heru Praptono	Bank Indonesia/Universitas Indonesia, Indonesia
Hideitsu Hino	Institute of Statistical Mathematics, Japan
Hidemasa Takao	University of Tokyo, Japan
Hiroaki Inoue	Kobe University, Japan
Hiroaki Kudo	Nagoya University, Japan
Hiromu Monai	Ochanomizu University, Japan
Hiroshi Sakamoto	Kyushu Institute of Technology, Japan
Hisashi Koga	University of Electro-Communications, Japan
Hiu-Hin Tam	City University of Hong Kong, Hong Kong
Hongbing Xia	Beijing Normal University, China
Hongtao Liu	Tianjin University, China
Hongtao Lu	Shanghai Jiao Tong University, China
Hua Zuo	University of Technology Sydney, Australia
Hualou Liang	Drexel University, USA
Huang Chaoran	University of New South Wales, Australia
Huang Shudong	Sichuan University, China
Huawen Liu	University of Texas at San Antonio, USA
Hui Xue	Southeast University, China
Hui Yan	Shanghai Jiao Tong University, China
Hyeyoung Park	Kyungpook National University, South Korea
Hyun-Chul Kim	Kyungpook National University, South Korea
Iksoo Shin	University of Science and Technology, South Korea
Indrabayu Indrabayu	Universitas Hasanuddin, Indonesia
Iqbal Gondal	RMIT University, Australia
Iuliana Georgescu	University of Bucharest, Romania
Iwan Syarif	PENS, Indonesia
J. Kokila	Indian Institute of Information Technology, Allahabad, India
J. Manuel Moreno	Universitat Politècnica de Catalunya, Spain
Jagdish C. Patra	Swinburne University of Technology, Australia
Jean-Francois Couchot	University of Franche-Comté, France
Jelita Asian	STKIP Surya, Indonesia
Jennifer C. Dela Cruz	Mapua University, Philippines
Jérémie Sublime	ISEP, France
Jiahuan Lei	Meituan, China
Jialiang Zhang	Alibaba, China
Jiaming Xu	Institute of Automation, Chinese Academy of Sciences
Jianbo Ning	University of Science and Technology Beijing, China
Jianyi Yang	Nankai University, China
Jiasen Wang	City University of Hong Kong, Hong Kong
Jiawei Fan	Australian National University, Australia
Jiawei Li	Tsinghua University, China

Jiaxin Li	Guangdong University of Technology, China
Jiaxuan Xie	Shanghai Jiao Tong University, China
Jichuan Zeng	Bytedance, China
Jie Shao	University of Science and Technology of China, China
Jie Zhang	Newcastle University, UK
Jiecong Lin	City University of Hong Kong, Hong Kong
Jin Hu	Chongqing Jiaotong University, China
Jin Kyu Kim	Facebook, USA
Jin Ren	Beijing University of Technology, China
Jin Shi	Nanjing University, China
Jinfu Yang	Beijing University of Technology, China
Jing Peng	South China Normal University, China
Jinghui Zhong	South China University of Technology, China
Jin-Tsong Jeng	National Formosa University, Taiwan
Jiri Sima	Institute of Computer Science, Czech Academy of Sciences, Czech Republic
Jo Plested	Australian National University, Australia
Joel Dabrowski	CSIRO, Australia
John Sum	National Chung Hsing University, China
Jolfaei Alireza	Federation University Australia, Australia
Jonathan Chan	King Mongkut's University of Technology Thonburi, Thailand
Jonathan Mojoo	Hiroshima University, Japan
Jose Alfredo Ferreira Costa	Federal University of Rio Grande do Norte, Brazil
Ju Lu	Shandong University, China
Jumana Abu-Khalaf	Edith Cowan University, Australia
Jun Li	Nanjing Normal University, China
Jun Shi	Guangzhou University, China
Junae Kim	DST Group, Australia
Junbin Gao	University of Sydney, Australia
Junjie Chen	Inner Mongolia Agricultural University, China
Junya Chen	Fudan University, China
Junyi Chen	City University of Hong Kong, Hong Kong
Junying Chen	South China University of Technology, China
Junyu Xuan	University of Technology, Sydney
Kah Ong Michael Goh	Multimedia University, Malaysia
Kaizhu Huang	Xi'an Jiaotong-Liverpool University, China
Kam Meng Goh	Tunku Abdul Rahman University College, Malaysia
Katsuhiro Honda	Osaka Prefecture University, Japan
Katsuyuki Hagiwara	Mie University, Japan
Kazushi Ikeda	Nara Institute of Science and Technology, Japan
Kazuteru Miyazaki	National Institution for Academic Degrees and Quality Enhancement of Higher Education, Japan
Kenji Doya	OIST, Japan
Kenji Watanabe	National Institute of Advanced Industrial Science and Technology, Japan

Kok Wai Wong	Murdoch University, Australia
Kitsuchart Pasupa	King Mongkut's Institute of Technology Ladkrabang, Thailand
Kittichai Lavangnananda	King Mongkut's University of Technology Thonburi, Thailand
Koutsakis Polychronis	Murdoch University, Australia
Kui Ding	Nanjing Normal University, China
Kun Zhang	Carnegie Mellon University, USA
Kuntpong Woraratpanya	King Mongkut's Institute of Technology Ladkrabang, Thailand
Kurnianingsih Kurnianingsih	Politeknik Negeri Semarang, Indonesia
Kusrini	Universitas AMIKOM Yogyakarta, Indonesia
Kyle Harrison	UNSW Canberra, Australia
Laga Hamid	Murdoch University, Australia
Lei Wang	Beihang University, China
Leonardo Franco	Universidad de Málaga, Spain
Li Guo	University of Macau, China
Li Yun	Nanjing University of Posts and Telecommunications, China
Libo Wang	Xiamen University of Technology, China
Lie Meng Pang	Southern University of Science and Technology, China
Liew Alan Wee-Chung	Griffith University, Australia
Lingzhi Hu	Beijing University of Technology, China
Linjing Liu	City University of Hong Kong, Hong Kong
Lisi Chen	Hong Kong Baptist University, Hong Kong
Long Cheng	Institute of Automation, Chinese Academy of Sciences, China
Lukman Hakim	Hiroshima University, Japan
M. Tanveer	Indian Institute of Technology, Indore, India
Ma Wanli	University of Canberra, Australia
Man Fai Leung	Hong Kong Metropolitan University, Hong Kong
Maram Mahmoud A. Monshi	Beijing Institute of Technology, China
Marcin Wozniak	Silesian University of Technology, Poland
Marco Anisetti	Università degli Studi di Milano, Italy
Maria Susan Anggreainy	Bina Nusantara University, Indonesia
Mark Abernethy	Murdoch University, Australia
Mark Elshaw	Coventry University, UK
Maruno Yuki	Kyoto Women's University, Japan
Masafumi Hagiwara	Keio University, Japan
Masataka Kawai	NRI SecureTechnologies, Ltd., Japan
Media Ayu	Sampoerna University, Indonesia
Mehdi Neshat	University of Adelaide, Australia
Meng Wang	Southeast University, China
Mengmeng Li	Zhengzhou University, China

Miaohua Zhang	Griffith University, Australia
Mingbo Zhao	Donghua University, China
Mingcong Deng	Tokyo University of Agriculture and Technology, Japan
Minghao Yang	Institute of Automation, Chinese Academy of Sciences, China
Minho Lee	Kyungpook National University, South Korea
Mofei Song	Southeast University, China
Mohammad Faizal Ahmad Fauzi	Multimedia University, Malaysia
Mohsen Marjani	Taylor's University, Malaysia
Mubasher Baig	National University of Computer and Emerging Sciences, Lahore, Pakistan
Muhammad Anwar Ma'Sum	Universitas Indonesia, Indonesia
Muhammad Asim Ali	Shaheed Zulfikar Ali Bhutto Institute of Science and Technology, Pakistan
Muhammad Fawad Akbar Khan	University of Engineering and Technology Peshawar, Pakistan
Muhammad Febrian Rachmadi	Universitas Indonesia, Indonesia
Muhammad Haris	Universitas Nusa Mandiri, Indonesia
Muhammad Haroon Shakeel	Lahore University of Management Sciences, Pakistan
Muhammad Hilman	Universitas Indonesia, Indonesia
Muhammad Ramzan	Saudi Electronic University, Saudi Arabia
Muideen Adegoke	City University of Hong Kong, Hong Kong
Mulin Chen	Northwestern Polytechnical University, China
Murtaza Taj	Lahore University of Management Sciences, Pakistan
Mutsumi Kimura	Ryukoku University, Japan
Naoki Masuyama	Osaka Prefecture University, Japan
Naoyuki Sato	Future University Hakodate, Japan
Nat Dilokthanakul	Vidyasirimedhi Institute of Science and Technology, Thailand
Nguyen Dang	University of Canberra, Australia
Nhi N. Y. Vo	University of Technology Sydney, Australia
Nick Nikzad	Griffith University, Australia
Ning Boda	Swinburne University of Technology, Australia
Nobuhiko Wagatsuma	Tokyo Denki University, Japan
Nobuhiko Yamaguchi	Saga University, Japan
Noor Akhmad Setiawan	Universitas Gadjah Mada, Indonesia
Norbert Jankowski	Nicolaus Copernicus University, Poland
Norikazu Takahashi	Okayama University, Japan
Noriyasu Homma	Tohoku University, Japan
Normaziah A. Aziz	International Islamic University Malaysia, Malaysia
Olarik Surinta	Mahasarakham University, Thailand

Olutomilayo Olayemi Petinrin	Kings University, Nigeria
Ooi Shih Yin	Multimedia University, Malaysia
Osamu Araki	Tokyo University of Science, Japan
Ozlem Faydasicok	Istanbul University, Turkey
Parisa Rastin	University of Lorraine, France
Paul S. Pang	Federation University Australia, Australia
Pedro Antonio Gutierrez	Universidad de Cordoba, Spain
Pengyu Sun	Microsoft
Piotr Duda	Institute of Computational Intelligence/Czestochowa University of Technology, Poland
Prabath Abeysekara	RMIT University, Australia
Pui Huang Leong	Tunku Abdul Rahman University College, Malaysia
Qian Li	Chinese Academy of Sciences, China
Qiang Xiao	Huazhong University of Science and Technology, China
Qiangfu Zhao	University of Aizu, Japan
Qianli Ma	South China University of Technology, China
Qing Xu	Tianjin University, China
Qing Zhang	Meituan, China
Qinglai Wei	Institute of Automation, Chinese Academy of Sciences, China
Qingrong Cheng	Fudan University, China
Qiufeng Wang	Xi'an Jiaotong-Liverpool University, China
Qiulei Dong	Institute of Automation, Chinese Academy of Sciences, China
Qiuye Wu	Guangdong University of Technology, China
Rafal Scherer	Częstochowa University of Technology, Poland
Rahmadya Handayanto	Universitas Islam 45 Bekasi, Indonesia
Rahmat Budiarto	Albaha University, Saudi Arabia
Raja Kumar	Taylor's University, Malaysia
Rammohan Mallipeddi	Kyungpook National University, South Korea
Rana Md Mashud	CSIRO, Australia
Rapeeporn Chamchong	Mahasarakham University, Thailand
Raphael Couturier	Université Bourgogne Franche-Comté, France
Ratchakoon Pruengkarn	Dhurakij Pundit University, Thailand
Reem Mohamed	Mansoura University, Egypt
Rhee Man Kil	Sungkyunkwan University, South Korea
Rim Haidar	University of Sydney, Australia
Rizal Fathoni Aji	Universitas Indonesia, Indonesia
Rukshima Dabare	Murdoch University, Australia
Ruting Cheng	University of Science and Technology Beijing, China
Ruxandra Liana Costea	Polytechnic University of Bucharest, Romania
Saaveethya Sivakumar	Curtin University Malaysia, Malaysia
Sabrina Fariza	Central Queensland University, Australia
Sahand Vahidnia	University of New South Wales, Australia

Saifur Rahaman	City University of Hong Kong, Hong Kong
Sajib Mistry	Curtin University, Australia
Sajib Saha	CSIRO, Australia
Sajid Anwar	Institute of Management Sciences Peshawar, Pakistan
Sakchai Muangsrinoon	Walailak University, Thailand
Salomon Michel	Université Bourgogne Franche-Comté, France
Sandeep Parameswaran	Myntra Designs Pvt. Ltd., India
Sangtae Ahn	Kyungpook National University, South Korea
Sang-Woo Ban	Dongguk University, South Korea
Sangwook Kim	Kobe University, Japan
Sanparith Marukatat	NECTEC, Thailand
Saptakatha Adak	Indian Institute of Technology, Madras, India
Seiichi Ozawa	Kobe University, Japan
Selvarajah Thuseethan	Sabaragamuwa University of Sri Lanka, Sri Lanka
Seong-Bae Park	Kyung Hee University, South Korea
Shan Zhong	Changshu Institute of Technology, China
Shankai Yan	National Institutes of Health, USA
Sheeraz Akram	University of Pittsburgh, USA
Shenglan Liu	Dalian University of Technology, China
Shenglin Zhao	Zhejiang University, China
Shing Chiang Tan	Multimedia University, Malaysia
Shixiong Zhang	Xidian University, China
Shreya Chawla	Australian National University, Australia
Shri Rai	Murdoch University, Australia
Shuchao Pang	Jilin University, China/Macquarie University, Australia
Shuichi Kurogi	Kyushu Institute of Technology, Japan
Siddharth Sachan	Australian National University, Australia
Sirui Li	Murdoch University, Australia
Sonali Agarwal	Indian Institute of Information Technology, Allahabad, India
Sonya Coleman	University of Ulster, UK
Stavros Ntalampiras	University of Milan, Italy
Su Lei	University of Science and Technology Beijing, China
Sung-Bae Cho	Yonsei University, South Korea
Sunu Wibirama	Universitas Gadjah Mada, Indonesia
Susumu Kuroyanagi	Nagoya Institute of Technology, Japan
Sutharshan Rajasegarar	Deakin University, Australia
Takako Hashimoto	Chiba University of Commerce, Japan
Takashi Omori	Tamagawa University, Japan
Tao Ban	National Institute of Information and Communications Technology, Japan
Tao Li	Peking University, China
Tao Xiang	Chongqing University, China
Teddy Mantoro	Sampoerna University, Indonesia
Tedjo Darmanto	STMIK AMIK Bandung, Indonesia
Teijiro Isokawa	University of Hyogo, Japan

Thanh Tam Nguyen	Leibniz University Hannover, Germany
Thanh Tung Khuat	University of Technology Sydney, Australia
Thaweesak Khongtuk	Rajamangala University of Technology Suvarnabhumi, Thailand
Tianlin Zhang	University of Chinese Academy of Sciences, China
Timothy McIntosh	Massey University, New Zealand
Toan Nguyen Thanh	Ho Chi Minh City University of Technology, Vietnam
Todsanai Chumwatana	Murdoch University, Australia
Tom Gedeon	Australian National University, Australia
Tomas Maul	University of Nottingham, Malaysia
Tomohiro Shibata	Kyushu Institute of Technology, Japan
Tomoyuki Kaneko	University of Tokyo, Japan
Toshiaki Omori	Kobe University, Japan
Toshiyuki Yamane	IBM, Japan
Uday Kiran	University of Tokyo, Japan
Udom Silparcha	King Mongkut's University of Technology Thonburi, Thailand
Umar Aditiawarman	Universitas Nusa Putra, Indonesia
Upeka Somaratne	Murdoch University, Australia
Usman Naseem	University of Sydney, Australia
Ven Jyn Kok	National University of Malaysia, Malaysia
Wachira Yangyuen	Rajamangala University of Technology Srivijaya, Thailand
Wai-Keung Fung	Robert Gordon University, UK
Wang Yaqing	Baidu Research, Hong Kong
Wang Yu-Kai	University of Technology Sydney, Australia
Wei Jin	Michigan State University, USA
Wei Yanling	TU Berlin, Germany
Weibin Wu	City University of Hong Kong, Hong Kong
Weifeng Liu	China University of Petroleum, China
Weijie Xiang	University of Science and Technology Beijing, China
Wei-Long Zheng	Massachusetts General Hospital, Harvard Medical School, USA
Weiqun Wang	Institute of Automation, Chinese Academy of Sciences, China
Wen Luo	Nanjing Normal University, China
Wen Yu	Cinvestav, Mexico
Weng Kin Lai	Tunku Abdul Rahman University College, Malaysia
Wenqiang Liu	Southwest Jiaotong University, China
Wentao Wang	Michigan State University, USA
Wenwei Gu	Chinese University of Hong Kong, Hong Kong
Wenxin Yu	Southwest University of Science and Technology, China
Widodo Budiharto	Bina Nusantara University, Indonesia
Wisnu Ananta Kusuma	Institut Pertanian Bogor, Indonesia
Worapat Paireekreng	Dhurakij Pundit University, Thailand

Xiang Chen	George Mason University, USA
Xiao Jian Tan	Tunku Abdul Rahman University College, Malaysia
Xiao Liang	Nankai University, China
Xiaocong Chen	University of New South Wales, Australia
Xiaodong Yue	Shanghai University, China
Xiaoqing Lyu	Peking University, China
Xiaoyang Liu	Huazhong University of Science and Technology, China
Xiaoyang Tan	Nanjing University of Aeronautics and Astronautics, China
Xiao-Yu Tang	Zhejiang University, China
Xin Liu	Huaqiao University, China
Xin Wang	Southwest University, China
Xin Xu	Beijing University of Technology, China
Xingjian Chen	City University of Hong Kong, Hong Kong
Xinyi Le	Shanghai Jiao Tong University, China
Xinyu Shi	University of Science and Technology Beijing, China
Xiwen Bao	Chongqing Jiaotong University, China
Xu Bin	Northwestern Polytechnical University, China
Xu Chen	Shanghai Jiao Tong University, China
Xuan-Son Vu	Umeå University, Sweden
Xuanying Zhu	Australian National University, Australia
Yanling Zhang	University of Science and Technology Beijing, China
Yang Li	East China Normal University, China
Yantao Li	Chongqing University, China
Yanyan Hu	University of Science and Technology Beijing, China
Yao Lu	Beijing Institute of Technology, China
Yasuharu Koike	Tokyo Institute of Technology, Japan
Ya-Wen Teng	Academia Sinica, Taiwan
Yaxin Li	Michigan State University, USA
Yifan Xu	Huazhong University of Science and Technology, China
Yihsin Ho	Takushoku University, Japan
Yilun Jin	Hong Kong University of Science and Technology, Hong Kong
Yiming Li	Tsinghua University, China
Ying Xiao	University of Birmingham, UK
Yingjiang Zhou	Nanjing University of Posts and Telecommunications, China
Yong Peng	Hangzhou Dianzi University, China
Yonghao Ma	University of Science and Technology Beijing, China
Yoshikazu Washizawa	University of Electro-Communications, Japan
Yoshimitsu Kuroki	Kurume National College of Technology, Japan
Young Ju Rho	Korea Polytechnic University, South Korea
Youngjoo Seo	Ecole Polytechnique Fédérale de Lausanne, Switzerland

Yu Sang	PetroChina, China
Yu Xiaohan	Griffith University, Australia
Yu Zhou	Chongqing University, China
Yuan Ye	Xi'an Jiaotong University, China
Yuangang Pan	University of Technology Sydney, Australia
Yuchun Fang	Shanghai University, China
Yuhua Song	University of Science and Technology Beijing
Yunjun Gao	Zhejiang University, China
Zeyuan Wang	University of Sydney, Australia
Zhen Wang	University of Sydney, Australia
Zhengyang Feng	Shanghai Jiao Tong University, China
Zhenhua Wang	Zhejiang University of Technology, China
Zhenqian Wu	University of Electronic Science and Technology of China, China
Zhenyu Cui	University of Chinese Academy of Sciences, China
Zhenyue Qin	Australian National University, Australia
Zheyang Shen	Aalto University, Finland
Zhihong Cui	Shandong University, China
Zhijie Fang	Chinese Academy of Sciences, China
Zhipeng Li	Tsinghua University, China
Zhiri Tang	City University of Hong Kong, Hong Kong
Zhuangbin Chen	Chinese University of Hong Kong, Hong Kong
Zongying Liu	University of Malaya, Malaysia

Contents – Part II

Theory and Algorithms

LSMVC:Low-rank Semi-supervised Multi-view Clustering for Special
Equipment Safety Warning. 3
    Fukang Zhang, Hongwei Yin, Xinmin Cheng, Wenhui Du,
    and Huangzhen Xu

Single-Skeleton and Dual-Skeleton Hypergraph Convolution Neural
Networks for Skeleton-Based Action Recognition 15
    Changxiang He, Chen Xiao, Shuting Liu, Xiaofei Qin, Ying Zhao,
    and Xuedian Zhang

A Multi-Reservoir Echo State Network with Multiple-Size Input Time
Slices for Nonlinear Time-Series Prediction 28
    Ziqiang Li and Gouhei Tanaka

Transformer with Prior Language Knowledge for Image Captioning 40
    Daisong Yan, Wenxin Yu, Zhiqiang Zhang, and Jun Gong

Continual Learning with Laplace Operator Based Node-Importance
Dynamic Architecture Neural Network. 52
    Zhiyuan Li, Ming Meng, Yifan He, and Yihao Liao

Improving Generalization of Reinforcement Learning for Multi-agent
Combating Games. 64
    Kejia Wan, Xinhai Xu, and Yuan Li

Gradient Boosting Forest: a Two-Stage Ensemble Method Enabling
Federated Learning of GBDTs 75
    Feng Wang, Jinxiang Ou, and Hairong Lv

Random Neural Graph Generation with Structure Evolution 87
    Yuguang Zhou, Zheng He, Tao Wan, and Zengchang Qin

MatchMaker: Aspect-Based Sentiment Classification
via Mutual Information 99
    Jing Yu, Yongli Cheng, Fang Wang, Xianghao Xu, Dan He,
    and Wenxiong Wu

PathSAGE: Spatial Graph Attention Neural Networks with Random
Path Sampling 111
    Junhua Ma, Jiajun Li, Xueming Li, and Xu Li

Label Preserved Heterogeneous Network Embedding.   121
  Xiangyu Li and Weizheng Chen

Spatio-Temporal Dynamic Multi-graph Attention Network for Ride-
Hailing Demand Prediction.   133
  Ya Chen, Wanrong Jiang, Hao Fu, and Guiquan Liu

An Implicit Learning Approach for Solving the Nurse
Scheduling Problem.   145
  Aymen Ben Said, Emad A. Mohammed, and Malek Mouhoub

Improving Goal-Oriented Visual Dialogue by Asking Fewer Questions   158
  Soma Kanazawa, Shoya Matsumori, and Michita Imai

Balance Between Performance and Robustness of Recurrent Neural
Networks Brought by Brain-Inspired Constraints on Initial Structure   170
  Yuki Ikeda, Tomohiro Fusauchi, and Toshikazu Samura

Single-Image Smoker Detection by Human-Object Interaction
with Post-refinement   181
  Hua-Bao Ling and Dong Huang

A Lightweight Multi-scale Feature Fusion Network for Real-Time
Semantic Segmentation   193
  Tanmay Singha, Duc-Son Pham, Aneesh Krishna, and Tom Gedeon

Multi-view Fractional Deep Canonical Correlation Analysis
for Subspace Clustering   206
  Chao Sun, Yun-Hao Yuan, Yun Li, Jipeng Qiang, Yi Zhu,
  and Xiaobo Shen

Handling the Deviation from Isometry Between Domains and Languages
in Word Embeddings: Applications to Biomedical Text Translation.   216
  Félix Gaschi, Parisa Rastin, and Yannick Toussaint

Inference in Neural Networks Using Conditional Mean-Field Methods.   228
  Ángel Poc-López and Miguel Aguilera

Associative Graphs for Fine-Grained Text Sentiment Analysis   238
  Maciej Wójcik, Adrian Horzyk, and Daniel Bulanda

k-Winners-Take-All Ensemble Neural Network.   250
  Abien Fred Agarap and Arnulfo P. Azcarraga

Performance Improvement of FORCE Learning for Chaotic Echo
State Networks   262
  Ruihong Wu, Kohei Nakajima, and Yongping Pan

Generative Adversarial Domain Generalization via Cross-Task Feature
Attention Learning for Prostate Segmentation   273
    Yifang Xu, Dan Yu, Ye Luo, Enbei Zhu, and Jianwei Lu

Context-Based Deep Learning Architecture with Optimal Integration Layer
for Image Parsing   285
    Ranju Mandal, Basim Azam, and Brijesh Verma

Kernelized Transfer Feature Learning on Manifolds   297
    R. Lekshmi, Rakesh Kumar Sanodiya, R. J. Linda, Babita Roslind Jose,
    and Jimson Mathew

Data-Free Knowledge Distillation with Positive-Unlabeled Learning   309
    Jialiang Tang, Xiaoyan Yang, Xin Cheng, Ning Jiang, Wenxin Yu,
    and Peng Zhang

Manifold Discriminative Transfer Learning for Unsupervised
Domain Adaptation   321
    Xueliang Quan, Dongrui Wu, Mengliang Zhu, Kun Xia,
    and Lingfei Deng

Training-Free Multi-objective Evolutionary Neural Architecture Search
via Neural Tangent Kernel and Number of Linear Regions   335
    Tu Do and Ngoc Hoang Luong

Neural Network Pruning via Genetic Wavelet Channel Search   348
    Saijun Gong, Lin Chen, and Zhicheng Dong

Binary Label-Aware Transfer Learning for Cross-Domain Slot Filling   359
    Gaoshuo Liu, Shenggen Ju, and Yu Chen

Condition-Invariant Physical Adversarial Attacks via Pixel-Wise
Adversarial Learning   369
    Chenchen Zhao and Hao Li

Multiple Partitions Alignment with Adaptive Similarity Learning   381
    Hao Dai

Recommending Best Course of Treatment Based on Similarities
of Prognostic Markers   393
    Sudhanshu, Narinder Singh Punn, Sanjay Kumar Sonbhadra,
    and Sonali Agarwal

Generative Adversarial Negative Imitation Learning from Noisy
Demonstrations   405
    Xin Cao and Xiu Li

Detecting Helmets on Motorcyclists by Deep Neural Networks
with a Dual-Detection Scheme   417
    Chun-Hong Li and Dong Huang

Short-Long Correlation Based Graph Neural Networks for Residential
Load Forecasting.   428
    Yiran Deng, Yingjie Zhou, and Zhiyong Zhang

Disentangled Feature Network for Fine-Grained Recognition   439
    Shuyu Miao, Shuaicheng Li, Lin Zheng, Wei Yu, Jingjing Liu,
    Mingming Gong, and Rui Feng

Large-Scale Topological Radar Localization Using Learned Descriptors.   451
    Jacek Komorowski, Monika Wysoczanska, and Tomasz Trzcinski

Rethinking Binary Hyperparameters for Deep Transfer Learning   463
    Jo Plested, Xuyang Shen, and Tom Gedeon

Human Centred Computing

Hierarchical Features Integration and Attention Iteration Network
for Juvenile Refractive Power Prediction   479
    Yang Zhang, Risa Higashita, Guodong Long, Rong Li, Daisuke Santo,
    and Jiang Liu

Stress Recognition in Thermal Videos Using Bi-directional Long-Term
Recurrent Convolutional Neural Networks   491
    Siyuan Yan and Abhijit Adhikary

StressNet: A Deep Neural Network Based on Dynamic Dropout Layers
for Stress Recognition   502
    Hao Wang and Abhijit Adhikary

Analyzing Vietnamese Legal Questions Using Deep Neural Networks
with Biaffine Classifiers.   513
    Nguyen Anh Tu, Hoang Thi Thu Uyen, Tu Minh Phuong,
    and Ngo Xuan Bach

BenAV: a Bengali Audio-Visual Corpus for Visual Speech Recognition   526
    Ashish Pondit, Muhammad Eshaque Ali Rukon, Anik Das,
    and Muhammad Ashad Kabir

Investigation of Different G2P Schemes for Speech Recognition
in Sanskrit   536
    C. S. Anoop and A. G. Ramakrishnan

GRU with Level-Aware Attention for Rumor Early Detection
in Social Networks 548
Yu Wang, Wei Zhou, Junhao Wen, Jun Zeng, Haoran He, and Lin Liu

Convolutional Feature-Interacted Factorization Machines for Sparse
Contextual Prediction 560
Ruoran Huang, Chuanqi Han, and Li Cui

A Lightweight Multidimensional Self-attention Network for Fine-Grained
Action Recognition 573
Hao Liu, Shenglan Liu, Lin Feng, Lianyu Hu, Xiang Li, and Heyu Fu

Unsupervised Domain Adaptation with Self-selected Active Learning
for Cross-domain OCT Image Segmentation...................... 585
Xiaohui Li, Sijie Niu, Xizhan Gao, Tingting Liu, and Jiwen Dong

Adaptive Graph Convolutional Network with Prior Knowledge for Action
Recognition 597
Guihong Lao, Lianyu Hu, Shenglan Liu, Zhuben Dong, and Wujun Wen

Self-adaptive Graph Neural Networks for Personalized
Sequential Recommendation.................................. 608
Yansen Zhang, Chenhao Hu, Genan Dai, Weiyang Kong, and Yubao Liu

Spatial-Temporal Attention Network with Multi-similarity Loss
for Fine-Grained Skeleton-Based Action Recognition................ 620
*Xiang Li, Shenglan Liu, Yunheng Li, Hao Liu, Jinjing Zhao, Lin Feng,
Guihong Lao, and Guangzhe Li*

SRGAT: Social Relational Graph Attention Network for Human
Trajectory Prediction 632
Yusheng Peng, Gaofeng Zhang, Xiangyu Li, and Liping Zheng

FSE: a Powerful Feature Augmentation Technique for Classification Task ... 645
Yaozhong Liu, Yan Yang, and Md Zakir Hossain

AI and Cybersecurity

FHTC: Few-Shot Hierarchical Text Classification in Financial Domain 657
Anqi Wang, Qingcai Chen, and Dongfang Li

JStrack: Enriching Malicious JavaScript Detection Based on AST Graph
Analysis and Attention Mechanism 669
*Muhammad Fakhrur Rozi, Tao Ban, Seiichi Ozawa,
Sangwook Kim, Takeshi Takahashi, and Daisuke Inoue*

Author Index 681

Theory and Algorithms

LSMVC:Low-rank Semi-supervised Multi-view Clustering for Special Equipment Safety Warning

Fukang Zhang[1], Hongwei Yin[1(✉)], Xinmin Cheng[1], Wenhui Du[2], and Huangzhen Xu[2]

[1] Huzhou University, Huzhou, Zhejiang, China
02713@zjhu.edu.cn
[2] Huzhou Special Equipment Inspection Center, Huzhou, Zhejiang, China

Abstract. In order to effectively prevent accidents of special equipment, numerous management platforms utilize the multi-source data of special equipment to predict the safety state of equipment. However, there is still a lack of methods to deal with noise when fusing different data sources. This paper proposes a novel low-rank semi-supervised multi-view clustering for special equipment safety warning (LSMVC). Which achieves robust multi-view clustering by using low rank representation (LRR) to reduce the impact of noise. To solve this non-smooth optimization problem, we propose an optimization procedure based on the Alternating Direction Method of Multipliers. Finally, experiments are carried out on six real datasets including the Elevator dataset, which is collected from the actual work. The results show that the proposed clustering method can achieve better clustering performance than other clustering method.

Keywords: Multi-view clustering · Semi-supervised learning · Low-rank representation · Non-negative matrix factorization · Special equipment

1 Introduction

Special equipment refers to the eight major types of equipment, including pressure vessels, pressure piping, elevators, lifting machinery, passenger ropeways, large amusement rides and field (plant) special motor vehicles. A lot of special equipment are used in daily life. If special equipment breakdown, it will cause serious safety accidents. Therefore safety warning for special equipment is very important.

In order to improve the accuracy of safety warning special equipment, previous methods manually set thresholds. However, they cannot take advantage of the special equipment features to realize special equipment safety warning. Once

Supported by National Natural Science Foundation of China Projects (U20A20228). Huzhou special equipment testing institute commissioned development projects (073–20201210-02).

T. Mantoro et al. (Eds.): ICONIP 2021, LNCS 13109, pp. 3–14, 2021.
https://doi.org/10.1007/978-3-030-92270-2_1

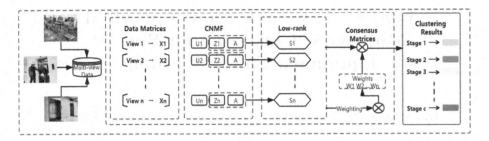

Fig. 1. LSMVC framework: X_i is multi-view matrix. A is label matrix. U_i and Z_i from the CNMF factorization. S_i is low-rank matrix. W_i is the weight of each view. Finally, different classters correspond to different risk stages. If a special equipment is assigned to a cluster with a high risk, this means that the risk of that sample may also be high.

there is multiple special equipment, these methods are not applicable. In addition, they did not take into account how noise affects the results. Noise is usually caused by network transmission problems, improper manual recording and other circumstances. In order to solve such two problems, this paper proposes a novel Low-rank Semi-supervised Multi-view Clustering for Special Equipment Safety Warning (LSMVC). The algorithm in this paper can cluster according to the latent features of special equipment. Low Rank Representation (LRR) [6] is embedded in our algorithm. The LRR can effectively reduce the impact of noise. In the process of collecting data, we can get a lot of labeled samples. We use a semi-supervised algorithm that can take advantage of the labeled information. The final clustering is obtained by the non-negative consensus matrix. The structure diagram is shown in Fig. 1. The main work of this paper includes:

1. A semi-supervised multi-view clustering model is proposed. Labels are added to our approach to enhance the discernment of data. Even in the new representation space, data of the same class that is marked up is allocated to the same representation space. Clustering capabilities for special equipment will be enhanced.
2. Data noise can affect the result of the algorithm. Therefore, low-rank representation is applied to our model to solve this problem. The important information of each view is extracted by the LRR. The experimental results show that the algorithm embedded with LRR can effectively reduce the impact of noise.
3. Different from other methods that predict the safety status of special equipment by setting thresholds, LSMVC can realize adaptive clustering through the hidden relationship of special equipment features. The low-rank representation is embedded to reduce the impact of noise. The final experiment shows that our LSMVC algorithm can realize the special equipment safety warning.

2   Related Work

The multi-source data can reflect the state of the special equipment. When special equipment breakdown, the data will deviate from normal. Therefore, the utilization of multi-source data can effectively realize the special equipment safety warning.

For example, [16] uses Dempster-Shafer (DS) evidence theory information to fuses data from multiple special equipment brakes to achieve fault location. [13] uses decision tree to analyze the main characteristics of historical data and predict the safety state of special equipment. [17] integrates UML, database, XML, mobile APP and other key technologies, and uses BP neural networks to build a safety evaluation model. [9] uses a Softmax Regression to analyze historical data of special equipment to predict possible failures in the coming week. There are many machine learning methods that use multi-source data to achieve special equipment safety warning.

However, based on the above works, there are mainly two problems. Firstly, they need manual set thresholds. That means every time you add a new type special equipment. You have to reset the parameters. It will bring great inconvenience. Secondly, due to the influence of the real environment, there will be a lot of noise in the process of data collection. Noise will affect the accuracy of the algorithm for special equipment safety warning.

The multi-view clustering algorithm takes into account the diversity of different views when fusing data. The algorithm in this paper can cluster by combining consistent and complementary information among multiple views. The LRR is embedded to effectively reduce the impact of noise. Among multi-view clustering algorithms, the Nonnegative matrix factorization (NMF) [14] is one of the most widely used frameworks. However, it is unsupervised. So Constrained nonnegative matrix factorization (CNMF) [12] is proposed by Liu et al. Which as a semi-supervised learning method and can use labeling information as an additional hard constraint. The data points in the same class will be fused into the same new representation.

2.1   Multi-view Non-negative Matrix Factorization

Multi-view non-negative matrix factorization (NMF) [8] has been developed to learn the latent representation from multi-view non-negative data in recent years. Given a multi-view data set $X^v = [X^1, \ldots, X^n]$, NMF aims to approximate $X^v \approx U^v(V^v)^T$ for each view, meanwhile learns a latent representation V^* via the following objective function:

$$O_F = \sum_{v=1}^{n_v} \left\| X^v - U^v(V^v)^T \right\|_F^2 + \sum_{v=1}^{n_v} \lambda_v \left\| V^v - V^* \right\|_F^2 \tag{1}$$

Where U can be considered as a basis matrix. V can be considered as a new representation of X. λ_v tunes the relative weight among different views. V^* is consensus matrix, which considered to reflect the latent clustering structure shared by different views.

2.2   Multi-view Semi-supervised Non-negative Matrix Factorization

A Multi-view semi-supervised non-negative matrix factorization [7] is proposed, which uses label information as an additional hard constraint. The data points within the same class will be fused into the same new representation. An index matrix C is introduced to construct the label matrix A:

$$A = \begin{pmatrix} C_{l \times c} & 0 \\ 0 & I_{n-1} \end{pmatrix} \tag{2}$$

Where $C_{ij} = 1$ means x_i belongs to the j th class, otherwise $C_{ij} = 0$; I_{n-1} is an identity matrix. Given a multi-view data set X^v , where the former l data points have label information and the remaining $n - l$ data points are not labeled. The matrix A^v is equal to A .The coefficient matrix V^v for each view can be defined as follows:

$$V^v = AZ^v \tag{3}$$

Where if x_i and x_j belong to the same class or label (for example,$A_{i:} = A_{j:}$), then $v_i = v_j$. In the new representation space, if the original data points belong to the same label, they must have the same representation in the new representation space. Therefore, its objective function can be rewritten as follows:

$$O_F = \left\| X^v - U^v (Z^v)^T A^T \right\|_F^2 + \sum_{v=1}^{n_v} \lambda_v \| Z^v - Z^* \|_F^2 \quad \text{s.t. } V^v = AZ^v \tag{4}$$

2.3   Low Rank Representation (LRR)

The Low rank representation (LRR) method [6] learns a low-rank subspace representation with better clustering performance, which can reduce noise interference. The data matrix X is given, LRR solves the self-representation problem by finding the lowest rank representation of all data as:

$$\min_Z \|Z\|_* + \lambda \|E\|_p \quad \text{s.t. } X = XZ + E \tag{5}$$

Where Z is the learned low-rank subspace representation of data X. E is the error matrix. $\| \cdot \|_*$ denotes the nuclear norm of a matrix. The learned subspace Z is an effective data representation method and provides a clear clustering structure.

3   Low-Rank Semi-supervised Multi-view Clustering for Special Equipment Safety Warning

In order to solve the problem that traditional methods can't adaptively classify special equipment and solve the noise problem, this paper proposes a novel Low-rank Semi-supervised Multi-view Clustering for Special Equipment Safety Warning (LSMVC). The algorithm can cluster according to the latent features of special equipment. That can take known label information as a constraint. The data points within the same class can be fused into a new representation. So the clustering capabilities for special equipment will be enhanced.

During the data collection process, data noise may be generated due to interference in the physical environment, poor network transmission environment and improper manual recording. Therefore, LRR [6] is embedded in our algorithm to reduce the impact of noise. In our model: The first, multi-view matrix data set X^v is decomposed into U^v and Z^v. A is the label matrix. Secondly, matrix S^v is obtained by LRR. Thirdly, the nonnegative consensus matrix is obtained. The non-negative consensus matrix fuses the information of all the views. Finally, the safety status of special equipment is obtained by spectral clustering method. The entire process is shown in Fig. 1.

3.1   Objective Function

In the objective function (6), the first term is CNMF. The data points from the same class can be compressed together in the new representation space V^v. To be specific, the row vector of V^v is a low-dimensional representation of the view v . However, due to the existence of data noise, it is not robust enough. We introduce the second item. The second item can effectively reduce the impact of noise by applying LRR. We can obtain low-rank matrix S^v from coefficient matrix V^v.

$$\mathcal{O}_F = \left\| X^v - U^v (V^v)^T \right\|_F^2 + \lambda_1 \sum_{v=1}^{n_v} (\| V^v - V^v S^v \|_F^2 + \| S^v \|_*)$$
$$\text{s.t. } U^v, Z^v \geq 0, V^v = AZ^v \tag{6}$$

The third item preserves not only the important information of each view but also the potential clustering structure. Inspired by [16] , we proposed a centroid-based nonnegative consensus matrix. The non-negative consensus matrix S^* fuses the information of low-rank matrix S^v. The final algorithm model is as follows:

$$\mathcal{O}_F = \left\| X^v - U^v (Z^v)^T A^T \right\|_F^2 + \lambda_1 \sum_{v=1}^{n_v} (\| V^v - V^v S^v \|_F^2 + \| S^v \|_*)$$
$$+ \lambda_2^v \sum_{v=1}^{n_v} \| S^v - S^* \|_F^2 \quad \text{s.t. } U^v, Z^v \geq 0, V^v = AZ^v \tag{7}$$

Where $X^v = [X^1, ..., X^n]$ represents multi-view data set. U^v and Z^v are the basis matrix and presentation matrix obtained by decomposition. A is the label matrix. $\lambda_1 > 0$ controls the low-rank constraint. λ_2^v is the weight for the view.

3.2   Optimization

The objective function contains multiple parameter variables: U^v, V^v, S^v and S^* .Therefore, the function is optimized by the method of exchange iteration.

Fixing V^v, S^v and S^*, Updating U^v: Retaining only the part of (7) that is relevant U^v, we introduce the Lagrange multiplier α^v for the constraint $U^v \geq 0$, and we obtain:

$$
\begin{aligned}
L_1 = \lambda_1 \mathrm{tr}((X^v)^T X^v &- (X^v)^T U^v (Z^v)^T A^T - A Z^v (U^v)^T X^v \\
&+ A Z^v (U^v)^T U^v (Z^v)^T A^T) + \mathrm{tr}(\alpha^v (U^v)^T)
\end{aligned}
\tag{8}
$$

Find the partial derivative of U^v with respect to L_1:

$$
\frac{\partial L_1}{\partial U_{ij}^v} = -2\lambda_1 \left(X^v A Z^v \right)_{ij} + 2\lambda_1 \left(U^v (Z^v)^T A^T A Z^v \right)_{ij} + \alpha_{ij}^v
\tag{9}
$$

Next, by setting $\partial L_1 / \partial U_{ij}^v = 0$ and using the KKT condition [1] $\alpha_{ij}^v u_{ij}^v = 0$, we obtain:

$$
-2\lambda_1 (X^v A Z^v)_{ij} u_{ij}^v + 2\lambda_1 \left(U^v (Z^v)^T A^T A Z^v \right)_{ij} u_{ij}^v = 0
\tag{10}
$$

Finally, the following rules can be updated from the above formula:

$$
u_{ij}^v \leftarrow u_{ij}^v \frac{(X^v A Z^v)_{ij}}{\left(U^v (Z^v)^T A^T A Z^v \right)_{ij}}
\tag{11}
$$

Fixing U^v, V^v and S^*, Updating Z^v: Similarly, we introduce the Lagrange multi-plier β^v for the constraint $Z^v \geq 0$ in (7) and we find the partial derivative of Z^v with respect to L_2:

$$
\frac{\partial L_2}{\partial z_{ij}^v} = -2\lambda_1 (A^T (X^v)^T U^v)_{ij} z_{ij}^v - (A^T A Z^v (U^v)^T U^v)_{ij} z_{ij}^v) + \beta_{ij}^v
\tag{12}
$$

Next, by setting $\partial L_2 / \partial Z_{ij}^v = 0$ and using the KKT condition [1] $\alpha_{ij}^v u_{ij}^v = 0$, we obtain:

$$
2\theta_v (-A^T (X^v)^T U^v + A^T A Z^v (U^v)^T U^v)_{ij} z_{ij}^v = 0
\tag{13}
$$

Finally, the following rules can be updated from the above formula:

$$
z_{ij} \leftarrow z_{ij} \frac{\left(A^T X^T U \right)_{ij}}{\left(A^T A Z U^T U \right)_{ij}}
\tag{14}
$$

Fixing U^v, V^v and S^*, Updating S^v: Retaining only the part of (7) that is relevant S^v, we introducing auxiliary variables $A^v, Y_1{}^v, Y_2{}^v$, we obtain:

$$
\frac{\lambda_1}{2} \|V^v - V^v S^v\|_F^2 + \lambda_1 \left\| Y_1{}^{(v)} \right\|_* + \lambda_2^v \sum_{v=1}^{V} \|S^v - S^*\|_F^2
\tag{15}
$$

$$
\text{s.t. } A^{(v)} = Y_1{}^{(v)}, A^{(v)} = Y_2{}^{(v)}
$$

Let us write it in Lagrangian form:

$$
\begin{aligned}
L3 = \frac{\lambda_1}{2} \|V^v - V^v A^v\|_F^2 &+ \lambda_1 \left\| Y_1{}^{(v)} \right\|_* + \frac{u}{2} \left\| A^{(v)} - Y_1{}^{(v)} + \Lambda_1 \right\|_F^2 \\
&+ \frac{u}{2} \left\| A^{(v)} - Y_2{}^{(v)} + \Lambda_2 \right\|_F^2 + \lambda_2{}^v \sum_{v=1}^{V} \left\| Y_2{}^{(v)} - S^* \right\|_F^2
\end{aligned}
\tag{16}
$$

$$
\text{s.t. } A^{(v)} = Y_1{}^{(v)}, A^{(v)} = Y_2{}^{(v)}
$$

Where u is the penalty parameter and Λ_i is the Lagrangian dual variable. ADMM method [2] is used to solve the convex optimization problem in (16).

Find the partial derivative of $A^{(v)}$ with respect to (16):

$$A^{(v)} = \left(\lambda_1 (V^v)^T V^v + 2\mu I\right)^{-1} \left(\lambda_1 (V^v)^T V^v + \mu \left(Y_1^{(v)} - \Lambda_1^{(v)}\right) + \mu \left(Y_2^{(v)} - \Lambda_2^{(v)}\right)\right)$$
(17)

According to [15], the $Y_1^{(v)}$ update is as follows:

$$Y_1^{(v)} = \Pi_{\frac{\lambda_1}{\mu}} \left(A^{(v)} + \frac{\Lambda_2^{(v)}}{\mu}\right)$$
(18)

Where $\Pi_\beta(Y_1^{(v)}) = U\pi_\beta(\Sigma)V^T$ performs a soft threshold operation on the singular values of $Y_1^{(v)}$, and $U\Sigma V^T$ is a singular value decomposition of $Y_1^{(v)}$.

Find the partial derivative of $Y_2^{(v)}$ with respect to (16):

$$Y_2^{(v)} = \left(2\lambda_2^{(v)} + \mu\right)^{-1} \left(2\lambda_2^{(v)} S^* + \mu(A^{(v)} + \Lambda_2^{(v)})\right)$$
(19)

Update rules for dual variables Y_1^v and Y_2^v:

$$\Lambda_1^{(v)} = \Lambda_1^{(v)} + \mu \left(A^{(v)} - Y_1^{(v)}\right); \Lambda_2^{(v)} = \Lambda_2^{(v)} + \mu \left(A^{(v)} - Y_2^{(v)}\right); \mu = \min\left(\rho\mu, \mu_{\max}\right)$$
(20)

Where μ_{max} is the upper limit of the multiplier and ρ is the positive coefficient of the multiplier. The detailed steps of solving (16) are shown in Algorithm 1.

Fixing U^v, V^v **and** S^v, **Updating** S^*: According to [16],we find the partial derivative of S^* with respect to (16):

$$S^* = \frac{\sum_{v=1}^{n_v} \lambda_2^{(v)} A^{(v)}}{\sum_{v=1}^{n_v} \lambda_2^{(v)}}$$
(21)

The nonnegative consensus matrix S^* is obtained by a centroid-based regularization scheme [16]. Finally, the cluster partition of the data is obtained by using spectral clustering on S^*. The detailed steps are shown in Algorithm 2.

3.3 Using LSMVC to Predict Special Equipment Safety

In practical, the multi-view data set of special equipment is taken as $X_{d \times n}^v$, where d is the number of features. n is the number of samples. v is the number of views. Label matrix $A_{n \times (n+c-l)}^v$ is generated with known special equipment information, where l is the number of labeled special equipment. c is the number of clusters. Each view first initializes the matrix $U_{d \times r}^v$ and $Z_{(n+c-l) \times r}^v$. Where r denotes the desired reduced dimension.

Then iterative optimization of $U_{d \times r}^v$, $Z_{(n+c-l) \times r}^v$ and $S_{n \times n}^v$ is started. $S_{n \times n}^v$ is a symmetric matrix obtained through low-rank processing of $V_{n \times r}^v$. We choose to

Algorithm 1: Solving the problem (18) with ADMM
Input:Lagrange multiplier penalty value μ, penalty value limit maximum $\mu_{\max}$, iteration parameters ε;   Initialize:$A^v = Y_1{}^v = Y_2{}^v = 0$ 1.While not converged do   2.Fix other, update A^v by (17);   3.Fix other, update $Y_1{}^v, Y_2{}^v$ by   (18), (19);   4.Update the dual variables by (20);   5.Judge the convergence condition $\|A^v - Y_1^v\| \leq \varepsilon$; $\|A^v - Y_2^v\| \leq \varepsilon$;   6.End while
Output:S^v

Algorithm 2: Low-rank Semi-supervised Multi-view Clustering (LSMVC)
Input: X^v, k, low rank parameters λ_1, view weights $\lambda_2{}^{(v)}$, proportion of labeled sample m;   1.For v=1:numview   2.   Initialize U^v, Z^v;   3.End for   4.While not converged do   5.For v=1:numview   6.By equation (11), Update U^v;   7.By equation (14), Update Z^v;   8.By Algorithm 1, Update S^v;   9.End for   10.End while   11.Update S^* by equation (21);
Output:Centroid matrix S^*

extract low-rank structure from $V_{n \times r}^v$, because $V_{n \times r}^v$ contains reduced dimension r and complete sample number n. After the convergence is complete, a consensus view integrated information for all views. We set a centroid-based nonnegative consensus matrix $S_{n \times n}^*$. The non-negative consensus matrix $S_{n \times n}^*$ is constructed to guarantee different views of information as complementary and robust as possible.

Finally, spectral clustering is performed. The spectral clustering algorithm divides the data points into multiple clusters. Each cluster contain high, medium and low risk samples, some more, some less. If a special equipment is classified into clusters with many high-risk samples, that means it is also high-risk. According to this, we can realize the special equipment safety warning.

4 Experiment

4.1 Datasets

To verify the effectiveness of the algorithm, in this paper, we used five public real datasets: uci-digit[1], 3-sources[2], MSRC_v1[3], BBCSport[4] and BBC[5] , and one real dataset Elevator. We make a briefly summary in Table 1. We describe in detail the source of Elevator dataset in the following. Elevator: This is a multi-view dataset taken from the real world. Includes elevator operation data in 2020 provided by Huzhou Special Equipment Inspection Institute, Zhejiang, China.

[1] http://archive.ics.uci.edu/ml/datasets.
[2] http://mlg.ucd.ie/datasets/3sources.html.
[3] https://www.cnblogs.com/picassooo/p/12890078.html.
[4] http://mlg.ucd.ie/datasets/segment.html.
[5] http://mlg.ucd.ie/datasets/segment.html.

Table 1. Multi-view datasets.

Dataset	Size	View	Cluster	View1	View2	View3	View4	View5
MSRC_v1	210	5	7	ms1(24)	ms2(576)	ms3(512)	ms4(256)	ms5(254)
Uci-digit	2000	3	10	fac(216)	fou(76)	kar(64)	/	/
Elevator	385	3	3	el01(12)	el02(16)	el03(307)	/	/
3-sources	169	3	6	bbc(3560)	reu(3631)	gua(3068)	/	/
BBCSport	544	2	5	bc1(3183)	bc2(3203)	/	/	/

Table 2. Comparison of clustering performance.

Method	ACC					
	BBCSport	uci-digit	3-sources	MSRC_v1	BBC	Elevator
MVGL	0.726	0.772	0.727	0.747	0.684	0.955
MCGC	0.855	0.776	0.733	0.742	0.743	0.924
RMSC	0.84	**0.846**	0.615	0.814	0.741	0.942
NMF(best)	0.416	0.628	0.528	0.647	0.46	0.976
CNMF(best)	0.636	0.771	0.636	0.702	0.553	0.966
GNMF(best)	0.462	0.657	0.467	0.579	0.467	0.974
LSMVC	**0.858**	0.838	**0.758**	**0.849**	**0.749**	**0.977**

The selected area was Wuxing District, Huzhou City, Zhejiang Province. Total of 358 samples were selected. Each sample has three views, the production view from the manufacturing information, the environment view from the installation environment of the special equipment, and the maintenance view from the regular manual inspection. And every sample has a risk level, it can be divided into three categories: high, medium and low.

4.2   Experimental Settings

The algorithm contains five main parameters: $lambda, m, r, wight^{(v)}, \mu$. λ is used to control the degree of low rank. m is used to control the percentage of true labels. We set $m = 0.5$. r is used to control the implicit function variables in the matrix decomposition. $wight^{(v)}$ is used to control the weights of each view. In the experiments, each view in the same dataset is set to the same weight. μ represents the iteration parameter, and we set $\mu = 10$. We first choose the optimal parameter $lambda$ from the interval [0.001,0.01, 0.1,1.0,10]. Then select the optimal parameter r and leave the rest of the parameters unchanged. The results are shown in Fig. 2.

4.3   Compared Methods

To validate the performance of our model (LSMVC), we compared it with other algorithms. Including three multi-view clustering algorithms and three matrix factorization-based algorithms. Multiview clustering algorithms include: Multiview Consensus Graph Clustering (MCGC) [4], Graph Learning for Multi-view

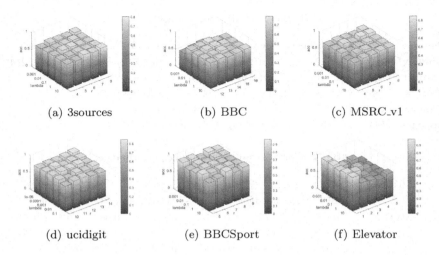

(a) 3sources   (b) BBC   (c) MSRC_v1

(d) ucidigit   (e) BBCSport   (f) Elevator

Fig. 2. Performance of LSMVC under different values of *lambda* and r .

Clustering (MVGL) [5], Robust Multi-view Spectral Clustering via Low-rank and Sparse Decomposition(RMSC) [11]. The algorithms based on matrix factorization include: Non-negative matrix factorization (NMF) [7], Graph regularized Non-negative Matrix Factorization (GNMF) [3], Constrained Non-negative Matrix Factorization for Image Representation (CNMF) [10], where the CNMF algorithm is a semi-supervised algorithm.

4.4 Results on Public Datasets

We use the ACC to evaluate the accuracy. In order to avoid errors caused by random initialization in the algorithm, each algorithm runs 10 times to take the average value. For single-view algorithms, we take the best value. The results are shown in Table 2: It is obvious that multi-view algorithms are superior to single-view algorithms because they combine more information. Compared with the multi-view clustering algorithm such as RMSC, which adopts low-rank strategy to control noise,the experimental results of LSMVC are usually 1–3% higher. This is because LSMVC retains more matrix information. Compared with the matrix factorization algorithms such as NMF and GNMF, LSMVC can get better performance. Because it combines the labeled information, which can enhance recognition ability. Compared with the Semi-supervised algorithms such as CNMF without noise processing, the experimental results of LSMVC are 13% higher. That indicate LSMVC can effectively reduce the impact of noise.

4.5 Results on Elevator Dataset

To verify the semi-supervised effect of the LSMVC algorithm in realistic dataset. Further experiments are conducted on Elevator dataset. Only the label proportion in [0.1,0.2,0.3,0.4] are changed during the experiments. The corresponding

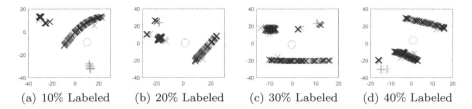

| (a) 10% Labeled | (b) 20% Labeled | (c) 30% Labeled | (d) 40% Labeled |

Fig. 3. Elevator's clustering results with different label proportion. (Color figure online)

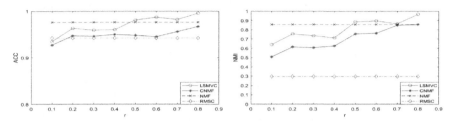

Fig. 4. Results of LSMVC and other algorithms with different label proportion.

clustering results are shown in Fig. 3: blue, green, and red shapes indicate low-risk, medium-risk, and high-risk devices. In general, special equipment at the same risk level tends to have similar features. The clustering algorithm will cluster the data points with similar features together.

The LSMVC is also compared with a semi-supervised algorithm (CNMF), Multi-view (RMSC) and single-view (NMF). The results are shown in Fig. 4. We use the ACC and NMI to evaluate algorithms. It can be seen that as the proportion of true label information increases. The corresponding index also has a significant increase with semi-supervised algorithms. It shows that the inclusion of label information is present to improve the clustering effect.

5   Concluding Remarks

The existing methods can't adaptively classify special equipment and solve the noise problem in the original data. We propose a novel LSMVC algorithm. The algorithm in this paper can cluster according to the latent features of special equipment. The LRR is embedded to effectively reduce the impact of noise. The effectiveness of the algorithm is demonstrated by experimental results on real datasets. In addition, the effect of label sample proportion is also studied in Elevator datasets. The percentage of labeled samples in the dataset increases. This will enhance the clustering performance of the semi-supervised algorithm.

The algorithm still has some shortcomings. Such as the effect of view weights for multi-view data is not considered in the model. The special equipment dataset is also still too small. In the future, we will further study the optimization of the algorithm under adaptive weights and collect more special equipment datasets. To achieve better results in terms of special equipment safety warning.

References

1. Boyd, S., Boyd, S.P., Vandenberghe, L.: Convex optimization. Cambridge University Press (2004)
2. Boyd, S., Parikh, N., Chu, E.: Distributed optimization and statistical learning via the alternating direction method of multipliers. Now Publishers Inc (2011)
3. Brbic, M., Kopriva, I.: Multi-view low-rank sparse subspace clustering **73**, 247–258 (2018)
4. Cai, H., Liu, B., Xiao, Y., Lin, L.: Semi-supervised multi-view clustering based on constrained nonnegative matrix factorization. Elsevier **182**, 104798 (2021)
5. Cai, J.F., Candès, E.J., Shen, Z.: A singular value thresholding algorithm for matrix completion **20**(4), 1956–1982 (2010)
6. Cai, D., He, X., Han, J., Huang, T.S.: Graph regularized nonnegative matrix factorization for data representation **33**(8), 1548–1560 (2010)
7. Lee, D.D., Seung, H.S.: Algorithms for non-negative matrix factorization. In: Proceedings of the 13th International Conference on Neural Information Processing Systems, pp. 535–541. NIPS'00, MIT Press, event-place: Denver CO (2000)
8. Liu, G., Lin, Z., Yu, Y.: Robust subspace segmentation by low-rank representation. In: Proceedings of the 27th International Conference on International Conference on Machine Learning, pp. 663–670. ICML'10, Omnipress, event-place: Haifa, Israel (2010)
9. Liu, H., Wu, Z., Cai, D., Huang, T.S.: Constrained nonnegative matrix factorization for image representation **34**(7), 1299–1311 (2011)
10. Liu, J., Jia, Z.: Study on elevator-maintenance cloud platform for intelligent management and control based on big data **030**(16), 39–42,64 (2019)
11. Liu, J., Wang, C., Gao, J., Han, J.: Multi-view clustering via joint nonnegative matrix factorization. In: Proceedings of the 2013 SIAM International Conference on Data Mining, pp. 252–260. Society for Industrial and Applied Mathematics (2013)
12. Liu, L.: Intelligent monitoring and diagnosis system of elevator brake based on information fusion. In: 2014 National Special Equipment Safety and Energy Conservation Academic Conference (2014)
13. Xia, R., Pan, Y., Du, L., Yin, J.: Robust multi-view spectral clustering via low-rank and sparse decomposition. In: Proceedings of the AAAI Conference on Artificial Intelligence, vol. 28, no. 1, p. 7 (2014)
14. Xu, B., Li, L., Zhong, L.: Design of intelligent elevator analysis pre-alarming platform based on big-data (41), 359–362 (2017)
15. Zeng, M.: Design and implementation of main factor analysis of elevator failure and early warning subsystem of national special equipment safety supervision system (2018)
16. Zhan, K., Nie, F., Wang, J., Yang, Y.: Multiview consensus graph clustering **28**(3), 1261–1270 (2019)
17. Zhan, K., Zhang, C., Guan, J., Wang, J.: Graph learning for multiview clustering **48**(10), 2887–2895 (2018)

Single-Skeleton and Dual-Skeleton Hypergraph Convolution Neural Networks for Skeleton-Based Action Recognition

Changxiang He[1]([✉])(iD), Chen Xiao[1], Shuting Liu[1], Xiaofei Qin[2], Ying Zhao[2], and Xuedian Zhang[2]

[1] College of Science, University of Shanghai for Science and Technology, Shanghai, China
[2] School of Optical-Electrical and Computer Engineering, University of Shanghai for Science and Technology, Shanghai, China

Abstract. In the last several years, the graph convolutional networks (GCNs) have shown exceptional ability on skeleton-based action recognition. Currently used mainstream methods often include identifying the movements of a single skeleton and then fusing the features. But in this way, it will lose the interactive information of two skeletons. Moreover, since there are some interactions between people (such as handshake, high-five, hug, etc.), the loss will reduce the accuracy of skeleton-based action recognition. To address this issue, we propose a two-stream approach (SD-HGCN). On the basis of single-skeleton stream (S-HGCN), a dual-skeleton stream (D-HGCN) is added to recognizing actions with interactive information between skeletons. The model mainly includes a multi-branch inputs adaptive fusion module (MBAFM) and a skeleton perception module (SPM). MBAFM can make the input features more distinguishable through two GCNs and an attention module. SPM may identify relationships between skeletons and build topological knowledge about human skeletons, through adaptive learning of the hypergraph distribution matrix based on the semantic information in the skeleton sequence. The experimental results show that the D-HGCN consumes less time and has higher accuracy, which meets the real-time requirements. Our experiments demonstrate that our approach outperforms state-of-the-art methods on the NTU and Kinetics datasets.

Keywords: Skeleton-based action recognition · Hypergraph convolution network · Adaptive fusion

1 Introduction

Action recognition is a relatively new area of study that has garnered considerable attention in academia in recent years. It is extensively utilized in human-computer

Supported by organization the Artificial Intelligence Program of Shanghai under Grant 2019-RGZN-01077.

T. Mantoro et al. (Eds.): ICONIP 2021, LNCS 13109, pp. 15–27, 2021.
https://doi.org/10.1007/978-3-030-92270-2_2

interaction, self-driving vehicles, intelligent medical treatment, and video surveillance. However, the technique of video-based [1] action recognition is computationally expensive, and the context of human activity is complex. When completing an action recognition task, it is frequently obstructed by a variety of factors, significantly increasing the complexity of the work. people can quickly and accurately obtain the 3D bone coordinates of the human body, so the skeleton-based action recognition task has been well developed.

Mercifully, with the development of visual sensors such as the Microsoft Kinect [1], people can quickly and accurately obtain the 3D bone coordinates of the human body, which are highly resistant to interference from the external environment, allowing for a well-developed skeleton-based action recognition task.

Section 2 will present the mostly skeleton-based action recognition technique. However, The existing mainly methods usually recognize single-skeleton actions separately and then fuse the features to fuse the features of the two skeletons to achieve the effect of recognition of the actions of the two skeletons. But in this way, it will lose the interactive information of two skeletons. Moreover, since there are some interactions between people (such as handshake, high-five, hug, etc.), the loss will reduce the accuracy of skeleton-based action recognition.

To overcome these problems, we propose a single-skeleton and dual-skeleton hypergraph convolution neural network (SD-HGCN), which contains a single-skeleton stream (S-HGCN) and a dual-skeleton stream (D-HGCN). By examining the relationships between various hierarchical information in the skeleton sequence and between numerous skeletons, we build a multi-scale hierarchical network. On the one hand, we propose a multi-branch inputs adaptive fusion module (MBAFM), which retains the underlying information for each branch input using two GCNs and then combines the information for many branch inputs using an attention module to make input characteristics more distinct. On the other hand, we offer a skeleton perception module (SPM) to represent the skeleton graph's topological information, which can discover the relationship between skeletons and construct the topological information of human skeletons through adaptive learning of the hypergraph distribution matrix according to the skeleton sequence's semantic information. Our work makes the following significant contributions: (1) We design D-HGCN to recognizing actions with interactive information between skeletons. The D-HGCN consumes less time and has higher accuracy, which meets the real-time requirements. (2) We propose MBAFM and SPM to make the input features more distinguishable and discover the relationship between skeletons. (3) Experiments show that our approach outperforms currently available state-of-the-art methods.

2   Related Works

2.1   Skeleton-Based Action Recognition

The majority of early research on skeleton-based action recognition relied on handmade features generated [2–4] by relative 3D rotation and translation of joints. Deep learning has transformed action recognition by providing methods for increasing robustness [1] and achieving previously unheard-of performance. The majority of methods in this category depend on various elements of skeleton data: (1) The recurrent neural network (RNN) approach [5–7] treats the input skeletal data as a time series by using the order of joint coordinates. (2) The convolutional neural network (CNN) [8,9] approach is complementary to RNN-based methods in that it makes use of spatial information. Indeed, the 3D skeleton sequence is translated to a pseudo-image that reflects the rows' and columns' temporal dynamics and skeleton joints, respectively. (3) The graph neural network (GNN) [10–14] approach makes use of the information stored in the inherent topological structure of the human skeleton, as well as geographical and temporal data. The last method has been shown to be the most expressive of the three. Among these, the Spatio-Temporal Graph Convolutional Network (ST-GCN) [10] is the first model that accurately captures the balance of spatiotemporal dependency (ST-GCN).

2.2 Graph Neural Networks

Graph deep learning [15] is an umbrella term that encompasses all new technologies that apply deep learning models to non-Euclidean domains such as graphs. The concept of GNN was initially outlined by [16] and further elaborated by [17]. The notion behind GNN is that nodes in the graph represent things or ideas, while edges reflect their connections. Convolution was subsequently expanded from grids to graph data as a result of the success of convolutional neural networks. GNN iteratively processes the graph, representing each node as a consequence of applying a transformation to the node's and its neighbors' characteristics. On the image, the first CNN formula is courtesy of [18], who use spectral construction to generalize the convolution to the signal. This method had computational shortcomings that were addressed in [19] and [20]. It was simplified and built upon further by [21]. A related method is the spatial method, which is described as information aggregation through graph convolution [22–24]. We use the spectral structure suggested by [21] in this study.

3 Method

This section covers our proposed SD-HGCN, MBAFM, and SPM.

3.1 Skeleton Sequences Representation

The joint data consists of the human joints' original coordinate sequences. We use the bone modeling method described in [13]. Each bone is treated as a vector; given a bone with a source joint $v_1 = (x_1, y_1, z_1)$ and a target joint $v_2 = (x_2, y_2, z_2)$, the bone's vector is computed as $e_{v1,v2} = (x_2 - x_1, y_2 - y_1, z_2 - z_1)$. However, since the number of bones is fewer than the number of joints, we add an empty bone with a value of 0 to the center joint. Thus, we may associate each bone with a distinct joint by concatenating the joint position sequence J_p and the bone position sequence B_p to get the position sequence $X_p = J_p||B_p$. The motion sequence X_m is the difference between each of the position sequence's two consecutive frames. To guarantee that the motion and position sequences have the same number of frames, we fill the first frame of the motion sequence with a series of zeros.

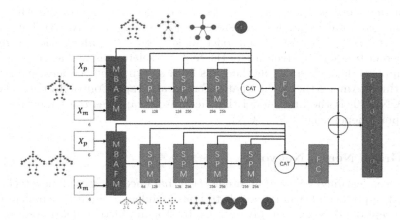

Fig. 1. We propose a description of the SD-HGCN architecture. MBAFM is a multi-branch inputs adaptive fusion module. SPM is a skeleton prediction module. Two numbers indicate the number of input and output channels, respectively. FC is a fully connected module. CAT refers to the concatenation of features from several layers in order to create a stable representation of the activity. $\oplus$ denotes adding the scores of the two streams and getting the final prediction.

3.2   SD-HGCN

Figure 1 depicts a high-level overview of our proposed SD-HGCN. The proposed SD-HGCN has two streams providing data on the human skeleton's single and dual skeletons. We input X_p and X_m as two branches into the network of each stream. For a single-skeleton stream, We use one MBAFM and three SPMs to learn action representation. Since the number of joints of the dual-skeleton is twice that of the single-skeleton, the dual-skeleton stream has one more SPM than the single-skeleton stream. We integrate the scores from the two streams into a weighted total throughout the experiment to get the final prediction during the test.

3.3   MBAFM

The input of multiple branches of the model is one method to enhance the model's impact. The previous method is to first perform two or three graph

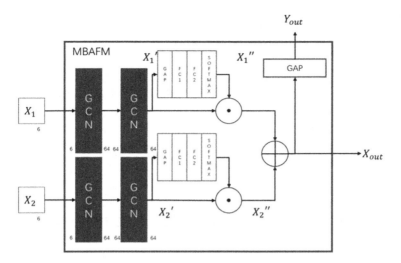

Fig. 2. The illustration of MBAFM. GCN is the module for spatial-temporal graph convolution. Two numbers indicate the number of input and output channels, respectively. GAP is an acronym for global average pooling. The symbol ⊙ indicates element-by-element multiplication. The symbol ⊕ indicates element-by-element addition.

convolutional layers on multiple branches respectively, and then concatenate or mean multiple branches to achieve the effect of multi-branch inputs fusion. We improve the operation of multi-branch inputs fusion and propose our multi-branch inputs adaptive fusion module. This module can complete the adaptive fusion of multiple branch inputs.

As shown in Fig. 2, we present the MBAFM for efficiently fusing multiple branch inputs. The module is composed of two components. One half component utilizes graph convolution to learn the characteristics of bodily parts, while the other component refines the features using the attention method. The first component takes into account the skeleton's input $X_i \in R^{C \times T \times V}$, $i \in \{1,2\}$, and the up-sample output $X'_i \in R^{C' \times T \times V}$ through two GCNs,

$$X'_i = GCN\left(GCN\left(X_i\right)\right).$$

The second component uses the Part-wise Attention (PartAtt) [25] to process the average feature of each group. PartAtt, based on the global context feature map, can determine the relative significance of various groups and improve the spatio-temporal feature downsampling. The second component may be expressed as follows:

$$X''_i = X'_i \odot Softmax\left(\sigma\left(GAP\left(X'_i\right)W_1\right)W_2\right), \tag{1}$$

where ⊙ is element-wise multiplication, σ represent the ReLU activation function, W_1, W_2 are trainable parameter matrices, and GAP is a global pooling module, which aggregates the graph into one node by mean function. Finally, we merge the features of multiple branches in a summation manner,

$$X_{out} = X_1'' \oplus X_2'' \quad and \quad Y_{out} = GAP\left(X_{out}\right).$$

The MBAFM module in our model combines several branches, thus expanding the spatial receptive field and extracting valuable discriminative information.

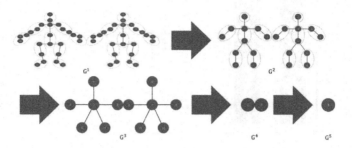

Fig. 3. A schematic diagram of the SPM of the NTU-RGB+D human dual-skeleton graph. There are a total of 5 scales of skeleton diagrams, from left to right representing G^1, G^2, G^3, G^4 and G^5. The vertices in the blue circle are combined into one vertex according to the weighted sum. Each joint has its own weight to merge vertices.

3.4   SPM

This module downsamples graph data to obtain subgraphs at various levels and is driven by the data according to the semantic information contained in the skeleton data to enable learning the non-physical connections between body parts, particularly the non-physical connections between different skeleton.

Assume that the graph of the human skeleton G is split into k components. That is, we subdivide the G skeleton network with n nodes into k subgraphs $g_i, i \in \{1, 2, \ldots, k\}$ corresponding to the k body components. Each subgraph gi is composed of a collection of joints Ji. Construct an assignment matrix S according to the subgraph's division, where Si, j is defined as:

$$S_{i,j} = \begin{cases} 1, & if \quad j \in J_i. \\ 0, & if \quad j \notin J_i. \end{cases}$$

As shown in Fig. 3, we perform three SPMs on a single-skeleton graph composed of 25 nodes, and obtain three new graphs, where the number of super nodes is 11, 6 and 1, respectively. Due to the fact that the dual-skeleton graph has twice the amount of nodes as the single-skeleton graph, we conduct four SPMs on the 50-node double-skeleton graph to produce four new graphs with a total of 22, 12, 2, and 1 super node. We may get a pooling matrix $S^p \in R^{V \times K}$ based on the partition methods to reflect how we manually arrange V nodes into K groups. In the dual-skeleton graph, we can see that $S^2 \in R^{50 \times 22}$, $S^3 \in R^{26 \times 12}$, $S^4 \in R^{12 \times 2}$ and $S^5 \in R^{2 \times 1}$. We use A^p to denote the adjacency matrix of the

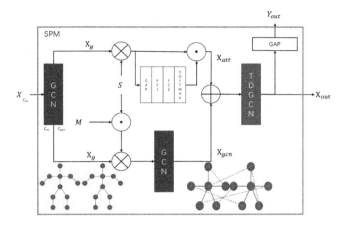

Fig. 4. The illustration of SPM. GCN is the module for spatial-temporal graph convolution. Two numbers indicate the number of input and output channels, respectively. TDGCM is a block of spatiotemporal graph convolution with time dilation. The symbol $\otimes$ indicates matrix multiplication. The symbol $\odot$ indicates element-by-element multiplication. The symbol $\oplus$ indicates element-by-element addition.

p-th hierarchical graph G^p, $A^1 \in R^{25 \times 25}$ is the adjacency matrix of predefined skeleton graph G^1 according to physical connections of human articulations. We can get the new adjacency matrix A^p for the pooled graph G^p using A^{p-1} and S^p in the following manner:

$$A^p = (S^p)^T A^{p-1} S^p,$$

we use $\hat{S}^p \in R^{K \times V}$ to denote the normalized matrix of pooling matrix $(S^p)^T$.

As shown in Fig. 4, we propose the SPM to realize the efficient merging operation of the multi-scale skeleton graph. In the first component, for a given features input $X \in R^{C_{in} \times T \times V}$, we can get the output $X_g \in R^{C_{out} \times T \times V}$ through one GCN,

$$X_g = GCN(X).$$

In the second component, given features X_g, We pool these V joints into K groups as follows:

$$X_{gcn} = \left(\hat{S}^p \odot M \right) X_g W,$$

where $\odot$ is element-wise multiplication, $M \in R^{K \times V}$ is the trainable weight for contribution of joint V in group K, $W \in R^{C_{in} \times C_{out}}$ is the weight of the convolution operation. We use PartAtt in the third component to work on the average characteristics of each group. The third component may be stated as follows:

$$X_{att} = X_g \hat{S}^p \odot Softmax \left(\sigma \left(GAP \left(X_g \hat{S}^p \right) W_1 \right) W_2 \right),$$

where $\odot$ is element-wise multiplication, σ represent the ReLU activation function, and W_1, W_2 are trainable parameter matrices. Finally, we merge the features of X_{gcn} and X_{att} in a summation manner, and we input W_1, W_2 into the TDGCN, which is the spatialtemporal graph convolution module with temporal dilated convolution,

$$X_{out} = TDGCN(X_{gcn} \oplus X_{att}) \quad and \quad Y_{out} = GAP(X_{out}).$$

The SPM can enlarge the spatial receptive field and extract valuable discriminative information in our model.

4   Experiments

4.1   Datasets

NTU-RGB+D Datasets. The NTU-RGB+D datasets [26] are one of the most comprehensive collections of indoor motion recognition data available. It includes 56,880 sequences divided into 60 action classes that were recorded concurrently by three Microsoft Kincet v2 cameras positioned at $-45°$, $0°$, and $45°$. These activities were carried out by forty distinct individuals ranging in age from ten to thirty-five years. The skeletal data is provided by the three-dimensional spatial coordinates of each object's 25 joints. This data collection is evaluated using two distinct protocols: (1) **Cross-Subject (X-Sub):** The training set has 40,320 samples taken from 20 individuals, whereas the test set contains 16,560 samples. (2) **Cross-View (X-View):** Divide the 37,290 samples gathered by the No. 2 and No. 3 cameras into a training set and the 18,960 samples gathered by the No. 1 camera into a training set.

NTU-RGB+D120 Datasets. The NTU-RGB+D120 datasets [27] are an extension of the NTU-RGB+D datasets. It includes 114,480 skeletal samples classified into 120 activity types. These samples were taken from 106 individuals using 32 different camera settings. As with NTU-RGB+D, it is suggested that the evaluation index for these datasets be run in two modes: (1) **Cross-Subject (X-Sub120):** The 53 individuals' samples are utilized for training, while the remaining samples are used for testing. The training set has 63,026 samples, whereas the test set contains 50,922 samples. (2) **Cross-Setup (X-Set120):** The samples acquired by camera sets with even IDs are utilized for training purposes, while the remaining samples are used for testing purposes. The training set has 54,471 and the test set contains 59,477 samples, respectively.

Kinetics Datasets. Kinetics [28] is a massive data collection on human behavior that includes 300,000 video clips from 400 different categories that were collected from YouTube. Due to the absence of skeletal data in the original video clip, [10] utilized the publicly accessible Open-Pose toolbox [29] to estimate the location of the clip's 18 joints each frame. We test our model using their provided data (Kinetics-Skeleton), which consists of 240,000 movies split into a training set and 20,000 videos used for validation.

4.2 Implementation Details

Our experiments are performed on the Pytorch deep learning framework with an NVIDIA GTX 3090 GPU. Use Stochastic Gradient Descent (SGD) with Nesterov momentum in our experiment as the optimization strategy of our method. The batch size is 64, the weight decay is 0.0001, and the initial learning rate is 0.1. The training process includes a total of 60 epochs. At the 25th, 35th, and 45th epochs, the learning rate is reduced to 0.02, 0.004, 0.0008. This cross entropy is used as the loss function.

4.3 Ablation Study

To validate the efficacy of the SD-HGCN module, we perform the following tests on the NTU RGB+D dataset using the Cross-View benchmark.

Table 1. Evaluating the effectiveness of proposed MBAFM module on the NTU-RGB+D dataset. (%)

Models	MBAFM	X-view
S-HGCN		95.4
D-HGCN		95.5
SD-HGCN		96.3
S-HGCN	✓	95.7
D-HGCN	✓	95.9
SD-HGCN	✓	**96.7**

Table 2. Evaluating the effectiveness of proposed SPM module on the NTU-RGB+D dataset. (%)

Models	Graph	X-view
S-HGCN	25,11,6,1	95.7
D-HGCN(3spm)	50,22,12,2	95.8
SD-HGCN		96.4
S-HGCN	25,11,6,1	95.7
D-HGCN(4spm)	50,22,12,2,1	95.9
SD-HGCN		**96.7**

Table 3. The average running time (ms) of per action sample on the NTU-RGB+D dataset.

Models	Graph	X-sub		X-view	
		Train	Val	Train	Val
S-HGCN	25,11,6,1	~19	~10	~19	~10
D-HGCN(3spm)	50,22,12,2	~14	~6	~14	~6
D-HGCN(4spm)	50,22,12,2,1	~15	~7	~15	~7

Evaluating of MBAFM. In order to analyze the effectiveness of MBAFM, we design the following test. As shown in Table 2, experimental data shows that by deleting the entire MBAFM we propose, the recognition accuracy is reduced by 0.4%, which shows that MBAFM are helpful for action recognition tasks.

Evaluating of SPM. In order to analyze the effectiveness of SPM, we design the following test. As shown in the Table 1, the experimental data shows that when S-HGCN and D-HGCN have the same SPM number, the accuracy of D-HGCN is 0.1% higher than that of S-HGCN. After adding an SPM module to D-HGCN, the accuracy of SD-HGCN has increased by 0.3%. This shows that SPM and interaction information between skeletons are helpful for action recognition tasks.

Time for Per Action Sample Recognition. S-HGCN is modified based on the [11], and the time to run each action sample is similar to that of SGCN. The dual-skeleton model D-HGCN improve accuracy relative to S-HGCN, and the time taken to run each action sample has also been reduced a lot, which shows that D-HGCN meets the real-time requirements, as shown in Table 3.

4.4   Comparisons with the State-of-the-Art Methods

As indicated in Table 4, our SD-HGCN is compared to state-of-the-art techniques. The accuracy of our method's Cross-Subject protocol and Cross-View protocol are 90.9% percent and 96.7% percent, respectively, for the NTU-RGB+D dataset. The accuracy of our method's Cross-Subject120 and Cross-Setup120 protocols is 87.0% percent and 88.2% percent, respectively, for the

Table 4. The action recognition performance comparison on the NTU-RGB+D datasets shows the classification accuracy of the X-sub and X-view benchmarks; the action recognition performance comparison on the NTU-RGB+D120 datasets shows the classification accuracy of X-sub120 and X-set120 benchmarks; the action recognition performance comparison on the Kinetics-Skeleton datasets shows the classification accuracy of top-1 and top-5 benchmarks.

Models	Years	NTU-RGB+D		NTU-RGB+D120		Kinetics	
		X-sub	X-view	X-sub120	X-set120	Top-1	Top-5
ST-GCN [10]	2018	81.5	88.3	70.7	73.2	30.7	52.8
AS-AGCN [30]	2019	86.8	94.2	77.9	78.5	34.8	56.5
2s-AGCN [13]	2019	88.5	95.1	82.5	84.2	36.1	58.7
RA-GCN [31]	2020	87.3	93.6	81.1	82.7	-	-
MSTGNN [32]	2021	**91.3**	95.9	87.4	87.6	-	-
2s-SGCN [11]	2021	90.1	96.2	-	-	37.1	60.0
S-HGCN(our)	-	89.2	95.7	85.8	86.4	36.3	59.2
D-HGCN(our)	-	89.8	95.9	86.3	86.9	36.8	59.6
SD-HGCN(our)	-	90.9	**96.7**	**87.0**	**88.2**	**37.4**	**60.5**

NTU-RGB+D120 dataset. The validation set's top-1 and top-5 accuracy percentages are 37.4% percent and 60.5% percent, respectively, for the kinetics-skeleton dataset. Although these are primarily methods for identifying multiple skeleton movements independently, they examine just the connection between the joints in each skeleton and disregard the interaction between the skeletons. We mix the two to get beneficial results.

5 Conclusion

Mainstream methods often include fusing the movements of a single skeleton and then fusing them together. But this approach loses interactive information of two skeletons, which is needed for skeleton-based action recognition. We propose a two-stream approach (SD-HGCN) on the basis of single-skeleton stream and dual-skeleton stream. In comparison to other mainstream methods, tests on large-scale skeleton-based action recognition datasets demonstrate that our suggested method offers a number of benefits. Additionally, based on the positive impact of the dual-skeleton model, we think that if the whole skeleton sequence is considered as a single graph, actions may be identified more effectively using a graph neural network.

References

1. Lei, W., Du, Q.H., Koniusz, P.: A comparative review of recent kinect-based action recognition algorithms. IEEE Trans. Image Process. **29**, 15–28 (2019)
2. Hu, J.F., et al.: Jointly learning heterogeneous features for RGB-D activity recognition. In: 2015 IEEE Conference on Computer Vision and Pattern Recognition (CVPR). IEEE, (2015)
3. Vemulapalli, R., Arrate, F., Chellappa, R.: Human action recognition by representing 3D skeletons as points in a lie group. In: 2014 IEEE Conference on Computer Vision and Pattern Recognition (CVPR) IEEE Computer Society (2014)
4. Hussein, M.E., et al.: Human action recognition using a temporal hierarchy of covariance descriptors on 3D joint locations. In: International Joint Conference on Artificial Intelligence (2013)
5. Lev, G., Sadeh, G., Klein, B., Wolf, L.: RNN fisher vectors for action recognition and image annotation. In: Leibe, B., Matas, J., Sebe, N., Welling, M. (eds.) Computer Vision. LNCS, vol. 9910. Springer, Cham (2016). https://doi.org/10.1007/978-3-319-46466-4_50
6. Wang, H., and L. Wang.: IEEE 2017 conference on computer vision and pattern recognition (CVPR) - Honolulu, HI. In: 2017 IEEE Conference on Computer Vision and Pattern Recognition (CVPR) - Modeling Temporal Dynamics and Spatial Configurations of Actions Usi, 3633–3642 (2017)
7. Liu, J., et al.: Global context-aware attention LSTM networks for 3D action recognition. In: IEEE Conference on Computer Vision and Pattern Recognition. IEEE (2017)

8. Chéron, G., Laptev, I., Schmid, C.: P-CNN: Pose-based CNN features for action recognition. IEEE (2015)
9. Simonyan, K., Zisserman, A.: Two-stream convolutional networks for action recognition in videos. Adv. Neural Inf. Process. Syst. (2014)
10. Yan, S., Xiong, Y.: Temporal graph convolutional networks for skeleton-based action recognition, Lin. (2018)
11. Yang, W. J., Zhang, J. L.: Shallow graph convolutional network for skeleton-based action recognition (2021)
12. Shi, L., et al.: Skeleton-based action recognition with directed graph neural networks. In: 2019 IEEE/CVF Conference on Computer Vision and Pattern Recognition (CVPR). IEEE (2020)
13. Shi, L., et al.: Two-stream adaptive graph convolutional networks for skeleton-based action recognition (2019)
14. Cheng, K., et al.: Skeleton-based action recognition with shift graph convolutional network. In: 2020 IEEE/CVF Conference on Computer Vision and Pattern Recognition (CVPR). IEEE (2020)
15. Bronstein, M.M., Bruna, J., LeCun, Y., Szlam, A., Vandergheynst, P.: Geometric deep learning: going beyond euclidean data. IEEE Signal Process. Mag. **34**(4), 18–42 (2017)
16. Gori, M., Monfardini, G., Scarselli, F.: A new model for learning in graph domains. In: 2005 Proceedings of IEEE International Joint Conference on Neural Networks, vol. 2, pp. 729–734 (2005)
17. Scarselli, F., Gori, M., Tsoi, A.C., Hagenbuchner, M., Monfardini, G.: The graph neural network model. IEEE Trans. Neural Netw. **20**(1), 61–80 (2008)
18. Bruna, J., Zaremba, W., Szlam, A., Lecun, Y.: Spectral networks and locally connected networks on graphs. In: International Conference on Learning Representations (ICLR2014), CBLS (2014)
19. Henaff, M., Bruna, J., LeCun, Y.: Deep convolutional networks on graph-structured data (2015). arXiv preprint arXiv:1506.05163
20. Defferrard, M., Bresson, X., Vandergheynst, P.: Convolutional neural networks on graphs with fast localized spectral filtering. Adv. Neural Inform. Process. Syst. 3844–3852 (2016)
21. Kipf, T.N., Welling, M.: Semi-supervised classification with graph convolutional networks. In: 5th International Conference on Learning Representations, ICLR (2017)
22. Micheli, A.: Neural network for graphs: a contextual constructive approach. IEEE Trans. Neural Netw. **20**(3), 498–511 (2009)
23. Niepert, M., Ahmed, M., Kutzkov, K.: Learning convolutional neural networks for graphs. In: International Conference on Machine Learning, pp. 2014–2023 (2016)
24. Such, F.P., et al.: Robust spatial filtering with graph convolutional neural networks. IEEE J. Sel. Top. Sign. Proces. (2017)
25. Song, Y. F., et al.: Stronger, faster and more explainable: a graph convolutional baseline for skeleton-based action recognition. ACM (2020)
26. Shahroudy, A., et al.: NTU-RGB+D: a large scale dataset for 3D human activity analysis. IEEE Comput. Soc. 1010–1019 (2016)
27. Liu, J., et al.: NTU-RGB+D 120: a large-scale benchmark for 3D human activity understanding (2019)
28. Kay, W., et al.: The kinetics human action video datasets (2017)
29. Cao, Z., et al.: Realtime multi-person 2D pose estimation using part affinity fields. IEEE Trans. Pattern Anal. Mach. Intell. (2018)

30. Li, M., Chen, S., Chen, X., Zhang, Y., Wang, Y., Tian, Q.: Actional-structural graph convolutional networks for skeleton-based action recognition. In: IEEE/CVF Conference on Computer Vision and Pattern Recognition (CVPR) (2019)
31. Song, Y.F., Zhang, Z., Shan, C., Wang, L.: Richly activated graph convolutional network for robust skeleton-based action recognition. IEEE Trans. Circuits Syst. Video Technol. (2020)
32. Feng, D., et al.: Multi-scale spatial temporal graph neural network for skeleton-based action recognition. IEEE Access (2021)

A Multi-Reservoir Echo State Network with Multiple-Size Input Time Slices for Nonlinear Time-Series Prediction

Ziqiang Li[1]([✉]) [iD] and Gouhei Tanaka[1,2]([✉]) [iD]

[1] Department of Electrical Engineering and Information Systems, Graduate School of Engineering, The University of Tokyo, Hongo, Bunkyo-ku, Tokyo 113-8656, Japan
{ziqiang-li,gtanaka}@g.ecc.u-tokyo.ac.jp
[2] International Research Center for Neurointelligence, The University of Tokyo, Hongo, Bunkyo-ku, Tokyo 113-0033, Japan

Abstract. A novel multi-reservoir echo state network incorporating the scheme of extracting features from multiple-size input time slices is proposed in this paper. The proposed model, Multi-size Input Time Slices Echo State Network (MITSESN), uses multiple reservoirs, each of which extracts features from each of the multiple input time slices of different sizes. We compare the prediction performances of MITSESN with those of the standard echo state network and the grouped echo state network on three benchmark nonlinear time-series datasets to show the effectiveness of our proposed model. Moreover, we analyze the richness of reservoir dynamics of all the tested models and find that our proposed model can generate temporal features with less linear redundancies under the same parameter settings, which provides an explanation about why our proposed model can outperform the other models to be compared on the nonlinear time-series prediction tasks.

Keywords: Reservoir computing · Echo state network · Multi-size input time slices

1 Introduction

Nonlinear Time-series Prediction (NTP) [19] is one of the classical machine learning tasks. The goal of this task is to make predicted values close to the corresponding actual values. Recurrent Neural Networks (RNNs) [12] are a subset of Neural Networks (NNs) and have been widely used in nonlinear time-series prediction tasks. Many related works have reported that RNNs-based methods outperform other NNs-based methods on some prediction tasks [9,13]. However, the classical RNN model and its extended models such as Long Short-Term Memory (LSTM) [5] and Gated Recurrent Unit (GRU) [1] often suffer from expensive computational costs along with gradient explosion/vanishing problems in the training process.

Reservoir Computing (RC) [6,17] is an alternative computational framework which provides a remarkably efficient approach for training RNNs. The most

© Springer Nature Switzerland AG 2021
T. Mantoro et al. (Eds.): ICONIP 2021, LNCS 13109, pp. 28–39, 2021.
https://doi.org/10.1007/978-3-030-92270-2_3

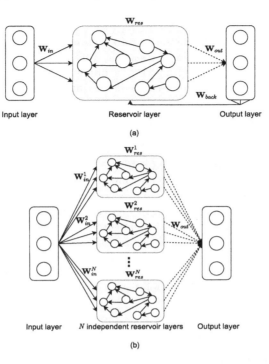

Fig. 1. (a) A standard ESN, (b) A standard GroupedESN.

important characteristic of this framework is that a predetermined non-linear system is used to map input data into a high-dimensional feature space. Based on this characteristic, a well-trained RNN can be built with relatively low computational costs.

As one of the important implementations of RC, Echo State Network (ESN), was first proposed in Ref. [6] and has been widely used to handle NTP tasks [8,14]. The standard architecture of ESN, including an input layer, a reservoir layer, and an output layer, is shown in Fig. 1(a). We can see that an input weight matrix, $\mathbf{W}_{in}$, represents the connection weights between the input layer and the reservoir layer. Moreover, a reservoir weight matrix denoted by $\mathbf{W}_{res}$ represents the connection weights between neurons inside the reservoir layer. The readout weight matrix, $\mathbf{W}_{out}$, represents the connection weights between the reservoir layer and the output layer. A feedback matrix from the output layer to the reservoir layer is denoted by $\mathbf{W}_{back}$. Typically, the element values of three matrices, $\mathbf{W}_{in}$, $\mathbf{W}_{res}$, and $\mathbf{W}_{back}$, are randomly drawn from certain uniform distributions and are kept fixed. Only $\mathbf{W}_{out}$ (dash lines) need to be trained by the linear regression.

Based on the above introduced ESN, we can quickly obtain a well-trained RNN. However, this simple architecture leads to a huge limitation in enhancing its representation ability and further makes the corresponding prediction performances on the NTP task hard to be improved. An effective remedy proposed in Ref. [4] is to feed the input time series into N independent randomly generated reservoirs for producing N different reservoir states, and then combine them to enrich the features used for training the output weight matrix. The authors of Ref. [4] called this architecture "GroupedESN" and reported that the prediction performances obtained by their proposed model is much better than those obtained by the standard ESN on some tasks. We show a schematic diagram of GroupedESN in Fig. 1(b).

The purpose of multi-reservoir ESN, including GroupedESN, inherited from the standard ESN is to extract features from each "data point" in a time series. In most of related works [4, 15], we found that they only used each sampling point in a time series as the "data point" and extracted the corresponding temporal feature from them in each reservoir. In fact, the scheme of extracting features from inseparable sampling points can capture the most fine-grained temporal dependency from the input series. However, this monotonous scheme unavoidably ignores some useful temporal information in the "time slices" composed of a period of continuous sampling points [20]. Figure 2 shows an example of transforming the original sampling points of a time series into several time slices of size two.

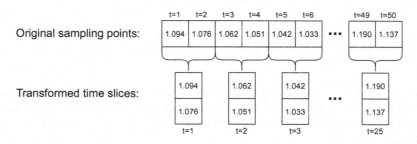

Fig. 2. An example of transforming original sampling points of a time series into time slices of size two.

In this paper, we propose a novel multi-reservoir ESN model, Multi-size Input Time Slice Echo State Network (MITSESN), which can extract various temporal features corresponding to input time slices of different sizes. We compare the proposed model with the standard ESN and the GroupedESN on the three NTP benchmark datasets and demonstrate the effectiveness of our proposed MITSESN. We provide an empirical analysis of richness in the reservoir-state dynamics to explain why our proposed model performs better than the other tested models on the NTP tasks.

The rest of this paper is organized as follows: We describe the details of the proposed model in Sect. 2. We report the experimental results, including results on three NTP benchmark datasets and corresponding analyses of richness in Sect. 3. We conclude this work in Sect. 4.

2   The Proposed Model: MITSESN

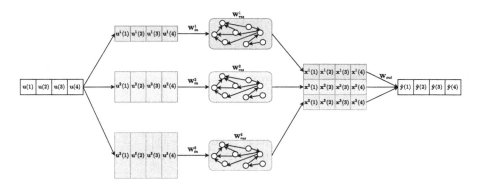

Fig. 3. An example of the proposed MITSESN with three independent reservoirs.

A schematic diagram of our proposed MITSESN is shown in Fig. 3. This is a case where an original input time series with length four is fed into the proposed MITSESN with three independent reservoirs. The original input time series is transformed into three time slices of different sizes. Then, each time slice is fed into the corresponding reservoir and the generated reservoir states are concatenated together. Finally, the concatenated state matrix is decoded to the desired values. Based on the above introduction, we can observe that our proposed MITSESN can be divided into three parts: the series-to-slice transformer, the multi-reservoir encoder, and the decoder. We introduce the details of these parts as below.

2.1   Series-to-Slice Transformer

We define the input vector and the target vector at time t as $\mathbf{u}(t) \in \mathbb{R}^{N_U}$ and $\mathbf{y}(t) \in \mathbb{R}^{N_Y}$, respectively. The length of input series and that of target series are denoted by N_T.

To formulate the transformation from the original input time-series points into input time slices of different sizes, we define the maximal size of the input slice used in the MITSESN as M. In our model, the maximal size of the input slice is equivalent to the number of different sizes. We denote the size of input slice by m, where $1 \le m \le M$. In order to keep the length of the transformed input time slice the same as those of the original input time series, we add zero paddings of length $(m-1)$ into the beginning of the original input series, which can be formulated as follows:

$$\mathbf{U}_{zp}^m = [\underbrace{\mathbf{0}, \dots, \mathbf{0}}_{m-1}, \mathbf{u}(1), \mathbf{u}(2), \dots, \mathbf{u}(N_T)], \tag{1}$$

where $\mathbf{U}_{zp}^m \in \mathbb{R}^{N_U \times (N_T + m - 1)}$ is the zero-padded input matrix. Based on the above settings, we can obtain the transformed input matrix corresponding to input time slices of size m as follows:

$$\mathbf{U}^m = [\mathbf{u}^m(1), \mathbf{u}^m(2), \dots, \mathbf{u}^m(N_T)], \qquad (2)$$

where $\mathbf{U}^m \in \mathbb{R}^{mN_U \times N_T}$ and $\mathbf{u}^m(t)$ is composed of the vertical concatenation of vectors from the t-th column to the $(t + m - 1)$-th column in $\mathbf{U}_{zp}^m$. We show an example of $\mathbf{U}^m$ when $m = 3$ as follows:

$$
\begin{aligned}
\mathbf{U}^3 &= \left[\mathbf{u}^3(1), \mathbf{u}^3(2), \dots, \mathbf{u}^3(N_T)\right] \\
&= \begin{bmatrix} \mathbf{0} & \mathbf{0} & \dots & \mathbf{u}(N_T - 2) \\ \mathbf{0} & \mathbf{u}(1) & \dots & \mathbf{u}(N_T - 1) \\ \mathbf{u}(1) & \mathbf{u}(2) & \dots & \mathbf{u}(N_T) \end{bmatrix}.
\end{aligned}
\qquad (3)
$$

2.2   Multi-reservoir Encoder

We adopt the basic architecture of GroupedESN in Fig. 1(b) to build the multi-reservoir encoder. However, the feeding strategy of the multi-reservoir encoder is different from that of GroupedESN. We assume that input time slices of size m are fed into the m-th reservoir. Therefore, there are totally M reservoirs in the multi-reservoir encoder. For the m-th reservoir, we define the input weight matrix and the reservoir weight matrix as $\mathbf{W}_{in}^m \in \mathbb{R}^{N_R^m \times mN_U}$ and $\mathbf{W}_{res}^m \in \mathbb{R}^{N_R^m \times N_R^m}$, respectively, where N_R^m represents the size of the m-th reservoir. The state of the m-th reservoir at time t, $\mathbf{x}^m(t)$, is calculated as follows:

$$\mathbf{x}^m(t) = (1 - \alpha)\mathbf{x}^m(t-1) + \alpha \tanh\left(\mathbf{W}_{in}^m \mathbf{u}^m(t) + \mathbf{W}_{res}^m \mathbf{x}^m(t-1)\right), \quad (4)$$

where the element values of $\mathbf{W}_{in}^m$ are randomly drawn from the uniform distribution of the range $[-\theta, \theta]$. The parameter θ is the input scaling. The element values of $\mathbf{W}_{res}^m$ are randomly chosen from the uniform distribution of the range $[-1, 1]$. To ensure the "Echo State Property" (ESP) [6], $\mathbf{W}_{res}^m$ should satisfy the condition described as follows:

$$\rho\left((1 - \alpha)\mathbf{E} + \alpha \mathbf{W}_{res}^m\right) < 1, \qquad (5)$$

where $\rho(\cdot)$ denotes the spectral radius of a matrix argument, the parameter α represents the leaking rate which is set in the range $(0, 1]$, and $\mathbf{E} \in \mathbb{R}^{N_R^m \times N_R^m}$ is the identity matrix. Moreover, we use the parameter η to denote the sparsity of $\mathbf{W}_{res}^m$.

We denote the reservoir-state matrix composed of N_T state vectors corresponding to the m-th reservoir as $\mathbf{X}^m \in \mathbb{R}^{N_R^m \times N_T}$. By concatenating M reservoir-state matrices in the vertical direction, we obtain a concatenated state matrix, $\mathbf{X} \in \mathbb{R}^{\sum_{m=1}^M N_R^m \times N_T}$, which can be written as follows:

$$\mathbf{X} = \left[\mathbf{X}^1; \mathbf{X}^2; \dots; \mathbf{X}^M\right]. \qquad (6)$$

2.3   Decoder

We use the linear regression for converting the concatenated state matrix into the output matrix, which can be formulated as follows:

$$\hat{\mathbf{Y}} = \mathbf{W}_{out}\mathbf{X}, \tag{7}$$

where $\hat{\mathbf{Y}} \in \mathbb{R}^{N_Y \times N_T}$ is the output matrix. The readout matrix $\mathbf{W}_{out}$ is given by the closed-form solution as follows:

$$\mathbf{W}_{out} = \mathbf{Y}\mathbf{X}^{\mathrm{T}} \left(\mathbf{X}\mathbf{X}^{\mathrm{T}} + \lambda\mathbf{I}\right)^{-1}, \tag{8}$$

where $\mathbf{Y} \in \mathbb{R}^{N_Y \times N_T}$ represents the target matrix, $\mathbf{I} \in \mathbb{R}^{\sum_{m=1}^{M} N_R^m \times \sum_{m=1}^{M} N_R^m}$ is an identity matrix, and the parameter λ symbolizes the Tikhonov regularization factor [18].

3   Numerical Simulations

In this section, we report the details and results of simulations. Specifically, three benchmark nonlinear time-series datasets and the corresponding task settings are described in Sec. 3.1, the evaluation metrics are listed in Sec. 3.2, the tested models and parameter settings are described in Sec. 3.3, the corresponding simulation results are presented in Sec. 3.4. The analyses of richness for all the tested models are given in Sec. 3.5.

3.1   Datasets Descriptions and Task Settings

We leverage three nonlinear time-series datasets, including the Lorenz system, MGS-17, and KU Leuven datasets, to evaluate the prediction performances of our proposed model. Glimpses of the above datasets are shown in Fig. 4. The partitions of the training set, the validation set, the testing set, and the initial transient set are listed in Table 1. We introduce the details of these datasets and task settings as below.

Lorenz System. The equation of Lorenz system [10] is formulated as follows:

$$\begin{aligned}
\frac{\mathrm{d}x}{\mathrm{d}t} &= \sigma(y - x), \\
\frac{\mathrm{d}y}{\mathrm{d}t} &= x(\delta - z) - y, \\
\frac{\mathrm{d}z}{\mathrm{d}t} &= xy - \beta z.
\end{aligned} \tag{9}$$

When $\delta = 28$, $\sigma = 10$, and $\beta = 8/3$, the system exhibits a chaotic behavior. In our evaluation, we used the chaotic Lorenz system and set the initial condition at $(x(0), y(0), z(0)) = (12, 2, 9)$. We adopted the sampling interval $\Delta t = 0.02$ and rescaled by the scaling factor 0.1, which is the same as those reported in [7]. We set a six-step-ahead prediction task on x values, which can be represented as $\mathbf{u}(t) = x(t)$ and $\mathbf{y}(t) = x(t + 6)$.

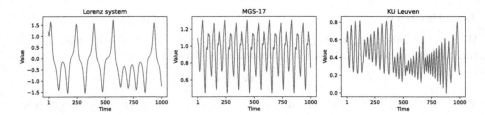

Fig. 4. Examples of three nonlinear time-series datasets.

MGS-17. The equation of Mackey-Glass system [11] is formulated as follows:

$$z(t+1) = z(t) + \delta \cdot \left(a \frac{z(t - \varphi/\delta)}{1 + z(t - \varphi/\delta)^n} - bz(t) \right), \tag{10}$$

where a, b, n, and δ are fixed at 0.2, -0.1, 10, and 0.1, respectively. The Mackey-Glass system exhibits a chaotic behavior when $\varphi > 16.8$. We kept the value of φ equal to 17 (MGS-17). The task on MGS-17 is to predict the 84-step-ahead value of z [7], which can be represented as $\mathbf{u}(t) = z(t)$ and $\mathbf{y}(t) = z(t + 84)$.

KU Leuven. KU Leuven dataset was first proposed in a time-series prediction competition held at KU Leuven, Belgium [16]. We set an one-step-ahead prediction task on this dataset for the evaluation.

Table 1. The partitions of Lorenz system, MGS-17, and KU Leuven datasets.

	Training set	Valiation set	Testing set	Initial transient set
Lorenz system	3000	1000	1000	500
MGS-17	3000	1000	1000	500
KU Leuven	2000	500	500	200

3.2   Evaluation Metrics

We use two evaluation metrics in this work, including Normalized Root Mean Square Error (NRMSE) and Symmetric Mean Absolute Percentage Error (SMAPE), to evaluate the prediction performances. These two evaluation metrics are formulated as follows:

$$\text{NRMSE} = \frac{\sqrt{\frac{1}{N_T} \sum_{t=1}^{N_T} (\hat{\mathbf{y}}(t) - \mathbf{y}(t))^2}}{\sqrt{\frac{1}{N_T} \sum_{t=1}^{N_T} (\mathbf{y}(t) - \bar{\mathbf{y}})^2}}, \tag{11}$$

$$\text{SMAPE} = \frac{1}{N_T} \sum_{t=1}^{N_T} \frac{|\hat{\mathbf{y}}(t) - \mathbf{y}(t)|}{(|\hat{\mathbf{y}}(t)| + |\mathbf{y}(t)|)/2}, \tag{12}$$

where $\bar{\mathbf{y}}$ denotes the mean of data values of $\mathbf{y}(t)$ from $t = 1$ to N_T.

3.3 Tested Models and Parameter Settings

In our simulation, we compared the prediction performances of our proposed model with those of ESN and GroupedESN. We denote the overall reservoir size $N_R = \sum_{m=1}^{M} N_R^m$ for all the models. Two architectures with $M = 2$ and $M = 3$ for GroupedESN and MITSESN were considered. We represent the architecture with M reservoirs as $N_R^1 - N_R^2 - \cdots - N_R^M$.

To make a fair comparison, we set N_R the same for each model. For simplicity, the size of each reservoir in the GroupedESN and the proposed MITSESN was kept the same. The parameter settings for all the tested models are listed in Table 2. The spectral radius, the sparsity of reservoir weights, and the Tikhonov regularization were set at 0.95, 90%, and 1E-06, respectively. The input scaling, the leaking rate and the overall reservoir size were searched in the ranges of [0.01, 0.1, 1], [0.1, 0.2, ..., 1], and [150, 300, ..., 900], respectively. For each setting, we averaged the results over 20 realizations.

Table 2. The parameter settings for all the tested models

Parameter	Symbol	Value
Spectral radius	ρ	0.95
Sparsity of reservoir weights	η	90%
Tikhonov regularization	λ	1E-06
Input scaling	θ	$[0.01, 0.1, 1]$
Leaking rate	α	$[0.1, 0.2, \ldots, 1]$
Overall reservoir size	N_R	$[150, 300, \ldots, 900]$

3.4 Simulation Results

We report the averaged prediction performances on the three datasets in Tables 3, 4 and 5. It is obvious that our proposed MITSESN with three reservoirs obtains the smallest NRMSE and SMAPE among all the tested models with the same overall reservoir size. By comparing the prediction performances of the GroupedESN with those of our proposed MITSESN, we can clearly find that the strategy of extracting temporal features from multi-size input time slices can significantly improve the prediction performances. Moreover, with the increase of the size of input time slices, the performance is obviously improved. Especially, our simulation results on MGS-17 show that only adding more reservoirs is not a universally effective method to improve prediction performances for the GroupedESN. Lastly, we observe that the best prediction performances of all the tested models are obtained under the maximal values in the searching range of input scaling and reservoir size, which indicates that all the models benefit from high richness [3]. We investigate how this important characteristic changes under different N_R for all the tested models in the following section.

Table 3. Average performances of the six-step-ahead prediction task on the Lorenz system.

Models	Architecture	NRMSE	SMAPE	Best parameters
ESN	900	7.44E-05	1.21E-04	$N_R = 900$, $\theta = 1$, $\alpha = 0.4$
GroupedESN ($M = 2$)	450-450	7.10E-05	1.04E-04	$N_R = 900$, $\theta = 1$, $\alpha = 0.4$
GroupedESN ($M = 3$)	300-300-300	7.04E-05	1.13E-04	$N_R = 900$, $\theta = 1$, $\alpha = 0.4$
MITSESN ($M = 2$)	450-450	6.29E-05	1.16E-04	$N_R = 900$, $\theta = 1$, $\alpha = 0.5$
MITSESN ($M = 3$)	300-300-300	**5.58E-05**	**9.23E-05**	$N_R = 900$, $\theta = 1$, $\alpha = 0.4$

Table 4. Average performances of the 84-step-ahead prediction task on the MGS-17.

Models	Architecture	NRMSE	SMAPE	Best parameters
ESN	900	1.87E-02	3.94E-03	$N_R = 900$, $\theta = 1$, $\alpha = 0.2$
GroupedESN ($M = 2$)	450-450	2.05E-02	4,14E-03	$N_R = 900$, $\theta = 1$, $\alpha = 0.3$
GroupedESN ($M = 3$)	300-300-300	2.16E-02	4.48E-03	$N_R = 900$, $\theta = 1$, $\alpha = 0.3$
MITSESN ($M = 2$)	450-450	1.49E-02	3.12E-03	$N_R = 900$, $\theta = 1$, $\alpha = 0.2$
MITSESN ($M = 3$)	300-300-300	**1.24E-02**	**2.41E-03**	$N_R = 900$, $\theta = 1$, $\alpha = 0.3$

3.5   Analysis of Richness

The richness is a desirable characteristic in the reservoir state as suggested by Ref. [2]. Typically the higher richness indicates the less redundancy held in the reservoir state. We leverage the Uncoupled Dynamics (UD) proposed in [3] to measure the richness of $\mathbf{X}$ for all the tested models. The UD of $\mathbf{X}$ is calculated as follows:

$$\arg\min_{d} \left\{ \sum_{k=1}^{d} R_k \mid \sum_{k=1}^{d} R_k \geq \mathcal{A} \right\}, \tag{13}$$

where $\mathcal{A}$ is in the range of $(0, 1]$ and represents the desired ratio of explained variability in the concatenated state matrix. We kept $\mathcal{A} = 0.9$ in the following evaluation. R_k denotes the normalized relevance of the i-th principal component, which can be formulated as follows:

$$R_i = \frac{\sigma_i}{\sum_{j=1}^{N_R} \sigma_j}, \tag{14}$$

where σ_i denotes the i-th singular value in the decreasing order. The higher the value of UD in Eq. (13) is, the less linear redundancy held in the concatenated state matrix $\mathbf{X}$ is. For the evaluation settings, we used a univariable time series of length 5000 and we randomly chose each value from the uniform distribution of the range $[-0.8, 0.8]$. We fixed the leaking rate $\alpha = 1$ and input scaling $\theta = 1$ in all the models.

The average UDs of all the tested models when varying N_R from 150 to 900 are shown in Fig. 5. It is obvious that the MITSESN ($M=3$) outperforms the other models when varying N_R from 300 to 900. With the increase of N_R (from

Table 5. Average performances of the one-step-ahead prediction task on the KU Leuven dataset.

Models	Architecture	NRMSE	SMAPE	Best parameters
ESN	900	2.31E-02	6.94E-03	$N_R = 900, \theta = 1, \alpha = 0.5$
GroupedESN ($M = 2$)	450-450	2.29E-02	7.18E-03	$N_R = 900, \theta = 1, \alpha = 0.4$
GroupedESN ($M = 3$)	300-300-300	2.28E-02	7.11E-03	$N_R = 900, \theta = 1, \alpha = 0.4$
MITSESN ($M = 2$)	450-450	2.25E-02	7.01E-03	$N_R = 900, \theta = 1, \alpha = 0.4$
MITSESN ($M = 3$)	300-300-300	**2.24E-02**	**6.64E-03**	$N_R = 900, \theta = 1, \alpha = 0.5$

$N_R = 450$), differences between UDs of our proposed MITSESN with those of ESN and GroupedESNs ($M = 2$ and 3) gradually become larger and larger, which indicates that our proposed MITSESN can generate less linear redundancy in the concatenated state matrix than the ESN and the GroupedESN under the case of the larger N_R. Moreover, we find that the larger size of input time slices is, the less linear redundancy in the concatenated state matrix of MITSESN is. The above analyses explain the reasons why our proposed MITSESN outperforms the ESN and the GroupedESNs, and the MITSESN ($M = 3$) has the best performances on the three prediction tasks.

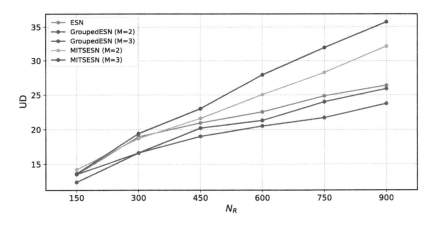

Fig. 5. UDs of all the tested models varying N_R from 150 to 900.

4 Conclusion

In this paper, we proposed a novel multi-reservoir echo state network, MITSESN, for nonlinear time-series prediction tasks. Our proposed MITSESN can extract various temporal features from multi-size input time slices. The prediction performances on three benchmark nonlinear time-series datasets empirically demonstrate the effectiveness of our proposed model. We provided an empirical analysis from the prospective of reservoir-state richness to show the superiority of MITSESN.

As future works, we will continue to evaluate the performances of the proposed model on the other temporal tasks such as time series classification tasks.

Acknowledgements. This work was partly supported by JSPS KAKENHI Grant Number 20K11882 and JST-Mirai Program Grant Number JPMJMI19B1, Japan (GT), and partly based on results obtained from Project No. JPNP16007, commissioned by the New Energy and Industrial Technology Development Organization (NEDO).

References

1. Cho, K., et al.: Learning phrase representations using RNN encoder-decoder for statistical machine translation. arXiv preprint arXiv:1406.1078 (2014)
2. Gallicchio, C., Micheli, A.: A Markovian characterization of redundancy in echo state networks by PCA. In: ESANN. Citeseer (2010)
3. Gallicchio, C., Micheli, A.: Richness of deep echo state network dynamics. In: Rojas, I., Joya, G., Catala, A. (eds.) IWANN 2019. LNCS, vol. 11506, pp. 480–491. Springer, Cham (2019). https://doi.org/10.1007/978-3-030-20521-8_40
4. Gallicchio, C., Micheli, A., Pedrelli, L.: Deep reservoir computing: a critical experimental analysis. Neurocomputing **268**, 87–99 (2017)
5. Hochreiter, S., Schmidhuber, J.: Long short-term memory. Neural Comput. **9**(8), 1735–1780 (1997)
6. Jaeger, H.: The "echo state" approach to analysing and training recurrent neural networks-with an erratum note. Ger. Natl. Res. Cent. Inf. Technol. GMD Tech. Rep. **148**(34), 13 (2001)
7. Li, Z., Tanaka, G.: Deep echo state networks with multi-span features for nonlinear time series prediction. In: 2020 International Joint Conference on Neural Networks (IJCNN), pp. 1–9. IEEE (2020)
8. Li, Z., Tanaka, G.: HP-ESN: echo state networks combined with hodrick-prescott filter for nonlinear time-series prediction. In: 2020 International Joint Conference on Neural Networks (IJCNN), pp. 1–9. IEEE (2020)
9. Liu, Y., Gong, C., Yang, L., Chen, Y.: DSTP-RNN: a dual-stage two-phase attentionbased recurrent neural network for long-term and multivariate time series prediction. Expert Syst. Appl. **143**, 113082 (2020)
10. Lorenz, E.N.: Deterministic nonperiodic flow. J. Atmos. Sci. **20**(2), 130–141 (1963)
11. Mackey, M.C., Glass, L.: Oscillation and chaos in physiological control systems. Science **197**(4300), 287–289 (1977)
12. Medsker, L., Jain, L.C.: Recurrent Neural Networks: Design and Applications. CRC Press, Boca Raton (1999)
13. Menezes, J.M., Barreto, G.A.: A new look at nonlinear time series prediction with narx recurrent neural network. In: 2006 Ninth Brazilian Symposium on Neural Networks (SBRN 2006), pp. 160–165. IEEE (2006)
14. Shen, L., Chen, J., Zeng, Z., Yang, J., Jin, J.: A novel echo state network for multivariate and nonlinear time series prediction. Appl. Soft Comput. **62**, 524–535 (2018)
15. Song, Z., Wu, K., Shao, J.: Destination prediction using deep echo state network. Neurocomputing **406**, 343–353 (2020)

16. Suykens, J.A., Vandewalle, J.: The KU leuven time series prediction competition. In: Suykens J.A.K., Vandewalle J. (eds) Nonlinear Modeling, pp. 241–253. Springer, Boston (1998). https://doi.org/10.1007/978-1-4615-5703-6_9
17. Tanaka, G., et al.: Recent advances in physical reservoir computing: a review. Neural Netw. **115**, 100–123 (2019)
18. Tikhonov, A.N., Goncharsky, A., Stepanov, V., Yagola, A.G.: Numerical Methods for the Solution of III-Posed Problems, vol. 328. Springer Science & Business Media, Berlin (2013)
19. Weigend, A.S.: Time Series Prediction: Forecasting The Future And Understanding The Past Routledge, Abingdon-on-Thames (2018)
20. Yu, Z., Liu, G.: Sliced recurrent neural networks. arXiv preprint arXiv:1807.02291 (2018)

Transformer with Prior Language Knowledge for Image Captioning

Daisong Yan[1], Wenxin Yu[1(✉)], Zhiqiang Zhang[2], and Jun Gong[3]

[1] Southwest University of Science and Technology, School of Computer Science and Technology, Sichuan, China
yuwenxin@swust.edu.cn
[2] Graduate School of Science and Engineering, Hosei University, Tokyo, Japan
[3] Beijing Institute of Technology, Beijing, China

Abstract. The Transformer architecture represents state-of-the-art in image captioning tasks. However, even the transformer uses positional encodings to encode sentences, its performance still not good enough in grammar. To improve the performance of image captioning, we present Prior Language Knowledge Transformer (PLKT)—a transformer-based model that can integrate learned a priori language knowledge for image captioning. In our proposal, when our model predicts the next word, it not only depends on the previously generated sequence but also relies on prior language knowledge. To obtain prior language knowledge, we embed a learnable memory vector inside the self-attention. Meanwhile, we use reinforcement learning to fine-tune the model in training. To prove the advancement and promising effectiveness of PLKT, we compare our approach with other recent image captioning methods in the experiments. Through objective results, our proposal increased the CIDEr score of the baseline by 0.6 points on the "Karpathy" test split when tested on COCO2014 dataset. In subjective results, our approach generated sentences is obviously better than baseline in grammar.

Keywords: Image caption · Transformer · Priori language knowledge · Self-attention

1 Introduction

The image caption is a cross-domain of computer vision (CV) and natural language processing (NLP). Specifically, it is the task of using a short sentence to describe the visual contents of an image. The image caption is a simple task for a human, but it is challenging for a machine. The reason is that it requires a model to detect the objects in an image, understand their relationships, and finally properly output a sentence to describe them.

Image caption task is similar to machine translation, and most image captioning models also adopt an encoder-decoder architecture. The image is encoded

---

D. Yan and W. Yu—These authors have contributed equally to this work.

T. Mantoro et al. (Eds.): ICONIP 2021, LNCS 13109, pp. 40–51, 2021.
https://doi.org/10.1007/978-3-030-92270-2_4

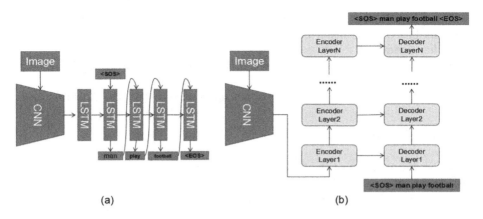

Fig. 1. (a) Architecture of the RNN (LSTM) for image caption. (b) Architecture of the Transformer for image caption (The last stage of caption generation).

into feature vectors by a convolutional neural network (CNN), and then the feature vector is decoded by a recurrent neural network (RNN) to generate the describe caption [1]. This architecture is still the dominant approach in the last few years. Currently, CNN models are capable of detecting objects accurately. However, the performance of the image captioning model is limited when generating sentences in RNN. Recently, Transformer and its variants have been introduced to the image captioning to replace the RNN in the traditional image caption, such as Fig. 1. It uses a self-attentive to build the model instead of recurrent structures [3]. Since its introduction in Transformer [2], Transformer and its variants have achieved satisfactory performance in the image caption, but it still exist some problems.

Since RNN uses a recurrent structure, it has a unique advantage in sentence encoding and can generate more consistent sentences with human grammar. The Transformer is based entirely on self-attention, which allows the network to run in parallel in training. However, there is no exact word order marking in the self-attention mechanism except for relative position embedding. This reliance on attention makes the Transformer's performance more unsatisfactory than RNN in grammar. To solve the above problem, we introduce learnable memory vectors in the decoder part of the Transformer. The input of the transformer decoder is both previously generated words and the output of the encoder. Our decoder can embed previously learned language knowledge by using learnable memory vectors when it generates the next words. In this way, the performance of the model in grammar has been improved. Also, in order to obtain better results, we used reinforcement learning in training. The experimental results show that our proposal improves the performance of the image caption model.

The contribution of this paper are as follows:

- To improve the model's language coding ability, we propose an image caption model based on Transformer, which innovatively introduces memory vectors to the decoder.
- When encoding sentences, our model can embed prior language knowledge and reduce reliance on the previously generated words.
- We demonstrate the effectiveness of our approach on the COCO2014 dataset.

2   Related Work

Since the development of the image caption, many methods have been proposed. The dominant approach use encoder-decoder structures, convolutional neural networks (CNN) as the encoder and recurrent neural networks (RNN) as the decoder [1]. In earlier research work, researchers have continued to improve the performance of CNN. Multiple CNN encoders are designed in [18] to generate features using complementary information between encoders. [19] uses a pre-trained CNN model to extract visual objects for each image and trains a visual analysis tree model to analyse visual objects relationships. In the encoder-decoder structure, the encoder encodes the input image into uniform semantic features. However, when the image size is too large, the uniform semantic features may not store enough information. To solve this problem, [4] uses faster-RCNN to extract multiple regions of the image as the input of the decoder, which significantly improves image captioning accuracy.

In the decoder of image captioning, [20] proposes a language model based on a convolutional neural network, which enables the model to be parallelized and offers performance near that of LSTM. [10] explicitly took the previous visual attention as context at each time step. [11] takes advantage of the scene graph model, integrates all the objects and the relationship between objects as the input. On the same line, [12] explicitly models semantic and geometric object interactions based on Graphic Convolutional Network (GCN) and makes full use of the alignment between linguistic words and visual semantic units for image captioning. Recently, the transformer has been introduced into the Image Caption field with great performance [2,21]. [13] proposes an object-relational transformer. Based on this approach, it merges information about the spatial relationships between objects detected by the input through geometric attention. [22] uses the Transformer framework to propose an object detection module in the image captioning task, and calculates a relational geometric weight between objects to scale the original attention weight. [14] introduces memory vectors in the transformer's image encoder. When encoding the region feature, the prior knowledge learned to be taken into account from the memory vector. In terms of training methods, [16,17] introduces reinforcement learning into the image captioning task, which enabled the model to be refined with non-differentiable caption metrics as optimization objectives.

Inspired by [14], to obtain the prior language knowledge, we introduce the memory vector into the language decoder. Our approach makes sentence decoding depend on previously generated words and relies on the previously learned language knowledge of memory vectors. To our knowledge, this is the first paper using memory vectors to improve the language decoding performance of Transformer. For our experiments, we used [14] as our baseline and the structure of the baseline in our approach. The structure of the baseline is shown in Fig. 2.

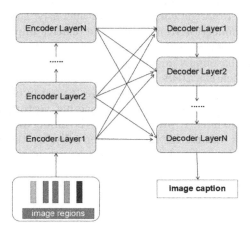

Fig. 2. This is our baseline model, and we follow the general structure of the baseline model. Image regions are regions of interest extracted by CNN. Multi-level encodings of encoders are connected to a language decoder through a meshed connectivity.

3 Preliminaries

3.1 Scaled Dot-Product Attention

We first review the self-attention mechanism, which called "Scaled Dot-Product Attention". Three sets of vectors implement the self-attention mechanism.

The three vectors are query vector Q, key vector K and value vector V. These three vectors are all obtained from the same vector through the linear transformation. Q is an n-dimensional vector, and the vectors K and V have the same dimension and are both d_m. First, we use Q and K to compute the energy scores E as:

$$E = QK^T \tag{1}$$

Where E is an $n \times m$ weight matrix. Then softmax function is applied to E to obtain the weight of the values. The output is the weighted sum of the weight matrix multiplied by the values as:

$$W = Attention(Q, K, V) = Softmax \left(\frac{E}{\sqrt{d}} \right) V \tag{2}$$

where d is a scaling factor, usually equal to the dimension of K.

3.2   Meshed-Memory Transformer for Image Captioning

In our work, we used [14] as our baseline. Figure 2 shows the architecture of the baseline. The model consists of a region encoder and language decoder and has the same number of layers. Baseline language decoders and encoders use mesh connections so that each layer of the language decoder can take advantage of different-level output from the encoder. Simultaneously, the learnable weight α is used in the mesh connection to modulate the relative importance between the outputs of different region encoding layers.

The input to the baseline's image encoder is region features extracted from an image, each of which is a 2048-dimensional feature vector. Before each input to the image decoder, the linear projection layer is used to change the dimension to the model dimension. Simultaneously, the one-hot vectors are used to encode each word, which is also linearly projected onto the model dimension. In addition, the language decoder cannot be input sequentially, so the fusion features of position encoding and word embedding encoding are used as the input at the bottom layer of the decoder.

4   Our Approach

Our model still uses the design of the baseline in the region encoder. After passing through the region encoder, each layer's output is respectively X_1 to X_n. Since the number of the region from each image are different, we use the mask operation to avoid the attention mechanism to operate the padding part.

In the language decoder, the attention mechanism is also used to encode the word sequence. The input of the decoder layers is the previously generated words Z and region encodings (X_1 to X_N). In this case, to get the query, key and value, linearly project the word sequences Z into the required dimension as:

$$
\begin{aligned}
S &= Attention(Q, K, V) \\
&= Attention(W_q Z, W_k Z, W_v Z)
\end{aligned}
\tag{3}
$$

Where W_Q, W_K and W_V are learnable weight matrices, and the number of elements in S is the same as that in Z. Each element in S can be thought of as a weighted sum of the elements in Z.

4.1   Prior Language Knowledge and Memory Vectors

When predicting the next word, the self-attention mechanism only depends on the previously generated words, but humans will also use the prior language knowledge. Therefore, we introduce the memory vectors into the self-attention mechanism in the decoder, which makes the decoder considering the prior language knowledge learned by the Scaled Dot-Product Memory Attention. In this way, it can improve the understanding ability of the language decoder and generate sentences that are more conform to human standards.

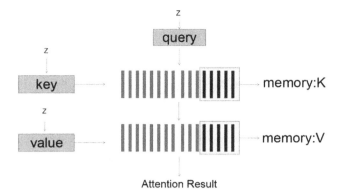

Fig. 3. Scaled dot-product memory attention. Memory vectors K and V are learnable vectors, and they connect with *key* and *value* respectively through concatenation.

4.2 Scaled Dot-Product Memory Attention

To combine memory vectors with self-attention operations, we use the approach in the baseline. The memory vector is combined with the K and V vectors in the self-attention through concatenation. Our method to make the memory vectors independent of the sentence sequence itself, and the memory vectors are learnable matrices. Simultaneously, in language decoding, the learned prior knowledge is embedded in the sentence sequence Z through self-attention operation, like Fig. 3. The formula is defined as:

$$
\begin{aligned}
S_m &= Attention(Q, K, V) \\
&= Attention(W_q Z, [W_k Z, M_k], [W_v Z, M_v])
\end{aligned}
\tag{4}
$$

Where is M_k and M_v represent memory vectors whose dimensionality is d_g, and $[,]$ indicates concatenation. Where the dimension of Q is d_n, and the dimension of K and V is d_m.

Like the self-attention mechanism, memory attention can be used in multi-head attention. In this case, we linearly project the different queries, keys, values and memory vectors for each head, and the memory attention is used h times. And then, these are concatenated from different heads and applied a linear projection to get the final output (Fig. 4).

4.3 Language Decoder

Mesh-Connection and Cross-Attention: The input of the language decoder includes the previously generated word sequence Z and region features encoded $(X_1$ to $X_n)$. To exploit the multi-level region features, we adopted the mesh connection and a cross-attention. Specifically, our operation is defined as:

$$
S_m^c(Z_m, X) = \sum_{i=1}^{N} \alpha_i \bigodot C(Z_m(Z), X_i)
\tag{5}
$$

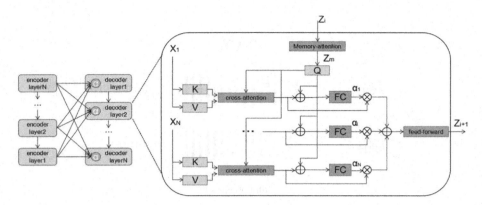

Fig. 4. Architecture of the PLKT. We use Memory-attention to encodes previously generated words with a priori knowledge. For clarity, AddNorm operations are not shown.

Where $Z_m()$ is the Scaled Dot-Product Memory Attention, $[,]$ indicates concatenation, and Z is previously generated words. $Z_m()$ is defined as:

$$Z_m(Z) = Attention(W_q Z, [W_k Z, M_k], [W_v Z, M_v]) \tag{6}$$

And $C(,)$ is the cross-attention operation, which uses Z_m to compute the query and X_i to get the value and key. Its internal calculation is the same as that of self-attention operations. $C(,)$ is defined as:

$$C(Z_m(Z), X_i) = Attention(W_q Z_m, W_k X_i, W_v X_i) \tag{7}$$

Moreover, α_i is a weight matrix used to modulate the relative importance of the different encoder layers for cross-attention. α_i is computed by computing the relevance between cross-attention and Z_m. α_i is defined as:

$$\alpha_i = sigmoid(W_i[Z_m(Z), C(Z_m(Z), X_i)] + b_i) \tag{8}$$

Where sigmoid is the activation function, and W_i is a weight matrix, b_i is a learnable vector. This calculation can be viewed as a linear projection operation that is applied the sigmoid activation function, and its result dimension is d_{model}.

Feed-Forward Layer: After the cross-attention operation, we get S_m^c. The S_m^c is applied to a position-wise feed-forward layer(FNN) which contains two linear transformation layers, and one of them uses activation functions. The FNN applied independently to each element of S_m^c. The formula is defined as:

$$F(S_m^c) = W_a ReLU(W_b S_m^c + a) + b \tag{9}$$

Where W_a and W_a are the learnable weight matrix, a and b are learnable bias vectors. Also, ReLU is the activation function.

Residual Connection and Layer Normalization: Like Transformer, we used residual connections and layer normalization for each sub-component (cross-attention and feed-forward layer). The residual connections and layer normalization is defined as:

$$R = AddNorm(O) \tag{10}$$

Where O is the output of sub-component, and addNorm represents a combination of residual connections and layer normalization.

In addition, we apply a multi-head fashion for each attention operation (meshed-cross-attention and Scaled dot-product Memory Attention) in the language decoder layer. Simultaneously, we use a Transformer-like architecture to stack multiple decoder layers on top of each other, which help the decoder to understand the input words sequence deep. When in training, to predict the next word only depending on the previously generated words, we add a mask to the input sequence. Finally, the operation of the decoder layer is defined as:

$$\hat{S} = AddNorm(S_m^c(Z_m(Mask(Z)), X))$$
$$\hat{Z} = AddNorm(F(\hat{S})) \tag{11}$$

Where Z are vectors of the words generated before, the $Mask$ represents the self-attention of the masking over time. After applying a linear projection and a softmax layer, it will output a probability of each word in the dictionary.

5   Experiments

5.1   Experiments Settings

Dateset. We used the COCO2014 dataset in our experiments, which is the most popular dataset for Image Caption. The COCO2014 dataset contains more than 120,000 images, each with five different captions. We used 'Karpathy' splits in our experiments, which contains 5000 pictures for validation and 5000 pictures for test and the rest for training. Our detection features are computed with the code provided by [4].

Metrics. Following the standard evaluation metrics, we employ automatic evaluation metrics to evaluate the quality of image captions: BLEU-4 [5], METEOR [6], ROUGE [7], CIDEr [8].

Implementation Details. In our model, we still set the hyperparameters like the transformer model. Specifically, the image feature dimension is 2048, and the dimension of the whole model is 512. The dimension inside the feed-forward operation is 2048. When we apply multi-head attention, the number of heads of multi-head attention is 8. For each attention layer and feed-forward layer, we applied the dropout layer with a keep probability of 0.9. We set the image encoder and the language decoder both have the same number of layers, and the number of layers is equal to 3. Our memory vector dimension is 40. The query, key, and value dimensions are all equal to 64. For training, word-level cross-entropy loss is used to pre-train the model, and then reinforcement learning is used to finetune the model. In the reinforcement learning phase, we used CIDEr-D as our reward [8]. We used a variant of the self-critical sequence training method [9] to sample the sequences using beam search. The beam size equal to 5. Pre-training follows the learning rate scheduling strategy [2], with a warm-up equal to 10,000 iterations. During the reinforcement learning, we used a fixed learning rate of 5×10^{-6}. We trained the entire model to use the Adam optimizer, and the batch size is 50.

5.2   Experimental Analysis

This section compares our method (PLKT) with other image caption methods. Our approach achieves better results (129.4 CIDEr) are shown in Table 1. In addition, it should be noted that the data of the baseline in the table is the result of our re-reproduction.

Table 1. Comparison with the state of the art approaches on the COCO2014 "Karpathy" test split.

Model	BLEU-4	METEOR	ROUGE	CIDEr*
Up-Down [4]	36.3	27.7	56.9	120.1
CAVP [10]	38.6	28.3	58.5	126.3
VSUA [12]	38.4	28.5	58.4	128.6
ORT [13]	38.6	28.7	**58.8**	128.3
Baseline [14]	**38.7**	29.0	58.4	128.8
PLKT	38.6	**29.2**	58.4	**129.4**

* CIDEr is the primary metrics to judge the performance of the image caption model.

We compare PLKT with the other approaches (Table 1), including Up-Down [4], CAVP [10], VSUA [12], ORT [13], and Baseline [14]. Up-Down, CAVP, and VSUA are based on Long Short-Term Memory (LSTM) networks. ORT and Baseline adopt Transformer-Based architecture.

In Table 1, we list our approach performance and other methods. Among all the evaluation metrics, the higher score, the better the performance of the model. For prediction captions and ground-truth captions, BLEU-4 computes the degree of overlap between them, and ROUGE calculates the longest common subsequence length of them. In short, for BLEU-4 and ROUGE, the more the same parts between the sentences, the higher the score. In consideration of the matching relationship of synonyms and definitions, Meteor uses WordNet to calculate the semantic similarity between prediction captions and ground-truth captions, which makes Meteor more relevant to human judgment stands. CIDEr is the closest to human judgment stands in all metrics, so it is the primary metrics to judge the performance of the image caption model. In Table 1, our approach surpasses other methods in METEOR and CIDEr.

GT: A motorbike sitting in front of a wine display case.

Baseline: A motorcycle parked on a counter in a room.

PLKT: An old motorcycle parked in front of a rack.

GT: A pinup-style photo of a woman sitting on a luggage trunk.

Baseline: A woman sitting on top of a suitcase.

PLKT: A woman in a dress sitting on a suitcase.

GT: A man on a bicycle riding next to a train.

Baseline: A man riding a bike next to a train on.

PLKT: A man riding a bike down a street next to a train.

Fig. 5. Example captions generated by the Baseline and our PLKT models. 'GT' denotes one of the five ground-truth captions.

Table 2. Comparison with the baseline of CIDEr scores during pre-training.

Epoch	Baseline [14]	**PLKT**
0	98.1	**100.8**
2	109.6	**110.3**
4	110.5	**111.0**
6	112.6	**113.7**
8	113.7	**114.0**

Our approach is slightly worse on BLEU-4 and ROUGE than baseline and ORT, respectively, which because the subjectivity of results becomes better, the objectivity inevitably becomes worse like Fig. 5. In Table 1, we increased the CIDEr score of the baseline by 0.6 points. The experimental results show that our method can generate more consistent captions with human judgment after integrating prior language knowledge. In Fig. 5, our approach is also better than baseline in grammar.

To further prove our method's effectiveness, we listed the changes of the CIDEr score in the Table 2. Because reinforcement learning is difficult to reproduce, we only list changes in CIDEr scores during the pre-training stage. It can be seen from Table 2, the CIDEr score of our method was higher than the baseline in all epochs.

6 Conclusion

In this paper, we propose a new image caption model based on Transformer. It introduces the memory vector that can learn prior knowledge into the language decoder of the Transformer. When the language decoder encodes the previously generated sentence, it embeds the previously learned language knowledge. To the best of our knowledge, this way of improving sentence decoding is unprecedented. We verify the effectiveness of our method on the COCO2014 dataset. Simultaneously, we also compare our method's performance with other methods, which shows our method's advance.

Acknowledgment. This research is supported by Sichuan Science and Technology Program (No. 2020YFS0307, No. 2020YFG0430, No. 2019YFS0146), Mianyang Science and Technology Program (2020YFZJ016).

References

1. Vinyals, O., Toshev, A., Bengio, S., Erhan, D.: Show and tell: lessons learned from the 2015 MSCOCO image captioning challenge. IEEE Trans. Pattern Anal. Mach. Intell. **39**(4), 652–663 (2016)
2. Vaswani, A., et al.: Attention is all you need. In: Advances in Neural Information Processing Systems (2017)
3. Devlin, J., Chang, M.-W., Lee, K., Toutanova, K.: BERT: pre-training of deep bidirectional transformers for language understanding. arXiv preprint arXiv: 1810.04805 (2018)
4. Anderson, P., et al.: Bottom-up and top-down attention for image captioning and VQA. arXiv preprint arXiv: 1707.07998 (2017)
5. Papineni, K., Roukos, S., Ward, T., Zhu, W.J.: Bleu: a method for automatic evaluation of machine translation. In: Meeting on Association for Computational Linguistics, pp. 311–318 (2002)
6. Denkowski, M., Lavie, A.: Meteor universal: language specific translation evaluation for any target language. In: The Workshop on Statistical Machine Translation, pp. 376–380 (2014)
7. Lin, C.-Y.: Rouge: a package for automatic evaluation of summaries. In: Text Summarization Branches Out (2004)
8. Vedantam, R., Lawrence Zitnick, C., Parikh, D.: Cider: consensus-based image description evaluation. In: Computer Science, pp. 4566–4575 (2015)
9. Rennie, S.J., Marcheret, E., Mroueh, Y., Ross, J., Goel, V.: Self-critical sequence training for image captioning. In: Proceedings of the IEEE Conference on Computer Vision and Pattern Recognition (2017)
10. Liu, D., Zha, Z.-J., Zhang, H., Zhang, Y., Wu, F.: Context-aware visual policy network for sequence-level image captioning. In: 2018 ACM Multimedia Conference on Multimedia Conference, pp. 1416–1424. ACM (2018)
11. Yang, X., Tang, K., Zhang, H., Cai, J.: Auto-encoding scene graphs for image captioning. In: Proceedings of the IEEE Conference on Computer Vision and Pattern Recognition, pp. 10685–10694 (2019)
12. Guo, L., Liu, J., Tang, J., Li, J., Luo, W., Lu, H.: Aligning linguistic words and visual semantic units for image captioning. In: ACM MM (2019)

13. Herdade, S., Kappeler, A., Boakye, K., Soares, J.: Image captioning: transforming objects into words. arXiv preprint arXiv: 1906.05963 (2019)
14. Cornia, M., Stefanini, M., Baraldi, L., Cucchiara, R.: Meshed-memory transformer for image captioning. In: 2020 IEEE/CVF Conference on Computer Vision and Pattern Recognition (CVPR). IEEE (2020)
15. Guo, L., Liu, J., Zhu, X., Yao, P., Lu, S., Lu, H.: Normalized and geometry-aware self-attention network for image captioning. In: 2020 IEEE/CVF Conference on Computer Vision and Pattern Recognition (CVPR) (2020)
16. Liu, S., Zhu, Z., Ye, N., Guadarrama, S., Murphy, K.: Improved image captioning via policy gradient optimization of SPIDEr. In: Proceedings of the International Conference on Computer Vision (2017)
17. Ranzato, M.A., Chopra, S., Auli, M., Zaremba, W.: Sequence level training with recurrent neural networks. In: Proceedings of the International Conference on Learning Representations (2015)
18. Jiang, W., Ma, L., Jiang, Y.-G., Liu, W., Zhang, T.: Recurrent fusion network for image captioning. In: Ferrari, V., Hebert, M., Sminchisescu, C., Weiss, Y. (eds.) ECCV 2018. LNCS, vol. 11206, pp. 510–526. Springer, Cham (2018). https://doi.org/10.1007/978-3-030-01216-8_31
19. Chen, F., Ji, R., Sun, X., Wu, Y., Su, J.: GroupCap: group-based image captioning with structured relevance and diversity constraints. In: 2018 IEEE/CVF Conference on Computer Vision and Pattern Recognition (CVPR). IEEE (2018)
20. Aneja, J., Deshpande, A., Schwing, A.G.: Convolutional image captioning. In: 2018 IEEE/CVF Conference on Computer Vision and Pattern Recognition (CVPR). IEEE (2018)
21. Huang, L., Wang, W., Chen, J., Wei, X.-Y.: Attention on attention for image captioning. In: Proceedings of the International Conference on Computer Vision (2019)
22. Herdade, S., Kappeler, A., Boakye, K., Soares, J.: Image captioning: transforming objects into words. arXiv preprint arXiv:1906.05963 (2019)

Continual Learning with Laplace Operator Based Node-Importance Dynamic Architecture Neural Network

Zhiyuan Li$^{(\boxtimes)}$, Ming Meng$^{(\boxtimes)}$, Yifan He, and Yihao Liao

School of Automation, Hangzhou Dianzi University, Hangzhou 310018, China
{hdulzy,mnming,hyfok,18051616}@hdu.edu.cn

Abstract. In this paper, we propose a continual learning method based on node-importance evaluation and a dynamic architecture model. Our method determines the important nodes according to the value of Laplace operator of each node. Due to the anisotropy of the important nodes, the sparse sub-networks for the specific task can be constructed by freezing the weights of the important nodes and splitting them with unimportant nodes to reduce catastrophic forgetting. Then we add new nodes in networks to prevent existing nodes from being exhausted after continuously learning many new tasks, and to lessen the negative transfer effects. We have evaluated our method on CIFAR-10, CIFAR-100, MNIST, Fashion-MNIST and CUB200 datasets and it achieves superior results when compared to other traditional methods.

Keywords: Continual learning · Node-importance · Dynamic architecture

1 Introduction

Continual learning is an open problem in the machine learning field. In this problem, the model is required to continuously learn new independent tasks without forgetting the previous tasks. Continual learning aims to overcome the stability-plasticity dilemma [6], if the model focuses on plasticity, it might suffers from the catastrophic forgetting problem [26] and if it focuses on stability too much, it will get poor transfer ability to new tasks. Study for continual learning was divided broadly into several categories: task-specific [8,19,33], replay memory [3,17,25], regularization-based method [1,7,14,15], etc.

In this paper, we will focus on both regularization-based and task-specific based dynamic architecture method. The regularization method will identify important learning weights for previous data and maintain them while learning new tasks. In order to achieve a structured sparsity model, several model compression methods [30,33] defined the group of weights to a node which referred as Lasso-like penalties. So, paying attention to the important nodes could lead to a sparse representative model and achieve better efficiency than simply focusing on weights importance. The dynamic architecture method focuses on the network structure, and uses additional layers or nodes to maintain the previous task

© Springer Nature Switzerland AG 2021
T. Mantoro et al. (Eds.): ICONIP 2021, LNCS 13109, pp. 52–63, 2021.
https://doi.org/10.1007/978-3-030-92270-2_5

structure while learning new tasks, and on this basis, reduces the interference with old knowledge.

We propose a new regularization-based node importance identification combine with dynamic architecture neural network continual learning method called Laplacian-based Node Importance Dynamic Architecture (LNIDA). To measure the node-importance, we introduce the Laplace Operator $\Delta\mathcal{L}$ in neural network gradient field. We notice that Laplace operator is a second-order differential operator in n dimensional Euclidean space, defined as the divergence of gradients. So, it calculates the gradient difference between the surrounding nodes and the specified node and obtains the total gain that may be obtained after a small perturbation to this node. Then, in order to generalize it to neural network, we propose that Laplace operator for neural network gradient field as

$$\Delta\mathcal{L}_i = \sum_{n=1}^{N} \frac{\partial^2 \mathcal{L}}{\partial w_{i,n}^2} \tag{1}$$

where $\mathcal{L}$ is the loss function, i represent the specified node in network and N is the nodes that connect to node i in previous layer.

Therefore, we could utilize the Laplace operator as a metric for how one node interfere with other nodes. Furthermore, it can also represent the importance of one node in previous tasks. Namely, we calculate the Laplace operator values of each node while learning groups of tasks as node-importance to identify important nodes for previous tasks and prevent the catastrophic forgetting via freezing the incoming weights between important nodes and the bias of important nodes, so that the important nodes group will form a sub-network which is related to the previous task. Moreover, after learning each task, we sever the weights that between important and unimportant nodes and re-initialize the remaining weights that pertain to unimportant nodes to minimize the negative interfere from previous tasks and maximize the plasticity of neural network (Fig. 1).

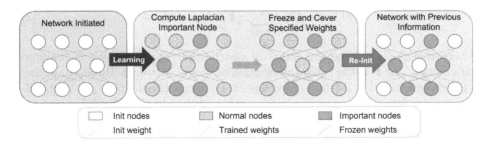

Fig. 1. There are summaries of the node-importance identify method. The init nodes and init weights mean that nodes and weights contain no information. After the learning process, important nodes can be identified by the Laplace operator. On this basis, we have the frozen sub-network consist of important nodes. Then, via re-init, we get the new network for new tasks and still maintain previous information

Based on the above methods, we considered another problem. After training with many tasks, the number of nodes allowed to be trained, or what we call "free nodes", will be reduced. Namely, with the free nodes decrease, the network-provided capacity for new tasks also decreases. That might lead to neural network with poor learning ability and negative transfer for all tasks. To prevent this problem, we applied dynamic architecture method to continuously increase additional nodes. While important nodes are frozen, we will add new additional nodes that have same connect architecture with important nodes to maintain the amount of free nodes, so that the negative transfer can be prevent and capacity of network will also remain stable.

The results show that our method efficiently mitigates the catastrophic forgetting while continuously learning new tasks. The primary contributions of this paper are as follows:

- Propose a new node-importance identify method for previous tasks and via freezing important nodes to mitigate catastrophic forgetting while continuously learning new tasks.
- Utilize dynamic architecture model to prevent negative transfer in node frozen regularization-based continual learning.

2   Related Work

The study of catastrophic forgetting in neural networks originated in 1980s [20]. With the revival of neural networks, catastrophic forgetting in continual learning has been reintroduced in 2016 [27]. The most famous method of this problem is Elastic Weight Consolidation (EWC) [14] which adds a weighted L2 penalty term on neural network's parameter. The weight of L2 term is defined as Fisher information metric which measures the importance of each parameter of previously distribution $\mathcal{D}_1$. So the loss function with $\mathcal{D}_2$ designed as:

$$\mathcal{L}(\theta) = \mathcal{L}_{\mathcal{D}_2}(\theta) + \sum_{i=1}^{N} \frac{\lambda}{2} \mathbf{F}_i^{\mathcal{D}_1} \cdot (\theta_i - \theta_{\mathcal{D}_1, i})^2 \tag{2}$$

Where the $\mathcal{L}_{\mathcal{D}_2}(\theta)$ is the loss value of distribution $\mathcal{D}_2$; The $\theta_{\mathcal{D}_1, i}$ is previous parameters; The $\mathbf{F}$ is the fisher information of neural networks. Besides, another type of regularization method encourages the current classifier output probabilities on old classes to approximate the output of the old classifier [15].

The replay strategy is initially proposed to relearn a subset of previously learned data when learning the current task [26]. Some recent works like large-scale incremental learning [31] storing a subset of old data fall into this category which called Real Data Replay(RDR). RDR only violates the continual learning requirement that old data are unavailable, but also is against the notion of bio-inspired design. According to complementary learning system theory [22], the hippocampus encodes experiences to help the memory in the neocortex consolidate. However, much evidence illustrates hippocampus works like the generative model, which has inspired the Deep Generative Replay (DGR) [28]. But the DGR may break down when it encounters complex data [11]. A remedy is to encode the raw data into features with a feature extractor pre-trained on datasets and replay features [32].

Another type of method called Task-specific that aim to prevent knowledge interference by establishing task-specific modules for different tasks. The task-specific modules can be designed as hidden units [19], network parameters [18], dynamically sub-networks [27] and dynamic architecture [8,33]. This type of method is designed for task incremental learning. But these methods require task identification to choose corresponding task-specific modules therefore they are not applicable for class incremental learning.

The subspace methods [34] retain previously learned knowledge by keeping the old mapping neural network induce fixed. In this goal, the gradients are projected to the subspace that is orthogonal for the inputs of previous tasks, such as Conceptor Aided Backprop (CAB) [10] and OWM algorithm [34]. These methods make the feature of previous tasks stable in the learning process.

3   Method

The previous study on continual learning has revealed that model drift [12] and negative transfer [9,12] are sources of catastrophic forgetting. The model drift problem is related to the changes in the weights of important nodes while learning new tasks. Meanwhile, the negative transfer is derived from the negative interference of unimportant nodes in the previous task to possible important nodes when learning each task. Hence, we propose the Laplace operator based Node-importance sub-network generates a method to eliminate the above source of catastrophic forgetting.

3.1   Laplace Operator Based Node-Improtance Sub-network

Notice that for the sake of clarity in the following paragraph, we assume that both the loss function and the activation function exist second derivatives. We denote the $\mathcal{L}$ as the loss function, a_i as the activate value of specified node i, $x_{i,j}$ as the j-th input value of node i and $w_{i,j}$ as the j-th weight of node i (corresponding to $x_{i,j}$).

To calculate the node-importance in the learning process, we establish the Laplace operator calculation equation in the neural network. For the single weight $w_{i,j}$ of node i in output layer in certain input, the second derivatives with loss function as

$$\frac{\partial^2 \mathcal{L}}{\partial w_{i,j}^2} = \left(\frac{\partial^2 \mathcal{L}}{\partial a_i^2} * \left(\frac{\partial a_i}{\partial z_i} \right)^2 + \frac{\partial^2 a_i}{\partial z_i^2} * \frac{\partial \mathcal{L}}{\partial a_i} \right) * x_{i,j}^2 \qquad (3)$$

where $z_i = \sum_{j=1}^{N} x_{i,j} \cdot w_{i,j} + b_i$. Corresponding to the error term δ in the back propagation, we define

$$\Omega_i = \frac{\partial^2 \mathcal{L}}{\partial a_i^2} * \left(\frac{\partial a_i}{\partial z_i} \right)^2 + \frac{\partial^2 a_i}{\partial z_i^2} * \frac{\partial \mathcal{L}}{\partial a_i} \qquad (4)$$

Moreover, we can deduce the second derivatives with weight $w_{i,j}$ of hidden node i in l layer as

$$\frac{\partial^2 \mathcal{L}}{\partial w_{i,j}^2} = \left(\Omega_i^{l+1} \cdot w_i^{l+1\,2} - 2\delta_i^{l+1} \cdot w_i^{l+1} a_i \right) * \left(1 - a_i^2 \right)^2 * x_{i,j}^2 \qquad (5)$$

where $l+1$ represent the parameters in forward layer of node i. For hidden nodes we let

$$\Omega_i = \left(\Omega_i^{l+1} \cdot w_i^{l+1\,2} - 2\delta_i^{l+1} \cdot w_i^{l+1} a_i \right) * \left(1 - a_i^2 \right)^2 \qquad (6)$$

Combining Eqs. 1, 5, and 6, we can get the Laplacian equation of node i as

$$\Delta \mathcal{L}_i = \Omega_i \sum_{j=1}^{N} x_{i,j}^2 \qquad (7)$$

On this basis, we can simply calculate the Laplace operator value for nodes in the neural network. For each layer in the current input, we have the Laplace operator calculation equation as

$$\Delta \mathcal{L}_l = \left[\left(\Omega_{l+1} w_{l+1}^{2}{}^{\top} - 2\delta_{l+1} w_{l+1}^{\top} * a_l \right) * \left(1 - a_l^2 \right)^2 \right] \cdot x_l^{2}{}^{\top} \qquad (8)$$

Hence, we calculate $\Delta \mathcal{L}_l^t$ in each epoch t and define the final Laplace operator value groups for the current learning task for nodes in each layer as $\Delta \mathcal{L}_l = \|\Delta \mathcal{L}_l^t\|_2$. Then, we naturally defined the threshold to identify important nodes in each layer as mean values of $\Delta \mathcal{L}_l$. As the neural network becomes stable after multiple trainings, the Laplacian of each node will show anisotropy [5, 21, 23]. So when the mean value is utilized as a threshold to identify important nodes, important nodes will also show sparseness. Therefore, we can construct a small sub-network of important nodes for the current task (Fig. 2).

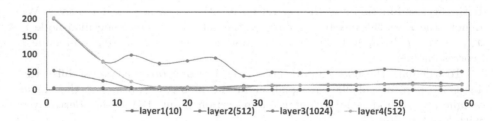

Fig. 2. This is a change of the important nodes number in 4 full connect layers. The horizontal axis is an epoch, and the vertical axis is the important nodes number. It can be seen that the number of important nodes in each layer is gradually decreasing and tends to be stable

After that, via frozen weights between important nodes and sever weights that from unimportant nodes to important nodes, we build the sub-network to maintain knowledge from previous task. The sub-network can also be considered as a certain function for a certain task roughly. Due to the variation of unimportant-nodes parameters, the accuracy of previous tasks still get lessen, but the catastrophic forgetting will be alleviated. However, we have to notice that even sub-network didn't interfere with new tasks, due to the data belonging to the same dataset all obey the approximate probability distribution [4,16,29], it still has contributions for tasks that belong to a same dataset. So the sub-networks might exist overlap.

3.2   Re-initialization and Addition of Unimportant Nodes

To decrease the negative interference from unimportant nodes of previous tasks to important nodes of future tasks, we utilized the re-initialization to release the weights of unimportant nodes. Base on that, the new learning process of future tasks will be stable and independent with the learning results of previous tasks. So the negative transfer is able to be minimized in our method.

After that, it's natural to consider an extreme condition that while the node-important based neural network learning multiple tasks, the nodes that able to release and re-training is depleted. In this case, there will be two problems. The neural network has been unable to learn new tasks, and due to the lack of auxiliary unimportant nodes, the sub-networks may face strong negative interference.

So, we introduce additional nodes to solve above problems. After the important nodes for the current task is determined, we add the new additional nodes to each layer and the number of additional nodes is equal to important nodes number. To maintain the original architecture before the current task and decrease the interference of architecture change to other tasks, the additional nodes will maintain the same connecting relationship as the important nodes (Fig. 3).

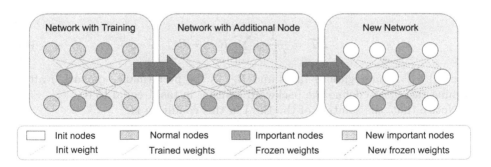

Fig. 3. There are summaries of additional nodes method. After a node is identified as an important node, the new initialized node will add and inherit the previous connect relationship of an important node to replace this node and maintain network architecture

Therefore, the summary of our method is following.

Algorithm 1: Laplacian based Node Importance Dynamic Architecture

---

Input: The set of training samples for current task; The model trained from the
   samples of the previous tasks;
Output: The model that contains the knowledge of the current task and the
   knowledge of the previous tasks, M;
1: Calculate Laplace operator for each layer in current learning task;
2: Identify important nodes by mean Laplacian value of each layer;
3: Add new additional nodes corresponding to important nodes base on the
   concatenate relationship;
4: Frozen and construct new sub-network;
5: Re-initialization neural network;
6: **return** M;

---

4   Experiments

We evaluated the performance of our method comparing with the representative
continual learning methods, EWC [14], SI [35], RWALK [7] and MAS [2] on multiple different vision datasets, {CIFAR-10/CIFAR-100/MNIST/CUB200/Fashion-
MNIST}. For experiments, we used multi-headed outputs and 5 different random
seeds to shuffle task sequences. After that, we deployed our method on VGG16
with a pre-trained model in ImageNet and replace the activation functions of full
connect layers as Tanh function. Our method was implemented by PyTorch and
Adam optimization method [13]. And we split each dataset into 5 tasks evenly.
Base on that, we consider the accuracy of each task and take the accuracy of task1
as baseline accuracy. We use the average accuracy of tasks after training each task
to evaluate the performance of continual learning. Meanwhile, we also use the backward transfer (Eq. 9) to characterize the degree of overall forgetting after learning
5 tasks.

$$BT = \frac{1}{T-1} \sum_{i=1}^{T-1} R_{T,i} - R_{i,i} \tag{9}$$

where the T is aggregate of tasks and $R_{i,j}$ is the performance on task j after
learning task i. We define the backward transfer in task1 equal to 0.

Notice that, the solution of the Eq. 4 in this condition is

$$\Omega_i = \frac{\partial a_i}{\partial z_i} \tag{10}$$

5   Results and Discussion

In this section, we present the results of experiments of continual learning in different datasets to evaluate the robustness of our method in different conditions.
The CIFAR-10, Fashion-MNIST and MNIST each contains 10 classes and have

been split into 5 tasks, each task contains 2 classes. These three datasets are used to evaluate the continual learning performance on regular conditions.

The CIFAR-100 contains 100 classes, each task in experiments contains 20 classes. It is used to prove the continual learning ability in a large number of classes dataset conditions. The CUB200 contains 200 classes and each task contains 40 classes. It can also be used to prove the continual learning ability in a large number of classes but due to that each class only have 30 samples. So its main purpose is to evaluate the method's continual learning performance on small samples conditions.

Table 1. Backward transfer results on datasets {CIFAR-10/CIFAR-100/MNIST/Fashion-MNIST/CUB200}

	Task	EWC	SI	RWALK	MAS	LNIDA(Ours)	Ablation
CIFAR-10	task1	0	0	0	0	0	0
	task2	−0.080	−0.060	−0.053	−0.048	**−0.020**	−0.089
	task3	−0.140	−0.119	**−0.090**	−0.137	−0.155	−0.159
	task4	−0.136	−0.128	−0.112	−0.142	**−0.110**	−0.392
	task5	−0.190	−0.157	**−0.130**	−0.166	−0.148	−0.457
CIFAR-100	task1	0	0	0	0	0	0
	task2	−0.092	−0.109	**−0.065**	−0.091	−0.085	−0.140
	task3	−0.119	−0.156	−0.200	**−0.095**	−0.113	−0.262
	task4	−0.137	−0.138	−0.170	−0.141	**−0.108**	−0.397
	task5	−0.187	−0.135	−0.155	−0.182	**−0.122**	−0.522
MNIST	task1	0	0	0	0	0	0
	task2	−0.071	−0.065	−0.102	**−0.038**	−0.100	−0.183
	task3	−0.122	−0.110	−0.128	−0.122	**−0.085**	−0.273
	task4	−0.120	−0.145	−0.168	−0.136	**−0.083**	−0.305
	task5	−0.150	−0.133	−0.183	−0.126	**−0.115**	−0.556
Fashion-MNIST	task1	0	0	0	0	0	0
	task2	−0.130	−0.079	−0.082	−0.065	**−0.062**	−0.153
	task3	−0.121	−0.117	−0.125	−0.116	**−0.115**	−0.287
	task4	−0.119	−0.132	**−0.103**	−0.139	−0.150	−0.353
	task5	−0.153	−0.141	−0.144	−0.145	**−0.138**	−0.502
CUB200	task1	0	0	0	0	0	0
	task2	−0.115	−0.082	−0.134	**−0.031**	−0.068	−0.168
	task3	−0.128	**−0.080**	−0.153	−0.101	−0.132	−0.352
	task4	−0.140	−0.119	−0.203	**−0.089**	−0.126	−0.443
	task5	−0.133	−0.134	−0.160	−0.153	**−0.129**	−0.513

Table 1 show the backward transfer results on each dataset. Our method showed a positive result on dataset CIFAR-10, MNIST and Fashion-MNIST. That result demonstrated our method can reduce catastrophic forgetting efficiently. Compared with other traditional methods, LNIDA has a higher average backward transfer value. However, for dataset CIFAR-100, the results show that our method also has positive continual learning ability on a large number of classes condition. That shows sub-networks based on important nodes can maintain information effectively. Meanwhile, after multiple tasks, our method also showed a high backward transfer results. That means, our method could efficiently lead to a lower forgetting rate. We also add the ablation test on the node-importance identify method. The result proves node-importance identify method has major effect in our continual learning method.

But we also should notice that experiment on CUB200 obtained a trivial result. That might means our method isn't fittable enough for small samples conditions. Base on the experiment results, the poor continual learning ability on small samples dataset may due to the limited samples and multiple classes make the important nodes become hard to identify. So the sub-network can not catch enough information for the current task. After several tasks, the results on CUB200 shown our method still keeping a high backward transfer as the results on other datasets, which indicate our method has a positive effect on long-range tasks.

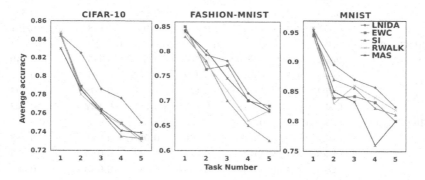

Fig. 4. Average accuracy results on CIFAR-10, Fashion-MNIST and MNIST datasets

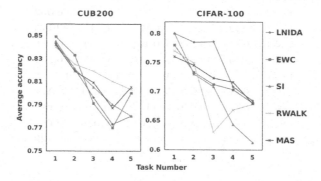

Fig. 5. Average accuracy results on CIFAR-100 and CUB200 datasets

Figure 4 and 5 show the average accuracy results on each dataset. Experiment results on CIFAR-10, Fashion-MNIST, MNIST and CIFAR-100 datasets show that on the same accuracy baseline, our method has a high average accuracy than other traditional methods. However, our method also has lower variance, which means it has better stability on each dataset.

But the advantage of CUB200 dataset is not obvious. The trend of average accuracy results on CUB200 have the same tendency as the backward transfer results. That indicates effective important nodes are difficult to identify on small samples datasets than regular datasets. This problem can be reduced by changing the important nodes identify criterion (LNIDA used the average Laplacian value) and expand important nodes identify the range. The wide important nodes range will lessen the confusion of sub-networks caused by small samples.

6 Conclusion

We proposed LNIDA, a new continual learning method based on Laplacian node-importance calculation and dynamic architecture neural network with the re-initialization trick to construct a sub-network for a certain task. The information of each task will be stored in each sub-network. To avoid the decrease in the number of free nodes as the sub-network increases, resulting in the inability to learn new tasks, we added new additional nodes to the network before constructing the sub-network.

The experiment results on 5 datasets showed our method has good robustness and generalization performance on regular datasets. Compared with other traditional methods, LNIDA shows a positive effect on long-range tasks. Although the results on the small samples dataset indicated our method may suffers confusion on small samples condition. But it still shows an effective long-range tasks continual learning ability.

Our future work will focus on continual learning algorithms on different conditions [24] and promote the learning ability on small samples datasets of LNIDA. Furthermore, the graph topology of the neural network also an untrivial impact on continual learning.

Acknowledgments. This work was supported by the National Natural Science Foundation of China (No. 61871427 and 61971168).

References

1. Ahn, H., Cha, S., Lee, D., Moon, T.: Uncertainty-based continual learning with adaptive regularization. arXiv preprint arXiv:1905.11614 (2019)
2. Aljundi, R., Babiloni, F., Elhoseiny, M., Rohrbach, M., Tuytelaars, T.: Memory aware synapses: learning what (not) to forget. In: Proceedings of the European Conference on Computer Vision (ECCV), pp. 139–154 (2018)
3. Ayub, A., Wagner, A.: Brain-inspired model for incremental learning using a few examples. arXiv preprint arXiv:2002.12411 (2020)

4. Bau, D., Zhu, J.Y., Strobelt, H., Lapedriza, A., Zhou, B., Torralba, A.: Understanding the role of individual units in a deep neural network. Proc. Natl. Acad. Sci. **117**(48), 30071–30078 (2020)
5. Belkin, M., Sun, J., Wang, Y.: Discrete laplace operator on meshed surfaces. In: Proceedings of the Twenty-Fourth Annual Symposium on Computational Geometry, pp. 278–287 (2008)
6. Carpenter, G.: Self organization of stable category recognition codes for analog input patterns. Appl. Opt. **3**, 4919–4930 (1987)
7. Chaudhry, A., Dokania, P.K., Ajanthan, T., Torr, P.H.: Riemannian walk for incremental learning: understanding forgetting and intransigence. In: Proceedings of the European Conference on Computer Vision (ECCV), pp. 532–547 (2018)
8. Golkar, S., Kagan, M., Cho, K.: Continual learning via neural pruning. arXiv preprint arXiv:1903.04476 (2019)
9. Gui, L., Xu, R., Lu, Q., Du, J., Zhou, Yu.: Negative transfer detection in transductive transfer learning. Int. J. Mach. Learn. Cybern. **9**(2), 185–197 (2017). https://doi.org/10.1007/s13042-016-0634-8
10. He, X., Jaeger, H.: Overcoming catastrophic interference using conceptor-aided backpropagation. In: International Conference on Learning Representations (2018)
11. Hu, W., et al.: Overcoming catastrophic forgetting for continual learning via model adaptation. In: International Conference on Learning Representations (2018)
12. Jung, S., Ahn, H., Cha, S., Moon, T.: Continual learning with node-importance based adaptive group sparse regularization. arXiv e-prints pp. arXiv-2003 (2020)
13. Kingma, D.P., Ba, J.: Adam: a method for stochastic optimization. arXiv preprint arXiv:1412.6980 (2014)
14. Kirkpatrick, J., et al.: Overcoming catastrophic forgetting in neural networks. Proc. Natl. Acad. Sci. **114**(13), 3521–3526 (2017)
15. Li, Z., Hoiem, D.: Learning without forgetting. IEEE Trans. Pattern Anal. Mach. Intell. **40**(12), 2935–2947 (2017)
16. Libardoni, A.G., Forest, C.E.: Sensitivity of distributions of climate system properties to the surface temperature dataset. Geophys. Res. Lett. **38**(22) (2011)
17. Lopez-Paz, D., Ranzato, M.: Gradient episodic memory for continual learning. arXiv preprint arXiv:1706.08840 (2017)
18. Mallya, A., Lazebnik, S.: Packnet: adding multiple tasks to a single network by iterative pruning. In: Proceedings of the IEEE Conference on Computer Vision and Pattern Recognition, pp. 7765–7773 (2018)
19. Masse, N.Y., Grant, G.D., Freedman, D.J.: Alleviating catastrophic forgetting using context-dependent gating and synaptic stabilization. Proc. Natl. Acad. Sci. **115**(44), E10467–E10475 (2018)
20. McCloskey, M., Cohen, N.J.: Catastrophic interference in connectionist networks: the sequential learning problem. In: Psychology of Learning and Motivation, vol. 24, pp. 109–165. Elsevier (1989)
21. Minakshisundaram, S., Pleijel, Å.: Some properties of the eigenfunctions of the laplace-operator on riemannian manifolds. Can. J. Math. **1**(3), 242–256 (1949)
22. O'Reilly, R.C., Norman, K.A.: Hippocampal and neocortical contributions to memory: Advances in the complementary learning systems framework. Trends Cogn. Sci. **6**(12), 505–510 (2002)
23. Ossandón, S., Reyes, C., Reyes, C.M.: Neural network solution for an inverse problem associated with the dirichlet eigenvalues of the anisotropic laplace operator. Comput. Math. Appl. **72**(4), 1153–1163 (2016)

24. Pan, L., Zhou, X., Shi, R., Zhang, J., Yan, C.: Cross-modal feature extraction and integration based RGBD saliency detection. Image Vis. Comput. **101**, 103964 (2020)
25. Rebuffi, S.A., Kolesnikov, A., Sperl, G., Lampert, C.H.: iCaRL: incremental classifier and representation learning. In: Proceedings of the IEEE Conference on Computer Vision and Pattern Recognition, pp. 2001–2010 (2017)
26. Robins, A.: Catastrophic forgetting in neural networks: the role of rehearsal mechanisms. In: Proceedings 1993 the First New Zealand International Two-Stream Conference on Artificial Neural Networks and Expert Systems, pp. 65–68. IEEE (1993)
27. Rusu, A.A., et al.: Progressive neural networks. arXiv preprint arXiv:1606.04671 (2016)
28. Shin, H., Lee, J.K., Kim, J., Kim, J.: Continual learning with deep generative replay. arXiv preprint arXiv:1705.08690 (2017)
29. Steele, M., Chaseling, J.: Powers of discrete goodness-of-fit test statistics for a uniform null against a selection of alternative distributions. Commun. Stat.-Simul. Comput. **35**(4), 1067–1075 (2006)
30. Wen, W., Wu, C., Wang, Y., Chen, Y., Li, H.: Learning structured sparsity in deep neural networks. arXiv preprint arXiv:1608.03665 (2016)
31. Wu, Y., et al.: Large scale incremental learning. In: Proceedings of the IEEE/CVF Conference on Computer Vision and Pattern Recognition, pp. 374–382 (2019)
32. Xiang, Y., Fu, Y., Ji, P., Huang, H.: Incremental learning using conditional adversarial networks. In: Proceedings of the IEEE/CVF International Conference on Computer Vision, pp. 6619–6628 (2019)
33. Yoon, J., Yang, E., Lee, J., Hwang, S.J.: Lifelong learning with dynamically expandable networks. arXiv preprint arXiv:1708.01547 (2017)
34. Zeng, G., Chen, Y., Cui, B., Yu, S.: Continual learning of context-dependent processing in neural networks. Nat. Mach. Intell. **1**(8), 364–372 (2019)
35. Zenke, F., Poole, B., Ganguli, S.: Continual learning through synaptic intelligence. In: International Conference on Machine Learning, pp. 3987–3995. PMLR (2017)

Improving Generalization
of Reinforcement Learning
for Multi-agent Combating Games

Kejia Wan[1], Xinhai Xu[2(✉)], and Yuan Li[2(✉)]

[1] Defence Innovation Institute, Beijing, China
[2] Academy of Military Science, Beijing, China
{xuxinhai,yuan.li}@nudt.edu.cn

Abstract. Multi-agent combating games have attracted great interest from the research community of reinforcement learning. Most work concentrates on improving the performance of agents, and ignores the generality of the trained model on different tasks. This paper aims to enhance the generality of reinforcement learning, which makes the trained model easily adapt to different combating tasks with a variable number of agents. We divide the observation of an agent to the common part which is related to the number of other agents, and the individual part which includes only its own information. The common observation is designed with a special representation of matrices irrelevant to the number of agents, and we use convolutional networks to extract valuable features from multiple matrices. The extracted features and the individual observation form new input for reinforcement learning algorithms. Meanwhile, the number of agents also changes during the combating process. We introduce a death mask technique to avoid the effects of the dead agents on the loss computation of multi-agent reinforcement learning. Finally, we conducted a lot of experiments in StarCraft-II on unit micromanagement missions. It turned out that the proposed method could significantly improve the generality and transferring ability of the model between different tasks.

Keywords: Multi agent reinforcement learning · Generalization · Combating games

1 Introduction

Combating games like unit micromanagement missions in StarCraft-II have been recognized as a standard testbed for multi-agent reinforcement learning (MARL) [10]. Various MARL algorithms have been put forward, and some of them perform very well in some difficult combating tasks [13]. MARL algorithms based on value decomposition have attacked much attention in recent years [16]. It follows the paradigm of centralized training with decentralized execution. The main idea is to decompose the global Q-value into individual Q-values for all agents. During the training, a sum function (see VDN in [8]) or a mixing network (see QMIX in [6]) is introduced to compute the global Q-value. During the execution, each agent

© Springer Nature Switzerland AG 2021
T. Mantoro et al. (Eds.): ICONIP 2021, LNCS 13109, pp. 64–74, 2021.
https://doi.org/10.1007/978-3-030-92270-2_6

computes the next action based on its own network model. Most work follows the structure of QMIX, and makes some different revisions to enhance the win rate on different maps, such as OWQMIX [5], ROMA [12] and QPLEX [11]. However, trained agents often overfit to a specific task, and the lack of generalization makes them unstable when they are applied for different tasks.

There have been some studies on the generalization of reinforcement learning on environments other than combating games. A natural strategy is to remove the input and output layers from the pre-trained QMIX network and fill the dimension of the observation with zero-vector that match the dimensions required of the scene, which includes the most number of the agents. The middle stack of the deep neural network is then fine-tuned with the RL loss [2]. Prior works have shown that such a transfer is effective in certain computer vision tasks [14]. However, it is not effective in multi-agent learning tasks, which will be analyzed in the experiment. However, a main character of combating tasks with different maps is the variable number of agents, which results in a dynamic input for the neural network. This is not taken into account for generalization techniques mentioned before.

In this paper we propose a method for improving the generalization of MARL models across different maps. The main idea is to construct a novel state representation which is invariant to the number of agents. We discrete the map into grids, which can be represented by a matrix. For each agent, we divide the observation into two parts: the common observation and the individual observation. The common observation contains information of all other agents, including enemies and allies. The individual information of one agent only contains its own information. The information of each agent may have many properties such as the position, health point, etc. The common observation is related to the number of agents while the individual observation is not. For the common observation, we design a special matrix representation which is irrelevant to the number of agents. Each matrix records a property of all other agents. The number of matrices is equivalent to the number of properties. Then we introduce the convolutional network to extract valuable features from multiple matrices. The extracted features and the individual observation form the new input for reinforcement learning algorithms. Meanwhile, the number of agents also varies during the combating process due to the death. We introduce a death mask technique to avoid the effects of the dead agents on the loss computation of multi-agent reinforcement learning. Although the death masking has been used in the work [15], it is mainly used to improve performance of the algorithm.

Another benefit of the designed observation representation is to improve the transference ability. Some transferring techniques have been studied. Paper [3] proposes the idea of policy distillation, where information from the teacher policy network is transferred to a student policy network. Some works [4] introduce techniques for knowledge transfer, which enables accelerating agent learning by leveraging either existing trained policies or using task demonstrations for imitation learning. Those methods always cost much computation or time resources to obtain previous experience. With the proposed model, the trained model could be reused for training on new tasks.

Finally, we conducted a lot of experiments in StarCraft-II on unit micromanagement missions over several state-of-the-art MARL methods. It turned out that the proposed way could not only effectively improve the generality of the model among different tasks, but also enhance the transferring ability of the trained model.

2   Background

2.1   Multi-agent Markov Decision Process

In the study we premeditate a combating game which can be modelled by a decentralized partially observable Markov decision process (Dec-POMDP). It is denoted by a tuple $<\mathcal{N}, \mathcal{S}, \mathcal{O}, \mathcal{A}, \mathcal{P}, \mathcal{R}, \lambda>$. $\mathcal{N}$ represents a set of all agents and $\mathcal{S}$ denotes the set of states. Because of a limited vision, the combating game is always a partially observable environment. Let $\mathcal{O} = \{O_1, O_2, ..., O_N\}$ denote observations of all agents, and $\mathcal{A} = \{A_1 \times A_2 \times ... \times A_N\}$ denote the set of actions for all agents. $\mathcal{P} : \mathcal{S} \times \mathcal{A} \times \mathcal{S}$ is the state transition function and $\mathcal{R}$ is the joint reward function;

$\pi_i : O_i \times A_i \rightarrow [0, 1]$ is defined as the policy for each agent. Each agent can get a local observation o_i^t and make an action choice $a_i^t \in A_i$ based on its own policy π_i at timestep t. The state transition follows the transition function $\mathcal{P}$ with the joint action of all agents which is combined by $\{a_1^t, a_2^t, ..., a_N^t\}$. And a reward r is returned by the environment following the reward function $\mathcal{R}$. To reach the optimal policy, the state value function $V^\pi(S)$ is introduced by cumulated rewards, see (1).

$$V^\pi(S) = \mathbb{E}\{\sum_{t=0}^{\infty}\lambda^t r^t | S_0 = S\} \tag{1}$$

where $\mathbb{E}$ is the expectation and $\lambda \in [0, 1]$ is the discount factor. The state-action value $Q^\pi(S, A)$ is defined as (2) based on $V^\pi(S)$.

$$Q^\pi(S^t, A^t) = \mathbb{E}\{r^{t+1} + \lambda V^\pi(S^{t+1})\} \tag{2}$$

2.2   MARL Algorithms Based on Value Decomposition

Recently, value-decomposition based MARL algorithms have been a hot research topic. It means that the global state-action value Q_{tot} is decomposed into the individual value Q_i. Akin ways have been proposed to study the relation between Q_{tot} and Q_i. Here we only introduce two typical methods, VDN and QMIX. Q_{tot} is expressed as a sum of Q_i in VDN. After this, a continuous monotonic function are adopted in QMIX in form of a mixing network to express this relation. And then each agent corresponds an independent neural network in QMIX, which is employed to compute Q_i based on its own local observation. We will train all independent neural networks and the mixing network at the same time. The loss is computed by Eq. (3), where $y_k^{tot} = r + max_{a'}Q_{tot}(o, a; \theta^-)$.

$$L(\theta) = \sum_{k=1}^{b}[(y_k^{tot} - Q_{tot}(o, a; \theta))^2] \tag{3}$$

θ represents parameters of eval networks while θ^- represents that of target networks. b is the number of samples used to train the neural network.

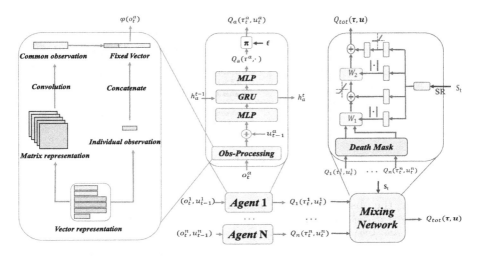

Fig. 1. The whole framework of the proposed method. The left-hand side (in orange) is the observation process modular. The middle part (in green) is the individual network for each agent. The right-hand side (in blue) is the structure of the mixing network, in which the death masking technique is applied. (Color figure online)

3   Methodology

In this section, we introduce the method for improving the generalization of the valued decomposition based MARL algorithms. The whole framework is shown in Fig. 1. We follow the structure of the QMIX algorithm [6], based on which we introduce two important techniques: matrix state representation and death masking.

As we can see, for each agent, we add a modular to process the observation before it is used for the input of the algorithm. The aim of this modular is to make the observation irrelative to the number of agent, which is beneficial for the generalization of MARL methods. Another modular, death masking, is imbedded in the mixing network. The two techniques are detailed introduced in the following subsections.

3.1   Multi-layer Matrix State Representation with Feature Extraction

In this subsection we introduce a multi-layer matrix state representation to deal with the dynamic number of agents in different combating tasks. The observation for an agent in a game is the partially observed state returned from the environment. It contains information for all observed agents. The variation of the number of agents often changes the dimension of the observation, which normally results in a dynamic input for the neural network.

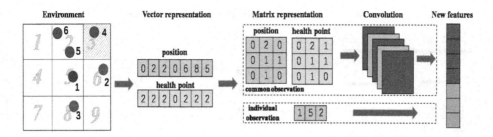

Fig. 2. A toy example of state representation.

Normally, the observation of an agent includes three parts: the state of enemies S^e, the state of allies S^a and the state of mine S^m. It can be represented by $o_i = (S_i^e, S_i^a, S_i^m)$ where o_i denotes the observation of agent i. For different combating tasks, the length of S^e and S^a will change with the number of agents while S^m will not. We call S^e and S^a as the common information and S^m the individual information.

To introduce the multi-layer matrix state representation, we firstly divide the map of the combating game into grids. All the following matrixes are constructed based on this divination. Let $\mathcal{G}$ denote the set of index of all grids. If an agent i locates in a grid $g \in \mathcal{G}$, then we have $i \in \delta(g)$. Let $\mathcal{C}$ represent the set of properties for each agent and v_i^c represent the value of the property c for agent i. For any property $c \in \mathcal{C}$, we construct a corresponding observation matrix D^c in which the value of each element is computed as $\sum_{i \in \delta(g)} v_i^c$.

Taking Fig. 2 as an example, there are 3 red agents and 3 blue agents. The map is divided into 9 grids and $v_i^1 = 1, v_i^2 = 2, v_i^3 = 3, \forall i \in \mathcal{N}$. Suppose each agent has two properties: the position and the health point. For agent 1, the vector representation of the observation $o_1 = (S_1^e, S_1^a, S_1^m)$ is shown in the figure. Note that the environment is partially observable, and agent 4 is invisible to agent 1. In the position vector, each element represents the index of the position of the agent, and $S_1^e = [0, 2, 2]$, $S_1^a = [0, 6, 8]$, $S_1^m = [5]$. Each element of the health point vector means the health value of each agent. Then we convert vector representation to the matrix representation, which consists of two parts: common observation and individual observation. In the common observation, each element of the position matrix is the number of agents in that grid, and the element of the health matrix is the sum of health points of agents. In the individual observation, the three elements represent the index, the position and the health point of agent 1. Note that there can be many matrixes in the common observation, and thus we introduce the convolution network which converts the matrixes into a vector. Combined the common vector and the individual vector together, we obtain new features which could be used for further learning.

Algorithm 1: MARL method with enhanced generalization

Initialize parameters θ of the network in MARL method;

while *the maximum number of episodes is not reached* **do**

    Start a new episode;

    while *the episode is not terminated* **do**

        for *each agent i in $\mathcal{N}$* **do**

            Get observation o_i^t from the environment;

            Divide o_i^t into common observation $\dot{o}_i^t$, individual observation $\ddot{o}_i^t$;

            (State modeling) Divide the map into grids with dimension $x \times y$;

            for *attribute j in each agent* **do**

                Construct a matrix m_j based on the common observation $\dot{o}_i^t$;

            Feature extraction Convert attribute matrices M_i to a feature vector $\bar{o}_i^t$ based on CNN;

            Concatenate $\bar{o}_i^t$ and $\ddot{o}_i^t$ into new observation $\hat{o}_i^t$;

            Calculate the individual Q value Q_i and next action u_i^t;

        Concaten u_i^t into joint action u^t;

        Receive global reward r^t and observe a new state s';

        Store$(s^t, o_i^t, u^t, r^t, s')$ in replay buffer $mathcalD$;

        if *UPDATE* **then**

            Sample a batch $\mathcal{D}'$ from $mathcalD$;

            For each sample in $\mathcal{D}'$, calculate Q_i by above steps;

            (Death masking) For each Q_i, if the agent i dies, Q_i equals to 0;

            Get the new global state s_{new}^t after state modeling and convolution network;

            Calculate Q_{tot} in the mixing network with s_{new}^t;

            Calculate the target value y_{tot};

            Calculate the loss and update θ by minimizing the TD loss;

            Update $\hat{\theta} : \hat{\theta} \leftarrow \theta$;

return θ;

3.2 Death Masking for Individual Q Values

In multi-agent combating games, an important character is that the number of agents varies due to the death event. For value decomposition based MARL algorithms, even if an agent dies, its corresponding Q value is still used for the value learning of the mixing network. Learning with the these informative states of dead agents will amplify the bias of the learned value function for the mixing network. Consequently, high value prediction errors in the mixing network will also be passed on to the individual network for each agent. And this error will also be accumulated during the computation, which would hinder finding a good policy.

Here we use the technique "death masking" to remove the effect of the dead agent on the value learning of the mixing network. Let function f represent the mixing network and $Q' = (Q_1, Q_2, ...Q_n)$. Then we have $Q_{tot} = f(Q')$. We introduce an extra death mask vector $M = (m_1, m_2, ...m_n)$, in which $m_i = 0$ when agent i dies during the game. We mask the Q values by $Q'' = M * Q'$ and then $Q_{tot} = f(Q'')$. In this way, the individual Q value of dead agent will be 0 when computing the total Q value.

In fact, the "death masking" technique could also improve the generality of MARL methods. During the combating process, the number of agents vary over the time steps, which also enhances the ability of the trained model to adapt to other tasks with different number of agents. This will be verified in Sect. 5.

Through this structure, we can train highly adaptive reinforcement learning algorithms that are not sensitive to the number of agents. The method is described in Algorithm 1.

4   Experiments

In this section, we performed experiments to quantify the efficacy of our algorithm for transfer learning in reinforcement learning. We address the following questions:

a) Can we do a successful transfer between two maps with different numbers of agents?
b) Which factors can affect generalization? For example, the similarity of the map or the type of agent.
c) What are the roles of state modeling and death masking, and which one is more effective?

4.1   Setup

Environments. All experiments use the default reward and observation settings of the SMAC benchmark. StarCraft II Micromanagement Challenge (SMAC) tasks were introduced [7]. In these tasks, decentralized agents must cooperate to defeat adversarial bots in various scenarios with a wide range of agent numbers. The original global state of SMAC contains information about all agents and enemies - this includes information such as the distance from each agent/enemy to the map center, the health of each agent/enemy, the shield status of each agent/enemy, and the weapon cooldown state of each agent. However, the local observation of each agent contain agent-specific information including agent id, agent movement options, agent attack options, relative distance to allies/enemies. Note that the local observation contains information only about allies/enemies within a sight radius of the agent. As described in Sect. 3, we utilize global information as input to the convolution network. This global vector augments the original global state provided by the SMAC environment by abstracting relevant features.

Baselines. In our setting, the target and the source networks have dissimilar input (the MDPs have different state-spaces because of the change of the number of agents). We take a classic transfer learning strategy (extended dimension of observation) [2] used by Deepmind in dota2 [1] as the baseline.

4.2   Comparison

For comparison, we define the following two methods: QMIX-SD: The QMIX method [6] with state modelling and death masking. QMIX-E: The QMIX method with extended dimension of observation. We consider several different transferring scenarios in SMAC environments, i.e., 3 m to 8 m (Map1-Up), 8 m to 8 m vs 9 m (Map2-Up), 3s5z to 1c3s5z (Map3-Up) and their rebellion (MapX-Down), which differs in the number or the type of agents (Marines, Colossus, Stalker, Zealots).

   In general, the purpose of generalization is to improve the test win rate in different scenarios, and the main goal of transfer is to accelerate learning. Whether or not a single-agent transfer algorithm is better than another is commonly evaluated through several of the following performance metrics, summarized by Taylor and Stone [9] and illustrated as follows: **Jumpstart (JS)**: Measures the improvement in the initial performance of the agent; **Time to threshold (TT)**: For domains in which the agents are expected to achieve a fixed or minimal performance level, the learning time taken for achieving it might be used as a performance metric. While both metrics are also generally applicable to MAS.

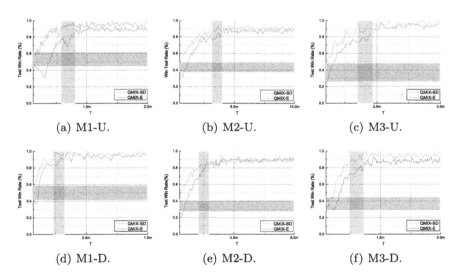

(a) M1-U.   (b) M2-U.   (c) M3-U.

(d) M1-D.   (e) M2-D.   (f) M3-D.

Fig. 3. Comparisons of the proposed method with QMIX-S and QMIX-E.

Figure 3 plots the win rate curves for the above main two methods in different transfer learning experiments. The green vertical stripes represent the gap between the two methods in TT, and the purple horizontal stripes represent the gap between the two methods in JS. For both, the wider the stripes, the higher the improvement of our algorithm relative to the benchmark. At test time, we measure the jumpstart on the test set, applying no fine-tuning to the agent's parameters. The first observation is that QMIX-SD gets larger values of the jumpstart in 6 cases. With the training iterations, the winning rate has dropped a little. It has risen back and reached the threshold albeit at a sluggish pace. Obviously, QMIX-SD performed the best again, with the shortest time to reach the threshold. We can observe that the performance of QMIX-SD improves greater at TT at the beginning, about 20% higher than QMIX-E in scenes of moving from a low dimension to a high dimension. In the scene of moving from a high dimension to a low dimension, QMIX-SD improves 15% speed than QMIX-E. Secondly, we find the gap between train and test performance determines the extent of overfitting if train and test data are drawn from the same distribution. If not, heterogeneous agents will bring lower jumpstart values and slower times to threshold (Fig. 3 (c, f)). As the number of agents grows, we expect the performance on the test set to improve. We can still get similar conclusions by combining with other algorithms(VDN [8], OW-QMIX [5], ROMA [12]) in the same scenario, as shown in Table 1.

The results of these experiments prove that firstly, these maps do have a structural commonality such that a policy trained in one task could be used advantageously to accelerate learning in a different task. And the more similar the map, the better the effect; Secondly, the effect of migration from low-dimensional to high-dimensional is more obvious in JS.

Table 1. Comparisons of the proposed method with the baseline in other algs.

Methods	JS						TT					
	M1-U	M1-D	M2-U	M2-D	M3-U	M3-D	M1-U	M1-D	M2-U	M2-D	M3-U	M3-D
VDN-E	31.3	29.3	26.7	21.1	23.6	20.5	1.23	3.51	3.92	2.75	2.69	1.77
VDN-SD	45.6	38.9	36.7	29.9	30.1	28.8	1.01	2.93	3.34	2.31	2.41	1.54
OWQMIX-E	45.3	44.9	40.7	30.5	35.6	33.5	0.78	2.78	3.22	2.15	2.26	1.07
OWQMIX-SD	65.5	60.1	52.5	44.3	52.9	44.2	0.62	2.31	2.77	1.78	2.03	0.98
ROMA-E	50.3	51.6	44.2	31.5	32.6	30.8	0.73	2.61	2.9	1.94	2.16	0.95
ROMA-SD	66.1	63.4	53.2	45.7	53.1	48.6	0.59	2.16	2.54	1.62	1.95	0.84

4.3 Ablation

In addition to exploring the adaptability, we also conducted experiments on the effects of the ability of each component. Based on the previous section, two new ablation experiments have been added: Algs-S: The QMIX method with state modelling. Algs-D: The QMIX method with death masking. As shown in Fig. 4, Algs-SD is a successful approach for achieving such a transfer. Without death masking, the results of Algs-S show that it gets less value of the jumpstart than QMIX-SD, not much difference. Compared with QMIX-E, the use of death masking can improve the versatility of QMIX-SD. Secondly, QMIX-E still outperforms VDN-E, implying the winning rate of the original algorithm also has a great influence on the result of generalization.

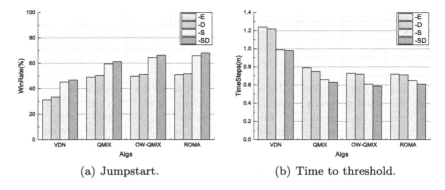

(a) Jumpstart.   (b) Time to threshold.

Fig. 4. Ablation results on Map1-Up.

Experiments have proved that SD is the best method on the two indicators of JS and TT. This works even when the source and target MDPs have different state and observation spaces, and is realized by learning the global state with convolution network. And death mask can also help accelerate the training and improve the performance of transferring on a small margin.

5   Conclusion

In this paper we propose an efficient method for improving the generalization of MARL algorithms. We design a matrix state modelling technique to deal with variable number of agents among different tasks. Meanwhile, we use the death masking technique to dispose the dynamic change of the number of agents in a combating task. The results show that the proposed method improve both the generalization and the transferring ability of MARL methods. It would be interesting to investigate the proposed method on more MARL algorithms.

References

1. Berner, C., et al.: Dota 2 with large scale deep reinforcement learning. arXiv preprint arXiv:1912.06680 (2019)
2. Chen, T., Goodfellow, I., Shlens, J.: Net2Net: accelerating learning via knowledge transfer. arXiv preprint arXiv:1511.05641 (2015)
3. Czarnecki, W.M., Pascanu, R., Osindero, S., Jayakumar, S., Swirszcz, G., Jaderberg, M.: Distilling policy distillation. In: The 22nd International Conference on Artificial Intelligence and Statistics, pp. 1331–1340. PMLR (2019)
4. Da Silva, F.L., Costa, A.H.R.: A survey on transfer learning for multiagent reinforcement learning systems. J. Artif. Intell. Res. **64**, 645–703 (2019)
5. Rashid, T., Farquhar, G., Peng, B., Whiteson, S.: Weighted QMIX: expanding monotonic value function factorisation for deep multi-agent reinforcement learning. In: NeurIPS (2020)

6. Rashid, T., Samvelyan, M., Schroeder, C., Farquhar, G., Foerster, J., Whiteson, S.: QMIX: monotonic value function factorisation for deep multi-agent reinforcement learning. In: International Conference on Machine Learning, pp. 4295–4304 (2018)
7. Samvelyan, M., et al.: The StarCraft multi-agent challenge. CoRR arXiv:1902.04043 (2019)
8. Sunehag, P., et al.: Value-decomposition networks for cooperative multi-agent learning based on team reward. In: Proceedings of the International Conference on Autonomous Agents and Multiagent Systems, pp. 2085–2087 (2018)
9. Taylor, M.E., Stone, P.: Transfer learning for reinforcement learning domains: a survey. J. Mach. Learn. Res. **10**(7), 1633–1685 (2009)
10. Vinyals, O., et al.: StarCraft II: a new challenge for reinforcement learning. arXiv preprint arXiv:1708.04782 (2017)
11. Wang, J., Ren, Z., Liu, T., Yu, Y., Zhang, C.: QPLEX: duplex dueling multi-agent Q-learning (2020)
12. Wang, T., Dong, H., Lesser, V., Zhang, C.: ROMA: multi-agent reinforcement learning with emergent roles. In: International Conference on Machine Learning, pp. 9876–9886. PMLR (2020)
13. Ye, D., et al.: Mastering complex control in MOBA games with deep reinforcement learning. In: Proceedings of the AAAI Conference on Artificial Intelligence, pp. 6672–6679 (2020)
14. Yosinski, J., Clune, J., Bengio, Y., Lipson, H.: How transferable are features in deep neural networks? arXiv preprint arXiv:1411.1792 (2014)
15. Yu, C., Velu, A., Vinitsky, E., Wang, Y., Bayen, A., Wu, Y.: The surprising effectiveness of MAPPO in cooperative, multi-agent games. arXiv preprint arXiv:2103.01955 (2021)
16. Zhang, T., et al.: Multi-agent collaboration via reward attribution decomposition (2020)

Gradient Boosting Forest: a Two-Stage Ensemble Method Enabling Federated Learning of GBDTs

Feng Wang[1] , Jinxiang Ou[1] , and Hairong Lv[1,2] (✉)

[1] Department of Automation, Tsinghua University, Beijing 100084, China
{wangf19,ojx19}@mails.tsinghua.edu.cn, lvhairong@tsinghua.edu.cn
[2] Fuzhou Institute of Data Technology, Fuzhou 350207, China

Abstract. Gradient Boosting Decision Trees (GBDTs), which train a set of decision trees in sequence with a gradient boosting strategy to fit the features of training data, has become very popular in recent years due to its strong capability in dealing with machine learning tasks. In many well-known machine learning competitions, GBDT even outperforms very complicate deep neural networks. Nevertheless, training such tree-based models requires accessing the whole dataset to find the split points on the features, which makes distributed training of GBDT models difficult. Particularly, in Federated Learning (FL), where training data is decentralized distributed and cannot be shared considering the privacy and security, training GBDT becomes challenging. To address this issue, in this paper, we propose a new tree-boosting method, named Gradient Boosting Forest (GBF), where the single decision tree in each gradient boosting round of GBDT is replaced by a set of trees trained from different subsets of the training data (referred to as a forest), which enables training GBDT in Federated Learning scenarios. We empirically prove that GBF outperforms the existing GBDT methods in both centralized (GBF-Cen) and federated (GBF-Fed) cases. In a series of experiments, GBF-Cen achieves 1.1% higher accuracy on HIGGS-1M dataset over XGBoost and GBF-Fed obtains 12.2%–48.0% lower RMSE loss over the state-of-the-art federated GBDT methods.

Keywords: Machine learning · GBDT · Federated learning

1 Introduction

Federated learning (FL) techniques [14,18,19] aim to train machine learning models on private date stored across a variety of devices without straightforwardly sharing them. Represented by FedAvg [18], the FL algorithms often include two steps—updating and aggregating, where the data owners train a

---

Supported by the National Nature Science Foundation of China under Grant No. 42050101 and U1736210.

T. Mantoro et al. (Eds.): ICONIP 2021, LNCS 13109, pp. 75–86, 2021.
https://doi.org/10.1007/978-3-030-92270-2_7

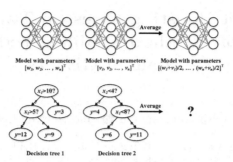

Fig. 1. Model aggregation for neural networks and decision trees. A simple way to aggregate neural networks is averaging their parameters. However, this approach is not applicable to decision trees since the parameters in decision trees are not additive.

shared model (e.g. a neural network) on their local data in the updating step and these local models are then averaged in the aggregating step. The alternated updating and aggregating steps in FL helps the model learn useful knowledge from different devices.

However, unlike in neural networks, most parameters in the tree models represent the split points on the features, which are difficult to aggregate (see Fig. 1). Therefore, most of the commonly used FL algorithms are not applicable to the tree-based models such as GBDT [8]. More importantly, the non-independently identically distribution (non-iid) over different devices of the training data may also bring about big challenge in training GBDT, since it is difficult to find globally optimal split points on the features without accessing all the training data over different devices. The existing literatures try to address these issues by training federated GBDT models in an Incremental Learning [5,15] manner, i.e., the data owners train the model by sequence and the model is trained only once on each device [13,23]. However, such methods suffer from poor performance with non-iid data, due to the so-called "catastrophic forgetting" (the model forgets the knowledge previously learned when trained on new data).

In this paper, we propose a two-stage ensemble algorithm named Gradient Boosting Forest (GBF), which trains a set of decision trees (the stage one) with different subsets of the training data, instead of the single tree in each gradient boosting (the stage two) round in GBDT. The set of trees in each round, which is referred to as a forest, brings about two merits of our method. Firstly, ensembling multiple decision trees trained on bootstrap or subsampled data helps improving accuracy and robustness [2,3], which is also empirically proved in our experiments (see Sect. 5.2). Secondly, since the decision trees in a forest can be trained on disjoint data, GBF enables federated learning of the tree-based models.

2   Related Work

There have been a number of literatures attempting to improve the original GBDT in terms of accuracy, robustness, and computational efficiency [4,11,20]. The approximation approaches to find the split points in these methods enable the distributed training of GBDTs. Nevertheless, these approaches still require the access to the entire dataset to compute the split point candidates, which violates the principles of privacy in FL.

To this end, a straightforward way is sharing the encrypted data or features [7,17,21]. However, even the encrypted data may lead to high risk of privacy leakage and the encryption algorithms such as homomorphic encryption [1,9,22] are computationally expensive, which prevents the practical use of these FL methods. Recently, people train federated GBDT models in the way of Incremental Learning [5,15], where the data owners train the GBDT model on their local data by sequence [23]. Further, literature [13] proposed SimFL, which first estimated the similarity of samples from different devices by locality-sensitive hashing [6] and then added the gradient terms of the similar samples from other devices while training a GBDT model.

These methods above are able to deal with some simple federated learning tasks. For the large scale and non-iid data, however, the federated GBDT methods with incremental manner [23] or encrypted data suffer from either low predictive performance or high additional computational cost. In contrast, our two-stage ensemble approach addresses these issues and achieves superior empirical performance.

3   Preliminaries

GBDT is an ensemble learning method based on decision trees, with the main idea to fit the residual loss of current output by training extra regression trees. Formally, given a dataset with N examples and d features $\mathcal{D} = \{(x_i, y_i)\}(|\mathcal{D}| = N, x_i \in \mathbb{R}^d, y_i \in \mathbb{R})$, GBDT predicts the output by T decision trees:

$$\hat{y}_i = \phi(x_i) = \sum_{t=1}^{T} f_t(x_i), \tag{1}$$

where $f_t(x_i)$ is the t-th regression tree in an ensemble. To efficiently train a GBDT model, [4] proposed a regularized loss function

$$\mathcal{L}(\phi) = \sum_i l(\hat{y}_i, y_i) + \sum_t \Omega(f_t(x_i)), \tag{2}$$

where

$$\Omega(f) = \delta_1 M + \frac{1}{2}\delta_2||\omega||^2$$

is a regularization term and l is a differentiable convex loss function which measures the difference between the predicted value and target value of a specific example. M denotes the number of leaf nodes in a tree and $\omega \in \mathbb{R}^M$ is leaf weights vector. δ_1 and δ_2 are hyper parameters controlling the degree of regularization.

GBDT optimizes the objective function in Eq. (2) in a greedily adding manner as minimizing

$$\mathcal{L}^{(t)} = \sum_{i=1}^{N} l(y_i, \hat{y}_i^{(t-1)} + f_t(x_i)) + \Omega(f_t) \tag{3}$$

in each training step, where $\hat{y}_i^{(t)}$ denotes the prediction of x_i in the t-th iteration. By applying second-order approximation and then omitting constant terms, Eq. (3) can be approximated as

$$\tilde{\mathcal{L}}^{(t)} = \sum_{i=1}^{N} [g_i f_t(x_i) + \frac{1}{2} h_i f_t^2(x_i)] + \Omega(f_t), \tag{4}$$

where the gradient statistics on loss function $g_i = \partial_{\hat{y}^{(t-1)}} l(y_i, \hat{y}^{(t-1)})$ and $h_i = \partial_{\hat{y}^{(t-1)}}^2 l(y_i, \hat{y}^{(t-1)})$ are applied.

4   Gradient Boosting Forest

4.1   An Overview of GBF

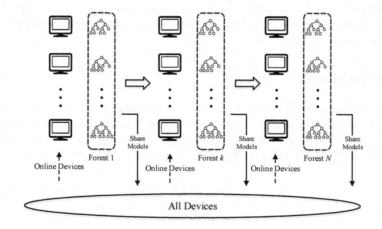

Fig. 2. Framework of GBF-Fed. In each iteration, online devices train decision trees which are then aggregated as a decision forest and shared globally.

Figure 2 shows the framework of GBF in federated learning, which is referred to as GBF-Fed. In each step of iteration, the online devices[1] train a decision tree

---

[1] Like most neural network based federated learning algorithms, our method allows part of the devices being offline and absent for training for a while, which improves the flexibility and saves communication cost, whereas the existing federated GBDT algorithms such as SimFL [13] requires the devices to be online all the time.

on their local data. These locally trained decision trees constitute a forest, which will be then shared globally for gradient boosting after every certain number of iterations.

Notably, GBF also supports centralized training, which is shown to be more effective and robust compared to the original GBDT algorithms (see Sect. 5.2). Here we refer to GBF with centralized data as GBF-Cen. In accordance with the claims in the existing literatures [2,3,16], ensembling decision trees with bootstrap sampling often outperforms the single tree in accuracy. Our method, GBF-Cen, is able to utilize this merit to improve the model performance, when the entire dataset is available. The main difference between GBF-Fed and GBF-Cen is that the training data is divided by their availability in GBF-Fed, while they are manually divided by bootstrap sampling in GBF-Cen.

The followings are the notations we use in this paper. Suppose that we have a training set with N instances and d features $\mathcal{D} = \{(x_i, y_i)\}(|\mathcal{D}| = N, x_i \in \mathbb{R}^d, y_i \in \mathbb{R})$, which is distributed in K devices and K' devices of them are online. In regression tasks, y_i can be a continuous value. In classification tasks, $y_i \in [0, M - 1]$ where M denotes number of classes. At the k-th node, there are n_k instances, where $\sum_{k=1}^{K} n_k = N$. Due to the privacy and security restrictions, training data is only locally available, and the goal of our method is to train a distributed GBDT in T communication rounds. In addition, we define the coefficient λ to represent the number of training iterations per communication.

4.2 GBF for Regression

Algorithm 1: Gradient Boosting Forest for Regression

Input: Datasets $\mathcal{D}_1, \mathcal{D}_2, ..., \mathcal{D}_K$; Shrinkage η; Weights of devices $[\alpha_1, ..., \alpha_K]$; Initial model $f^{(0)} : \mathbb{R}^d \to \mathbb{R}$

Output: An ensemble model $f : \mathbb{R}^d \to \mathbb{R}$

1 **for** $t \leftarrow 1$ **to** T **do**

2   Choose K' devices;

3   **for** $k \leftarrow 1$ **to** K' **do**

4   **for** *instances* $i \leftarrow 1$ ***to*** n_k **do**

5        $g_i \leftarrow \frac{\partial l(y_i, f^{(t-1)}(x_i))}{\partial f^{(t-1)}(x_i)}$;

6        $h_i \leftarrow \frac{\partial^2 l(y_i, f^{(t-1)}(x_i))}{\partial (f^{(t-1)}(x_i))^2}$;

7   **end**

8   Train a regression tree $f_k^{(t)}$ locally by Eq.(8);

9   **end**

10    $f^{(t)} \leftarrow f^{(t-1)} + \eta \sum_{k=1}^{K'} \alpha_k f_k^{(t)} / \sum_{k=1}^{K'} \alpha_k$;

11 **end**

12 $f \leftarrow f^{(T)}$;

GBF is able to deal with regression tasks, where the training process is summarized in Algorithm 1. Given a decentralized dataset, the objective of our method is to train a shared predictor $f : \mathbb{R}^d \to \mathbb{R}$. The model can be initialized by zero initialization or ensembling the trees trained by the online devices for one iteration. In each gradient boosting round, the K' online devices train a CART tree [12] in parallel by the local loss function (Eq. (8)). After local updates, the CART trees are aggregated as a forest and shared globally. Formally, the output of one decision forest is calculated by

$$f(x) = \sum_{k=1}^{K} \alpha_k f_k(x), \tag{5}$$

where $f_k(x)$ is the decision tree trained on $\mathcal{D}_k$, the training data in the k-th device. α_k denotes the weight of the k-th device, which is here set to n_k/N. With the two-stage ensemble design in our method, the model output in Eq. (1) is transformed into

$$\hat{y}_i = \sum_{k=1}^{K} \alpha_k \phi_k(x_i) = \sum_{t=1}^{T} \sum_{k=1}^{K} \alpha_k f_{tk}(x_i). \tag{6}$$

Similarly, the loss function in the t-th iteration is formulated as

$$\mathcal{L}^{(t)} = \sum_{k=1}^{K} (\sum_{i=1}^{n_k} l(y_i, \hat{y}_i^{(t-1)} + \sum_{k=1}^{K} \alpha_k f_{tk}(x_i)) + \Omega(f_{tk})). \tag{7}$$

In each iteration, we split the total loss into K parts and approximate $\sum_{k=1}^{K} \alpha_k f_{tk}(x_i)$ with $f_{tk}(x_i)$ in the k-th device. Thus, the approximated loss function in the k-th device is

$$\tilde{\mathcal{L}}_k^{(t)} = \sum_{i=1}^{n_k} [g_i f_{tk}(x_i) + \frac{1}{2} h_i f_{tk}^2(x_i)] + \Omega(f_{tk}). \tag{8}$$

Therefore, the GBF model is trained by minimizing $\mathcal{L}^{(t)}$ locally and aggregating the local updates every λ iterations.

4.3   GBF for Classification

GBF is also competent at classification tasks. We need to first transform y_i into one-hot labels and separately train regression trees for each class to predict the probability. In multi-class tasks, we try to minimize the loss

$$\mathcal{L} = -\sum_{m=1}^{M} y_m \log p_m(x), \tag{9}$$

where M denotes the total number of classes and y_m denotes the m-th term in one-hot label of a training sample. $p_m(x)$ is calculated by softmax function which denotes the probability that label of x equals m. Similarly, GBF for classification greedily adds new regression trees to reduce the bias between y_m and $p_m(x)$. We train regression trees by the gradient terms $\tilde{y}_m = -\partial \mathcal{L} / \partial F_m$ to fit this bias. [8] has proven that for each device, the optimal leaf node value in step t under loss function (9) is

$$\gamma_{jm}^{(t)} = \frac{M}{M-1} \frac{\sum_{x_i \in S_{jm}^{(t)}} \tilde{y}_{mi}}{\sum_{x_i \in S_{jm}^{(t)}} |\tilde{y}_{mi}|(1 - |\tilde{y}_{mi}|)}, \tag{10}$$

Algorithm 2: Gradient Boosting Forest for Classification

Input: Datasets $\mathcal{D}_1, \mathcal{D}_2, ..., \mathcal{D}_K$; Shrinkage η; device weight α_k,
$(k \in [1, K])$; Initial model $f^{(0)} = \left[p_m^{(0)} : \mathbb{R}^d \to \mathbb{R} \right]_{m=1}^{M}$

Output: An ensemble model $f : \mathbb{R}^d \to \mathbb{R}^M$

1 **for** $t \leftarrow 1$ **to** T **do**
2   Choose K' devices;
3   **for** $k \leftarrow 1$ **to** K' **do**
4   **for** $i \leftarrow 1$ **to** n_k, $m \leftarrow 1$ **to** M **do**
5       $\tilde{y}_{mi}^{(t)} = y_{mi} - p_m^{(t-1)}(x_i)$;
6   **end**
7   **for** $m \leftarrow 1$ **to** M **do**
8       $\{S_{jkm}^{(t)}\}_1^J \leftarrow J$-leaves tree($\{x_i, \tilde{y}_{mi}^{(t)}\}_{i=1}^{n_k}$);
9   get $\gamma_{jkm}^{(t)}$ by Eq. (10);
10   **end**
11   **end**
12    $\Delta p_{km}^{(t)} \leftarrow \frac{\exp\left(\sum_{j=1}^{J} \gamma_{jkm}^{(t)} I(x \in S_{jkm}^{(t)}) \right)}{\sum_{m=1}^{M} \exp\left(\sum_{j=1}^{J} \gamma_{jkm}^{(t)} I(x \in S_{jkm}^{(t)}) \right)}$;
13    $p_m^{(t)} \leftarrow p_m^{(t-1)} + \eta \sum_{k=1}^{K'} \alpha_k \Delta p_{km}^{(t)} / \sum_{k=1}^{K'} \alpha_k$;
14 **end**
15 $f \leftarrow \left[p_m^{(T)} \right]_{m=1}^{M}$;

---

where the feature space is divided into J parts by the CART tree trained by m-th label in t-th iteration. $S_{jm}^{(t)}$ denotes the j-th part of feature space, and $\gamma_{jm}^{(t)}$ is the optimal value of j-th leaf node. The training process of classification is shown in Algorithm 2, which is similar to that in GBF regression tasks, whilst the key difference between GBF classification and regression are the loss function and leaf node outputs.

4.4 Analyses of Privacy

In most federated learning literatures [18,19], people assume that sharing trained models is safe and acceptable. Here we follow this assumption. Figure 3 shows the information flow of GBF and the baseline methods. GBF and the sequential training method [23], referred to as SeqFL, only share the trained models, which

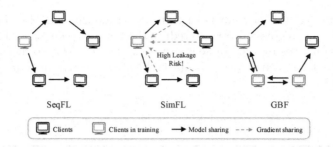

Fig. 3. Information flow of SeqFL, SimFL and GBF's training process, where SeqFL and GBF only share models. However, SimFL shares gradient terms, which may results in high risk of privacy leakage.

is safe under our security assumption. However, in SimFL [13], the gradient terms of other devices are required for training, which may result in higher risk of privacy leakage.

5 Experiments

5.1 Experiment Settings

In this section, we evaluate the effect of GBF and compare it to the competitive GBDT baselines, including:

- XGBoost [4], also referred to as Centralized Training in this paper, which is the state-of-the-art GBDT method for centralized data.
- SimFL, a federated GBDT method proposed by [13], which improves the performance by sharing some gradient information among devices.
- SeqFL, a distributed GBDT method proposed by [23], which is able to train federated GBDT in an Incremental Learning manner.

Specifically, we compare our method, GBF-Cen, with XGBoost since they only support centralized training, i.e., they both require the access to the entire dataset. We compare GBF-Fed with SimFL and SeqFL since they support federated training. Generally, a way to demonstrate the effect of federated learning algorithms is to show the margin of performance between these methods and centralized training. Therefore, we also take XGBoost as a baseline of federated GBDT methods and refer to it as Centralized Training. To achieve optimal performance of SeqFL, we omit the differential privacy in its prototype. In SimFL, we adopted the same method to generate LSH functions as [13] that set $L = \min(40, d - 1)$ hash functions where d is the dimension of features.

All the experiments were conducted on a Linux machine with two Intel(R) Xeon(R) Gold-6148 CPUs and 512 GB main memory. We use eight public datasets from LIBSVM[2] and scikit-learn[3] for regression and classification. The

---

[2] https://www.csie.ntu.edu.tw/~cjlin/libsvm/.
[3] https://scikit-learn.org.

Table 1. The datasets used in our experiments.

Dataset	# of examples	# of features	# of classes	Task
abalone	4,177	8	–	Regression
cadata	20,640	8	–	Regression
cpusmall	8,192	12	–	Regression
space_ga	3,107	6	–	Regression
YearPredictionMSD	515,345	90	–	Regression
digits	1,797	64	10	Classification
letter	20,000	16	26	Classification
satimage	6,435	36	6	Classification
Sensorless	58,509	48	11	Classification
HIGGS	1,000,000	28	2	Classification

detailed information about the datasets is shown in Table 1. We randomly select 75% of a dataset for training and 25% for evaluating. The maximum depth of a decision tree is set to 4–8 based on the size of dataset. The minimum number of samples per leaf is set to 10. We also adopt shrinkage in training, which is set to 0.1 in GBF-Cen and XGBoost, while it is set $\min(K, 10)$ times smaller in the federated methods.

To simulate non-iid, we split training sets in a specialized way [10]. For regression, we first divide the training set into K parts $\mathcal{D}_1, ..., \mathcal{D}_K$, then randomly sample $\beta|\mathcal{D}_k|$ instances without replacement from $\mathcal{D}_k$ and $(1 - \beta)|\mathcal{D}_k|/(K - 1)$ instances from other parts as the k-th subset, where $0 < \beta < 1$ measures the level of non-iid. For classification, we generate non-iid data by Dirichlet distribution. In this method, the classes proportion q subjects to $Dir(\theta p)$, where p is the prior distribution of total data and $\theta \geq 0$ tunes the degree of non-iid. As the smaller θ is, the generated data is more heterogeneous. Specifically, it is equivalent to a uniform distribution when $\theta \to \infty$, and it approximates to a one-class problem when $\theta = 0$.

5.2   Evaluation Results

Table 2. Test RMSE loss on the regression datasets. The best results are **bolded**.

Datasets	abalone	cadata	cpusmall	space_ga	YearPredictionMSD
XGBoost	**2.20**	4.85×10^4	2.76	0.105	9.12
GBF-Cen	2.21	$\mathbf{4.76 \times 10^4}$	**2.69**	**0.101**	**8.98**

We summarize the performance on the centralized data in Table 2 and 3. In most cases, GBF outperforms XGBoost on both regression and classification

Table 3. Test accuracy on the classification datasets. The best results are **bolded**.

Datasets	Digits	Letter	Satimage	Sensorless	HIGGS
XGBoost	97.5	98.2	94.6	**99.8**	70.3
GBF-Cen	**97.6**	**98.6**	**95.1**	**99.8**	**71.4**

tasks. Particularly, GBF-Cen achieves 8.98 RMSE on YearPredictionMSD and 71.4 accuracy on HIGGS.

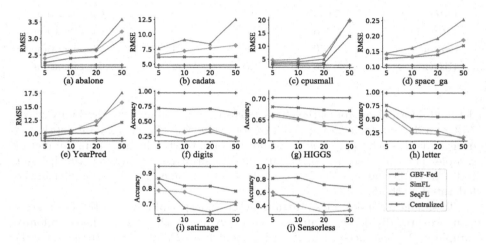

Fig. 4. Performance under different numbers of devices. (a)–(e) are regression tasks. (f)–(j) are classification tasks. The x-axis denotes the number of devices. In (b), the RMSE loss has been scaled into 1/10000 times.

We further study the effect of data partitioning on the performance of the federated models. For a given dataset, we split it into different numbers of parts to observe the change of performance. In the experiments, we set $\beta = 0.8$ for regression and $\theta = 0.1$ for classification tasks, respectively. We range the number of devices from $\{5, 10, 20, 50\}$ to examine the impact on RMSE or accuracy. Figure 4 demonstrates that GBF-Fed has low susceptibility to data decentralization and outperforms the other two methods in most cases.

We also examine the three federated methods' robustness to the degree of non-iid. To figure out the impact of β or θ, we set the number of devices to 20 in regression tasks and 50 in classification tasks. We range β from $\{0, 0.5, 0.6, 0.7, 0.8\}$, θ from $\{+\infty, 1.0, 0.5, 0.2, 0.1\}$, where $\beta = 0$ or $\theta = +\infty$ denotes iid partitioning. As is illustrated in Fig. 5, GBF-Fed outperforms the other two approaches in any degree of non-iid. On each of the ten examined datasets, GBF-Fed reduces the testing RMSE by 12.22%–48.00% and raises testing accuracy by up to 56.97%. These experimental results indicate that GBF-Fed is less susceptible to non-iid data than the existing methods.

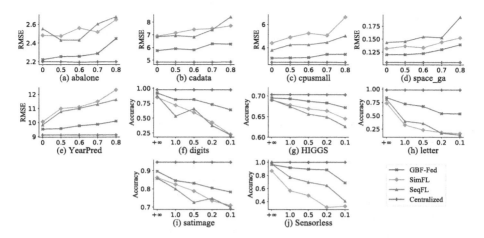

Fig. 5. Performance under different degrees of non-iid. (a)–(e) are regression tasks. (f)–(j) are classification tasks. The x-axis denotes β in (a)–(e) and θ in (f)–(j). In (b), the RMSE loss has been scaled into $1/10000$ times.

6 Conclusion

In this paper, we propose a novel tree-boosting method, Gradient Boosting Forest for both centralized and federated learning. Our new design, replacing the single decision tree by a decision forest, brings about significant improvements on GBDT's predictive performance. Also, Gradient Boosting Forest enables efficient and flexible federated training of tree-boosting models, with GBF-Fed achieving state-of-the-art performance. We present an effective solution for model aggregation in federated learning by a two-stage ensemble, which may put Gradient Boosting Machines to a broader use and inspire related studies on GBDT and federated learning.

References

1. Acar, A., Aksu, H., Uluagac, A.S., Conti, M.: A survey on homomorphic encryption schemes: theory and implementation. ACM Comput. Surv. (CSUR) **51**(4), 1–35 (2018)
2. Breiman, L.: Bagging predictors. Mach. Learn. **24**(2), 123–140 (1996)
3. Buhlmann, P., Yu, B., et al.: Analyzing bagging. Ann. Stat. **30**(4), 927–961 (2002)
4. Chen, T., Guestrin, C.: XGBoost: a scalable tree boosting system. In: Proceedings of the 22nd ACM SIGKDD International Conference on Knowledge Discovery and Data Mining, pp. 785–794 (2016)
5. Chen, Z., Liu, B.: Lifelong Machine Learning. Synthesis Lectures on Artificial Intelligence and Machine Learning, vol. 12, no. 3, pp. 1–207 (2018)
6. Datar, M., Immorlica, N., Indyk, P., Mirrokni, V.S.: Locality-sensitive hashing scheme based on p-stable distributions. In: Proceedings of the Twentieth Annual Symposium on Computational Geometry, pp. 253–262 (2004)

7. Feng, Z., et al.: SecureGBM: secure multi-party gradient boosting. In: 2019 IEEE International Conference on Big Data (Big Data), pp. 1312–1321. IEEE (2019)
8. Friedman, J.H.: Greedy function approximation: a gradient boosting machine. Ann. Stat. **29**, 1189–1232 (2001)
9. Gentry, C.: Fully homomorphic encryption using ideal lattices. In: Proceedings of the Forty-first Annual ACM Symposium on Theory of Computing, pp. 169–178 (2009)
10. Hsu, T.M.H., Qi, H., Brown, M.: Measuring the effects of non-identical data distribution for federated visual classification. arXiv preprint arXiv:1909.06335 (2019)
11. Ke, G., et al.: LightGBM: a highly efficient gradient boosting decision tree. In: Advances in Neural Information Processing Systems, pp. 3146–3154 (2017)
12. Lewis, R.J.: An introduction to classification and regression tree (CART) analysis. In: Annual Meeting of the Society for Academic Emergency Medicine in San Francisco, California, vol. 14 (2000)
13. Li, Q., Wen, Z., He, B.: Practical federated gradient boosting decision trees. In: AAAI, pp. 4642–4649 (2020)
14. Li, T., Sahu, A.K., Talwalkar, A., Smith, V.: Federated learning: challenges, methods, and future directions. IEEE Sig. Process. Mag. **37**(3), 50–60 (2020)
15. Li, Z., Hoiem, D.: Learning without forgetting. IEEE Trans. Pattern Anal. Mach. Intell. **40**(12), 2935–2947 (2017)
16. Liaw, A., Wiener, M., et al.: Classification and regression by random forest. R News **2**(3), 18–22 (2002)
17. Liu, Y., Ma, Z., Liu, X., Ma, S., Nepal, S., Deng, R.: Boosting privately: privacy-preserving federated extreme boosting for mobile crowd sensing. arXiv preprint arXiv:1907.10218 (2019)
18. McMahan, B., Moore, E., Ramage, D., Hampson, S., Arcas, B.A.: Communication-efficient learning of deep networks from decentralized data. In: Artificial Intelligence and Statistics, pp. 1273–1282. PMLR (2017)
19. Mirhoseini, A., Sadeghi, A.R., Koushanfar, F.: CryptoML: secure outsourcing of big data machine learning applications. In: 2016 IEEE International Symposium on Hardware Oriented Security and Trust (HOST), pp. 149–154. IEEE (2016)
20. Wen, Z., Shi, J., He, B., Li, Q., Chen, J.: ThunderGBM: fast GBDTS and random forests on GPUs (2019)
21. Yang, M., Song, L., Xu, J., Li, C., Tan, G.: The tradeoff between privacy and accuracy in anomaly detection using federated XGBoost. arXiv preprint-arXiv:1907.07157 (2019)
22. Yang, Q., Liu, Y., Chen, T., Tong, Y.: Federated machine learning: concept and applications. ACM Trans. Intell. Syst. Technol. (TIST) **10**(2), 1–19 (2019)
23. Zhao, L., et al.: InPrivate digging: enabling tree-based distributed data mining with differential privacy. In: IEEEINFOCOM 2018-IEEE Conference on Computer Communications, pp. 2087–2095. IEEE (2018)

Random Neural Graph Generation with Structure Evolution

Yuguang Zhou[1], Zheng He[1], Tao Wan[2,3(✉)], and Zengchang Qin[1,4(✉)]

[1] Intelligent Computing and Machine Learning Lab, School of Automation Science and Electrical Engineering, Beihang University, Beijing, China
{yuguangzhou,hz1998,zcqin}@buaa.edu.cn
[2] School of Biological Science and Medical Engineering,
Beihang University, Beijing, China
taowan@buaa.edu.cn
[3] Beijing Advanced Innovation Center for Biomedical Engineering,
Beihang University, Beijing, China
[4] Codemao AI Research, Codemao Inc., Shenzhen, China

Abstract. In deep learning research, typical neural network models are multi-layered architectures, and weights are tuned while optimizing a carefully designed loss function. In recent years, studies of randomized neural networks have been extended towards deep architectures, opening a new research direction to the design of deep learning models. However, how the structure of the network can influence the model performance still remains unclear. In this paper, we move a further step to investigate the relation between network topology and performance via a structure evolution algorithm. Experimental results show that the graph would evolve towards a more small-world topology at the beginning of the training session along with gaining accuracy, and would also evolve towards a structure with more scale-free property in the following periods. These conclusions could help explain the effectiveness of the randomly connected networks, as well as give us insights in new possibilities of network architecture design.

Keywords: Random neural graph · Network topology · Small-world · Scale-free

1 Introduction

Early perspectives of the connectionism [1,2] suggest the internal topology of a computational network is essential for building intelligent machines. These years, some deep neural networks achieved promising performance with manually elaborated architectures [3,4], and attempts like ResNet [5] and DenseNet [6] which challenge the layered architecture paradigm also acquired success. In recent studies, graph theory get new applications in network design: Xie *et al.*

This research is partially supported by the NSF of China (No. 61876197) and the Beijing NSF (Grant No. 7192105).

[7] show that constructing neural networks with random graph algorithms often outperforms a manually engineered architecture; Javaheripi *et al.* [8] reconstruct layered networks into a small-world topology and achieve superior performance. These advances seem to prove that using certain topology to construct neural networks could be an effective way of architecture design. Given this perspective, we would like to ask the following questions: *What if we let a network choose its own preferred architecture during training? What kind of network topology would it evolve in the processing of loss minimization?*

Methods about evolving the topology of computational networks can be backtracked to the use of genetic algorithm to train the structure of Turing's *Unorganised Machines* [2,9]. Recently, Wortsman *et al.* [10] propose using gradient descent methods to jointly learn model parameters and discover network inner connections. However, as it still emphasises on the weights updating process, this approach cannot help us to perceive how structural optimization can directly bring better performance: to study the structure evolution of networks requires de-emphasizing the weights updating process. Responding to this, some approaches are developed to find competitive sub-networks within an overparameterized layered network with fixed weights [11–13]. Work followed up by Ramanujan *et al.* [14] introduces a structural searching algorithm of such subnetworks without explicit weights training. Nevertheless, these works are still confined to layered architectures, without paying attention to topological evolution of networks.

Our work builds upon the structural searching method [14] mentioned above. By adapting it to random neural graphs, we could perform structural evolution of neural graphs independently from traditional weights updating process. **Our main contribution is three-folds:** (1) By breaking the layered constrains of neural networks, we explore the possibility of self-evolving networks with graph structures. (2) To our best knowledge, we are the first to simultaneously investigate the evolution of topological properties and the model performance of a neural graph during training. (3) The experimental results on the image recognition benchmark dataset CIFAR-10 [15] show that the performance of model (validation accuracy) is strongly correlated with its small-world topological property on the early training stage, and the evolution of the inner connectivity also shows that the neural graph evolves towards a structure with stronger scale-free property. More importantly, this work could help explain the effectiveness of neural networks generated with random graph theories, and could also give new possibilities on network architecture design.

2   Related Work

2.1   Neural Networks in Graph Structure

The original *Unorganised Machine* proposed by Turing [2] was built out of a crowd of neuron-like binary elements, connecting with each other in a random manner. In 2001, Teuscher [9] re-explores this idea further and applies genetic algorithm to evolve the architecture of unorganized machines. These machines are then able to perform very simple digit recognition tasks, and their inner

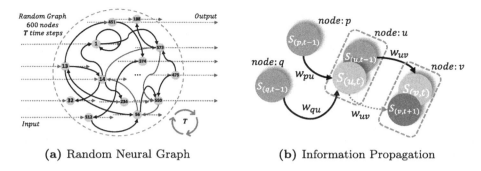

(a) Random Neural Graph   (b) Information Propagation

Fig. 1. Illustration of a random neural graph (a); Information propagation on node-level (b).

structure still presents randomness. On the contrary, recent research including RandWire [7] and SWNet [8] pay more attention on building random but fixed network structure before training, which, in another way, has also been proven successful. To examine the effectiveness of RandWire, You *et al.* [16] investigate on a huge number of modern networks the relationship between the graph-based representation of network structure and model performance, and find that the latter is approximately a smooth function of the clustering coefficient and average path length of its network's graph representation.

2.2 Neural Architecture Search and Lottery Tickets Hypothesis

Structure evolution inside a fully connected space is in fact a Neural Architecture Search (NAS) problem. However, traditional NAS approaches are hardly to be able to apply in structure learning, which needs to de-emphasis the updating process of connection weights. Recent advances in Lottery Tickets Hypothesis (LTH) [11–13] bring us the possibility to discover new pruning strategies inside networks with fixed weights. Followed up work by Ramanujan *et al.* [14] gives a structural searching algorithm without explicit weight training, which aims to find effective sub-structure with fixed number of connections within a large layered network with fixed weights. Our work builds upon this structural searching method while adapting it to evolving the structure of a randomly connected graph inside a pairwise connected dense space. Our objective differs from theirs as we pay attention to the topological properties that a randomly wired network may acquire during the learning of architecture rather than weight values.

3 Methodology

3.1 Information Propagation Inside Random Neural Graphs

A random neural graph is a directed weighted graph $\mathcal{G} = (\mathcal{V}, \mathcal{E})$ consisting of a nodes set $\mathcal{V}$ and an edges set $\mathcal{E} \subset \mathcal{V} \times \mathcal{V}$, where an edge from u to v is associated with a random connection weight w_{uv}. The inner connection of a n nodes graph can then be represented by a weighted connection matrix $C_{n \times n}$. An illustration of a random graph is presented in Fig. 1a. The information inside the neural graph is propagated through a maximum propagation time T (also called time steps), and the information is passed one-edge-long every single time step. For any $t \in [\![0, T]\!]$, the information held on a node v (i.e., the node state) is represented by $S_{(v,t)}$. At each node v, a node operation function f_v can be applied (in deep learning, this function could be a composition of activation, normalisation and other node-level operation such as single channel convolution). The state of node v at time $t \in [\![1, T]\!]$ computes as:

$$S_{(v,t)} = \sum_{(u,v) \in \mathcal{E}} w_{uv} f_v(S_{(u,t-1)}) + S_{(v)}^{\text{init}}, \qquad (1)$$

where $S_{(v)}^{\text{init}}$ is the initial information that is fed constantly for any $t \in [\![0, T]\!]$ to node v in the graph. The information sent outside are the node states at $t = T$. Figure 1b illustrates a zoom of the information propagation mechanism: the states at $t - 1$ on nodes p and q are used to update the state at t for node u, and so on. If the nodes states are written in vector form, i.e. $\vec{S}_t = [S_{(1,t)}, S_{(2,t)}, ..., S_{(n,t)}]^{\text{T}}$, then the nodes states are updated as:

$$\vec{S}_t = C_{n \times n} \cdot f_v(\vec{S}_{t-1}) + \vec{S}_0. \qquad (2)$$

3.2 Model Structure of Networks with Random Neural Graph

Our experimental model is designed to perform image recognition task. It has 3 main components: firstly 3 convolution layers to down-sample the original image, then a random neural graph, and at last other layers used to generate the final output. As we intend to evolve the network structure from the finest level, for the graph, each node receives a single channel of the output from previous convolution layer, and each potential edge is mapped to an 1×1 convolution kernel. The node operation is composed of a ReLU [17] activation followed by instance normalization [18] and a 3×3 convolution (the three play together the role of f_v above). After the graph operation, the node states at the last time step are fed to pooling and fully-connected operations for the classification.

3.3 Structural Evolution of Random Neural Graphs

During the graph generation process, any connection between two nodes could be activated. Each potential edge is initialised with a fixed weight w_{uv} following the Kaiming normal distribution [19]. These weights form the fixed weight tensor W_{init}. To generate a sparse neural graph inside this densely connected space, another trainable parameter m_{uv} is associated to each edge uv. These parameters then form a trainable mask tensor M. During the training, only M is updated, and once the absolute value of a trainable parameter passes the pre-defined activation threshold thr, the corresponding connection will be activated.

The activation process is noted as $h(\cdot)$, which maps the tensor M to another binary tensor $h(M)$, with the 1's being activated. The masked weighted connection matrix is then obtained as $C_{n \times n} = W_{\text{init}} \odot h(M)$. In the forward stage, information propagates with $C_{n \times n}$. To stay simple, f_v is set as identity operation in this section. Formally, the input (the state) of node v at time t is:

$$I_{(v,t)} := S_{(v,t)} = \sum_{u \in \mathcal{V}^{\text{pred}}} w_{uv} S_{(u,t)} h(m_{uv}) + S_{(v,0)}. \tag{3}$$

In the backward stage, instead of training and updating the weights w_{uv}, we choose to update the mask tensor M, using the straight-through gradient estimator (h is treated as the identity operation) to compute the estimated gradient:

$$\hat{g}_{m_{uv}} = \frac{\partial \mathcal{L}}{\partial I_{(v,T)}} \cdot \frac{\partial I_{(v,T)}}{\partial h(m_{uv})} \cdot \frac{\partial h(m_{uv})}{\partial m_{uv}} = \frac{\partial \mathcal{L}}{\partial I_{(v,T)}} w_{uv} S_{(u,T)}, \tag{4}$$

where $\mathcal{L}$ represents the final loss of the model, and the corresponding node states use the final values at T [14]. The value of each m_{uv} is then updated by $m_{uv} \leftarrow m_{uv} - \alpha \frac{\partial \mathcal{L}}{\partial I_{(v,T)}} w_{uv} S_{(u,T)}$ [14] where α the learning rate. The internal edges initially activated by the neural graph depend on the trainable parameters following Kaiming uniform distribution [19], which is independent of the weights distribution. During the structural evolution, whether an edge is activated will only depend on the value of m_{uv}. In other words, training m_{uv} allow us to evolve the structure of the graph, to a certain extent, independently to the weights.

3.4 Metrics for Topological Properties

Metrics for Small-Worldness. We firstly consider the connectivity between nodes, using the σ scalar measure to estimate the small-worldness of the graph topology: this is the ratio between the average clustering coefficient C and the average shortest path length L, normalized to an equivalent graph generated by the classical Erdős-Rényi model [20]: $\sigma = \frac{\Gamma}{\Lambda}, \Gamma = \frac{C_G}{C_{ER}}, \Lambda = \frac{L_G}{L_{ER}}$. Meanwhile, information encoded in edge weights might also reflect topological features, such as, larger weights may signify smaller topological distances [21], a weighted version of σ is then proposed in [22]. As the σ is normalized in the same way, we only report the score $SW = C_G/L_G$ (and similarly WSW for the weighted version) of the graph after each train epoch.

Metrics for Scale-Freedom. Another topological property is that most of the nodes have very small degrees and a small portion of nodes possesses the most of the connections: this is the scale-free property [23] which is represented by the power law distribution for its nodes degrees: $P(d) \sim d^{-b}$, where d stands for degree, $P(d)$ is the frequency of a certain degree d and b is the power coefficient. The power coefficient indicates how imbalance the degree distribution is, and is used in this work as the scale-freed score of a graph. Besides, the scale-freedom is estimated respectively for the in-degrees and out-degrees (we call them respectively the SFI score and the SFO score in the following).

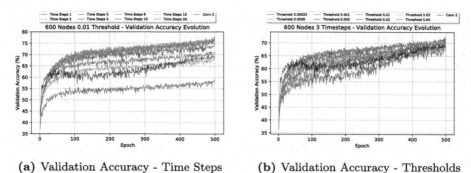

(a) Validation Accuracy - Time Steps   (b) Validation Accuracy - Thresholds

Fig. 2. Evolution in 500 epochs of validation accuracy.

4   Experimental Studies

4.1   Experiment Setup

We perform experiments on image recognition benchmark dataset CIFAR-10 [15]. To illustrate the validity of our model, we also compare with one simple layered model used in [13,14] (called *Conv 2*): it is composed of two 64-channel convolution layers each followed by a pooling operation, plus three fully-connected layers with 256, 256, 10 neurons, respectively. The structure evolution of *Conv 2* consists in finding an efficient sub-network within an overparameterized space with fixed weights, and only half of the total potential connections are activated (as layered structures could attain their optimum around this *sparsity* of 50% in these previous works). The convolution and fully-connected layers in graph models are also trained as in *Conv 2* with the sparsity of 50%. The batch normalisation [24] and the instance normalisation [18] are also set as non-affine.

In the following, we perform structure evolution experiments on graph models of 600 nodes, as in this way they have similar number of total trainable parameters ($560K$) as *Conv 2* ($580K$). Our experiments are performed in two directions: we study the evolution of model performance (validation accuracy) and report topological metrics for different maximum propagation time T (in this case the threshold is set as 0.01) and different threshold *thr* (in this case the time steps is set as 3). As the calculation of a densely connected component might be very costly, we only calculate the topological metrics of the sub-component with the absolutely highest 5% trainable parameters M as an proxy of the graph. The train process is performed by SGD using cosine learning rate decay, with weight decay 1e−4, momentum 0.9, batch size 512, and initial learning rate 0.1.

4.2   Model Performance

Figure 2a illustrates the evolution of validation accuracy over 500 epochs of *Conv 2* and graph models with different time steps ranging from 1 to 20. As we can see, the validation accuracy firstly increases fast during the first 100 epochs then

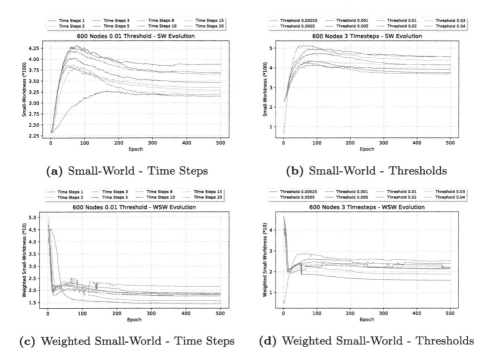

(a) Small-World - Time Steps

(b) Small-World - Thresholds

(c) Weighted Small-World - Time Steps

(d) Weighted Small-World - Thresholds

Fig. 3. Evolution in 500 epochs of small-worldness.

grows slowly in the rest of the training session. And the final accuracy score generally augments with T and stagnates starting from $T = 10$. For $T \geq 3$, the graph model generally outperforms the layered counterpart. The situation for different thresholds in Fig. 2b is more interesting: only when $0.001 \leq thr \leq 0.02$ the graph model achieves better performance than *Conv 2*.

4.3   Topological Properties

Small-Worldness. Figure 3a and Fig. 3b shows respectively the evolution of the small-world score for models with different time steps and different thresholds. The SW score increases sharply during about the first 80 epochs, which has simultaneous tendency with the validation accuracy in previous figures. After that, the validation accuracy keeps slowly growing, while the SW score slightly decreases then keeps quasi-unchanged without obvious fluctuation. This indicates that the rapid performance improvement at the beginning stage may come from the evolution of the network structure towards a small-worldish topology, and that in the following stages, this topology is kept so that the network performance is maintained at a stable level. Further, we notice that in Fig. 3a, the SW score increases from $T = 1$ to $T = 5$, then decrease from $T = 8$ to $T = 20$. This seems to indicate that models with higher propagation time need less small-world property to obtain better performance.

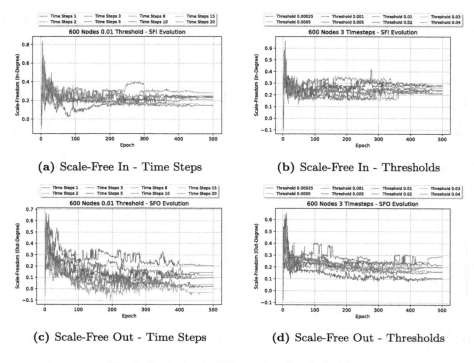

Fig. 4. Evolution in 500 epochs of scale-freedom.

Figure 3c and Fig. 3d shows the evolution of the weighted small-world score. We notice that, except the case $T = 1$ in the left and the case $thr = 0.04$ in the right, the WSW firstly falls sharply in the first 20 epochs, then rises slightly again and gradually diminishes during the rest training epochs. A reasonable explanation to this interesting behaviour would be: during the very first stages, the neural graph evolves itself towards a small-world topology without taking into account the weights information associated with the connections. Then from about the 20th epoch, the graph starts to be aware of the weighted topology and chooses to slightly evolve itself towards this. This supposition could be further established by observing the situation of $thr = 0.04$ in Fig. 3d: the behaviour of the WSW score acts similarly as the SW score. A possible reason for this could be: when the threshold is set too high, the number of activated connections are too few so that the graph needs to rely more on the weighed topology.

Scale-Freedom. The four figures in Fig. 4 show the evolution of the scale-freedom estimated on the in-degree and out-degree distribution of the graph for different time steps and different thresholds. We notice that these scores present similar behaviour over the entire training session: the SFI and SFO scores firstly jump sharply during the very first epochs, then fall gradually in the first 100 epochs. During the following stages, these scores present obvious

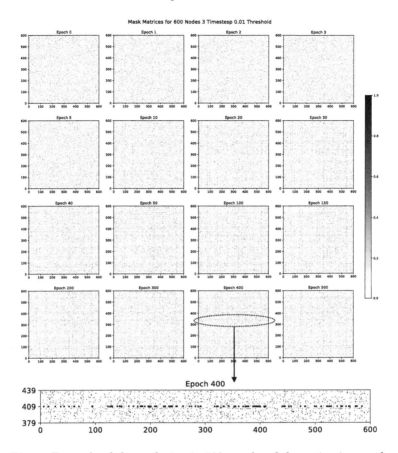

Fig. 5. Example of the evolution in 500 epochs of the activation mask.

fluctuations with a trend to diminish. Moreover, the SFO score seems to be more correlated to the evolution of the validation accuracy: in Fig. 4c, the Pearson correlation coefficient between SFO and the validation accuracy is generally between -0.61 for $T = 8$ and -0.88 for $T = 15$ (except for $T = 10$ is -0.51); in Fig. 4d, this correlation is between -0.71 and -0.84 for $0.0005 \leq thr \leq 0.01$ and for $thr = 0.04$. These observations imply that the fluctuation of the scale-free score may help explain the model performance improvement. However, the quantitative metrics SFI and SFO may not be good enough to reflect whether the structure of the graph evolves towards a scale-free topology.

To verify this fact, we show in Fig. 5 the exact representation of the mask tensor $h(M)$ at the end of some training epoch of one of the previous used models (others are similar). A black dot (value 1) represents an activated connection. We notice that, at the initial training period, the activation distribution is uniform. Starting from the 10th epoch, the mask matrix begins to show certain patterns: there are more observable black horizontal and vertical dashed lines.

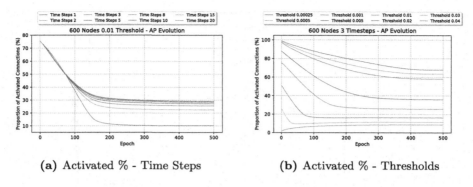

(a) Activated % - Time Steps   (b) Activated % - Thresholds

Fig. 6. Evolution in 500 epochs of activated percentage.

This indicates that some nodes in the graph become more important and thus have more connections. We also show a zoom of the node that has the most in-connections at the end of the 400th epoch. This connectivity pattern is exactly the characteristic of a classical scale-free network. Therefore, we could conclude that the graph indeed evolves towards a scale-free topology.

Activated Percentage. Figure 6a and Fig. 6b shows the evolution of activated percentage for different time steps and thresholds. In Fig. 6a, the activated percentage of all models decreases by at least 45%. Figure 6b shows similar behaviour of the activation percentage, except when $thr = 0.04$, the activation percentage increases instead as initially activated connections are too few.

5   Conclusion

In this research, we applied quasi constraint-free structure evolution on randomly connected neural graphs without using explicit updating process for connection weights, in order to let the inner-structure evolves freely towards the most efficient topological situation. We studied the evolution of some topological properties of the graph structure alongside its training, in order to explain how the performance improvement is achieved during the evolution of network structure. We find that, at early stage of the training, the most efficient way to gain performance would be to move towards a more small-worldish topology. At the same time, during the following stages, it could also be confirmed that the inner connectivity evolves towards a structure with more scale-free feature. Some parts of the performance improvement still remain unexplained and require further investigation. This work presents new possibilities of efficient neural network architecture design using random graphs theory without human over-intervention.

References

1. Fodor, J.A., Pylyshyn, Z.W.: Connectionism and cognitive architecture: a critical analysis. Cognition **28**, 3–71 (1988)
2. Turing, A.: Intelligent machinery. The Essential Turing, pp. 395–432 (1948)
3. Krizhevsky, A., Sutskever, I., Hinton, G.E.: ImageNet classification with deep convolutional neural networks. In: Advances in Neural Information Processing Systems, vol. 25, no. 1097–1105 (2012)
4. Simonyan, K., Zisserman, A.: Very deep convolutional networks for large-scale image recognition. In: Conference Track Proceedings of ICLR (2015)
5. He, K., Zhang, X., Ren, S., Sun, J.: Deep residual learning for image recognition. In: Proceedings of CVPR, vol. 2016, no. 770–778 (2016)
6. Huang, G., Liu, Z., Van Der Maaten, L., Weinberger, K.Q.: Densely connected convolutional networks. In: Proceedings of CVPR, vol. 2017, no. 4700–4708 (2017)
7. Xie, S., Kirillov, A., Girshick, R., He, K.: Exploring randomly wired neural networks for image recognition. In: Proceedings of ICCV, vol. 2019, no. 1284–1293 (2019)
8. Javaheripi, M., Rouhani, B.D., Koushanfar, F.: SWNet: small-world neural networks and rapid convergence. arXiv preprint arXiv:1904.04862 (2019)
9. Teuscher, C., Sanchez, E.: A revival of Turing's forgotten connectionist ideas: exploring unorganized machines. In: French, R.M., Sougné, J.P. (eds.) Connectionist Models of Learning, Development and Evolution, pp. 153–162 (2001). Springer, London. https://doi.org/10.1007/978-1-4471-0281-6_16
10. Wortsman, M., Farhadi, A., Rastegari, M.: Discovering neural wirings. In: Advances in Neural Information Processing Systems, vol. 32, no. 2680–2690 (2019)
11. Rosenfeld, A., Tsotsos, J.K.: Intriguing properties of randomly weighted networks: generalizing while learning next to nothing. In: Proceedings of CRV, vol. 2019, no. 9–16 (2019)
12. Frankle, J., Carbin, M.: The lottery ticket hypothesis: finding sparse, trainable neural networks. In: Proceedings of ICLR (2019)
13. Zhou, H., Lan, J., Liu, R., Yosinski, J.: Deconstructing lottery tickets: zeros, signs, and the supermask. In: Advances in Neural Information Processing Systems, vol. 32, pp. 3592–3602 (2019)
14. Ramanujan, V., Wortsman, M., Kembhavi, A., Farhadi, A., Rastegari, M.: What's hidden in a randomly weighted neural network? In: Proceedings of CVPR, vol. 2020, no. 11893–11902 (2020)
15. Krizhevsky, A., Hinton, G.: Learning multiple layers of features from tiny images. Technical report, University of Toronto (2009)
16. You, J., Leskovec, J., He, K., Xie, S.: Graph structure of neural networks. In: Proceedings of the ICML, vol. 2020, no. 10881–10891 (2020)
17. Nair, V., Hinton, G.E.: Rectified linear units improve restricted Boltzmann machines. In: Proceedings of ICML, vol. 2010, no. 807–814 (2010)
18. Ulyanov, D., Vedaldi, A., Lempitsky, V.: Instance normalization: the missing ingredient for fast stylization. arXiv preprint arXiv:1607.08022 (2016)
19. He, K., Zhang, X., Ren, S., Sun, J.: Delving deep into rectifiers: surpassing human-level performance on ImageNet classification. In: Proceedings of ICCV, vol. 2015, no. 1026–1034 (2015)
20. Erdős, P., Rényi, A.: On the evolution of random graphs. Publ. Math. Inst. Hung. Acad. Sci. **5**, 17–60 (1960)

21. Bassett, D.S., Bullmore, E.T.: Small-world brain networks revisited. Neuroscientist **23**, 499–516 (2017)
22. Onnela, J.-P., Saramäki, J., Kertész, J., Kaski, K.: Intensity and coherence of motifs in weighted complex networks. Phys. Rev. E **71**, 065103 (2005)
23. Barabási, A.-L., Bonabeau, E.: Scale-free networks. Sci. Am. **288**, 60–69 (2003)
24. Ioffe, S., Szegedy, C.: Batch normalization: accelerating deep network training by reducing internal covariate shift. In: Proceedings of ICML, vol. 2015, no. 448–456 (2015)

MatchMaker: Aspect-Based Sentiment Classification via Mutual Information

Jing Yu[1], Yongli Cheng[1,2]($\boxtimes$), Fang Wang[2], Xianghao Xu[3], Dan He[4], and Wenxiong Wu[1]

[1] College of Computer and Data Science, FuZhou University, FuZhou, China
`chengyongli@fzu.edu.cn`
[2] Wuhan National Laboratory for Optoelectronics, Huazhong University of Science and Technology, Wuhan, China
[3] School of Computer Science and Engineering, Nanjing University of Science and Technology, Nanjing, China
[4] Information Engineering School of Nanchang Hangkong University, Nanchang, China

Abstract. Aspect-based sentiment classification (ABSC) aims to determine the sentiment polarity toward a specific aspect. In order to finish this task, it is difficult to match a specific aspect with its opinion words since there are usually multiple aspects with different opinion words in a sentence. Many efforts have been made to address this problem, such as graph neural networks and attention mechanism, however come at the cost of the introduced extraneous noise, leading to mismatches of the aspect with its opinion words. In this paper, we propose a Mutual Information-based ABSC model, called MatchMaker, which introduces Mutual Information estimation to strengthen the correlations between a specific aspect and its opinion words without introducing any extraneous noise, thus significantly improving the accuracy when determining the sentiment polarity toward a specific aspect. Experimental results show that our method with Mutual Information is effective. For example, MatchMaker obtains a significant improvement of accuracy over ASGCN model by 3.1% on the Rest14.

Keywords: Sentiment analysis · Aspect-based sentiment analysis · Mutual information

1 Introduction

In the field of Natural Language Processing (NLP), aspect-based sentiment classification (ABSC) is a fine-grained subtask of sentiment analysis. It aims to identify the sentiment polarity (e.g. positive, negative or neutral) of a specific aspect in a given sentence [1]. For instance, considering the sentence of *"The Japanese*

Supported by the Open Project Program of Wuhan National Laboratory for Optoelectronics NO. 2018WNLOKF006, and Natural Science Foundation of Fujian Province under Grant No. 2020J01493. This work is also supported by NSFC No. 61772216 and 61862045.

© Springer Nature Switzerland AG 2021
T. Mantoro et al. (Eds.): ICONIP 2021, LNCS 13109, pp. 99–110, 2021.
https://doi.org/10.1007/978-3-030-92269-2_9

sushi is great but the price is too expensive!", *Japanese sushi* and *price* are the two aspects with opposite sentiment polarities in the sentence.

There are two challenges during the process of ABSC task. One is to effectively model the relatedness between the context words and aspect words. The other one is to further match a specific aspect with its opinion words correctly by strengthening their correlations. Usually, the latter is more challenging since there are multiple aspects and opinion words within a regular sentence [2].

Many efforts have been made to tackle the two challenges. Prevalent approaches combine attention mechanism with the deep neural networks (e.g. RNNs and CNNs) to capture the semantic information of the context and aspects [3–6]. Graph neural networks (GNNs) are investigated to learn the syntactical dependencies of the sentence over dependency tree [2,7,8]. These works have yielded good results, however come at the cost of producing extraneous noise, reducing the total effectiveness of the ABSC models. The reasons for the noise caused by the attention mechanism and the syntactic methods are that the former may design inappropriate attention weights to aspect-irrelevant words, and the latter may suffer from parsing errors [2]. Furthermore, since the two methods cannot distinguish the difference of opinion words, they are unable to efficiently solve the second challenge.

In this paper, we propose a Mutual Information-based ABSC model, called MatchMaker, which helps ABSC task determine the sentiment polarity to a specific aspect more accurately by introducing two key modules. One is the well-designed Feature Extraction Module, which first uses the attention mechanism and graph convolutional networks to model the semantic and syntactic relationship of the sentence, and then integrates the double gate mechanism with the two components to filter the inherent noise of them. The three efforts above help our MatchMaker effectively extract aspect-relevant sentiment features. The other is the Mutual Information Maximizing Module, which first models the correlations between the aspect and corresponding opinion words by using the Mutual Information estimation, and then strengthens the correlations by maximizing the Mutual Information objective values. By combining with the two modules, our MatchMaker learns better representation of the sentence to obtain higher prediction accuracy, and outperforms existing ABSC models significantly.

The rest of this paper is organized as follows. We first introduce the preliminary knowledge of Mutual Information in Sect. 2. The proposed model MatchMaker is given in Sect. 3. Experimental evaluation of the MatchMaker is presented in Sect. 4. We finally conclude this paper in Sect. 5.

2   Mutual Information

Mutual Information $MI(X, Y)$ can measure the degree of interdependence between two random variables X and Y, which can be defined precisely as follow:

$$MI(X, Y) = D_{KL}(p(X, Y) \parallel p(X)p(Y)) \tag{1}$$

where D_{KL} is the Kullback-Leibler (KL) divergence between the joint distribution $p(X, Y)$ and the product of marginals $p(X)p(Y)$.

Yeh et al. [9] propose a QAInfomax model, which takes the MI estimation method as a regularizer, aiming to tackle the question answering task. The main idea of the estimation is that it uses adversarial learning to train a classifier, which distinguishes among positive samples (x, y) from the joint distribution and negative samples $(x, \bar{y})$ and $(\bar{x}, y)$ from the product of marginals. QAInfomax maximizes the mutual information objective function to improve the robustness. The objective formula is listed as follow:

$$MI(X,Y) \geq \mathbb{E}_{\mathbb{P}}[\log(g(x,y))] + \frac{1}{2}\mathbb{E}_{\mathbb{N}}[\log(1 - g(x,\bar{y}))] + \frac{1}{2}\mathbb{E}_{\mathbb{N}}[\log(1 - g(\bar{x},y))] \quad (2)$$

where $\mathbb{E}_{\mathbb{P}}$ and $\mathbb{E}_{\mathbb{N}}$ denote the expectations of positive and negative samples respectively, and $g(x,y) = x^{\top}W_g y$ is a discriminator function modeled by a neural network, W_g is the learnable matrix.

In spite of many efforts, it is a challenge that how to further strengthen the correlations between the aspect and its opinion words. In the case of multiple aspects, the challenge becomes tricky. Meanwhile, Mutual Information can be used to measure the dependent correlation of two random variables. Therefore, we introduce the Mutual Information estimation in ABSC area to strengthen the correlations of the aspect with its opinion words, aiming to improve the accuracy of sentiment polarity prediction.

3 MatchMaker

The overall architecture of MatchMaker is shown in Fig. 1. The Input Module is a fundamental module of the model, turning the word into contextual embedding. Feature Extraction(FE) Module extracts features effectively with the help of double gate mechanism, which is employed to reduce the inherent noise of attention mechanism and graph convolutional networks. Mutual Information Maximizing(MIMax) Module is the key module of our model that learns the deep information by the Mutual Information maximization. We update the shared parameters of the FE module and MIMax module in the backpropagation by optimizing the total loss function, integrating the benefits from them. Details are described below.

3.1 Input Module

For a given sentence with n context words $\{w_1, w_2, ..., w^a_{i+1}, ..., w^a_{i+m}, ..., w_n\}$, there is a subsequence $\{w^a_{i+1}, ..., w^a_{i+m}\}$ consisting of the given aspect words, where m denotes the length of the given aspect. We map the context words and aspect words respectively, into the context vectors $S = \{e_1, e_2, ..., e_n\}$ and aspect vectors $T = \{e^a_1, e^a_2, ..., e^a_m\}$, according to the pre-trained GloVe [10] embedding matrix $E \in \mathbb{R}^{|V| \times d_e}$, where $|V|$ is the vocabulary size and d_e is the dimensionality of the word embeddings. We further input S and T into the Bidirectional LSTM (BiLSTM) to obtain the contextual hidden states of context words $H^c = \{h^c_1, h^c_2, ...h^c_n\}$ and aspect words $H^a = \{h^a_1, h^a_2, ..., h^a_m\}$, where $H^c \in \mathbb{R}^{n \times 2d_h}$, $H^a \in \mathbb{R}^{m \times 2d_h}$, and d_h is the dimensionality of a hidden state learnt by the unidirectional LSTM.

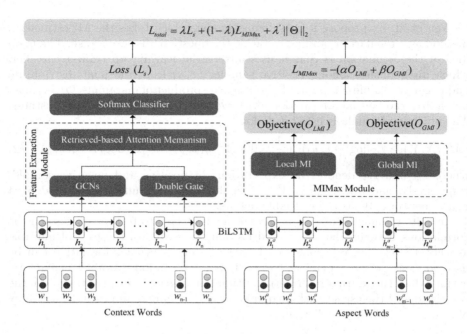

Fig. 1. The overall architecture of MatchMaker.

3.2   Feature Extraction Module

Graph Convolutional Networks (GCNs) Layer: We construct a syntactic graph G over the syntactical dependency tree of a given sentence, where each node represents a word in the sentence and each edge is the dependency between two given nodes. An adjacency matrix A is used to represent graph G. Specifically, $A_{ij} = 1$ denotes that there is an edge between node i and node j, and $A_{ij} = 0$ otherwise. We then employ a two-layer GCNs over the graph G. The advantage of this method is that the representation of each node can be updated iteratively by convolution operation, to aggregate information from its neighbor nodes [8]. The input of the first layer is the hidden states of context words obtained from the BiLSTM (i.e. $H^0 = H^c$). The final output $H^L \in \mathbb{R}^{n \times 2d_h}$ can be obtained by formula:

$$h_i^l = ReLU(\sum_{j=1}^{n} \frac{1}{d_i} A_{ij}(W_l h_j^{l-1} + b_l)) \tag{3}$$

where $h_i^l \in \mathbb{R}^{2d_h}$ is the hidden state of node i at the l-th layer and d_i denotes the degree of node i shown in the adjacency matrix. W_l and b_l are the learnable parameters.

Double Gate Layer: We introduce the double gate mechanism to help alleviate the issue of parsing errors caused by the GCNs layer. In this layer, for a given aspect, a shrinking gate is used to filter the irrelevant features, and an enlarging gate then is employed to magnify the relevant features. The input of this layer also comes from the output of BiLSTM. The shrinking gate converts the initial context representations H^c and the average aspect representation H^a_{mean} into an intermediate vector M, according to Eqs. (4–6). The enlarging gate then amplifies the features by imposing the forward bias, as shown in Eq. (8). The middle vector M and H^a_{mean} are converted into the final output vector $F \in \mathbb{R}^{n \times 2d_h}$ by Eq. (7):

$$m_t = s_t \odot o_t \tag{4}$$

$$s_t = \sigma(W_s \cdot [h_t^c; H^a_{mean}] + b_s) \tag{5}$$

$$o_t = ReLU(W_v \cdot h_t^c + b_v) \tag{6}$$

$$h_t^f = m_t \odot z_t \tag{7}$$

$$z_t = \sigma(W_z \cdot [o_t; H^a_{mean}] + b_z) + 1 \tag{8}$$

where h_t^c is the context hidden state vector at time step t, $m_t \in \mathbb{R}^{2d_h}$ and $z_t \in \mathbb{R}^{2d_h}$ are the outputs of the shrinking gate and the enlarging gate respectively. We obtain the final output $h_t^f \in \mathbb{R}^{2d_h}$ learnt by the two gates jointly. $\odot$ is element-wise multiplication. σ is the sigmoid function. If the variables are relevant, the function value will reach the upper bound but strictly zero otherwise. ReLU function, as a nonlinear function, is applied to generate better representations since it prevents overfitting. Where W_s, W_v, W_z, b_s, b_v and b_z are the learnable parameters of the model.

Retrieval-Based Attention Layer: We combine the output vectors of the GCNs layer and double gate layer to obtain the highlight and rich representations $H^{LF} = H^L + H^F$. Based on $H^{LF} = \{h_1^{lf}, \cdots, h_i^{lf}, \cdots, h_{i+m}^{lf}, \cdots, h_n^{lf}\}$, we mask the representations of non-aspect words to obtain the aspect-specific representations $H^{LF}_{mask} = \{0, \cdots, h_i^{lf}, \cdots, h_{i+m}^{lf}, \cdots, 0\}$. By feeding the amplified vectors H^{LF}_{mask} into the retrieval-based attention mechanism, this layer can retrieve features related to the aspect, and assign more proper retrieval-based weights μ for the context words, to extract relevant features more effectively. The final sentence representation r for prediction is produced by:

$$\mu_t = softmax(\sum_{i=1}^{n} h_t^{c\top} h_i^{lf}) \tag{9}$$

$$r = \sum_{t=1}^{n} \mu_t h_t^c \tag{10}$$

3.3 Mutual Information Maximizing Module

We introduce two units in this module, called Local Mutual Information (LMI) and Global Mutual Information (GMI), to learn the deep interaction information between context words and aspect words by maximizing their mutual information, from the local and global perspectives. To estimate mutual information, we take another sentence as a distractor sentence, which is randomly chosen from training samples. We then use the target sentence and distractor sentence to construct positive samples and negative samples. The hidden states of context words in the two sentences provided by BiLSTM are $H_T^c = \{h_1^c, \cdots, h_{i+1}^c, \cdots, h_{i+m}^c, \cdots, h_n^c\}$ and $H_D^c = \{h_1^c, \cdots, h_{j+1}^c, \cdots, h_{j+p}^c, \cdots, h_q^c\}$, the corresponding aspect hidden states are $H_T^a = \{h_1^a, \cdots, h_m^a\}$ and $H_D^a = \{h_1^a, \cdots, h_p^a\}$, where m and p represent the lengths of the two aspects, n and q are the lengths of the two sentences.

Local MI: Many researchers hold that the context words close to the given aspect and the aspect itself may have more relevant sentiment features [6,11]. Hence, we not only maximize the average MI between each aspect word vector and its local context word vectors, but also among the aspect word vectors in the given aspect term, thus preventing the model from losing information. Specifically, given a sentence and an aspect, we construct positive samples (x, y) by pairing one aspect token representation $x \in H_T^a$ with its surrounding tokens in the preset window and all other aspect tokens, i.e. $y \in v_T^c = \{h_{i+1-C}^c, \cdots, h_{i+m+C}^c\} \backslash \{x\}$, where C denotes the window size. We take the corresponding tokens from the distractor sentence to construct the negative samples $(x, \bar{y})$ and $(\bar{x}, y)$, where $\bar{x} \in H_D^a$ is the fake aspect token and $\bar{y} \in v_D^c = \{h_{j+1-C}^c, \cdots, h_{j+p+C}^c\} \backslash \{\bar{x}\}$ is the fake local context token. The estimation objective of LMI is listed as follow:

$$O_{LMI} = \frac{1}{|2C + m - 1|} \sum_{y_i \in v_T^c} log(g(x, y_i)) + \frac{1}{2|2C + p - 1|} \sum_{\bar{y}_j \in v_D^c} log(1 - g(x, \bar{y}_j))$$

$$+ \frac{1}{2|2C + m - 1|} \sum_{\bar{y}_j \in v_T^c} log(1 - g(\bar{x}, y_j))$$

$$(11)$$

Global MI: We further maximize the average MI between the summarized aspect word vector $x = S(H_T^a)$ and the whole context word vectors except for the aspect words, i.e. $y \in v_T^c = \{H_T^c\} \backslash \{H_T^a\}$, to learn the global deep interaction information. We choose $S(H_T^a) = sigmoid(\frac{1}{m} \sum h_i^a)$ as the summarization function to consider the whole aspect term. Similarly, we match the processed aspect vector x with the context vector y as the positive samples (x, y). For negative samples $(x, \bar{y})$ and $(\bar{x}, y)$, we pair the summarized vector x with the fake context word vector $\bar{y} \in v_D^c = \{H_D^c\} \backslash \{H_D^a\}$, and pair the fake summarized aspect vector $\bar{x} = S(H_D^a)$ with context word vector y, where $\bar{x}$ and $\bar{y}$ are both from the distractor sentence. The estimation objective of GMI is listed as follow:

$$O_{GMI} = \frac{1}{|n-m|} \sum_{y_i \in v_T^c} log(g(x, y_i)) + \frac{1}{2|q-p|} \sum_{\bar{y}_j \in v_D^c} log(1 - g(x, \bar{y}_j))$$
$$+ \frac{1}{2|n-m|} \sum_{y_j \in v_T^c} log(1 - g(\bar{x}, y_j)) \tag{12}$$

3.4 Sentiment Classification Layer

The output vector r of the attention layer is projected into a fully-connected layer to calculate the probability distribution for classifying the sentiment polarity:

$$\hat{y} = softmax(W_y \cdot r + b_y) \tag{13}$$

where $\hat{y} \in \mathbb{R}^{d_p}$ is the predicted value, $W_y \in \mathbb{R}^{d_p \times 2d_h}$ and $b_y \in \mathbb{R}^{d_p}$ are the weight matrix and bias of this layer.

3.5 Loss Function

We feed each training batch that contains training samples into the model, and shuffle the whole batch. The feature extraction task is optimized by the standard gradient descent algorithm with the cross-entropy loss L_s:

$$L_s = -\frac{1}{B} \sum_{i=1}^{B} \sum_{j=1}^{d_p} y_j log\hat{y}_j \tag{14}$$

where B is the batch size, d_p represents the number of sentiment polarity categories and y denotes the ground truth.

For the mutual information maximizing task, we combine the objectives of LMI and GMI to obtain the loss function L_{MIMax} of this task, which is listed as follow:

$$L_{MIMax} = -\frac{1}{B} \sum_{i=1}^{B} (\alpha O_{LMI} + \beta O_{GMI}) \tag{15}$$

where α and β are hyperparameters.

Eventually, the total loss function of our MatchMaker model is formulated as follow:

$$L_{total} = \lambda L_s + (1 - \lambda)L_{MIMax} + \lambda'||\Theta||_2 \tag{16}$$

where $\lambda \in (0, 1)$ is a hyperparameter that balances the proportions of the two tasks in the model, λ' is the L_2 regulation term and Θ denotes all parameters of the model.

4   Experiments

4.1   Dataset and Setting

As shown in Table 1, we conduct extensive experiments on five common used benchmark datasets: Rest 14, Rest 15, Rest 16, Lap14 and Twitter. Rest 14, Rest 15 and Rest 16 are retrieved from the SemEval 2014, 2015 and 2016 respectively [12–14], consisting of reviews from the domain of restaurant. Lap14 consists of reviews from the laptop domain, which is also retrieved from the SemEval 2014. We further evaluate the performance of MatchMaker on the Twitter dataset [15]. In the preprocessing stage, we remove the conflicting samples in Rest15 and Rest16.

Table 1. Statistics in different datasets.

Dataset	Positive		Neural		Negative	
	Train	Test	Train	Test	Train	Test
Rest14	2164	728	637	196	807	196
Lap14	994	341	464	169	870	128
Rest15	912	326	36	34	256	182
Rest16	1240	469	69	30	439	117
Twitter	1561	173	3127	346	1560	173

In our experiments, GloVe embedding dimensionality and the hidden dimensionality of unidirectional LSTM are set to 300. The batch size is set to 32. We choose Adam as the optimizer with the learning rate of 0.001 and the L2-regularization weight λ' of 10^5, and the dropout of 0.5. All weights of the model are initialized with a uniform distribution of U(−0.01,0.01). Moreover, we set α and β to 0.5 and 0.5 respectively.

4.2   Baselines

We compare our model with several popular baselines proposed in recent years, which can be divided into two groups as follows.

Semantic-Based Models:

ATAE-LSTM [3]: It combines the attention mechanism and LSTM to capture the concerned aspect information comprehensively.

IAN [4]: It applies two LSTMs to generate representations for context and aspect respectively, and then utilizes the attention mechanism to learn the relation between them interactively.

MGAN [5]: It captures the word-level interaction between context and aspect by using fine-grained attention networks.

GCAE [16]: It effectively extracts the relevant sentiment features by designing the novel Gated Tanh-ReLU unit.

BERT-QA [17]: It fine-tunes the pre-trained model from BERT by constructing an auxiliary sentence.

FDN [11]: It applies double gate mechanism to reduce noise and distill aspect-related sentiment features.

Syntactic-Based Models:

ASGCN [8]: It firstly introduces GCNs to generate the aspect-oriented representations, and then uses retrieval-based attention mechanism to retrieve significant features for aspect-based sentiment classification.

RepWalk [2]: It constructs an aspect-aware subtree of the dependency tree by performing a replicated random walk on a syntactic graph, to learn better representations of the aspect term.

Table 2. Comparison with baselines on five benchmark datasets. Accuracy (%) and Macro-F1 (%) are the evaluation metrics, and the best performances are bold-typed. The symbol – denotes the unrepoted results.

Model	Rest14		Lap14		Twitter		Rest15		Rest16	
	Acc	F1	Acc	F1	Acc	F1	Acc	F1	Acc	F1
LSTM	74.49	59.32	66.51	59.44	69.22	66.52	75.40	53.30	80.67	54.53
ATAE-LSTM	78.60	67.02	68.88	63.93	68.64	66.60	78.48	62.84	83.77	61.71
IAN	79.26	70.09	72.05	67.38	72.50	70.81	78.54	52.65	84.74	55.21
MGAN	81.25	71.94	75.39	72.47	72.54	70.81	–	–	–	–
GCAE	79.35	70.50	73.30	70.10	71.80	69.60	–	–	–	–
BERT-QA	81.96	73.29	**78.21**	**73.56**	74.28	72.38	81.89	65.80	84.87	66.00
FDN	82.30	75.00	76.80	72.50	73.70	72.20	–	–	–	–
ASGCN	80.77	72.02	75.55	71.05	72.15	70.40	79.89	61.89	88.99	67.48
RepWalk	83.80	76.90	**78.20**	**74.30**	74.40	72.60	–	–	89.60	71.20
Ours w/o DG	82.50	73.86	76.18	72.33	72.98	70.96	80.81	64.56	89.77	68.36
Ours w/o MIMax	82.23	75.21	75.55	71.15	73.31	71.53	80.44	63.23	89.12	71.32
Ours w/o LMI	83.13	75.12	75.86	71.86	74.13	72.53	80.81	63.02	89.61	72.88
Ours w/o GMI	83.30	76.07	76.01	71.96	74.28	72.95	80.99	63.24	89.44	72.34
Our MatchMaker	**83.84**	**77.07**	76.95	73.47	**74.86**	**73.9**	**81.90**	**66.41**	**90.26**	**74.31**

4.3 Results and Discussions

We conduct experiments to compare our MatchMaker with baseline models on five datasets. As shown in Table 2, the experimental results indicate that our MatchMaker can obtain a significant performance improvement over all baselines on the datasets of Rest14, Rest15, Rest16 and Twitter. Moreover, our model outperforms baselines on the Lap14, except for the BERT-QA and RepWalk. The reason for the discrepancy is that the Lap14 consists of more implicit samples, making it difficult to learn the deep information. These experimental results indicate that our method of Mutual Information is effective, which makes our MatchMaker model obtain higher performance over most models in terms of each key metric. For example, MatchMaker outperforms ASGCN by 3.1%, 2.71%, 2% on Rest14, Rest15 and Twitter, in terms of the metrics of accuracy. From Table 2, we also can see that our MatchMaker obtain a significant improvement of accuracy rate over ASGCN by 1.4% and 1.3%, in terms of Lap14 and Rest16. This benefit comes from the Mutual Information maximization that strengthens the correlations of the aspect with its opinion words, which helps the feature extraction task to extract more exact aspect-relevant sentiment feature.

4.4   Ablation Study

We further conduct ablation experiments on five datasets to investigate the effect of Mutual Information Maximizing (MIMax) module and double gate layer in our MatchMaker model. In Table 2, the notation *Ours w/o DG* denotes a model from which the double gate layer has been removed, and the performance of the ablated model drops on all datasets. This indicates that the double gate mechanism is helpful to our model.

Meanwhile, we conduct many ablation experiments to demonstrate the effectiveness of Mutual Information estimation. Experimental results, as shown in Table 2, the accuracy is reduced significantly when the MIMax module has been removed from our MatchMaker model. There are two units (i.e. Global MI unit and Local MI unit) in the MIMax module. To further study the individual role of each unit, experiments are also conducted by moving one unit at a time. There are two observations from the results of ablation experiments. First, we find that the ablated model with any one of two units is better than the case of the entire MIMax module removed. These results show that each unit is beneficial to our MatchMaker model. Furthermore, the entire MIMax case is better than the case of only one unit employed. This indicates that the two-unit design of our MIMax is reasonable. Since one is designed to focus on the local information and the other is designed to capture the global information. Second, the ablated model that only contains the LMI unit is better than the GMI case on most datasets. This observation indicates that the LMI unit plays a more significant role than GMI.

4.5   Hyperparameter Analysis

Effects of Hyperparameter λ: We also study the optimal hyperparameter λ, which balances the loss of feature extraction task and MIMax task in our model. Experiments are conducted on the Rest14 by using different values of λ. The results indicate that our model achieves the peak performance at the point of $\lambda = 0.4$.

Effects of Hyperparameter C: In the Local MI unit, the span size C of local context words impacts the performance of our model. Therefore, we explore the optimal hyperparameter C through experiments on the Rest14. Experiment results show that when C = 3, our model can take the most advantage of the LMI unit (Fig. 2).

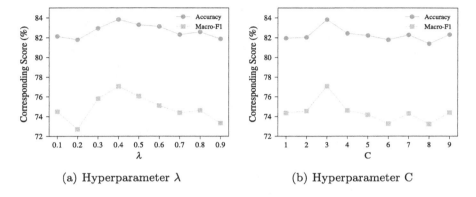

(a) Hyperparameter λ (b) Hyperparameter C

Fig. 2. Effects of hyperparameters.

5 Conclusion

In this work, we propose an ABSC model called MatchMaker, which introduces Mutual Information estimation to effectively strengthen the correlations of the given aspect with its opinion words, improving the precision rate for each aspect in a sentence to match its sentiment polarity. This benefit comes from the improved representation of the sentence, which employs Mutual Information estimation to well model the correlations between aspects and their opinion words without the introduction of extraneous noise. Experimental results demonstrate that our MatchMaker outperforms various baseline models.

References

1. Habimana, O., Li, Y., Li, R., Gu, X., Yu, G.: Sentiment analysis using deep learning approaches: an overview. SCIENCE CHINA Inf. Sci. **63**(1), 1–36 (2019). https://doi.org/10.1007/s11432-018-9941-6
2. Zheng, Y., Zhang, R., Mensah, S., Mao, Y.: Replicate, walk, and stop on syntax: an effective neural network model for aspect-level sentiment classification. In: Proceedings of the AAAI Conference on Artificial Intelligence, vol. 34, pp. 9685–9692 (2020)
3. Wang, Y., Huang, M., Zhu, X., Zhao, L.: Attention-based LSTM for aspect-level sentiment classification. In: Proceedings of the 2016 Conference on Empirical Methods in Natural Language Processing, pp. 606–615 (2016)

4. Ma, D., Li, S., Zhang, X., Wang, H.: Interactive attention networks for aspect-level sentiment classification. In: Proceedings of the 26th International Joint Conference on Artificial Intelligence, pp. 4068–4074 (2017)
5. Fan, F., Feng, Y., Zhao, D.: Multi-grained attention network for aspect-level sentiment classification. In: Proceedings of the 2018 Conference on Empirical Methods in Natural Language Processing, pp. 3433–3442 (2018)
6. Zeng, B., Yang, H., Xu, R., Zhou, W., Han, X.: LCF: a local context focus mechanism for aspect-based sentiment classification. Appl. Sci. **9**(16), 3389 (2019)
7. Sun, K., Zhang, R., Mensah, S., Mao, Y., Liu, X.: Aspect-level sentiment analysis via convolution over dependency tree. In: Proceedings of the 2019 Conference on Empirical Methods in Natural Language Processing and the 9th International Joint Conference on Natural Language Processing (EMNLP-IJCNLP), pp. 5683–5692 (2019)
8. Zhang, C., Li, Q., Song, D.: Aspect-based sentiment classification with aspect-specific graph convolutional networks. In: Proceedings of the 2019 Conference on Empirical Methods in Natural Language Processing and the 9th International Joint Conference on Natural Language Processing (EMNLP-IJCNLP), pp. 4560–4570 (2019)
9. Yeh, Y.T., Chen, Y.N.: QAInfomax: learning robust question answering system by mutual information maximization. In: Proceedings of the 2019 Conference on Empirical Methods in Natural Language Processing and the 9th International Joint Conference on Natural Language Processing (EMNLP-IJCNLP), pp. 3361–3366 (2019)
10. Pennington, J., Socher, R., Manning, C.D.: Glove: global vectors for word representation. In: Proceedings of the 2014 Conference on Empirical Methods in Natural Language Processing (EMNLP), pp. 1532–1543 (2014)
11. Shuang, K., Yang, Q., Loo, J., Li, R., Gu, M.: Feature distillation network for aspect-based sentiment analysis. Inf. Fusion **61**, 13–23 (2020)
12. Pontiki, M., Galanis, D., Pavlopoulos, J., Papageorgiou, H., Androutsopoulos, I., Manandhar, S.: SemEval-2014 task 4: aspect based sentiment analysis. In: Proceedings of the 8th International Workshop on Semantic Evaluation (SemEval 2014), pp. 27–35 (2014)
13. Pontiki, M., Galanis, D., Papageorgiou, H., Manandhar, S., Androutsopoulos, I.: SemEval-2015 task 12: aspect based sentiment analysis. In: Proceedings of the 9th International Workshop on Semantic Evaluation (SemEval 2015), pp. 486–495 (2015)
14. Pontiki, M., et al.: SemEval-2016 task 5: aspect based sentiment analysis. In: International Workshop on Semantic Evaluation, pp. 19–30 (2016)
15. Dong, L., Wei, F., Tan, C., Tang, D., Zhou, M., Xu, K.: Adaptive recursive neural network for target-dependent twitter sentiment classification. In: Proceedings of the 52nd Annual Meeting of the Association for Computational Linguistics (volume 2: Short papers), pp. 49–54 (2014)
16. Xue, W., Li, T.: Aspect based sentiment analysis with gated convolutional networks. In: Proceedings of the 56th Annual Meeting of the Association for Computational Linguistics (Volume 1: Long Papers), pp. 2514–2523 (2018)
17. Sun, C., Huang, L., Qiu, X.: Utilizing BERT for aspect-based sentiment analysis via constructing auxiliary sentence. In: Proceedings of the 2019 Conference of the North American Chapter of the Association for Computational Linguistics: Human Language Technologies, Volume 1 (Long and Short Papers), pp. 380–385 (2019)

PathSAGE: Spatial Graph Attention Neural Networks with Random Path Sampling

Junhua Ma[1], Jiajun Li[2], Xueming Li[1(✉)], and Xu Li[1]

[1] College of Computer Science, Chongqing University, Chongqing 400044, China
{majunhua,lixuemin}@cqu.edu.cn
[2] School of Information Technology and Electrical Engineering, The University of Queensland, St Lucia, Qld 4072, Australia
jiajun.li1@uqconnect.edu.au

Abstract. Graph Convolutional Networks (GCNs) achieve great success in non-Euclidean structure data processing recently. In existing studies, deeper layers are used in CCNs to extract deeper features of Euclidean structure data. However, for non-Euclidean structure data, too deep GCNs will confront with problems like "neighbor explosion" and "over-smoothing", it also cannot be applied to large datasets. To address these problems, we propose a model called PathSAGE, which can learn high-order topological information and improve the model's performance by expanding the receptive field. The model randomly samples paths starting from the central node and aggregates them by Transformer encoder. Path-SAGE has only one layer of structure to aggregate nodes which avoid those problems above. The results of evaluation shows that our model achieves comparable performance with the state-of-the-art models in inductive learning tasks.

Keywords: Neural network models · Path · GCNs · Transformer · Random

1 Introduction

Convolutional Neural Networks (CNNs) have been successfully used in various tasks with Euclidean structure data in recent years. For non-Euclidean structure datasets, graph convolutional networks (GCNs) use the same idea to extract the topological structure information.

[1] firstly proposed two GCN models with spectral and spatial construction respectively. The spectral model using a Laplacian matrix to aggregate neighborhood information of each node in a graph. The spatial model partition graph into clustering and update them by aggregating function. In order to extract deeper features, model based on CNNs usually deepen the model's layers. While in GCNs, Deepening layers will cause a lot of problems. In spectral construction method, too many layers lead to **"over smoothing"** [13,25]: the features of nodes in graph will tend to be the same. In spatial construction method, it

© Springer Nature Switzerland AG 2021
T. Mantoro et al. (Eds.): ICONIP 2021, LNCS 13109, pp. 111–120, 2021.
https://doi.org/10.1007/978-3-030-92270-2_10

will cause exponential growth of the number of sampled neighbor nodes, called **"neighbor explosion"**. Node sampling and layer sampling [3,6,7,9,20] were proposed to handle this problem, but due to incomplete sampling, the inaccuracy of nodes' representation accumulates errors between layers.

To address these two problems, in this paper, we propose a model called PathSAGE, which can learn high-order topological information by expanding the receptive field. Firstly, We design a path sampling technique based on random walk to sample paths starting from central nodes with different lengths, then the sequences of paths are fed into Transformer encoder [17], which can extract the semantic and distance information in sequence effectively. As shown in Fig. 1, We view the sequences of paths from the tail nodes to the central node as the central node' neighbors. Secondly, following this idea, we take the average of paths of the same length as the representation of the central node in this level of reception field. Finally, after concatenating the aggregated features of paths with different lengths, the final representation of the central node are used for downstream tasks.

For the two problems mentioned earlier, on the one hand, the aggregation of the central node only perform once in training of a sample, which never cased "over-smoothing". On the other hand, all the paths were sampled with a fixed number for representing the central node in our model, instead of recursively sampling exponentially growing neighbor nodes. And each path only contributes to the central node, we do not need to calculate and store the temporary representations of nodes from the middle layer. Furthermore, it prevents the error propagation caused by incomplete sampling.

Our contribution can be summarized in three points:

- We utilize the path sampling to take place of the node sampling to avoid error accumulation caused by incomplete node sampling.
- We propose and evaluate our model Path-SAGE to solve the existing "neighbour explosion" and "over-smoothing" problems. The model can capture richer and more diverse patterns around the central node with only one layer of structure to the central node.
- We evaluate our model on three inductive learning tasks, and it reaches the state-of-the-art performance on two of them. We analyze the attention weights of Transformer encoder and detect some patterns in the attention mechanism, which can further illustrate how the model works.

2   Related Work

GNNs model was initially proposed by [1], and the convolution operation in the traditional Euclidean structure was introduced into the graph network with the non-Euclidean structure in this article. They [1] divided GNN into two construction methods: spectral construction and spatial construction. Subsequently, many studies are carried out around these two aspects.

In spectral construction, [5] used Chebyshev polynomials with learnable parameters to approximate a smooth filter in the spectral domain, which improves computation efficiency significantly. [11] further reduced the computational cost through local first-order approximation. In spatial construction, MoNet was proposed in [15], developing the GCN model by defining a different weighted sum of the convolution operation, using the weighted sum of the nodes as the central node feature instead of the average value. [7] attempted various aggregator to gather the features of neighbor nodes. [18,23] defined the convolution operation with a self-attention mechanism between the central node and neighbor nodes, [14] brought the LSTM (Long Short Term Memory networks) [8] from NLP (Natural Language Processing) to GNN, and built an adaptive depth structure by applying the memory gates of LSTM. [19] sampled the shortest path based on their attention score to the central node and combined them in a mixed way. PPNP and APPNP [12] used the relationship between GCN and PageRank to derive an improved communication scheme based on personalized PageRank.

With the scale of graph data increasing, the full-batch algorithm is no longer applicable, and the mini-batch algorithm using stochastic gradient descent is applied. In recent years, some studies based on different sampling strategies have been proposed. [7] tried random neighbor node sampling for the first time to limit large amounts of nodes caused by recursive node sampling. Layer-wise sampling techniques was applied in [3,6,9,20], which only consider fixed number of neighbors in the graph to avoid "neighbor explosion". [2,3] used a control variate-based algorithm that can achieve good convergence by reducing the approximate variance. Besides, subgraph sampling was first introduced by [21], which used a probability distribution based on degree. [4] performed clustering decomposition on the graph before the training phase and randomly sampled a subgraph as a training batch at every step. [22] sampled subgraph during training based on node and edge importance. All these sampling techniques are based on GCNs, how to extract deeper topology information from large graph datasets is still a problem.

3   Proposed Method

In this section, we present the PathSAGE. Firstly, the sampling algorithm is introduced in Sect. 3.1. Secondly, we detail the aggregator in Sect. 3.2. Finally, we discuss the difference between our model and the related models.

3.1   Path Sampling

Except deepening model, another way to expand receptive field of CNNs is to increase the size of convolution kernel. Following this idea, we sample the node sequences starting from central node and regard these paths as the context of the central node. Therefore, the receptive field can be expanded by extending these paths.

We use a straightforward random sampling algorithm based on random walk, shown in **Algorithm** 1: for a central node, a sampling starts from it and randomly selects a neighbor node each time until reaching the preset path length,

---

Algorithm 1: Random Path Sampling

Input: garph $G(V, E)$ central node c
sample depth s
sample num each length $L = \{n_1, n_2, \dots, n_s\}\}$
Output: path sequences with different length
$\{\boldsymbol{P_1}, \boldsymbol{P_2}, \dots, \boldsymbol{P_s}\}$

1 Random Path Sampling(G, c, s, L)
2 **foreach** $l = 0 \rightarrow s$ **do**
3 $\quad$ $i \leftarrow 0;$
4 $\quad$ **while** $i < n_l$ **do**
5 $\quad\quad$ $u \leftarrow c;$
6 $\quad\quad$ $P \leftarrow \{u\};$
7 $\quad\quad$ **for** $j = 0 \rightarrow l$ **do**
8 $\quad\quad\quad$ $u \leftarrow$ Node randomly selected from u's neighbors;
9 $\quad\quad\quad$ $P \leftarrow P \cup \{u\}$
10 $\quad\quad$ **end**
11 $\quad\quad$ $\boldsymbol{P_l} \leftarrow \boldsymbol{P_l} \cup \{P\};$
12 $\quad\quad$ $i \leftarrow i + 1;$
13 $\quad$ **end**
14 **end**
15 **return** $\{\boldsymbol{P_1}, \boldsymbol{P_2}, \dots, \boldsymbol{P_s}\}$

---

and multiple path sequences for the corresponding central node of various lengths can be obtained in this way, constituting a training sample (Fig. 2).

3.2 Aggregator

There are two aggregations in the aggregator: the first one aggregate path sequences; the second one aggregate different paths as the final representation of the central node. For the first aggregation, We formulate each path sequence as:

$$\boldsymbol{P}_{ij} = \{a_{ij}^1, a_{ij}^2, a_{ij}^3, \dots, a_{ij}^j\} \tag{1}$$

To utilize the position information of path sequence, we define positions of nodes in paths as their distances to the central nodes. As same as Transformer, we add nodes' features and their positional vectors together as the input of the structure:

$$\tilde{\boldsymbol{P}}_{ij} = \{a_{ij}^1 + pos\_emb\,(1), \dots, a_{ij}^j + pos\_emb\,(j)\} \tag{2}$$

For the $pos\_emb(\cdot)$, we use *sine* and *cosine* functions of different frequencies to generate the positional vectors. In each dimension, the positional embedding is:

$$pos\_emb(p)_{2i} = sin(\frac{p}{10000^{\frac{2i}{d}}}) \tag{3}$$

$$pos\_emb(p)_{2i+1} = cos(\frac{p}{10000^{\frac{2i}{d}}}) \tag{4}$$

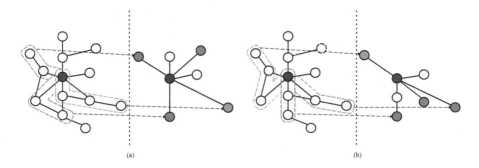

Fig. 1. Mechanism of sampling and aggregation. (a) and (b) are two different possible training samples with the same central node. Paths with same colors are with same lengths same length and share the same aggregators.

Where p is the position in the sequences, $2i$ and $2i + 1$ mean the odd and even dimensions of position embedding, and d is the number of features.

After that, we apply the Transformer encoder on each path:

$$\tilde{\boldsymbol{P}}^k_{ij} = transformer\_block^k(\tilde{\boldsymbol{P}}^{k-1}_{ij}) \tag{5}$$

Where k means k-th Transformer encoder layer. Noted that, in a m-layer Transformer encoder, the output of the last layer $\tilde{\boldsymbol{P}}^m_{ij}$ is a sequence of features. We only take the output at position 0 as the final representation of the path sequence.

$$\tilde{P}'_{ij} = \left[\tilde{\boldsymbol{P}}^m_{ij} \right]_0 \tag{6}$$

Following these Eqs. (2)–(6), we can obtain representations of paths with different length to the central node, as shown in Fig. 1. In the second aggregation, we apply an average pooling layer to aggregate the central node paths. Then we concatenate all the path representation and apply a feed-forward layer with nonlinearity to fuse these features. The final output of a central node C' is computed as following:

$$C = concat\,(C_1,\ ...,\ C_s)$$
$$\text{(7)}$$
$$where\ C_i\ =\ Average\left(\tilde{P}'_{i1},\ ...,\ \tilde{P}'_{in} \right)$$

$$C'\ =\ max\,(0,\ C\boldsymbol{W}_1\ +\ b_1)\,\boldsymbol{W}_2\ +\ b_2 \tag{8}$$

where s denotes the sample depth, n denote the number of paths sampled in

3.3   Comparisons to related Work

– We introduce a sequence transduction model into our structure, but it is distinct from related work based on these models. LSTM [8] was also used in GraphSAGE [22] to aggregate node's features, it is very sensitive to the order

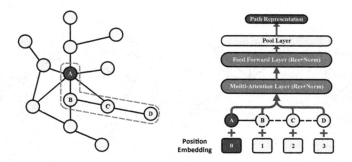

Fig. 2. Structure of first aggregation in aggregator. A specific sampled path's features adds the position embedding, through Transformer encoder layers to fuse the information.

of the input sequence. In contrast, neighbor nodes are disordered, and the authors have rectified it by consistently feeding randomly-ordered sequences to the LSTM. In our model, paths to the central node already have orders, which are naturally suitable for sequence processing model.

- The difference between our attention mechanism and GAT is that we collect all the mutual attention information of each node in the path. The output of our model is integrated information of entire sequence, instead of only attend to the central node.

- Path sampling method is also used in SPAGAN [19]. But each sampled path in SPAGAN is the shortest one to the central node, this sample technique will leads to a high computational overhead and limits the ability of the model to be applied to large graph datasets. By contrast, we sample paths randomly, which greatly save the computation overhead. Besides, in our sample algorithm, the same node may have a different path to the central node, which may help the model acquire more diverse patterns around the central node while saving the computational overhead.

4   Experiments

In this section, we introduce the datasets and the experiment setting in Sects. 4.1 and 4.2 respectively. We present the results of evaluation in Sect. 4.3.

4.1   Dataset

Three large-scale datasets are used to evaluate our model: 1) **Reddit**[7]: a collection of monthly user interaction networks from the year 2014 for 2046 subreddit communities from Reddit that connect users when one has replied to the other, 2) **Flickr**: a social network built by forming links between images sharing

Table 1. Summary of inductive learning tasks' statistics.

	Reddit	Yelp	Flickr
Type	Single-label	Multi-label	Multi-label
# Node	232,965	716,847	89,250
# Edges	11,606,919	6,977,410	899,756
# Features	602	300	500
# Classes	41	100	7
Train/Val /Test	66%/10% /24%	75%/10% /15%	50%/25% /25%

common metadata from Flickr. 3) **Yelp** [22]: a network that links up businesses with the interaction of its customers. Reddit is a multiclass node-wise classification task; Flickr and Yelp are multilabel classification tasks. The detail statistics of these datasets are shown in Table 1.

4.2   Experiment Setup

We build our model on Pytorch framework [16] and construct the Transformer encoder based on UER [24]. For all tasks, We train the model with Adam SGD optimizer [10] with learning rate $1e-3$ and a learning rate scheduler with 0.1 warmup ratio. We use two layers Transformer, each of which has 8 attention heads to gather the features of paths. The batch size is 32. The dropout layers are applied between each sub-layer in Transformer layer. We use different dropout rates in the output layer and Transformer encoder, which are 0.3 and 0.1 respectively. We set the sampling length of the path ranging from 1 to 8 (depth $s = 8$), and the number of paths sampled in each length are [5, 5, 5, 5, 5, 10, 10, 10]. For multi-label classification task Flickr and Yelp, the final output is obtained through a *sigmoid* activation, and in Reddit, the final output is obtained through a *softmax* activation. The hidden dimension is 128 for Reddit and Flickr, 512 for Yelp.

4.3   Result

We compare our model with seven state-of-the-art model: GCN [11], Graph-SAGE [7], FastGCN [3], S-GCN [2], AS-GCN [9], ClusterGCN [4], GraphSAINT [22]. GraphSAGE uses a random node sampling and LSTM aggregator. Fast-GCN and S-GCN use a control variate-based algorithm that can achieve good convergence by reducing the sampling variance. ClusterGCN performs clustering decomposition on the graph before the training phase and randomly sampling a subgraph as a training-batch at every step. GraphSAINT samples subgraph while training with several sampling methods based on node and edge importance. The evaluation on these tasks uses the $Micro - F1$ metric, and report the mean and confidence interval of the metrics by five runs.

Table 2. Performance on inductive learning tasks (Micro-F1).

Model	Reddit	Yelp	Flickr
GCN	0.933 ± 0.000	0.378 ± 0.001	0.492 ± 0.003
GraphSAGE	0.953 ± 0.001	0.634 ± 0.006	0.501 ± 0.013
FastGCN	0.924 ± 0.001	0.265 ± 0.053	0.504 ± 0.001
S-GCN	0.964 ± 0.001	0.640 ± 0.002	0.482 ± 0.003
AS-GCN	0.958 ± 0.001	–	0.504 ± 0.002
ClusterGCN	0.954 ± 0.001	0.609 ± 0.005	0.481 ± 0.005
GraphSAINT	0.966 ± 0.001	**0.653 ± 0.003**	0.511 ± 0.001
Ours	**0.969 ± 0.002**	0.642 ± 0.005	**0.511 ± 0.003**

The results of inductive learning experiments are shown in Table 2. As we can see from the table, For Reddit, our model outperforms all the baseline models. For Flickr, we achieve a comparable F1-score with the top-performing model. For Yelp, we surpass most GCN models, second only to GraphSAINT. One hypothesis to explain the difference of the results is: Reddit and Flickr have more training samples and number of nodes features, which makes the attention mechanism have enough data to capture the relationships between nodes.

5   Attention Analysis

We observed the attention weight of the trained model in the PubMed test set. At some attention heads, we find that nodes with the same labels get very high attention scores to each other. We visualize an example in Fig. 3. Central node

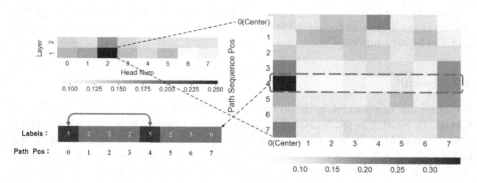

Fig. 3. Detection of attention weights in Transformer that node have the same label in the path sequence may receive higher weights at some attention heads. The heatmap in the upper-left corner is the weight of position 4 of different layers attend to position 2 and attention heads, the right one is the complete heatmap of the selected head. The lower-left picture shows the labels of the nodes in the sequence.

0 and node 4 have the same label and receive extremely high attention scores on second attention heads of the first layer. By observing the attention scores for the whole sequence, we can see that this score also occupies a significant share in the sequence.

This observation proves that the attention mechanism can successfully capture the information that is helpful to the downstream tasks in the aggregation of paths.

6   Conclusion

In this paper, we propose a model—PathSAGE, which can expand the size of receptive field without stacking layers which solved the problem of "neighbor explosion" and "over smoothing". Regarding all paths as the neighbor nodes of the central node, PathSAGE samples paths starting from the central node based on random walk and aggregates these features by a strong encoder—Transformer. Consequently, the model can obtain more features of neighbor nodes and more patterns around the central node. In our experiment, PathSAGE achieves the state-of-the-art on inductive learning tasks. Our model provides a novel idea to handle large-scale graph data.

Acknowledgments. This work is supported by National Key R&D Program of China (No. 2017YFB1402405-5), and the Fundamental Research Funds for the Central Universities (No.2020CDCGJSJ0042). The authors thank all anonymous reviewers for their constructive comments.

References

1. Bruna, J., Zaremba, W., Szlam, A., LeCun, Y.: Spectral networks and locally connected networks on graphs (2014)
2. Chen, J., Zhu, J., Song, L.: Stochastic training of graph convolutional networks with variance reduction (2018)
3. Chen, J., Ma, T., Xiao, C.: FastGCN: fast learning with graph convolutional networks via importance sampling (2018)
4. Chiang, W.L., Liu, X., Si, S., Li, Y., Bengio, S., Hsieh, C.J.: Cluster-GCN: an efficient algorithm for training deep and large graph convolutional networks. In: Proceedings of the 25th ACM SIGKDD International Conference on Knowledge Discovery and Data Mining, pp. 257–266 (2019)
5. Defferrard, M., Bresson, X., Vandergheynst, P.: Convolutional neural networks on graphs with fast localized spectral filtering (2017)
6. Gao, H., Wang, Z., Ji, S.: Large-scale learnable graph convolutional networks. In: Proceedings of the 24th ACM SIGKDD International Conference on Knowledge Discovery and Data Mining, pp. 1416–1424 (2018)
7. Hamilton, W.L., Ying, R., Leskovec, J.: Inductive representation learning on large graphs (2018)
8. Hochreiter, S., Schmidhuber, J.: Long short-term memory. Neural Comput. **9**(8), 1735–1780 (1997)

9. Huang, W., Zhang, T., Rong, Y., Huang, J.: Adaptive sampling towards fast graph representation learning. arXiv preprint arXiv:1809.05343 (2018)
10. Kingma, D.P., Ba, J.: Adam: A method for stochastic optimization. arXiv preprint arXiv:1412.6980 (2014)
11. Kipf, T.N., Welling, M.: Semi-supervised classification with graph convolutional networks (2017)
12. Klicpera, J., Bojchevski, A., Günnemann, S.: Predict then propagate: Graph neural networks meet personalized pagerank. arXiv preprint arXiv:1810.05997 (2018)
13. Li, Q., Han, Z., Wu, X.M.: Deeper insights into graph convolutional networks for semi-supervised learning. In: Proceedings of the AAAI Conference on Artificial Intelligence, vol. 32 (2018)
14. Liu, Z., Chen, C., Li, L., Zhou, J., Li, X., Song, L., Qi, Y.: Geniepath: graph neural networks with adaptive receptive paths. In: Proceedings of the AAAI Conference on Artificial Intelligence, vol. 33, pp. 4424–4431 (2019)
15. Monti, F., Boscaini, D., Masci, J., Rodola, E., Svoboda, J., Bronstein, M.M.: Geometric deep learning on graphs and manifolds using mixture model CNNs. In: Proceedings of the IEEE Conference on Computer Vision and Pattern Recognition, pp. 5115–5124 (2017)
16. Paszke, A., et al.: Automatic differentiation in Pytorch (2017)
17. Vaswani, A., et al.: Attention is all you need (2017)
18. Veličković, P., Cucurull, G., Casanova, A., Romero, A., Lio, P., Bengio, Y.: Graph attention networks. arXiv preprint arXiv:1710.10903 (2017)
19. Yang, Y., Wang, X., Song, M., Yuan, J., Tao, D.: SPAGAN: shortest path graph attention network (2021)
20. Ying, R., You, J., Morris, C., Ren, X., Hamilton, W.L., Leskovec, J.: Hierarchical graph representation learning with differentiable pooling. arXiv preprint arXiv:1806.08804 (2018)
21. Zeng, H., Zhou, H., Srivastava, A., Kannan, R., Prasanna, V.: Accurate, efficient and scalable graph embedding. In: 2019 IEEE International Parallel and Distributed Processing Symposium (IPDPS), pp. 462–471. IEEE (2019)
22. Zeng, H., Zhou, H., Srivastava, A., Kannan, R., Prasanna, V.: GraphSaint: graph sampling based inductive learning method (2020)
23. Zhang, J., Shi, X., Xie, J., Ma, H., King, I., Yeung, D.Y.: GaAN: gated attention networks for learning on large and spatiotemporal graphs (2018)
24. Zhao, Z., et al.: UER: an open-source toolkit for pre-training models. EMNLP-IJCNLP **2019**, 241 (2019)
25. Zhou, J., et al.: Graph neural networks: A review of methods and applications. arXiv preprint arXiv:1812.08434 (2018)

Label Preserved Heterogeneous Network Embedding

Xiangyu Li[1(✉)] and Weizheng Chen[2(✉)]

[1] School of Software Engineering, Beijing Jiaotong University, Beijing, China
lixiangyu@bjtu.edu.cn
[2] Beijing, China

Abstract. Recently, the heterogeneous network embedding (HNE for short) methods have been attracting increasing attention due to their simplicity, scalability, and effectiveness. However, the rich node label information is not considered by these HNE methods, which leads to suboptimal node embeddings. In this paper, we propose a novel **L**abel **P**reserved **H**eterogeneous **N**etwork **E**mbedding (LPHNE) method to tackle this problem. Briefly, for each type of the nodes, LPHNE projects these nodes and their labels into a same low-dimensional hidden space by modeling the interactive relationship between the labels and the contexts of the nodes. Thus, the discriminability of node embedding is improved by utilizing the label information. The extensive experimental results demonstrate that our semi-supervised method outperforms the various competitive baselines on two widely used network datasets significantly.

Keywords: Heterogeneous network · Network embedding · Semi-supervise learning

1 Introduction

With the tremendous development of the online social media, heterogeneous information network which is made up of multiple types of nodes and edges has become more common. For example, the photo-sharing site Flickr can be viewed as a heterogeneous network composed of users and photos, the bibliography site DBLP can be regarded as a heterogeneous network composed of authors and papers. It is a great challenge to directly apply machine learning techniques to analyze the heterogeneous networks since their high-dimensional structure is highly complex. A promising solution, heterogeneous network embedding, whose target is to map nodes to a continuous low-dimensional space, has received a lof of attention recently. The heterogeneous network embedding based approaches provide a new point of view to many important tasks, such as ride matching [14] and product recommendation [11].

There have been quite a few models to learn heterogeneous network embedding, such as Esim [10]. However, without considering the supervised label

W. Chen—Independent Researcher.

T. Mantoro et al. (Eds.): ICONIP 2021, LNCS 13109, pp. 121–132, 2021.
https://doi.org/10.1007/978-3-030-92270-2_11

information, the node embeddings learned in these unsupervised models are suboptimal for the node classification task. Note that in real life, nodes generally have rich labels. For instance, in Flickr, users are belonging to various interest groups, and photos are labeled with different tags by the users. Overall, the existing heterogeneous network embedding methods could not benefit from the heterogeneous structure and label information simultaneously. As far as we know, the valuable label-context relationship has not been considered to solve the heterogeneous network embedding problem. In fact, the label-context relationship is crucial to increase the prediction ability of the node embeddings. For example, as shown in the left side of Fig. 1, a partially labeled information network has 6 users, including two politicians ("Obama" and "Hillary"), a hoopster ("LeBron") and three other users ("user1", "user2" and "user3"). Only two users have labels, i.e. the label of user2 is "politics" and the label of user3 is "sport". Now we want to predict the label of user1. Although we have known that user1, user2 and "Hillary" share the same context (C_{obama}), the relation between context and label is unknown. Thus, it's difficult to predict the label of user1 without the supervised label information. However, if we have the label-context network, just like the right side of Fig. 1, the context C_{obama} is highly relevant to the label "politics", then we can predict that the label of user1 is "politics" because its context is C_{obama}.

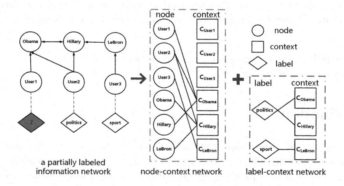

Fig. 1. Demonstration of how to convert a partially labeled information network to separaed bipartite networks. The left side is a partially labeled toy Twitter network which has six users (nodes) and two of them are labeled. We first decompose this network into a node-context bipartite network and a label-context bipartite network. The node-context network and the label-context network are used to retain the unsupervised structure information and the unsupervised structure information respectively. The label-context network is generated by gathering the label-level context co-occurrences. **Here, we let the neighbouring nodes of a labeled node serve as the common contexts of this node and its label.**

Motivated by the above observation, we propose a Label Preserved Heterogeneous Network Embedding (LPHNE) method to overcome the shortcoming of the existing HNE methods, which can incorporate the label information. LPHNE first extracts multiple overlapping node-context and label-context bipartite networks from a heterogeneous network. Then, LPHNE embed these bipartite networks jointly via a unified neural matrix embedding approach. Furthermore, LPHNE can intergrate the text information of the nodes if available. Finally, the effectiveness of LPHNE model is verified by conducting node classification experiment on largescale, real-world heterogeneous network datasets.

2 Related Work

According to whether the label information is used, the existing network embedding models can be roughly classified into two groups.

The first group is the unsupervised network embedding methods. In recent years, distributed representation learning, has become a fundamental technology in the field of network data mining. DeepWalk is the first model to adopt neural network approaches to learn node embeddings. Similar to DeepWalk, several structure based network embeddings models, LINE [13], node2vec [3], are all proven to be equivalent to a specific matrix factorization. SDNE [16] adopts a deep autoencoder to learn node embeddings from the structure information. All the aforementioned models are not applicable for the homogeneous networks. Thus, HNE [1] and EOE [17] are proposed to learn unsupervised node embeddings from the heterogeneous networks.

The second group is the semi-supervised network embedding methods. A semi-supervised network embedding model is usually a linear combination of a structure based homogeneous network embedding method and a classification method trained on the nodes whose label is known. For example, MMDW [15] and DDRW [7] are both a transductive extension of DeepWalk. NLSTNE [2] is a transductive extension of LINE. Recently, graph neural networks like GCN [5] have been extensively studied in the homogeneous network embedding area. The main contribution of GCN is to fuse the network topology structure and local node features in a deep convolutional neural network. All the aforementioned models are not applicable for the heterogeneous networks since their network embedding part is designed for the homogeneous network. LSHM [4] is a semi-supervised HNE model, which trains several max-margin classifiers for different types of nodes. Semi-supervised heterogeneous graph neural networks [8] usually use a softmax classifier as the output layer to predict the node labels.

3 Problem Formulation

In this section, our research problem is formally defined. We use a capital non-bold italic letter to denote a set (e.g. V). A matrix is represented as a capital bold letter (e.g. $\mathbf{M}$). All vectors are column vectors (e.g. $\vec{z}$).

Definition 1 (Partially Labeled Heterogeneous Information Network).
For simplicity, the partially labeled heterogeneous information network is defined as $G = (V, A, L, U, C, Y, W, D)$. $V = \{V_1, ..., V_t\}$ is composed of t disjoint node sets, where $V_i = \{v_{i1}, ..., v_{in_i}\}$ is the set of ith type nodes and t is the number of

node types. If $t = 1$, the network G will degenerate into a homogeneous network. where $\mathbf{A}_{ij}$ is a $n_i \times n_j$ adjacency matrix. For the node $v_{ip} \in V_i$ and the node $v_{jq} \in V_q$, $\mathbf{A}_{ij,pq}$ is the weight of the edge between v_{ip} and V_{ip}. If there is no edge between v_{ip} and v_{jq}, we set $\mathbf{A}_{ij,pq} = 0$. $L = \{L_1, ..., L_t\}$ is composed of t disjoint labeled node sets, where $L_i = \{v_{i1}, ..., v_{i|L_i|}\}$ is the set of ith type labeled nodes. $U = \{U_1, ..., U_t\}$ is composed of t disjoint unlabeled node sets, where $U_i = V_i - L_i$ is the set of ith type unlabeled nodes. $C = \{C_1, ..., C_t\}$ is composed of t class label sets, where $C_i = \{c_{i1}, ..., c_{im_i}\}$ is the set of the possible class labels for the ith type nodes. $Y = \{\mathbf{Y}_1, ..., \mathbf{Y}_t\}$ is a set of t label matrices, where $\mathbf{Y}_i$ is a $n_i \times m_i$ matrix which encodes the label information for the ith type nodes. If a node v_{ip} has a label c_{iq}, we set $\mathbf{Y}_{i,pq} = 1$, otherwise $\mathbf{Y}_{i,pq} = 0$. Note that the label information of any nodes that belongs to U_i is unknown, so the last $|U_i|$ rows of $\mathbf{Y}_i$ are all zeros. $W = \{w_1, ..., w_r\}$ is the set of r text attributes, namely words. $D = \{\mathbf{D}_1, ..., \mathbf{D}_t\}$ is a set of t text content matrices, we use the tf.idf matrix in our paper.

The goal of network embedding is to acquire a low-dimensional vector $\overrightarrow{z_{v_{ip}}} \in \mathbb{R}^d$ for a node v_{ip}, where $d \ll |V|$. The ideal node vectors should preserve the structure and label information simultaneously. When a node v_{ip} is served as a context, it is represented as a context vector $\overrightarrow{h_{v_{ip}}} \in \mathbb{R}^d$. Similarly, when a word w_p is served as a context, it is represented as a context vector $\overrightarrow{h_{w_p}} \in \mathbb{R}^d$. Since we embed the labels and the nodes to the same space, each label c_{ip} is associated with a vector $\overrightarrow{z_{c_{ip}}} \in \mathbb{R}^d$.

4 The Proposed Model

4.1 Neural Matrix Embedding

Most real-world data are matrices. Thus, inspired by [6], we generalize the idea of LINE [13], an unsupervised homogeneous network embedding model, to learn distributed representations for an arbitrary matrix. For ease of presentation, the task of matrix embedding is defined as follow:

Definition 2 (Matrix Embedding). Considerin a sample set $S = \{s_1, ..., s_{|S|}\}$ and a feature set $F = \{f_1, ..., f_{|F|}\}$ and a sample-feature matrix $\mathbf{M}$ in which $\mathbf{M}_{ij}$ is the value of feature f_j in the sample s_i, the target of matrix embedding is to get a meaningful $|S| \times d$ matrix $\mathbf{Z}$ whose ith row is the transpose of input vector $\overrightarrow{z_{s_i}} \in \mathbb{R}^d$ and a $|F| \times d$ matrix $\mathbf{H}$ whose jth row is the transpose of output context vector $\overrightarrow{h_{f_i}} \in \mathbb{R}^d$.

Next, we adopt the neural network approach to solve the matrix embedding problem. Given a sample s_i, the following softmax function is adopted to model the probability that the feature f_j is generated:

$$\Pr(f_j|s_i) = \frac{\exp(\overrightarrow{z_{s_i}}^T \overrightarrow{h_{f_j}})}{\sum_{k=1}^{|F|} \exp(\overrightarrow{z_{s_i}}^T \overrightarrow{h_{f_k}})}. \tag{1}$$

The corresponding empirical probability of $\Pr(f_j|s_i)$ can be defined according to the observed information in $\mathbf{M}$:

$$\widehat{\Pr}(f_j|s_i) = \frac{\mathbf{M_{ij}}}{\sum_{k=1}^{|F|} \mathbf{M_{ik}}}. \tag{2}$$

If the input sample vectors and the output feature vectors are meaningful, the two probability distributions $\Pr(\cdot|s_i)$ and $\widehat{\Pr}(\cdot|s_i)$ should be close. So we adopt Kullback-Leibler divergence as the distance metric and minimize the following objective loss function:

$$O_{SF} = \sum_{i=1}^{|S|} \lambda_{s_i} D_{KL}(\widehat{\Pr}(\cdot|s_i)||\Pr(\cdot|s_i)), \tag{3}$$

where $\lambda_{s_i} = \sum_{k=1}^{|F|} \mathbf{M}_{ik}$ is the importance of s_i in S. After removing some constants, O_{sf} is simplified as:

$$O_{SF} = -\sum_{i=1}^{|S|} \sum_{j=1}^{|F|} \mathbf{M}_{ij} \log \Pr(f_j|s_i). \tag{4}$$

Finally, the sample input vectors and the feature context vectors can be obtained by minimizing O_{sf}. The assumption of the above model is that two samples should have close representations if they share similar features. We call the above method **N**eural **M**atrix **E**mbedding (NME).

4.2 Heterogeneous Structure Information Modeling

When we consider a adjacency matrix $\mathbf{A}_{ij}$ independently, we treat the nodes in the node set V_i as samples and the nodes in the node set V_j as contexts (features). Naturally, by applying the NME method to $\mathbf{A}_{ij}$, the objective loss function is as follow:

$$O_{V_i V_j} = -\sum_{p=1}^{n_i} \sum_{q=1}^{n_j} \mathbf{A}_{ij,pq} \log \Pr(v_{jq}|v_{ip}). \tag{5}$$

Then, to model the completely unsupervised heterogeneous structure information of the heterogeneous information network G consistently, we learn the node input vectors and the node context vectors from all the adjacency matrices jointly. So the objective loss function of modeling the network structure information is:

$$O_{net} = \sum_{i=1}^{t} \sum_{j=1}^{t} O_{V_i V_j}. \tag{6}$$

4.3 Heterogeneous Label Information Modeling

When we consider the adjacency matrix $\mathbf{A}_{ij}$ and the label information of V_i contained in $\mathbf{Y}_i$ jointly, we let the nodes belongs to V_j serve as the contexts of the labels in C_i. More specifically, as illustrated in the right side of Fig. 1, we build immediate connections between the labels and the contexts. So we define a $m_i \times n_j$ label-context matrix $\mathbf{B}_{ij}$ to encode the label-context relationship between the label set C_i and the node set V_j:

$$\mathbf{B}_{ij} = \mathbf{Y}_i^T \mathbf{A}_{ij}. \tag{7}$$

In a similar way, the loss function for $\mathbf{B}_{ij}$ is defined as follow:

$$O_{C_i V_j} = -\sum_{p=1}^{m_i}\sum_{q=1}^{n_j} \mathbf{B}_{ij,pq} \log \Pr(v_{jq}|c_{ip}). \tag{8}$$

Then, we can define a set of t^2 label-context matrices $B = \{\mathbf{B}_{11}, ..., \mathbf{B}_{tt}\}$. To model the supervised heterogeneous label-context information of G consistently, we learn the label input vectors and the node context vectors from all the label-context matrices jointly. So the objective loss function of modeling the label information is:

$$O_{label} = \sum_{i=1}^{t}\sum_{j=1}^{t} O_{C_i V_j}. \tag{9}$$

4.4 Heterogeneous Text Information Modeling

When we consider a text content matrix $\mathbf{D}_i$, we treat the nodes in V_i as samples and the words in the word set W as contexts (features). Naturally, by applying the NME method to $\mathbf{D}_i$, the objective loss function is defined as follow:

$$O_{V_i W} = -\sum_{p=1}^{n_i}\sum_{q=1}^{r} \mathbf{D}_{i,pq} \log \Pr(v_{jq}|v_{ip}). \tag{10}$$

Then, to model the heterogeneous text information of the heterogeneous information network G, we can learn the node input vectors and the word vectors from all the text content matrices collectively. Thus, the objective loss function of modeling the text information is:

$$O_{text} = \sum_{i=1}^{t} O_{V_i W}. \tag{11}$$

4.5 LPHNE

To combine structure, label and text information into a joint embedding framework, the objective loss function of the LPHNE model is a linear combination of O_{net}, O_{label} and O_{text}:

$$\begin{aligned}
O_{LPHNE} &= O_{net} + \alpha O_{label} + \beta O_{text} \\
&= \sum_{i=1}^{t}\sum_{j=1}^{t}\{O_{V_i V_j} + \alpha O_{C_i V_j}\} + \sum_{i=1}^{t}\beta O_{V_i W},
\end{aligned} \tag{12}$$

where α and β are two tunable tradeoff parameters whose values both default to 1. When the text information is not available, we set $\beta = 0$. When the label information is not available, we set $\alpha = 0$.

In LPHNE, we model the label-context relationship directly, which in turn improves the discriminative power of the input node vectors due to the sharing of the output context vectors among $O_{V_i V_j}$ and $O_{C_i V_j}$. As we discussed earlier in Sect. 2, the way of incorporating the label information in LPHNE is totally different from existing semi-supervised network embedding methods.

4.6   Training and Complexity Analysis

Like the term O_{SF} defined in the Eqn. (4), $O_{V_i V_j}$, $O_{C_i V_j}$ and $O_{V_i W}$ have the same format. So we first introduce how to optimize O_{SF} separately. Directly calculating the probability term $\Pr(f_j|s_i)$ in O_{SF} is computationally expensive since we need to iterate all the features in F. Therefore we adopt the negative sampling method [9] to reduce the computational complexity. Finally O_{sf} is rewritten as:

$$O_{SF} = -\sum_{i=1}^{|S|}\sum_{j=1}^{|F|}\mathbf{M}_{ij}\left\{\log\sigma(\overrightarrow{z_{s_i}}^T\overrightarrow{h_{f_j}}) + K\mathbb{E}_{f_n \sim P_{SF}(\cdot)}\left[\log\sigma(-\overrightarrow{z_{s_i}}^T\overrightarrow{h_{f_n}})\right]\right\}, \quad (13)$$

where $\sigma(x) = \frac{1}{1+\exp(-x)}$ is the sigmoid function, K is the number of negative features for an observed sample-feature pair, $\mathbb{E}$ indicates the mathematical expectation and $P_{SF}(\cdot)$ is the noise feature distribution over the feature set F that can be calculated as:

$$P_{SF}(j) = \frac{\lambda_{f_j}}{\Lambda} \quad (14)$$

where $\lambda_{f_j} = \left(\sum_{i=1}^{|S|}\mathbf{M}_{ij}\right)^{0.75}$ is the importance of f_j in F, $\Lambda = \sum_{k=1}^{|F|}\lambda_{f_k}$ is the normalization constant.

For clarity, we set $O_1^{s_i f_j} = -\log\sigma(\overrightarrow{z_{s_i}}^T\overrightarrow{h_{f_j}})$ and $O_2^{s_i f_n} = -\log\sigma(-\overrightarrow{z_{s_i}}^T\overrightarrow{h_{f_n}})$ to represent the loss function for an observed sample-feature pair (s_i, f_j) and a negative noisy sample-feature pair (s_i, f_n) respectively. In a similar way, we can rewrite $O_{V_i V_j}$, $O_{C_i V_j}$ and $O_{V_i W}$ as follows:

$$O_{V_i V_j} = \sum_{p=1}^{n_i}\sum_{q=1}^{n_j}\mathbf{A}_{ij,pq}\left\{O_1^{v_{ip} v_{jq}} + K\mathbb{E}_{v_{jn} \sim P_{V_i V_j}(\cdot)}O_2^{v_{ip} v_{jn}}\right\}, \quad (15)$$

$$O_{C_i V_j} = \sum_{p=1}^{m_i}\sum_{q=1}^{n_j}\mathbf{B}_{ij,pq}\left\{O_1^{c_{ip} v_{jq}} + K\mathbb{E}_{v_{jn} \sim P_{C_i V_j}(\cdot)}O_2^{c_{ip} v_{jn}}\right\}, \quad (16)$$

$$O_{V_i W} = \sum_{p=1}^{n_i}\sum_{q=1}^{r}\mathbf{D}_{i,pq}\left\{O_1^{v_{ip} w_q} + K\mathbb{E}_{w_n \sim P_{V_i W}(\cdot)}O_2^{v_{ip} w_n}\right\}, \quad (17)$$

where $P_{V_i V_j}(\cdot)$ and $P_{C_i V_j}(\cdot)$ are two different noise context distributions that are determined by $\mathbf{A}_{ij}$ and $\mathbf{B}_{ij}$ respectively, $P_{V_i W}(\cdot)$ is a noise word distributions which is determined by $\mathbf{D}_i$. The definitions of them are similar to the Eq. (12).

As shown in Algorithm 1, we use a stochastic gradient descent based joint training approach to optimize the Eq. (12). We define a function $g(\mathbf{M})$ to represent the number of non-zero elements in $\mathbf{M}$. In each iteration, Algorithm 2 is called to handle the node-context, label-context and node-word matrices sequentially. In Algorithm 2, we sample an observed sample-feature pair and K negative noise features, then we update the corresponding input sample vectors and output feature vectors by adopting the given updating formulas. When the number of iterations I is large enough, we can finally get converged node embeddings.

By adopting the alias method, sampling a pair whose weight is greater than 0 from a matrix or sampling a negative noise feature from a noise distribution only takes $O(1)$ time. So the time complexity of Algorithm 2 is $O(dK)$. In practice, we find that the number of iterations I should be proportional to $X = max(\{g(M)|M \in A \cup B \cup D\})$. Therefore, the overall time complexity of LPHNE model is $O(dKX)$ and does not depend on the number of nodes in G.

Algorithm 1: Joint training for LPHNE

---

    input   : number of iterations I, number of negative samples K, tradeoff parameters α and β, learning rate η, a partially labeled heterogeneous information network $G = (V, A, L, U, C, Y, W, D)$

    output: node embeddings $\overrightarrow{z_v}$

1   initialize all vectors randomly from the uniform distribution [-1,1]

2   **while** $iter \leq I$ **do**

3   **for** $i \leftarrow 1$ **to** t **do**

4   **for** $j \leftarrow 1$ **to** t **do**

5   **if** $g(A_{ij}) > 0$ **then**

6   UpdateNME(V_i, V_j, $\mathbf{A}_{ij}$, η)

7   UpdateNME(C_i, V_j, $\mathbf{B}_{ij}$, $\eta\alpha$)

8   **end**

9   **end**

10   **if** $g(D_i) > 0$ **then**

11   UpdateNME(V_i, W, $\mathbf{D}_i$, $\eta\beta$)

12   **end**

13   **end**

14 **end**

Algorithm 2: UpdateNME(S, F, $\mathbf{M}$, η)

---

1   sample a pair (s_i, f_j) according to $\mathbf{M}_{ij} > 0$

2   $\overrightarrow{z_{s_i}} \leftarrow \overrightarrow{z_{s_i}} - \eta\dfrac{\partial O_1^{s_i f_j}}{\partial \overrightarrow{z_{s_i}}} = \overrightarrow{z_{s_i}} + \eta\sigma(-\overrightarrow{z_{s_i}}^T \overrightarrow{h_{f_j}})\overrightarrow{h_{f_j}}$

3   $\overrightarrow{h_{f_i}} \leftarrow \overrightarrow{h_{f_j}} - \eta\dfrac{\partial O_1^{s_i f_j}}{\partial \overrightarrow{h_{f_j}}} = \overrightarrow{h_{f_j}} + \eta\sigma(-\overrightarrow{z_{s_j}}^T \overrightarrow{h_{f_j}})\overrightarrow{z_{s_i}}$

4   **for** $k \leftarrow 1$ **to** K **do**

5   sample a negative noise feature f_n from $P_{sf}(\cdot)$

6      $\overrightarrow{z_{s_i}} \leftarrow \overrightarrow{z_{s_i}} - \eta\dfrac{\partial O_2^{s_i f_n}}{\partial \overrightarrow{z_{s_i}}} = \overrightarrow{z_{s_i}} - \eta\sigma(\overrightarrow{z_{s_i}}^T \overrightarrow{h_{f_n}})\overrightarrow{h_{f_n}}$

7      $\overrightarrow{h_{f_n}} \leftarrow \overrightarrow{h_{f_n}} - \eta\dfrac{\partial O_2^{s_i f_n}}{\partial \overrightarrow{h_{f_n}}} = \overrightarrow{h_{f_n}} - \eta\sigma(\overrightarrow{z_{s_i}}^T \overrightarrow{h_{f_n}})\overrightarrow{z_{s_i}}$

8   **end**

---

5   Experiments

5.1   Datasets and Experiment Settings

In the network embedding area, the multi-class node classification task is widely adopted to evaluate the quality of node embeddings quantitatively [18]. We also follow these works, and conduct the classification experiments on two real-world information networks, i.e., DBLP and Flickr.

DBLP [4] is a bibliography network heterogeneous information network composed of 14,475 authors and 14,376 papers. The titles and abstracts are used as text contents of the paper. For each author, the text contents of the papers written by him or her are used as the text contents of the author. The size of the word vocabulary is 8,920. There is only the author-paper relationship in the DBLP network. The authors are categorized into 4 different research areas. Each paper is labeled with the conference name that it is published in (20 conferences). Each paper or author only has one label.

Flickr [4] is a photo-sharing heterogeneous information network composed of 4,760 users and 46,926 photos. There is no text attribute information in this network. Flickr have two types of relationship, the *following* relation between the users, the *authorship* between the users and the photos. The number of the user-user edges is 175,779 and the number of the user-photo edges is 46,926. The 42 subscribed interest groups are treated as the user labels. The 21 possible tags are regarded as the photo labels. Each user or photo can have more than one label.

For DBLP, we adopt Accuracy as the evaluation metric. For Flickr, each node can have multiple labels, so we report Micro-F1. The performance of our LPHNE model is compared with the following baselines:

- PTE [12]. A classical semi-supervised document embedding method. It takes the average vector of all words in the text contents to represent a node.
- LINE [13]. We use LINE with second-order proximity in our paper.
- LSHM [4]. A competitive semi-supervised heterogeneous network embedding method, which utilizes the label information, the structure information and the text information. Words are also treated as nodes in LSHM.

Among the above baselines, LINE is only applicable for the homogeneous networks. Therefore, we treat DBLP and Flickr as homogeneous networks and feed them to LINE directly. Because we have no text content in the Flickr dataset, PTE is not compared on Flickr.

For a fair comparison, we set the vector length $d = 200$ for all the models. The parameter setting of LPHNE is the same for the two datasets, $\alpha = 1$, $K = 5$ and $I = 4$ million. We set $\beta = 1$ for DBLP and $\beta = 0$ for Flickr. The rest parameters of other baselines are set to the suggested values according to their original papers since the same datasets are used in [4].

We follow the previous works [18] and adopt the one-vs-the-rest linear SVM for classification. More specifically, we random sample a certain proportion of nodes as labeled data to train the SVM, the rest nodes are used for test. We repeat the process 20 times and report the averaged results.

Table 1. Accuracy (%) of author classification on DBLP

Labeled Nodes	1%	2%	3%	4%	5%	6%	7%	8%	9%
LINE	36.43	37.75	39.03	41.64	43.32	44.07	44.59	46.93	48.21
PTE	52.61	68.43	76.35	76.65	76.96	76.92	77.31	77.17	77.38
LSHM	37.85	39.98	40.85	42.08	54.03	59.26	63.25	64.57	66.36
PLHNE	**59.77**	**67.07**	**70.71**	**71.74**	**74.08**	**76.32**	**77.92**	**79.42**	**80.28**

Table 2. Accuracy (%) of paper classification on DBLP

Labeled Nodes	1%	2%	3%	4%	5%	6%	7%	8%	9%
LINE	15.85	16.12	16.61	16.65	17.60	17.78	17.91	18.03	18.80
PTE	19.04	21.07	22.13	22.60	22.72	23.02	23.19	23.65	23.97
LSHM	18.89	21.17	25.61	26.17	26.51	26.91	27.75	29.41	29.95
LPHNE	**25.88**	**27.18**	**27.98**	**29.33**	**29.66**	**30.13**	**30.38**	**31.77**	**32.24**

Table 3. Micro-F1 (%) of user classification on Flickr

Labeled Nodes	1%	2%	3%	4%	5%	6%	7%	8%	9%
LINE	21.38	23.75	24.83	26.32	27.87	30.07	31.91	33.06	34.51
LSHM	37.85	39.98	40.85	42.08	44.03	45.20	45.82	45.83	45.98
PLHNE	**46.64**	**46.76**	**49.05**	**49.45**	**49.62**	**50.42**	**51.46**	**51.61**	**51.71**

Table 4. Micro-F1 (%) of photo classification on Flickr

Labeled Nodes	1%	2%	3%	4%	5%	6%	7%	8%	9%
LINE	15.32	17.08	17.85	18.36	19.29	20.63	21.90	22.91	24.04
LSHM	40.02	42.76	43.42	44.02	44.81	45.23	45.57	46.66	47.56
LPHNE	**39.19**	**40.21**	**44.78**	**48.19**	**50.37**	**51.45**	**54.05**	**54.27**	**56.74**

5.2 Results of Classification

The performance of different models with varied proportion of labeld nodes ranging from 1% to 9% is shown in Tables 1, 2, 3 and 4. From these results, we can obtain:

(1) On both two datasets, LINE is the worst-performing model due to the lack of considering the content and the label information. Furthermore, treating a heterogeneous network as a homogeneous network is oversimplified. The performance of PTE is unsatisfactory on DBLP-paper classification task, because the author vectors and the paper vectors could not mutually enhance each other without modeling the structure information.

(2) LPHNE generally outperforms other models by a notable margin. Compared with LHSM, LPHNE achieves nearly 20% and 3.1% improvement in

the measure of Accuracy on DBLP-author and DBLP-paper classification task respectively when the proportion of labeled nodes is 0.05. On Flickr, LPHNE achieves nearly 5.6% and 5.5% improvement over LSHM in the measure of Micro-F1 on user and photo classification task respectively when the proportion of labeled nodes is 0.05. This result validates the superiority of our joint neural matrix embedding approach over the transductive LSHM model.

5.3   Parameter Sensitivity

In this part, the parameter sensitivity is studied to show the stability of our model. Concretely speaking, we report the classification performance of LPHNE and LSHM w.r.t. different values of d when the proportion of training data is 0.05. When d is under test, we alter d with the others fixed to their default values.

The dimension sensitivity of different models with $d \in [10, 1280]$ is shown in Fig. 2. When d increases from 10, the performance of two semi-supervised models, LPHNE and LSHM, are both relatively stable. However, LPHNE always outperforms LSHM by a noticeable margin. Even when $d = 10$, LPHNE also achieve its optimal performance on both two datasets. In other words, our model can reduce the memory requirements to store the node vectors significantly.

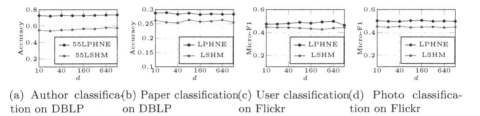

(a) Author classifica-tion on DBLP (b) Paper classification on DBLP (c) User classification on Flickr (d) Photo classification on Flickr

Fig. 2. Dimension sensitivity of LPHNE and LSHM

6   Conclusion

In this paper, the problem of learning heterogeneous network embeddings is explored by modeling the label-context relationship. We propose LPHNE, which encodes the heterogeneous structure, label and text information into the unified node embeddings. Experimental results on the real-world datasets demonstrate that our model outperforms the competitive methods significantly.

Acknowledgments. This work is supported by Natural Science Foundation of China [62003028]. We thank the anonymous reviewers for their valuable comments.

References

1. Chang, S., Han, W., Tang, J., Qi, G.-J., Aggarwal, C.C., Huang, T.S.: Heterogeneous network embedding via deep architectures. In: SIGKDD, pp. 119–128. ACM (2015)

2. Chen, W., Zhang, X., Wang, J., Zhang, Y., Yan, H., Li, X.: Non-linear smoothed transductive network embedding with text information. In: ACML, pp. 1–16 (2016)
3. Grover, A., Leskovec, J.: Node2vec: scalable feature learning for networks. In: SIGKDD, KDD '16, pp. 855–864. ACM, New York, NY, USA (2016)
4. Jacob, Y., Denoyer, L., Gallinari, P.: Learning latent representations of nodes for classifying in heterogeneous social networks. In: Proceedings of the 7th ACM International Conference on Web Search and Data Mining, pp 373–382. ACM (2014)
5. Kipf, T.N., Welling, M.: Semi-supervised classification with graph convolutional networks. arXiv preprint arXiv:1609.02907 (2016)
6. Levy, O., Goldberg, Y.: Neural word embedding as implicit matrix factorization. In: Advances in Neural Information Processing Systems, pp. 2177–2185 (2014)
7. Li, J., Zhu, J., Zhang, B.: Discriminative deep random walk for network classification. In: Association for Computational Linguistics (ACL), pp. 1004–1013 (2016)
8. Linmei, H., Yang, T., Shi, C., Ji, H., Li, X.: Heterogeneous graph attention networks for semi-supervised short text classification. In: Proceedings of the 2019 Conference on Empirical Methods in Natural Language Processing and the 9th International Joint Conference on Natural Language Processing (EMNLP-IJCNLP), pp. 4821–4830 (2019)
9. Mikolov, T., Sutskever, I., Chen, K., Corrado, G.S., Dean, J.: Distributed representations of words and phrases and their compositionality. In: Advances in Neural Information Processing Systems, pp. 3111–3119 (2013)
10. Shang, J., Qu, M., Liu, J., Kaplan, L.M., Han, J., Peng, J.: Meta-path guided embedding for similarity search in large-scale heterogeneous information networks. arXiv preprint arXiv:1610.09769 (2016)
11. Shi, C., Hu, B., Zhao, W.X., Philip, S.Y.: Heterogeneous information network embedding for recommendation. IEEE Trans. Knowl. Data Eng. **31**(2), 357–370 (2018)
12. Tang, J., Qu, M., Mei, Q.: Pte: Predictive text embedding through large-scale heterogeneous text networks. In: SIGKDD, pp. 1165–1174. ACM (2015)
13. Tang, J., Qu, M., Wang, M., Zhang, M., Yan, J., Mei, Q.: Line: Large-scale information network embedding. In: Proceedings of the 24th International Conference on World Wide Web, WWW '15, pp. 1067–1077. ACM, New York, NY, USA (2015)
14. Tang, L., Liu, Z., Zhao, Y., Duan, Z., Jia, J.: Efficient ridesharing framework for ride-matching via heterogeneous network embedding. ACM Trans. Knowl. Discovery Data (TKDD) **14**(3), 1–24 (2020)
15. Tu, C., Zhang, W., Liu, Z., Sun, M.: max-margin deepwalk: discriminative learning of network representation. In: Proceedings of the Twenty-Fifth International Joint Conference on Artificial Intelligence (IJCAI-16), pp. 3889–3895 (2016)
16. Wang, D., Cui, P., Zhu, W.: Structural deep network embedding. In: Proceedings of the 22Nd ACM SIGKDD International Conference on Knowledge Discovery and Data Mining, KDD '16, pp. 1225–1234. ACM, New York, NY, USA (2016)
17. Xu, L., Wei, X., Cao, J., Yu, P.S.: Embedding of embedding (eoe): Joint embedding for coupled heterogeneous networks. In: WSDM, WSDM '17, pp. 741–749. ACM, New York, NY, USA (2017)
18. Yang, C., Liu, Z., Zhao, D., Sun, M., Chang, E.Y.: Network representation learning with rich text information. In: Proceedings of the 24th International Conference on Artificial Intelligence, IJCAI'15, pp. 2111–2117. AAAI Press (2015)

Spatio-Temporal Dynamic Multi-graph Attention Network for Ride-Hailing Demand Prediction

Ya Chen, Wanrong Jiang, Hao Fu, and Guiquan Liu[✉]

School of Computer Science and Technology, University of Science and Technology of China, Hefei, China
{chenya88,jwr,hfu}@mail.ustc.edu.cn, gqliu@ustc.edu.cn

Abstract. Accurate ride-hailing demand prediction is of great importance in traffic management and urban planning. Meanwhile, this is a challenging task due to the complicated Spatio-temporal correlations. Existing methods mainly focus on modeling the Euclidean correlations among spatially adjacent regions and modeling the non-Euclidean correlations among distant regions through the similarities of features such as points of interest (POI). However, due to these invariable regional characteristics, the spatial correlations obtained from them are static. These approaches all ignore the real-time dynamic correlations which change over time such as passenger flow between regions. Dynamic correlations can reflect the travel status of residents in real time and is the important factor for an accurate demand forecasting. In this paper, we propose Spatio-temporal dynamic multi-graph attention network (STDMG) to solve this problem. First, we encode the feature similarity and passenger flow between regions into multiple static and dynamic graphs at each time step. Then, the dynamic multi-graph fusion module is proposed to capture spatial dependencies by modeling these graphs. Finally, we design a temporal attention module which consisting of ConvLSTM layer and attention layer, to capture the influence of adjacent Spatio-temporal dependencies by combining the global context information. Experiments on three real-world datasets demonstrate the effectiveness of our approach over state-of-the-art methods.

Keywords: Demand prediction · Spatio-temporal · Deep learning

1 Introduction

The goal of region-level ride-hailing demand prediction is to predict the future demand of urban regions under the given historical observation, help cities allocate resources in advance to meet travel demand, reduce the waste of vehicle resources and alleviate traffic congestion [22]. Due to the complex Spatio-temporal correlations, composed of multiple spatial dependencies among regions and nonlinear temporal dependencies among observed data, predicting ride-hailing demand is generally challenging [21].

© Springer Nature Switzerland AG 2021
T. Mantoro et al. (Eds.): ICONIP 2021, LNCS 13109, pp. 133–144, 2021.
https://doi.org/10.1007/978-3-030-92270-2_12

The development of deep learning makes it possible to model the complex Spatio-temporal correlation of demand forecasting [27]. For Euclidean correlations between spatially adjacent regions, most studies utilize convolutional neural network (CNN) for modeling [22], and utilize recursive neural network (RNN) to model the temporal correlation [5]. For non-Euclidean correlations between distant regions, most recent studies model them as graph. For example, taking each region as a node and the passenger demand as an attribute of nodes, the spatial correlations are captured by Graph Convolutional Network (GCN) [1,17,30]. However, these models mainly construct graphs based on static correlations such as POI similarity and passenger demand patterns similarity [1,5,26], and these graphs have fixed topology due to invariant correlations. They ignore real-time dynamic correlations such as the passenger flow between departure region and destination region, and the similarity of short-term passenger demand patterns in the same day or several hours. The static correlations can describe the long-term fixed spatial dependence between regions, while the dynamic correlations can reflect the real-time travel conditions of residents, and capture the real-time spatial dependence between regions.

We propose STDMG to solve the above problems, it contains three modules: Dynamic Multi-graph Fusion Module (DMFM), Temporal Attention Module (TAM) and Prediction Module. First, we describe the non-Euclidean spatial correlations between regions as POI similarity, passenger flow, long-term and short-term passenger demand pattern similarity, and encode them as dynamic multi-graph. Then, DMFM based on Graph Attention Networks (GAT) [18] is proposed to capture spatial dependencies by modeling these graphs. Next, by integrating ConvLSTM and attention mechanism into TAM, we combine global context information and allocate adjacent time importance to model Spatio-temporal correlation. Finally, prediction module makes the final prediction.

To sum up, the main contributions of paper are as follows: (1) We propose a novel ride-hailing prediction model that can simultaneously capture a variety of dynamic and static Spatio-temporal correlations. (2) We identify non-Euclidean correlations among regions and encode them in dynamic multi-graphs, then further propose DMFM to model these graphs. (3) We propose TAM to capture the spatial dependence of adjacent regions and the temporal dependence of adjacent times. (4) We conduct extensive experiments on three real-world datasets, and we compare it with state-of-the-art models. Experimental results show that our model can outperform all baselines.

2   Related Work

2.1   Traffic Spatio-Temporal Prediction

Traffic Spatio-temporal prediction has been studied extensively over the past few decades. Early methods mainly utilized traditional time series models (e.g., auto-regressive integrated moving average (ARIMA) [19]) and machine learning models (e.g., regression tree [13], clustering model [25]). But time series models only consider time information and machine learning model cannot self-adaptively extract Spatio-temporal features. Later, in order to construct the complex Spatio-temporal relationship, deep learning methods with better performance and adaptive capturation of Spatio-temporal features were gradually adopted. CNN is widely used to extract spatial features of traffic status images [24], long short-term memory (LSTM) is used to capture temporal characteristics [29], and some studies joint CNN and LSTM to capture the complex and nonlinear Spatio-temporal relationships [20,22]. In recent years, many studies have processed traffic data as graph data, using Graph Neural Network (GNN) to capture non-Euclidean spatial correlations between distant regions. A large number of models based on GNN have achieved obvious results in various traffic Spatio-temporal prediction problems [1,8,11]. Yu et al. [23], modeling multi-scale traffic networks based on GCN, can capture comprehensive Spatio-temporal correlations more effectively, and Song et al. [17] proposed a Spatio-temporal synchronous modeling mechanism consisting of a group of graph convolutional operations, which can effectively capture the heterogeneities in localized Spatio-temporal graphs.

2.2 Deep Learning on Graph

In recent years, the excellent performance of deep learning in the fields of acoustics, image, natural language processing, and the ubiquity of graph structures in real life, make the generalization of neural networks to graph structure data into an emerging topic [14]. A line of studies generalized CNN to model arbitrary graphs on spectral [4,6] or spatial [3,12,31] data. Spectral domain convolution mainly uses Discrete Fourier Transform to map topological graph to spectral domain for convolution operation, and the most classic model is GCN [10]. In the process of spectral domain convolution, the network structure is fixed and there are great limitations in processing dynamic graphs. Spatial domain convolution mainly extracts the spatial features of the topology graph by finding the neighbor nodes of each vertex, such as GAT, which is a variant of GCN defined in the vertex domain. Different from GCN, it uses the attention layer to dynamically adjust the importance of neighbor nodes and thus it can handle dynamic graphs.

3   Preliminaries

Dynamic Multi-graphs. To start We explain how to use dynamic multi-graphs to represent the complex spatial correlations between urban areas and passenger demand. First, the urban area is divided into $N = d_w \times d_h$ regions, we represent each region as a node of the graph, and the relationships between regions are edges of the graph. Then, we consider six non-Euclidean spatial dependencies and model them as static and dynamic graphs respectively.

Static graphs: As shown in Fig. 1(a) and (c), they are POI similarity graph G_P, long-term pattern similarity graph of inflow G_{Li} and demand G_{Lo}. These

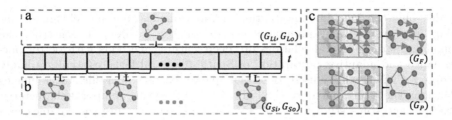

Fig. 1. Dynamic Multi-Graphs. (a) G_{Li}, G_{Lo}. (b) G_{Si}, G_{So}. (c) G_F, G_P.

graphs are constructed in a similar way. Lets take $G_P = (V, A_P)$ as an examples, where V is the set of nodes and A_P is a matrix that records neighbor nodes. If $Sim(Q_i, Q_j) > \epsilon$, then $A_{P,ij} = 1$, otherwise $A_{P,ij} = 0$, where Sim is Pearson Correlation Coefficient, and the threshold $\epsilon = 0.9$. The difference is that Q_i, Q_j of G_{Li} and G_{Lo} are the overall historical t time inflow or demand data at i, j region, and the Q_i, Q_j of G_P are POI features of i, j region.

Dynamic graphs: As shown in Fig. 1(b) and (c), they are passenger flow graph G_F, short-term pattern similarity graph of inflow G_{Si} and demand G_{So}. At t time step, $G_F^t = (V, A_F^t)$, and if $Flow_{ij}^t > 0$ then $A_{F,ij}^t = 1$, otherwise $A_{F,ij}^t = 0$, where $Flow_{ij}^t$ is the passenger flow between i, j region. G_{Si}^t and G_{So}^t have the same construction method as G_P, but Q_i, Q_j of G_{Si}^t and G_{So}^t are the inflow or demand data for the $\lfloor \frac{t}{L} \rfloor$-th length of L time steps at i, j region.

Problem Definition. Ride-hailing demand prediction problem, that is, how to forecast the demand in the future by historical demand data. In our framework, $I_t = (D_t, \mathcal{G}_t)$ denotes the input at t time step, where $\mathcal{G}_t = (G_{Li}^t, ..., G_{So}^t)$ and the number of graphs in $\mathcal{G}_t$ is S ($S = [4, 6]$). The demand data observed are expressed as $D_t \in \mathbb{R}^{N \times F}$, where F denotes the feature dimension of each regions. Given most recent K time steps history data $X = (I_{t-K+1}, I_{t-K+2}, .., I_t)$, the target is to learn a prediction function Γ that predicts the ride-hailing demand $\hat{Y}_{t+1}$ at $t + 1$ time step. The specific formula can be described as follows:

$$\hat{Y}_{t+1} = \Gamma(I_{t-K+1} + I_{t-K+2}, ..., I_t) \tag{1}$$

4 Proposed Model

4.1 Overview

Figure 2 illustrates the framework of our proposed STDMG, which has three parts: DMFM, TAM and Prediction Module. As shown in Fig. 2(a), we use multiple static graphs and dynamic graphs to describe various spatial correlations. As shown in Fig. 2(b), the Graph Attention (GAtt) layer and Multi-Graphs Fusion (MGF) layer included in DMFM are applied to capture non-Euclidean spatial correlation between regions, we add extra features such as weather to

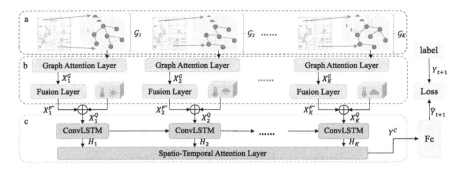

Fig. 2. Architecture of STDMG. (a) Dynamic Multi-Graphs. (b) Dynamic Multi-Graph Fusion Module. (c) Temporal Attention Module.

their output. As shown in Fig. 2(c), TAM is composed of ConvLSTM layer and Spatio-Temporal Attention (STatt) layer, which is used to capture temporal correlation, neighbor spatial relationship and obtain dynamic influence of adjacent time. Prediction Module is a fully connected (FC) layer for final prediction.

4.2 Dynamic Multi-graph Fusion Module

Graph Attention Layer. By determining the weight of each node's neighbors, GAtt layer can combine neighbor features with its own features to obtain new features. The input of GAtt layer at time t is $I_t = (D_t, \mathcal{G}_t)$, where $D_t = (q_1, q_2, ..., q_N)$ and $q_i \in \mathbb{R}^F$. For any graph in $\mathcal{G}_t$, take G_P as an examples, the attention coefficient α_{ij} between q_i and q_j is:

$$\alpha_{ij} = \frac{exp(\text{LeakyReLU}(a^T[Wq_i \| Wq_j]))}{\sum_{k \in \mathcal{K}_i} exp(\text{LeakyReLU}(a^T[Wq_i \| Wq_k]))} \quad (2)$$

where $W \in \mathbb{R}^{F' \times F}$ is a weight matrix, $a \in \mathbb{R}^{2F'}$ is a weight vector, $\|$ denotes the concatenation operation, LeakyReLU is activation function, $\mathcal{K}_i$ denotes the set of neighbor nodes of q_i (if $A_{P,ik} > 0, k \in \mathcal{K}_i$). The output of the node i is:

$$q_i' = \sigma(\sum_{k \in \mathcal{K}_i} \alpha_{ik} W q_k) \quad (3)$$

where $q_i' \in \mathbb{R}^{F'}$, F' is the output dimension of GAtt layer and is set to 8, σ denotes nonlinear activation function. The output of GAtt layer is $X_t^G = (x_{t,P}^G, .., x_{t,So}^G)$, where $x_{t,P}^G = (q_1', ..., q_N')$.

Multi-graphs Fusion Layer. At time t, $X_t^G \in \mathbb{R}^{S \times N \times F'}$ is the input of MGF layer. As shown in Fig. 3, MGF layer is mainly composed of function F_{concat} and a convolution operation. F_{concat} is used to weight and concatenate the X_t^G. Its formula is defined as $X_t^F = W_C \circ X_t^G$, where $X_t^F \in \mathbb{R}^{S \times N \times F'}$, $W_C \in \mathbb{R}^{S \times N \times F'}$ is

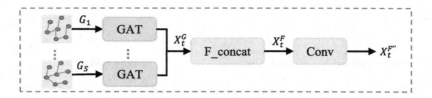

Fig. 3. The architecture of the Dynamic Multi-Graph Fusion Module.

the weight matrix, $\circ$ denotes the Hadamard product. The convolution operation is used for the fusion and dimensionality reduction of X_t^F, and the output is $X_t^{F''} \in \mathbb{R}^{N \times F''}$, where F'' is output dimension. Finally, $X_t^{F''}$ adds extra features such as weather and time meta, and the output of MGF layer is $X_t^Q \in \mathbb{R}^{C \times d_w \times d_h}$.

4.3 Temporal Attention Module

ConvLSTM Layer. The ConvLSTM introduces convolution operation on the basis of LSTM, so that the model could not only describe long short-time memory like LSTM, but also extract local spatial features like CNN. For input X_t^Q at time t , the key equations of ConvLSTM Layer are defined as follows:

$$
\begin{aligned}
i_t &= \sigma(W_{xi} * X_t^Q + W_{hi} * H_{t-1} + W_{ci} \circ C_{t-1} + b_i) \\
f_t &= \sigma(W_{xf} * X_t^Q + W_{hf} * H_{t-1} + W_{cf} \circ C_{t-1} + b_f) \\
C_t &= f_t \circ C_{t-1} + i_t \circ tanh(W_{xc} * X_t^Q + W_{hc} * H_{t-1} + b_c) \\
o_t &= \sigma(W_{xo} * X_t^Q + W_{ho} * H_{t-1} + W_{co} \circ C_{t-1} + b_o) \\
H_t &= o_t \circ tanh(C_t)
\end{aligned}
\tag{4}
$$

where the $*$ denotes the convolution operator, C_t is cell output, $H_t \in \mathbb{R}^{C' \times d_w \times d_h}$ is hidden state, and gates i_t, f_t, o_t are 3D tensors whose last two dimensions are d_w and d_h. We adopt two-layer ConvLSTM networks in each ConvLSTM Layer, where kernel size is $(3, 3)$, hidden layer's dimension is $(64, 64)$.

Spatio-Temporal Attention Layer. Through a sequence composed of K ConvLSTM layers, we get $X^C = (H_1, ..., H_K)$ as the input. Coordinate Attention (CoordAtt) [7] can encode channel relationships and long-term dependencies with precise location information, STatt layer, which is based on CoordAtt, allows us to capture the importance of each feature channel on the temporal dimension while maintaining location information, thus allocating attention.

For input X^C, we shape it into $X^A \in \mathbb{R}^{P \times d_w \times d_h}$, where $P = K \times C'$, by using the pooled cores of the dimensions $(d_h, 1)$ and $(1, d_w)$ to code each channel along the horizontal and vertical coordinates, and the output of the m-th channel at height h and the output of the m-th channel at width w can be formulated as:

$$
z_m^h = \frac{1}{d_w} \sum_{0 \le i < d_w} X_m^A(h, i), \quad z_m^w = \frac{1}{d_h} \sum_{0 \le j < d_h} X_m^A(j, w)
\tag{5}
$$

Concatenating z^h, z^w and sending them to a shared 1×1 convolutional transformation function F_1: $f = \delta(F_1(z^h||z^w))$, where $||$ denotes the concatenation operation along the spatial dimension, δ is a non-linear activation function. Using two 1×1 convolutional transformations F_h and F_w separately transform f^h and f^w to tensors with the same channel number to the input X^A: $g^h = \sigma(F_h(f^h)), g^w = \sigma(F_w(f^w))$. The output of STatt layer is $Y^C \in \mathbb{R}^{P \times d_w \times d_h}$:

$$Y_m^C(i,j) = X_m^A(i,j) \times g_m^h(i) \times g_m^w(j) \tag{6}$$

4.4   Prediction Module

Our goal is predicting the demand at time step $t + 1$, through the historical demand data of the most recent K time steps and other related information (e.g. weather). The prediction module is a fully connected layer, and the formula is $\hat{Y}_{t+1} = W_p Y^C + b_p$, where $\hat{Y}_{t+1} \in \mathbb{R}^{N \times 1}$, W_p and b_p are learnable parameters.

Loss Function. In our experiments, we adopted the joint loss function. The loss function consists of mean square loss and mean absolute percentage loss:

$$L(\theta, \hat{y}_{t+1}) = \frac{1}{N} \sum_{i=1}^{N} (\alpha(\hat{y}_{t+1}^i - y_{t+1}^i)^2 + \beta \left| \frac{\hat{y}_{t+1}^i - y_{t+1}^i}{y_{t+1}^i} \right|) \tag{7}$$

where θ represents all the learnable parameters, $\hat{Y}_{t+1} = (\hat{y}_{t+1}^1, ..., \hat{y}_{t+1}^N)$ denotes the prediction of the model, $Y_{t+1} = (y_{t+1}^1, ..., y_{t+1}^N)$ denotes the ground truth, α and β are hyperparameters, which values are 3 and 12. We use Adam [9] to optimize and PyTroch [15] to implement our model.

5   Experiments

Datasets. We evaluate the proposed method on three real-world datasets:

- Hefei: The dataset Hefei is the ride-hailing order data which collected in HeFei from November 1 to 30, 2019. It mainly includes: the latitude and longitude of departure place, the latitude and longitude of the destination, the departure time, the arrival time, the price and so on. After experimental processing, we obtained inflow data, demand data and passenger flow data between regions with time intervals of 15 min, 30 min and 1 h in the 15×15 regions, and the last 20% are used for testing.
- TaxiBJ [24]: The public dataset TaxiBJ was collected in Beijing from November 1, 2015 to January 21, 2016. It contains passenger demand, passenger inflow, time meta, meteorological data, and each time step is 30 min. Consistent with Hefei, after processing, we obtained two datasets with time intervals of 30 min and 1 h in the 15×15 regions, and the last 20% are used for testing.
- BikeNYC [24]: Different from Hefei and TaxiBJ, the public dataset BikeNYC contains new-flow, end-flow and time meta in the 16×8 regions. The dataset BikeNYC was collected from Apr 1, 2014 to Sept 30, 2014 in New York. Each time step is 1h, and the last ten days' data are used for testing.

Experimental Settings. We evaluate the performance of the proposed model based on three evaluation metrics: Root Mean Squared Error (RMSE), Mean Absolute Error (MAE) and Mean Absolute Percentage Error (MAPE). The baselines are as follows.

We compare our model with the following baseline methods:(1) Historical Average (HA); (2) Support vector regression (SVR); (3) Convolutional LSTM Network (ConvLSTM) [16]; (4) Spatio-temporal graph convolutional networks (STGCN) [23], which combines graph convolutional layers and convolutional sequence learning layers; (5) Spatial-temporal graph to sequence (STG2Seq) [1], which uses multiple gated graph convolutional module and seq2seq architecture with attention mechanisms to make multi-step prediction; (6) Spatial-Temporal Synchronous Graph Convolutional Networks (STSGCN) [17], capturing Spatio-temporal correlations through Spatio-temporal synchronous modeling mechanism which consists of graph convolutional operations.

6   Experimental Results

6.1   Performance Comparison

First we processed the Hefei order data into three demand datasets with time intervals of 15 min, 30 min and 1 h. In the experiment, we described Spatio-temporal correlations by six graphs, including $G_{Li}, G_{Lo}, G_{Si}, G_{So}, G_P, G_F$. Table 1 shows the comparison between STDMG and baselines on the three datasets. We can observe the following phenomena: (1) Compared with other baselines, deep learning based methods, including ConvLSTM, STGCN, STG2Seq, STSGCN and STDMG, all have better performance, which shows the advantage of deep learning based methods in capturing complex Spatio-temporal correlations. (2) On the three datasets at different time intervals, our model achieves the best performance on the three metrics, and is at least 4.5% better than baselines on RMSE, indicating the effectiveness of STDMG model for Spatio-temporal correlation modeling.

Table 1. Comparison with different baselines on Hefei.

Model	Hefei		
	15 min RMSE/MAE/MAPE	30 min RMSE/MAE/MAPE	1 h RMSE/MAE/MAPE
HA	5.423/3.304/0.465	10.338/6.313/0.417	19.923/12.247/0.452
SVR	5.158/3.107/0.445	10.344/6.327/0.420	19.977/12.291/0.454
ConvLSTM	2.568/1.469/0.230	4.207/2.384/0.221	7.386/4.008/0.212
STGCN	2.621/1.493/0.231	4.275/2.402/0.220	7.508/4.168/0.213
STG2Seq	2.574/1.443/0.233	4.077/2.254/0.211	7.136/3.817/0.208
STSGCN	3.278/2.226/0.378	5.156/3.251/0.320	8.691/5.023/0.221
STDMG	**2.475/1.412/0.219**	**3.883/2.146/0.208**	**6.486/3.553/0.199**

Table 2. Comparison with different baselines on TaxiBJ and BikeNYC.

Model	TaxiBJ		BikeNYC	
	30 min RMSE/MAE/MAPE	1 h RMSE/MAE/MAPE	New_flow RMSE/MAE/MAPE	End_flow RMSE/MAE/MAPE
HA	80.24/53.86/0.920	168.08/113.64/0.897	14.92/7.48/0.476	14.87/7.42/0.470
SVR	31.22/18.48/0.157	58.65/34.12/0.183	6.50/2.78/0.270	6.70/2.82/0.273
ConvLSTM	26.25/15.75/0.157	52.24/31.51/0.160	4.92/2.42/0.226	5.37/2.62/0.243
STGCN	26.17/15.08/0.146	54.97/28.89/0.173	4.57/2.30/0.212	4.92/2.41/0.220
STG2Seq	24.16/14.27/0.145	50.29/29.53/0.146	5.47/2.77/0.224	**4.66**/2.33/0.215
STSGCN	26.78/15.84/0.338	51.70/29.97/0.162	5.64/2.70/0.227	5.89/2.73/0.229
STDMG	**23.49/14.20/0.137**	**47.38/27.46/0.136**	**4.55/2.25/0.211**	4.67/**2.27/0.215**

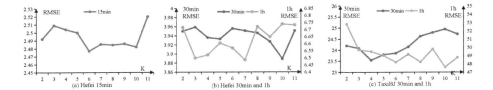

Fig. 4. Effect of the number of ConvLSTM layer K.

Second, to check the applicability of STDMG on other datasets, we conducted experiments on two public datasets TaxiBJ and BikeNYC. Like Hefei, we processed the TaxiBJ into two datasets with time intervals of 30 min and 60 min respectively. Because of data limitations, we describe Spatio-temporal correlations by four graphs in these experiments, including $G_{Li}, G_{Lo}, G_{Si}, G_{So}$. Table 2 shows the comparison between STDMG and baselines on TaxIBJ and BikeNYC. We can observe that: (1) Among all evaluation indicators, STDMG has the second best RMSE value in end_flow, and the others are the best. Its RMSE indicator is at least 2.7% better than all baselines on TaxiBJ. (2) The STDMG model also has excellent prediction effect on these datasets, which proves the effectiveness of STDMG on other datasets.

6.2 Parameter Studies

To study the effects of hyperparameters of the STDMG, we evaluate models on Hefei and TaxiBJ by varying two of the most important hyperparameters.

The Number of ConvLSTM Layers K: As Fig. 4, K takes the value in {2, 3,..., 11}. We observe that K has an optimal value within this range, and when K value is greater than the optimal value, the accuracy of the model will decrease due to overfitting.

Short-Term Similarity Time Length L: If L is too large or too small, the results get worse, because the time is too short or too long which is not enough to reflect the real-time status of passengers. According to the above reasons, as can be seen in Fig. 5, L takes the value in {6, 12, 24}. We observe that L has significant effect on the experimental results, indicating the validity of the similarity of short-term passenger demand patterns for a certain length of time.

6.3   Effect of Each Component

In order to study the influence of each component in STDMG, we conducted experiments on Hefei and TaxiBJ datasets from two aspects : (1) Remove the DMFM and STatt layer from STDMG respectively to verify the influence of each module. (2) Delete each graph from dynamic multi-graphs for experiment to verify the validity of each graph.

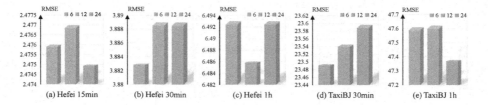

Fig. 5. Effect of the short-term similarity time length L.

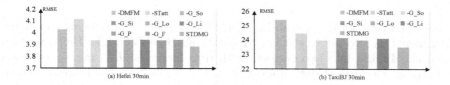

Fig. 6. Effect of each component of STDMG.

Figure 6 shows the RMSE value of the above experiments. We can observe that: (1) STDMG always has better performance, and the lack of two modules reduces the prediction accuracy by at least 3%. It shows the importance and effectiveness of each STDMG module, and the STatt layer have obvious effects on dynamic time capture. (2) The lack of any graph reduces the accuracy of the model, indicating the necessity of each graph in dynamic multi-graphs, and modeling multiple dynamic and static correlations are more suitable for capturing complex spatial correlations.

7   Conclusion

In this paper, we proposed a novel Spatio-temporal dynamic model denoted STDMG for the ride-hailing demand prediction. We use dynamic multi-graphs to describe the non-Euclidean dynamic spatial correlations between regions, and propose DMFM to capture spatial dependencies by modeling these graphs. Then we use TAM to capture Spatio-temporal features by combining adjacent time and neighbor space. Among them, our model considers the dynamic influence of temporal correlation by using the STatt layer based on the attention mechanism. Finally, we evaluated the model on three large real-world datasets to verify the STDMG's prediction accuracy is higher than state-of-the-art methods.

Acknowledgement. This paper has been supported by the National Key Research and Development Program of China (No.2018YFB1801105).

References

1. Bai, L., Yao, L., Kanhere, S., Wang, X., Sheng, Q., et al.: Stg2seq: spatial-temporal graph to sequence model for multi-step passenger demand forecasting. arXiv preprint arXiv:1905.10069 (2019)
2. Chai, D., Wang, L., Yang, Q.: Bike flow prediction with multi-graph convolutional networks. In: Proceedings of the 26th ACM SIGSPATIAL International Conference on Advances in Geographic Information Systems, pp. 397–400 (2018)
3. Chen, J., Ma, T., Xiao, C.: Fastgcn: fast learning with graph convolutional networks via importance sampling. arXiv preprint arXiv:1801.10247 (2018)
4. Defferrard, M., Bresson, X., Vandergheynst, P.: Convolutional neural networks on graphs with fast localized spectral filtering. arXiv preprint arXiv:1606.09375 (2016)
5. Geng, X., et al.: Spatiotemporal multi-graph convolution network for ride-hailing demand forecasting. In: Proceedings of the AAAI Conference on Artificial Intelligence, vol. 33, pp. 3656–3663 (2019)
6. Hammond, D.K., Vandergheynst, P., Gribonval, R.: Wavelets on graphs via spectral graph theory. Appl. Comput. Harmonic Anal. **30**(2), 129–150 (2011)
7. Hou, Q., Zhou, D., Feng, J.: Coordinate attention for efficient mobile network design. arXiv preprint arXiv:2103.02907 (2021)
8. Huang, R., Huang, C., Liu, Y., Dai, G., Kong, W.: Lsgcn: Long short-term traffic prediction with graph convolutional networks (2020)
9. Kingma, D.P., Ba, J.: Adam: a method for stochastic optimization. In: Bengio, Y., LeCun, Y. (eds.) 3rd International Conference on Learning Representations, ICLR 2015, San Diego, CA, USA, May 7–9, 2015, Conference Track Proceedings (2015). http://arxiv.org/abs/1412.6980
10. Kipf, T.N., Welling, M.: Semi-supervised classification with graph convolutional networks. arXiv preprint arXiv:1609.02907 (2016)
11. Li, M., Tong, P., Li, M., Jin, Z., Huang, J., Hua, X.S.: Traffic flow prediction with vehicle trajectories. In: Proceedings of the AAAI Conference on Artificial Intelligence, vol. 35, pp. 294–302 (2021)
12. Li, Y., Yu, R., Shahabi, C., Liu, Y.: Diffusion convolutional recurrent neural network: Data-driven traffic forecasting. arXiv preprint arXiv:1707.01926 (2017)
13. Li, Y., Zheng, Y., Zhang, H., Chen, L.: Traffic prediction in a bike-sharing system. In: Proceedings of the 23rd SIGSPATIAL International Conference on Advances in Geographic Information Systems, pp. 1–10 (2015)
14. Pareja, A., et al.: Evolvegcn: evolving graph convolutional networks for dynamic graphs. In: Proceedings of the AAAI Conference on Artificial Intelligence, vol. 34, pp. 5363–5370 (2020)
15. Paszke, A., et al.: Pytorch: An imperative style, high-performance deep learning library. arXiv preprint arXiv:1912.01703 (2019)
16. Shi, X., Chen, Z., Wang, H., Yeung, D.Y., Wong, W.K., Woo, W.C.: Convolutional lstm network: A machine learning approach for precipitation nowcasting. arXiv preprint arXiv:1506.04214 (2015)

17. Song, C., Lin, Y., Guo, S., Wan, H.: Spatial-temporal synchronous graph convolutional networks: A new framework for spatial-temporal network data forecasting. In: Proceedings of the AAAI Conference on Artificial Intelligence, vol. 34, pp. 914–921 (2020)
18. Veličković, P., Cucurull, G., Casanova, A., Romero, A., Lio, P., Bengio, Y.: Graph attention networks. arXiv preprint arXiv:1710.10903 (2017)
19. Williams, B.M., Hoel, L.A.: Modeling and forecasting vehicular traffic flow as a seasonal arima process: theoretical basis and empirical results. J. Transp. Eng. **129**(6), 664–672 (2003)
20. Wu, Y., Tan, H.: Short-term traffic flow forecasting with spatial-temporal correlation in a hybrid deep learning framework. arXiv preprint arXiv:1612.01022 (2016)
21. Yao, H., Tang, X., Wei, H., Zheng, G., Li, Z.: Revisiting spatial-temporal similarity: a deep learning framework for traffic prediction. In: Proceedings of the AAAI Conference On Artificial Intelligence, vol. 33, pp. 5668–5675 (2019)
22. Yao, H., et al.: Deep multi-view spatial-temporal network for taxi demand prediction. In: Proceedings of the AAAI Conference on Artificial Intelligence, vol. 32 (2018)
23. Yu, B., Yin, H., Zhu, Z.: Spatio-temporal graph convolutional networks: A deep learning framework for traffic forecasting. arXiv preprint arXiv:1709.04875 (2017)
24. Zhang, J., Zheng, Y., Qi, D.: Deep spatio-temporal residual networks for city-wide crowd flows prediction. In: Proceedings of the AAAI Conference on Artificial Intelligence, vol. 31 (2017)
25. Zhang, K., Feng, Z., Chen, S., Huang, K., Wang, G.: A framework for passengers demand prediction and recommendation. In: 2016 IEEE International Conference on Services Computing (SCC), pp. 340–347. IEEE (2016)
26. Zhang, X., He, L., Chen, K., Luo, Y., Zhou, J., Wang, F.: Multi-view graph convolutional network and its applications on neuroimage analysis for parkinson's disease. In: AMIA Annual Symposium Proceedings, vol. 2018, p. 1147. American Medical Informatics Association (2018)
27. Zhang, Z., Cui, P., Zhu, W.: Deep learning on graphs: a survey. IEEE Transactions on Knowledge and Data Engineering (2020)
28. Zhao, L., et al.: T-gcn: a temporal graph convolutional network for traffic prediction. IEEE Trans. Intell. Transp. Syst. **21**(9), 3848–3858 (2019)
29. Zhao, Z., Chen, W., Wu, X., Chen, P.C., Liu, J.: Lstm network: a deep learning approach for short-term traffic forecast. IET Intell. Transp. Syst. **11**(2), 68–75 (2017)
30. Zheng, C., Fan, X., Wang, C., Qi, J.: Gman: a graph multi-attention network for traffic prediction. In: Proceedings of the AAAI Conference on Artificial Intelligence, vol. 34, pp. 1234–1241 (2020)
31. Zhou, J., et al.: Graph neural networks: a review of methods and applications. AI Open **1**, 57–81 (2020)

An Implicit Learning Approach for Solving the Nurse Scheduling Problem

Aymen Ben Said[1], Emad A. Mohammed[2], and Malek Mouhoub[1(✉)]

[1] Department of Computer Science, University of Regina, Regina, SK, Canada
{aymenbensaid,mouhoubm}@uregina.ca
[2] Department of Electrical and Software Engineering, University of Calgary,
Calgary, AB, Canada
eamohamm@ucalgary.ca

Abstract. The Nurse Scheduling Problem (NSP) is one of the challenging combinatorial optimization problems encountered in the healthcare sector. Solving the NSP consists in building weekly schedules by assigning nurses to shift patterns, such that workload constraints are satisfied, while nurses' preferences are maximized. In addition to the difficulty to tackle this NP-hard problem, extracting the problem constraints and preferences from an expert can be a tedious task. Moreover, the NSP is a highly dynamic optimization application that is often affected by unpredictable and unforeseen events, such as outbreaks, accidents, and nurses call in sick. This dynamic nature may force the automatic scheduling system to adapt to the first possible solution that only satisfies work conditions without considering the competing interest of nurse satisfaction. To overcome this limitation, we propose an alternative approach that relies on automatically and implicitly learning NSP constraints and preferences from available historical data, and without any prior knowledge. To evaluate the performance of our proposed approach, we have conducted comparative experiments against COUNT-OR, a solving method that explicitly learns constraints. We measure the closeness of the generated solutions to the input data to quantify the quality of using our implicit approach in capturing the embedded constraints and preferences in historical data. The experiment results, reported in this paper, show an improvement in the computed average errors based on Frobenius Norm.

Keywords: Nurse Scheduling Problem · Combinatorial optimization · Machine learning

1 Introduction

The Nurse Scheduling Problem (NSP) also known as the Nurse Rostering Problem (NRP), is an NP-hard combinatorial optimization problem [5,19,22]. The NSP consists in appointing feasible shift patterns to nurses while ensuring a fair consideration of preferences between all the nurses. Several methods were developed to solve the NSP problem using exact and heuristic approaches which focused on satisfying the workload requirements without paying attention to the

© Springer Nature Switzerland AG 2021
T. Mantoro et al. (Eds.): ICONIP 2021, LNCS 13109, pp. 145–157, 2021.
https://doi.org/10.1007/978-3-030-92270-2_13

competing interest of nurse satisfaction [12,13]. Manual scheduling [8] involves a human scheduler relying on predefined constraints and preferences to solve the NSP using a constraint solver. However, due to the dynamic nature of the problem, this method may trigger challenges such as dealing with unpredictable and unforeseen events, such as outbreaks, accidents, and nurses call in sick. Maintaining the satisfaction of workload constraints, as well as fairness of preferences distribution among nurses, will be very difficult when the problem requirements change over time [8]. Moreover, in case of no knowledge about the constraints and preferences, it is unpractical to manually examine the schedules and extract useful information due to the massive amount of data available for nurses' workload. This motivates the use of automatic learning methods from historical data.

Recently, there has been considerable interest in machine learning and data mining techniques to solve combinatorial optimization problems, as constraint solving can benefit from the learning and mining framework [16]. Machine learning methods such as active learning [2,3,15,18,21] can be used to explicitly learn the NSP where the learner interacts with an expert during the learning process, however, the procedure is iterative and may be time consuming as discussed in [11]. The later motivates the use of implicit solving approaches where constraints and preferences are captured and represented in the learned frequent patterns from historical data. Even though implicit methods do not guarantee the optimality of the solution returned, they are more practical compared to explicit solving methods. Furthermore, implicit methods provide a way to facilitate the manual solving task where the human operator can rely on the generated association rules to assign nurses in critical situations. Given the limitations of the explicit NSP problem solving methods, we adopt an approach that implicitly learns constraints and preferences from historical data. The proposed method has several advantages, including solving this dynamic problem in a reasonable time-frame guaranteeing an overall preference fulfillment among all nurses. The proposed method comes at the following cost; availability of sufficient data and the data pre-processing overhead. To evaluate the performance of our proposed approach, we have conducted comparative experiments against COUNT-OR, a solving method that explicitly learns constraints. The comparison criteria is the quality of the generated schedules based on the Frobenius Norm. The results of these experiments demonstrate a slight improvement in terms of closeness to the input schedule.

2   Literature Review

Kumar, M. et al. [12], proposed an explicit constraint learning method known as COUNT-OR to solve the NSP problem using tensors representation of past schedules. COUNT-OR works well for real-world problems as it first learns the bounds for some pre-defined quantities of interest (e.g., minimum numbers of working nurses per day), then used these bounds to elicit the constraints and generate new schedules. ARNOLD [13] is another proposed explicit learning method based on tensor representation of past schedules. This method partially automates the modelling process by learning integer programs that capture polynomial constraints from past feasible solutions and represents them using a constraint language. ARNOLD slightly differs from COUNT-OR since it assumes integers such as tensor bounds to be provided manually instead of automatically learning them. Constantino et al. [7] proposed a hybrid heuristic algorithm that tackles specific objectives of the NSP, which ensures a balanced preference satisfaction between the nurses by maximizing the preference satisfaction for each nurse individually (which is different in our case). The proposed algorithm has two phases, the first one consists in producing initial shift patterns, and the second phase involves reassignments of shifts to improve the quality of the solution in terms of balancing the preferences of nurses. Constantino et al. [4] proposed an extension of this work, where instead of using local search in the second phase, they used the Variable Neighborhood Search (VNS) to improve their initial solution. Jingpeng L, and Uwe Aickelin [1,14] proposed a Bayesian Optimization Algorithm that involves an explicit learning of the best scheduling rule from a set of four building rules (corresponding to a selection method of shift patterns) that contribute to the building of a Bayesian Network that represents the joint distribution of solutions. A possible solution is defined as a sequence of rules from the first nurse to the last one. The objective function is about minimizing the total preference cost of all nurses and the fitness function is associated with a penalty cost for uncovered shifts. Snehasish Karmakar, et al. [10] implemented four meta-heuristic algorithms based on explicit constraint learning to solve the NSP problem; namely Firefly Algorithm, Particle Swarm Optimization (PSO), Simulated Annealing and Genetic Algorithm using predefined constraints and preferences, and initial set of shift patterns. These four nature-inspired techniques are then compared in terms of solution quality in terms of minimizing the overall cost that involves penalty weights of both hard and soft constraints (nurse' preferences). The conducted experiments showed that the Firefly Algorithm performed better in terms of average final cost, and PSO performed poorly overall. This result is claimed to be due to the randomness nature of the Firefly algorithm which triggers more exploitation in the search space. Most of the methods presented in the literature [4,7,10,14] are based on explicitly learning constraints and preferences, and verifying that the generated solutions satisfy the learned constraints and optimizes the learned preferences. These methods are approximation methods based on meta-heuristics and may not guarantee the optimality of the returned solution. Unlike explicit learning and solving techniques, implicit solving methods can be explored to generate new solutions that satisfy the embedded constraints and optimizes the preferences through the learned patterns among data. Some of the constraints and preferences are visible in historical data (e.g., specific nurse working in particular shift). However, historical data may also reflect other types of constraints that are not visible (e.g., financial constraints, conflicts between staff, etc.). Thus, we use implicit learning methods based on machine learning algorithms to generate realistic schedules specific to the utilized data and further maintain the integrity and confidentiality of the data.

3   General Problem Formulation

The NSP consists in creating weekly schedules by assigning nurses to shift patterns, such that the hospital workload constraints are satisfied, while nurses' preferences are maximized. The following is a formulation of the NSP.

Parameters and Indices:
n = Number of nurses
m = Number of possible shift patterns
p_{ij} = Preference cost of nurse i working shift pattern j
h_d = Hospital's minimum shift coverage requirements on day d
x_i = Maximum number of shifts for nurse i per week
i = $\{1,..,n\}$ is the nurse index
j = $\{1,...,m\}$ is the index of the weekly shift pattern
d = $\{1,..,7\}$ is the day index
s = $\{1,..,4\}$ is the shift index (morning, afternoon, evening, night)

Decision variables:

$$w_{ij} = \begin{cases} 1 & \text{if nurse i is assigned shift pattern j} \\ 0 & \text{Otherwise} \end{cases} \tag{1}$$

$$c_{jsd} = \begin{cases} 1 & \text{if shift pattern j covers shift s on day d} \\ 0 & \text{Otherwise} \end{cases} \tag{2}$$

Target Function: *Minimizing the total preference cost for all nurses*

$$\sum_{i=1}^{n} \sum_{j=1}^{m} p_{ij} w_{ij} \to min \tag{3}$$

Hard Constraints:
Every nurse must work exactly one feasible shift pattern

$$\sum_{j=1}^{m} w_{ij} = 1 \quad \forall i \tag{4}$$

Each schedule must satisfy the hospital's minimum daily shift coverage requirements

$$\sum_{i=1}^{n} \sum_{j=1}^{m} \sum_{s=1}^{4} w_{ij} c_{jsd} \geq h_d \quad \forall d \tag{5}$$

Each nurse may work a maximum of two shifts in a single day

$$\sum_{j=1}^{m} \sum_{s=1}^{4} w_{ij} c_{jsd} \leq 2 \quad \forall i, \forall d \tag{6}$$

No nurse may work a night shift followed immediately by a morning shift

$$\sum_{i=1}^{n} \sum_{j=1}^{m} w_{ij} c_{j4d} + w_{ij} c_{j1d+1} \leq 1 \quad \forall d = 1,..,6 \tag{7}$$

Each nurse may not exceed the maximum number of shifts per week

$$\sum_{j=1}^{m}\sum_{d=1}^{7}\sum_{s=1}^{4} w_{ij}c_{jsd} \leq x_i \quad \forall i \tag{8}$$

4   Proposed Methodology

To compare different methodologies used to solve the NSP problem, Vanhoucke, M. et al. [19] proposed an extensive benchmark library namely NSPLib[1], a repository containing thousands of data instances arranged through different classes using multiple complexity indicators. We use historical data from [19] to create nurses' weekly schedules and assume that these schedules satisfy the working constraints as the constraints and preferences are already embedded. The purpose of using machine learning algorithms is to identify the patterns and produce new schedules with the same properties as those of original data. We expect the generated schedule to have the minimum average error to historical data computed using the Frobenius Norm.

Association rules mining [23] is an unsupervised learning method for finding co-occurrences, frequent patterns, associations among items in a set of transactions or a database. Association rules mining methods are applied to various transaction databases to extract rules in the form of "item1 & item2 → item3" which means that if item1 and item2 occur together, there is a high probability that item3 also occurs. In the NSP problem context, association rules mining methods are applied on previously defined schedules to extract nurses' frequent assignments with interestingness parameters above the user-defined threshold. The generated rules will represent the assignments that some constraints or preferences could enforce to simulate new scheduling scenarios using these rules. The Apriori algorithm [23] is used for association rules learning in transaction databases based upon predefined support and confidence thresholds set to select the significant rules that capture the association between the data variables. Let us consider two variables A and B with the item-set, $A \Rightarrow B$ indicates an association rule and T is a set of transactions in a database. The support measures the percentage of occurrence of an item-set and is computed as follows. $Support(A) = \frac{|\{A \subseteq T\}|}{|T|}$. Confidence, $Confidence(A \Rightarrow B) = \frac{Support(A \cup B)}{Support(B)}$, is the probability of multiple items that occur together, indicating an estimation of how many times a given rule is valid. In the Apriori experiment, we fix the support at 0.25 because it is suggested to use relatively low min-support value for sparse data [20], then we create multiple scheduling scenarios based on the generated rules with the following rounded confidence ranges; [0.6 to 0.7], [0.7 to 0.8], [0.8 to 1]. We further compare the obtained schedules against the original schedules. The pseudo-code listed in Algorithm 1 shows the different steps to find the best schedule from the confidence ranges; the rules are generated according to the support and confidence. A schedule is created using these rules,

---

[1] https://www.projectmanagement.ugent.be.

then the average error is computed using the Frobenius Norm. The best schedule is selected based on the minimum value.

Algorithm 1: Selecting the best schedule using the Apriori Algorithm

---

$Best\_schedule(Data\_schedule):$
$rules \leftarrow Apriori(Data\_schedule, Support, Confidence)$
$New\_schedule \leftarrow Fill(rules)$
$frob\_norm \leftarrow Frobenious\_Norm(Data\_schedule, New\_schedule)$
while $frob\_norm > min\_frob\_norm$ **do**
 | $Best\_schedule(Data\_schedule)$
end
return $New\_schedule$

High Utility Item-set Mining (HUIM) [23] is a method derived from the association rules mining framework. In this method, the rules are learned from a transaction database through the number of occurrences of the item-sets (See Sect. 5.1 Data). It is assumed that the user is only interested in the high frequency item-sets, however, the frequency may not be enough to determine the relevance of a given association rule because it only indicates the number of occurrences of the item-set without revealing its utility. Thus, frequent pattern mining with utility may be used to extract more interesting rules to build more accurate solutions. In the scope of NSP, The items are the nurses and the utility is the average preference cost of a specific nurse working in a particular shift which we compute from the preference matrix provided by the NSPLib instances [19]. The minimum utility threshold is set to select the rules with the maximum-profits/highest-utilities [23]. We use a Two-Phase algorithm for this method where the inputs are; the workload coverage during a month and the nurses' average preferences with min-utility threshold of 120 [23]. Setting a standard minimum utility is very difficult as stated in literature [20], for simplicity, we fix a high min-utility value since we are interested in item-sets of high utilities. A pseudo-code of the process is presented in Algorithm 2. The rules with utility greater or equal to min-utility are generated, then the schedule is created with these rules, finally, the average error is computed using the Frobenius Norm.

Algorithm 2: Generating the schedule using a Two-Phase Algorithm

---

$HUIM\_schedule():$
$rules \leftarrow TwoPhase(Coverage\_table, Utility\_table, min\_utility)$
$New\_schedule \leftarrow Fill(rules)$
$frob\_norm \leftarrow Frobenious\_Norm(Data\_schedule, New\_schedule)$
return $New\_schedule$

Naive Bayes classifier are supervised machine learning algorithm used to classify new observations of data using labeled data-set with the conditional independence assumption for all the features given the class label. Naive Bayes assumes that all the features contribute independently to the predicted outcome, which means that a particular attribute in a given class is irrelevant to the presence of other attributes. These classifiers are derived from the Bayes theorem, a mathematical formula used to compute the probability of an event occurring based on prior observations. In the NSP problem context, Naive Bayes classifiers are used to predict remaining shift assignments based on newly observed data. So the input schedules are expected to have partial assignments of shifts and the goal is to fill in the missing ones based on prior shift observations that could be enforced by the constraints and preferences. The following is a formulation of the NSP problem according to the Bayes theorem. Let $i = \{1, .., n\}$ be the nurse index, $j = \{1, .., m\}$ is the shift index.

$$P(N_{i+1j'}|N_{ij}) = \frac{P(N_{ij}|N_{i+1j'})\ P(N_{i+1j'})}{P(N_{ij})} \qquad (9)$$

In the Naive Bayes method, we split the data into 70% and 30% for training and testing data-sets respectively. We train the model with the training set and use the testing set to validate the predicted class labels as shown in Fig. 1. Note that in our experiment, we predict the assignments of $Nurse1$ in $shift4$ and we use Laplace add-1 smoothing [17] in our computation for missing observation in the training data that can be found in the testing data.

A graphical model is a method of representing joint probability distribution over a set of random variables in a compact and intuitive form which can be used to represent a causal relationship between these random variables. A Bayesian Network can be described as a Directed Acyclic Graph (DAG) where the nodes represent random variables, and the directed edges are the causal relationship between these variables. According to Markov's assumption [9], only the children nodes are conditioned by their parent nodes, which means that the features are independent and contribute independently to the result. The joint probability distribution is calculated as depicted in Eq. (10). Let $X = \{X_1, ..., X_n\}$ be a set of variables, x_i designate values of the variable X_i, and $Parents(X_i)$ designate the values for the parents of X_i in the Bayesian Network.

$$P(x_1, ..., x_n) = \prod_{i=1}^{n} P(x_i \mid Parents(X_i)) \qquad (10)$$

Bayesian Networks are used to model multinomial data with both discrete and continuous variables by encoding the relationship between the variables contained in the nurse scheduling data. The generation of new instances of the variables (nurse scheduling scenarios) is achieved by learning joint probability distributions over the data variables simultaneously using Bayesian Network inference. The motivation of using a Bayesian Network is to have a stochastic distribution to allow us to sample and produce potentially better estimation because the Naive Bayes method is limited to the data distribution of the used schedules. Moreover, the Naive Bayes model is the product of the prior and likelihood parameters which follows a specific distribution themselves. Note that our input schedules follows a Bernoulli distribution where 1 means a specific nurse work in a particular shift and 0 if not, and the probability of assigning a nurse is equally likely to guarantee fairness. Thus, the data generating process of our data is of Bernoulli likelihood and Uniform probability. The flowchart of generating the schedules using Bayesian Network is presented in Fig. 1.

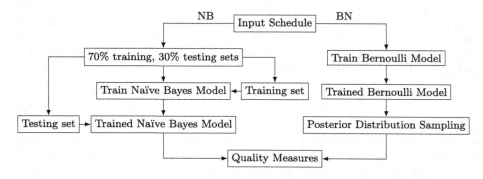

Fig. 1. Flowchart of Class Prediction using Naive Bayes (NB), and Sampling from the Posterior Distribution using the Bayesian Network (BN)

5   Experimentation

5.1   Data

Multiple data-set settings are provided in the NSPLib benchmark library [19], such as the workload coverage of the different number of nurses (e.g., 25, 75, 100) throughout a different number of days (e.g., 7, 28), more details can be found in [19]. Note that in our experimentation, we use instances from the following setting; 25 nurses, 7 d, and 4 shifts per day. For simplicity, we reduce the number of nurses to 5 and distribute the coverage requirements accordingly. Note that for the testing of our methods, we collect the coverage matrices from 52 data instances which represent 52 weeks of workload coverage, then distribute these coverage requirements randomly to introduce fairness among nurses. To create a schedule for a given day of the week, we gather all the corresponding shift assignments of nurses in that specific day from all the 52 weeks. Thus, for our test setting (5 nurses, 4 shifts), the input schedule is of dimensions; 52 rows of $day1$ gathered from the instances and 20 columns representing the nurses assigned in shifts. In the High Utility Item-set Mining [23] (see Sect. 4 Proposed Methodology) we multiply the workload by 4 to reflect the number of shift assignments of nurses in a period of a month because the input for this method is assumed to show the quantity of interest. A utility table is another input required in this method which basically reflect the average preference cost attached to the items (nurse working in a specific shift), and it is mainly needed to extract frequent patterns of high interest.

5.2   Evaluation Matrix and Quality Measures

The objective of our methods is to automatically generate schedules with similar properties to the original schedule, which means that we want to capture the frequent patterns from the input schedules and generate new solutions that conserves the learned patterns. We measure the performance of the obtained schedules by computing the distance between the original and the generated schedules using the Frobenius Norm [6], which is a matrix norm that calculates the average error between two matrices element-wise, as depicted in Eq. 11.

$$\|M - N\|_F = \sqrt{\sum_{i=1}^{x} \sum_{j=1}^{y} (m_{ij} - n_{ij})^2} \tag{11}$$

M and N are two matrices with m_{ij} and n_{ij} their respective entries. For the Naive Bayes method, we measure the accuracy by comparing the matched and unmatched predicted class labels with the testing set.

5.3   Results and Discussion

The experimental results, reported in Table 1, show that the average error values found in all the methods are very close to each other.

Table 1. Experiment results of our proposed methods

Method	Settings	Quality measure
Apriori	0.25 Support and [0.6 to 0.7] confidence range 0.25 Support and [0.7 to 0.8] confidence range 0.25 Support and [0.8 to 1] confidence range	17.94 20 17.94
Two-Phase	Minimum Utility = 120	21.40
Naïve Bayes	Training: 70% Testing: 30%	Positive hits: 11 Negative hits: 5 Accuracy: 68%
Bayesian Network	Probability: Uniform, Likelihood: Bernoulli	20.83

Note that choosing the confidence ranges in the Apriori method is due to the fact that association rules mining has a very sparse nature. This results in rules with close values of support and confidence, as shown in Fig. 2 (darker shade indicates higher number of rules).

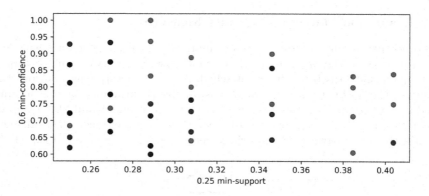

Fig. 2. Rules distribution

The Apriori algorithm with confidence ranges [0.6 to 0.7] and [0.8 to 1] performed better than Two-Phase and the Bayesian Network with a minimum average error value of 17.94. This value reflects the average error element-wise between the input and the generated schedules. Ideally, this value is preferred to be closer to 0 to show that the obtained schedules are very close to the original one. However, the results of our methods reported in Table 1 are acceptable because we do not learn any explicit preference and use limited data (we will conduct a rigorous external validation by consulting with scheduler officer at hospitals in our future work). The schedule generation using our methods is an iterative process because it relies on specific parameters (e.g., Support, Confidence) so it is expected to obtain relatively better results if we relax these interestingness measures for Apriori and Two-Phase, or by more sampling from Bayesian Network. Thus, the next phase of our methods may concern the learning of how to find the optimal/sub-optimal solutions without iterations by investigating the correlation between the quality of the generated solutions and the iterations.

The ROC curve presented in Fig. 3 outlines the trade-off between the sensitivity and the specificity of our Naive Bayes model for the binary class labels 1 (assignment of shift) and 0 (non assignment of shift). Ideally the ROC curve should be closer to the true positive rate axis indicating that the higher probabilities are assigned to the correct class label. The accuracy of our Naive Bayes model is 68%, however, Naive Bayes is known as instance based learning which means that the accuracy of the model is dependent on the training set; the better quality of data the better accuracy. Figure 4 shows the posterior distribution from Markov Chain Monte Carlo (MCMC) sampling which represent the convergence of the prior probabilities after seeing more evidence in the data. The plot reveals that 94% of the probability is assigned within 0.28 and 0.34 probabilities which means that the likelihood values in this interval yields the highest accuracy.

The comparison results of our methods against COUNT-OR [12] regarding the quality of the generated schedules measured by the Frobenius Norm is as

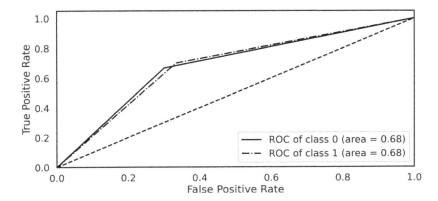

Fig. 3. ROC curve for class 0 and 1

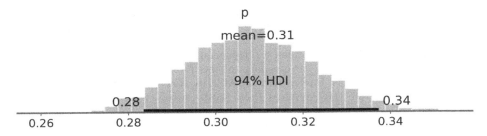

Fig. 4. Highest Density Interval (HDI) of Posterior Distribution

follows; Apriori: 5.83, Two-Phase: 6.63, Bayesian Network: 6.70, COUNT-OR: 6.78. Note that, we exclude the Naive Bayes method from this comparison due to the different quality measure scope. Overall, our methods performed slightly better because it is known that machine learning algorithms capture and learn the patterns and generate similar historical data instances. Moreover, COUNT-OR works by explicitly learning the constraints of the input schedules against predefined quantities of interest (obtained using tensor operations) that represent the constraints, so if there is any inconsistency in the input schedules (the schedules do not satisfy the predefined quantities of interest), COUNT-OR may fail in producing a solution. Contrarily, our methods learns the constraints and preferences implicitly, and if there is any inconsistency in the data, it will be disregarded simply because our methods are used to extract highly interesting frequent patterns. Though, our methods may fail in providing an optimal solution, they guarantee a solution compared to the given data.

6   Conclusion and Future Work

We have proposed an NSP solving approach, relying on implicitly learning constraints and preferences from historical data, and without any prior knowledge. The experiments we conducted show that our approach produce slightly better results than COUNT-OR. One advantage of our learning method is the ability to capture relevant shift assignments when using association rules mining, and to predict unknown assignments from partially observed ones, when relying on Naive Bayes. Based on this fact, we plan to apply our methodology to other dynamic combinatorial optimization applications, including reactive planning and scheduling, vehicle routing and timetabling.

References

1. Aickelin, U., Li, J.: An estimation of distribution algorithm for nurse scheduling. Ann. Oper. Res. **155**(1), 289–309 (2007). https://doi.org/10.1007/s10479-007-0214-0
2. Alanazi, E., Mouhoub, M., Zilles, S.: The complexity of learning acyclic cp-nets. In: Kambhampati, S. (ed.) Proceedings of the Twenty-Fifth International Joint Conference on Artificial Intelligence, IJCAI 2016, New York, NY, USA, 9–15 July 2016, pp. 1361–1367. IJCAI/AAAI Press (2016). http://www.ijcai.org/Abstract/16/196
3. Alanazi, E., Mouhoub, M., Zilles, S.: The complexity of exact learning of acyclic conditional preference networks from swap examples. Artif. Intell. **278**, 103182 (2020). https://doi.org/10.1016/j.artint.2019.103182
4. Aparecido Constantino, A., Tozzo, E., Lankaites Pinheiro, R., Landa-Silva, D., Romão, W.: A variable neighbourhood search for nurse scheduling with balanced preference satisfaction. In: Proceedings of the 17th International Conference on Enterprise Information Systems - Volume 1, pp. 462–470. SCITEPRESS - Science and Technology Publications, Lda (2015). https://doi.org/10.5220/0005364404620470
5. Bagheri, M., Gholinejad Devin, A., Izanloo, A.: An application of stochastic programming method for nurse scheduling problem in real word hospital. Comput. Ind. Eng. **96**, 192–200 (2016). https://doi.org/10.1016/j.cie.2016.02.023, https://www.sciencedirect.com/science/article/pii/S036083521630050X
6. Böttcher, A., Wenzel, D.: The frobenius norm and the commutator. Linear Algebra Appl. **429**(8), 1864–1885 (2008)
7. Constantino, A.A., de Melo, E.L., Landa-Silva, D., Romão, W.: A heuristic algorithm for nurse scheduling with balanced preference satisfaction. In: 2011 IEEE Symposium on Computational Intelligence in Scheduling (SCIS), pp. 39–45 (2011)
8. El Adoly, A.A., Gheith, M., Nashat Fors, M.: A new formulation and solution for the nurse scheduling problem: a case study in egypt. Alexandria Eng. J. **57**(4), 2289–2298 (2018). https://doi.org/10.1016/j.aej.2017.09.007
9. Friedman, N.: A qualitative markov assumption and its implications for belief change. In: In Proceedings of the Twelfth Conference on Uncertainty in Artificial Intelligence (UAI '96), pp. 263–273. Morgan Kaufmann (1996)

10. Karmakar, S., Chakraborty, S., Chatterjee, T., Baidya, A., Acharyya, S.: Meta-heuristics for solving nurse scheduling problem: a comparative study. In: 2016 2nd International Conference on Advances in Computing, Communication, & Automation (ICACCA) (Fall), pp. 1–5 (2016)
11. Kottke, D., Schellinger, J., Huseljic, D., Sick, B.: Limitations of assessing active learning performance at runtime. arXiv e-prints (2019). http://arxiv.org/abs/1901.10338
12. Kumar, M., Teso, S., De Causmaecker, P., De Raedt, L.: Automating personnel rostering by learning constraints using tensors. In: 2019 IEEE 31st International Conference on Tools with Artificial Intelligence (ICTAI), pp. 697–704 (2019)
13. Kumar, M., Teso, S., De Raedt, L.: Acquiring integer programs from data. In: Proceedings of the Twenty-Eighth International Joint Conference on Artificial Intelligence, IJCAI-19, pp. 1130–1136 (2019). https://doi.org/10.24963/ijcai.2019/158
14. Li, J., Aickelin, U.: A bayesian optimization algorithm for the nurse scheduling problem. In: The 2003 Congress on Evolutionary Computation, 2003. CEC '03, vol. 3, pp. 2149–2156 (2003). https://doi.org/10.1109/CEC.2003.1299938
15. Mouhoub, M., Marri, H.A., Alanazi, E.: Learning qualitative constraint networks. In: Alechina, N., Nørvåg, K., Penczek, W. (eds.) 25th International Symposium on Temporal Representation and Reasoning, TIME 2018, Warsaw, Poland, October 15–17, 2018. LIPIcs, vol. 120, pp. 19:1–19:13. Schloss Dagstuhl - Leibniz-Zentrum für Informatik (2018). https://doi.org/10.4230/LIPIcs.TIME.2018.19
16. Passerini, A., Tack, G., Guns, T.: Introduction to the special issue on combining constraint solving with mining and learning. Artif. Intell. **244**, 1–5 (2017). https://doi.org/10.1016/j.artint.2017.01.002
17. Setyaningsih, E., Listiowarni, I.: Categorization of exam questions based on bloom taxonomy using naïve bayes and laplace smoothing. In: 2021 3rd East Indonesia Conference on Computer and Information Technology (EIConCIT), pp. 330–333 (2021). https://doi.org/10.1109/EIConCIT50028.2021.9431862
18. Tsouros, D.C., Stergiou, K., Bessiere, C.: Omissions in constraint acquisition. In: Simonis, H. (ed.) CP 2020. LNCS, vol. 12333, pp. 935–951. Springer, Cham (2020). https://doi.org/10.1007/978-3-030-58475-7_54
19. Vanhoucke, M., Maenhout, B.: Nsplib-a nurse scheduling problem library: a tool to evaluate (meta-) heuristic procedures (2007)
20. Vu, L., Alaghband, G.: An efficient approach for mining association rules from sparse and dense databases. In: 2014 World Congress on Computer Applications and Information Systems (WCCAIS), pp. 1–8 (2014)
21. Vu, X.H., O'Sullivan, B.: A unifying framework for generalized constraint acquisition. Int. J. Artif. Intell. Tools **17**, 803–833 (2008)
22. Wang, T., Meskens, N., Duvivier, D.: Scheduling operating theatres: mixed integer programming vs. constraint programming. Eur. J. Oper. Res. **247**(2), 401–413 (2015). https://doi.org/10.1016/j.ejor.2015.06.008, https://www.sciencedirect.com/science/article/pii/S0377221715005226
23. Yao, H., Hamilton, H., Butz, C.: A foundational approach to mining itemset utilities from databases. In: Proceedings of the Fourth SIAM International Conference on Data Mining, Lake Buena Vista, Florida, USA, April 22–24, 2004, vol. 4 (2004). https://doi.org/10.1137/1.9781611972740.51

Improving Goal-Oriented Visual Dialogue by Asking Fewer Questions

Soma Kanazawa$^{(\boxtimes)}$, Shoya Matsumori, and Michita Imai

Graduate School of Science and Technology, Keio University, 3-14-1 Hiyoshi,
Kohoku-ku, Yokohama, Kanagawa 223-8522, Japan
{kanazawa,shoya,michita}@ailab.ics.keio.ac.jp
https://www.ailab.ics.keio.ac.jp

Abstract. An agent who adaptively asks the user questions to seek information is a crucial element in designing a real-world artificial intelligence agent. In particular, goal-oriented visual dialogue, which locates an object of interest from a group of visually presented objects by asking verbal questions, must be able to efficiently narrow down and identify objects through question generation. Several models based on Guess-What?! and CLEVR Ask have been published, most of which leverage reinforcement learning to maximize the success rate of the task. However, existing models take a policy of asking questions up to a predefined limit, resulting in the generation of redundant questions. Moreover, the generated questions often refer only to a limited number of objects, which prevents efficient narrowing down and the identification of a wide range of attributes. This paper proposes Two-Stream Splitter (TSS) for redundant question reduction and efficient question generation. TSS utilizes a self-attention structure in the processing of image features and location features of objects to enable efficient narrowing down of candidate objects by combining the information content of both. Experimental results on the CLEVR Ask dataset show that the proposed method reduces redundant questions and enables efficient interaction compared to previous models.

Keywords: Goal-oriented visual dialogue · Attention mechanism ·
Visual state estimation · GuessWhat?! · CLEVR Ask

1 Introduction

The acquisition of information by agents through interaction with humans is one of the most vital abilities for artificial intelligence. Real-world agents often experience unknown situations that do not exist at the time of task design/learning. Particularly in situations where the user and agent collaborate to accomplish a specific task, the agent needs to understand which parts of the environment the user is paying attention to and which objects the user is referring to in order to understand the user's instructions accurately. However, in scenes where expressions are abbreviated, or there are multiple candidates for relevant objects

© Springer Nature Switzerland AG 2021
T. Mantoro et al. (Eds.): ICONIP 2021, LNCS 13109, pp. 158–169, 2021.
https://doi.org/10.1007/978-3-030-92270-2_14

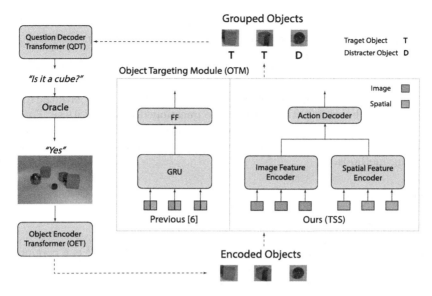

Fig. 1. Comparison of previous model [6] and Two-Stream Splitter (TSS). TSS replaces UniQer's Object Targeting Module (OTM) and assigns target objects which are to be set apart from distracter objects in Question Decoder Transformer (QDT).

in the environment, the user's reference may not be uniquely determined, thus requiring the agent not only to receive instructions passively but also to seek information by asking the user questions. In this paper, we tackle Goal-Oriented Visual Dialogue (GOVD), a task to locate the object referred to by the opponent through a question from a set of visually presented objects.

GuessWhat?! [1] is a prominent example of a GOVD task. Several models for solving GuessWhat?! tasks based on reinforcement learning have been trained to maximize the success reward by generating question word tokens as actions [10, 11]. Unified Questioner Transformer (UniQer) [6] enables the question generation with referring expressions that can identify individual objects in an image on the CLEVR Ask framework [6] by dividing the multiple roles previously performed by a single module into different modules.

Nevertheless, existing models, including UniQer, are still too irritating for users to easily interact with agents, particularly because 1) the questioner often generates redundant questions, increasing the number of questions required to identify an object, and 2) the questioner tends to generate questions that only refer to a few objects, making it inefficient for narrowing down candidate objects. In order to identify an object in an image, the model generates questions over multiple turns. However, some of the questions generated may be redundant, making the number of questions to be asked before the task succeeds too large. For example, there are cases where the model generates questions with exactly the same content repeatedly, cases where the model continues to ask questions up to the limit even if the object can be logically identified, and cases where the model generates meaningless questions that do not make any progress in narrowing down the candidate objects.

In this paper, we propose Two-Stream Splitter (TSS), which decides the next action to take to replace the Object Targeting Module of UniQer (Fig. 1). TSS is a Transformer-based model that extracts one object in the image with the highest confidence and determines the object to be questioned based on its similarity to other objects. Experimental results on the CLEVR Ask dataset show that our model reduces redundant questions and enhances efficient interaction compared to existing methods.

2   Related Works

2.1   GuessWhat?!

A typical test-bed for GOVD is GuessWhat?! [1]. GuessWhat?! is a cooperative two-player game in which the agent guesses through a series of questions and answers between the agent and Oracle. Initially, Oracle selects one object in the image, and the agent asks Oracle yes/no questions about the image to locate the object selected by Oracle through a dialogue. The agent asks the next question in response to Oracle's answer, and if the agent can finally identify the object, the task is considered successful. The GuessWhat?! dataset consists of 155K dialogues with 822K questions and 67K natural images. Images containing three to 20 objects are collected from the MSCOCO dataset [5]. The number of images is 67K, and the total number of objects in the dataset is 134K. For GuessWhat?!, studies have been conducted in the framework of supervised learning since its inception. However, in recent years, several additional methods have been incorporated into the framework of reinforcement learning [11]. Many studies have examined ways to obtain more accurate models by devising reward functions in the policy gradient method [12] of reinforcement learning, such as evaluating whether a useful question has been asked [10]. In most of these methods, the questioner consists of a question generator that understands the current state of the dialogue and asks questions, and a guesser that guesses which objects are likely according to the responses. The question generator performs both understanding of the current situation and question generation, and the input to the Oracle is not images but only object categories and location information. In research on the improvement of the guesser, a model [7] that uses the probability distribution of each object in the dialogue has been proposed, contributing to the improvement of accuracy. As for the improvement of the question generator, there is a model [8] that improves the accuracy and reduces the repetition of questions by taking the difference of image features between objects.

Other research on GuessWhat?! has focused on not only improving the task success rate but also reducing the number of questions [9]. However, most of these methods require an external module to determine whether a question should be continued or ended and need a separate dataset for the end-of-question decision and modular training.

2.2 UniQer

UniQer [6] is a GOVD system architecture that achieves a higher task success rate than conventional methods thanks to two unique features. First, it enables question generation using referring expressions, which the previous methods cannot do. Second, it solves the problem of the low correct answer rate of Oracle and enables accurate answers. In GuessWhat?! task, the models are trained on the assumption that Oracle's answers are always accurate, so if the answers are not accurate, the model will not train correctly. Using the CLEVR Ask dataset [6], which is based on the artificial synthetic image dataset CLEVR [3], UniQer can give accurate yes-no answers, which solves the problem of low accuracy of Oracle answers in GuessWhat?!. UniQer consists of the following three modules.

- Object Encoder Transformer (OET)
- Object Targeting Module (OTM)
- Question Decoder Transformer (QDT)

OET provides an understanding of the current situation and is implemented using a Transformer [13] encoder architecture. It encodes the image features of each object in the scene and the history of questions and answers during the dialogue to infer which object has the highest probability of being Oracle's reference object, and then outputs a probability distribution.

OTM is a module that decides which objects to focus on in generating questions by dividing each object into three groups: masked object property, targeted object property, and distracted object property. The masked object property is a group of objects that are considered unimportant and is excluded from consideration in question generation. The objects that are judged to have a probability of being Oracle's reference objects are assigned to two groups called targeted object property and distracted object property. The next question is generated to distinguish between the targeted and distracted groups.

QDT's role is to generate questions that distinguish between the target object and the distracter object according to group IDs provided by OTM. QDT is a Transformer decoder-based model that takes embeddings of group ID and object features. QDT generates questions with a tokenized word in an auto-regressive manner.

3 The CLEVR Ask Framework

We briefly explain here the CLEVR Ask task that will serve as a task for our dialogue system. An example dialogue of the CLEVR Ask task is shown in Fig. 2. CLEVR Ask is an object identification task with two players, Oracle and questioner. Oracle first selects an object from the scene at random, and the questioner's goal is to locate the object selected by Oracle through questions. The questioner has to guess the process of getting an answer from Oracle by asking a yes-no question, and if it judges that it can identify the object, it terminates the question and submits the guess result. If the guess matches the object selected

Questioner		Oracle
Q1. Is it a cube?	Oracle's reference	A1. Yes
Q2. Is it a blue metal cube?		A2. No
Q3. Is it in front of a blue cube?		A3. Yes
[Submit]		Result: [Success]

Fig. 2. An example dialogue for the CLEVR Ask task, where the questioner needs to ask questions to guess Oracle's reference object and submit the result at the appropriate time. The red circle represents the Oracle's reference object.

by Oracle, the task is counted as a success; if it does not, it is counted as a failure. The CLEVR Ask dataset consists of an image containing a scene file and a set of QAs based on the image. For problems that were not answered correctly by Oracle in GuessWhat?!, Oracle in CLEVR Ask can provide correct answers based on the scene files describing the environment of CLEVR. There are two datasets in CLEVR Ask, Ask3 and Ask4, each consisting of 85K images, balanced by the respective number of attributes of the objects.

4   Model

Our proposed Two-Stream Splitter (TSS) model is shown in Fig. 3. We focus on OTM to improve the model.   OTM is a module that determines the next action to be taken by the questioner. OTM's action is to decide which object to focus the question on or to terminate the question and submit the prediction of the Oracle's reference object. OTM only pays attention to top-k high-scored candidate objects calculated from the object probability, and ignores the others.

There are two problems in the existing OTM: 1) it is difficult to choose the submit action and 2) it often generates a split action that groups a narrow range of objects. In order to solve these problems, we made the following architectural design. 1) We assume that the most probable object should always be addressed in the question, and fix the object with the highest probability to be classified into the target object group in the model. As a result, the number of possible actions was reduced from 3^k in UniQer's OTM to 3^{k-1}, making it relatively easier to select the submit action. 2) We divide the stream that takes the self-attention of the image and spatial features, and calculate the similarity of the image features between the object with the highest probability and each object. This structure makes it easier to select an action that groups a wide range of objects with similar attributes. This is because objects with similar attributes are likely to have smaller differences in features, and with such kind information, the model can put them into the same group.

TSS is composed of three sub-modules. 1) Spatial Feature Encoder, 2) Image Feature Encoder, and 3) Action Decoder. The Spatial Feature Encoder obtains

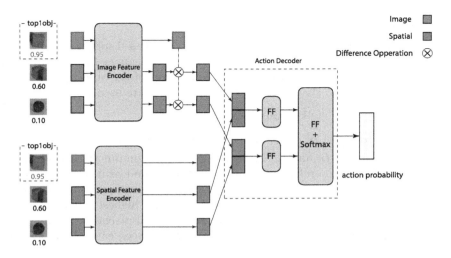

Fig. 3. The structure of our Two-Stream Splitter. Our model processes each piece of information with a Transformer encoder and generates a split focusing on the difference from top1obj.

the positional relationship between objects in a self-attention structure, and the Image Feature Encoder measures the relationship and similarity of the image features of objects.

4.1   Spatial Feature Encoder

The Spatial Feature Encoder is a sub-module for capturing the interrelationship of location information of each object. The spatial feature of each object is a concatenation of the 5D coordinates of its position in the environment and the 5D coordinates of its positional relationship with other objects [15]. The spatial feature is represented by $o_s = \{o_s^1, ..., o_s^k\}$, a set of top-k 5D geometric features stored for each object. The object probability $P_{\hat{o}}$ contains the probability distribution of how well each object matches the current dialogue history as $P_{\hat{o}} = \{P_{\hat{o}}^1, ..., P_{\hat{o}}^k\}$, output from the OET. A Transformer encoder-based sub-module that captures and encodes the positional relationship between objects is described as

$$\tilde{\mathcal{X}}_S = \mathcal{F}_S([f_A(o_s)f_p(P_{\hat{o}})]),\qquad(1)$$

where $\mathcal{F}_S$ is a Transformer encoder, f_A and f_p are linear transformation functions. By calculating the positional information of each object using Self-Attention with other objects, a vector $\tilde{\mathcal{X}}_S$ containing the mutual positional relationships with other objects is output from the Spatial Feature Encoder.

4.2   Image Feature Encoder

The Image Feature Encoder is a Transformer encoder-based sub-module for capturing the interrelationships between objects using their image features. In order to obtain the relationship between the image features of the objects, image features such as the color, shape, and size of each object are used, as

$$\tilde{\mathcal{X}}_I = \mathcal{F}_I([f_B(o_{img}), f_p(P_{\hat{o}})]), \tag{2}$$

where $o_{img} = \{o^1_{img}, ..., o^k_{img}\}$ is an image feature value of each object, $\mathcal{F}_I$ is a Transformer encoder, f_B and f_p are linear transformation functions, and $\tilde{\mathcal{X}}_I$ is a vector that captures the relationship between the image features of objects.

Let the object that the OET judges to have the highest probability of being the goal object (in other words, the object with the highest value of $P_{\hat{o}}$) be the top1obj, and let the vector after calculating the self-attention by the image feature be $\tilde{\mathcal{X}}_I^1$. The similarity of the image feature between each object and the top1obj is then computed by Difference Operation [8] as

$$v = \text{softmax}([\tilde{\mathcal{X}}_I^1 \odot (\tilde{\mathcal{X}}_I^1 - \tilde{\mathcal{X}}_I^i)]_{i \in \mathcal{K}, i \neq 1}), \tag{3}$$

where i is the index of an object and $\mathcal{K}$ is a set of indices of top-k objects. Equation 3 calculates the similarity by taking the difference between the top1obj and every other object and then obtaining the element product with the top1obj. By using this value, Two-Stream Splitter can explicitly obtain the feature values of which objects are similar and which are not and then use them to generate actions.

4.3   Action Decoder

The Action Decoder determines the action to be taken by the questioner based on the values calculated by the Spatial Feature Encoder and Image Feature Encoder through the self-attention structure. To obtain the vector for each object, a feed forward layer is used, as

$$O(i) = \mathcal{F}_f([\tilde{\mathcal{X}}_S^i, \tilde{\mathcal{X}}_I^i, v(i)]), \tag{4}$$

where $O(i)$ represents the vector of each object and $\mathcal{F}_f$ is a representation of the feed forward layer for each object.

In the final layer of the Action Decoder, the vectors of the top-k objects are concatenated, and the probability distribution of the action ID is obtained through linear transformation and activation, as

$$P_{\text{RL}} = \text{softmax}(\mathcal{F}_l([\{O(i)\}_{i \in \mathcal{K}}])), \tag{5}$$

where P_{RL} represents the probability distribution of all possible actions the questioner can take. In UniQer, the top-k objects were split into three groups, so 3^k probability distributions of actions were output. Since Two-Stream Splitter always classifies the top1obj into the Target Object Group and calculates the actions for $k - 1$ objects, Eq. 5 outputs the probability distributions for 3^{k-1} actions.

5   Experiments

Dataset. We used the CLEVR Ask3 dataset containing 70K training, 7.5K validation, and 7.5K test images. The number of objects is 455,632, 48,458, and 48,900, for training, validation and test sets respectively.

Metrics. In order to compare the quality of dialogue generated by questions from the model, we examined the number of questions that met the following criteria. The number of useless questions (*useless q*) represents the average number of times a questioner asks a redundant question without submitting a prediction, even though the oracle's reference object can be logically identified from the dialogue. The number of narrow down failure questions (*nd fail*) indicates the number of questions that failed to make progress in narrowing down the candidate objects. The details of each metric is described as follows. The condition of the object corresponding to the dialogue history between Oracle and the questioner at time t in the scene image is described as $D_t(o)$, and the set of candidate objects in the image at turn t is denoted as $O_t = \{o|D_t(o)\}$. For instance, if Oracle's reply to the question "Is it a red object?" in turn t is "Yes", then all the red objects in the scene are included in O_t. Therefore, the state in which the object candidates in the scene are narrowed down to one in the turn t and Oracle's reference object can be determined is represented as $|O_t| == 1$. If the question at t results in $|O_t| == 1$ and the questioning is continued without selecting submit in the action at $t + 1$, the question at $t + 1$ is considered redundant and counted as *useless q*. In the case where $|O_t| == |O_{t+1}|$ is true, the questions generated at $t + 1$ are counted as *nd fail* because the questions fail to narrow down the objects due to redundant or inefficient questions.

Ablation Study. We conducted ablation experiments to evaluate the individual contribution of components in TSS. The following three models were tested for comparison:

- **Ours (single-stream)**: The model that substitutes UniQer's GRU with Transformer encoder architecture. Image features and spatial features of each object are concatenated and fed into a single Transformer encoder. This model is equivalent to a single-stream version of TSS.
- **Ours (w/o top1fix)**: The model that does not fix the top1obj to target object group. Therefore, the number of possible actions of this model is 3^k.
- **Ours (w/o diff-op)**: The model that does not perform the difference operation in TSS's Image Feature Encoder.

5.1   Training Details

We pre-trained the Object Encoder Transformer in UniQer with binary cross-entropy as a loss function in multi-label classification and pre-trained the Question Decoder Transformer with a negative loss likelihood. During the training

Table 1. Comparison results and ablation studies of OTM on the task success rate and number of questions. The bold numbers represent the best performance with regard to Num of Questions, which is the primary metric of this study.

Model	Task success (%)		Num of questions	
	New Img↑	New Obj↑	New Img↓	New Obj↓
UniQer (GRU)	$80.90_{\pm 2.56}$	$80.90_{\pm 2.61}$	$3.75_{\pm 0.38}$	$3.75_{\pm 0.38}$
Ours (single-stream)	$82.85_{\pm 3.42}$	$82.41_{\pm 3.42}$	$3.80_{\pm 0.40}$	$3.80_{\pm 0.40}$
Ours (w/o top1fix)	$65.69_{\pm 33.06}$	$65.19_{\pm 32.79}$	$3.34_{\pm 0.83}$	$3.36_{\pm 0.82}$
Ours (w/o diff-op)	$81.73_{\pm 3.71}$	$81.59_{\pm 3.70}$	$3.16_{\pm 0.43}$	$3.17_{\pm 0.42}$
Ours	$82.79_{\pm 3.31}$	$82.22_{\pm 2.87}$	$\mathbf{2.98}_{\pm 0.20}$	$\mathbf{2.95}_{\pm 0.08}$

of TSS, we freeze both OET and QDT. All experiments were conducted using six Quadro RTX 8000 GPUs and implemented in PyTorch. As an image feature extractor, we used ImageNet pre-trained ResNet34 [2]. We used REINFORCE [14], a policy gradient method [12], as our reinforcement learning algorithm. For the reinforcement learning, all experiments were trained for 140 epochs using the Adam [4] optimizer with the learning rate 5e-4 and a batch size of 1024.

5.2 Quantitative Evaluations

We tested our model in two experimental settings [1,6]: new image and new object. In New Image, the image and the target object were completely new, and both were tested on data that did not appear during training. In New Object, the image itself existed at the time of training, but the target object was set to a new one. We conducted five experimental runs across different seeds.

The comparison between Task Success and Number of Questions is presented in Table 1. The task success rate is defined as the rate of correct predictions submitted by a questioner. The number of questions indicates the average number of times the questioner asked the question before submitting the prediction result for the target object.

As we can see, our model achieved higher Task Success than the baseline GRU model, and at the same time, it was able to submit predictions for the target object with fewer questions. In addition, even with simple scenes, the baseline model tended to ask up to the specific number of questions, which were either 3 or 4. The baseline often took the strategy of submitting the prediction result after asking a consistent number of questions for any image, even when the number of objects in the image was small. TSS, on the other hand, was able to terminate the question and submit the prediction result flexibly depending on the given image.

As shown in Table 2, the *useless q* was 0.83 for Ours (single-stream) and 0.07 for TSS. Thus, in Ours (single-stream), questions were often continued even though the object had been logically narrowed down by the question, while TSS was able to determine the end of the questioning appropriately in many cases.

Table 2. A comparison on number of low quality questions generated by different models. The bold numbers indicate the best performance results. Our model successfully reduced the number of redundant questions. *useless q* indicates the number of questions that continue to ask even though the answer can be determined, and *nd fail* indicates the number of questions that fail to narrow down the object candidates.

Model	Num of useless q		Num of nd fail	
	New Img↓	New Obj↓	New Img↓	New Obj↓
UniQer (GRU)	$0.73_{\pm 0.23}$	$0.73_{\pm 0.23}$	$0.50_{\pm 0.18}$	$0.50_{\pm 0.18}$
Ours (single-stream)	$0.83_{\pm 0.32}$	$0.82_{\pm 0.32}$	$0.45_{\pm 0.06}$	$0.44_{\pm 0.06}$
Ours (w/o top1fix)	$0.31_{\pm 0.50}$	$0.31_{\pm 0.50}$	$0.47_{\pm 0.33}$	$0.47_{\pm 0.32}$
Ours (w/o diff-op)	$0.22_{\pm 0.29}$	$0.21_{\pm 0.29}$	$0.43_{\pm 0.21}$	$0.44_{\pm 0.21}$
Ours	$\mathbf{0.07}_{\pm 0.08}$	$\mathbf{0.07}_{\pm 0.08}$	$\mathbf{0.28}_{\pm 0.04}$	$\mathbf{0.30}_{\pm 0.03}$

In addition, while Ours (w/o top1fix) had an average *useless q* of 0.31, TSS reduced to 0.07, suggesting that both fixing the group of top1obj and limiting the number of actions are effective in submitting the prediction results at the right turn. Even in the case of *nd fail*, UniQer generated about 0.50 questions per session that could not narrow down the object candidates, while TSS succeeded in reducing the number to 0.28. Our model could reduced the *nd fail* by more than 1.5 on average for Ours (w/o diff-op), suggesting that calculating the difference operation of image features can help generate questions that can efficiency narrow down object candidates.

5.3 Qualitative Evaluations

A comparison of the dialogue examples generated by the proposed method and baseline is shown in Fig. 4. In the upper example, the baseline and TSS both submitted their predictions and eventually succeed in identifying the Oracle's reference object, but TSS was more efficient at submitting with fewer questions. The baseline tended to take the strategy of asking four questions before submitting a prediction regardless of scenes, resulting in generating more redundant questions than TSS. For example, the second question in the baseline asked about the attribute of the object already identified in the first question. In contrast, TSS was able to submit prediction results based on the answer to the second question without generating redundant questions.

In the lower example, the baseline failed the task with four questions, while TSS succeeded with three questions. The baseline only asked questions about a narrow range of attributes, whereas TSS asked questions about a wide range of attributes in the first question to narrow down the object more effectively. As a result, TSS succeeded in identifying the object with only three questions. This implies that the model can group objects with similar attributes, leading to the generation of efficient questions that refers to the wide range of objects.

	Baseline		Ours	
	(1) Is it a blue metal cylinder?	Yes	(1) Is it a blue metal cylinder?	Yes
	(2) Is it a rubber blue cylinder?	No	(2) Is it behind a blue metal cylinder?	No
	(3) Is it to the right of a blue metal cylinder?	No	[Submit]	Success
	(4) Is it a blue metal cylinder?	Yes		
	[Submit]	Success		
	(1) Is it a red glass cube?	No	(1) Is it a cube?	No
	(2) Is it a green rubber cube?	No	(2) Is it a blue metal cylinder?	No
	(3) Is it to the left of a red glass cube?	No	(3) Is it a blue rubber sphere?	No
	(4) Is it a blue rubber sphere?	No	[Submit]	Success
	[Submit]	Fail		

Fig. 4. Examples where our model achieves task success by asking fewer questions than the baseline. The Oracle's reference object is annotated in red circle.

6   Conclusion

In this study, we proposed Two-Stream Splitter, which selects the next action in CLEVR Ask based on the difference of image features between images and the self-attention of spatial location information. Experimental results showed that Two-Stream Splitter achieved a higher task success rate than existing models, and also achieved task success with a smaller number of questions. Although we worked only on CLEVR Ask in this study, in future work we will apply the model to dialogue tasks using natural images such as GuessWhat?! so that it can be used to select appropriate actions in a variety of highly general dialogue tasks.

Acknowledgments. This work was supported by JSPS KAKENHI Grant Number JP21J13789 and JST CREST Grant Number JPMJCR19A1, Japan.

References

1. De Vries, H., Strub, F., Chandar, S., Pietquin, O., Larochelle, H., Courville, A.: GuessWhat?! visual object discovery through multi-modal dialogue. In: CVPR, pp. 5503–5512 (2017). https://hal.inria.fr/hal-01549641
2. He, K., Zhang, X., Ren, S., Sun, J.: Deep residual learning for image recognition. In: CVPR, pp. 770–778 (2016)
3. Johnson, J., Hariharan, B., van der Maaten, L., Fei-Fei, L., Zitnick, C.L., Girshick, R.: Clevr: a diagnostic dataset for compositional language and elementary visual reasoning. In: CVPR, pp. 1988–1997 (2017). https://doi.org/10.1109/CVPR.2017.215
4. Kingma, D.P., Ba, J.: Adam: A method for stochastic optimization. arXiv preprint arXiv:1412.6980 (2014)

5. Lin, T.Y., et al.: Microsoft COCO: common objects in context. In: Fleet, D., Pajdla, T., Schiele, B., Tuytelaars, T. (eds.) ECCV 2014. LNCS, vol. 8693, pp. 740–755. Springer, Cham (2014). https://doi.org/10.1007/978-3-319-10602-1_48
6. Matsumori, S., Shingyouchi, K., Abe, Y., Fukuchi, Y., Sugiura, K., Imai, M.: Unified questioner transformer for descriptive question generation in goal-oriented visual dialogue. arXiv preprint arXiv:2106.15550 (2021)
7. Pang, W., Wang, X.: Guessing state tracking for visual dialogue. In: Vedaldi, A., Bischof, H., Brox, T., Frahm, J.-M. (eds.) ECCV 2020. LNCS, vol. 12361, pp. 683–698. Springer, Cham (2020). https://doi.org/10.1007/978-3-030-58517-4_40
8. Pang, W., Wang, X.: Visual dialogue state tracking for question generation. In: AAAI, vol. 34, pp. 11831–11838 (2020)
9. Shekhar, R., Baumgärtner, T., Venkatesh, A., Bruni, E., Bernardi, R., Fernandez, R.: Ask no more: deciding when to guess in referential visual dialogue. In: COLING, pp. 1218–1233. Association for Computational Linguistics, Santa Fe, New Mexico, USA (2018). https://www.aclweb.org/anthology/C18-1104
10. Shukla, P., Elmadjian, C., Sharan, R., Kulkarni, V., Turk, M., Wang, W.Y.: What should I ask? using conversationally informative rewards for goal-oriented visual dialog. In: ACL, pp. 6442–6451. Association for Computational Linguistics, Florence, Italy, July 2019. https://doi.org/10.18653/v1/P19-1646
11. Strub, F., de Vries, H., Mary, J., Piot, B., Courville, A., Pietquin, O.: End-to-end optimization of goal-driven and visually grounded dialogue systems. In: IJCAI, pp. 2765–2771 (2017). https://doi.org/10.24963/ijcai.2017/385
12. Sutton, R.S., McAllester, D.A., Singh, S.P., Mansour, Y.: Policy gradient methods for reinforcement learning with function approximation. In: NeurIPS, pp. 1057–1063 (2000)
13. Vaswani, A., et al.: Attention is all you need. In: NeurIPS, pp. 5998–6008 (2017)
14. Williams, R.J.: Simple statistical gradient-following algorithms for connectionist reinforcement learning. Mach. Learn. 8(3–4), 229–256 (1992)
15. Yu, L., Tan, H., Bansal, M., Berg, T.L.: A joint speaker-listener-reinforcer model for referring expressions. In: CVPR, pp. 7282–7290 (2017)

Balance Between Performance and Robustness of Recurrent Neural Networks Brought by Brain-Inspired Constraints on Initial Structure

Yuki Ikeda, Tomohiro Fusauchi, and Toshikazu Samura[✉]

Graduate School of Sciences and Technology for Innovation, Yamaguchi University,
2-16-1, Tokiwadai, Ube-shi, Yamaguchi 755-8611, Japan
{b055vgv,samura}@yamaguchi-u.ac.jp

Abstract. The advantage of brain computation is not only the performance but also the robustness in its network. A biological neural network can avoid dysfunction due to natural damage. In this study, we focus on the robustness of artificial recurrent neural networks (RNNs) where the initial network structure is constrained by brain's anatomical properties. RNNs with inappropriate constraints cause a tradeoff between performance and robustness, although the constraints are derived from the brain's anatomical structures. We found that RNNs, which are composed of excitatory and inhibitory neurons, with two brain-inspired constraints overcame the tradeoff. The constraints tended to improve the performance of trained RNNs and maintained the improvement after the destruction of connections. The first constraint was an excitatory/inhibitory balance constraint on a single neuron. The second was a partial connection constraint, limiting the number of connections. Consequently, we proposed the brain-inspired constraints as RNN initializers to achieve a balance between performance and robustness of RNNs.

Keywords: Recurrent neural network · Brain-inspired constraints · Robustness

1 Introduction

An artificial neural network models the network of neurons in the brain. Neurons connect with other neurons and send signals to each other. They implement parallel distributed computation on the network. One of the advantages of brain computation is its robustness in a network. In the brain, when a part of a network is damaged, it decreases the performance of a certain function, but its dysfunction is avoided, e.g., spatial learning in the hippocampus [1]. However, many studies focus on the functional performance of an artificial neural network that mimics the brain computation, e.g., convolutional neural network mimicking vision [2]. The mutual connectivity of the brain network is replicated by recurrent neural networks (RNNs). We had introduced the properties of brain

© Springer Nature Switzerland AG 2021
T. Mantoro et al. (Eds.): ICONIP 2021, LNCS 13109, pp. 170–180, 2021.
https://doi.org/10.1007/978-3-030-92270-2_15

connectivity to RNNs as constraints on an initial network structure —brain-inspired constraints— and improved the performance of RNNs on a sequential prediction task [3, 4]. Moreover, fault tolerance in feed-forward neural networks have been evaluated, and mitigation methods for the failure have been proposed in previous studies [5]. However, the fault tolerance of RNNs, which mimics the mutual connectivity of the brain network, has been less focused on. Therefore, in this study, we evaluate RNNs' robustness with the brain-inspired constraints. Balances between excitatory and inhibitory neurons are crucial in RNN performance improvement [3]. In this study, we focus on the excitatory/inhibitory (E/I) balance on a single neuron and elucidate the relationship between robustness and brain-inspired constraints.

2 Methods

2.1 RNNs

We used a simple RNN where the initial connectivity among neurons is constrained by the properties of brain's anatomical structures [3, 4]. Figure 1 shows the structure of a vanilla RNN that is unconstrained by the brain's anatomical properties. In this study, we assumed that an RNN solves a sequential prediction task. An RNN receives a single value in every step and predicts the next single value. The input and output layers comprised single neurons, respectively. The hidden layer comprised 100 neurons. The input neuron was connected with 90 hidden neurons. The output neuron was connected with 10 hidden neurons. In the hidden layer, each neuron was also connected with all other neurons. The synaptic weights between neurons with connections were derived from the Gaussian distribution $N(0, 0.1)$ in the input and hidden layers. The synaptic weights from the hidden to output layers were initialized by the Glorot normal initializer [6]. In all layers, the synaptic weights between neurons without connections were set to 0. The activation functions of the input and output neurons were linear and that of the hidden neurons was hyperbolic tangent.

2.2 RNNs with Brain-Inspired Constraints

We introduced four constraints into the hidden layer of the vanilla RNN to define five models: Models 1–5 (Table 1). The initial structure of each model is defined according to the combination of constraints (Fig. 2). The first three constraints, neuron type, partial connection, and small world [7], were the brain's anatomical properties introduced in a previous study [3].

The neuron type constraint was introduced to all models to focus on the differences of E/I balance. The neuron type constraint introduced to Models 1, 2, 4, and 5 divides neurons into excitatory and inhibitory neurons in the hidden layer of the vanilla RNN (Figs. 2 (a, b, d, and e)). The ratio of excitatory neurons to inhibitory neurons was 8:2, which was similar to the ratio observed in the brain [8]. The synaptic weights from neurons were defined as the absolute

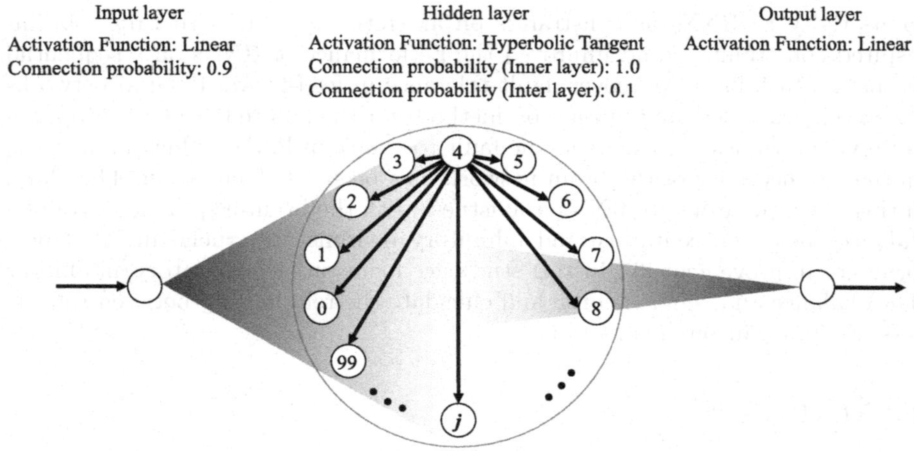

Input layer	Hidden layer	Output layer
Activation Function: Linear	Activation Function: Hyperbolic Tangent	Activation Function: Linear
Connection probability: 0.9	Connection probability (Inner layer): 1.0	
	Connection probability (Inter layer): 0.1	

Fig. 1. Structure of a vanilla RNN. Some parts of connections from fourth neuron in the hidden layer are illustrated.

values of the initial weights of the hidden neurons. Further, the signs of synaptic weights only from the inhibitory neurons were inverted. All synaptic weights from the inhibitory neurons were negative. In Model 3 (Fig. 2 (c)), although the ratio of positive weight (excitatory) connections to negative weight (inhibitory) connections was the same as that of Model 2, the signs of synaptic weights from a neuron was not uniform in Model 3.

A partial connection constraint introduced into Models 1, 4, and 5 removed the connections. An excitatory neuron was connected with 10% of excitatory neurons and 5% of inhibitory neurons, whereas an inhibitory neuron was connected with 20% of both the excitatory and inhibitory neurons (Figs. 2 (a, d, e)) [3,4]. In Models 1 and 4, the connections between the excitatory neurons were removed according to the distance between pre- and post-neurons in the hidden layer (Figs. 2 (a, d)). We assumed that excitatory neurons were positioned on a ring according to their neuron ID (Fig. 1).

The small world constraint introduced into Models 1, 4 rewired 10% connections between the excitatory neurons regardless of the distance (Fig. 2 (a, d)). A part of the excitatory connections became a long-distance connection. The network only among excitatory neurons organized the small world network as similar to the hippocampal CA3 in the brain [9].

Additionally, we introduce an E/I balance on a single neuron as a novel constraint. In Ref [3], it has been suggested that an global E/I balance on the number of connections is crucial in RNNs' performance improvement of RNNs. The global E/I balance brings the E/I balance on a single neuron at a certain level. Therefore, we focused on the local E/I balance on a single neuron. The E/I balance on the ith neuron can be calculated as follows:

$$B_i = w_i^{\text{inh}}/w_i^{\text{ex}}, \tag{1}$$

Table 1. The differences in constraints among Models 1–5.

Models	Brain-inspired initial constraints			
	Neuron type	Partial connection	Small world	Excitatory/inhibitory balance
Model 1	Yes			Balance 1
Model 2	Yes	No		Balance 2
Model 3	Yes (connection)	No		Balance 3
Model 4	Yes			Balance 2
Model 5	Yes		No	Balance 2

(a) Model 1   (b) Model 2   (c) Model 3

(d) Model 4   (e) Model 5

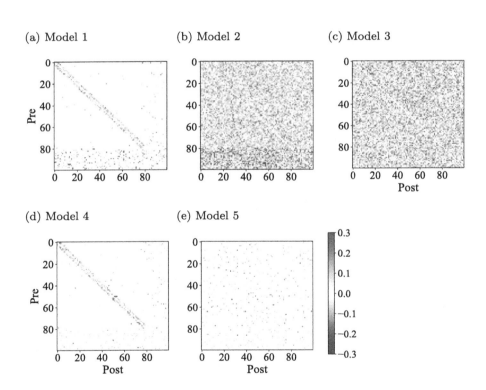

Fig. 2. Initial synaptic weights of hidden layer in Model 1–5.

where $w_i^{ex,\ inh}$ is the total weights of excitatory or inhibitory connections on the ith neuron. We defined the distribution of B_i obtained from Models 1, 2, and 3 as Balances 1, 2, and 3 (Fig. 3 (a, b, c)), respectively. In Models 4 and 5, the distribution of B_2, i.e., Balance 2 derived from the global E/I balance in the brain was reproduced by adjusting the synaptic weights of inhibitory connections to the ith neuron as follows (Fig. 4). The adjusting rate a_i of ith neuron is given by Eq. (2).

$$a_i = B_{\mathrm{rand}}^2 / B_i, \tag{2}$$

where B_{rand}^2 is a random value obtained from the Balance 2. The synaptic weights of inhibitory connections onto ith neuron were adjusted by a_i.

$$w_{ij}^{\mathrm{new}} = a_i w_{ij}^{\mathrm{old}}, \tag{3}$$

where w_{ij} is the synaptic weight from the jth neuron that is, i.e., a negative value.

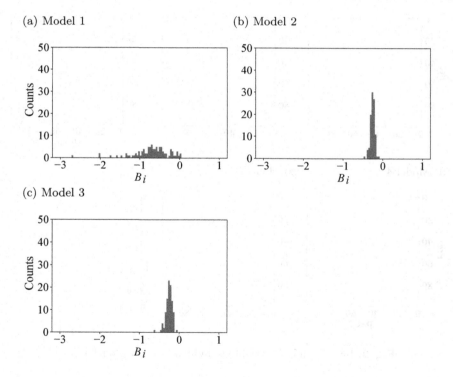

(a) Model 1   (b) Model 2

(c) Model 3

Fig. 3. E/I balance on a single neuron.

(a) Model 4   (b) Model 5

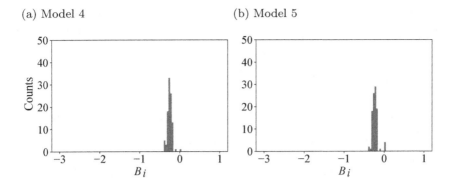

Fig. 4. Reproduction of the E/I balance of Model 2.

2.3 Timeseries Prediction Task

In Ref [3], a dataset that causes overfitting in vanilla RNNs was used. Similarly, we obtained five datasets from the logistic map ($x_0 = 0.8, a = 3.9$) for training and evaluation of RNN's performances. Each dataset comprised 200 training and 9971 test data. The training and test data were obtained from different parts of the logistic map. Each data comprised of 31 consecutive values obtained from the logistic map with stride 1. The last value was defined as teaching signal for the data. We defined the loss function as the mean squared error (MSE) between the predicted value from 30 values and the teaching signal. We used the Adam optimizer for training. An RNN was trained using each dataset for 1000 epochs 10 times. We obtained 50 RNNs for each model after training.

2.4 RNN Evaluation

We evaluated 50 trained RNNs using the two criteria prediction performance and robustness.

The prediction performance was defined as MSE between the RNN prediction and teaching signal in data. If RNNs in a model acquire a low MSE in several trials, the model has a high prediction performance. Robustness is defined as the modulation index (MI) of prediction errors for the test data before and after faults on a trained RNN as follows.

$$MI = \frac{E_{\text{after}} - E_{\text{before}}}{E_{\text{after}} + E_{\text{before}}}, \tag{4}$$

where E_{before} and E_{after} are the prediction errors for the test data before and after faults, respectively. The faults were implemented as the zero replacement of synaptic weights from selected neurons. Some neurons in the hidden layer were randomly selected according to the failure rate (1%–5%) and the synaptic weights of their output connections were replaced by 0. We selected the top 25 trained RNNs with high prediction performance from each model and induced fault four times for each selected RNN. We calculated MI from 100 RNNs after faults. If the RNNs in a model acquire a low MI in several trials, the model has the high robustness.

3   Results

3.1   Tradeoff Between Prediction Performance and Robustness

Figure 5 shows the MSEs for training and test data for Models 1–3. Models 1 and 3 had high prediction performance for the training data. However, the RNNs in Models 1 and 3 caused overfitting. The prediction error for the test data tended to be larger than that of the training data in Models 1 and 3. Meanwhile, RNNs in Model 2 reduced overfitting. The prediction error for the test data was at the same level with that of the training data. Consequently, Model 2 exhibited the best prediction performance by reducing overfitting, although the prediction performance for the training data was lower than those of Models 1 and 3.

Figure 6 shows the robustness of RNNs with errors for each model. When a RNN acquires high robustness, the prediction performance does not change before and after failure. The absolute MI tended to be small. Although MI increased with the increase in failure rate for all Models, MI tended to be larger in Models 2 and 3 than in Model 1. Therefore, there was a tradeoff between prediction performance and robustness in the models.

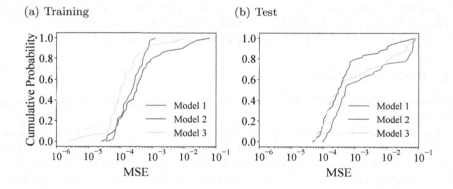

Fig. 5. Prediction performance for Models 1–3.

3.2   Overcoming Tradeoff Between Prediction Performance and Robustness

Figure 7 shows the MSEs for the training and test data for Models 4 and 5. Models 4 and 5 showed high prediction performance for training data at the same level with Model 1. Although the prediction error for test data decreased in Models 4 and 5, the change was moderate compared to Model 1 (Fig. 7 (b)). Models 4 and 5 also reduced overfitting. The difference in constraint between Models 4, 5 and Model 1 was the E/I balance constraint. Thus, the E/I balance on a single neuron based on the brain's anatomy contributed to reduce the

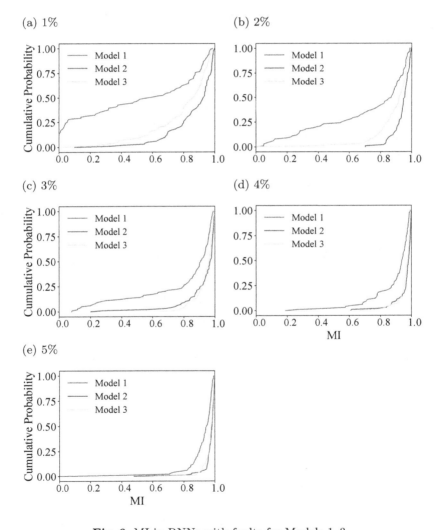

Fig. 6. MI in RNNs with faults for Models 1–3.

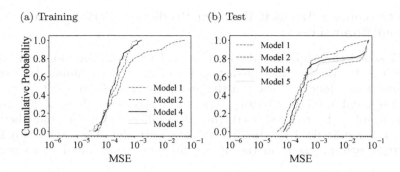

Fig. 7. Prediction performance for Models 4 and 5. The prediction performances for Models 1 and 2 shown in Fig. 5 are reprint.

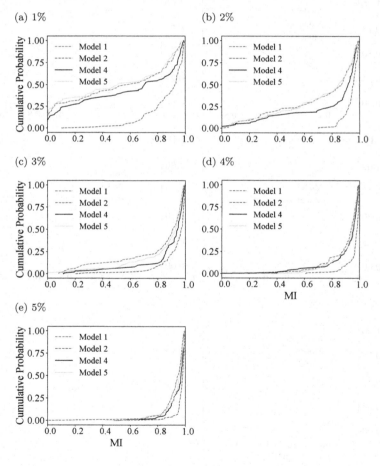

Fig. 8. Modulation indexes in RNNs with faults for Models 4 and 5. The modulation indexes for Models 1 and 2 shown in Fig. 6 are reprint.

overfitting. Figure 8 shows the robustness of RNNs for Models 4 and 5. The robustness for Models 4 and 5 were higher than that for Model 2 for 1% or 2% failure (Fig. 8 (a, b)). The robustness improved compared to Model 2 in all cases. The difference in the constraints between Models 2 and 5 was the presence/absence of the partial connection constraint. The RNNs in Model 2 had all-to-all connected structures in the hidden layer at the initial stage. Meanwhile, the RNNs in Model 5 had partially connected structures. The partial connection constraint was common among Models 1, 4, and 5, which showed high robustness. Therefore, the partial connection constraint contributed to the improvement of RNN's robustness. From the differences in constraints between Models 4 and 5, the small world constraint improved the prediction performance slightly for the test data when RNNs were not damaged (Fig. 7). However, the small world constraint tended to decrease the robustness (Fig. 8).

4 Conclusion

In this study, we introduced brain-inspired constraints to RNNs and evaluated their performance and robustness. Brain-inspired constraints in the previous studies could not overcome tradeoff between prediction performance and robustness. We proposed a novel constraint that is an E/I balance on a single neuron. We showed that the constraints that organize RNNs as partially connected structures with the E/I balance on a single neuron achieved a balance between prediction performance and robustness in RNNs. The partial connection constraint and E/I balance constraint contribute to the improvement of robustness and prediction performance, respectively.

Moreover, the small world constraint improved RNN predict performance without faults but decreased the robustness. In the brain, hub neurons are connected with each other and form a rich club [8]. A rich club constraint may balance between prediction performance and robustness in RNNs.

The brain-inspired constraints define RNN's initial structures. These constraints work as the initializer of RNNs. Although conventional initializing methods improved RNN prediction performance [10–13], the brain-inspired constraints can improve the prediction performance and the robustness. We introduced E/I balance constraint into Long short-term memory (LSTM) [14]. The LSTM with E/I balance improved the predict performance (data not shown). Furthermore, the robustness improvement is achieved by refining the learning algorithm but the repetitive computations during training increase the cost [5]. However, the initializing method is calculated at once at the initial. The brain-inspired constraints have the advantage of low computation cost. These mean that the proposed constraints have a potential of wide application.

Consequently, the brain-inspired constraints are RNN initializers to achieve a balance between performance and robustness of RNNs at a low computation cost.

Acknowledgment. A part of this study is supported by JSPS KAKENHI Grant Numbers 18K11527.

References

1. Moser, M.B., Moser, E.I., Forrest, E., Andersen, P., Morris, R.G.: Spatial learning with a minislab in the dorsal hippocampus. Proc. Natl. Acad. Sci. USA **92**, 9697–9701 (1995)
2. LeCun, Y.: Backpropagation applied to handwritten zip code recognition. Neural Comput. **1**, 541–551 (1989)
3. Fusauchi, T., Toshikazu, S.: Suppression of overfitting in a recurrent neural network by excitatory-inhibitory initializer. In: Proceedings of the 2019 International Symposium on Nonlinear Theory and its Applications (NOLTA 2019), pp. 196–199 (2019)
4. Samura, T., Fusauchi, T.: Improvement on performance of recurrent neural network through initializing of input and output structures similar to partial connection. In: Proceedings of NCSP 2021, pp. 345–348 (2021)
5. Torres-Huitzil, C., Girau, B.: Fault and error tolerance in neural networks: a review. IEEE Access. **5**, 17322–17341 (2017)
6. Glorot, X., Bengio, Y.: Understanding the difficulty of training deep feedforward neural networks. In: Proceedings of the Thirteenth International Conference on Artificial Intelligence and Statistics, PMLR, vol. 9, pp. 249–256 (2010)
7. Watts, D.J., Strogatz, S.H.: Collective dynamics of "small-world" networks. Nature **393**, 440–442 (1998)
8. Gal, E., et al.: Rich cell-type-specific network topology in neocortical microcircuitry. Nat. Neurosci. **20**, 1004–1013 (2017)
9. Takahashi, N., Sasaki, T., Matsumoto, W., Matsuki, N., Ikegaya, Y.: Circuit topology for synchronizing neurons in spontaneously active networks. Proc. Natl. Acad. Sci. USA **107**, 10244–10249 (2010)
10. Le, Q.V., Jaitly, N., Hinton, G.E.: A simple way to initialize recurrent networks of rectified linear units. arXiv:1504.00941 (2015)
11. Talathi, S.S., Vartak, A.: Improving performance of recurrent neural network with relu nonlinearity. arXiv:1511.03771 (2015)
12. Arjovsky, M., Shah, A., Bengio, Y.: Unitary evolution recurrent neural networks. arXiv:1511.06464 (2016)
13. Li, S., Li, W., Cook, C., Zhu, C., Gao, Y.: Independently recurrent neural network (IndRNN): building a longer and deeper RNN. arXiv:1803.04831 (2018)
14. Gers, F.A., Schmidhuber, J., Cummins, F.: Learning to forget: continual prediction with LSTM. In: 1999 Ninth International Conference on Artificial Neural Networks ICANN 1999 (Conf. Publ. No. 470), pp. 850–855 (1999)

Single-Image Smoker Detection by Human-Object Interaction with Post-refinement

Hua-Bao Ling[1,2] and Dong Huang[1,2(✉)]

[1] College of Mathematics and Informatics, South China Agricultural University, Guangzhou, China
hbling@stu.scau.edu.cn, huangdonghere@gmail.com
[2] Guangzhou Key Laboratory of Intelligent Agriculture, Guangzhou, China

Abstract. This paper addresses the problem of smoker detection in a single image. Previous smoker detection works usually focus on cigarette detection, yet often neglect the rich information of smoking behavior (especially the interaction information between smoker and cigarette). Though some attempts have been made to detect the smoking behavior, they typically rely on the temporal information in videos and are not suitable for single images. To tackle these issues, this paper proposes a novel smoker detection framework based on human-object interaction (HOI) and post-refinement. In particular, based on deep neural networks, we develop a one-stage HOI module to identify the interaction between smoker and cigarette, and exploit an additional fine-grained detector to further improve the HOI accuracy in the post-refinement module. Remarkably, we present a new benchmark dataset named SCAU smoker detection (SCAU-SD), which, to the best of our knowledge, is the first benchmark dataset for the specific task of smoker detection in single images with HOI annotations. Extensive experimental results demonstrate the superior single-image smoker detection performance of the proposed framework.

Keywords: Smoker detection · Object detection · Human-object interaction · Post-refinement

1 Introduction

Detecting smoking behaviors (or smokers) in public places has been a critical task due to the increasing concerns of public health and fire prevention. For dealing with this task, the recent development in computer vision has provided a variety of promising techniques, among which the object detection technique plays an important role [12,13].

Object detection is one of the most fundamental and challenging problems in computer vision [12,13]. In recent years, several attempts have been made to address the smoker detection problem by different object detection models [3,9,12,15]. Han et al. [9] proposed a cigarette detection method based on

© Springer Nature Switzerland AG 2021
T. Mantoro et al. (Eds.): ICONIP 2021, LNCS 13109, pp. 181–192, 2021.
https://doi.org/10.1007/978-3-030-92270-2_16

(a)   (b)

(c)   (d)

Fig. 1. Illustration of our smoker detection procedure. Red boxes indicate the smoker. Green boxes indicate the cigarette. Each colored line linking the smoker and the cigarette involved in the same color indicates an HOI class. Given an input image (a) and the conventional object detection results (b), our method can identify the interaction between smoker-cigarette pairs by the HOI module (c). Then a post-refinement module is exploited to refine the final result (d). (Color figure online)

Faster R-CNN [12], which takes into consideration both face detection and color segmentation. Chien et al. [3] used the YOLOv2 [11] model for driver's cigarette detection. Although they can achieve cigarette detection through object detectors, they didn't actually detect the smoking behaviors. For example, if there is a cigarette on the table, they may also identify it as smoking in their models [3,12]. Further, Wu and Chen [15] proposed a smoking behavior detection method based on the human face analysis. This method first locates the position of the mouth through the existing facial detector, and then utilizes white balance processing and HSV color space to segment the possible location of the cigarette. This method still cannot detect the action interaction of smoking behavior, and also fails to take into account the situation that the cigarette is sometimes in the hand and sometimes in the mouth. In view of the above issues, Danilchenko and Romanov [4] proposed to train two cigarette detectors to detect the cigarette in the hand and the cigarette in the mouth, respectively. Besides, Wu et al. [14] took into consideration the interaction between human and cigarette to better depict the smoking behavior. However, both of above two models [4,14] rely on the context information of the video, which require a larger computational cost and are not feasible for smoker detection in single images.

Human-object interaction(HOI) detection, as an emerging technique for behavior recognition with a deeper image semantic understanding, has received

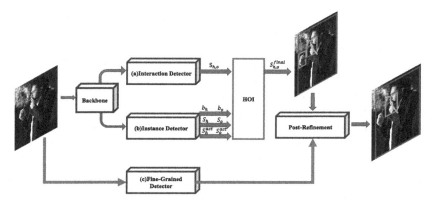

Fig. 2. Overview of the proposed framework. Our smoker detection model consists of three streams, namely, (a) the interaction detector for identifying the interaction between smoker and cigarette, (b) the instance detector for detecting human and object, which shares the same feature map from backbone with interaction detector, and (c) the fine-grained detector for locating cigarette in the post-refinement module.

increasing attention recently [2,6,7]. The purpose of HOI is to locate a person and an object, and identify the interaction between the person and the object (e.g., eating an apple, riding a bike, etc.). Conventional HOI methods typically employed a two-stage paradigm [2,6,7]. In the first stage, a pre-trained object detector localizes both humans and objects within single images, and obtains the answer of what is where. In the second stage, an action classification network with multi-stream architectures recognizes the interaction categories for each human-object pair, and obtains the answer of what is happening. Besides the two-stage structures, several one-stage HOI detection methods have also been proposed in the past two years. Liao et al. [10] reformulated HOI detection as a keypoint detection and matching problem and proposed a one-stage model by parallel point detection and matching. Fang et al. [5] proposed a new concept called interaction regions and thus devised a one-stage model by dense interaction region voting (DIRV). Despite the significant progress, on the one hand, these methods are often designed for generic HOI detection and cannot well address the specific task of smoker detection. On the other hand, these HOI detection methods often rely on one-stage interactive detection, and cannot integrate fine-grained information to further refine the detected interaction results.

In this paper, to deal with the above-mentioned issues, we propose a novel single-image smoker detection method based on human-object interaction and post refinement. Specifically, a trainable human-object interaction module is presented to learn and identify the interaction region for smoker detection. Besides, an accurate and effective cigarette detector for the more accurate location of cigarette detection is utilized in the post-refinement module. To the best of our knowledge, this work is the first attempt to apply the human-object interaction to smoker detection in single images. As shown in Fig. 1, given an image, our proposed framework can effectively detect the smoking behavior with both the human-object interaction and the post-refinement considered. Furthermore, in the field of HOI detection, the existing datasets, such as HICO-DET [2] and V-COCO [8], are mostly designed for general HOI detection, and not suitable for the specific smoker detection task. In light of this, this paper introduces a new dataset termed SCAU Smoker Detection (SCAU-SD), which is the first benchmark dataset for the specific smoker detection task with HOI annotations. To summarize, the main contributions of this paper are listed as follows:

1. This paper for the first time exploits the human-object interaction for smoker detection in single images.
2. This paper presents an effective post-refinement strategy to incorporate fine-grained information for further improving the smoker detection accuracy.
3. A new benchmark dataset with HOI annotations is presented for the single-image smoker detection task.

The rest of the paper is organized as follows. Section 2 describes the proposed smoker detection framework. Section 3 introduces the SCAU-SD dataset. Section 4 reports the evaluation analysis and ablation study, followed by the conclusion in Sect. 5.

2   Proposed Framework

In this section, we introduce the proposed smoker detection framework. An overview of the framework is given in Sect. 2.1. Then, the human-object interaction module is described in Sect. 2.2. Next, the post-refinement module is developed in Sect. 2.3. Finally, the inference procedure is provided in Sect. 2.4.

2.1   Framework Overview

In this paper, the smoker detection aims to identify the HOI categories, which are based the smoker boxes, the cigarette boxes and the interaction class between the smoker-cigarette pairs. We break up this task into three steps: (i) person/cigarette detection, (ii) HOI interactive recognition, and (iii) post-refinement. Particularly, given the dataset, we first adopt the EfficientDet [13] to train the person/cigarette object detectors. Let b_h denote the detected bounding box for a person and b_o denote the detected bounding box for a cigarette instance. For HOI detection, the HOI classification scores $S_{h,o}^{final}$ can be achieved by the dense interaction region voting [5]. To further improve the accuracy of the HOI detection, a post-refinement mechanism is devised based on an additional fine-grained detector. An overview of the proposed framework is illustrated in Fig. 2.

2.2   Person-Cigarette Interaction Module

To identify the person-cigarette interaction, the first task is to detect human and objects with high accuracy. In this paper, we adopt EfficientDet-D4 [13] as the backbone of our HOI module. The backbone is trained on our SCAU-SD dataset (which will be described in Sect. 3), and loaded into the interactive recognition network as pre-trained weight, which is frozen during training. To identify the interactive action, we use a one-stage HOI detection model dense interaction region voting [5] as our main structure.

Our HOI module has two parallel branches, i.e., an instance detector and an interaction detector. On the one hand, the instance detector is responsible for the person/cigarette detection, which reuses the features extracted by Efficient-Det and thus its computation is lightweight. Specifically, given an input image, the instance detector outputs a set of human boxes (i.e., b_h), or cigarette boxes (i.e., b_o) and a class score for each box of human or object (denoted as s_h and s_o, respectively). In addition, based on the features of instance appearance, the instance detector also outputs the action scores of humans and objects (denoted as s_h^{act} and s_o^{act}, respectively), as additional information for the association of person-cigarette pair. On the other hand, the interaction detector is responsible for the interactive action recognition, which shares the same feature map from the EfficientDet backbone. In this branch, it directly outputs the interaction score (i.e., $s_{h,o}$) of each person-cigarette pair based on the elaborate visual features in interaction regions and a voting strategy [5].

2.3   Post-refinement Module

The HOI module mainly focuses on the interaction between human and object, yet lack the ability to exploit additional fine-grained detection to refine the HOI result. It is noteworthy that, instead of detecting a single cigarette, our instance detector typically detect the instance of cigarette-in-hand and that of cigarette-in-mouth. Though the instance detector of cigarette-in-hand or cigarette-in-mouth is helpful in the HOI module (as it takes the spatial context of the cigarette into account), we also empirically find that the fine-grained single-cigarette information can be exploited to further refine the smoker detection result *after* the HOI module, which is in fact the main motivation of our post-refinement module.

To train an accurate and efficient cigarette detector to post-refine the HOI outputs, we first annotate the cigarette in single images from the SCAU-SD training set. Then we train the fine-grained detector and evaluate its effectiveness. Through empirical comparison, we observe that the YOLOv4 [1] model achieves higher accuracy than the EfficientDet-D4 in our task. More experimental details can be found in Sect. 4.5. Based on empirical comparison, we adopt the state-of-the-art YOLOv4 as the main model in the post-refinement stage. Given the input image, the fine-grained detector outputs the cigarette box (denoted as box_1), and we indicate that the predicted cigarette (associated with cigarette-in-hand or cigarette-in-mouth) box from person-cigarette interaction detection results as box_2. The final identified category C_{final} is decided as follows:

$$C_{final} = \begin{cases} True\,Positive, & \text{if } box_1 \cap box_2 > 0, \\ False\,Positive, & \text{otherwise.} \end{cases} \tag{1}$$

When the cigarette box of the HOI-predicted result is estimated to be false positive, we will choose the next prediction result of the top-3 predictions and repeat the above decision process. Due to the high accuracy of the fine-grained detector, the post-refinement module can effectively reduce false detections, and locate the area of true-positive cigarette, which is also be demonstrated by our ablation analysis (as shown in Table 3).

2.4   Inference Procedure

Based on the person-cigarette interaction detection module, we predict scores of action classification in a similar manner to other HOI models. Consequently, the final interactive classification scores $S_{h,o}^{final}$ for a person-cigarette pair (b_h, b_o) is represented as:

$$S_{h,o}^{final} = s_h \cdot s_o \cdot (s_h^{act} + s_o^{act}) \cdot s_{h,o} \tag{2}$$

Where s_h, s_o, s_h^{act}, s_o^{act}, $s_{h,o}$ have been introduced in Sect. 2.2. For our task, all the action classes correspond to the persons and the cigarette objects, so there is no need to consider the classes which do not involve any objects (e.g. walking, smile, etc.).

In the training process, the design of the loss is similar to that in [5], which is the sum of loss functions from several branches. The final loss L_{final} is computed as:

$$L_{final} = L_{reg,h} + L_{reg,o} + L_{cls}^{inter} + L_{cls}^{inst} \tag{3}$$

Where $L_{reg,h}$ and $L_{reg,o}$ denote the smooth L_1 losses for human and object branches, respectively, and L_{cls}^{inter} denotes the ignorance loss [5]. Besides, L_{cls}^{inst} is the standard binary cross-entropy loss function for the instance detector branch.

3   SCAU-SD Dataset

In this section, we present the SCAU-SD dataset for the smoker detection task. The construction of the dataset is first described in Sect. 3.1. Then the statistics of the dataset is provided in Sect. 3.2.

3.1   Dataset Construction

In this paper, we build a new smoker detection dataset named SCAU-SD with HOI annotations to facilitate the research of this field. The details of the constructing the SCAU-SD dataset will be described below.

Table 1. Sample annotations of our SCAU-SD dataset.

In hand and mouth	In hand	In mouth

* Red and green boxes indicate smoker and cigarette, respectively. Each linking line denotes the smoker-cigarette pair and each column denotes the interactive action.

Data Collection and Division. We collect smoking images from three sources. First, we randomly select 232 smoking images from an open-source HOI dataset called HOI-A [10]. Second, we collect 581 smoking images from the Kaggle dataset[1]. Third, we have crawled 747 smoking images from the Internet. Thereby, our final SCAU-SD dataset contains 1560 smoking images. Then we randomly select 360 images as the test set, leaving the rest as the training set. There is no overlap between the training set and the test set. As shown in Table 1, we exhibit some samples of the SCAU-SD dataset. Note that the SCAU-SD dataset has a variety of different scenarios, which reveal challenges of various human poses, illumination situations, incomplete objects, and complicated backgrounds of the real scenes. In addition, as shown in Table 1, we further divide the smoking behavior into three sub-categories, which include cigarette-in-hand-and-mouth, cigarette-in-hand, and cigarette-in-mouth.

Data Annotation. The annotation task for smoker detection with HOI, is not as trivial as drawing bounding boxes around the smoker and the cigarette. We also need to link up their relations. Therefore, the process of annotation consists of two steps, i.e., box annotation and interaction annotation. First, we define two categories of objects, i.e., person and cigarette. All objects in the pre-defined categories need to be annotated with a box and the corresponding label. For the annotation cigarette, we include the additional region of hand or mouth/nose around the cigarette, which serves as spatial context to help identify the cigarette. The experimental results show that this strategy is effective compared to the area where only cigarettes are marked. Second, we annotate the interaction label between each person-cigarette pair. As show in Table 1, colored lines are used to connect the interaction between person and cigarette, and the corresponding interaction labels are added to corresponding annotation files.

---

[1] https://www.kaggle.com/vitaminc/cigarette-smoker-detection.

3.2   Data Statistics

To our knowledge, the SCAU-SD dataset is the first benchmark for smoker detection with HOI annotations. It contains 1560 annotated smoking images, 2 kinds of objects and 3 action categories. It is divided into a training set with 1200 images and a test set with 360 images. Table 2 shows the statistics of the division of the dataset. Different from other interactive actions, for this task, we only consider the case where there is a person interacting with a cigarette in single images.

Table 2. Statistics of our SCAU-SD dataset.

	#images
Train	1200
Test	360
Total	1560

4   Experiments

In this section, we conduct experiments on the SCAU-SD dataset to evaluate our proposed approach for smoker detection. First, we describe the evaluation metric in Sect. 4.1 and implementation details in Sect. 4.2. Then, we report the results by quantitative and qualitative analysis in Sect. 4.3 and 4.4, respectively. Finally, we analyze the contributions of different components to the final performance in Sect. 4.5.

4.1   Evaluation Metric

For smoker detection in single images, the goal is to detect pairs of a person and a cigarette with an interaction class label connecting them. Following the standard evaluation metric for object detection and human-object interaction, we use the mean average precision (mAP) as the evaluation metric. Similar to other HOI tasks, a prediction is judged as a true positive if two conditions are satisfied, i.e., (i) the person-cigarette interaction recognition result is correct and (ii) the bounding boxes of the person and the cigarette both have the IoUs that exceed 0.5 with ground truth. Formally, it can be expressed as $min(IoU_h, IoU_o) > 0.5$.

4.2   Implementation Details

For HOI detection, we use EfficientDet-D4 [13] as our backbone, which loads a pre-trained weight that is trained on the training set of SCAU-SD. We apply Adam optimizer to optimize the loss function. In addition, we use a learning rate decay strategy that starts with an initialized learning rate of 1e-3, and reduces the learning rate when the loss stop downing for 5 consecutive epochs.

Table 3. Comparison results of our model with(w) or without(w/o) post-refinement.

	Model w/o post-refinement		Model w post-refinement	
	AP (%)	Recall (%)	AP (%)	Recall (%)
In hand and mouth	63.81	69.17	**66.39**	**71.67**
In hand	68.28	77.50	**75.24**	**81.67**
In mouth	63.92	73.33	**67.03**	**75.83**
Mean	65.33	73.33	**69.55**	**76.39**

We train the network for 9000 iterations on the training set, which takes 4 h on a single NVIDIA RTX2080Ti GPU. As for the post-refinement module, the YOLOv4 [1] is used for the fine-grained cigarette detection. We apply mosaic data augmentation approach and cosine decay strategy for learning rate on the SCAU-SD training set.

4.3 Quantitative Analysis

This section reports the quantitative results in terms of *AP* and *Recall* on the SCAU-SD test set. The person-cigarette interaction detection is based on the dense interaction region voting [5] and the result of the experiment is shown in Table 3. As shown in Table 3, our model without post-refinement has an mAP of 65.33 and a mean recall of 73.33. By adding the post-refinement module, our model achieves an mAP of 69.55 and a mean recall of 76.39. It can be observed that the post-refinement module is effective for improving the smoker detection performance.

4.4 Qualitative Analysis

In this section, we visualize the smoker detection results of several image samples on the SCAU-SD test set. We highlight the detected smoker and cigarette with red and green bounding boxes, respectively. Besides, we use the colored line to connect the center points of the person and the interacted cigarette. The prediction of interactive category is shown in the upper left corner of each person's box. As shown in Table 4, our model can accurately locate the bounding boxes of person and cigarette, and identify the interactive categories. In these examples, our proposed method can deal with various situations very well. For instance, our model can accurately identify the interactive action category in the dark scene of Table 4(c), the complicated background of Table 4(a) and Table 4(d), the incomplete person of Table 4(b), the low-resolution image of Table 4(g), and the illumination variance of Table 4(f) and Table 4(h).

4.5 Ablation Study

In this section, we analyze the contributions of different components to the final performance.

Table 4. Visualization results for smoker detection based on our method. (**Class 1**: cigarette in hand and mouth. **Class 2**: cigarette in hand. **Class 3**: cigarette in mouth.)

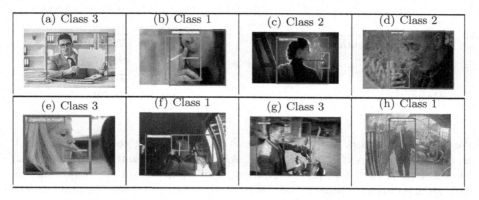

(a) Class 3	(b) Class 1	(c) Class 2	(d) Class 2
(e) Class 3	(f) Class 1	(g) Class 3	(h) Class 1

Table 5. Ablation study for backbone.

	Params	mAP (%)
EfficientDet-D0	3.9M	56.8
EfficientDet-D1	6.6M	60.8
EfficientDet-D2	8.1M	70.0
EfficientDet-D3	12M	85.1
EfficientDet-D4	21M	**88.0**

Backbone. We first perform an ablation analysis on the backbone of person/cigarette object detection. Experimental comparison with EfficientDet-D0, D1, D2, D3 and D4 [13] is carried out. As shown in Table 5, EfficientDet-D0 only obtains an mAP of 56.8, while EfficientDet-D4 achieves an mAp of 88.0. For the requirement of the accuracy of object location, in this paper, we adopt the EfficientDet-D4 as the backbone in our model.

Annotated Area of Cigarette. As mentioned in Sect. 3.1, we add additional mouth or hand/nose area around the cigarette in annotation. To verify the effectiveness of this strategy, we design a comparative experiment, whose results are shown in Table 6. In Table 6, the contextual box refers to the bounding box around cigarette with hand or mouth/nose, while the tight box refers to the bounding box around the exact area of the cigarette. According to the results in Table 6, our model with the contextual box performs significantly better than that with tight box.

Post-refinement. In terms of the post-refinement module mentioned in Sect. 2.3, we compare our basic model without post-refinement against that with

Table 6. Ablation study for annotated area of cigarette. (**contextual box**: the bounding box around cigarette with the area of hand or mouth/nose. **tight box**: the exact bounding box of the cigarette.

Method	mAP (%)
Our HOI model with contextual box	**69.55**
Our HOI model with tight box	53.35

Table 7. Ablation study for post-refinement module. (**Our HOI model with E**: refers to the HOI model with post-refinement by EfficientDet-D4 cigarette detector. **Our HOI model with Y**: refers to the HOI model with post-refinement by YOLOv4 cigarette detector.

	Post-refinement	mAP (%)
(a) Our HOI model	✗	65.33
(b) Our HOI model with E	✓	66.88
(c) Our HOI model with Y	✓	**69.55**

the post-refinement module on the SCAU-SD test set. In the post-refinement module, two types of cigarette detectors, based on EfficientDet-D4 and YOLOv4, respectively, are tested. As reported in Table 7, the model with the post-refinement module can improve the performance from 65.33% mAP to 66.88% mAP, which demonstrates the effectiveness of the post-refinement module. In addition, we further investigate different object detectors in the post-refinement module, which shows that YOLOv4 performs better than EfficientDet-D4 in post-refinement.

5   Conclusion

In this paper, we present a new single-image smoker detection framework that consists of a human-object interaction module and a post-refinement module. Specifically, an HOI module is exploited to locate and recognize the interactions between smoker-cigarette pairs. After that, we take advantage of a fine-grained detector in the post-refinement module to refine the HOI outputs, which further improves the smoker detection performance. Furthermore, in this paper, we present the first single-image smoker detection dataset with HOI annotations, which is termed SCAU-SD. Extensive experimental results have demonstrated the superiority of the proposed framework in detecting the smoking behavior in single images.

Acknowledgement. This work was supported by the Project of Guangzhou Key Laboratory of Intelligent Agriculture (201902010081), the NSFC (61976097), and the Natural Science Foundation of Guangdong Province (2021A1515012203).

References

1. Bochkovskiy, A., Wang, C.Y., Liao, H.Y.M.: YOLOv4: optimal speed and accuracy of object detection. arXiv preprint arXiv:2004.10934 (2020)
2. Chao, Y.W., Liu, Y., Liu, X., Zeng, H., Deng, J.: Learning to detect human-object interactions. In: Proceedings of the IEEE Winter Conference on Applications of Computer Vision, pp. 381–389 (2018)
3. Chien, T.C., Lin, C.C., Fan, C.P.: Deep learning based driver smoking behavior detection for driving safety. J. Image Graph., 15–20 (2020)
4. Danilchenko, P., Romanov, N.: Neural networks application to detect the facts of smoking in video surveillance systems. J. Phys., 1794 (2021)
5. Fang, H.S., Xie, Y., Shao, D., Lu, C.: DIRV: dense interaction region voting for end-to-end human-object interaction detection. In: Proceedings of the AAAI Conference on Artificial Intelligence, pp. 1291–1299 (2021)
6. Gao, C., Zou, Y., Huang, J.B.: iCAN: instance-centric attention network for human-object interaction detection. In: Proceedings of the British Machine Vision Conference, p. 41 (2018)
7. Gkioxari, G., Girshick, R., Dollár, P., He, K.: Detecting and recognizing human-object interactions. In: Proceedings of the IEEE Conference on Computer Vision and Pattern Recognition, pp. 8359–8367 (2018)
8. Gupta, S., Malik, J.: Visual Semantic Role Labeling. arXiv preprint arXiv:1505.04474 (2015)
9. Han, G., Li, Q., Zhou, Y., Duan, J.: Rapid cigarette detection based on faster R-CNN. In: Proceedings of the IEEE Symposium Series on Computational Intelligence, pp. 2759–2765 (2019)
10. Liao, Y., Liu, S., Wang, F., Chen, Y., Qian, C., Feng, J.: PPDM: parallel point detection and matching for real-time human-object interaction detection. In: Proceedings of the IEEE Conference on Computer Vision and Pattern Recognition, pp. 482–490 (2020)
11. Redmon, J., Farhadi, A.: YOLO9000: better, faster, stronger. In: Proceedings of the IEEE Conference on Computer Vision and Pattern Recognition, pp. 7263–7271 (2017)
12. Ren, S., He, K., Girshick, R., Sun, J.: Faster R-CNN: towards real-time object detection with region proposal networks. In: Proceedings of the Annual Conference on Neural Information Processing Systems, pp. 91–99 (2015)
13. Tan, M., Pang, R., Le, Q.V.: EfficientDet: scalable and efficient object detection. In: Proceedings of the IEEE Conference on Computer Vision and Pattern Recognition, pp. 10781–10790 (2020)
14. Wu, P., Hsieh, J.W., Cheng, J.C., Cheng, S.C., Tseng, S.Y.: Human smoking event detection using visual interaction clues. In: Proceedings of the International Conference on Pattern Recognition, pp. 4344–4347 (2010)
15. Wu, W.C., Chen, C.Y.: Detection system of smoking behavior based on face analysis. In: Proceedings of the IEEE on International Conference on Genetic and Evolutionary Computing, pp. 184–187 (2011)

A Lightweight Multi-scale Feature Fusion Network for Real-Time Semantic Segmentation

Tanmay Singha[1]([✉]), Duc-Son Pham[1], Aneesh Krishna[1], and Tom Gedeon[2]

[1] School of Electrical Engineering, Computing, and Mathematical Sciences, Curtin University, Bentley, WA 6102, Australia
tanmay.singha@postgrad.curtin.edu.au
[2] Research School of Computer Science, The Australian National University, Canberra, Australia

Abstract. Recently, semantic segmentation has become an emerging research area in computer vision due to a strong demand for autonomous vehicles, robotics, video surveillance, and medical image processing. To address this demand, several real-time semantic segmentation models have been introduced. Relying on existing Deep Convolution Neural networks (DCNNs), these models extract contextual features from the input image and construct the output at the decoder end by simply fusing deep features with shallow features which causes a large semantic gap. However, this large gap causes boundary degeneration and noisy feature effects in the output. To address this issue, we propose a novel architecture, called Feature Scaling Feature Fusion Network (FSFFNet) which alleviates the gap by successively fusing features at consecutive levels in multiple directions. For better dense pixel-level representation, we also employ a feature scaling technique which helps the model assimilate more contextual information from the global features and improves model performance. Our proposed model achieves 71.8% validation accuracy (mIoU) on the Cityscapes dataset whilst having only 1.3M parameters.

Keywords: Semantic segmentation · Feature scaling · Feature fusion · Deep learning · Deep neural networks · Real-time applications

1 Introduction

Semantic scene segmentation is an important task in many applications such as medical image processing, autonomous vehicles, and damage detection. Previous studies [5,7–9,17] have shown the wide application of Deep Convolutional Neural Networks (DCNNs) in different computer vision tasks. However, semantic segmentation is still a challenging task, partly due to objects of various scales in a complex scene. Whilst filters of varying sizes can be used to create multiple receptive fields to process these objects, they can lead to an exponential growth of parameters and computational cost. To address this issue, PSPNet

© Springer Nature Switzerland AG 2021
T. Mantoro et al. (Eds.): ICONIP 2021, LNCS 13109, pp. 193–205, 2021.
https://doi.org/10.1007/978-3-030-92270-2_17

[22] introduced a Pyramid Pooling Method (PPM) in which multiple parallel pooling branches with different pool sizes and strides are deployed in order to capture multi-scale contextual information from the scene. Although, multiple pooling branches create receptive fields of different sizes, the pooling operation causes a loss of neighboring information of each object in the scene. To address this problem, the authors of [2] presented a new approach, called Atrous Spatial Pyramid Pooling (ASPP). It uses multiple dilated convolution branches with different dilation rates for multiple receptive fields. A higher dilation rate enlarges the field of view without contributing any extra parameters and GFLOPs. However, it uses dilated convolutions which are sensitive to input image resolution. Moreover, getting a trade-off between dilation rates and input size is a challenging task in ASPP. Similar to feature scaling, object positioning is also an important factor for better scene representation. The literature has shown that global features, produced by the encoder, are highly sensitive to the entire objects whereas local features mainly focus on the boundaries and edges of the object [2, 22]. Therefore, fusing local feature with rich global feature is essential for accurate object localization. Existing off-line [2,3,22] and real-time [9,11–13,16,21] semantic segmentation models mainly focus on feature extraction and contextual representation. For object localization, deep features at low resolution are simply fused with shallow features at higher resolution, creating a large semantic gap between the feature maps. This gap produces semantic inconsistency due to the background noisy features. The study [10] also shows that the fusing of global and local features directly is less effective. Therefore, our work introduces an optimised multi-stage feature fusion module at the decoder side for better object localization. Our key contributions are as follows:

- We design a lightweight backbone using MobileNetV2 residual blocks, capable of handling high-resolution input images in real-time environments;
- We introduce a multi-scale Feature Scaling Module (FSM), inspired by [2], and obtain the best trade-off between input size and dilation rates;
- We introduce a multi-stage Feature Fusion Module (FFM) to bridge the semantic gap and improve semantic performance; and
- The proposed model produces state-of-the-art results on Cityscapes among all the real-time models having less than 5 million parameters.

2   Related Work

Traditionally, a semantic segmentation design typically follows an image pyramid structure, which is inefficient for real-time applications as it increases training and inference time. Later on, [5,9,20] introduce an encoder-decoder architecture which involves both a pyramid structure to create semantic features and upsampling layers to produce segmentation. However, many of these models, for example DeepLab [1,2], PSPNet [22], HANet [3], are not suitable for real-time applications as they use a large encoder, such as ResNet [18].

To address real-time requirements, several approaches (Bayesian SegNet [5], ENet [11], ICNet [21], BiSeNet [20], DFANet [6]) have been proposed using a simpler variant of ResNet, but their parameters and GFLOP counts of these models are still high. More recent models such as ContextNet [12], FAST-SCNN [13], FANet [15], ESPNet [16] have achieved further reduction by using MobileNet bottleneck residual block (MBConv) whilst still maintaining good segmentation performance.

Feature Scaling. Using multiple scales is necessary for better contextual representation in semantic segmentation [2, 4, 22]. There are three notable approaches to feature scaling: 1)PPM [22] achieves better contextualization but lacks fine details and is still expensive; 2) DMNet [4] provides a dynamic scaling whilst being expensive at high resolutions; and 3) ASPP [2] provides a robust scaling and more controllability through varying dilation rates.

Feature Fusion. Traditionally, high-level features are upsampled and then fused with lower-level features in the deconvolution process [7]. Many offline [2, 3] and some real-time [6, 9, 12, 13, 21] semantic segmentation models skipped the intermediate stages and upsampled semantic features directly by between 2^3 to 2^5 times, which causes a large semantic gap while fusing features and loses object localization. To address this issue, PAN [8] has introduced a new bottom-up path for accurate signal propagation from lower layers to higher layers for instance segmentation, and this bi-directional propagation has been utilised in FANet [15] and DSMRSeg [19].

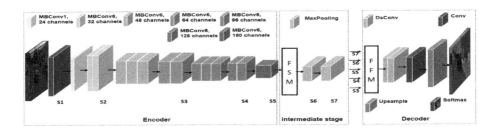

Fig. 1. Complete architecture of the proposed model

3   Proposed Method

Figure 1 displays the complete pipeline of our proposed model. Our work addresses the above challenges by appropriately using multi-scale feature representation and multi-stage feature fusion together with a slim backbone.

3.1   Network Architecture

Encoder. In our design, we exploit MobileNetV2 bottleneck residual blocks (MBConv) to design the backbone of our proposed model as they are much more efficient than other residual blocks. The layer architecture of the encoder network is shown in Table 1. Two types of MBConv blocks are used- MBConv1 with expansion ratio 1 and MBConv6 with expansion ratio 6.

While passing through a typical bottleneck architecture, the channel c of the feature gets expanded based on the expansion ratio t and becomes tc. To reduce the complexity, we employ a Depth-wise Convolution (DwConv) layer at the expansion stage. The layout of the bottleneck residual block is exhibited in Table 2. Variables h, w denote the spatial dimensions of input feature map and s represents stride. ReLU non-linearity is deployed in the first two layers, however it is skipped at the last stage of each MBConv block to preserve meaningful information of the input feature map.

Table 1. Layer architecture of encoder

Stage (i)	Input	Operators	Stride	Layers (n)	Output
1	$1024 \times 2048 \times 3$	Conv, k3 $\times$ 3	2	1	$512 \times 1024 \times 32$
2	$512 \times 1024 \times 32$	MBConv1, k3 $\times$ 3	2	1	$256 \times 512 \times 24$
	$256 \times 512 \times 24$	MBConv6, k3 $\times$ 3	1	2	$256 \times 512 \times 32$
3	$256 \times 512 \times 32$	MBConv6, k3 $\times$ 3	2	3	$128 \times 256 \times 48$
	$128 \times 256 \times 48$	MBConv6, k3 $\times$ 3	1	2	$128 \times 256 \times 64$
4	$128 \times 256 \times 64$	MBConv6, k3 $\times$ 3	2	3	$64 \times 128 \times 96$
	$64 \times 128 \times 96$	MBConv6, k3 $\times$ 3	1	2	$64 \times 128 \times 128$
5	$64 \times 128 \times 128$	MBConv6, k3 $\times$ 3	2	1	$32 \times 64 \times 160$

Table 2. Bottleneck residual block

Input	Operator	Output
$h \times w \times c$	1×1 Conv,1/1, Relu	$h \times w \times tc$
$h \times w \times tc$	3×3 DwConv, 3/s, Relu	$h/s \times w/s \times tc$
$h/s \times w/s \times tc$	1×1 Conv,1/1, -	$h/s \times w/s \times c'$

We employ 14 MBConv blocks to design the backbone. The filter size of each block is controlled by a tunable hyper-parameter, called width multiplier. Following the suggestion in [14], the width multiplier is set between 0.35 and 0.5 to obtain a better trade-off between the model's accuracy and performance. The encoder processes the input through 5 stages, each stage reduces the input feature by half. It is suggested in [2] that an output stride of 2^3 or 2^4 is optimal for an input of 512×512px. As we target a higher resolution (1204×2048), we have found that an output stride of 2^5 provides a better trade-off between model accuracy and performance whilst not overlooking small objects. The complete layout of each stage is displayed in Table 1.

Intermediate Stage. This stage addresses multi-scale representation, which is crucial for complex scene analysis [2,4,22]. Motivated by ASPP [2], we develop a Feature Scaling Module (FSM) targeting real-time applications. Our FSM employs three scaling branches with different dilation rates and one feature pooling branch. In contrast to ASPP, we use Depth-wise Separable Convolution (DsConv) in each branch. The dilation rate in each branch is sensitive to the input size. At smaller dilation rates, the size of the receptive field is small and it takes more number of operations to filter the input, whereas a higher dilation rate enlarges model's field-of-view whilst potentially causing artifacts. To obtain the best trade-off among the dilation rates and input sizes, we conducted an ablation study which shows that a dilation rate of {8,16,24} gives the best result for input of size 1024×2048. The layout of each branch is shown in Table 3. At the end of FSM, we concatenate all four branches with the input feature. After FSM, we deploy two successive MaxPooling layers to create two additional stages for feature fusion. The MaxPooling layer does not contribute any parameters, hence the model's real-time performance is not hampered.

Table 3. Layer architecture of FSM

Branch	Input	Operator	Filter	Dilation rate	Output
Dilated branch 1	$h \times w \times c$	DsConv, Bn, f	3×3	r_1	$h \times w \times c'$
Dilated branch 2	$h \times w \times c$	DsConv, BN, f	3×3	r_2	$h \times w \times c'$
Dilated branch 3	$h \times w \times c$	DsConv, BN, f	3×3	r_3	$h \times w \times c'$
Feature pooling	$h \times w \times c$	AveragePooling2D Conv, BN, f UpSampling2D	1×1	-	$h \times w \times c'$

Decoder. The proposed decoder has two modules: multi-stage feature fusion (FFM) and classifier. FFM is required for identifying the region and localizing the objects in the scene. In a pyramid encoder design, neurons at top levels strongly respond to entire objects while neurons at lower level more likely capture local texture and patterns. Motivated by this idea, we introduce an effective multi-stage feature fusion module at the decoder side. It takes five rich semantic features from five different levels and fuses it through three different paths. The operation of FFM is illustrated in Fig. 2.

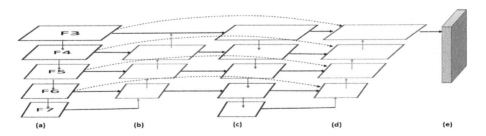

Fig. 2. Multi-stage Feature Fusion Module: (a) Features F_3-F_7 generated by encoder, (b) Top-down path for feature fusion, (c) Bottom-up path for object localization, (d) Top-down path for better contextual assimilation. Dotted lines mean skip connections from the encoder.

Traditionally, a high-level rich semantic feature map F_i $(C_i \times H_i \times W_i)$ is up-sampled and fused with the former low-level features F_{i-1} $(C_{i-1} \times H_{i-1} \times W_{i-1})$ [5,7] to regenerate the scene. This one direction (top-down) FFM may lead to localization issues in the scene. Moreover, each upsampling method contributes to the loss of neighboring information. This phenomenon clearly manifests the necessity of deploying another bottom-up path for accurate object localization in the entire feature hierarchy and also the need for lateral connections in the encoder to prevent the loss of neighboring details. For that reason, we introduce a new bottom-up path where local feature maps are fused with global features in order to achieve better localization. Finally, we introduce another top-down path for better contextual assimilation using some skip connections from the different stages of the encoder (dotted lines in Fig. 2).

Finally, we deploy a simple, yet effective classifier consisting of two DsConv, one point-wise Conv, one Upsampling and one softmax layer. The activation function softmax is used to assign a class to each pixel of the image. We also use one Dropout layer to avoid overfitting.

4   Experiment

4.1   Datasets

Cityscapes. It is a large-scale data set for semantic understanding of urban street scenes. It provides 5,000 fine-tune and 20,000 coarse annotated images. The fine-tune images are divided into three parts: a training set with 2,975 images, a validation set with 500 images, and a test set with 1,525 images, all at 1024×2048. This data set has 33 classes, 19 of which are used for training.

BDD100K. It is a recent data set developed to meet the growing demand in the field of autonomous car industry. It is the largest driving video dataset with 100K videos. Compared to Cityscapes, it is more challenging due to its diverse nature. It provides 8,000 fine-grained, pixel-level annotations, 7,000 of which are used for training and 1,000 for validation. The class labelling of this benchmark is compatible with Cityscapes (see our Github for further detail). Each image in this dataset has 720×1280 pixels.

4.2   Implementation Details

All our experiments are conducted in a dual Nvidia TITAN RTX GPUs system, each GPU having 24 GB of memory. Our environment includes CUDA 10.2, tensorflow 2.1.0, keras 2.3.1., and Horovod. For training, we set a batch size of 2 for full input resolution and 4 for low input resolution. For model optimizer, we employ stochastic gradient decent (SGD) with a momentum of 0.9. Following [13,22], we use the 'poly' learning rate policy which computes the current learning rate ($LR_{current}$) in each epoch as $LR_{current} = LR_{base} \times (1 - \text{iter}/\text{maxiter})^{\text{power}}$, where iter defines current iteration and maxiter defines maximum number of iterations in each epoch. We set LR_{base} to 0.045 and power to 0.9.

To overcome the limited samples of the data sets, we implement several data augmentation techniques, such as random horizontal flip, random crop, resizing of image, adjust the brightness of images. We also employ few regularization techniques such as ℓ_2 regularization and Dropout in the classifier module. We set the ℓ_2 regularization hyper-parameter to 0.00004 and dropout rate to 0.35. We utilize the categorical cross-entropy function for the model loss.

4.3   Ablation Study

At the initial stage, without using any FSM and FFM modules, we evaluated model performance on the Cityscapes data set. In the next stage, we deployed the FSM module on top of the backbone and reported the results. Table 4 clearly displays that the use of FSM module enhances model performance. We exploit different feature-scaling techniques. In the final stage, we introduce one multi-stage feature fusion module on top of FSM to exploit the benefits of FFM. We also report the model's performance by utilizing other existing feature fusion module and compare the results with our designs. Table 4 demonstrates that with the use of our FSM and FFM modules, the proposed model produces better results on the Cityscapes validation set. To get a best trade-off between dilation rate and input size, we also trained our model with different rates. It was shown in [2] that ASPP performed better at dilation rates {6,12,18,24}, whereas this study shows that our model attains best performance at dilation rates {8,16,24} for

Table 4. Segmentation performance evaluation

Backbone	FSM	FFM	mIoU (%)	Number of parameters (Million)	GFLOPs
14MBConv	–	–	60.3	1.11	50.9
14MBConv	PPM	–	63.2	1.32	53.4
14MBConv	ASPP	–	64.1	1.73	53.9
14MBConv	Ours	–	64.6	1.21	51.3
14MBConv	Ours	FPN	66.4	1.25	50.3
14MBConv	Ours	PAN	67.2	1.36	49.4
14MBConv	Ours	Bi-FPN	67.9	1.27	48.7
14MBConv	Ours	Ours	68.3	1.29	50.8

Table 5. Segmentation performance evaluation

Input size	output stride	Global feature size	Dilation rates	mIoU (%)
1024 × 2048	32	32 × 64	4,8,12	67.4
1024 × 2048	32	32 × 64	6,12,18	68
1024 × 2048	32	32 × 64	**8,16,24**	**68.3**
1024 × 2048	32	32 × 64	12,24,36	67.9

an input at 1024×2048. The difference is because the proposed model handles higher resolution input than [2]. As the stride is fixed, the size of the global feature map is large for higher input resolution. Therefore, larger dilation rates are required for better receptive fields. However, it is noted that the model performance drops when the dilation rate is beyond 8,16,24. Table 5 displays the performance at different rates. To reduce the training time in the ablation study, we only used the fine-tune set and stopped at epoch 500, which is suitable for its purpose. In the main experiments, we use all relevant sets and extend the number of epochs to 1000.

4.4   Model Evaluation

Model performance on Cityscapes. This section demonstrates model performance on Cityscapes dataset and compares its performance with other existing off-line and real-time semantic segmentation models. Table 6 reports its performance over 19 classes of Cityscapes validation and test sets. All classes of Cityscapes dataset are divided into 7 categories. Table 7 displays model performance on each category of Cityscapes dataset. The class-based result demonstrates that our model attains an accuracy of above 90% for 5 classes. Similarly, its accuracy is more than 90% in 5 categories.

Table 6. Class-wise FSFFNetperformance on Cityscapes validation and test sets

Dataset	Road	S. walk	Build.	Wall	Fence	Pole	T.light	T.sign	Veg.	Terrain
Validation set	**96.4**	77.7	**90.6**	57.0	52.1	58.3	63.5	72.7	**91.0**	62.2
	Sky	Person	Rider	Car	Truck	Bus	Train	M.cycle	Bicycle	**mIoU**
	93.2	75.9	51.2	**93.3**	67.8	79.1	64.0	47.5	70.8	**71.8**
Dataset	Road	S.walk	Build.	Wall	Fence	Pole	T.light	T.sign	Veg.	Terrain
Test set	**97.4**	78.5	**90.7**	41.8	46.1	57.8	65.3	68.5	**92.0**	63.9
	Sky	Person	Rider	Car	Truck	Bus	Train	M.cycle	Bicycle	**mIoU**
	94.4	79.2	56.9	**93.9**	55.4	65.7	54.4	50.4	65.8	**69.4**

Table 7. Category-wise FSFFNetperformance on Cityscapes validation and test set

Dataset	Flat	Construction	Object	Nature	Sky	Human	Vehicle	mIoU
Validation set	**96.0**	**90.3**	65.0	**91.4**	**93.2**	77.9	**91.2**	**86.4**
Test set	**96.8**	**91.5**	64.1	**94.4**	**90.2**	79.8	**92.7**	**87.1**

Table 8. Performance evaluation of different models on Cityscapes validation set

Type	Model	Input Size	Class mIoU(%)	Category mIoU(%)	Parameters (Million)	GFLOPs
Off-line	DeepLabV3+ [2]	1024 × 2048	64.5	82.6	54.8	344.9
	PSPNet [22]	713 × 713	81.2	91.2	250.8	516
	HANet [3]	768 × 768	80.9	-	65.4	2138.02
Real-time	Bayesian SegNet* [5]	1024 × 2048	63	82.1	30	2729.2
	BiseNet [20]	360 × 640	69	-	5.8	2.9
	ContextNet* [12]	1024 × 2048	60.4	81.5	1.0	37.5
	DFANet-B [6]	360 × 640	68.4	-	4.9	2.1
	FAST-SCNN* [13]	1024 × 2048	63.3	82.2	1.2	14.9
	FANet* [15]	1024 × 2024	65.9	83.6	1.1	11.4
	ICNet [21]	1024 × 2048	67.7	-	6.68	58.5
Real-time	**FSFFNet***	**1024 × 2048**	**71.8**	**86.4**	**1.3**	**50.8**

We also trained some existing semantic segmentation models under the same system configuration and presented results in Table 8. Models trained by us are marked with '*' sign in Table 8. Results of other models in Table 8 are obtained from either the literature or Cityscapes leaderboard. Note that the authors of DeeplabV3+ [2] deployed a new Xception (X-65) model on top of DeepLab previous version [1] to make a deeper encoder for semantic segmentation. However, we only used X-65 as feature extractor. We also replaced standard Conv layers of DeepLab and Bayesian SegNet [5] by DsConv layers to make the models computationally efficient in our system. We also incorporated ASPP on top of X-65 as a dense feature extractor.

Without utilizing image-level features, we attained 64.5% validation mean Intersection over Union (mIoU) on Cityscapes validation set. Among other trained models, we achieved 63.3% and 60.5% mIoU for FAST-SCNN [13] and ContextNet [12] respectively. They are different from the original claim of 68.6% and 65.9% mIoU on the validation set. We conjecture that the existing models might have been pre-trained with other datasets or some post-processing techniques might have used to boost their performance. In this study, we trained the model with the fine-tune set first, then followed by the coarse set. After that, we again trained the model with the fine-tune set. Table 8 clearly demonstrates that among all the real-time scene segmentation models, our proposed model produces the best validation accuracy (71.8%) on Cityscapes while having only 1.3 million parameters. Similar observation can be drawn from Table 9. Our proposed model achieves 69.3% test accuracy, setting a new state-of-the-art result among the existing real-time semantic segmentation models having less than 5 million parameters. While comparing model parameter and GFLOPs, we noticed that existing models reported their GFLOPs count at lower input resolution. With the increase of input resolution, GFLOPs increases exponentially. Therefore, comparing GFLOPs at different input resolutions is not an appropriate approach. We can compare model parameters as it does not depend on

Table 9. Performance evaluation of different models on Cityscapes test set

Model	Input size	Class mIoU (%)	Category mIoU (%)	parameters (Million)	GFLOPs
ENet [11]	360 × 640	58.3	80.4	0.4	3.8
ICNet [21]	1024 × 2048	69.5	-	6.68	-
FCN 8S [9]	512 × 1024	65.3	85.7	57	-
BiseNet [20]	360 × 640	68.4	-	5.8	2.9
ContextNet [12]	1024 × 2048	66.1	82.8	1.0	37.5
DFANet-B [6]	360 × 640	67.1	-	4.9	2.1
FAST-SCNN [13]	1024 × 2048	68.0	84.7	1.2	14.9
FANet [15]	1024 × 2048	64.1	83.1	1.1	11.4
FSFFNet	**1024 × 2048**	**69.4**	**87.1**	**1.3**	**50.8**

Table 10. Performance evaluation on validation set of BDD100K dataset

Type	Model	Input size	Parameters (Million) (%)	GFLOPs	Class mIoU (%)
Off-line	HANet (MobileNetV2) [3]	608 × 608	14.8	142.7	58.9
	HANet (ResNet-101)	608 × 608	64.2	2137.8	64.8
Real-time	ContextNet*	768 × 1280	1.0	37.5	44.5
	FAST-SCNN*	768 × 1280	1.2	14.9	47.9
	FANet*	768 × 1280	1.1	11.4	50.0
	FSFFNet*	**768 × 1280**	**1.3**	**50.8**	**55.2**

input size. Compared to previous state-of-the-art performance by ICNet [21], FSFFNetis 5 times smaller, however it produces similar test accuracy and more than 3% higher validation accuracy.

Model Performance on BDD100K. We also trained our model with the BDD100K dataset and presented the results in Table 10. From the literature, we found only one off-line model (HANet [3]) evaluated on BDD100K. The remaining models presented in Table 10 were trained by us under the same settings. As HANet [3] is an off-line model, it was expected to have better accuracy than our proposed real-time model.

To better see the effect of large models, we implemented a shallow variant of HANet which uses MobileNetV2 [14] as the backbone instead, and it produced 58.9% mIoU on BDD100K validation set. Our method (55.2%) is only few percent behind this variant whilst having 11 times smaller number of parameters. Compared to other models reported in Table 10, our proposed model FSFFNet-consistently achieves a better segmentation accuracy.

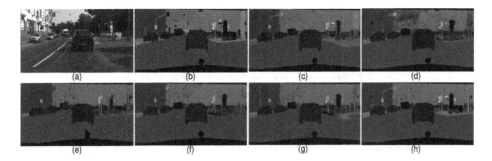

Fig. 3. Cityscapes val. set: (a) RGB input, (b) Color annotation, (c) DeepLabV3, (d) Bayesian SegNet, (e) ContextNet, (f) FAST-SCNN, (g) FANet, (h) FSFFNet

Fig. 4. Output by FSFFNeton Cityscapes test set

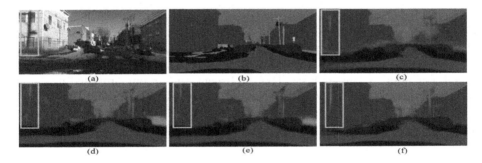

Fig. 5. Output by different models on BDD100K validation set: (a) RGB, (b) Colored Annotation, (c) ContextNet, (d) FAST-SCNN, (e) FANet, (f) FSFFNet

Qualitative Performance Analysis. We present samples of the output generated by all trained models in Figs. 3, 4, 5, 6. It can be clearly seen that the quality of the output generated by FSFFNetis better than the other models: boundary degeneration, overlapping classes, noisy feature effects can be observed in the output generated by other models whereas in FSFFNet, sharp boundaries of

Fig. 6. Output by FSFFNeton BDD100K test set

each object and the presence of tiny objects clearly demonstrate performance superiority in semantic segmentation. Due to the complex nature of BDD100K, noisy feature effects, wrong classification can be observed in the output generated by all models. However, inline with the quantitative results, the segmented images by FSFFNetare much better than others.

5   Conclusion

This study presents a computationally efficient real-time semantic segmentation model based on a light-weighted backbone, capable of handling high-resolution input images. The performance of the model is evaluated by two publicly available benchmarks and the results clearly demonstrate that our proposed model sets a new state-of-the-art results on Cityscapes benchmark in real-time semantic segmentation. Our feature fusion module helps reduce the semantic gap between the features, whereas our feature scaling module assimilates more contextual information for better scene representation. In the future, we plan to extend the model for indoor scene analysis. The implementation of our proposed model is publicly available at https://github.com/tanmaysingha/FSFFNet.

References

1. Chen, L.C., Papandreou, G., Kokkinos, I., Murphy, K., Yuille, A.L.: DeepLab: semantic image segmentation with deep convolutional nets, atrous convolution, and fully connected CRFs. IEEE TPAMI **40**(4), 834–848 (2017)
2. Chen, L.C., Zhu, Y., Papandreou, G., Schroff, F., Adam, H.: Encoder-decoder with atrous separable convolution for semantic image segmentation. In: Proceedings ICCV, September 2018
3. Choi, S., Kim, J.T., Choo, J.: Cars can't fly up in the sky: improving urban-scene segmentation via height-driven attention networks. In: Proceedings CVPR, pp. 9373–9383 (2020)
4. He, J., Deng, Z., Qiao, Y.: Dynamic multi-scale filters for semantic segmentation. In: Proceedings ICCV, pp. 3562–3572 (2019)

5. Kendall, A., Badrinarayanan, V., Cipolla, R.: Bayesian SegNet: model uncertainty in deep convolutional encoder-decoder architectures for scene understanding. arXiv preprint arXiv:1511.02680 (2015)
6. Li, H., Xiong, P., Fan, H., Sun, J.: DfaNet: deep feature aggregation for real-time semantic segmentation. In: Proceedings CVPR, pp. 9522–9531 (2019)
7. Lin, T.Y., Dollár, P., Girshick, R., He, K., Hariharan, B., Belongie, S.: Feature pyramid networks for object detection. In: Proceedings CVPR, pp. 2117–2125 (2017)
8. Liu, S., Qi, L., Qin, H., Shi, J., Jia, J.: Path aggregation network for instance segmentation. In: Proceedings CVPR, pp. 8759–8768 (2018)
9. Long, J., Shelhamer, E., Darrell, T.: Fully convolutional networks for semantic segmentation. In: Proceedings CVPR, pp. 3431–3440 (2015)
10. Pang, Y., Li, Y., Shen, J., Shao, L.: Towards bridging semantic gap to improve semantic segmentation. In: Proceedings ICVV, pp. 4230–4239 (2019)
11. Paszke, A., Chaurasia, A., Kim, S., Culurciello, E.: ENet: a deep neural network architecture for real-time semantic segmentation. arXiv preprint arXiv:1606.02147 (2016)
12. Poudel, R.P., Bonde, U., Liwicki, S., Zach, C.: ContextNet: exploring context and detail for semantic segmentation in real-time. arXiv preprint arXiv:1805.04554 (2018)
13. Poudel, R.P., Liwicki, S., Cipolla, R.: Fast-SCNN: fast semantic segmentation network. arXiv preprint arXiv:1902.04502 (2019)
14. Sandler, M., Howard, A., Zhu, M., Zhmoginov, A., Chen, L.C.: MobileNetv 2: inverted residuals and linear bottlenecks. In: Proceedings CVPR, pp. 4510–4520 (2018)
15. Singha, T., Pham, D.S., Krishna, A.: FaNet: feature aggregation network for semantic segmentation. In: Proceedings DICTA, pp. 1–8. IEEE (2020)
16. Singha, T., Pham, D.-S., Krishna, A., Dunstan, J.: Efficient segmentation pyramid network. In: Yang, H., Pasupa, K., Leung, A.C.-S., Kwok, J.T., Chan, J.H., King, I. (eds.) ICONIP 2020. CCIS, vol. 1332, pp. 386–393. Springer, Cham (2020). https://doi.org/10.1007/978-3-030-63820-7_44
17. Tan, M., Le, Q.V.: EfficientNet: rethinking model scaling for convolutional neural networks. arXiv preprint arXiv:1905.11946 (2019)
18. Targ, S., Almeida, D., Lyman, K.: ResNet in ResNet: generalizing residual architectures. arXiv preprint arXiv:1603.08029 (2016)
19. Yang, M., Shi, Y.: DSMRSeg: dual-stage feature pyramid and multi-range context aggregation for real-time semantic segmentation. In: Gedeon, T., Wong, K.W., Lee, M. (eds.) ICONIP 2019. CCIS, vol. 1142, pp. 265–273. Springer, Cham (2019). https://doi.org/10.1007/978-3-030-36808-1_29
20. Yu, C., Wang, J., Peng, C., Gao, C., Yu, G., Sang, N.: BiSeNet: bilateral segmentation network for real-time semantic segmentation. In: Ferrari, V., Hebert, M., Sminchisescu, C., Weiss, Y. (eds.) ECCV 2018. LNCS, vol. 11217, pp. 334–349. Springer, Cham (2018). https://doi.org/10.1007/978-3-030-01261-8_20
21. Zhao, H., Qi, X., Shen, X., Shi, J., Jia, J.: ICNet for real-time semantic segmentation on high-resolution images. In: Ferrari, V., Hebert, M., Sminchisescu, C., Weiss, Y. (eds.) ECCV 2018. LNCS, vol. 11207, pp. 418–434. Springer, Cham (2018). https://doi.org/10.1007/978-3-030-01219-9_25
22. Zhao, H., Shi, J., Qi, X., Wang, X., Jia, J.: Pyramid scene parsing network. In: Proceedings CVPR, pp. 2881–2890 (2017)

Multi-view Fractional Deep Canonical Correlation Analysis for Subspace Clustering

Chao Sun[1], Yun-Hao Yuan[1,2(✉)], Yun Li[1(✉)], Jipeng Qiang[1], Yi Zhu[1],
and Xiaobo Shen[3]

[1] School of Information Engineering, Yangzhou University, Yangzhou, China
{yhyuan,liyun}@yzu.edu.cn
[2] School of Computer Science and Technology, Fudan University, Shanghai, China
[3] School of Computer Science and Engineering,
Nanjing University of Science and Technology, Nanjing, China

Abstract. Canonical correlation analysis (CCA) is a classic unsupervised dimensionality reduction approach, but it has difficulty to analyze the nonlinear relationship and learn from more than two views. In addition, real-world data sets often contain much noise, which makes the performance of machine learning algorithms degraded. This paper presents a multi-view fractional deep CCA (MFDCCA) method for representation learning and clustering tasks. The proposed MFDCCA method utilizes fractional order embedding to reduce the adverse effect of noise in multi-view data and deep neural networks to explore the nonlinear relationship. Experimental results on three popular datasets have demonstrated the effectiveness of the proposed MFDCCA.

Keywords: Multi-view learning · Canonical correlation analysis · Deep learning · Subspace clustering

1 Introduction

The development of data collecting technologies has led to an increase in information diversity. A large amount of data are presented in various forms, generating the so-called multi-view data [1]. For example, an object can be described by the picture, text, sound, and video features. Multi-view data contain more feature information than single view data. Because these features describe the same objects from different angles, different view data can complement each other. At present, more and more researchers are interested in multi-view learning [2]. However, multi-view data not only have complex associations, but are highly

Supported by the National Natural Science Foundation of China under Grant Nos. 61402203 and 61703362, the China Postdoctoral Science Foundation under Grant No. 2020M670995, and Yangzhou Science Project Foundation under Grant No. YZ2020173. It is also sponsored by Excellent Young Backbone Teacher (Qing Lan) Project and Scientific Innovation Project Fund of Yangzhou University under Grant No. 2017CXJ033.

T. Mantoro et al. (Eds.): ICONIP 2021, LNCS 13109, pp. 206–215, 2021.
https://doi.org/10.1007/978-3-030-92270-2_18

dimensional. Directly dealing with multi-view data may encounter the problem of "curse of dimensionality". This means that multi-view learning faces more challenges [3] than single view learning.

Dimension reduction (DR) is an effective way to reduce the dimensionality of the data. Principal component analysis (PCA) [4] and locality preserving projections (LPP) [5] are two typical single view DR methods. PCA extracts features by maximizing the variance of the projected data, while LPP by minimizing the local scatter. In addition, kernel PCA (KPCA) [6] was presented to deal with nonlinear scenario. For two-view data, canonical correlation analysis (CCA) [7] is an available method for DR. It aims to find two projection matrices that project two-view samples into low-dimensional spaces, such that the projected data are maximally correlated.

In recent years, there have been a lot of CCA-related variants. For example, kernel CCA (KCCA) [8] is a famous nonlinear extension of CCA. It uses kernel functions to project the original data into the Hilbert space for learning nonlinear representations. However, KCCA has an obvious problem that fixed kernels restrict its generalization ability and applicability. Moreover, it is difficult for KCCA to choose proper parameters for a good performance in actual scenarios. Recently, deep CCA (DCCA) [9] was proposed to learn deep nonlinear representations through deep neural networks (DNNs), which is more flexible than KCCA. DCCA has been successfully applied to many high-dimensional scenarios with complex nonlinear relationship. But, DCCA is only able to be used for two views. To deal with three or more views, deep multi-view CCA (DMCCA) [10] was presented to learn nonlinear latent multi-view representations.

In reality, it is hardly possible for CCA and related methods to obtain the true within- and between-set covariance matrices in advance. Therefore, their computation relies heavily on training samples. When training samples have much noise, the sample covariance matrix is not a good estimation of true one, which may not enable CCA to get optimal representations for learning tasks. As a result, fractional-order embedding CCA (FECCA) [11] was proposed by introducing fractional parameters into the sample covariance matrices and reconstructing them through spectral decomposition. This paper borrows the idea of fractional-order embedding and propose a multi-view fractional deep CCA (MFDCCA) method for clustering tasks. MFDCCA first learns nonlinear projections and then uses spectral clustering algorithm to cluster multi-view data. Experimental results are encouraging.

2   Related Work

2.1   Multiset CCA

Assume m views of samples are given as $\{X^i = [x_1^i, x_2^i, \cdots, x_n^i] \in R^{d_i \times n}\}_{i=1}^m$, where n and d_i are respectively the number of samples and dimension in i-th view. The goal of multiset CCA (MCCA) [12] is to learn a set of vectors that project the original data into low-dimensional subspaces, such that the correlation coefficient of the projected data is maximized. MCCA can be expressed by the following optimization problem:

$$\max_{\omega_1,\cdots,\omega_m} \text{corr}(\{\omega_i^T X^i\}_{i=1}^m) = \sum_{i=1}^m \sum_{j\neq i}^m \omega_i^T S_{ij}\omega_j$$
$$s.t. \ \sum_{i=1}^m \omega_i^T S_{ii}\omega_i = 1 \tag{1}$$

where corr($\cdot$) denotes the correlation measure, $\{\omega_i \in R^{d_i}\}_{i=1}^m$ are the projection vectors, S_{ij} ($i \neq j$) is the between-set covariance matrix of X^i and X^j, and S_{ii} is the within-set covariance matrix of X^i.

Through the Lagrange multiplier method, the problem in (1) can be solved by the generalized eigenvalue problem:

$$\sum_{j=1,j\neq i}^m S_{ij}\omega_j = \lambda S_{ii}\omega_i, \ i = 1, 2, \cdots, m, \tag{2}$$

or concisely

$$A\omega = \lambda B\omega, \tag{3}$$

where λ is the eigenvalue associated with the generalized eigenvector ω,

$$\omega = [\omega_1^T, \omega_2^T, \cdots, \omega_m^T]^T,$$

$$A = \begin{bmatrix} 0 & S_{12} & \cdots & S_{1m} \\ S_{21} & 0 & \cdots & S_{2m} \\ \vdots & \vdots & \ddots & \vdots \\ S_{m1} & S_{m2} & \cdots & 0 \end{bmatrix},$$

$$B = \begin{bmatrix} S_{11} & & & \\ & S_{22} & & \\ & & \ddots & \\ & & & S_{mm} \end{bmatrix}.$$

2.2 Deep MCCA

An obvious deep extension of MCCA is to use DNNs for multi-view data transformation. Deep MCCA gets the projection vectors of multiple views through multi-layer network transformations such that the correlation among multiple projections is maximized. Let $f_i(\cdot)$ ($i = 1, 2, \cdots, m$) represent DNNs and their final outputs be $f_i(X^i, \theta_i)$, where θ_i denotes the network parameters of i-th DNN. Deep MCCA is expressed as the following maximization problem:

$$\max_{\substack{\theta_1,\cdots,\theta_m \\ \omega_1,\cdots,\omega_m}} \text{corr}(\{\omega_i^T f_i(X^i, \theta_i)\}_{i=1}^m) = \sum_{i=1}^m \sum_{j\neq i}^m \omega_i^T f_i(X^i,\theta_i)f_j(X^j,\theta_j)^T\omega_j$$
$$s.t. \ \sum_{i=1}^m \omega_i^T f_i(X^i,\theta_i)f_i(X^i,\theta_i)^T\omega_i = 1. \tag{4}$$

The optimization problem in (4) can be computed by backpropagation.

2.3   Spectral Clustering

In the area of machine learning, clustering is a major task in unsupervised learning. Different from conventional clustering methods, spectral clustering (SC) combines k-means algorithm and graph Laplacian by the assumption that sample points are the vertices of a graph and the edge weights are expressed by the similarity of sample points.

For any two sample points x_i and x_j, the adjacency matrix $W = [w_{ij}]$ can be constructed by

$$w_{ij} = \exp\left(-\frac{\|x_i - x_j\|_2^2}{2\sigma^2}\right), \tag{5}$$

where $\|\cdot\|_2$ is the 2-norm operator and σ is the parameter of the Gaussian function that controls the size of neighborhood. With (5), the graph Laplacian matrix L can be computed by

$$L = D - W, \tag{6}$$

where D is a diagonal matrix whose i-th diagonal element is the i-th row (column) sum of W. By using (6), the normalized graph Laplacian matrix $\hat{L}$ can be constructed by the following

$$\hat{L} = D^{-1/2} L D^{-1/2} \tag{7}$$

Calculating the first k smallest eigenvectors of $\hat{L}$ in (7) to form an $n \times k$ matrix F. Using traditional k-means algorithm to cluster all rows of F. More details of spectral clustering can be found in [13].

3   Multi-view Fractional Deep CCA

Multi-view samples in real world often have some noise. This leads to a negative impact on machine learning algorithms. FECCA uses fractional order embedding to adjust the eigenvalues and singular values of intra-set and inter-set sample covariance matrices. Experiments show that this idea is able to effectively reduce the influence of noise and enhance the algorithm performance. Thus, we introduce this idea of fractional order embedding into deep MCCA.

For view X^i, its singular value decomposition can be obtained by $X^i = U^i \Sigma^i V^{iT}$, where U^i and V^i consist of the left and right singular vectors of X^i, respectively, and Σ^i is a diagonal matrix with diagonal elements as singular values in descending order. The fractional order form of X^i is expressed as:

$$(X^i)^{\alpha_i} = U^i(\Sigma^i)^{\alpha_i} V^{iT}, \ i = 1, 2, \cdots, m, \tag{8}$$

where α_i is a fractional order parameter and $0 \leq \alpha_i \leq 1$. Through using (8), the optimization model of multi-view fractional deep CCA (MFDCCA) can be built by the following

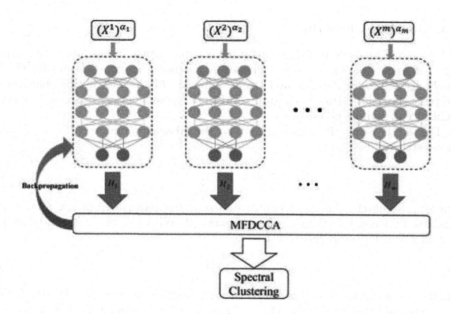

Fig. 1. Overall flowchart of our MFDCCA method.

$$\max_{\substack{\theta_1,\cdots,\theta_m \\ \omega_1,\cdots,\omega_m}} \mathrm{corr}(\{\omega_i^T H_i\}_{i=1}^m) = \sum_{i=1}^m \sum_{j \neq i}^m \omega_i^T \hat{S}_{ij} \omega_j$$
$$s.t. \sum_{i=1}^m \omega_i^T \hat{S}_{ii} \omega_i = 1. \tag{9}$$

where $H_i = f_i\big((X^i)^{\alpha_i}, \theta_i\big)$ is the final output of DNN f_i, $\hat{S}_{ij} = H_i H_j^T$ $(i \neq j)$, and $\hat{S}_{ii} = H_i H_i^T$, $i = 1, 2, \cdots, m$. Figure 1 shows the overall flowchart of our MFDCCA method.

The optimization problem in (9) can use the Lagrange multiplier technique to obtain the following generalized eigenequation:

$$P\omega = \eta Q\omega \tag{10}$$

where η is the eigenvalue corresponding to the generalized eigenvector ω,

$$P = \begin{bmatrix} 0 & H_1 H_2^T & \cdots & H_1 H_m^T \\ H_2 H_1^T & 0 & \cdots & H_2 H_m^T \\ \vdots & \vdots & \ddots & \vdots \\ H_m H_1^T & H_m H_2^T & \cdots & 0 \end{bmatrix},$$

$$Q = \begin{bmatrix} H_1 H_1^T & & & \\ & H_2 H_2^T & & \\ & & \ddots & \\ & & & H_m H_m^T \end{bmatrix}.$$

The inverse propagation algorithm based on small-batch samples is used to optimize MFDCCA's parameters continuously until the maximum correlation between views is achieved.

4   Experiments

To test the effectiveness of the proposed MFDCCA, we compare it with related methods including MCCA, DMCCA, and labeled multiple CCA (LMCCA) [14] on three popular datasets. After representation learning, spectral clustering algorithm is used for clustering tasks. In addition, we also perform SC algorithm as a baseline.

4.1   Data Preparation

Experiments are carried out on three real datasets, i.e., COIL-100[1], Fashion MNIST[2], and MNIST[3]. To get three different views, original images are first downsampled to low-resolution images and then they are restored to the original resolution, which are used as the first view. We use the Symlets wavelet transform to extract wavelet features from original images for the second view. The third view is obtained through a mean filter with a window size of 3×3. The accuracy and rand index (RI) are used to evaluate different methods.

4.2   Experiment on COIL-100

The COIL-100 dataset contains 7,200 color images of 100 objects. Each object has 72 images. On this dataset, we respectively select 1,200, 1,800, and 2,400 images of the first 20, 30, and 40 objects to form three subsets, denoted as subset-20, subset-30, and subset-40. Note that in each subset, each object has 60 images selected randomly from 72 images. For each subset, 30 images per object are selected randomly to generate the training set, and the rest are used for the testing set. Five tests are independently performed and the performances of different methods are evaluated by calculating the average results. Table 1 shows the average results of SC, MCCA, LMCCA, DMCCA, and our MFDCCA. Figure 2 shows the accuracy of MFDCCA when two fractional order parameters α_1 and α_2 vary and α_3 is fixed as 0.3.

It can be seen from Table 1 that our MFDCCA is superior to the other four methods on all cases. MCCA achieves the worst performance. LMCCA and DMCCA achieve comparable results with MFDCCA. These results suggest that our MFDCCA method is an effective technique for clustering tasks.

---

[1] https://www1.cs.columbia.edu/CAVE/software/softlib/coil-100.php.
[2] https://github.com/zalandoresearch/fashion-mnist.
[3] http://yann.lecun.com/exdb/mnist/.

Table 1. The results of SC, MCCA, LMCCA, DMCCA, and our MFDCCA methods on the COIL-100 dataset. In MFDCCA, $\alpha_1 = 0.1$, $\alpha_2 = 0.1$, and $\alpha_3 = 0.3$.

Method	subset-20		subset-30		subset-40	
	Accuracy	Rand index	Accuracy	Rand index	Accuracy	Rand index
SC	0.583	0.949	0.601	0.969	0.604	0.976
MCCA	0.506	0.929	0.479	0.951	0.498	0.964
LMCCA	0.621	0.948	0.623	0.965	0.599	0.973
DMCCA	0.670	0.959	0.659	0.973	0.644	0.978
MFDCCA	**0.693**	**0.962**	**0.673**	**0.974**	**0.658**	**0.979**

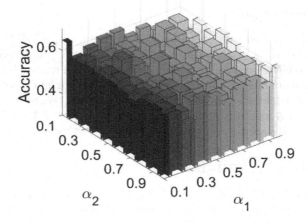

Fig. 2. Accuracy of MFDCCA when fractional order parameters α_1 and α_2 vary and $\alpha_3 = 0.3$ on the COIL-100 dataset.

4.3 Experiment on Fashion-MNIST

The Fashion-MNIST dataset consists of 10 categories of 70,000 grayscale images with size of 28 × 28. Each category has 7,000 images. This dataset is divided into two parts in advance. The first part contains 60,000 images for training and the second part includes 10,000 images for testing. To reduce the computational cost, 60, 80, and 100 images per class are randomly selected from the first part for training, while 60, 80, and 100 images per class from the second part for testing. Table 2 summarizes the average results of different methods with five runs. Figure 3 shows the accuracy of MFDCCA when two fractional order parameters α_1 and α_2 vary and α_3 is fixed as 0.3. From Table 2, it is able to be seen that the proposed MFDCCA is superior to the other approaches, indicating that our method is more effective.

Table 2. The results of SC, MCCA, LMCCA, DMCCA, and our MFDCCA methods on the Fashion-MNIST dataset. In MFDCCA, $\alpha_1 = 0.8$, $\alpha_2 = 0.6$, and $\alpha_3 = 0.3$.

Method	60 Train.		80 Train.		100 Train.	
	Accuracy	Rand index	Accuracy	Rand index	Accuracy	Rand index
SC	0.236	0.358	0.221	0.509	0.234	0.512
MCCA	0.416	0.849	0.428	0.850	0.393	0.846
LMCCA	0.573	0.889	0.565	0.886	0.590	0.889
DMCCA	0.649	0.904	0.643	0.903	0.623	0.899
MFDCCA	**0.673**	**0.907**	**0.651**	**0.904**	**0.639**	**0.900**

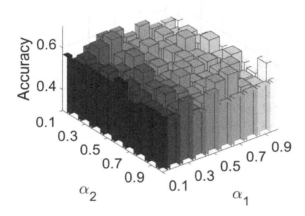

Fig. 3. Accuracy of MFDCCA when fractional order parameters α_1 and α_2 vary and $\alpha_3 = 0.3$ on the Fashion-MNIST dataset.

4.4   Experiment on MNIST

The MNIST dataset consists of 10-class handwritten digits. This dataset is divided into two parts in advance. That is, the first part is the training set including 60,000 samples and the second part is the testing set including 10,000 samples. We randomly choose 100, 400, and 600 samples per class from the first part for training, respectively, and 100, 400, and 600 samples per class from the second part for testing. Table 3 shows the average results of different methods with five runs. As seen, the MFDCCA method is again superior to other approaches.

Table 3. The results of SC, MCCA, LMCCA, DMCCA, and our MFDCCA methods on the MNIST dataset. In MFDCCA, $\alpha_1 = 0.3$, $\alpha_2 = 0.4$, and $\alpha_3 = 0.8$.

Method	100 Train.		400 Train.		600 Train.	
	Accuracy	Rand index	Accuracy	Rand index	Accuracy	Rand index
SC	0.297	0.685	0.180	0.428	0.169	0.422
MCCA	0.328	0.835	0.301	0.837	0.341	0.843
LMCCA	0.567	0.875	0.593	0.884	0.615	0.888
DMCCA	0.579	0.886	0.574	0.885	0.589	0.885
MFDCCA	**0.597**	**0.888**	**0.602**	**0.888**	**0.627**	**0.891**

5   Conclusion

In this paper, we propose a new representation learning method called MFD-CCA for subspace clustering. The proposed MFDCCA approach makes use of fractional order embedding to reduce the adverse effect of noise in multi-view data and deep neural networks to explore the nonlinear relationship. We use an inverse propagation algorithm based on small-batch samples to optimize the model. Experimental results on three famous benchmark datasets show that MFDCCA is superior to other related approaches.

References

1. Zheng, T., Ge, H., Li, J., Wang, L.: Unsupervised multi-view representation learning with proximity guided representation and generalized canonical correlation analysis. Appl. Intell. **51**(1), 248–264 (2021)
2. Zhao, J., Xie, X., Xu, X., Sun, S.: Multi-view learning overview: recent progress and new challenges. Inf. Fusion **38**, 43–54 (2017)
3. Li, Y., Yang, M., Zhang, Z.: A survey of multi-view representation learning. IEEE Trans. Knowl. Data Eng. **31**(10), 1863–1883 (2018)
4. Martinez, A.M., Kak, A.C.: PCA versus LDA. IEEE Trans. Pattern Anal. Mach. Intell. **23**(2), 228–233 (2001)
5. He, X., Niyogi, P.: Locality preserving projections. Adv. Neural. Inf. Process. Syst. **16**(16), 153–160 (2004)
6. Schölkopf, B., Smola, A., Müller, K.R.: Nonlinear component analysis as a kernel eigenvalue problem. Neural Comput. **10**(5), 1299–1319 (1998)
7. Hardoon, D.R., Szedmak, S., Shawe-Taylor, J.: Canonical correlation analysis: an overview with application to learning methods. Neural Comput. **16**(12), 2639–2664 (2004)
8. Lai, P.L., Fyfe, C.: Kernel and nonlinear canonical correlation analysis. Int. J. Neural Syst. **10**(5), 365–377 (2000)
9. Andrew, G., Arora, R., Bilmes, J., Livescu, K.: Deep canonical correlation analysis. In: Proceedings of the 30th International Conference on Machine Learning, pp. 1247–1255. JMLR.org, Atlanta, USA (2013)

10. Somandepalli, K., Kumar, N., Travadi, R., Narayanan, S.: Multimodal representation learning using deep multiset canonical correlation. arXiv preprint, arXiv:1904.01775 (2019)
11. Yuan, Y.-H., Sun, Q.-S., Ge, H.-W.: Fractional-order embedding canonical correlation analysis and its applications to multi-view dimensionality reduction and recognition. Pattern Recogn. **47**(3), 1411–1424 (2014)
12. Nielsen, A.A.: Multiset canonical correlations analysis and multispectral, truly multitemporal remote sensing data. IEEE Trans. Image Process. **11**(3), 293–305 (2002)
13. Von Luxburg, U.: A tutorial on spectral clustering. Stat. Comput. **17**, 395–416 (2007)
14. Gao, L., Zhang, R., Qi, L., Chen, E., Guan, L.: The labeled multiple canonical correlation analysis for information fusion. IEEE Trans. Multimedia **21**(2), 375–387 (2018)

Handling the Deviation from Isometry Between Domains and Languages in Word Embeddings: Applications to Biomedical Text Translation

Félix Gaschi[1,2]([✉]), Parisa Rastin[2], and Yannick Toussaint[2]

[1] SAS Posos, 55 rue de la Boétie, 75008 Paris, France
felix@posos.fr
[2] LORIA, UMR 7503, BP 239, 54506 Vandoeuvre-lès-Nancy, France
{felix.gaschi,parisa.rastin,yannick.toussaint}@loria.fr

Abstract. Previous literature has shown that it is possible to align word embeddings from different languages with unsupervised methods based on a distance-preserving mapping, with the assumption that the embeddings are isometric. However, these methods seem to work only when both embeddings are trained on the same domain. Nonetheless, we hypothesize that the deviation from isometry might be reduced between relevant subsets of embeddings from different domains, which would allow to partially align them. To support our hypothesis, we leverage the Bottleneck distance, a topological data analysis tool used to approximate the deviation from isometry. We also propose a cross-domain and cross-lingual unsupervised alignment method based on a proxy embedding, as a first step towards new cross-lingual alignment methods that generalize to different domains. Results of such a method on translation tasks show that unsupervised alignment methods are not doomed to fail in a cross-domain setting. We obtain BLEU-1 scores ranging from 0.38 to 0.50 on translation tasks, where previous fully unsupervised alignment methods obtain near-zero scores in cross-domain settings.

Keywords: Machine learning · Natural language processing · Biomedical information · Multilingual representations · Domain adaptation

1 Introduction

Word embeddings provide useful representations for many downstream tasks [13] and have been generalized in the cross-lingual context to the concept of Unsupervised Cross-lingual Embedding (UCE) [2,6]. UCEs learn a distance-preserving transformation, or isometry, mapping one language to the other. This kind of method was shown to work well in some cases, but fails when the embeddings of each language come from a different domain [19]. Our work is motivated by a need for effective UCEs in cross-domain settings. Domain-specific data,

© Springer Nature Switzerland AG 2021
T. Mantoro et al. (Eds.): ICONIP 2021, LNCS 13109, pp. 216–227, 2021.
https://doi.org/10.1007/978-3-030-92270-2_19

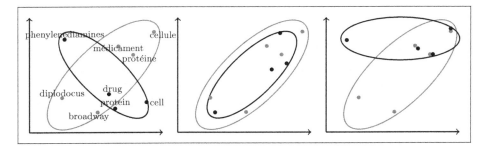

Fig. 1. Toy example showing different alignment of a domain (grey) with another (black), the initial unaligned embeddings (left) do not align well when considering all words (center), we aim to align them partially (right)

such as scientific publications, can be rare in languages other than English. Our goal is to show that UCEs are not doomed to fail in a setting where we have a domain-specific English embedding (e.g. trained on PubMed) and a general-domain non-English embedding (e.g. trained on French Wikipedia).

Methods based on distance-preserving transformations seem to fail in such a cross-domain setting [19]. This suggests that the approximate isometry between embeddings, which is necessary for such methods to work, is not verified in this setting. However, we noticed that alignment methods usually try to align the whole embeddings together, or more precisely a large set of the most frequent words of each embedding, typically 20k [2]. Yet, between two different domains, the distribution of the vocabulary may vary. Some words might be more frequent in one domain and less in the other, or even absent.

We hypothesize that it is possible to improve unsupervised cross-lingual embedding in a cross-domain setting by trying to align well-chosen subsets of each vocabulary. This "partial alignment" of embeddings could be suitable for certain domain-specific tasks. We provide a toy example in Fig. 1 to illustrate this.

Our contribution is threefold: (1) we measure reduced deviation from isometry between parallel vocabularies for embeddings from different domains; (2) we propose an unsupervised alignment method based on a partial alignment outperforming other unsupervised methods in a cross-domain setting; (3) we visualize this partial alignment with t-SNE, a dimensionality-reduction technique [12].

Before detailing our approach in Sect. 3, we describe the context of our research with isometry-based embedding alignment methods (Sect. 2.1) and the Bottleneck distance as a measure of the deviation from isometry (Sect. 2.2). Then we introduce our proposed method for aligning cross-domain embeddings (Sect. 3), before detailing experiments and results (Sect. 4).

2   Context

2.1   Isometry-Based Alignment Methods

Shortly after dense word embeddings were introduced [13], it was proposed to map embeddings from distinct languages to a shared space [14]. To preserve the shape of monolingual embeddings, the mapping applied to an embedding must be distance-preserving [6,14]. Such a transformation is called isometry.

Definition 1. *Let $\mathcal{X}$ and $\mathcal{Y}$ be two metric spaces. A map $f : \mathcal{X} \to \mathcal{Y}$ is an isometry if for any $(x_i, x_j) \in \mathcal{X}^2$, we have:*

$$d(x_i, x_j) = d(f(x_i), f(x_j)) \tag{1}$$

A linear finite-dimensioned isometry produces an orthogonal matrix, a square matrix whose transpose is its inverse. Supervised methods for cross-lingual embeddings learn an orthogonal mapping between representations of a bilingual dictionary [14]. Following the formalism of [6], we have:

$$W^* = \arg \min_{W \in \mathcal{O}_d} ||AW - B||_F^2 \tag{2}$$

where $A \in \mathbb{R}^{N \times d}$ and $B \in \mathbb{R}^{N \times d}$ are the representations of the N entries of the dictionary in the source and target embeddings of dimension d and W^* is the learned mapping in $\mathcal{O}_d$, the set of orthogonal matrices. Procrustes analysis [18] gives us $W^* = UV^\top$, with the singular value decomposition (SVD) $A^\top B = USV^\top$.

But to learn a mapping in an unsupervised way, given two embeddings X and Y, we also need to learn a dictionary as a permutation matrix P, which is also an isometry (permutation matrices are orthogonal matrices):

$$W^*, P^* = \arg \min_{P \in \mathcal{P}_n, W \in \mathcal{O}_d} ||XW - PY||_F^2 \tag{3}$$

With the advent of iterative self-learning [1] which alternates between learning the mapping W and the dictionary P, alignments required fewer and fewer training pairs of words. It led to fully unsupervised methods with adversarial learning [6] and initialization heuristics. VecMap [2] is one of those self-learning methods which introduces decreasing random noise in the process, inspired by simulated annealing, for more robust alignment. To account for local variations of the density of embeddings, a Cross-domain Similarity Local Scaling (CSLS) criterion [6] is often used [2,6,8], leveraging the average distance d of k nearest neighbors[1]:

$$CSLS(u, v) = d(u, v) - \frac{1}{k} \sum_{x \in \mathcal{N}_k(u)} d(x, u) - \frac{1}{k} \sum_{y \in \mathcal{N}_k(v)} d(y, v) \tag{4}$$

---

[1] Usually $k = 10$.

But for X and Y to align correctly with such methods, they must be approximately isometric. Unfortunately, it was shown that UCEs relying on an orthogonal mapping need three conditions to give accurate results [19]: (1) languages must be morphologically similar; (2) the monolingual training corpora must be from the same domain; and (3) the same model must be used (CBOW Spanish embeddings did not align with Skip-gram English). These drawbacks eventually led to several methods featuring a weak orthogonality constraint [16]. But we hypothesize that, when we have embeddings from different domains, the isometry assumption must not be completely abandonned, as there might still be approximate isometry between relevant subsets of those embeddings.

2.2 Measuring Deviation from Isometry

To verify our hypothesis, we need a way to measure the deviation from isometry of two metric spaces. First, we must be able to evaluate to what extent two aligned sets coincide. For that we can rely on the Hausdorff distance.

Definition 2. *Let $\mathcal{X}$ and $\mathcal{Y}$ be two compact subsets of a metric space $(\mathcal{Z}, d_Z)$. The Hausdorff distance is defined by:*

$$d_H^{\mathcal{Z}}(\mathcal{X}, \mathcal{Y}) = \max\left(\sup_{x \in \mathcal{X}} \inf_{y \in \mathcal{Y}} d_Z(x, y), \sup_{y \in \mathcal{Y}} \inf_{x \in \mathcal{X}} d_Z(x, y)\right) \tag{5}$$

Intuitively, the Hausdorff distance is the maximum distance between pairs of nearest neighbors. From that we can build the Gromov-Hausdorff distance, which gives a theoretical measure of the deviation from isometry.

Definition 3. *Let $(\mathcal{X}, d_X)$ and $(\mathcal{Y}, d_Y)$ be two metric spaces. The Gromov-Hausdorff distance is defined by:*

$$d_{GH}((\mathcal{X}, d_X), (\mathcal{Y}, d_Y)) = \inf_{\mathcal{Z}, f, g} d_H^{\mathcal{Z}}(f(\mathcal{X}), g(\mathcal{Y})) \tag{6}$$

With $f : \mathcal{X} \to \mathcal{Z}$ and $g : \mathcal{Y} \to \mathcal{Z}$ isometries matching both metric spaces to a single metric space $(\mathcal{Z}, d_Z)$.

The Gromov-Hausdorff distance is the minimum over all isometric transformations of the Hausdorff distance. It measures how well two metric spaces can be aligned without deforming them. This distance, intractable to compute in practice, needs to be approximated. Several works use diverse metrics. A similarity based on a spectral analysis of neighborhood graphs is correlated with performances of alignment methods [19]. Earth Mover's Distance between embeddings was linked with typological similarity between languages [20], showing why distant language pairs are more difficult to align. Another metric is the Bottleneck distance between the persistence diagrams of the Rips complex filtrations of each metric space and it was shown to be a tight lower-bound for the Gromov-Hausdorff distance [5], and was found to better correlate with the ability to align embedding with an orthogonal mapping than previously mentioned metrics [16].

For a more formal definition of persistence diagram, Rips complex and filtrations, the reader might refer to relevant literature [5]. In our case, we create a parameter t that varies from 0 to $+\infty$, and compute a graph for each embedding such that two points of an embedding that are at a distance smaller than $2t$ are linked by an edge. This graph is a Rips complex, or rather a simplified version of it because we only look at connected components, hence only edges, not higher-dimensional simplexes. We start with as many connected components as elements in the embedding ($t = 0$) and gradually decrease their number by merging them. This sequence of Rips complexes is called a filtration, on which we can compute a persistence diagram for each embedding: a list of points ($t_{\text{birth}}, t_{\text{death}}$) for each connected component that appears during the filtration recording the $t = t_{\text{birth}}$ at which it appears and the $t = t_{\text{death}}$ at which it is merged with another.

Comparing the persistence diagrams for two embeddings allows us to measure to what extent they differ topologically. This is the Bottleneck distance.

Definition 4. *Given two multi-sets A and B in $\overline{\mathbb{R}}^2$, the Bottleneck distance is defined by:*

$$d_B^\infty(A, B) = \min_\gamma \max_{p \in A} ||p - \gamma(p)||_\infty \tag{7}$$

With γ ranging over bijections between A and B.

The Bottleneck distance between Rips filtration of metric spaces gives us a lower bound for the Gromov-Hausdorff distance.

Theorem 5. *From [5]. For any finite metric spaces $(\mathcal{X}, d_X)$ and $(\mathcal{Y}, d_Y)$ and for any $k \in \mathbb{N}$:*

$$d_B^\infty(\mathcal{D}_k\mathcal{R}(\mathcal{X}, d_X), \mathcal{D}_k\mathcal{R}(\mathcal{Y}, d_Y)) \leq d_{GH}((\mathcal{X}, d_X), (\mathcal{Y}, d_Y)) \tag{8}$$

To make it simpler, in the following, when we mention the "Bottleneck distance" between two sets of embeddings, we will be actually referring to the Bottleneck distance between the persistence diagrams of the filtrations of Rips complexes built over those embeddings. We will use this Bottleneck distance in our first experiment to measure the deviation from isometry between diverse subsets of embeddings from different languages and domains, showing that some subsets of embeddings from different domains are more topologically similar than the whole embeddings themselves.

3   Proposed Approach

The proposed hypothesis states that embeddings from different domains could still be partially aligned. So the biggest challenge is to find the relevant subsets that align well. We propose here a method based on a proxy embedding of the same domain as one embedding and the same language as the other.

The proposed approach can be summarized as follows: we align the source embedding and the proxy embedding, which are from the same domain, with an isometry-based alignment method. We build a dictionary from it that takes into account the target vocabulary and use this dictionary to find a mapping between the source and target. The alignment between source and proxy should work since both embeddings are from same domain and the filtered dictionary used to align the source and the target should allow this partial alignment and avoid aligning subsets of vocabulary that are not relevant to each other.

More formally, as shown in Algorithm 1, we have X and Z two embeddings of distinct languages and domains (e.g. French Wikipedia and English PubMed). Let Y be a proxy embedding of same domain as X and same language as Z (English Wikipedia in our example). The proposed method aligns X and Y together (of same domain) by solving Eq. 3, giving aligned embeddings $\widetilde{X}$ and $\widetilde{Y}$. We use the VecMap algorithm [2], a state-of-the-art unsupervised method based on self-learning as in Sect. 2.1, which performs well on same-domain embeddings. We then compute a dictionary by nearest-neighbor search between elements of $\widetilde{X}$ and $\widetilde{Y}$ using the CSLS criterion. We filter the dictionary by keeping only the pairs for which the entry word for the target language is in the vocabulary of Z. And finally, we align X and Z by learning the orthogonal mapping matching the embeddings of the entries A and B of the filtered dictionary (Eq. 2). Using Procrustes analysis, it can be obtained as $W^* = UV^\top$ with the result of the SVD $A^\top B = USV^\top$.

We have XW^* and Z the aligned cross-domain and cross-lingual embeddings.

This cross-domain and cross-lingual alignment method could serve as initialization for unsupervised translation models [3,11] and we will evaluate it on two domain-specific translation tasks in the following section.

Algorithm 1: Proposed method of alignment with proxy

Input : source embedding X, target Z and proxy Y.
Output: aligned source and target XW^* and Z.
Create aligned $\widetilde{X}$ and $\widetilde{Y}$ with VecMap [2] on X and Y.
Initialize $D = \{\}$
for *word w in $\widetilde{X}$* **do**
    Find w' nearest-neighbor of w in $\widetilde{Y}$.
    if *w' in vocabulary of Z* **then**
       | Add (w, w') to dictionary D.
    end
end
for *word w' in $\widetilde{Y}$ and vocabulary of Z* **do**
    Find w nearest-neighbor of w' in $\widetilde{X}$.
    Add (w, w') to dictionary D.
end
Build $A \in \mathbb{R}^{N \times d}$ and $B \in \mathbb{R}^{N \times d}$ embeddings entries of D in X and Z.
Perform SVD: $A^\top B = USV^\top$.
Solve Eq. 2: $W^* = UV^\top$.
Compute aligned embeddings XW^* and Z.

---

4   Experimental Validation

To assess our hypothesis we perform three experiments: (1) showing that some vocabulary subsets of embeddings from different domains are topologically similar with Bottleneck distance; (2) evaluating the performance of the proposed partial alignment method on a translation task; (3) visualizing the learned partial alignment with a dimensionality reduction technique.

4.1   Datasets

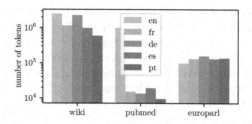

Fig. 2. Number of distinct tokens in each embedding by corpus and by language

For all experiments, we leverage embeddings built on three different corpora (Wikipedia, PubMed, EuroParl) with five different languages (English, French, German, Spanish and Portuguese). PubMed is a collection of approximately 21 million biomedical abstracts[2] mainly written in English. EuroParl [10] is a parallel corpus built from proceedings of the European Parliament. All embeddings were built with FastText [4] with 300 dimensions. Those from Wikipedia were obtained directly from the FastText website and we trained FastText ourselves on PubMed and EuroParl using the official implementation. We show on Fig. 2 the vocabulary size for each embedding. Wikipedia embeddings as well as the English PubMed embeddings have a vocabulary size of the same order of magnitude. The other corpora are not comparable. This allows us to emphasize on why we need cross-domain alignment methods as domain-specific data is sometimes lacking in languages other than English.

4.2   Measuring the Deviation from Isometry

We use the Bottleneck distance (Sect. 2.2) to measure the deviation from isometry between various subsets of different pairs of embeddings. We limit the subsets to 5k words for computability reasons and because previous works [16] have used the same constant and shown correlation with the ability to align embeddings. We compute the Bottleneck distance between three kinds of subsets:

---

[2] Courtesy of the U.S. National Library of Medicine https://www.nlm.nih.gov/databases/download/pubmed_medline.html.

Table 1. Bottleneck distance between the 5k most frequent words for different subsets and different domain pairs averaged over all language pairs with English (standard deviation between parenthesis)

Most frequent words of each embedding			English		
			wiki (1)	Pubmed (2)	Europarl (3)
fr/de/es/pt	wiki	(a)	0.08 (0.03)	0.12 (0.02)	0.15 (0.05)
	Pubmed	(b)	0.21 (0.08)	0.17 (0.06)	0.23 (0.08)
	Europarl	(c)	0.19 (0.03)	0.16 (0.01)	0.05 (0.02)
Most frequent words in PubMed parallel corpus			English		
			wiki	Pubmed	Europarl
fr/de/es/pt	wiki	(d)	0.08 (0.02)	0.08 (0.01)	0.07 (0.03)
	Pubmed	(e)	0.20 (0.06)	0.16 (0.07)	0.23 (0.08)
	Europarl	(f)	0.14 (0.03)	0.10 (0.03)	0.07 (0.01)
Most frequent words in Europarl parallel corpus			English		
			wiki	Pubmed	Europarl
fr/de/es/pt	wiki	(g)	0.09 (0.04)	0.10 (0.03)	0.13 (0.04)
	Pubmed	(h)	0.18 (0.06)	0.16 (0.07)	0.21 (0.06)
	Europarl	(i)	0.17 (0.02)	0.16 (0.03)	0.05 (0.02)

- The 5k most frequent words for each embedding, which was shown to correlate with the ability to align whole embeddings [16];
- The 5k most frequent words in each language of a parallel Pubmed corpus;
- The 5k most frequent words in each language of a parallel Europarl corpus.

We measure the Bottleneck distance for all domain pairs for language pairs involving English only, as we were able to obtain only parallel copora between English and other languages. Results are reported on Table 1. We average over the other languages to summarize, as results were consistent across languages as shown by the low standard deviations, except for pairs including non-English Pubmed embeddings for which the small size of those embeddings (Fig. 2) might explain the high Bottleneck distance and standard deviation. The lines corresponding to the latter situation (b,e,h) are discarded from further analysis.

The Bottleneck distance for same-domain pairs and most frequent words for Wikipedia and Europarl (a1,c3) indicates that topologically similar subsets have a Bottleneck distance below 0.09 whereas disimilar ones, such as cross-domain pairs have Bottleneck distances above 0.10. This verifies that embedding as a whole cannot be aligned if not from the same domain.

For each cross-domain pairs, the Bottleneck distance for parallel subsets is lower than for frequent words or at least equal. The most striking cases are when comparing non-English Wikipedia with English Europarl or Pubmed on the Pubmed parallel vocabulary (d2,d3), where the Bottleneck distance becomes comparable to those of same-domain pairs. For other domain-pairs, the Bottleneck distance is higher but we can still observe decreases with respect to the first subset (comparing c1 with f1,i1 and c2 with f2,i2). This suggests that the embeddings of parallel vocabularies for cross-domain pairs are topologically close.

With this first experiment, we can support our hypothesis that the quasi-isometry assumption might still hold between well-chosen subsets of embeddings from different domains. Indeed, the Bottleneck distance for cross-domain pairs was systematically smaller for parallel vocabularies. However, in an unsupervised method, such subsets may not be straightforward to find. That is why we devised the method described in Sect. 3, which we evaluate in the following section.

4.3   Unsupervised Cross-Domain and Cross-Lingual Alignement

The method proposed in Sect. 3 is evaluated on the Europarl parallel corpora and on the test set of the Biomedical WMT19 dataset[3]. This latter dataset consists of non-English Pubmed abstracts (around 100 by language pair) and their English translation. We proceed token-by-token by nearest-neighbor search with cosine distance and CSLS criterion. We use the BLEU-1 score [15] to evaluate translations:

$$\text{BLEU-1}(r, h) = \min\left(1, e^{1-\frac{|r|}{|h|}}\right) \frac{\sum_{w \in h} \min(\text{count}_h(w), \text{count}_r(w))}{\sum_{w \in h} \text{count}_h(w)} \quad (9)$$

It is a modified precision measure on the tokens with an additional term penalizing short translations and a clipping on the count of occurences to avoid giving high scores to candidate translations (h) which repeat words from the reference translation (r).

We choose translation over bilingual lexicon induction, which does not account for morphological variations of words [7] and gives too much importance to specific words such as proper nouns [9]. Moreover, cross-lingual embeddings can be used as initialization models in unsupervised translation models [3,11].

Table 2. BLEU-1 scores on Biomedical WMT19 and Europarl

Cross-domain	Biomedical WMT19				Europarl			
	fr-en	es-en	de-en	pt-en	fr-en	es-en	de-en	pt-en
MUSE	0.05	0.06	0.06	0.07	0.00	0.03	0.01	0.01
WP	0.08	0.08	0.05	0.05	0.01	0.01	0.01	0.01
VecMap (unsupervised)	0.09	0.06	0.07	0.07	0.02	0.02	0.03	0.02
VecMap (weakly supervised)	0.30	0.37	0.25	0.28	0.33	0.39	0.31	0.33
Proposed method	**0.38**	**0.50**	**0.31**	**0.47**	**0.40**	**0.44**	**0.37**	**0.44**
Same-domain	Biomedical WMT19				Europarl			
	fr-en	es-en	de-en	pt-en	fr-en	es-en	de-en	pt-en
MUSE	0.43	0.58	**0.40**	0.53	0.43	0.47	**0.41**	0.43
WP	0.45	0.57	0.36	0.51	0.45	0.47	0.40	0.46
VecMap (unsupervised)	**0.46**	**0.58**	0.37	**0.56**	**0.46**	**0.49**	0.41	**0.47**
UCAM (supervised)	–	*0.71*	*0.61*	–	–	–	–	–

---

[3] http://www.statmt.org/wmt19/biomedical-translation-task.html.

Results for translation on the Biomedical WM19 and Europarl tasks are shown on Table 2. The proposed method is compared to the unsupervised methods VecMap [2], MUSE [6] and Wasserstein-Procrustes (WP) [8] applied to cross-domain embeddings, a weakly supervised method based on identical words and VecMap applied to cross-domain embeddings as well, the same unsupervised methods (VecMap, MUSE, WP) applied to same-domain embeddings, and the UCAM submission [17] to the Biomedical WMT19 challenge, a supervised deep learning model which gives an upper-bound baseline.

In all languages and domains, the proposed method outperforms any other that tries to align embeddings from different domains. In such cross-domain settings, fully unsupervised methods obtain near-zero translation scores, whereas the proposed unsupervised method gives fair results, even outperforming the weakly-supervised method based on identical words. The proposed method demonstrates that it is possible to align embeddings from different domains in an unsupervised manner, although it is outperformed by same-domain alignments.

Nevertheless, the fact that we were able to obtain fair translation scores with a strictly isometric mapping suggests that our hypothesis of isometry between subsets might hold. To further validate this hypothesis and the intuition behind our method, we show in the following section with a visualization method that our alignment is indeed partial.

4.4   Visualization with t-SNE

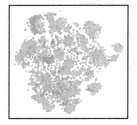 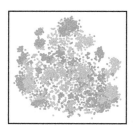 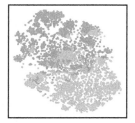

(a) VecMap, same domain. (b) VecMap, cross-domain. (c) ours (cross-domain).

Fig. 3. t-SNE for alignment between French Wikipedia and English Wikipedia (same domain) or English Pubmed (cross-domain)

We use a dimensionality-reduction technique, t-SNE [12], to confirm that the proposed method performs a partial alignment. Results for the French-English pair are shown on Fig. 3. Blue points are t-SNE representations of embeddings of French words from Wikipedia embedding and orange points are for English words either from Wikipedia (Fig. 3(a)) or Pubmed (Fig. 3(b), 3(c)) embedding, represented after being aligned by VecMap (Fig. 3(a), 3(b)) or the proposed method (Fig. 3(a)). A classic alignment method like VecMap aligns the embedding as a whole. In the same-domain setting (Fig. 3(c)) it works well as suggested by our previous results in the translation tasks and the seemingly local alignment of small clusters. In the cross-domain setting (Fig. 3(b)), we do not observe the same local alignment, which is corroborated by the near-zero BLEU-1 score in the previous experiment. Finally, t-SNE for the proposed method (Fig. 3(c)) seems to show that there is a partial alignment. And even if we do not observe local alignment of clusters as in VecMap for same domain, we can confirm with this visualization that our filtered dictionary allows a partial alignment as hypothesized in introduction (Fig. 1).

5   Discussion and Conclusion

We proposed a novel method for aligning embeddings from different domains, a setting where other unsupervised methods failed. By aligning only a well-chosen subsets instead of the whole embeddings, we showed that unsupervised isometry-based alignment were not doomed to fail in such setting. However, our method needs additional data. Further work could try to improve on the proposed method by removing this need for additional data or to use it as initialization of an unsupervised neural machine translation model as is already done with same-domain alignments [3,11]. Another limitation of our work is that the Bottleneck distance is only a lower bound of the Gromov-Hausdorff distance. Although this bound is expected to be tight [5], to the best of our knowledge, there is no formal proof of this tightness. Nevertheless, there is empirical evidence of correlation between Bottleneck distance and the ability to align embeddings with an orthogonal mapping [16], although this was only shown for different pairs of languages in the same domain. Further work would try to demonstrate the same correlation for different domains.

We showed with Bottleneck distance that some subsets of embeddings were topologically similar, despite being from different languages and domains. We demonstrated that orthogonal mapping alignments are not doomed to fail thanks to the proposed method which outperforms any other unsupervised distance-preserving alignment applied to embeddings from different domains. Finally, we confirmed that the proposed method was performing the hypothesized partial alignment with a dimensionality-reduction technique. With these three pieces of evidence, we are confident that the approximate isometry assumption can still hold between well-chosen subsets of cross-domain embeddings.

References

1. Artetxe, M., Labaka, G., Agirre, E.: Learning bilingual word embeddings with (almost) no bilingual data. In: Proceedings of the 55th Annual Meeting of ACL (2017). https://doi.org/10.18653/v1/P17-1042
2. Artetxe, M., Labaka, G., Agirre, E.: A robust self-learning method for fully unsupervised cross-lingual mappings of word embeddings. In: Proceedings of the 56th Annual Meeting of ACL (2018). https://doi.org/10.18653/v1/P18-1073
3. Artetxe, M., Labaka, G., Agirre, E.: An effective approach to unsupervised machine translation. In: Proceedings of the 57th Annual Meeting of ACL (2019). https://doi.org/10.18653/v1/P19-1019

4. Bojanowski, P., Grave, E., Joulin, A., Mikolov, T.: Enriching word vectors with subword information. Trans. ACL **5**, 135–146 (2017). https://www.aclweb.org/anthology/Q17-1010
5. Chazal, F., Cohen-Steiner, D., Guibas, L.J., Mémoli, F., Oudot, S.Y.: Gromov-hausdorff stable signatures for shapes using persistence. In: Proceedings of the Symposium on Geometry Processing, SGP 2009, pp. 1393–1403. Eurographics Association, Goslar, DEU (2009)
6. Conneau, A., Lample, G., Ranzato, M., Denoyer, L., Jégou, H.: Word translation without parallel data (2017). http://arxiv.org/abs/1710.04087
7. Czarnowska, P., Ruder, S., Grave, E., Cotterell, R., Copestake, A.: Don't forget the long tail! a comprehensive analysis of morphological generalization in bilingual lexicon induction. In: Proceedings of EMNLP-IJCNLP (2019). https://doi.org/10.18653/v1/D19-1090
8. Joulin, A., Bojanowski, P., Mikolov, T., Jégou, H., Grave, E.: Loss in translation: Learning bilingual word mapping with a retrieval criterion. In: Proceedings of the 2018 Conference of EMNLP (2018). https://doi.org/10.18653/v1/D18-1330
9. Kementchedjhieva, Y., Hartmann, M., Søgaard, A.: Lost in evaluation: misleading benchmarks for bilingual dictionary induction. In: Proceedings of EMNLP-IJCNLP (2019). https://doi.org/10.18653/v1/D19-1328
10. Koehn, P.: Europarl: a parallel corpus for statistical machine translation. In: Conference Proceedings: the tenth Machine Translation Summit. AAMT, AAMT, Phuket, Thailand (2005). http://mt-archive.info/MTS-2005-Koehn.pdf
11. Lample, G., Conneau, A., Denoyer, L., Ranzato, M.: Unsupervised machine translation using monolingual corpora only. In: International Conference on Learning Representations (2018). https://openreview.net/forum?id=rkYTTf-AZ
12. van der Maaten, L., Hinton, G.: Visualizing data using t-SNE. J. Mach. Learn. Res. **9**(86), 2579–2605 (2008)
13. Mikolov, T., Chen, K., Corrado, G., Dean, J.: Efficient estimation of word representations in vector space (2013). http://arxiv.org/abs/1301.3781
14. Mikolov, T., Le, Q.V., Sutskever, I.: Exploiting similarities among languages for machine translation (2013). http://arxiv.org/abs/1309.4168
15. Papineni, K., Roukos, S., Ward, T., Zhu, W.J.: BLEU: a method for automatic evaluation of machine translation. In: Proceedings of the 40th Annual Meeting of ACL (2002). https://doi.org/10.3115/1073083.1073135
16. Patra, B., Moniz, J.R.A., Garg, S., Gormley, M.R., Neubig, G.: Bilingual lexicon induction with semi-supervision in non-isometric embedding spaces. In: Proceedings of the 57th Annual Meeting of ACL (2019). https://doi.org/10.18653/v1/P19-1018
17. Saunders, D., Stahlberg, F., Byrne, B.: UCAM biomedical translation at WMT19: transfer learning multi-domain ensembles. In: Proceedings of the Fourth Conference on Machine Translation (2019). http://www.aclweb.org/anthology/W19-5421
18. Schönemann, P.H.: A generalized solution of the orthogonal procrustes problem. Psychometrika **31**(1), 1–10 (1966). https://doi.org/10.1007/BF02289451
19. Søgaard, A., Ruder, S., Vulić, I.: On the limitations of unsupervised bilingual dictionary induction. In: Proceedings of the 56th Annual Meeting of ACL (2018). https://doi.org/10.18653/v1/P18-1072
20. Zhang, M., Liu, Y., Luan, H., Sun, M.: Earth mover's distance minimization for unsupervised bilingual lexicon induction. In: Proceedings of the 2017 Conference on EMNLP (2017). https://doi.org/10.18653/v1/D17-1207

Inference in Neural Networks
Using Conditional Mean-Field Methods

Ángel Poc-López[1]([✉])[iD] and Miguel Aguilera[1,2][iD]

[1] ISAAC Lab, I3A Engineering Research Institute of Aragón, University of Zaragoza,
Zaragoza, Spain
696295@unizar.es
[2] Department of Informatics Sussex Neuroscience, University of Sussex, Brighton, UK
sci@maguilera.net

Abstract. We extend previous mean-field approaches for non-
equilibrium neural network models to estimate correlations in the system.
This offers a powerful tool for approximating the system dynamics, as
well as a fast method for inferring network parameters from observations. We develop our method for the asymmetric kinetic Ising model
and test its performance on 1) synthetic data generated by an asymmetric Sherrington Kirkpatrick model and 2) recordings of in vitro neuron
spiking activity from the mouse somatosensory cortex. We find that our
mean-field method outperforms previous ones in estimating networks
correlations and successfully reconstructs network dynamics from data
near a phase transition showing large fluctuations.

Keywords: Mean-field · Fluctuations · Neural network · Ising model ·
Inference · Spike train

1 Introduction

Biological and neural networks generally exhibit out-of-equilibrium dynamics
[9]. Resulting physiological rhythms and emerging patterns are in continuous
and asymmetrical interactions within and between networks and are often found
to self-organize near critical regimes with large fluctuations [15]. Although new
data acquisition technologies provide descriptions of the dynamics of hundreds
or thousands of neurons in different animals [3,12], these properties make it challenging to analyze them assuming an asymptotic equilibrium state or standard
approximation methods. This problem demands mathematical tools for capturing and reproducing the types of non-equilibrium fluctuations found in large
biological systems.

The kinetic Ising model with asymmetric couplings is a prototypical model for
such non-equilibrium dynamics in biological systems [10]. The model is described
as a discrete-time Markov chain of interacting binary units, resembling the nonlinear dynamics of recurrently connected neurons. A popular application of the

---

Á. Poc-López and M. Aguilera—Contributed equally to this work.

© Springer Nature Switzerland AG 2021
T. Mantoro et al. (Eds.): ICONIP 2021, LNCS 13109, pp. 228–237, 2021.
https://doi.org/10.1007/978-3-030-92270-2_20

model (known as the 'inverse Ising problem' [1]) involves inference of the model parameters to capture the properties of observed data from biological systems [17]. Unfortunately, exact solutions for describing network dynamics and inference often become computationally too expensive due to combinatorial explosion of patterns in large systems, limiting applications using sampling methods to around a hundred of neurons [15, 16]. In consequence, analytical approximation methods are necessary for large networks. To this end, mean-field methods are powerful tools to track down otherwise intractable statistical quantities. The standard mean-field approximations to study equilibrium Ising models are the classical naive mean-field (nMF) and the more accurate Thouless-Anderson-Palmer (TAP) approximations [13]. In non-equilibrium networks, however, it is not obvious how to apply mean-field methods. Alternatives involve the use of information geometric approaches [2, 7] or Gaussian approximations of the network effective fields [8]. In this work, we will expand the latter to explicitly address fluctuations for network simulation and inference.

2   Kinetic Ising Model

We model neural network activation using a kinetic Ising model, i.e. a generalized linear model with binary states and pairwise couplings. The network consists of a system of N interacting neurons $\mathbf{s}_t$. Neuron i at a time t can take values $s_{i,t} \in \{+1, -1\}, i = 1, 2, \ldots, N, t = 0, 1, \ldots, T$. At time t, the activation probability is defined by a nonlinear sigmoid function,

$$P(s_{i,t}|\mathbf{s}_{t-1}) = \frac{e^{s_{i,t}h_{i,t}}}{2\cosh h_{i,t}}. \tag{1}$$

Activation is driven by effective fields $\mathbf{h}_t$, composed of a bias term $\mathbf{H} = \{H_i\}$ and couplings to units at the previous time step $\mathbf{J} = \{J_{ij}\}$,

$$h_{i,t} = H_i + \sum_j J_{ij} s_{j,t-1}. \tag{2}$$

When the couplings are asymmetric (i.e., $J_{ij} \neq J_{ji}$), the system is away from equilibrium because the process is irreversible with respect to time. In this article, we are interested in estimating first and second-order statistical moments of the system. That is, the components of the mean activation of a system and the fluctuations around this mean. Thus, we will calculate the activation rates $\mathbf{m}_t$, correlations between pairs of units (covariance function) $\mathbf{C}_t$, and delayed correlations $\mathbf{D}_t$ defined as

$$m_{i,t} = \sum_{\mathbf{s}_t} s_{i,t} P(\mathbf{s}_t), \tag{3}$$

$$C_{ik,t} = \sum_{\mathbf{s}_t} s_{i,t} s_{k,t} P(\mathbf{s}_t) - m_{i,t} m_{k,t}, \tag{4}$$

$$D_{il,t} = \sum_{\mathbf{s}_t, \mathbf{s}_{t-1}} s_{i,t} s_{l,t-1} P(\mathbf{s}_t, \mathbf{s}_{t-1}) - m_{i,t} m_{l,t-1}. \tag{5}$$

3   Gaussian Mean-Field Method

In [8], the authors proposed that, in some cases, the second term of Eq. 2 is a sum of a large number of weakly coupled components. Assuming weak and asymmetric couplings, given the Central Limit Theorem, they approximate this term by a Gaussian distribution $P(h_{i,t}) \approx \mathcal{N}(g_{i,t}, \Delta_{i,t})$, with mean and variance:

$$g_{i,t} = H_i + \sum_j J_{ij} m_{j,t-1}, \tag{6}$$

$$\Delta_{i,t} = \sum_j J_{ij}^2 (1 - m_{j,t-1}^2). \tag{7}$$

Yielding mean-field activation of neuron $s_{i,t}$ at time t as:

$$m_{i,t} \approx \int D_z \tanh(g_{i,t} + z\Delta_{i,t}), \tag{8}$$

where $D_z = \frac{dz}{\sqrt{2\pi}} exp(-\frac{1}{2}z^2)$ describes a Gaussian integral term with mean zero and unity variance. As well, the method provides a relation between $\mathbf{D}_t$ and $\mathbf{C}_{t-1}$ [8]. Alternatively, these equations can be derived by defining a mean-field problem using path integral methods [4] or information geometry [2]. This approximation is exact in the thermodynamic limit for fully asymmetric methods [8]. However, in [2] it was shown that this method (extended to add calculations of same-time correlations) fails to approximate the behaviour of fluctuations near a ferromagnetic phase transition for networks of hundreds of neurons.

4   Conditional Gaussian Mean-Field Method

The motivation of this article is to explore extensions of the Gaussian mean-field method in [8] to accurately capture fluctuations in non-equilibrium systems, even in the proximity of critical dynamical regimes.

Time-Delayed Correlations. In order to better estimate correlations, instead of using a Gaussian approximation to compute $m_{i,t}$ (which results in a fully independent model), we propose the use of multiple conditional Gaussian distributions, aimed to capture conditional averages $m_{i,t}(s_{l,t-1})$ for a fixed neuron l at the previous time-step $s_{l,t-1}$. This conditional average can be approximated using a similar mean-field assumption:

$$m_{i,t}(s_{l,t-1}) = \sum_{s_{t-1}} \tanh(h_i) P(s_{t-1}|s_{l,t-1})$$

$$\approx \int D_z \tanh(g_{i,t}(s_{l,t-1}) + z\Delta_{i,t}(s_{l,t-1})), \tag{9}$$

where the statistical moments of the Gaussian distribution are computed as

$$g_{i,t}(s_{l,t-1}) = H_i + \sum_j J_{ij} m_{j,t-1}(s_{l,t-1}), \tag{10}$$

$$\Delta_{i,t}(s_{l,t-1}) = \sum_j J_{ij}^2 (1 - m_{j,t-1}^2(s_{l,t-1})). \tag{11}$$

Here, $m_{j,t-1}(s_{l,t-1})$ are now conditional averages of two spins at time $t-1$. As a pairwise distribution $P(s_{j,t-1}, s_{l,t-1})$ is completely determined by its moments $m_{j,t-1}, m_{l,t-1}, C_{jl,t-1}$. We derive the equivalence

$$m_{j,t-1}(s_{l,t-1}) = m_{j,t-1} + \frac{s_{l,t-1} - m_{l,t-1}}{1 - m_{l,t-1}^2} C_{jl,t-1}. \tag{12}$$

Once $m_{i,t}(s_{l,t-1})$ is known (Eq. 9), computing the marginal over $s_{l,t-1} \in \{1, -1\}$ we calculate $m_{i,t}$ as

$$m_{i,t} = \sum_{s_{l,t-1}} m_{i,t}(s_{l,t-1}) \frac{1 + s_{l,t-1} m_{l,t-1}}{2}. \tag{13}$$

Finally, having the values of the conditional magnetizations we compute time-delayed correlations $D_{il,t}$ as

$$D_{il,t} = \sum_{s_{l,t-1}} m_{i,t}(s_{l,t-1}) \frac{s_{l,t-1} + m_{l,t-1}}{2} - m_{i,t} m_{l,t-1}. \tag{14}$$

This sequence approximates the values of $\mathbf{m}_t, \mathbf{D}_t$ knowing the values of $\mathbf{C}_{t-1}$. In order to recursively apply this method, we need to complement our equations with a method for computing $\mathbf{C}_t$ from $\mathbf{m}_t, \mathbf{D}_t$.

Equal-Time Correlations. We follow a similar procedure for equal-time correlations. First, we calculate the conditional average $m_{i,t}(s_{k,t})$, now conditioned on a neuron at time t:

$$m_{i,t}(s_{k,t}) \approx \int Dz \, \tanh(g_{i,t}(s_{k,t}) + z\Delta_{i,t}(s_{k,t})), \tag{15}$$

$$g_{i,t}(s_{k,t}) = H_i + \sum_j J_{ij} m_{j,t-1}(s_{k,t}), \tag{16}$$

$$\Delta_{i,t}(s_{k,t}) = \sum_j J_{ij}^2 (1 - m_{j,t-1}^2(s_{k,t})). \tag{17}$$

Here, we see that the Gaussian integral depends on averages $m_{j,t-1}(s_{k,t})$, conditioned on the next time step. We determine these quantities from the delayed correlations computed by Eq. 14 at the previous step

$$m_{j,t-1}(s_{k,t}) = m_{j,t-1} + \frac{s_{k,t} - m_{k,t}}{1 - m_{k,t}^2} D_{kj,t}. \tag{18}$$

Once computed this conditional magnetization value, and having obtained the magnetizations from Eq. 13, correlations are computed as:

$$C_{ik,t} = \sum_{s_{k,t}} m_{i,t}(s_{k,t}) \frac{s_{k,t} + m_{k,t}}{2} - m_{i,t} m_{k,t}. \qquad (19)$$

5   Results

In this section, we test our method 1) in an asymmetric version of the well-known Sherrington-Kirkpatrick (SK) model, and 2) in vitro recordings of neuron spiking activity from the mouse somatosensory cortex [6].

5.1   Sherrington-Kirkpatrick Model

We compare the performance of our method with respect of two widely used methods: the TAP equations [11] and the Gaussian mean-field method [8] (implemented as in [2] to account for same-time correlations). We use a dataset with simulations of an asymmetric kinetic version of the SK model with $N = 512$ neurons [2], known to have a phase transition between ordered and disordered phases (although the spin glass phase is absent for fully asymmetric models). This critical point maximizes fluctuations of the system, thus being challenging for mean-field methods for finite sizes. Approximating network behaviour near criticality is highly relevant as many biological systems, like neural networks, are believed to be poised near critical points [15]. External fields $\mathbf{H}_i$ are sampled from independent uniform distributions $\mathcal{U}(-\beta H_0, \beta H_0)$, $H_0 = 0.5$, and coupling terms J_{ij} from independent Gaussian distributions $\mathcal{N}(\beta \frac{J_0}{N}, \beta^2 \frac{J_\sigma^2}{N})$, $J_0 = 1, J_\sigma = 0.1$, where β is a scaling parameter (i.e., an inverse temperature). The model displays a ferromagnetic phase transition, which takes place at $\beta_c \approx 1.1108$ [2].

Network Dynamics. First, we examine the performance of the different methods for computing the statistics of the model, i.e., $\mathbf{m}$, $\mathbf{C}$, and $\mathbf{D}$. We simulate the behaviour of the model for $T = 128$ steps from $\mathbf{s}_0 = \mathbf{1}$, comparing exact and mean-field behaviour for different values of the inverse temperature $\beta \in [0.7\beta_c, 1.3\beta_c]$. In Fig. 1A, B, C we observe the average evolution of the magnetizations and equal-time and delayed correlations. Figure 1D, E, F shows a direct comparison between approximated and real values. We observe that our method makes the best approximation at the critical point. The approximations of both $\mathbf{m}$ and $\mathbf{D}$ show small errors, although these are accumulated resulting in a less accurate prediction of $\mathbf{C}$. Besides, Fig. 1G, H, I shows that our method performs better than the others at all inverse temperatures, including near the critical point.

Inference. Second, we compare the performance of the different methods inferring the model parameters from data. Starting from $\mathbf{H} = 0$ and $\mathbf{J} = 0$, a gradient ascent on the log-likelihood is performed, approximating $\mathbf{m}$ and $\mathbf{D}$ using

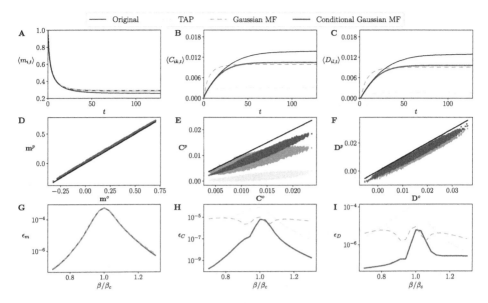

Fig. 1. Approximation of neural dynamics in the SK model. Top: evolution of average magnetizations (A), equal-time correlations (B) and delayed correlations (C) found by different mean-field methods for $\beta = \beta_c$. Middle: comparison of magnetizations (D), equal-time correlations (E) and delayed correlations (F) found by the different mean-field approximations (ordinate, p superscript) with the original values (abscissa, o superscript) for $\beta = \beta_c$ and $t = 128$. Black lines represent the identity line. Bottom: mean Squared Error (MSE) of the magnetizations $\epsilon_\mathbf{m} = \langle\langle (m_{i,t}^o - m_{i,t}^p)^2\rangle_i\rangle_t$ (G), equal-time correlations $\epsilon_\mathbf{C} = \langle\langle (C_{ik,t}^o - C_{ik,t}^p)^2\rangle_{ik}\rangle_t$ (H), and delayed correlations $\epsilon_\mathbf{D} = \langle\langle (D_{ik,t}^o - D_{ik,t}^p)^2\rangle_{il}\rangle_t$ (I) for 21 values of β in the range $[0.7\beta_c, 1.3\beta_c]$.

the mean-field equations (see [2]). The learning algorithm is run a maximum of $R = 10^6$ trials per step. Figure 2A, B displays the inferred external fields (**H**) and couplings (**J**) against the real ones. We observe how the results for the TAP are displaced away from the identity line and how the results for the Gaussian mean-field method and ours are very similar. However, in Fig. 2C, D we observe that our method obtains a lower Mean Squared Error (ε) for all inverse temperatures.

Phase Transition Reconstruction. Finally, we reconstruct the phase transition in the model by combining the inverse and forward Ising problem. We use the **H** and **J** inferred in the inverse problem to calculate the systems' statistical moments and determine if the learned model is able to reproduce the original behaviour of the system in unobserved conditions. In order to reproduce the phase transition, the learned **H** and **J** in the previous section are multiplied by a fictitious temperature in the range $[0.7\beta_c, 1.3\beta_c]$. In Fig. 3 we display the averaged statistical moments after simulation. Our method achieves a better adjustment

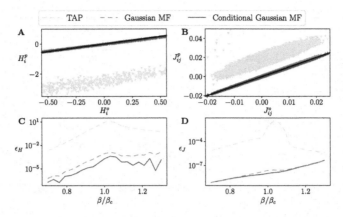

Fig. 2. Network inference in the SK model. Top: inferred external fields (A) and couplings (B) found by different mean-field models, plotted versus the real ones for $\beta = \beta_c$. Black lines represent the identity line. Bottom: mean Squared Error of inferred external fields $\epsilon_{\mathbf{H}} = \langle (H_i^o - H_i^p)^2 \rangle_i$ (C) and couplings $\epsilon_{\mathbf{J}} = \langle (J_{ij}^o - J_{ij}^p)^2 \rangle_{ij}$ (D) for 21 values of β in the range $[0.7\beta_c, 1.3\beta_c]$.

of the averages. While the Gaussian mean-field method from [8] achieves a good approximation far from the critical point, our method achieves a close approximation at all inverse temperatures. We not only reduce the difference between the approximation and the expected values for the system's statistics, but also preserve the shape around the critical point where other methods flatten out. Our method could help to better characterize non-equilibrium systems of high dimensionality poised near a phase transition.

5.2   In Vitro Neuronal Spike Train Data

Finally, we test the performance of our conditional Gaussian mean-field method on in vitro neural dynamics. To this end, we selected a dataset containing neural spiking activity from mouse somatosensory cortex in organotypic slice cultures [6]. Additional information about this dataset can be consulted in [5]. Specifically, we selected dataset 1, which contained 166 neurons. In order to adjust the dynamics of the dataset and the model, we binned spikes in discrete time windows of length δt and we extended the model to introduce asynchronous updates [18] in which each neuron $s_{i,t}$ is updated with Eq. 1 with a probability γ.

Inference and Network Dynamics. We infer the model parameters that best fit the data applying the learning algorithm in [18] with mean-field approximations starting from $\mathbf{H} = 0$ and $\mathbf{J} = 0$. Different learning hyperparameters were manually selected and tested, resulting in $\delta t = 70\,\text{ms}$ and $\gamma = 0.77$. Early stopping was used, taking the minimum MSE of one-step-estimation of $\mathbf{C}$ as the stopping criterion. After learning, we generated new data simulating the

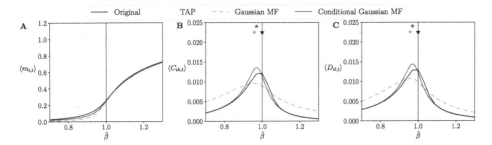

Fig. 3. Phase transition reconstruction of the SK model. Average of the Ising model's magnetizations (A), non-diagonal equal-time correlations (B), and non-diagonal delayed correlations (C), at the last step $t = 128$ of a simulation, found by different mean-field methods using the reconstructed network $\mathbf{H}, \mathbf{J}$ by solving the inverse Ising problem at $\beta = \beta_c$ and multiplying the estimated parameters by a fictitious inverse temperature $\tilde{\beta}$. The stars indicate the values of $\tilde{\beta}$ with maximum fluctuations.

model. In Fig. 4 we compare the statistics of the inferred model with respect to the original values. As we observe from the figure, almost all the statistics for each neuron lie near the identity line, leading to a MSE of $\epsilon_{\mathbf{m}} = 1.18 \cdot 10^{-06}$, $\epsilon_{\mathbf{C}} = 6.96 \cdot 10^{-06}$ and $\epsilon_{\mathbf{D}} = 6.26 \cdot 10^{-06}$ at the last step of the simulation.

Phase Transition Reconstruction. Finally, we explore if the inferred model presents signatures of a phase transition. We multiply $\mathbf{H}$ and $\mathbf{J}$ by a fictive inverse temperature $\tilde{\beta}$ in the range $[0, 2]$ and simulate every neural system, reaching a steady state. Figure 5 displays the average statistics at the different temperatures. As in the SK model (Fig. 3) we observe a peak in correlations around the operating temperature (i.e. $\tilde{\beta} = 1$), suggesting the presence of a continuous phase transition. Further analysis (e.g. testing of different network sizes, entropy estimations [14]) should be performed to confirm this result.

6 Discussion

Many biological networks are found to self-organize at points of their parameter space, maximizing fluctuations in the system and showing non-equilibrium dynamics. Although mean-field methods have been successfully proposed as a tool to approximate complex network phenomena and transitions, successfully capturing fluctuations in non-equilibrium conditions is a challenging open problem. Here, we extend a previous method proposing a Gaussian mean-field estimation of the average activation of a system [8]. This method is known to accurately capture average activations in fully asymmetric networks, but capturing fluctuations or transitions in networks presenting different degrees of symmetry is still challenging. We have shown how an extension based in computing Gaussian estimations of the conditional input field offers a good approximation of pairwise

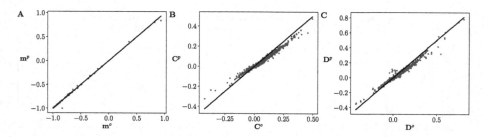

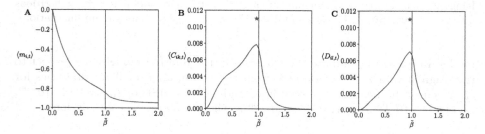

Fig. 4. Inference and neural dynamics approximation of in vitro neural observations. Comparison of magnetizations (A), equal-time correlations (B) and delayed correlations (C) found after solving the inverse Ising problem (ordinate, p superscript) with the original values (abscissa, o superscript) for $\beta = 1.0$ and $t = 128$. Black lines represent the identity line.

Fig. 5. Phase transition of in vitro neural observations. Average of the Ising model's magnetizations (A), non-diagonal equal-time correlations (B), and non-diagonal delayed correlations (C), by solving the inverse Ising problem at $\beta = 1.0$ and multiplying a fictitious inverse temperature $\tilde{\beta}$ to the estimated parameters at time $t = 128$. The stars are marked at the values of $\tilde{\beta}$ that yield maximum fluctuations.

correlations even in the proximity of a ferromagnetic phase transition in well-known neural network theoretical models. This is important as it allows not only to simulate network dynamics and fluctuations, but also as a fast method to infer neural network models from data. Our approach can be useful for studying the connectivity of biological neural circuits displaying complex dynamics. As well, we show in a preliminary test how the method is able to successfully infer a model reproducing the statistics of neural spike trains recorded from a sensori-motor cortex culture, suggesting that it operates near a critical phase transition. This is expected to foster useful tools to efficiently analyse large-scale properties of neural network dynamics.

Acknowledgements. M.A. was funded by the European Union's Horizon 2020 research and innovation programme under the Marie Skłodowska-Curie grant agreement No. 892715.

References

1. Ackley, D.H., Hinton, G.E., Sejnowski, T.J.: A learning algorithm for Boltzmann machines. Cogn. Sci. **9**(1), 147–169 (1985)
2. Aguilera, M., Moosavi, S.A., Shimazaki, H.: A unifying framework for mean-field theories of asymmetric kinetic Ising systems. Nat. Commun. **12**(1), 1–12 (2021)
3. Ahrens, M.B., Orger, M.B., Robson, D.N., Li, J.M., Keller, P.J.: Whole-brain functional imaging at cellular resolution using light-sheet microscopy. Nat. Methods **10**(5), 413–420 (2013)
4. Bachschmid-Romano, L., Battistin, C., Opper, M., Roudi, Y.: Variational perturbation and extended Plefka approaches to dynamics on random networks: the case of the kinetic Ising model. J. Phys. A: Math. Theor. **49**(43), 434003 (2016)
5. Ito, S., et al.: Large-scale, high-resolution multielectrode-array recording depicts functional network differences of cortical and hippocampal cultures. PLOS ONE **9**(8), 1–16 (2014)
6. Ito, S., Yeh, F.C., Timme, N.M., Hottowy, P., Litke, A.M., Beggs, J.M.: Spontaneous spiking activity of hundreds of neurons in mouse somatosensory cortex slice cultures recorded using a dense 512 electrode array. CRCNS. org (2016)
7. Kappen, H.J., Spanjers, J.J.: Mean field theory for asymmetric neural networks. Phys. Rev. E **61**(5), 5658–5663 (2000)
8. Mézard, M., Sakellariou, J.: Exact mean-field inference in asymmetric kinetic Ising systems. J. Stat. Mech.: Theory Exp. **2011**(07), L07001 (2011)
9. Nicolis, G., Prigogine, I.: Self-Organization in Nonequilibrium Systems: From Dissipative Structures to Order through Fluctuations. Wiley, New York (1977)
10. Roudi, Y., Dunn, B., Hertz, J.: Multi-neuronal activity and functional connectivity in cell assemblies. Curr. Opin. Neurobiol. **32**, 38–44 (2015)
11. Roudi, Y., Hertz, J.: Dynamical TAP equations for non-equilibrium Ising spin glasses. J. Stat. Mech.: Theory Exp. **2011**(03), P03031 (2011)
12. Stringer, C., Pachitariu, M., Steinmetz, N., Carandini, M., Harris, K.D.: High-dimensional geometry of population responses in visual cortex. Nature **571**, 361–365 (2019)
13. Thouless, D.J., Anderson, P.W., Palmer, R.G.: Solution of 'Solvable model of a spin glass'. Philos. Mag.: J. Theor. Exp. Appl. Phys. **35**(3), 593–601 (1977)
14. Tkačik, G., Marre, O., Amodei, D., Schneidman, E., Bialek, W., Ii, M.J.B.: Searching for collective behavior in a large network of sensory neurons. PLOS Comput. Biol. **10**(1), e1003408 (2014)
15. Tkačik, G., et al.: Thermodynamics and signatures of criticality in a network of neurons. Proc. Natl. Acad. Sci. **112**(37), 11508–11513 (2015)
16. Tyrcha, J., Roudi, Y., Marsili, M., Hertz, J.: The effect of nonstationarity on models inferred from neural data. J. Stat. Mech.: Theory Exp. **2013**(03), P03005 (2013)
17. Witoelar, A., Roudi, Y.: Neural network reconstruction using kinetic Ising models with memory. BMC Neurosci. **12**(1), P274 (2011)
18. Zeng, H.L., Alava, M., Aurell, E., Hertz, J., Roudi, Y.: Maximum likelihood reconstruction for Ising models with asynchronous updates. Phys. Rev. Lett. **110**, 210601 (2013)

Associative Graphs for Fine-Grained Text Sentiment Analysis

Maciej Wójcik, Adrian Horzyk$^{(\boxtimes)}$, and Daniel Bulanda

AGH University of Science and Technology in Krakow, 30-059 Krakow, Poland
mawojcik@student.agh.edu.pl, horzyk@agh.edu.pl, daniel@bulanda.net

Abstract. Due to social media's ubiquitousness, most advertising campaigns take place on such platforms as Facebook, Twitter, or Instagram. As a result, Natural Language Processing has become an essential tool to extract information about the users: their personality traits, brand preferences, distinctive vocabulary, etc. Such data can be further used to create text adverts profiled to engage with users who share a certain set of features. While most of the algorithms capable of processing the text are neural network-driven, associative graphs serve the same purpose, attaining usually similar or better accuracy, but being more explainable than black box-like models based on neural networks. This paper presents an associative graph for natural language processing and fine-grained sentiment analysis. The ability of associative graphs to represent complex relations between phrases can be used to create a model capable of classifying the input data into many categories simultaneously with high accuracy and efficiency. This approach enabled us to acquire a model performing similarly or better than the state-of-the-art solutions while being more explicit and easier to create and explain.

Keywords: Associative graph data structures · Natural language processing · Knowledge graphs · Sentiment analysis · Matching algorithms

1 Introduction

Natural language processing (NLP) is a discipline that focuses on the analysis of human language with the use of computers. It includes semantic analysis, next sentence word prediction, topic extraction, part-of-speech tagging, machine translation, etc. [1]. Currently, many applications are based on various artificial intelligence approaches, especially on deep neural networks, which supply us with the outstanding results, using the increasing computational power of new hardware solutions [2, 18, 20, 21]. Sentiment analysis and the capability of extracting the information about the text's context and characteristics have many appliances in many branches of modern industry, e.g. social media marketing.

Artificial intelligence (AI) and NLP in marketing can be used to create chatbots to improve the customers' experience, and on trend extraction [19] - analyzing

Supported by Native Hash LTD and the AGH University of Science and Technology.

T. Mantoro et al. (Eds.): ICONIP 2021, LNCS 13109, pp. 238–249, 2021.
https://doi.org/10.1007/978-3-030-92270-2_21

the user's actions and reactions, e.g. a content posted on social media, shopping habits, reviews of certain services. It can also be further used to tailor advertising campaigns to resemble the user's character traits and their needs. Creating well-matched campaigns is beneficial for companies since it may result in an increase in sale and decrease of advertising costs.

The most popular solutions used in sentiment analysis are based on convolutional neural networks (CNN) [11] and recurrent neural networks, especially Long-Short Term Memory (LSTM) or Gated Recurrent Unit (GRU) networks of various kinds [16] as well as Transformers like BERT [10]. Separate problem is fine-grained classification, where the algorithm does not only classify the text to one of the defined categories but also calculates a *scale* of text's similarity to a class. Furthermore, the most complex models face the issue of both fine-grained and coarse classification, where the data can be a part of one or more classes, differing in the similarity per class.

This paper proposes a multi-label multi-class classification model based on Associative Graph Data Structures (AGDS) [7], which allows replicating the dependencies between phrases appearing in texts. Applying certain mathematical operations to the structure and preprocessed text enables us to acquire a vectorized set of features, describing the input data in terms of similarity to a predefined list of character traits and brand archetypes. Using these graph structures is highly effective because of low computational costs, accurate representation and maintaining good generalization.

The main task of this paper is to create a model capable of associating the input text with 25 personality traits [9] and 12 brand archetypes [13], while some of the traits are not exclusive and can appear in conjunction with the other personality types. Therefore, it is a multi-class and multi-label classification problem, different from the typical situation of selecting a single class out of a defined list.

We also present the current state-of-the-art solutions to compare their efficiency, benefits and disadvantages to our proposed solution for the described NLP problem. The AGDS-based solution can act as a base for more complex solutions, i.e. encoding and representing a training dataset for neural networks, therefore boosting the whole training process due to the fact that the graph already resembles the dataset. This solution resembles a fine-tuning process instead of training a neural network from scratch.

2   Sentiment Analysis with Coarse and Fine-Grained Classification

Sentiment analysis is one of the most popular tasks of natural language processing since it is used in a vast area of fields: from hate speech detection [4] to user recommendation systems [12]. Due to its popularity and complexity, the task has been approached with the use of neural networks because of their ability to process the dataset and learn the relations between the phrases and words. Such approaches were highly successful throughout the years [20,21], acquiring the accuracy higher than 90%.

Current state-of-the-art models used for sentiment analysis are based on BERT [10] - Bidirectional Encoder Representations from Transformers, a language representation model presented in 2019 by Devlin et al. [5]. BERT-driven models are analyzing the context of each word from both sides of the word, which results in extracting contextual information from the network's input. The information about the words' neighborhood is immensely beneficent for the model's accuracy; hence, the best performing models acquire accuracy as high as 97.5% on SST-2 dataset [10].

The aforementioned models serve the purpose of general semantic analysis, without trying to estimate the classification more specifically - to acquire the degree of affiliation between the input and each class. It cannot be mistaken with the classification confidence, which defines how probable is the assumption made by the model. The fine-grained semantic analysis is a more complex task, which is shown by an extreme drop in the accuracy of the models [17], compared to the ones that perform the binary or multi-class semantic analysis. While the fine-grained analysis being a continuous domain task, it can be discretized to a multi-class classification - at the cost of losing some information details because of the finite number of class similarity degrees, we get the ability to numerically describe how much is the data similar to the given class. Such an approach is shown in this paper. Our model is capable of classifying the input data to each of the classes simultaneously, meaning that one class can be described with one of five degrees of similarity (subclasses), while not interfering or even prohibiting the other classes from being classified.

3 Trait Classification

The problem of trait classification is to create a system capable of classifying the input text - consisting of a few sentences, varying in length - to each of the available character traits and brand archetypes. The character traits are placed on an extrovert-introvert spectrum, allowing to precisely describe the person's characteristics in relation to a specified set of 10 pairs of opposite traits, i.e. maximalist (possessive, keen on extreme activities) versus minimalist (easy to satisfy, reluctant to take an effort) [9]. Hence being a descriptive feature set, character traits can be used to profile and target the content published in social media by companies to appeal to a certain set of people. Targeted advertisement campaigns have substantial effect on the target's interest in the brand's activity [15]. Introduced in [13], the brand archetypes are a construct that associate the personality traits with a given archetype of a person, such as the rebellious Outlaw or the brave Hero. An archetype, being a general model representing certain traits, can be used as a base for a marketing company to build the brand's DNA - since archetypes are not exclusive, it is easy to identify with them. In this paper, both personality traits and archetypes will be defined as "traits", for the sake of simplicity of the following description.

Based on the aforementioned description of personality traits and brand archetypes, the dataset consisting of 685 unique influencer Instagram accounts has been created. Every dataset entry is a list of 15 most popular posts of a single influencer and a list of annotations: every trait has been given a score from 0 to 4, where 0 means no association between the account and the trait, 1 is little to no trait-account similarity, 2 means that the relation is unknown, 3 shows some degree of resemblance between the trait and influencer, and 4 is the maximum similarity between account and trait. Therefore, for every entry there are 37 features, each of them has five subclasses that define how strong is the account associated with a given personality trait.

Having described the dataset, the main goal is to create a model capable of associating the input text (list of sentences, as mentioned in the beginning of this chapter), with the personality traits, and acquiring influencers from the database that are the most similar to the input. This problem is more complex than the binary classification because the model must be able to classify the input text to the multiple classes simultaneously, making it a multi-class and multi-label classification. Such a model can be further used as a validation tool, to enhance the quality of the marketing campaign - having designed the advertisement's text, it can be analyzed in terms of similarity to the desired traits.

To be able to use the dataset in this task, the features of the text have to be generated - to emphasize the importance of a phrase in a given trait subclass, the Term Frequency is used - calculating the occurrence of each term per single class can show the distinct words for each class, what can be further used to differentiate the final class. Afterwards, the frequencies are normalized for each phrase and trait, highlighting the differences between usage of the phrase per every subclass. Processed phrases and their weights are now in a form of a modified Bag-of-Words format that will be used to generate an associative graph model.

Such a task differs from the usual process of defining the belonging of the input to a given class by a binary manner - each input text writer can belong to one or more of the available classes. What is more, the similarity grade may not be identical and it has intermediate states, meaning that the input can be associated with the class only to some extent. In this scenario, binary classification is insufficient since the fine-grained similarity information is lost during the process. The problem becomes a multi-class and multi-label classification because the continuous scale describing the similarity grade can be discretized enough to keep the required level of specificity. The main obstacle is to acquire a model capable of holding the information that can be further used to discriminate the input text's class belonging, while maintaining a high classification accuracy. Therefore, we have proposed the use of associative graphs for the purpose of creating an efficient representation of the dataset that can be used to predict the classes of the input text.

The currently available methods used for multi-label and multi-class classification require using deep structures [20] that reduces the model's explainability. Because of an emerging trend of Explainable AI (XAI) [6], the possibility to describe the model's decision-making process is crucial for respecting the social right to explanation. Using the graph structures, which can represent the complex relations in an understandable manner, it is possible to acquire a model capable of precisely describing the associations between the nodes (in this paper - between phrases and traits) while maintaining a high degree of transparency - the weights can have a strict mathematical definition. The associative graph-based model proposed in this paper solves both of the encountered models - high efficacy rate is present in conjunction with the ability to show the non-convoluted relations between the nodes, phrases, and traits.

4   Associative Graph Data Structures

The approach to multi-label and multi-class text classification presented in this paper is based on creating an associative graph structure and training it as a specific associative neural network on the gathered dataset in order to make it resemble the data and enable the model to associate the user's input (text) with the database containing text content published by influencers in social media, as described in Sect. 3. The Associative Graph Data Structures (AGDS), introduced by Adrian Horzyk [7,8], are able to easily represent any tabular data in a sparse graph structure, linking all similar values of each attribute, removing duplicated values, and associating represented objects (table records) according to all similarities of all attributes simultaneously. They also reduce the allocated memory size and the computational complexity when searching for similar objects or values.

Our adaptation of the associative graph data structure (AGDS) used to solve the defined problem creates an AGDS structure that consists of two custom layers of specific nodes. The first layer represents the phrases appearing in the whole corpus, and the second layer represents different personality traits - selected character types and brand archetypes [9,13]. This graph structure can be defined as

$$G = (V, E, W) \tag{1}$$

where V are the nodes - phrases appearing in the dataset and the personality traits that the phrases can be associated with, and E being the edges, carrying the information about the relation between the phrase and a given trait, which strength is represented by the weights W. The process of weights calculation is described in Sect. 5.

The main advantage of this graph structure is the aggregated representation of the duplicated elements, i.e. every node in this graph represents a unique value. The same unique value can be used to define various traits, using connections between nodes. The traits are defined by annotating the given input text with one of five degrees of similarity per each trait. Therefore, the trait consists of five nodes, and the weighted edges between the phrase and trait nodes give information about how much is the given phrase node associated with the trait class node. The trait class nodes are connected with influencer nodes, which represent the social media accounts that the source texts have been gathered from. The graph structure is presented on Fig. 1 - the information flow is defined by the arrows in graph edges. Representing data in tabular format is insufficient

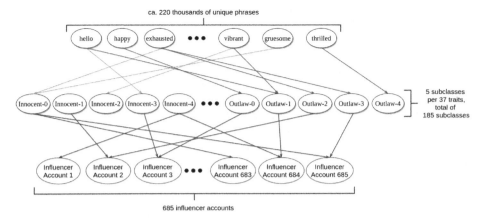

Fig. 1. Associative graph structure - example phrases, traits, and influencer accounts.

for this task - many of the phrases occur in texts associated with multiple traits, creating redundant data table rows. Since the graph nodes are unique, graph structures are more efficient in presenting the highly repetitive data. Duplicate removal reduces the memory usage - a common issue for computer-powered natural language processing. As already mentioned, browsing the graph structure is also highly efficient since all of the connections between nodes are already defined, and the computations required to get all of the nodes associated with the source node are limited to acquiring the nodes' indices or pointers from the edges.

5   Graph-Based Sentiment Analysis

The graph structure proposed in Sect. 4 consists of two sparsely connected layers, representing the relationships between words and personality traits, including the relation types and the association strengths. This graph is then treated similarly to a neural network - the word layer is treated as the input to the model, and the model's output is the vector representing the input's similarity to each of the available class. The output vector from the graph structure consists of 185 elements, which is then divided into 37 sets of 5 classes, where each set describes one feature. The softmax function is then applied to each set, giving as a result the probability of the word to belong to each of the feature's subclasses - how strong the word is associated with the trait. Afterwards, the arguments of the maxima of each set's probability are selected - the final form of the output, a class that describes the similarity of the input to each of the available personality traits and archetypes. The whole process is presented in Fig. 2 and could be defined in the following steps:

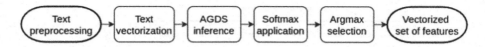

Fig. 2. Associative model flow diagram.

The model starts with text preprocessing - the input text is cleared of non-alphanumeric symbols, the stopwords are removed, and all the sentences are tokenized at the end to enable a smooth transition from textual space to a numerical one.

In the second stage, the term frequencies of all features and their subclasses are calculated - it allows to show the basic links between the phrases and the personality traits. The weights are calculated as follows:

$$w_i = \frac{TF_i}{\sum TF} \tag{2}$$

where TF is a term frequency of a phrase for a given trait and $\sum TF$ is a sum of all phrase occurrences - such kind of weight normalization results in the possibility to compare the classes with each other because of the same scale of values in each subclass.

Having created a matrix representing the connections between the phrases and the subclasses, the weights need to be normalized per each feature - even the basic type of normalization, such as *min-max* normalization (3), amplifies the differences between the subclasses of a single feature, making it easier to select the most similar class.

$$w_{i,c} = \frac{w_{i,c} - w_{cmin}}{w_{cmax} - w_{cmin}} \tag{3}$$

where $w_{i,c}$ is a weight for phrase i per class c, with w_{cmax} and w_{cmin} being maximum and minimum weight values for a given class.

Next, the normalized weights are fine-tuned using the backpropagation algorithm. Because of the graph's nature of AGDS, it is easy to convert it to a trainable form. For this purpose, the weights have been modified using Keras package [3] with the RMSProp optimizer and categorical crossentropy as the loss function.

Using the weights calculated in the previous step (when developing AGDS structure) as initial weights, the training process converges very quickly and with much higher accuracy than when using a similar network without AGDS associative initialization. This is because the weights (calculated using AGDS) already preserve the information about the relationships between the phrases, traits, and classes together with their strengths.

The next step is to extract the input text's features using an AGDS model - the tokenized input is converted to a vector describing the term frequency of phrases that appear both in corpus and in the source text - the result is a vector of integers V_{input}, representing the input text in the numerical domain.

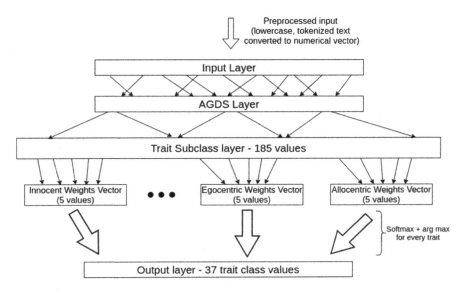

Fig. 3. AGDS-based model structure.

To calculate the similarity between the V_{input} and the personality traits of graph A_{traits}, the dot product of the input vector and each of the subclass weights has to be calculated (Fig. 3) - the result is a 185-element long vector, where each element describes the similarity between the V_{input} and the subclasses.

$$V_{out} = V_{input} \cdot A_{traits} \qquad (4)$$

The trait label have been described in Sect. 3. After acquiring the input in all-class representation V_{out}, the output vector has to be grouped in five element sets X_{trait} (Fig. 3), one set for each personality trait, and the Softmax function is applied to this set in order to obtain the probability of the input being of a given subclass:

$$X_{prob} = Softmax(X_{trait}) \qquad (5)$$

The last step is to select the argument of the maxima of each feature set and, as a result, acquire a 37-element long vector, representing all the defined traits:

$$X_{out} = argmax(X_{prob}) \qquad (6)$$

The acquired vector consists of 37 elements, each of being a class label describing the similarity of the input text to each of the traits. Such output can be used to evaluate the input text's compliance with the assumptions made about its character.

6   Results and Comparison

The whole model was trained on a dataset consisting of 685 unique Instagram accounts, which have posted in total about 33,500 posts, varying in length. In general, the opposite traits (e.g. dominant vs submissive) are exclusive in the dataset, while there are some erroneous annotations where such traits have been similarly annotated (i.e. the text is both allocentric and egocentric). As mentioned in Sect. 3, the traits that are not archetypes are placed on an extrovert-introvert spectrum, making it impossible for excluding traits to be annotated with the same class. If such situation occurs, the sample should be removed from the dataset. The archetypes are a more convoluted form of describing the personality, nevertheless the exclusion rule also applies to them, e.g., a person cannot be associated with outlaw and innocent at the same time.

Table 1. The accuracy of AGDS-based, HerBERT and GRU models achieved for all traits.

Trait	Accuracy AGDS	Accuracy HerBERT	Accuracy GRU	Trait	Accuracy AGDS	Accuracy HerBERT	Accuracy GRU
Innocent	91.3%	63.0%	62.32%	Sage	84.06%	48.0%	43.48%
Explorer	92.75%	25.0%	39.13%	Caregiver	95.65%	28.0%	40.58%
Outlaw	94.2%	31.0%	31.88%	Magician	91.3%	31.0%	44.93%
Hero	95.65%	35.0%	49.28%	Ruler	89.86%	35.0%	50.72%
Lover	95.65%	30.0%	36.23%	Jester	94.2%	28.0%	40.58%
Everyman	88.41%	62.0%	59.42%	Creator	95.65%	31.0%	42.03%
Dominant	89.86%	51.0%	47.83%	Submissive	88.41%	54.0%	36.23%
Maximalist	94.2%	63.0%	52.17%	Minimalist	91.3%	56.0%	44.93%
Inspiring	94.2%	64.0%	68.12%	Systematic	86.96%	56.99%	42.03%
Discovering	89.86%	52.0%	43.48%	Conservative	88.41%	56.99%	37.68%
Verifying	92.75%	60.0%	53.62%	Overlooking	89.86%	56.99%	39.13%
Sharpening	92.75%	50.0%	40.58%	Harmonic	94.2%	60.0%	66.67%
Empathic	91.3%	59.0%	50.72%	Matter of fact	81.16%	39.0%	37.68%
Brave	85.51%	39.0%	33.33%	Protective	95.65%	50.0%	36.23%
Generous	94.2%	68.0%	59.42%	Thrifty	91.3%	54.0%	42.03%
Favourable	85.51%	43.0%	39.13%	Balanced	91.3%	59.0%	47.83%
Sensuality	89.86%	59.0%	46.38%	Intelligent	89.86%	48.0%	47.83%
Believe	89.86%	49.0%	47.83%	Egocentric	94.2%	34.0%	21.74%
Allocentric	92.75%	50.0%	40.58%				

We performed many tests on the dataset divided into training and testing sets in the ratio of 90% to 10%. The best average accuracy on the testing dataset was 91.19%, with the average accuracy on the training dataset being 88.83%. The accuracy per trait has been shown in Table 1, varying in values from 85.51% to 95.65% - the difference is due to the difference in class representation - the traits have different class distributions and word counts, which can take a toll on the accuracy. While validating the model by the means of 10-fold cross-validation, the average accuracy was fluctuating between 44% and 91%. This issue will be addressed by the authors in the future since it requires additional work on the model's robustness and generalization.

Defining the problem as the multi-class classification allows us to compare the performance of the associative graph-based model with the neural network-driven solutions using our dataset. We compared our solution to two models: first is the classifier based on HerBERT model, introduced in [14], which is a state-of-the-art BERT-powered classifier, and the second on uses two Gated Recurrent Unit (GRU) layers - such kind of approach preceded the BERT models in terms of the best performance in text classification. Both models have been trained on our dataset, to compare the models' performances in similar conditions. While HerBERT model has only been fine-tuned, instead of training it from scratch, the GRU model has been trained on our dataset from scratch with random weight initialization. The deep models' results are presented in Table 1.

The best accuracy of HerBERT-based model was 48.05% and of GRU model was 44.97% on the test set, which is significantly lower than the score achieved by the AGDS-based model, which was 91.19%. Since the deep models' performance is inferior to AGDS-based model, several operations have been undertaken in order to improve the models, such as shuffling the dataset, changing the train-test split ratio or modification of the models' structure. None of those actions resulted in an accuracy improvement.

Therefore, it can be claimed that the associative sparse graph with weights initialized from term frequency reproducing real relations performs better in terms of multi-label and multi-class classification. Sequential models are harder to train since they require a vast amount of data to learn the relations between the sequence's elements. Texts presented as a bag-of-words are easier to classify, especially for smaller text corpora. AGDS-based models use the bag-of-words approach and fine-tune it with the backpropagation algorithm what works as a training booster since the weights already carry information about relations between text and trait.

7   Conclusions and Summary

In this paper, we have proposed the AGDS-based model for multi-label and multi-class classification. The associative graph structure was created by calculating the term frequency of the phrases appearing in our dataset and using them as initial weights for a neural network model. The trained weights were used to classify the input text to the available traits by taking the dot product of the weights and input vector, grouping the product's elements into a sets of five subclasses, and extracting the argument with the maximum probability. Such an approach allows us to boost the training process since the model is to be fine-tuned using the backpropagation algorithm, not trained from scratch. As AGDS weights already carry the information about the relations in the dataset, it is easier to meet the global minimum of the loss function; therefore, acquiring a highly accurate model capable of classifying the input data to multiple classes simultaneously. All of the aforementioned issues create a complex problem of acquiring the similarity between the input text and the available character traits.

The AGDS-based model achieved high 91.19% accuracy on our dataset, significantly higher than deep models, such as HerBERT or GRU, which achieved 48.05% and 44.97%. The AGDS training process itself was not very time consuming, apart from generating a graph structure from tabular data (computational complexity of $O(n * logn)$, where n is the number of table rows).

Our proposal is to use associative graphs as a booster structure for neural networks since the interoperability between those types of algorithms is hassle-free - neural networks, in some degree, derive from the graphs. The AGDS-based model is able to classify the multiple labels simultaneously while maintaining high accuracy. Given the quite basic assumptions - term frequency used as the main association metric, *min-max* normalization, there is still some space for further improvement for graph-powered association algorithms, and exploration of this area may result in a rise of the accuracy of multi-label classification models in the future.

Furthermore, the associative graph-based solutions are beneficial to use because of their reduced complexity in comparison with deep learning methods, which results in a decrease in training time and model size. The number of parameters in our proposed associative graph structure was ca 40 million, while there are 110 million of parameters in BERT base model [5]. What is more, the total training time of AGDS for our dataset is about 60 min on a 6th generation Intel CPU, compared to 5 h of fine-tuning a HerBERT-based model with Nvidia Tegra P100 GPU. The resulting associative graph model has a size of about 310 MB, while the HerBERT model is an order of magnitude larger - having over 3 GB.

The approach of creating an associative graph and using its weights as initial ones for the neural network model can be used in other branches of machine learning - for example, in image classification, the image features can be extracted and saved in a graph structure, which can be further used as a base for training a shallow neural network model, capable of classifying the images to classes with high accuracy.

References

1. Bates, M.: Models of natural language understanding. Proc. Natl. Acad. Sci. U. S. A. **92**(22), 9977–9982 (1995). https://doi.org/10.1073/pnas.92.22.9977
2. Bohnet, B., McDonald, R., Simoes, G., Andor, D., Pitler, E., Maynez, J.: Morphosyntactic tagging with a meta-bilstm model over context sensitive token encodings (2018)
3. Chollet, F., et al.: Keras (2015). https://github.com/fchollet/keras
4. Davidson, T., Warmsley, D., Macy, M., Weber, I.: Automated hate speech detection and the problem of offensive language (2017)
5. Devlin, J., Chang, M., Lee, K., Toutanova, K.: BERT: pre-training of deep bidirectional transformers for language understanding. In: Proceedings of the 2019 Conference of the North American Chapter of the Association for Computational Linguistics: Human Language Technologies, vol. 1 (Long and Short Papers), pp. 4171–4186. Association for Computational Linguistics, Minneapolis (June 2019). https://doi.org/10.18653/v1/N19-1423

6. Gunning, D., Stefik, M., Choi, J., Miller, T., Stumpf, S., Yang, G.Z.: XAI-explainable artificial intelligence. Sci. Robot. **4**(37), eaay7120 (2019). https://doi.org/10.1126/scirobotics.aay7120

7. Horzyk, A.: Associative graph data structures with an efficient access via avb+trees. In: 2018 11th International Conference on Human System Interaction (HSI), pp. 169–175 (2018)

8. Horzyk, A., Gadamer, M.: Associative text representation and correction. In: Rutkowski, L., Korytkowski, M., Scherer, R., Tadeusiewicz, R., Zadeh, L.A., Zurada, J.M. (eds.) ICAISC 2013, Part I. LNCS (LNAI), vol. 7894, pp. 76–87. Springer, Heidelberg (2013). https://doi.org/10.1007/978-3-642-38658-9_7

9. Horzyk, A., Tadeusiewicz, R.: A psycholinguistic model of man-machine interactions based on needs of human personality. In: Cyran, K.A., Kozielski, S., Peters, J.F., Stańczyk, U., Wakulicz-Deja, A. (eds.) Man-Machine Interactions, Advances in Intelligent and Soft Computing, vol. 59. Springer, Heidelberg (2009). https://doi.org/10.1007/978-3-642-00563-3_5

10. Jiang, H., He, P., Chen, W., Liu, X., Gao, J., Zhao, T.: Smart: robust and efficient fine-tuning for pre-trained natural language models through principled regularized optimization. In: Proceedings of the 58th Annual Meeting of the Association for Computational Linguistics (2020). https://doi.org/10.18653/v1/2020.acl-main.197

11. Johnson, R., Zhang, T.: Effective use of word order for text categorization with convolutional neural networks (2015)

12. Liu, H., et al.: NRPA. In: Proceedings of the 42nd International ACM SIGIR Conference on Research and Development in Information Retrieval (2019). https://doi.org/10.1145/3331184.3331371

13. Mark, M., Pearson, C.: The Hero and the Outlaw: Building Extraordinary Brands Through the Power of Archetypes. McGraw-Hill Education, New York (2001)

14. Mroczkowski, R., Rybak, P., Wróblewska, A., Gawlik, I.: HerBERT: efficiently pretrained transformer-based language model for Polish. In: Proceedings of the 8th Workshop on Balto-Slavic Natural Language Processing, pp. 1–10. Association for Computational Linguistics, Kiyv (April 2021)

15. Nguyen, B., Li, M., Chen, C.H.: The targeted and non-targeted framework: differential impact of marketing tactics on customer perceptions. J. Target. Meas. Anal. Mark. **20**(2), 96–108 (2012). https://doi.org/10.1057/jt.2012.7

16. Sachan, D.S., Zaheer, M., Salakhutdinov, R.: Revisiting LSTM networks for semi-supervised text classification via mixed objective function. In: Proceedings of the AAAI Conference on Artificial Intelligence, vol. 33, pp. 6940–6948 (2019). https://doi.org/10.1609/aaai.v33i01.33016940

17. Sun, Z., et al.: Self-explaining structures improve NLP models (2020)

18. Takase, S., Kiyono, S.: Lessons on parameter sharing across layers in transformers (2021)

19. Verma, S., Sharma, R., Deb, S., Maitra, D.: Artificial intelligence in marketing: systematic review and future research direction. Int. J. Inf. Manag. Data Insights **1**(1), 100002 (2021). https://doi.org/10.1016/j.jjimei.2020.100002

20. Wu, Z., Ong, D.C.: Context-guided BERT for targeted aspect-based sentiment analysis (2020)

21. Yang, Z., Dai, Z., Yang, Y., Carbonell, J., Salakhutdinov, R., Le, Q.V.: Xlnet: generalized autoregressive pretraining for language understanding (2020)

k-Winners-Take-All Ensemble Neural Network

Abien Fred Agarap[✉] and Arnulfo P. Azcarraga

College of Computer Studies, De La Salle University, 2401 Taft Ave, Malate, Manila, 1004 Metro Manila, Philippines
{abien_agarap,arnulfo.azcarraga}@dlsu.edu.ph
https://dlsu.edu.ph

Abstract. Ensembling is one approach that improves the performance of a neural network by combining a number of independent neural networks, usually by either averaging or summing up their individual outputs. We modify this ensembling approach by training the sub-networks concurrently instead of independently. This concurrent training of sub-networks leads them to cooperate with each other, and we refer to them as "cooperative ensemble". Meanwhile, the mixture-of-experts approach improves a neural network performance by dividing up a given dataset to its sub-networks. It then uses a gating network that assigns a specialization to each of its sub-networks called "experts". We improve on these aforementioned ways for combining a group of neural networks by using a k-Winners-Take-All (kWTA) activation function, that acts as the combination method for the outputs of each sub-network in the ensemble. We refer to this proposed model as "kWTA ensemble neural networks" (kWTA-ENN). With the kWTA activation function, the losing neurons of the sub-networks are inhibited while the winning neurons are retained. This results in sub-networks having some form of specialization but also sharing knowledge with one another. We compare our approach with the cooperative ensemble and mixture-of-experts, where we used a feedforward neural network with one hidden layer having 100 neurons as the sub-network architecture. Our approach yields a better performance compared to the baseline models, reaching the following test accuracies on benchmark datasets: 98.34% on MNIST, 88.06% on Fashion-MNIST, 91.56% on KMNIST, and 95.97% on WDBC.

Keywords: Theory and algorithms · Competitive learning · Ensemble learning · Mixture-of-experts · Neural network models

1 Introduction and Related Works

We use artificial neural networks in a myriad of automation tasks such as classification, regression, and translation among others. Neural networks would approach these tasks as a function approximation problem, wherein given a dataset of input-output pairs $D = \{(x_i, y_i) | x_i \in \mathcal{X}, y_i \in \mathcal{Y}\}$, their goal is to learn the mapping $\mathcal{F} : \mathcal{X} \mapsto \mathcal{Y}$. They accomplish this by optimizing their parameters θ with some

© Springer Nature Switzerland AG 2021
T. Mantoro et al. (Eds.): ICONIP 2021, LNCS 13109, pp. 250–261, 2021.
https://doi.org/10.1007/978-3-030-92270-2_22

modification mechanism, such as the retro-propagation of output errors [14]. We then deem the parameters to be optimal if the neural network outputs are as close as possible to the target outputs in the training data, and if it can adequately generalize on previously unseen data. This can be achieved when a network is neither too simple (has a high bias) nor too complex (has a high variance).

Through the years, combining a group of neural networks is among the simplest and most straightforward ways to achieve this feat. The two basic ways to combine neural networks are by ensembling [1,3,4,15], and by using a mixture-of-experts (MoE) [5,6]. In an ensemble, a group of independent neural networks is trained to learn the entire dataset. Meanwhile in MoE, each network is trained to learn their own and different subsets of the dataset.

In this work, we use a group of neural networks for a classification task on the following benchmark datasets: MNIST [8], Fashion-MNIST [18], Kuzushiji-MNIST (KMNIST) [2], and Wisconsin Diagnostic Breast Cancer (WDBC) [17]. We introduce a variant of ensemble neural networks that uses a k-Winners-Take-All (kWTA) activation function to combine the outputs of its sub-networks instead of using averaging, summation, or voting schemes to combine such outputs. We then compare our approach with an MoE and a modified ensemble network on a classification task on the aforementioned datasets.

1.1   Ensemble of Independent Networks

We usually form an ensemble of networks by independently or sequentially (in the case of boosting) training them, and then by combining their outputs at test time usually by averaging [1] or voting [4]. In this work, we opted to use the averaging scheme for ensembling.

That is, we have a group of neural networks $f_1, \ldots, f_M$ parameterized by $\theta_1, \ldots, \theta_M$, and we compute its final output as,

$$o = \frac{1}{M} \sum_{m=1}^{M} f_m(x; \theta_m) \tag{1}$$

Each sub-network is trained independently to minimize their own loss function, e.g. cross entropy loss for classification, $\ell_{ce}(y, o) = -\sum y \log(o)$. Then Eq. 1 is used to get the model outputs at test time.

1.2   Mixture of Experts

The Mixture-of-Experts (MoE) model consists of a set of M "expert" neural networks $E_1, \ldots, E_m$ and a "gating" neural network G [5]. The experts are assigned by the gating network to handle their own subset of the entire dataset. We compute the final output of this model using the following equation,

$$o = \sum \arg \max G(x) E_m(x) \tag{2}$$

where $G(x)$ is gating probability output to choose E_m for a given input x. The gating network and the expert networks have their respective set of parameters.

Then, we compute the MoE model loss by using the following equation,

$$\mathcal{L}_{MoE}(x, y) = \frac{1}{M} \sum_{m=1}^{M} \left[\frac{1}{n} \sum_{i=1}^{n} \arg\max G(x_i) \cdot \ell_{ce}(y_i, E_m(x_i)) \right] \quad (3)$$

where ℓ_{ce} is the cross entropy loss, $G(x)$ is the weighting factor to choose E_m.

In this system, each expert learns to specialize on the cases where they perform well, and they are imposed to ignore the cases on which they do not perform well. With this learning paradigm, the experts become a function of a sub-region of the data space, and thus their set of learned weights highly differ from each other as opposed to traditional ensemble models that result to having almost identical weights for their learners.

1.3   Cooperative Ensemble Learning

We refer to the ensemble learning we described in Sect. 1.1 as traditional ensemble of independent neural networks. However, in our experiments, we trained the ensemble sub-networks concurrently instead of independently or sequentially. In Algorithm 1, we present our modified version of the traditional ensemble, and we call it "cooperative ensemble" (CE) for the rest of this paper.

Algorithm 1: Cooperative Ensemble Learning

Input   : Dataset $D = \{(x_i, y_i)|x_i \in \mathbb{R}^d, y_i = 1, \ldots, k\}$, randomly initialized
networks $f_1, \ldots, f_M$ parameterized by $\theta_1, \ldots, \theta_M$
Output : Ensemble of M trained networks $f_1, \ldots, f_M$
1 Initialization;
2 Sample mini-batch $B \subset D$;
3 **for** $t \leftarrow 0$ **to** *convergence* **do**
4   **for** $m \leftarrow 1$ **to** M **do**
5   # Forward pass: Compute model outputs for mini-batch
6        $\hat{y}_{m,1}, \ldots, \hat{y}_{m,B} = f_m(x_B)$
7   **end**
8     $o = \frac{1}{M} \sum_{m}^{M} \hat{y}_m$
9   # Backward pass: Update the models
10     $\theta_m^* = \theta_m - \alpha \nabla \ell(y, o)$
11 **end**

First, in a training loop, we compute each sub-network output $\hat{y}_{m,B}$ for mini-batches of data B (line 6). Then, similar to a traditional ensemble, we compute the output of this model o by averaging over the individual network outputs (line 8). Finally, we optimize the parameters of each sub-network in the ensemble based on the gradients of the loss between the ensemble network outputs o and the target labels y (line 10).

In contrast, a traditional ensemble of independent networks train each sub-network independently before ensembling, thus not allowing an interaction among the members of the ensemble and not allowing a chance for each member to contribute to the knowledge of one another.

Cooperative ensemble may have already been used in practice in the real world, but we take note of this variant for it presents itself as a more competitive baseline for our experimental model. This is because cooperative ensemble introduces some form of interaction among the sub-networks during training since there is an information feedback from the combination stage to the sub-network weights, thus giving each sub-network a chance to share their knowledge with one another [9].

The contributions of this study are as follows,

1. The conceptual introduction of cooperative ensembling as a modification to the traditional ensemble of independent networks. The cooperative ensemble is a competitive baseline model for our experimental model (see Sect. 3).
2. We introduce an ensemble network that uses a kWTA activation function to combine its sub-network outputs (Sect. 2). Our approach presents better classification performance on the MNIST, Fashion-MNIST, KMNIST, and Wisconsin Diagnostic Breast Cancer (WDBC) datasets (see Sect. 3).

2   Competitive Ensemble Learning

We take the cooperative ensembling approach further by introducing a competitive layer as a way to combine the outputs of the sub-networks in the ensemble.

We propose to use a k-Winners-Take-All (kWTA) activation function for a fully connected layer which combines the sub-network outputs in the ensemble, and we call the resulting model "kWTA ensemble neural network" (kWTA-ENN). As per Majani et al. (1989) [10], the kWTA activation function admits $k \geq 1$ winners in a competition among neurons in a hidden layer of a neural network (see Eq. 4 for the kWTA activation function).

$$\phi_k(z)_j = \begin{cases} z_j & z_j \in \{\max_k z\} \\ 0 & z_j \notin \{\max_k z\} \end{cases} \tag{4}$$

where z is an activation output, and k is the percentage of winning neurons we want to get. We set $k = 0.75$ in all our experiments, but it could still be optimized as it is a hyper-parameter. This kWTA activation function that we used is the classical one [10] as we are only inhibiting the losing neurons in the competition while retaining the values of the winning neurons. Due to competition, the winning neurons gain the right to respond to particular subsets of the input data, as per Rumelhart and Zipser (1985) [13].

We have seen the training algorithm for our cooperative ensemble in Algorithm 1, wherein we train the sub-networks concurrently instead of independently or sequentially. We incorporate the same manner of training in kWTA-ENN, and we lay down our proposed training algorithm in Algorithm 2.

Our model first computes the sub-network outputs $f_m(x_B)$ for each mini-batch of data B (line 6) but as opposed to cooperative ensemble, we do not

---

Algorithm 2: k-Winners-Take-All Ensemble Network

    Input   : Dataset $D = \{(x_i, y_i)|x_i \in \mathbb{R}^d, y_i = 1, \ldots, k\}$, randomly initialized
                 networks $f_1, \ldots, f_M$ parameterized by $\theta_1, \ldots, \theta_M$
    Output: Ensemble of M trained networks $f_1, \ldots, f_M$

1 Initialization;
2 Sample mini-batch $B \subset D$;
3 **for** $t \leftarrow 0$ **to** *convergence* **do**
4   **for** $m \leftarrow 1$ **to** M **do**
5   # Forward pass: Compute model outputs for mini-batch
6         $\hat{y}_{m,1}, \ldots, \hat{y}_{m,B} = f_m(x_B)$
7   **end**
8     $\hat{Y} = \hat{y}_{1,B}, \ldots, \hat{y}_{M,B}$
9     $z = \theta_z \hat{Y} + b_z$
10    $o = \phi_k(z)$
11   # Backward pass: Update the models
12    $\theta_m^* = \theta_m - \alpha \nabla \ell(y, o)$
13 **end**

---

use a simple averaging of the sub-network outputs. Instead, we concatenate the sub-network outputs $\hat{Y}$ (line 8) and use it as an input to a fully connected layer (line 9). We then pass the fully connected layer output z to the kWTA activation function (line 10). Finally, we update our ensemble based on the gradients of the loss between the kWTA-ENN outputs o and the target labels y (line 12).

To further probe the effect of the kWTA activation function in the combination of sub-network outputs, we add a competition delay parameter d. We define this delay parameter as the number of initial training epochs where the kWTA activation function is not yet used on the fully connected layer output that combines the sub-network outputs. We set $d = 0; 3; 5; 7$.

3   Experiments

To demonstrate the performance gains using our approach, we used four benchmark datasets for evaluation: MNIST [8], Fashion-MNIST [18], KMNIST [2], and WDBC [17]. We ran each model ten times, and we report the average, best, and standard deviation of test accuracies for each of our model. Then, we ran a Kruskal-Wallis H test on the test accuracy results from ten runs of the baseline and experimental models.

3.1   Datasets Description

We evaluate and compare our baseline and experimental models on three benchmark image datasets and one benchmark diagnostic dataset. We list the dataset statistics in Table 1.

Table 1. Dataset statistics.

Dataset	# Samples	Input dimension	# Classes
MNIST	70,000	784	10
Fashion-MNIST	70,000	784	10
KMNIST	70,000	784	10
WDBC	569	30	2

All the *MNIST datasets consist of 60,000 training examples and 10,00 test examples each – all in grayscale with 28×28 resolution. We flattened each image pixel matrix to a 784-dimensional vector.

- **MNIST**. MNIST is a handwritten digit classification dataset [8].
- **Fashion-MNIST**. Fashion-MNIST is said to be a more challenging alternative to MNIST that consists of fashion articles from Zalando [18].
- **KMNIST**. Kuzushiji-MNIST (KMNIST) is another alternative to the MNIST dataset. Each of its classes represent one character representing each of the 10 rows of Hiragana [2].
- **WDBC**. The WDBC dataset is a binary classification dataset where its 30-dimensional features were computed from a digitized image of a fine needle aspirate of a breast mass [17]. It consists of 569 samples where 212 samples are malignant and 357 samples are benign. We randomly over-sampled the minority class in the dataset to account for its imbalanced class frequency distribution, thus increasing the number of samples to 714. We then splitted this dataset to 70% training set and 30% test set.

We randomly picked 10% of the training samples for each of the dataset to serve as the validation dataset.

3.2 Experimental Setup

The code implementations for both our baseline and experimental models are found in https://gitlab.com/afagarap/kwta-ensemble.

Hardware and Software Configuration. We used a laptop computer with an Intel Core i5-6300HQ CPU with Nvidia GTX 960M GPU for training all our models. Then, we used the following arbitrarily chosen 10 random seeds for reproducibility: 42, 1234, 73, 1024, 86400, 31415, 2718, 30, 22, and 17. All our models were implemented in PyTorch 1.8.1 [11] with some additional dependencies listed in the released source code.

Training Details. For all our models, we used a feed-forward neural network with one hidden layer having 100 neurons as the sub-network, and then we vary the number of sub-networks per model from 2 to 5. The sub-network weights were initialized with Kaiming uniform initializer [7].

Table 2. Classification results on the benchmark datasets (bold values represent the best results) in terms of average, best, and standard deviation of test accuracies (in %). Our kWTA-ENN achieves better test accuracies than our baseline models with statistical significance. * denotes at $p < 0.05$, ns denotes not significant.

# nets	Acc	kWTA-ENN $d = 0$	$d = 3$	$d = 5$	$d = 7$	MoE	CE
		MNIST					
2	AVG	98.16	**98.18**	**98.18**	**98.18**	96.43	97.90
	MAX	98.28	98.28	98.28	98.28	96.66	97.96
	STD	0.08	0.08	0.08	0.08	0.25	0.05
		* $H = 41.51$, $p = 7.39 \times 10^{-8}$					
3	AVG	98.24	**98.26**	**98.26**	**98.26**	94.67	97.62
	MAX	98.36	98.39	98.39	98.39	96.33	97.71
	STD	0.06	0.08	0.08	0.08	0.99	0.05
		* $H = 41.19$, $p = 8.61 \times 10^{-8}$					
4	AVG	**98.30**	98.27	98.27	98.27	92.349	97.33
	MAX	98.43	98.39	98.39	98.39	95.02	97.39
	STD	0.07	0.08	0.08	0.08	1.30	0.05
		* $H = 41.60$, $p = 7.11 \times 10^{-8}$					
5	AVG	98.33	**98.34**	**98.34**	**98.34**	90.63	97.02
	MAX	98.52	98.42	98.42	98.42	91.94	97.13
	STD	0.08	0.05	0.05	0.05	1.25	0.06
		* $H = 41.58$, $p = 7.17 \times 10^{-8}$					

# nets	Acc	kWTA-ENN $d = 0$	$d = 3$	$d = 5$	$d = 7$	MoE	CE
		Fashion-MNIST					
2	AVG	87.53	87.54	87.54	87.54	86.59	**87.84**
	MAX	87.78	87.70	87.70	87.70	87.54	88.00
	STD	0.16	0.12	0.12	0.12	0.40	0.11
		* $H = 36.75$, $p = 6.72 \times 10^{-7}$					
3	AVG	87.73	**87.81**	**87.81**	**87.81**	85.54	87.69
	MAX	88.01	88.10	88.10	88.10	87.15	87.86
	STD	0.18	0.15	0.15	0.15	0.58	0.09
		* $H = 28.32$, $p = 3.16 \times 10^{-5}$					
4	AVG	87.88	**87.93**	**87.93**	**87.93**	84.47	87.40
	MAX	88.22	88.15	88.15	88.15	86.69	87.54
	STD	0.14	0.15	0.15	0.15	1.20	0.09
		* $H = 42.04$, $p = 5.78 \times 10^{-8}$					
5	AVG	87.99	**88.06**	**88.06**	**88.06**	82.89	87.15
	MAX	88.22	88.27	88.27	88.27	85.80	87.27
	STD	0.15	0.15	0.15	0.15	2.18	0.05
		* $H = 42.26$, $p = 5.22 \times 10^{-8}$					

# nets	Acc	kWTA-ENN $d = 0$	$d = 3$	$d = 5$	$d = 7$	MoE	CE
		KMNIST					
2	AVG	**90.64**	90.53	90.53	90.53	85.23	89.94
	MAX	91.11	90.74	90.74	90.74	87.14	90.19
	STD	0.29	0.12	0.12	0.12	0.99	0.12
		* $H = 41.63$, $p = 6.99 \times 10^{-8}$					
3	AVG	91.16	**91.17**	**91.17**	**91.17**	81.12	89.47
	MAX	91.4	91.51	91.51	91.51	87.59	89.61
	STD	0.14	0.19	0.19	0.19	2.77	0.12
		* $H = 41.09$, $p = 9.00 \times 10^{-8}$					
4	AVG	**91.39**	91.31	91.31	91.31	77.55	88.72
	MAX	91.68	91.54	91.54	91.54	83.04	88.94
	STD	0.18	0.15	0.15	0.15	2.89	0.13
		* $H = 41.67$, $p = 6.88 \times 10^{-8}$					
5	AVG	**91.56**	91.52	91.52	91.52	74.17	87.87
	MAX	91.82	91.76	91.76	91.76	79.99	88.02
	STD	0.16	0.18	0.18	0.18	3.47	0.09
		* $H = 41.28$, $p = 8.24 \times 10^{-8}$					

# nets	Acc	kWTA-ENN $d = 0$	$d = 3$	$d = 5$	$d = 7$	MoE	CE
		WDBC					
2	AVG	95.43	95.36	95.36	95.36	94.49	**95.79**
	MAX	98.62	98.62	98.62	98.62	98.57	99.05
	STD	1.98	2.48	2.48	2.48	2.37	2.13
		(ns) $H = 1.40$, $p = 9.24 \times 10^{-1}$					
3	AVG	94.76	**95.64**	**95.64**	**95.64**	92.68	95.35
	MAX	98.15	99.07	99.07	99.07	95.45	98.17
	STD	1.92	2.33	2.33	2.33	2.36	2.45
		(ns) $H = 9.20$, $p = 1.02 \times 10^{-1}$					
4	AVG	94.98	**95.97**	**95.97**	**95.97**	91.79	95.65
	MAX	98.62	98.62	98.62	98.62	96.67	98.15
	STD	2.87	2.20	2.20	2.20	4.20	2.00
		* $H = 12.56$, $p = 2.78 \times 10^{-2}$					
5	AVG	95.03	**95.40**	**95.40**	**95.40**	90.93	95.04
	MAX	98.61	99.05	99.05	99.05	96.33	98.61
	STD	2.73	2.83	2.83	2.83	2.60	2.47
		* $H = 12.16$, $p = 3.27 \times 10^{-2}$					

We trained our baseline and experimental models on the MNIST, Fashion-MNIST, and KMNIST datasets using mini-batch stochastic gradient descent (SGD) with momentum [12] of 9×10^{-1}, a learning rate of 1×10^{-1} decaying to 1×10^{-4}, and weight decay of 1×10^{-5} on a batch size of 100 for 10,800 iterations (equivalent to 20 epochs). As for the WDBC dataset, we used the same hyper-parameters except we trained our models for only 249 iterations (equivalent to 20 epochs). All these hyper-parameters were arbitrarily chosen since we did not perform hyper-parameter tuning for any of our models. This makes the comparison fair for our baseline and experimental models, and we also did not have the computational resources to do so, which is why we chose a simple architecture as the sub-network.

We recorded the accuracy and loss during both the training and validation phases. We then used the validation accuracy as the basis to checkpoint the best model parameters θ so far in the training. By the end of each training epoch, we load the best recorded parameters to be used by the model at test time.

3.3 Classification Performance

We evaluate the performance of our proposed approach in its different configurations as per the competition delay parameter d and compare it with our baseline models: Mixture-of-Experts (MoE) and Cooperative Ensemble (CE). The empirical evidence shows our proposed approach outperforms our baseline models on the benchmark datasets we used. However, we are not able to observe a proper trend in performance with respect to the varying values of d, and thus it may warrant further investigation.

For the full classification performance results of our baseline and experimental models, we refer the reader to Table 2, from where we can observe the following:

1. MoE performed the least among the models in our experiments, which may be justified with our choice of mini-batch size of 100. MoE performs better on larger datasets and/or larger batch sizes [5,16].
2. CE is indeed a competitive baseline as we can see from the performance margins when compared to our proposed model.
3. Our model in its different variations has consistently outperformed our baseline models in terms of average test accuracy (with the exception of two sub-networks for Fashion-MNIST and WDBC).
4. Our model has higher margins on its improved test accuracy on the KMNIST dataset, which we find appealing since the said dataset is also supposed to be more difficult than the MNIST dataset and thus it better demonstrates the performance gains using our model.
5. Finally, we can observe that there is a statistical significance among the differences in performance of the baseline and experimental models at $p < 0.05$ (on WDBC, for $M = 4, 5$ sub-networks), which indicates that the performance gains through our proposed approach are statistically significant.

3.4 Improving Cooperation Through Competitive Learning

In the context of our work, we refer to *cooperation* in a group of neural networks as the phenomenon when the members of the group contribute to the overall group performance. For instance, in CE, all the sub-networks contribute to the knowledge of one another as opposed to the traditional ensemble, where there is no interaction among the ensemble members [9]. Meanwhile, *specialization* is when members of a group of neural networks are tasked to a specific subset of the input data, which is the intention behind the design of MoE [5]. In this respect, *competition* can be thought of leading to specialization since it is when the winning units gain the right to respond to a particular subset of the dataset.

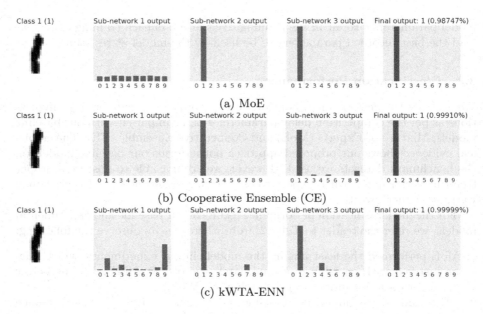

Fig. 1. Predictions of each sub-network on a sample MNIST data and their respective final outputs. In (a) we can infer that MoE sub-networks 2 and 3 are specializing on class 1. In (b) all CE sub-networks have high probability outputs for class 1. In (c) all kWTA-ENN sub-networks contributed but with the kWTA activation function, the neurons for other classes were most likely inhibited at inference, thus its higher probability output than MoE and CE.

We argue that with our proposed approach, we employ the notion of all three: competition, specialization, and cooperation.

kWTA-ENN uses a kWTA activation function so that the neurons from its sub-networks could compete for their right to respond to a particular subset of the dataset. We demonstrate this in Figs. 1 and 2. Let us recall that kWTA-ENN gets its outputs by computing a linear combination of the outputs of its sub-networks, and then passing the linear combination results to a kWTA activation function. As per the referred figures, even though each kWTA-ENN sub-network is not providing high probability output per class as compared to MoE and CE sub-networks, the final kWTA-ENN output is on par with the MoE and CE probability outputs.

We can then infer two things from this: (1) the kWTA activation function inhibits the neurons of the losing kWTA-ENN sub-networks, and (2) the probability outputs of the winning sub-network neurons enable the sub-networks to help one another. For instance, in Fig. 1c, we can observe a probability output for class 1 from sub-network 1, however minimal, and a higher probability output for class 1 from sub-networks 2 and 3, but then their final output has even higher probability output for the same class when compared to MoE and CE probability outputs. The same could be observed in Fig. 2c. This is because

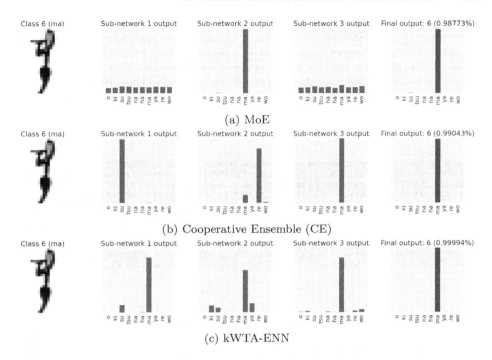

(a) MoE

(b) Cooperative Ensemble (CE)

(c) kWTA-ENN

Fig. 2. Predictions of each sub-network on a sample KMNIST data and their respective final outputs. In (a) we can infer that MoE sub-network 2 is specializing on class 6 ("ma"). In (b) CE sub-network 3 was assisted by sub-network 2. In (c) all kWTA-ENN sub-networks contributed but with the kWTA activation function, the neurons for other classes were most likely inhibited at inference, thus its higher probability output than MoE and CE.

the losing neurons are inhibited in the competition process while retaining the winner neurons, thus improving the final probability output of the model.

In Table 3, we further support this by showing the per-class accuracy of each kWTA-ENN sub-network with varying number of sub-networks. We can see that there is some apparent division of classes among the sub-networks even without pre-defining such divisions, but the final per-class accuracies of the entire model are even better than the per-class accuracies of the sub-networks, thus suggesting that there is indeed a sharing of responsibility among the sub-networks due to the inhibition of losing sub-network neurons and retention of the winning sub-network neurons, even with the competition in place.

260   A. F. Agarap and A. P. Azcarraga

Table 3. Classification results of each kWTA-ENN sub-network and kWTA-ENN itself on MNIST (a) and KMNIST (b) datasets. The tables show the test accuracy of each sub-network on each dataset class, indicating a degree of specialization among the sub-networks. Furthermore, the final model accuracy on each class shows that combining the sub-network outputs have stronger predictive capability. These divisions were in no way pre-determined but they show how cooperation by specialization can be done through competitive ensemble.

(a)

n = 2

	Net-1	Net-2	Final
0	93.65	93.07	98.38
1	95.97	85.42	99.03
2	84.15	87.29	98.55
3	80.95	55.70	97.64
4	87.01	88.22	98.47
5	88.11	85.79	98.87
6	97.50	93.12	98.43
7	74.34	82.96	98.06
8	70.23	93.52	97.44
9	95.61	79.95	97.80

n = 3

	Net-1	Net-2	Net-3	Final
0	96.44	90.38	94.49	98.78
1	65.52	96.13	75.91	99.30
2	62.92	94.89	83.01	98.26
3	45.22	90.98	91.16	98.42
4	53.23	87.55	72.75	98.27
5	92.61	69.26	67.32	98.65
6	83.32	93.76	90.21	98.85
7	85.11	74.32	90.80	98.25
8	57.64	77.35	72.58	97.44
9	51.96	72.86	45.49	97.42

n = 4

	Net-1	Net-2	Net-3	Net-4	Final
0	71.55	84.36	99.56	69.15	98.68
1	54.44	91.07	100.00	52.46	99.47
2	82.80	74.53	35.96	77.51	97.88
3	36.96	63.28	70.62	62.89	98.32
4	76.58	42.58	68.86	80.73	98.07
5	59.88	56.59	68.49	70.17	98.87
6	91.64	50.00	78.41	82.21	98.53
7	60.91	72.37	57.07	48.87	98.64
8	70.26	50.69	51.66	54.81	97.64
9	59.16	32.77	81.57	74.63	98.01

n = 5

	Net-1	Net-2	Net-3	Net-4	Net-5	Final
0	96.10	93.11	85.83	94.91	75.07	99.29
1	87.96	71.03	NaN	58.35	71.60	99.03
2	75.66	56.49	37.32	80.63	95.86	98.44
3	38.42	84.77	37.93	44.11	82.18	97.85
4	53.01	81.43	62.47	82.61	51.95	97.97
5	40.99	78.57	50.75	25.00	53.43	99.20
6	84.51	76.62	90.84	93.66	74.83	98.54
7	76.30	78.04	76.23	72.87	50.41	98.73
8	30.41	34.74	62.99	33.42	74.02	97.64
9	49.32	88.89	34.84	59.78	93.09	97.35

(b)

n = 2

	Net-1	Net-2	Final
o	76.26	92.15	92.61
ki	73.97	84.29	91.93
su	48.96	57.00	86.77
tsu	62.89	68.20	90.31
na	67.75	58.87	88.86
ha	78.82	85.22	95.81
ma	79.54	73.49	88.42
ya	69.87	57.35	93.95
re	72.54	83.40	90.61
wo	68.23	71.51	94.10

n = 3

	Net-1	Net-2	Net-3	Final
o	66.90	77.87	57.51	93.36
ki	36.16	66.43	64.12	90.91
su	64.27	63.46	43.32	86.82
tsu	81.94	72.38	79.35	92.96
na	61.84	76.86	51.20	90.96
ha	70.00	85.87	85.66	96.11
ma	47.10	61.70	65.72	86.67
ya	78.68	78.20	78.64	94.36
re	79.91	58.84	63.71	90.89
wo	76.15	40.68	56.52	92.71

n = 4

	Net-1	Net-2	Net-3	Net-4	Final
o	75.75	86.90	81.82	47.54	93.47
ki	50.00	44.32	42.90	58.70	90.62
su	30.37	44.03	48.54	46.96	86.40
tsu	56.40	49.27	58.38	74.32	93.52
na	54.63	78.43	44.91	72.18	90.65
ha	50.40	60.07	92.27	63.64	95.80
ma	77.54	43.15	69.14	55.65	88.95
ya	68.37	56.80	70.08	73.91	94.68
re	40.09	78.33	70.06	44.14	89.86
wo	65.73	81.18	36.86	63.54	92.88

n = 5

	Net-1	Net-2	Net-3	Net-4	Net-5	Final
o	89.81	89.66	57.80	66.26	46.32	93.89
ki	32.40	44.29	42.84	59.19	33.91	91.58
su	44.73	35.02	34.72	49.94	29.66	85.88
tsu	70.08	51.31	81.44	81.71	53.56	92.85
na	66.21	50.35	39.13	48.31	54.84	91.38
ha	52.29	57.13	81.25	53.85	78.43	95.51
ma	41.11	56.41	48.51	40.66	37.87	87.75
ya	55.13	54.11	58.63	70.67	57.61	94.62
re	45.38	49.66	42.62	43.62	49.06	91.00
wo	58.00	75.00	55.14	48.84	68.35	94.15

4 Conclusion and Future Works

We introduce the k-Winners-Take-All ensemble neural network (kWTA-ENN) which uses a kWTA activation function as the means to combine the sub-network outputs in an ensemble as opposed to the conventional way of combining sub-network outputs through averaging, summation, or voting. Using a kWTA activation function induces competition among the sub-network neurons in an ensemble. This in turn leads to some form of specialization among them, thereby improving the overall performance of the ensemble.

Our comparative results showed that our proposed approach outperforms our baseline models, yielding the following test accuracies on benchmark datasets: 98.34% on MNIST, 88.06% on Fashion-MNIST, 91.56% on KMNIST, and 95.97% on WDBC. We intend to pursue further exploration into this subject by comparing the performance of our baseline and experimental models with respect to varying mini-batch sizes, by training on other benchmark datasets, and finally, by using a more rigorous statistical treatment for a more formal comparison between our proposed model and our baseline models.

References

1. Breiman, L.: Stacked regressions. Mach. Learn. **24**(1), 49–64 (1996)
2. Clanuwat, T., et al.: Deep Learning for Classical Japanese Literature. arXiv preprint arXiv:1812.01718 (2018)
3. Freund, Y., Schapire, R.E.: Experiments with a new boosting algorithm. In: ICML, vol. 96 (1996)
4. Hansen, L.K., Salamon, P.: Neural network ensembles. IEEE Trans. Pattern Anal. Mach. Intell. **12**(10), 993–1001 (1990)
5. Jacobs, R.A., et al.: Adaptive mixtures of local experts. Neural Comput. **3**(1), 79–87 (1991)
6. Jordan, M.I., Jacobs, R.A.: Hierarchies of adaptive experts. In: Advances in Neural Information Processing Systems (1992)
7. He, K., et al.: Delving deep into rectifiers: surpassing human-level performance on imagenet classification. In: Proceedings of the IEEE International Conference on Computer Vision (2015)
8. LeCun, Y.: The MNIST database of handwritten digits (1998). http://yann.lecun.com/exdb/mnist/
9. Liu, Y., Yao, X.: A cooperative ensemble learning system. In: 1998 IEEE International Joint Conference on Neural Networks Proceedings. IEEE World Congress on Computational Intelligence (Cat. No. 98CH36227), vol. 3. IEEE (1998)
10. Majani, E., Erlanson, R., Abu-Mostafa, Y.: On the K-winners-take-all network, pp. 634–642 (1989)
11. Paszke, A., et al.: PyTorch: an imperative style, high-performance deep learning library. In: Advances in Neural Information Processing Systems, vol. 32, pp. 8026–8037 (2019)
12. Qian, N.: On the momentum term in gradient descent learning algorithms. Neural Netw. **12**(1), 145–151 (1999)
13. Rumelhart, D.E., Zipser, D.: Feature discovery by competitive learning. Cogn. Sci. **9**(1), 75–112 (1985)
14. Rumelhart, D.E., Hinton, G.E., Williams, R.J.: Learning representations by back-propagating errors. Nature **323**(6088), 533–536 (1986)
15. Schapire, R.E.: The strength of weak learnability. Mach. Learn. **5**(2), 197–227 (1990)
16. Shazeer, N., et al.: Outrageously large neural networks: the sparsely-gated mixture-of-experts layer. arXiv preprint arXiv:1701.06538 (2017)
17. Wolberg, W.H., Nick Street, W., Mangasarian, O.L.: Breast cancer Wisconsin (diagnostic) data set. UCI Machine Learning Repository (1992). http://archive.ics.uci.edu/ml/
18. Xiao, H., Rasul, K., Vollgraf, R.: Fashion-MNIST: a Novel Image Dataset for Benchmarking Machine Learning Algorithms. arXiv preprint arXiv:1708.07747 (2017)

Performance Improvement of FORCE Learning for Chaotic Echo State Networks

Ruihong Wu[1], Kohei Nakajima[2], and Yongping Pan[1(✉)]

[1] School of Computer Science and Engineering, Sun Yat-sen University,
Guangzhou 510006, China
wurh5@mail2.sysu.edu.cn, panyongp@mail.sysu.edu.cn
[2] Graduate School of Information Science and Technology, University of Tokyo,
Tokyo 113-8656, Japan
k_nakajima@mech.t.u-tokyo.ac.jp

Abstract. Echo state network (ESN) is a kind of recurrent neural networks (RNNs) which emphasizes randomly generating large-scale and sparsely connected RNNs coined reservoirs and only training readout weights. First-order reduced and controlled error (FORCE) learning is an effective online training approach for chaotic RNNs. This paper proposes a composite FORCE learning approach enhanced by memory regressor extension to train chaotic ESNs efficiently. In the proposed approach, a generalized prediction error is obtained by using regressor extension and linear filtering operators with memory to retain past excitation information, and the generalized prediction error is applied as additional feedback to update readout weights such that partial parameter convergence can be achieved rapidly even under weak partial excitation. Simulation results based on a dynamics modeling problem indicate that the proposed approach largely improves parameter converging speed and parameter trajectory smoothness compared with the original FORCE learning.

Keywords: Echo state network · Chaotic neural network · Composite learning · Memory regressor extension · Dynamics modeling

1 Introduction

Theoretically, recurrent neural networks (RNNs) can be applied to approximate arbitrary dynamical systems, which makes it a powerful modeling tool, especially for nonlinear dynamical systems [1]. However, due to the recurrent connectivity, RNNs is hard to be trained effectively and efficiently compared with feedforward neural networks. Several traditional training methods of RNNs suffer from some problems such as local minima, slow convergence, and bifurcation that may cause gradient information to become invalid, and these problems are difficult to be solved or even debugged [2–5]. As a result, RNNs were prevented from being widely used in practical applications.

© Springer Nature Switzerland AG 2021
T. Mantoro et al. (Eds.): ICONIP 2021, LNCS 13109, pp. 262–272, 2021.
https://doi.org/10.1007/978-3-030-92270-2_23

Echo state network (ESN) was proposed as an efficient paradigm for designing and training RNNs with simplicity [6]. Several similar RNN training approaches, such as ESN, liquid state machine [7], and backpropagation decorrelation learning rule [8], are unified to be a reservoir computing framework, which emphasizes randomly generating large-scale and sparsely connected RNNs coined reservoirs and only training readout weights [9–11]. In the ESN, the reservoir is driven by external inputs and neuron activations to obtain very rich echo states so that the training of readout weights can always be treated as a simple linear regression problem. In this way, the difficulties of traditional RNN training methods can be largely avoided. ESN has been used in many applications and outperforms fully trained RNNs in many tasks, e.g., see [12–22].

Based on the ESN framework, an online training approach termed first-order reduced and controlled error (FORCE) learning was proposed for chaotic RNNs in [23]. FORCE learning emphasizes a strong and rapid way to adjust network weights such that the network output keeps up with a target output. The network output is fed back to the reservoir through an external feedback loop to suppress the chaos such that the training becomes possible. Besides, FORCE learning can take advantage of the initial activity of spontaneous chaos in the reservoir to make the training more rapid, accurate and robust. Some recent developments of FORCE learning can be referred to [24–27]. While recursive least square (RLS) is an alternative approach to achieve FORCE learning, there is a least mean squares (LMS)-based FORCE learning approach that only uses a scalar time-varying learning rate rather than a matrix one in RLS [23]. Although the LMS-based FORCE learning approach is more biological and computationally efficient, it suffers from slow parameter convergence compared with the RLS method.

From the viewpoint of classical linear regression, the convergence of parameter estimation is ensured by a well-known condition termed persistency of excitation (PE) [28]. However, PE requires sufficient training data containing rich spectrum information, which is too stringent to be satisfied in practice. This requirement is more difficult for ESNs that have large-scale and sparsely connected reservoirs. Even if PE exists, the rate of parameter convergence in gradient-based parameter estimation depends heavily on the excitation strength, which usually results in a slow learning speed [29]. This is a major reason why the LMS-based FORCE learning in [23] exhibits slow parameter convergence. In recent years, several innovative parameter estimation techniques, such as concurrent learning [30], composite learning [31], and dynamic regressor extension and mixing [32], have been proposed to guarantee parameter convergence without the PE condition. The common features of these techniques include regressor extension using algebraic operations and memory data exploitation using integral-like operations, which are useful to relax the strict requirements on data richness and learning duration for parameter convergence. Note that the composite learning was named to be memory regressor extension (MRE) in [33]. Please refer to [34] for the comparison among the above parameter estimation techniques.

This paper presents a composite FORCE learning approach enhanced by the MRE technique to train chaotic ESNs efficiently. The design procedure of the proposed approach is as follows: First, a chaotic ESN with an external feedback loop is introduced for dynamics modeling; second, the original regression equation is multiplied by a regression vector to obtain a new regression equation with a regression matrix; third, linear filtering operators with memory are applied to construct a generalized prediction error; last, this prediction error is incorporated to the LMS-based FORCE learning approach to enhance the learning ability of the chaotic ESN. A modeling problem of nonlinear dynamics using chaotic ESNs is considered for simulations, where the proposed composite FORCE learning approach is compared with the LMS-based FORCE learning.

The rest of the paper is organized as follows. Section 2 describes a basic ESN architecture; Sect. 3 revisits the original FORCE learning approach and presents the composite FORCE learning approach enhanced by MRE; Sect. 4 provides an illustrative example; Sect. 5 draws some conclusions. Throughout this paper, $\mathbb{R}$, $\mathbb{R}^+$, $\mathbb{R}^N$ and $\mathbb{R}^{N \times M}$ denote the spaces of real numbers, positive real numbers, real N-vectors and real $N \times M$-matrices, respectively, $\mathbb{N}^+$ denotes the set of positive natural numbers, $\|x\|$ denotes the Euclidean norm of $x \in \mathbb{R}^N$, x_i denotes the ith elements of x, and $i, N, M \in \mathbb{N}^+$.

2   Neural Network Architecture

As shown in Fig. 1, consider a basic ESN with $N \in \mathbb{N}^+$ reservoir nodes, one input node and one output node, where $W \in \mathbb{R}^{N \times N}$, $W_{\text{in}} \in \mathbb{R}^{N \times 1}$, $W_{\text{fb}} \in \mathbb{R}^{N \times 1}$ and $\hat{W}_{\text{o}} \in \mathbb{R}^{N \times 1}$ denote internal, input, output feedback, and output connection weights, respectively, $u(k) \in \mathbb{R}$ and $z(k) \in \mathbb{R}$ denote network input and output, respectively, $x(k) \in \mathbb{R}^N$ and $\phi(k) \in \mathbb{R}^N$ denote a state and an activation of internal nodes, respectively, $p \in (0, 1)$ and $g \in \mathbb{R}^+$ are connectivity and chaotic factors of the reservoir, respectively, and $k \in \mathbb{N}^+$ is a time step. It is worth noting that the only parameters that need to be specified in the reservoir are the network size N, the connectivity factor p, and the chaotic factor g, where concrete values of W, W_{in}, and W_{fb} are randomly generated and then keep fixed.

The state x and the activation ϕ of the reservoir are updated by [23]

$$x(k) = (1 - \alpha)x(k-1) + \alpha W_{\text{in}}u(k) + \alpha(W\phi(k-1) + W_{\text{fb}}z(k-1)), \tag{1}$$

$$\phi(k) = \tanh(x(k)) \tag{2}$$

respectively, where $\alpha \in \mathbb{R}^+$ is a leaky rate, and $\tanh(x) = [\tanh(x_1), \tanh(x_2),$ $\cdots, \tanh(x_N)]^T$ is an activation function which works in the element-wise manner (other types of activation functions can also be chosen). The network output z is a weighted sum of the activation values in the reservoir as follows:

$$z(k) = \phi^T(k)\hat{W}_{\text{o}}(k). \tag{3}$$

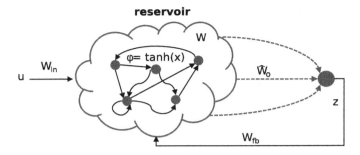

Fig. 1. The basic ESN architecture used in this study, where red dashed arrows denote trainable connections, and black solid arrows denote fixed connections.

The above ESN model is similar to a leaky-integrator ESN model in [15] which can be utilized to accommodate the network to temporal characteristics of a learning task. The differences between the two ESN models lie in the position of the leaky rate α and the information transmitted to the output layer to generate the network output z. Please refer to [15] for more details.

By properly setting the parameters N, p and g, a large-scale, sparsely connected, and spontaneously chaotic reservoir can be obtained to build a chaotic ESN. As shown in Fig. 1, an external feedback loop is set between the output layer and the reservoir with the output feedback weight W_{fb} such that the chaotic activity of the ESN can be suppressed by *i)* training the output connection weight $\hat{W}_o$ only, and *ii)* driving the reservoir with the feedback signal z and the activation ϕ to generate a target output. In contrast to those traditional RNN training methods that are computationally demanding and difficult to be used and debugged, the ESN training of this study is restricted to a linear output layer, largely simplifying the whole training procedure.

3   Enhanced FORCE Learning Design

3.1   Introduction to FORCE Learning

FORCE learning, an online training approach for chaotic RNNs, aims at keeping modeling errors sufficiently small even from the initial training phase so that the network output z is almost equal to a target output [23]. During the FORCE learning process, sampled fluctuations can well solve the stability problem that is hard to overcome when using a teacher forcing approach that feeds back a target output instead of the network output z.

Consider a dynamics modeling problem that aims to make the network output z track a target output $f(k) \in \mathbb{R}$ such that the ESN can produce target output correctly after training [23]. In this case, the network input u is not needed so that the external driving force for the reservoir is only z. At the time step k, define a prediction error as follows:

$$e(k) = f(k) - \phi^T(k)\hat{W}_o(k-1) \tag{4}$$

FORCE learning allows feeding back the output error e but keeps it small enough in a strong and rapid way. There are several learning approaches meeting the requirements of FORCE learning. In this study, the LMS-based simpler learning method in the supplemental material of [23] is considered as follows:

$$\hat{W}_o(k) = \hat{W}_o(k - 1) + n(k)e(k)\phi(k) \tag{5}$$

where $n(k) \in \mathbb{R}^+$ is a scalar time-varying learning rate updated by

$$n(k) = n(k - 1) - n(k - 1)(|e(k)|^\gamma - n(k - 1)) \tag{6}$$

in which $\gamma \in \mathbb{R}^+$ is a constant and $\gamma \in [1, 2]$ can be effective. It follows from (6) that as the prediction error e decreases, the learning rate n becomes smaller.

The chaotic RNN with FORCE learning is powerful because [23]: $i)$ It can generate an extensive range of outputs such as periodic signals, chaotic attractors, one-shot non-repeating signals, and other complex patterns; $ii)$ it can achieve quicker training and generate a more accurate and robust network output than its non-chaotic counterpart. The FORCE learning rule composed of (5) and (6) is a basic delta rule which is computationally simpler but has the following main problems: $i)$ the weight convergence depends on the PE condition; $ii)$ the converging speed is much slower than RLS estimation [23].

3.2   Composite FORCE Learning Design

To solve the above-mentioned problems, a composite FORCE learning approach enhanced by the MRE technique is proposed for training chaotic ESNs as follows. The main idea of MRE is to generate new extended regression equations via linear filtering operators with memory. Multiplying (4) by $\phi(k)$, one gets an extended regression equation as follows:

$$\phi(k)e(k) = \phi(k)f(k) - \phi(k)\phi^T(k)\hat{W}_o(k - 1). \tag{7}$$

Applying a linear filtering operator $L(z)$ to (7), one gets a new extended regression equation with memory as follows:

$$\boldsymbol{\xi}(k) = \boldsymbol{\eta}(k) - \Phi(k)\hat{W}_o(k - 1) \tag{8}$$

in which $\boldsymbol{\xi}(k) = L\{\phi(k)e(k)\}$, $\boldsymbol{\eta}(k) = L\{\phi(k)f(k)\}$, $\Phi(k) = L\{\phi(k)\phi^T(k)\}$, z is a Z-transform operator, and $\lambda \geq 0$ is a filtering parameter. If we choose $L(z) = \frac{\lambda}{1-(1-\lambda)z^{-1}}$ with $\lambda > 0$ and an integral interval $[0, t]$, then it will be a linear filter with forgetting memory; if we choose $L(z) = \frac{1}{1-z^{-1}}$ with an integral interval $[0, t]$, then it will be a normal integral with long-term memory; if we choose $L(z) = \frac{1}{1-z^{-1}}$ with an integral interval $[t - \tau, t]$ and $\tau \in \mathbb{R}^+$, then it will be a interval integral with short-term memory. As a result, a MRE-based composite FORCE learning algorithm of $\hat{W}_o$ is given as follows:

$$\hat{W}_o(k) = \hat{W}_o(k - 1) + n(k)(e(k)\phi(k) + \kappa \boldsymbol{\xi}(k)) \tag{9}$$

where $\kappa \in \mathbb{R}^+$ is a weight parameter.

As the procedure shown above, the major intention of using the MRE technique is to take advantage of the memory function of linear filtering operators such that past values of the extended regressor $\phi\phi^T$ can be retained. According to existing results [35–37], if the regressor ϕ meets a partial interval-excitation (IE) condition which is much weaker than the PE condition, the MRE-based estimator (9) makes partial elements of $\hat{W}_o$ exponentially converge to their optimal values, where the rate of parameter convergence can be made arbitrarily fast by increasing the learning rate n. In this way, the proposed composite FORCE learning can relax the strict requirements on data richness and learning duration for parameter convergence, which leads to a more efficient learning approach for training chaotic ESNs with respect to both data and computation. The computational complexity of the proposed composite FORCE learning is strongly related to the computation of $\xi(k)$, which heavily relies on the number of reservoir nodes N. The choice of the linear filtering operator $L(z)$ and the weight parameter κ influence the actual convergence performance of the proposed method. In this paper, the linear filtering operator used in simulations is $L(z) = \frac{\lambda}{1-(1-\lambda)z^{-1}}$ with an integral interval $[0, t]$, providing forgetting memory. Theoretically, a smaller λ means a weaker forgetting ability, and thus, it should be determined by the dynamical property of the considered modeling problem.

4   Numerical Verification

Consider a modeling problem of nonlinear dynamics which has the following function of the target output [23]:

$$f(t) = 0.87\sin\left(\frac{\pi t}{20}\right) + 0.43\sin\left(\frac{\pi t}{10}\right) + 0.29\sin\left(\frac{3\pi t}{20}\right).$$

Our objective is to train the chaotic ESN in Sect. 2 using the approaches in Sec. 3 such that it ends up producing a periodic wave output autonomously. For this case, the proposed composite FORCE learning with $N \in [400, 1000]$, $\lambda \in [0.01, 1]$, $\kappa \in [1, 5]$ and $\gamma \in [1, 2]$ can take effect, and other hyper-parameters should be well tuned by trial-and-error.

The generation of a chaotic ESN has the following steps: First, the internal weight W is generated with a connectivity p, where nonzero elements are drawn from a Gaussian distribution with zero mean and variance $1/(pN)$; second, the internal weight W is scaled by the chaotic factor g to make the reservoir chaotic; last, the feedback weight W_{fb} is drawn from a uniform distribution between $[-1, 1]$. For the proposed composite FORCE learning, we set $N = 400$, $p = 0.1$, $g = 1.5$, $\alpha = 0.1$, $n(0) = 0.02$, $\kappa = 5$, $\lambda = 0.01$, $\hat{W}_o(0) = \mathbf{0}$, and $\gamma = 1.8$. The proposed method is compared with the LMS-based FORCE learning during simulations, where all shared parameters of the two learning methods keep identical to ensure a fair comparison. In our case, we simply set $\kappa = 0$ in the proposed algorithm to get the LMS-based FORCE learning algorithm. Simulations are carried out in the MATLAB software environment.

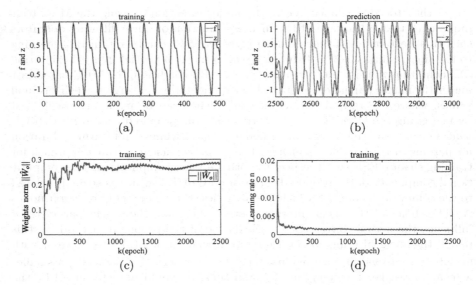

Fig. 2. Modeling results using the chaotic ESN with the LMS-based FORCE learning in [23]. (a) Training performance, where only the first 500 epochs of the total 2500 epochs are displayed for clear illustration. (b) Prediction performance. (c) Norm of the readout weight $\hat{W}_o$ during training. (d) Evolution of learning rate during training.

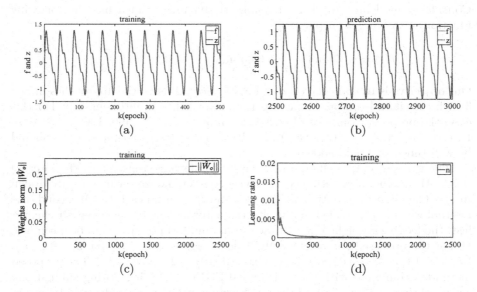

Fig. 3. Modeling results using the chaotic ESN with the proposed composite FORCE learning. (a) Training performance, where only the first 500 epochs of the total 2500 epochs are displayed for clear illustration. (c) Norm of the readout weight $\hat{W}_o$ during training. (d) Evolution of learning rate during training.

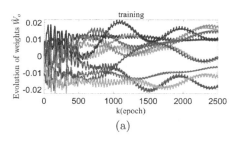 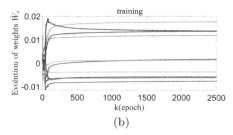

(a)   (b)

Fig. 4. Evolving trajectories of the readout weight $\hat{W}_o$, where only 10 randomly chosen components of the output weight $\hat{W}_o$ are displayed for clear illustration. (a) By the LMS-based FORCE learning in [23]. (b) By the proposed composite FORCE learning.

The training procedure is introduced as follows: First, the chaotic ESN is randomly generated according to the above steps; second, the output weight $\hat{W}_o$ is updated online, and the modeling performance of the ESN, the evolution of the output weight $\hat{W}_o$, and the weight norm $\|\hat{W}_o\|$ are observed in real-time; last, when the training time is up to the specified duration (60 cycles of the target wave in this example), online learning will be turned off, and the network keeps generating the target wave on its own.

Simulation results are provided in Figs. 2, 3, 4 and 5. For the chaotic ESN with the LMS-based FORCE learning, the network output z matches the target wave rapidly as soon as the training starts [see Fig. 2(a)], but deviates from the target wave quickly after training, which means that the network cannot predict the target output well [see Fig. 2(b)]. The norm of the output weight $\hat{W}_o$ oscillates violently and does not converge to any certain constant within 60 cycles of the target wave in this example [see Fig. 2(c)]. Besides, the learning rate n decreases during training from an initial value of 0.02 to a final value of 0.0015, at which point learning is stopped [see Fig. 2(d)]. Actually, it takes more that 6000 epochs for the LMS-based FORCE learning to achieve the convergence and to learn the example function in this case.

For the chaotic ESN trained with the proposed composite FORCE learning, the network output z matches the target wave rapidly while the learning starts [see Fig. 3(a)], and still keeps generating the target wave accurately after learning [see Fig. 3(b)]. It is shown that the proposed learning method is much more efficient as the norm of the readout weight $\hat{W}_o$ converges to a certain constant within only 12 cycles of the wave. Besides, the learning rate n decreases rapidly during training from an initial value of $n = 0.02$ to a final value of 0.00012, at which point learning is stopped [see Fig. 3(d)].

It is clearly shown in Fig. 4 that the parameter convergence of the proposed composite FORCE learning is significantly improved in terms of converging speed and parameter trajectory smoothness. In addition, the firing rates of four sample nodes from the reservoir shown in Fig. 5 implies that even the network activity exhibits chaos before learning, it turns out to be periodic as soon as the learning

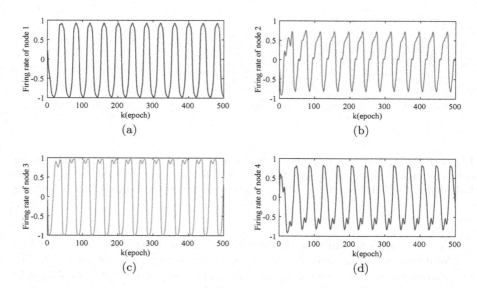

Fig. 5. Activation strengths of some internal nodes in the reservoir by the proposed composite FORCE learning, where only 4 randomly chosen internal nodes with the first 500 epochs of the total 2500 epochs are displayed for clear illustration. (a) Activation strength of node 1. (b) Activation strength of node 2. (c) Activation strength of node 3. (d) Activation strength of node 4.

starts. Such a phenomenon appears in both the learning methods, indicating the power of the feedback loop in driving the reservoir to an appropriate dynamic space that matches the requirement of certain tasks. In addition, the proposed method also works better than the RLS-based FORCE learning in terms of convergence speed, where it takes about 200 epochs to achieve the convergence compared to over 800 epochs for the RLS-FORCE learning.

5   Conclusions

In this paper, a composite FORCE learning approach enhanced by the MRE technique has been proposed to train chaotic ESNs efficiently. By the exploitation of regressor extension and linear filtering operators with memory, the proposed approach effectively enhances converging speed and parameter trajectory smoothness compared with the LMS-based FORCE learning. Simulation results based on a dynamics modeling problem have validated the effectiveness and superiority of the proposed approach. The extension of this study to robot control would be interested and some preliminary work has been done in [38].

Acknowledgments. This work was supported in part by the Guangdong Pearl River Talent Program of China under Grant No. 2019QN01X154.

References

1. Funahashi, K.I., Nakamura, Y.: Approximation of dynamical systems by continuous time recurrent neural networks. Neural Netw. **6**(6), 801–806 (1993)
2. Werbos, P.J.: Backpropagation through time: what it does and how to do it. Proc. IEEE **78**(10), 1550–1560 (1990)
3. Doya, K.: Bifurcations in the learning of recurrent neural networks. In: IEEE International Symposium on Circuits and Systems, San Diego, CA, pp. 2777–2780 (1992)
4. Jaeger, H.: Tutorial on training recurrent neural networks, covering BPPT, RTRL, EKF and the echo state network approach. Technical report, GMD Report 159, German National Research Center for Information Technology, Bonn, Germany (2002)
5. Chen, G.: A gentle tutorial of recurrent neural network with error backpropagation. arXiv preprint, arXiv:1610.02583 (2016)
6. Jaeger, H.: The "echo state" approach to analysing and training recurrent neural networks. Technical report, GMD Report 148, German National Research Center for Information Technology, Bonn, Germany (2001)
7. Maass, W., Natschläger, T., Markram, H.: Real-time computing without stable states: a new framework for neural computation based on perturbations. Neural Comput. **14**(11), 2531–2560 (2002)
8. Steil, J.J.: Backpropagation-decorrelation: online recurrent learning with o(n) complexity. In: IEEE International Joint Conference on Neural Networks, Budapest, Hungary, pp. 843–848 (2004)
9. Schrauwen, B., Verstraeten, D., Van Campenhout, J.: An overview of reservoir computing: theory, applications and implementations. In: European Symposium on Artificial Neural Networks, pp. 471–482 (2007)
10. Lukoševičius, M., Jaeger, H.: Reservoir computing approaches to recurrent neural network training. Comput. Sci. Rev. **3**(3), 127–149 (2009)
11. Nakajima, K.: Physical reservoir computing-an introductory perspective. Jpn. J. Appl. Phys. **59**(6), Article ID 060501 (2020)
12. Jaeger, H., Haas, H.: Harnessing nonlinearity: predicting chaotic systems and saving energy in wireless communication. Science **304**(5667), 78–80 (2004)
13. Jaeger, H.: Adaptive nonlinear system identification with echo state networks. In: International Conference on Neural Information Processing Systems, Cambridge, MA, USA, pp. 609–616 (2002)
14. Jaeger, H.: Short term memory in echo state networks. Technical report, GMD Report 152, German National Research Center for Information Technology, Bonn, Germany (2002)
15. Jaeger, H., Lukoševičius, M., Popovici, D., Siewert, U.: Optimization and applications of echo state networks with leaky-integrator neurons. Neural Netw. **20**(3), 335–352 (2007)
16. Bush, K., Anderson, C.: Modeling reward functions for incomplete state representations via echo state networks. In: IEEE International Joint Conference on Neural Networks, Montreal, Canada, pp. 2995–3000 (2005)
17. Sun, X., Li, T., Li, Q., Huang, Y., Li, Y.: Deep belief echo-state network and its application to time series prediction. Knowl.-Based Syst. **130**, 17–29 (2017)
18. Chen, Q., Shi, H., Sun, M.: Echo state network-based backstepping adaptive iterative learning control for strict-feedback systems: an error-tracking approach. IEEE Trans. Cybern. **50**(7), 3009–3022 (2019)

19. Waegeman, T., Schrauwen, B., et al.: Feedback control by online learning an inverse model. IEEE Trans. Neural Netw. Learn. Syst. **23**(10), 1637–1648 (2012)
20. Deihimi, A., Showkati, H.: Application of echo state networks in short-term electric load forecasting. Energy **39**(1), 327–340 (2012)
21. Ishu, K., van Der Zant, T., Becanovic, V., Ploger, P.: Identification of motion with echo state network. In: Oceans 2004 MTS/IEEE Techno-Ocean 2004 Conference, Kobe, Japan, pp. 1205–1210 (2004)
22. Skowronski, M.D., Harris, J.G.: Automatic speech recognition using a predictive echo state network classifier. Neural Netw. **20**(3), 414–423 (2007)
23. Sussillo, D., Abbott, L.F.: Generating coherent patterns of activity from chaotic neural networks. Neuron **63**(4), 544–557 (2009)
24. Nicola, W., Clopath, C.: Supervised learning in spiking neural networks with FORCE training. Nat. Commun. **8**, Article ID 2208 (2017)
25. DePasquale, B., Cueva, C.J., Rajan, K., Escola, G.S., Abbott, L.: Full-FORCE: a target-based method for training recurrent networks. PLoS ONE **13**(2), Article ID e0191527 (2018)
26. Inoue, K., Nakajima, K., Kuniyoshi, Y.: Designing spontaneous behavioral switching via chaotic itinerancy. Sci. Adv. **6**(46), Article ID eabb3989 (2020)
27. Tran, Q., Nakajima, K.: Higher-order quantum reservoir computing. arXiv preprint arXiv:2006.08999 (2020)
28. Sastry, S., Bodson, M.: Adaptive Control: Stability, Convergence and Robustness. Prentice Hall, New Jersey (1989)
29. Pan, Y., Yu, H.: Composite learning robot control with guaranteed parameter convergence. Automatica **89**, 398–406 (2018)
30. Chowdhary, G., Johnson, E.N.: Theory and flight-test validation of a concurrent-learning adaptive controller. J. Guid. Control. Dyn. **34**(2), 592–607 (2011)
31. Pan, Y., Yu, H.: Composite learning from adaptive dynamic surface control. IEEE Trans. Autom. Control **61**(9), 2603–2609 (2016)
32. Aranovskiy, S., Bobtsov, A., Ortega, R., Pyrkin, A.: Performance enhancement of parameter estimators via dynamic regressor extension and mixing. IEEE Trans. Autom. Control **62**(7), 3546–3550 (2017)
33. Ortega, R., Nikiforov, V., Gerasimov, D.: On modified parameter estimators for identification and adaptive control. A unified framework and some new schemes. Ann. Rev. Control **50**, 278–293 (2020)
34. Pan, Y., Bobtsov, A., Darouach, M., Joo, Y.H.: Learning from adaptive control under relaxed excitation conditions. Int. J. Adapt. Control Signal Process. **33**(12), 1723–1725 (2019)
35. Pan, Y., Sun, T., Liu, Y., Yu, H.: Composite learning from adaptive backstepping neural network control. Neural Netw. **95**, 134–142 (2017)
36. Pan, Y., Yang, C., Pratama, M., Yu, H.: Composite learning adaptive backstepping control using neural networks with compact supports. Int. J. Adapt. Control Signal Process. **33**(12), 1726–1738 (2019)
37. Huang, D., Yang, C., Pan, Y., Cheng, L.: Composite learning enhanced neural control for robot manipulator with output error constraints. IEEE Trans. Industr. Inf. **17**(1), 209–218 (2021)
38. Wu, R., Li, Z., Pan, Y.: Adaptive echo state network robot control with guaranteed parameter convergence. In: International Conference on Intelligent Robotics and Applications, pp. 587–595. Yantai, China (2021)

Generative Adversarial Domain Generalization via Cross-Task Feature Attention Learning for Prostate Segmentation

Yifang Xu[1], Dan Yu[1], Ye Luo[2(✉)], Enbei Zhu[1], and Jianwei Lu[2(✉)]

[1] Shanghai Maritime University, Shanghai, China
[2] Tongji University, Shanghai, China
{yeluo,jwlu}@tongji.edu.cn

Abstract. In real-life clinical deployment, domain shift on multi-site datasets is a critical and crucial problem for the generalization of prostate segmentation deep neural networks. To tackle this challenging problem, we propose a new Generative Adversarial Domain Generalization (GADG) network, which can achieve the domain generalization through the generative adversarial learning on multi-site prostate MRI images. Specifically, we compose a cascaded generation-segmentation "U-Net" shaped network, first we generate a new image with same structural but different style information to the source image via the generation path (i.e. one branch of "U"), then the domain discriminator is used to determine whether the generated image is true or false via the domain label classification. To make the prostate segmentation network (i.e. the other branch of "U") learned from the source domains still have good performance in the target domain, a Cross-Task Attention Module (CTAM) is designed to transfer the main domain generalized features from the generation branch to the segmentation branch. To maintain the credibility of style transferred images, the whole network is warmed-up first. To evaluate the proposed method, we conduct extensive experiments on prostate T2-weighted MRI images from six different domains with distribution shift, and the experimental results validate that our approach outperforms other state-of-the-art prostate segmentation methods with good domain generalization capability.

Keywords: Domain generalization · Generative adversarial learning · Prostate segmentation · Cross-task attention

1 Introduction

Deep learning has been proven effective in many image segmentation tasks [7], and early deep learning models assumed that the training and testing data come

---

Y. Xu, D. Yu and E. Zhu—These authors contributed equally to this work.
This work was supported by the General Program of National Natural Science Foundation of China (NSFC) (Grant No. 61806147).

© Springer Nature Switzerland AG 2021
T. Mantoro et al. (Eds.): ICONIP 2021, LNCS 13109, pp. 273–284, 2021.
https://doi.org/10.1007/978-3-030-92270-2_24

from the same data distribution, but this assumption is usually nonsensical. In the real world, particularly, in medical image segmentation tasks, the data distributions are often inconsistent across medical centers as they using different imaging protocols or scanner vendors. This makes it difficult for deep learning networks to obtain good generalization performance, which is a long-standing challenge in medical image processing. On the other hand, prostate disease is one of the most common diseases in men. In the past, it was usually segmented manually by radiologists, but this process is expensive and time-consuming, thus automatic segmentation of prostate MRI images is necessary. However, automatically segmenting multi-site prostate MRI images is always a challenging task in that images from different data source usually have various visual appearances, and the area of the gland itself is too small and lack of distinct boundary with the surrounding, the biggest challenge of multi-site prostate MRI image segmentation is the domain shift among data from different medical centers.

Recently, scholars have proposed various methods to solve the domain shift problem. The most common approach is transfer learning [13], where a portion of the data on the target domain is acquired and labeled by a physician to fine-tune the segmentation model. However, the target domain data are often expensive to label. Ganin et al. [15] proposed the unsupervised domain adaptive method that can free physicians from labeling. But in real-life medical image processing tasks, it is difficult to acquire MRI images from new medical centers in advance. Therefore, the domain generalization method is increasingly important and meaningful recently, i.e. train a deep learning model with good generalization performance which can directly deploy to the target domain.

Some recent domain generalization methods with better results are broadly classified into three types: 1) Model-independent meta-learning algorithm [7]. 2) Learning domain-invariant feature representations of cross-domain data, and matching them with the prior distribution of the target domain [8]. 3) Data augmentation, which envisages a broad transformation of the source data can help to adapt the model to a more diverse data distribution.

Inspired by [9], we propose a new Generative Adversarial Domain Generalization (GADG) network to solve the domain generalization problem for prostate segmentation on multi-site MRI images. The network architecture of the proposed method is under multi-task framework (i.e. image generation and image segmentation), in which the segmentation network and the generation network share one encoder but with respective decoders, and the two networks are cascaded into a "U-Net" shape. We conducted extensive experiment on the multi-site prostate dataset [7], and the results validated our method outperforms other state-of-the-art domain generalization prostate segmentation methods.

Our main contributions are three-fold: 1) A novel GADG network is proposed to solve the domain shift problem in prostate segmentation on multi-site MRI images. To our best knowledge, our GADG network is the first to simulate the domain shift between the source domains and the unseen domain by the data distribution shift between the input images and the generated images. 2) A cross-task attention module is designed to transfer the domain generalized features into the segmentation branch, and improve the prostate segmentation performance obviously. 3) Extensive experimental results validate the effectiveness of our method on the prostate segmentation with good domain generalization. Our method can also be applied to any other medical segmentation tasks besides prostate.

2 Related Work

2.1 Prostate Segmentation

Early scholars used some traditional algorithms, including modeling the shape of the prostate, and regressing the shape of the prostate edges. Nowadays, deep learning methods are more commonly for automatic prostate MRI segmentation. For example, Yu *et al.* [10] designed a novel volumetric convolutional neural network for prostate segmentation using a mixture of long-residual and short-residual connections to perform efficient. Nie *et al.* [11] proposed a semi-supervised learning strategy based on region attention to overcome the challenge of lacking sufficient training data using unlabeled data. To reduce the effect of noise and suppress the tissue around the prostate with similar intensity, Wang *et al.* [12] developed a novel deep neural network that selectively utilizes multilevel features for prostate segmentation using an attention mechanism. These works not only effectively solved the difficult problem of segmentation of prostate MRI images, but also well solved the problem of insufficient training data.

2.2 Domain Generalization

Although CNNs have been successfully applied to the automatic segmentation of medical images, the domain shift still lead the pre-trained model to drastic performance degradation when deployed to a new dataset. The domain generalization problem is gradually attracting scholars because it needs neither images nor masks from target domains, which is more common in real-life deployment scenarios. Some recent works such as, Zhou *et al.* [8] proposed a new method based on domain adversarial image generation (DDAIG) by designing a neural network including an image classifier, a domain classifier and an image generator to generate new images by mapping the source domain data to the unknown domain through a domain conversion network, aiming to generate new images that can deceive over the domain classifier while the image classifier can correctly classify whether true or false the generated images are, thus augment the source domains via the generated data with different distributions; Dou *et al.* [7] proposed a novel shape aware meta-learning (SAML) scheme to promote robust optimization of segmentation models by employing meta-train set and meta-test set, so as to simulate domain shifts during network training, they also proposed a new loss function for constraining the segmentation shape completeness and smoothness.

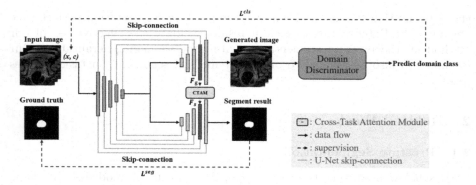

Fig. 1. Overview of our proposed Generative Adversarial Domain Generalization (GADG) network for prostate segmentation on multi-site MRI images.

3   The Proposed Method

The overview of the proposed framework is shown in Fig. 1. To tackle the problem of domain shift, we proposed a Generative Adversarial Domain Generalization (GADG) network, aiming to simulate the domain shift between the source domains and the target domain by the data distribution shift between the original inputs of the GADG network and the generated new images. In order to help the segmentation network improve its domain generalization performance, we proposed a Cross-task Attention Module (CTAM) to extract features from the generated images, which contain domain shift with the input images. We implement the image generation and the segmentation tasks into a shared encoder by a cascaded "U-Net" shaped network. Further, a shared encoder for image generation and image segmentation is helpful to extract structural information via shallow feature learning especially when the boundary of the gland in the prostate MRI image is blurred.

3.1   The Generative Adversarial Domain Generalization Network

The foundation of our method is a multi-task generative adversarial network [1], which includes the prostate segmentation task and the generative adversarial learning task. As Fig. 1 shows, the upper branch is a generative adversarial network, which consists of an image generator and a domain discriminator. The skip connections in the generator are useful to merge the shallow and deep features. As the generative adversarial network gradually converged, the discriminator is expected to classify the generated images into any other domains except the input one (i.e. the distance between the one-hot label of input images and the generated images is expected as far as possible). Thus we can get new images that contain the same structural information but different style information from the original ones. We assume that the distribution shift between original images and generated images can simulate the domain shift in the real world. The lower part of the network is the segmentation branch, which has the same structure as the generation network. It is consists of four down-samplings, four up-samplings, and skip-connections between encoder and decoder. Finally, the segmentation result is supervised by the ground truth of the input image.

Task1. Multi-site Prostate Segmentation Let $D = \{D_1, D_2, ..., D_k\}$ be the sets of data from k different domains. As Fig. 1 shows, our purpose is to construct a segmentation network $f(x)$, train it with images x from $k - 1$ source domains, and test it with images from the k_{th} domain. In this way, we expect the constructed segmentation network has well generalization result on prostate segmentation even on data from different domains. Here, without loss of the generality, we adopt the standard U-Net as the backbone of the segmentation network [20]. That's to say, we obtained the feature F_s at the second to last layer of the segmentation decoder, which is subsequently used to get and visualized the segmentation result y' as the final output of the segmentation decoder. However, directly training and testing the segmentation network above can not solve the domain shift problem, because the domain shift among different data domains did not take into account in the network learning.

Task2. Domain Shift Modeling by Generative Adversarial Learning In order to model the domain shift, the generative adversarial network (i.e. the upper branch of the cascade "U-Net") is proposed, which is consists of a generator and a domain discriminator as Fig. 1 shows. We expect that the new image generated by the generator has the same structure as the input but with different domain class label thus can cheat the domain discriminator. By this kind of generative adversarial learning, the domain shift (e.g. the distance between the true domain one-hot label of input and the generated domain one-hot label of the new image) can be learned by the network implicitly.

The generation network shares the same encoder as the segmentation network, followed by a decoder to generate a new image. Once obtained the generated image, the domain discriminator can perform the classification. Roughly, the shape of the generator network is also a U-Net. And the train of this branch of the network is as following. Randomly pick three continuous slices of an MRI images the input, and the middle slice of its corresponding three ground-truth masks as the mask x, and the domain label of the input is represented as a one-hot vector c. After the second to last layer of the decoder, we proposed a Cross-task Attention Module (CTAM) to extract main information from the feature map F_g, which is detailed in Sect. 3.2. F_g is used to get the generated image $\hat{x}$.

3.2 The Cross-Task Attention Module (CTAM)

The Cross-task Attention Module is very important in that it can effectively transfer the key features of the generated images as auxiliary information to the segmentation network, so as to improve the generalization performance of the prostate segmentation task. The structure of the CTAM module can be referred to Fig. 2. Different from previous methods which usually perform feature attention learning within a single task, we propose to learn the attention cross two different tasks based on the convolutional block attention module [6]. In other words, the learned feature contributed to the image generation task is re-calibrated via the channel attention and the spatial attention sequentially and then added to the feature learned into the segmentation task, aiming to obtain good segmentation performance across different data domains.

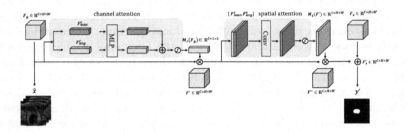

Fig. 2. Cross-task Attention Module adopted in proposed method.

As detailed above, F_g and F_s are features extracted from the second to last layer of the generation decoder and the segmentation decoder, respectively. Denote $F_g \in \mathbb{R}^{C \times H \times W}$ and $F_s \in \mathbb{R}^{C \times H \times W}$. To produce the channel attention map M_c, the 1D channel attention block aims to exploit the inter-channel relationship of features, by using both average-pooling and max-pooling operations, generating the average-pooled features F_{avg}^c and the max-pooled features F_{max}^c. Both features are then forward to a shared MLP with one hidden layer to generate M_c, which is defined as follow,

$$M_c(F) = \sigma(W_1(W_0(F_{avg}^c)) + W_1(W_0(F_{max}^c))), \tag{1}$$

where σ denotes the sigmoid function. Here, $W_0 \in R^{C/r \times C}$ and $W_1 \in R^{C \times C/r}$ represent the MLP weights. W_0 and W_1, are shared for both inputs and the ReLU activation function is followed by W_0. Then broadcast the channel attention value along the spatial dimension, and do element-wise multiplication between F_g and M_c as Eq. 2 shows.

$$F^{'} = M_c(F_g) \otimes F_g. \tag{2}$$

To utilize the inter-spatial information of features, we first apply max-pooling and average-pooling operations along the channel axis, generating two 2D maps F_{avg}^s and F_{max}^s, then concatenate them together, and apply a convolution operation with the filter size of 7×7,

$$M_s(F) = \sigma(f^{7 \times 7}([F_{avg}^s; F_{max}^s])), \tag{3}$$

where σ denotes the sigmoid function. Then broadcast the spatial attention value along the channel dimension, and do element-wise multiplication between

F' and M_s. Finally, we merge the output feature and segmentation feature F_s using element-wise summation. The recalibrated feature is defined as follows:

$$F'_s = F_s + M_s(F') \otimes F'. \tag{4}$$

3.3   End-to-End Loss Function

In order not to increase the complexity of the network while ensuring the credibility of the generated images, we set 50 epochs as the warm-up phase during the training process, and the generated images in this part are not involved in the update of the segmentation network parameters, but only used to update the parameters of the domain discriminator. After the warm-up phase, when the generated images are of high quality, we begin to adopt generated images to assist the segmentation network. Two sets of supervision are required during training, using the ground truth of original images as the supervision of the segmentation network, i.e. L^{seg}, and the domain class of the original images as the supervision of the domain discriminator, i.e. L^{cls}.

Considering that the prostate area only accounts for a small proportion of the entire image and the boundary is always blurred, which makes the segmentation difficult, the focal loss can tackle this unbalanced to a certain extent by dynamically changing the proportion of the foreground and background area when backpropagation is performed. The loss L^{seg} is, *i.e.*,

$$L^{seg} = -\frac{1}{N} \sum_{i}^{N} (\alpha g_i (1 - p_i)^{\gamma} log p_i + (1 - \alpha)(1 - g_i) p_i^{\gamma} log(1 - p_i)). \tag{5}$$

Specifically, according to the number of source domains, we assign a domain class c for each input image, for example, there are 5 source domains, a batch of input images respectively from 1st, 4th, 3rd domain, the class vector of input images is $c = [1, 4, 3]$. As we aim to transfer image style into $\hat{x}$ that cannot be identified by domain discriminator, so we adopt FCN to classify batches of $\hat{x}$. The loss L^{cls} is cross-entropy, *i.e.*,

$$L^{cls} = -\frac{1}{N} \sum_{i}^{N} (g_i log p_i + (1 - (1 - g_i) log(1 - p_i)) \tag{6}$$

As above all indicated, the end-to-end network intends to get better segmentation results and worse classification results, which means the generator can successfully generate images contain the same style information but different structural information with origin images. Therefore, the total loss function can be written as L, *i.e.*,

$$L = L^{seg} - L^{cls}. \tag{7}$$

Considering that at the beginning of the training process, the generated images are not good enough to contain different structural information from the original images, so the domain discriminator can easily classify the generated image into the same class as the original one. In other words, the L^{cls} could be large at the beginning, and the L value could be negative. After carefully analyzing the ranges of L^{seg} and L^{cls}, we finally adjust our loss function into, *i.e.*,

$$L = L^{seg} + \frac{1}{e^{L^{cls}}}. \tag{8}$$

4   Experiment

4.1   Dataset and Implementation Details

We adopt the Multi-site Dataset for Prostate MRI Segmentation from Chinese University of Hong Kong [7], which provides 116 prostate T2-weighted MRI from 6 different data sources with distribution shift. For pre-processing, we resized each sample to 384 × 384 in axial plane, and we extract three slices of each sample as input and normalized them to zero mean and unit variance.

We implement 3D U-Net as a backbone, the number of the domains of source and target are set as 5 and 1. The basic feature extraction network was resnet 18. The segmentation network was trained with Adam optimizer and the learning rate was 1e−4. We trained 400 epochs with batch size of 2 for each target domain.

4.2   Comparison with State-of-the-Arts

Table 1 presents the quantitative results of different methods on this Multi-site Dataset for Prostate MRI Segmentation in terms of Dice. The compared methods are all implemented in the same dataset with us, including a classifier regularization based method (Epi-FCR) [3], a data-augmentation based method (BigAug) [2], a meta-learning based method (MASF) [5] and a latent space regularization method (LatReg) [4]. Additionally, we compared with *DeepAll* baseline (i.e., aggregating data from all source domains for training a U-Net model) and 'Intra-site' setting (i.e., training and testing on the same domain, with some outlier cases excluded to provide general internal performance on each site). As Table 1 illustrates, Intra-site can achieve quite a good result, as training and testing data are from the same domain with the similar distribution. Naturally, *DeepAll* got the relatively worst result because this method did not do any processing on the domain shift, but simply combined data from different domains to train the network. The other approaches of LatReg, BigAug, MASF are more significantly better than *DeepAll*, with SAML method yielding the best results among them in previous work. Notably, our approach (cf. the last row) achieves higher performance over all these state-of-the-art methods on five sites, and performs second best on site C, it is worth noting that our method outperforms the *DeepAll* model by 2.46% simultaneously outperforms the SAML model by 0.31%, demonstrating the capability of our scheme to deal with domain generalization problem.

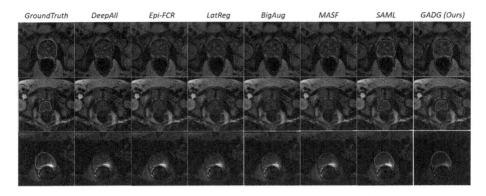

GroundTruth   DeepAll   Epi-FCR   LatReg   BigAug   MASF   SAML   GADG (Ours)

Fig. 3. Qualitative comparison on the generalization results of different methods, with three cases drawn from different unseen domains respectively.

We infer the reason why our method slightly drops on site C is the MRI data distribution of site C is significantly different compared to several other domains, and the prostate information in the picture is blurred, which shows that our method needs to be improved in the performance of difficult samples. Figure 3 shows the segmentation results of the comparison with the above methods with ours. It could be seen that our method can accurately segment the prostate margin, and our segmentation contour is more smoothly shown by the segmentation result. As Fig. 5 shown, it is worth noting that the latest work via federated learning (FedDG) [18] on this dataset improves the performance. Compared with the FedDG method, our network has increased by 1.38% on average of six domains, and still outperform significantly on site B (1.49%), D (2.97%) and E (6.75%). But FedDG has made improvement on site A (−0.46%), C (−1.08%), F (−1.35%). It can be seen that our method still improves the generalization performance on average and most domains.

Table 1. Generalization performance of various methods on Dice (%) ↑.

Method	Site A	Site B	Site C	Site D	Site E	Site F	Average
Intra-site	89.27	88.17	88.29	83.23	83.67	85.43	86.34
DeepAll	87.87	85.37	82.94	86.87	84.48	85.58	85.52
Epi-FCR [3]	88.35	85.83	82.56	86.97	85.03	85.66	85.74
LatReg [4]	88.17	86.65	83.37	87.27	84.68	86.28	86.07
BigAug [2]	88.62	86.22	83.76	87.35	85.53	85.83	86.21
MASF [5]	88.70	86.20	84.16	87.43	86.18	86.57	86.55
SAML [7]	89.66	87.53	**84.43**	88.67	87.37	88.34	87.67
GADG (**Ours**)	**89.73**	**88.66**	84.18	**91.20**	**89.77**	**89.12**	**88.78**

4.3   Ablation Experiment

Effect of CTAM Block. As mentioned above, the CTAM block helps to effectively extract the main features of generated images. We first conducted two sets of control experiments to study the contribution of attention block in our model. One group investigated the influence of the presence or absence of an attention block on the experimental results, and the other group investigated whether adding the attention block to the segmentation or generation decoder branch will affect the result.

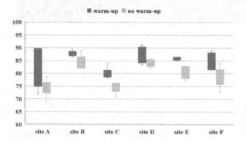

Fig. 4. Generalization performance of our method with or without the warm-up scheme.

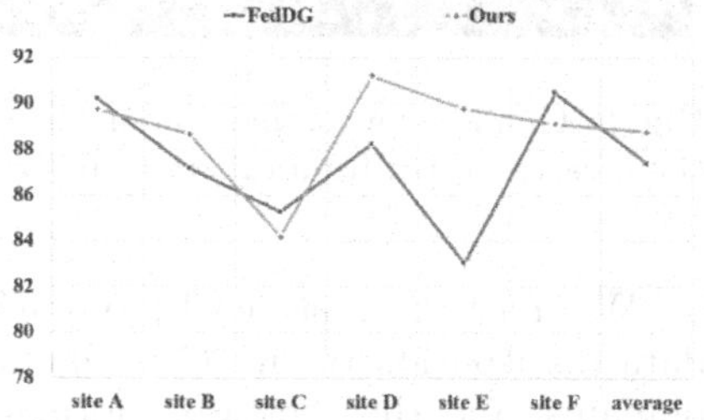

Fig. 5. Curves of generalization performance on six unseen domains of newest State-of-the-Art and our method.

As shown in Table 2, we adopt SE-Net [19] as a comparison with our Cross-Task Attention Module (CTAM), G_{ctam} represent using our CTAM to extract the main feature of generated images and fusion with segmentation feature, which has been detailed in Sect. 3.4. On the contrary, G_{senet} replaced the attention block with SE-Net. While S_{ctam} and S_{senet} represent the attention block is used from the segmentation path to the generation path. It can be seen that such multi-task generative adversarial network without attention block already outperformed SAML on Site D, Site E and Site F, though slightly lower on Site A, Site B and Site C. Simultaneously, G_{senet} outperformed No Attention method on last three domains but drop heavily on former three domains, it may because that in Site A-C, the main generated feature extracted by SE-Net are rather worse to contain key information. In conclusion, the CTAM attention block adding after the generation decoder significantly improved the performance, i.e. our G_{ctam} method outperformed any other methods on all domains except Site F (dropped 0.02%). **Effects of Network Warm-Up.** As Fig. 4 shown, the warm-up scheme can improve the performance across all six domains by improving the quality of generated images.

Ablation on Segmentation Loss. Considering that the prostate target is quite small in the entire image, we compare the contribution of focal loss and

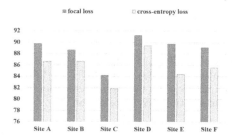

Fig. 6. Ablation results to analyze the effect of the loss function

cross-entropy loss, as Fig. 6 shown that focal loss is helpful for improving the performance on all domains.

Table 2. Results to the effect of the Cross-task Attention Module on Dice (%)↑.

Method	Site A	Site B	Site C	Site D	Site E	Site F	Average
No Attention	87.93	87.32	84.18	90.68	89.11	88.98	88.03
$G_{senet} + S_{senet}$	80.64	80.16	74.27	87.35	86.01	86.24	82.45
S_{senet}	77.72	76.51	61.47	88.47	87.21	87.39	79.79
G_{senet}	82.12	81.36	64.02	90.85	89.14	**89.14**	82.61
$G_{ctam} + S_{ctam}$	66.30	66.10	64.27	68.95	67.53	67.41	66.76
S_{ctam}	84.95	87.28	70.50	87.46	87.08	85.01	83.71
G_{ctam}	**89.73**	**88.66**	**84.18**	**91.20**	**89.77**	89.12	**88.78**

5 Conclusion

We provide a new generative adversarial domain generalization network via cross-task Feature attention learning for prostate segmentation on MRI images. We further introduce the warm-up scheme to maintain the credibility of transferred images. Extensive experiments demonstrate the effectiveness of our method. To our best knowledge, it is the first time incorporating a multi-task GAN-based method for domain generalization in the multi-site dataset for prostate MRI segmentation. Our method can be extended to various medical image segmentation scenarios that suffer from domain shift.

References

1. Li, L., et al.: Random style transfer based domain generalization networks integrating shape and spatial information. In: Puyol Anton, E., et al. (eds.) STACOM 2020. LNCS, vol. 12592, pp. 208–218. Springer, Cham (2021). https://doi.org/10.1007/978-3-030-68107-4_21

2. Zhang, L., Wang, X., Yang, D., Sanford, T., Harmon, S., et al.: Generalizing deep learning for medical image segmentation to unseen domains via deep stacked transformation. IEEE TMI (2020)
3. Li, D., Zhang, J., Yang, Y., Liu, C., Song, Y.Z., Hospedales, T.M.: Episodic training for domain generalization. In: ICCV, pp. 1446–1455 (2019)
4. Aslani, S., et al.: Scanner invariant multiple sclerosis lesion segmentation from MRI. In: 2020 IEEE 17th International Symposium on Biomedical Imaging (ISBI). IEEE (2020)
5. Dou, Q., de Castro, D.C., Kamnitsas, K., Glocker, B.: Domain generalization via model-agnostic learning of semantic features. In: NeurIPS, pp. 6450–6461 (2019)
6. Woo, S., Park, J., Lee, J.Y., et al.: Cbam: convolutional block attention module. In: Proceedings of the ECCV (2018)
7. Liu, Q., Dou, Q., Heng, P.A.: Shape-aware meta-learning for generalizing prostate MRI segmentation to unseen domains. In: Martel, A.L., et al. (eds.) MICCAI 2020. LNCS, vol. 12262, pp. 475–485. Springer, Cham (2020). https://doi.org/10.1007/978-3-030-59713-9_46
8. Zhou, K., et al.: Deep domain-adversarial image generation for domain generalisation. In: Proceedings of the AAAI Conference on Artificial Intelligence, vol. 34, no. 07 (2020)
9. Li, L., Zimmer, V.A., Ding, W., et al.: Random style transfer based domain generalization networks integrating shape and spatial information. arXiv preprint arXiv:2008.12205 (2020)
10. Yu, L., Yang, X., Chen, H., Qin, J., Heng, P.A.: Volumetric convnets with mixed residual connections for automated prostate segmentation from 3D MR images. In: AAAI (2017)
11. Nie, D., Gao, Y., Wang, L., Shen, D.: ASDNet: attention based semi-supervised deep networks for medical image segmentation. In: International Conference on MICCAI (2018)
12. Wang, Y., et al.: Deep attentional features for prostate segmentation in ultrasound. In: International Conference on MICCAI (2018)
13. Ghafoorian, M., et al.: Transfer learning for domain adaptation in MRI: application in brain lesion segmentation. In: International Conference on MICCAI (2017)
14. Dou, Q., Liu, Q., Heng, P.A., Glocker, B.: Unpaired multi-modal segmentation via knowledge distillation. IEEE TMI **39**, 2415-2425 (2020)
15. Ganin, Y., Lempitsky, V.: Unsupervised domain adaptation by backpropagation. In: International Conference on Machine Learning. PMLR (2015)
16. Guo, Y., Gao, Y., Shen, D.: Deformable MR prostate segmentation via deep feature learning and sparse patch matching. IEEE Trans. Med. Imaging **35**(4), 1077–1089 (2015)
17. Zhu, Q., Du, B., Turkbey, B., Choyke, P., Yan, P.: Exploiting interslice correlation for MRI prostate image segmentation, from recursive neural networks aspect. Complexity **2018**, 10 (2018)
18. Liu, Q., Chen, C., Qin, J., et al.: Feddg: federated domain generalization on medical image segmentation via episodic learning in continuous frequency space. In: Proceedings of the IEEE cconference on CVPR (2021)
19. Hu, J., Shen, L., Sun, G.: Squeeze-and-excitation networks. In: Proceedings of the IEEE Cnference on CVPR (2018)
20. Ronneberger, O., Fischer, P., Brox, T.: U-net: convolutional networks for biomedical image segmentation. In: International Conference on MICCAI (2015)

Context-Based Deep Learning Architecture with Optimal Integration Layer for Image Parsing

Ranju Mandal$^{(\boxtimes)}$, Basim Azam, and Brijesh Verma

Centre for Intelligent Systems, School of Engineering and Technology, Central Queensland University, Brisbane, Australia
{r.mandal,b.azam,b.verma}@cqu.edu.au

Abstract. Deep learning models have been proved to be promising and efficient lately on image parsing tasks. However, deep learning models are not fully capable of incorporating visual and contextual information simultaneously. We propose a new three-layer context-based deep architecture to integrate context explicitly with visual information. The novel idea here is to have a visual layer to learn visual characteristics from binary class-based learners, a contextual layer to learn context, and then an integration layer to learn from both via genetic algorithm-based optimal fusion to produce a final decision. The experimental outcomes when evaluated on benchmark datasets show our approach outperforms existing baseline approaches. Further analysis shows that optimized network weights can improve performance and make stable predictions.

Keywords: Deep learning · Scene understanding · Semantic segmentation · Image parsing · Genetic algorithm

1 Introduction

Image parsing is a popular topic in the computer vision research domain with a broad range of applications. Scene understanding, medical image analysis, robotic perception, hazard detection, video surveillance, image compression, augmented reality, autonomous vehicle navigation are some specific applications where image parsing is essential. Because of the performance of deep learning models on benchmark problems in a vast range of computer vision applications, there has been a volume of works focused on designing image parsing models using deep learning. Recent deep CNN-based architectures have the remarkable capability for precise semantic segmentation as well as pixel-wise labeling, however, the context presentations are implicit, and still scene parsing task remains a hard problem.

Two important aspects that make scene parsing challenging: i) clutter objects and shadows in scene images have high intra-class variability ii) presence of low inter-class variability due to texture similarity among objects. Recently published architectures [1–4] have made progress in pixel-wise labeling using CNN and context information

© Springer Nature Switzerland AG 2021
T. Mantoro et al. (Eds.): ICONIP 2021, LNCS 13109, pp. 285–296, 2021.
https://doi.org/10.1007/978-3-030-92270-2_25

at local and multiple levels. Semantic segmentation processes are mostly enhanced by rich global contextual information [5–7]. However, a lack of systematic investigation to utilize local and global contextual information parallelly in single network architecture is visible. Context plays a vital role in complex real-world scene image parsing by acquiring the critical surrounding information, and this captured statistical property can be utilized for predicting class label inference in many applications.

Contextual information can be categorized into two basic types: i) a global context that contains the statistical information derived from a holistic view and ii) local context captures the statistical information of adjacent neighbors. In general, the class label of a pixel in an image should maintain a high uniformity with the local context conveyed by adjacent pixels; match the global context of the entire scene semantics and layout too. Existing approaches are neither able to capture global contextual information from the images nor consider the labels of neighboring object categories. However, deep learning models have the potential to improve performance for semantic segmentation if assimilate such features explicitly. In fact, global and local context representations in deep learning models remain a challenging problem. Few methods have shown misclassification can be reduced by scene context at the class label. Conditional Random Fields (CRFs)-based graphical model [8], relative location [9], and object co-occurrence statistics-based model [10] have achieved significant performance improvement by the modeling of contextual information in natural scenes.

In this paper, we propose a novel architecture to integrate contextual and visual information in a new way to improve scene parsing. The proposed architecture extracts the visual features and the contextual features to produce precise superpixel-wise class labels and integrates them via a Genetic Algorithm (GA)-based optimization. Both global and local contextual features are represented by computing object occurrence priors of each superpixel. We introduce novel learning methods where multiple weak learners are fused to achieve high performance. The class-specific model classifier considers both; many weak classifiers and boosting algorithms to achieve improved accuracy of class labeling. The final integration layer combines three sets of class-wise probabilities acquired from the classification of superpixels, local context, and global context. A neural network optimized with the GA instead of gradient descent to avoid being trapped into 'local minima' and computes the final class label. The primary contributions of this paper can be summarized as follows:

a) The proposed new robust architecture incorporates both visual features and contextual features and unified them into a single network model. To leverage the binary classification, we use the one vs. all classification technique to obtain a class-wise probability of superpixels. Based on the number of target classes, our one vs. all solution consists of an equal number of binary classifiers for each possible outcome.

b) The model is trained with object co-occurrence and spatial context simultaneously from the training image samples in an unsupervised fashion. The proposed Context Adaptive Voting (CAV) features conserve relative and absolute spatial coordinates of objects by capturing Objects Co-occurrence Priors (OCPs) among spatially divided image blocks.

c) We optimize the weights of the integration layer using a novel fusion method to obtain the final classification label. GA-based optimization is proposed in this layer to overcome the 'local minima' problem.

d) We conduct exhaustive experimentations with benchmark datasets, namely Stanford Background [11] and CamVid [12] dataset and compare the results with state-of-the-art methods.

2 Related Works

An up-to-date review of the lately published state-of-the-arts works in image parsing is presented here. We critically reviewed multi-scale and spatial pyramid-based architectures, encoder-decoder networks, models based on visual attention, pixel-labeling CNN, and adversarial settings generative models. Besides, as the GA is incorporated in our integration layer, we explore few recent GA-based optimization techniques on the hyper-parameter solution space to optimize the CNN architectures' weights.

The fundamental attributes of the contemporary methods comprise of CNN-based model and multi-scale context. In contrast to the standard convolution operator, DenseA-SPP's [1] atrous convolution secure a larger receptive field without compromising spatial resolution and keeps the same number of kernel parameters. It encodes higher-level semantics by acquiring a feature map of the input shape with every output neuron computing a bigger receptive field. RPPNet [2] analyzes preceding successive stereo pairs for parsing result that efficiently learns the dynamic features from historical stereo pairs to ensure the temporal and spatial consistency of features to predict the representations of the next frame. STC-GAN [3] combines a successive frame generation model with a predictive image parsing model. It takes previous frames of a video and learns temporal representations by computing dynamic features and high-level spatial contexts. Zhang et al. [4] proposed perspective adaptive convolutions automatically adapt the sizes and shapes of receptive fields corresponding to objects' perspective deformation that attenuates the issue of inconsistent large and small objects, due to the fixed shaped receptive fields of standard convolutions.

Preliminary approaches obtain pixel-wise class labels using several low-level visual features extracted at pixel-level [13] or a patch around each pixel [14]. Few early network models on image parsing include feature hierarchies and region proposal-based architectures [15], which consider region proposals to produce class label accuracy. However, features that acquire global context achieve better as features at pixel-level fail to capture robust appearance statistics of the local region, and patch-wise features are susceptible to background noise from objects. PSPNet [16] exploits the global context of an image using the spatial pyramid pooling technique to improves performance. A hybrid DL-GP network [17] segments a scene image into two major regions namely lane and background. It integrates a convolutional encoder-decoder network with a hierarchical GP classifier. The network model outperforms SegNet [18] in both visual and quantitative comparisons, however, considers only the pedestrian lane class for evaluation.

RAPNet [20] proposes importance-aware street frame parsing to include the significance of various object classes and the feature selection module extracts salient features for label predictions, and the spatial pyramid module further improves labeling in a residual refinement method. UPSNet [21] tackles the panoptic segmentation task. A residual network model followed by a semantic segmentation module and a mask R-CNN-based [22] segmentation module, solves subtasks concurrently, and eventually, a panoptic module performs the panoptic segmentation using pixel classification. A deep gated attention network was proposed by Zhang et al. [23] in the context of pixel-wise labeling tasks which captures the scene layout and multi-level features of images. Cheng et al. [24] propose scene parsing using Panoptic-DeepLab architecture based on a fully convolutional method that simultaneously aims the semantic and instance segmentation in a single-shot, bottom-up fashion. RefineNet [26] enables high-resolution prediction by adopting a method based on long-range residual connections. It uses features obtained at multiple levels available along the down-sampling process. The chained residual pooling encapsulates the robust background context, and labeling was further refined by dilated convolutions, multi-scale features, and refining boundaries.

3 Proposed Approach

The proposed network architecture is constructed by pipelined stacked layers. As presented in Fig. 1, the first step is involved in computing superpixels level visual features. The class probabilities of superpixels are predicted in the visual feature prediction layer using the visual features and the class-semantic supervised classifiers. The contextual voting layer attains local and global contextual voting features of each superpixel. These contextual voting features are computed using the corresponding most probable class and the Object Co-occurrence Priors (OCPs). Subsequently, all three features are fed into our final integration layer to fuse optimally and produce the final target class. The optimal weights are learned using a small network with a single hidden layer, and GA is incorporated to find the best solution which holds all parameters that help to enhance the classification results. Three main steps involved in the proposed architecture for scene parsing are described below.

3.1 Superpixel Feature Computation and Classification

Let $I(v) \in R^3$ be an image defined on a set of pixels v, the goal of scene parsing is to set each pixel v into one of the predefined class labels $C = \{c_i | i = 1, 2, ..., M\}$ where M represents the number of attributes. For superpixel-based scene parsing, let $S(v) = \{s_j | j = 1, 2, ..., N\}$ represents the set of superpixels obtained from our segmentation process from an image I, and N represents the number of all superpixels, $F^V = \{f_j{}^V | j = 1, 2, ...N\}$, $F^l = \{f_j{}^l | j = 1, 2, ...N\}$ represents, and $F^g = \{f_j{}^g | j = 1, 2, ...N\}$ represent corresponding visual features, local contextual features, and global contextual features respectively, the goal is then set to labeling all image pixels v in s_j into a class $c_i \in C$ and $v \in s_j$, and the conditional probability of giving a correct label c_i for s_j can be formulated as:

$$P(c_i|s_j, W) = P\left(c_i|f_j^v, f_j^l, f_j^g; W_v, W_l, W_g\right) \tag{1}$$

$s.t. \Sigma_{1 \leq i \leq M} P(c_i|s_j) = 1$, where, $W = \{W_v, W_l, W_g\}$ denotes weight parameters for all the three features F_v, F_l, and F_g respectively, and W is learned during the training process. The final goal is to get a trained model that is able to maximize the sum of conditional probabilities of giving appropriate labels for all superpixels:

$$P(C|S, \theta) = max_{s_j \epsilon S \& c_i \epsilon C} P(c_i|s_j, W) \tag{2}$$

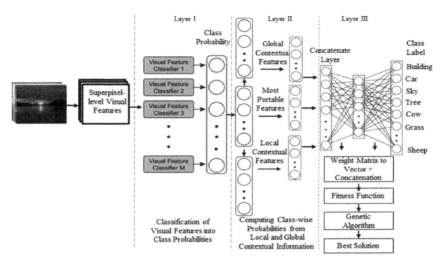

Fig. 1. The three-layered deep architecture predicts probabilities of every superpixel by class semantic 'One vs. all' classifiers in Layer I. In Layer II, the global and local Context Adaptive Voting (CAV) features are extracted by casting votes from superpixels in spatial blocks and neighboring superpixels respectively using the most probable class and object co-occurrence priors (OCPs). Finally, three class-wise probability vectors obtained from the visual feature, local, and global CAV features are integrated into Layer III to predict the final class label.

3.2 Prediction of Visual Features

The visual prediction layer comprises multiple binary (one vs. all) classifiers that take visual features as input to predict the probabilities approximately for all superpixels belonging to each class. It provides a way to leverage binary classification over a multi-class classifier. The resultant probabilities vector serves as an early prediction and becomes a foundation for computing local and global contextual features based on image context. For each superpixel, we obtained a probability vector of size equal to the total class number. A set of binary (one-vs-all) classifiers are developed and trained for an individual class instead of training a single multi-class classifier for all classes. Given a multi-class classification problem with M feasible solutions, a one-vs.-all classification model contains M independent binary classifiers, one binary classifier for every possible outcome. The use of class-specific classifiers provides three benefits: a) class-specific classifier-based architecture mitigates the imbalanced training data problem among classes, especially for an organic dataset containing a batch of sparsely occurring but crucial classes. In public benchmark scene parsing datasets, the distribution of the pixels is presumably heavy-tailed to many common attributes whereas sparse pixels can be available for uncommon classes. In such circumstances, training a multi-class model has the probability of totally disregard uncommon classes and being influenced towards prevalent classes. b) Reinforce class-specific feature selection, which permits the use of the exclusive feature sets for every class. c) Training of multiple class-specific classifiers that proved to be more potent contrarily to train a single multi-class classifier, especially for datasets containing vast class numbers.

3.3 Local and Global Context-Based Features

The contextual voting layer plays a pivotal role in the proposed architecture as it aims to learn the context-adaptive voting (CAV) features that incorporate both short- and long-range label connections among attributes in images and are robust to the local characteristic of images. The contextual voting layer computes CAV features of each superpixel at local and global context levels. Superpixel-wise predicted class labels and the corresponding object co-occurrence priors (OCPs) are used to compute CAV features. The contextual voting layer first computes global OCPs and local OCPs from the training dataset. Finally uses these two priors with the most probable class to computes global and local CAV features. Subsequently, these three (i.e. global CAV, local CAV, and class probability) vectors are fed to the final integration layer. The main objective of the contextual exploitation layer is to consider the contextual attributes of an object, computing the co-occurrence priors of object classes from the training images, and extract context-adaptive voting features for each superpixel. This layer receives the most probable class information by the initial prediction of visual features from layer I and uses it with OCPs to obtain CAV features. The local and global context priors that are learned are reflected in the contextual features, and we obtain these features by casting votes for image superpixels during testing as presented in Eqs. (3–5).

$$V^l(C|s_j) = \psi^l(P^v(C|s_j), OCP) \tag{3}$$

$$V^g(C|s_j) = \psi^g(P^v(C|s_j), OCP) \tag{4}$$

$$V^{con}(C|s_j) = \zeta\left(\overbrace{(V^l(C|s_j)}^{local}, \overbrace{(V^g(C|s_j)}^{global}\right) \tag{5}$$

where, $\psi^l and \psi^g$ indicate the voting functions for local and global context, respectively, and ζ is the function that fuses them. The contextual features adaptively obtain the superpixels level dependencies within an image.

It was observed that both visual and contextual features reveal crucial information in the scene parsing problem. Contextual features are regularized to class probabilities to integrate with the class probabilities obtained from the visual classifier. The unification layer models the correlation of class probabilities obtained from visual and contextual features.

$$P(C|s_j) = \mathcal{H}\left(\overbrace{(P^{vis}(C|s_j)}^{\text{visual features}}, \overbrace{(P^{con}(C|s_j)}^{\text{contextual features}}\right) \qquad (6)$$

$$P^T(C|s_j) = U_{1 \le i \le M} P\left(c_i | f_i^T ; W^T\right) \qquad (7)$$

where $\mathcal{H}$ indicates the joint modeling function for both class probabilities $P^T(C|s_j)$ of s_j using features f_i^T and the weights W^T, $T \epsilon \{vis, con\}$.

1) Visual features extracted from superpixels in the visual feature layer are utilized to train several class-specific classifiers to get a superpixel-wise approximate probability of belonging to each class. For the jth superpixel s_j, and its class probability for i^{th} class is c_j:

$$P^{vis}(c_j|s_j) = \phi_i(f_{i,j}^{vis}) = fn\left(w_{1,i} f_{i,j}^{vis} + b_{1,i}\right) \qquad (8)$$

where $f_{i,j}^{vis}$ represents visual features of s_j obtained for the ith class c_i, ϕ_i is trained binary classifier at the initial stage for c_i, fn represents prediction function of ϕ_i, trainable weights, constants parameters are $w_{1,i}$ and $b_{1,i}$ respectively. For all classes M, the probability vector we obtained for s_j:

$$P^{vis}(C|s_j) = \left[P^{vis}(c_1|s_j), \ldots, P^{vis}(c_i|s_j), \ldots, P^{vis}(c_M|s_j)\right] \qquad (9)$$

The P vector holds the likelihood of superpixel s_j belonging to all classes C, and s_j is assigned to the class that holds max probability:

$$s_j \epsilon \hat{c} \ if \ P^{vis}(\hat{c}|s_j) = \max_{1 \le i \le M}\left(P^{vis}(c_i|s_j)\right) \qquad (10)$$

2) The votes for the contextual label of all classes for s_j are computed using the contextual features:

$$V^{con}(C|s_j) = \left[V^{con}(c_1|s_j), \ldots, V^{con}(c_i|s_j), \ldots, V^{con}(c_M|s_j)\right] \qquad (11)$$

3.4 Integration Layer

The main objective of our integration layer is to determine a class label for each superpixel through learning optimal weights. The optimal weights seamlessly integrate class probabilities derived from the most probable class, local CAV features, and global CAV features. The final layer facilitates learning a set of class-specific weights to best represent the correlations among three predicted class probabilities and the class label. These weights are learned to inherently account for contributions of visual and contextual features to determine superpixel-wise class labels for testing images. The integration layer in the proposed approach has one hidden layer. The GA-based method [19, 27, 28] is applied on hyper-parameter solution space to optimize the network architectures' weights. As gradient descent-based weight optimization of Artificial Neural Networks is susceptible to multiple local minima, here we applied a standard GA to train the weights to overcome the 'local minima' problem. In GA, a single solution contains the full network weights of the proposed integrated layer, in the final step of our system architecture, and the steps are repeated for many generations to obtain the optimal weights. GA process is shown below in Algorithm 1.

Algorithm 1. GA Process for Network Optimization

Input: Class probability vectors: visual (F_v) local (F_l) and global (F_g), generations T, population N, mutation and crossover probabilities p_M and p_c, mutation parameter q_M, and crossover parameter q_C.

Initialization: generate a set of randomized individuals and recognition accuracies

Output: Class labels

for $t = 1, 2, \ldots, T$ do

    Selection: produce a new generation using the Russian roulette process

    Crossover: for each pair perform crossover with probability p_C and parameter q_C;

    Mutation: do mutation with probability p_M and parameter q_M for every non-crossover individual

    Evaluation: compute the recognition accuracy for each new individual

end for

Output: N sets of individuals in the final optimal weight matrix $W= \{W_1, W_2, \ldots, W_N\}$ with their recognition accuracies, where N presents the number of attributes or classes.

The iteration of our genetic optimization iteration starts from this original population, and a fitness evaluation is performed at first. The fitness evaluation checks where the models are on the optimization surface and determining which of the models perform best. The proposed network learns an optimized set of weights for all its layers. Indeed, the estimation of a globally optimized weight set is often very challenging for a deep architecture with each layer comprising multiple sub-layers. The model parameters are trained layer by layer. The proposed network learns weights in each layer independently to generate an approximation of the best weights.

4   Results and Discussion

A detailed comparative study with state-of-the-art scene parsing models is presented in this section, and here we draw a comparison from a performance perspective on two popular benchmark datasets.

4.1   Dataset

To evaluate the proposed approach, we adopt the widely used Stanford Background Dataset (SBD) [11] and the CamVid dataset [12]. The SBD contains 715 outdoor scene images and 8 object classes. Each image contains at least one foreground object and 320×240 pixels dimension. 572 images are randomly selected for training and 143 images for testing in each fold. The CamVid dataset [12] contains pixel-level ground truth labels. The original resolution of images was downsampled to 480×360, and semantic classes were reduced to 11 for our experiments to follow previous works.

4.2 Training Details

The proposed architecture is implemented in a MATLAB environment. Experiments are conducted on an HPC cluster, and we assigned a limited resource of 8 CPUs and 64 GB of memory for the experiment. The initial learning rate is 10-4, and 0.1 is the exponential rate of decays after 30 epochs. We initialized our initial population with random initialization and the number of resultant solutions to 8. The number of solutions to be selected as parents in the mating pool was set to 4, the number of generations was set to 1000, and the percentage of genes to mutate was 10. We considered single-point crossover operator, random mutation operator, and kept all parents in the next population. We observed our model convergence up to 1000 generations to find out the optimal solutions.

4.3 Evaluation

Table 1 shows the accuracies of the proposed architecture along with the accuracies reported by previous methods. The proposed network achieved global accuracies of 86.2% and a class accuracy of 85.5%. Table 2 shows the class accuracy we obtained on the SBD [11]. We obtained 73% and 73.6% mIOU (see Table 3, mIOU is calculated to compare with previous approaches) using MLP and SVM respectively as a classifier at the first layer on the CamVid dataset. The qualitative results obtained on the SBD are presented in Fig. 2. The results demonstrate that our approach successfully predicts object pixels with high precision.

Table 1. Performance (%) comparisons with previous state-of-the-art approaches on **Stanford Background Dataset**

Method	Pixel Acc.	Class Acc.
Gould et al. [11] (2009)	76.4	NA
Kumar et al. [30] (2010)	79.4	NA
Lempitsky et al. [31] (2011)	81.9	72.4
Farabet et al. [8] (2013)	81.4	76.0
Sharma et al. [32] (2015)	82.3	79.1
Chen et al. [33] (2018)	87.0	75.9
Zhe et al. [29] (2019)	87.7	79.0
Proposed approach	86.2	85.5

4.4 Comparative Study with Previous Approaches

Table 2 shows that our approach outperformed previous methods on average class-wise accuracy on the SBD. It has also achieved balanced accuracy across all classes. On the

CamVid dataset, we obtained 73.6% mIoU, which is comparable to the contemporary accuracies, and the comparison is presented in Table 3.

Table 2. Class wise performance (%) comparisons with previous approaches on **Stanford Background Dataset**

Method	Global	Avg.	Sky	Tree	Road	Grass	Water	Bldg.	Mt.	Frgd.
Gould et al. [11]	76.4	65.5	92.6	61.4	89.6	82.4	47.9	82.4	13.8	53.7
Munoz et al. [34]	76.9	66.2	91.6	66.3	86.7	83.0	59.8	78.4	5.0	63.5
Ladicky et al. [35]	80.9	70.4	94.8	71.6	90.6	88.0	73.5	82.2	10.2	59.9
Proposed method	**86.2**	**85.5**	**95.4**	**80.4**	**91.6**	**85.1**	**80.2**	**86.8**	**86.7**	**73.9**

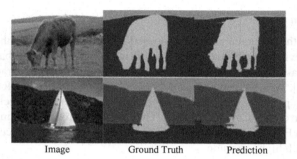

Image   Ground Truth   Prediction

Fig. 2. Qualitative results obtained on Stanford Background Dataset (best view in color mode). Original, ground truth, and predicted result images are presented column-wise.

Table 3. Performance (%) comparisons with recently proposed approaches on the **CamVid** dataset

Method	Pretrained	Encoder	mIoU (%)
SegNet [18]	ImageNet	VGG16	60.1
Dilate8 [25]	ImageNet	Dilate	65.3
BiSeNet [7]	ImageNet	ResNet18	68.7
PSPNet [16]	ImageNet	ResNet50	69.1
DenseDecoder [6]	ImageNet	ResNeXt101	70.9
VideoGCRF [11]	Citscapes	ResNet101	75.2
Proposed approach (MLP)	NA	NA	73.0
Proposed approach (SVM)	NA	NA	73.6

5 Conclusion and Future Works

The proposed end-to-end deep learning architecture computes object co-occurrence priors from the training phase and taking advantage of the CAV features. It captures both short- and long-range label correlations of object entities in the whole image while capable to acclimate to the local context. The GA optimization finds the optimized set of parameters for the integration layer. We demonstrate that overall performance is comparable with the recent methods. The integrated layer optimization will also be further investigated by incorporating multi-objective optimization.

Acknowledgements. This research was supported under Australian Research Council's Discovery Projects funding scheme (project number DP200102252).

References

1. Yang, M., Yu, K., Zhang, C., Li, Z., Yang, K.: DenseASPP for semantic segmentation in street scenes. In: CVPR, pp. 3684–3692 (2018)
2. Zhou, L., Zhang, H., Long, Y., Shao, L., Yang, J.: Depth embedded recurrent predictive parsing network for video scenes. IEEE Trans. ITS **20**(12), 4643–4654 (2019)
3. Qi, M., Wang, Y., Li, A., Luo, J.: STC-GAN: spatio-temporally coupled generative adversarial networks for predictive scene parsing. IEEE Trans. IP **29**, 5420–5430 (2020)
4. Zhang, R., Tang, S., Zhang, Y., Li, J., Yan, S.: Perspective-adaptive convolutions for scene parsing. IEEE Trans. PAMI **42**(4), 909–924 (2019)
5. Zhang, H., et al.: Context encoding for semantic segmentation. In: CVPR, pp. 7151–7160 (2018)
6. Heitz, G., Gould, S., Saxena, A., Koller, D.: Cascaded classification models: combining models for holistic scene understanding. NIPS **21**, 641–648 (2008)
7. Yu, C., Wang, J., Peng, C., Gao, C., Yu, G., Sang, N.: BiSeNet: bilateral segmentation network for real-time semantic segmentation. In: Ferrari, V., Hebert, M., Sminchisescu, C., Weiss, Y. (eds.) ECCV 2018. LNCS, vol. 11217, pp. 334–349. Springer, Cham (2018). https://doi.org/10.1007/978-3-030-01261-8_20
8. Farabet, C., Couprie, C., Najman, L., LeCun, Y.: Learning hierarchical features for scene labeling. IEEE Trans. PAMI **35**(8), 1915–1929 (2013)
9. Gould, S., Rodgers, J., Cohen, D., Elidan, G., Koller, D.: Multi-class segmentation with relative location prior. IJCV **80**(3), 300–316 (2008)
10. Micušlík, B., Košecká, J.: Semantic segmentation of street scenes by superpixel co-occurrence and 3D geometry. In: ICCV Workshops, pp. 625–632 (2009)
11. Gould, S., Fulton, R., Koller, D.: Decomposing a scene into geometric and semantically consistent regions. In: ICCV, pp. 1–8 (2009)
12. Brostow, G.J., Fauqueur, J., Cipolla, R.: Semantic object classes in video: a high-definition ground truth database. Pattern Recognit. Lett. **30**(2), 88–97 (2009)
13. Shotton, J., Winn, J., Rother, C., Criminisi, A.: Textonboost for image understanding: multi-class object recognition and segmentation by jointly modeling texture, layout, and context. IJCV **81**(1), 2–23 (2009)
14. Fulkerson, B., Vedaldi, A., Soatto, S.: Class segmentation and object localization with superpixel neighborhoods. In: ICCV, pp. 670–677 (2009)
15. Girshick, R., Donahue, J., Darrell, T., Malik, J.: Rich feature hierarchies for accurate object detection and semantic segmentation. In: CVPR, pp. 580–587 (2014)

16. Zhao, H., Shi, J., Qi, X., Wang, X., Jia, J.: Pyramid Scene Parsing Network. In: CVPR, pp. 2881–2890 (2017)
17. Nguyen, T.N.A., Phung, S.L., Bouzerdoum, A.: Hybrid deep learning-Gaussian process network for pedestrian lane detection in unstructured scenes. IEEE Trans. NNLS 31(12), 5324–5338 (2020)
18. Badrinarayanan, V., Kendall, A., Cipolla, R.: SegNet: a deep convolutional encoder-decoder architecture for image segmentation. IEEE Trans. PAMI 39(12), 2481–2495 (2017)
19. Ghosh, R., Verma, B.: A hierarchical method for finding optimal architecture and weights using evolutionary least square based learning. IJNS 13(1), 13–24 (2003)
20. Zhang, P., Liu, W., Lei, Y., Wang, H., Lu, H.: RAPNet: residual atrous pyramid network for importance-aware street scene parsing. IEEE Trans. IP 29, 5010–5021 (2020)
21. Xiong, Y., et al.: UPSNet: a unified panoptic segmentation network. In: CVPR, pp. 8818–8826 (2019)
22. He, K., Gkioxari, G., Dollár, P., Girshick, R.: Mask R-CNN. PAMI 42(2), 386–397 (2020)
23. Zhang, P., Liu, W., Wang, H., Lei, Y., Lu, H.: Deep gated attention networks for large-scale street-level scene segmentation. Pattern Recognit. 88, 702–714 (2019)
24. Cheng, B., et al.: Panoptic-DeepLab: a simple, strong, and fast baseline for bottom-up panoptic segmentation. In: CVPR, pp. 12475–12482 (2020)
25. Yu, F., Koltun, V.: Multi-scale context aggregation by dilated convolutions. In: ICLR, pp. 1–13 (2016)
26. Lin, G., Liu, F., Milan, A., Shen, C., Reid, I.: Refinenet: multi-path refinement networks for dense prediction. PAMI 42(5), 1228–1242 (2019)
27. Ding, S., Li, H., Su, C., Yu, J., Jin, F.: Evolutionary artificial neural networks: a review. Artif. Intell. Rev. 39(3), 251–260 (2013)
28. Azam, B., Mandal, R., Zhang, L., Verma, B. K.: Class probability-based visual and contextual feature integration for image parsing. In IVCNZ, pp. 1–6 (2020)
29. Zhu, X., Zhang, X., Zhang, X.-Y., Xue, Z., Wang, L.: A novel framework for semantic segmentation with generative adversarial network. JVCIR 58, 532–543 (2019)
30. Kumar, M.P., Koller, D.: Efficiently selecting regions for scene understanding. In: CVPR, pp. 3217–3224 (2010)
31. Lempitsky, V., Vedaldi, A., Zisserman, A.: Pylon model for semantic segmentation. In: NIPS, pp. 1485–1493 (2011)
32. Sharma, A., Tuzel, O., Jacobs, D.W.: Deep hierarchical parsing for semantic segmentation. In: CVPR, pp. 530–538 (2015)
33. Chen, L.-C., Papandreou, G., Kokkinos, I., Murphy, K., Yuille, A.L.: Deeplab: semantic image segmentation with deep convolutional nets, Atrous convolution, and fully connected CRFs. PAMI 40(4), 834–848 (2017)
34. Munoz, D., Bagnell, J.A., Hebert, M.: Stacked hierarchical labeling. In: Daniilidis, K., Maragos, P., Paragios, N. (eds.) ECCV 2010. LNCS, vol. 6316, pp. 57–70. Springer, Heidelberg (2010). https://doi.org/10.1007/978-3-642-15567-3_5
35. Ladický, L., Russell, C., Kohli, P., Torr, P.H.S.: Associative hierarchical random fields. PAMI 36(6), 1056–1077 (2014)

Kernelized Transfer Feature Learning on Manifolds

R. Lekshmi[1]([✉]), Rakesh Kumar Sanodiya[2]([✉]), R. J. Linda[1],
Babita Roslind Jose[1], and Jimson Mathew[3]

[1] Cochin University of Science and Technology, Kochi, Kerala, India
lekshmir@cusat.ac.in
[2] Indian Institute of Information Technology Sri City, Chittoor, Sri City, India
rakesh.s@iiits.in
[3] Indian Institute of Technology, Patna, Patna, India

Abstract. In the past few years in computer vision and machine learning, transfer learning has become an emerging research for leveraging richly labeled data in the source domain to construct a robust and accurate classifier for the target domain. Recent work on transfer learning has focused on learning shared feature representations by minimizing marginal and conditional distributions between domains for linear data sets only. However, they produce poor results if they deal with non-linear data sets. Therefore, in this paper, we put forward a novel framework called Kernelized Transfer Feature Learning on Manifold (KTFLM). KTFLM aims to align statistical differences and preserve the intrinsic geometric structure between the labeled source domain data and unlabeled target domain data. More specifically, we consider Maximum Mean Discrepancy for statistical alignment and Laplacian Regularization term for incorporating manifold structure. We experimented using benchmark data sets such as the PIE Face Recognition and the Office-Caltech (DeCAF features) object recognition dataset to discourse the limitations of the existing classical machine learning and domain adaptation methods. The performance comparison indicates that our model gave splendid accuracy of 79.41% and 91.97% for PIE and Office-Caltech data sets using linear and Gaussian kernels, respectively.

Keywords: Transfer learning · Kernelization · Unsupervised domain adaptation · Laplacian regularization · Visualization

1 Introduction

It is very difficult or even not possible to induce a supervised classifier for domain without any labeled information [5]. In the real world, this situation is very common. For example, a newly emerging domain (usually called target domain) may contain little and completely unlabeled data, but some other related domains (usually called source domains) may contain a large amount of labeled data. Therefore, it is impossible to induce target domain classifiers due to poorly labeled data or completely unlabeled data. To understand more deeply, consider

© Springer Nature Switzerland AG 2021
T. Mantoro et al. (Eds.): ICONIP 2021, LNCS 13109, pp. 297–308, 2021.
https://doi.org/10.1007/978-3-030-92270-2_26

an example shown in Fig. 1, where for each domain there are four images of four categories (bicycle, television, calculator, coffee-mug) i.e. source and target domains. In Fig. 1, the objects shown in the source and target domains are similar, but their distribution is different due to different origins. Recall the assumption for the target domain that the labeled information is not given, so it may be difficult for the classifier to train the target domain. Nevertheless, since objects in both domains are the same, if we try to train the target domain classifier using the source domain labeled images, it is not easy to train the target domain classifier due to the difference in distribution.

Fig. 1. Illustration of source domain and target domain images for four identical objects

The main contributions of this paper can be listed as follows:

1. We propose a novel Kernelized Transfer Feature Learning on Manifold (KTFLM) framework for unsupervised domain adaptation, in which we minimize the marginal distribution and conditional distribution, and preserve the intrinsic geometric structure of the data simultaneously.
2. In our proposed framework, we have incorporated linear, Gaussian (radial basis function), and polynomial kernel functions to deal with both linear and non linear datasets.

2   Previous Work

Unsupervised DA acquires a k-NN or SVM classifier with the available source data to predict and classify the unlabeled target domain [11]. Initially, dimensionality reduction algorithms like PCA, LDA does the feature transformation, reduces error in parameter estimation, and avoids overfitting. There are many types of distance measures like KL divergence, MMD that the researchers predefine to lessen the distribution distance between the domains [10]. Arthur Gretton et al. suggested a distribution-free distance estimate, maximum mean discrepancy (MMD) [4]. Some of the existing DA algorithms adopt MMD in a Reproducing Kernel Hilbert Space $\mathcal{H}$ (RKHS) to calculate the separation between the domains by taking their means of the two distributions. Pan et al. proposed TCA and SSTCA, which studies some of the transfer components in RKHS so that the domain gap is reduced while projecting the original input feature onto a common feature subspace by the transfer components [10].

Long et al. presented JDA method, which estimated marginal and conditional distribution simultaneously and reduced the distribution shift by constructing a new latent space [6]. M. Long et al. in TJM did instance reweighting and feature matching using marginal distribution adaptation across domains [7]. J. Zhang proposed JGSA, and he learns joined projections for reducing the variation of distributions statistically and geometrically among the two domains into low dimensional subspaces [15]. In SCA, M. Ghifary et al. introduced "scatter" to build a linear transformation using the variance of the subset of data and removes the unwanted distinctions within the labels and between the domains, but retains the required distinctions between labels [2]. L. Luo et al. in their algorithm DGA-DA discusses the discriminative instances and geometrical structure of data when considering the unlabeled target domain [8]. In the algorithm MEDA, J. Wang uses Grassman manifold for the dynamic distribution alignment of the domain adaptation classifier [14]. N. Courty et al. proposed a transportation model known as OTGL for aligning the data for both domains by using probability distribution functions that grouped the same class samples within the source domain throughout transport [1]. GFK method integrates the infinite number of subspaces that specify statistical and geometric change from known to unknown domain [3]. R. Sanodiya et al., in their work KUFDA, applied kernelization of source and target domain manifold and aligned the data geometrically and statistically [12]. So it is obvious by adding regularization or target selection or domain invariant clustering in the objective function leads to the improvement of JDA [5]. In our proposed work, we adopted Laplacian regularization, which exploits the geometrical similarity of the nearest points [14]. So by manifold assumption, we utilize the locality preserving property of the laplacian regularizer [10].

3   Kernelized Transfer Feature Learning on Manifolds

This section presents the Kernelized Transfer Feature Learning on Manifolds (KTFLM) approach.

3.1   Definitions and Notation

We first define the domain and task, and finally the domain adaptation problem.

Domain: A domain $\mathbb{D}$ consists of a feature space $\mathbb{X}$ and a marginal probability distribution $P(x)$, i.e. $\mathbb{D} = \{\mathbb{X}, P(x)\}$, where $x \in \mathbb{X}$.

Task: Given domain $\mathbb{D}$, a task $\mathbb{T}$ is basically composed of a classifier $f(x)$ and c- classes labeled set $\mathbb{Y}$, i.e. $\mathbb{T} = \{f(x), \mathbb{Y}\}$, where the classifier function $f(x)$ predicts the label $y \in \mathbb{Y}$ for input data $x \in \mathbb{X}$.

Problem Definition-Domain Adaptation: Given a source domain labeled data $\mathbb{D}_s = \{(x_1, y_1), (x_2, y_2), \ldots, (x_{n_s}, y_{n_s})\}$ and an unlabeled target domain $\mathbb{D}_t = \{x_{n_s+1}, x_{n_s+2}, \ldots, x_{n_s+n_t}\}$ under the assumption that $\mathbb{X}_s = \mathbb{X}_t$ and $\mathbb{Y}_s = \mathbb{Y}_t$. However, due to dataset shift, $P_s(x_s) \neq P_t(x_t)$ and $Q_s(y_s|x_s) \neq Q_t(y_t|x_t)$, where P and Q are the marginal and conditional distributions respectively.

3.2   KTFLM Formulation

To overcome the limitations of existing non-DA and DA approaches, the proposed framework KTFLM reduces the domain divergence statistically and preserves the intrinsic geometric structure by exploiting both the shared and domain specific features of the source and the target domains. More specifically, in KTFLM a common projection vector matrix V is calculated inorder to obtain a new domain feature representation in such a way that 1) variance of both domains is maximized, 2) minimize the marginal distribution between both domains, 3) minimize the conditional distribution between both domains, and 4) preserve intrinsic geometric structure between the source domain labeled data and the target domain unlabeled data.

Variance Maximization: To learn transformed features, we need to formulate an optimization problem by maximizing the variance of the domain. To maximize the variance, we consider a standard dimensionality reduction technique called principal component analysis (PCA). Basically, the PCA preserve most information of the data in its principal components that are orthogonal to each other in its vector space.

Let the input data matrix be $\mathbb{X} = \{x_1, x_2, \ldots x_n\} \in \mathbb{R}^{m \times n}$ where m represents the dimension of the input data space and $n = n_s + n_t$, where n_s is the number of samples in the source domain, n_t is the number of samples in the target domain. H_c is data centric matrix and computed by $H_c = I - \frac{1}{n}1$, where I is an identity matrix and 1 is $n \times n$ matrix of ones. Thus, co-varaince of input data (i.e., $\mathbb{X}H_c\mathbb{X}^T$) with respect to orthogonal transformation matrix $Z \in \mathbb{R}^{n \times p}$ is maximized as follows:

$$\max_{Z^T Z = I} \operatorname{tr}(Z^T \mathbb{X} H_c \mathbb{X}^T Z) \tag{1}$$

where tr(.) is the trace of a matrix.

To deal with linear or non-linear datasets, we consider some kernel functions (as shown in Table 1) such as linear, Gaussian, and polynomial into our proposed framework. Let Φ be a kernel function that transforms the inseparable data samples $\Phi : x \rightarrow \Phi(x)$, or $\Phi(x) = [\Phi(x_1), \Phi(x_2), \ldots, \Phi(x_n)]$, and kernel matrix $K = \Phi(\mathbb{X})^T \Phi(\mathbb{X})$. Finally we use the Representer Theorem $Z = \Phi(\mathbb{X})V$, then the objective function becomes,

$$\max_{V^T V = I} \operatorname{tr}(V^T K H_c K^T V) \tag{2}$$

This optimization problem can be solved by eigen decomposition $K H_c K^T V = V\sigma$, where $\sigma = \operatorname{diag}(\sigma_1, \sigma_2, \ldots, \sigma_p) \in \mathbb{R}^{n \times n}$ are the largest eigenvalues. $V \in \mathbb{R}^{n \times p}$ is the transformation matrix. However, p-leading eigenvector (i.e., corresponding to highest p eigen values) are selected. Thus, the subspace embedding becomes $A = V^T K$.

Table 1. Different kernel functions

Kernel function	Kernel matrix
Linear	$K(x, x^T) = \{x.x^T\}$
Gaussian	$K(x, x^T) = exp^{-\gamma|x-x^T|^2} \ \gamma > 0$
Polynomial	$K(x, x_i) = \sum_{i=1}^{m}(1 + xx_i^T)^m$, m is the degree of the polynomial

Matching Features with Marginal Distribution Adaptation: Even after projecting the initial space kernelized data K onto the learned feature vector V, the distribution difference between the source and target domains is not minimal. We consider Maximum Mean Discrepancy (MMD) to reduce the marginal distribution between the source domain and the target domain [10]. Basically, MMD is non parametric approach and computes the distance between the empirical expectations of both domains data with p dimensional embedding:

$$\min_{V} \left\| \frac{1}{n_s}\sum_{i=1}^{n_s} V^T k_i - \frac{1}{n_t}\sum_{j=n_s+1}^{n_s+n_t} V^T k_j \right\|_{\mathcal{H}}^2 = \text{tr}(V^T KMK^T V) \tag{3}$$

where M is the MMD matrix and can be computed as follows:

$$M_{ij} = \begin{cases} \frac{1}{n_s n_s} & k_i, k_j \in D_s \\ \frac{1}{n_t n_t} & k_i, k_j \in D_t \\ \frac{-1}{n_s n_t} & \text{otherwise} \end{cases} \tag{4}$$

Matching Features with Conditional Distribution Adaptation: If the data is distributed class-wise, minimizing the marginal probability distributions of the source $(P_s(x_s))$ and the target $(P_t(x_t))$ is not enough. Therefore, we need to minimize the conditional distributions between the source and the target domains i.e. $Q_s(y_s|x_s)$ and $Q_t(y_t|x_t)$. It is not an easy task to reduce the conditional distribution due to dealing with the problem of unsupervised domain adaptation(DA), where no labeled information is available. However, we utilize the pseudo label concept proposed by Long et al. [5], where initially the k-NN classifier is trained with the source domain labeled data and subsequently tested on the target domain features to generate the target domain pseudo-labels. Thus, by using the classes $(c \in \{1, \ldots, C\})$ available in pseudo labels of target domain and true labels of source domain, the conditional distribution between the source and the target domains (i.e., $Q_s(x_s|y_s = c)$ and $Q_t(x_t|y_t = c)$) can be matched by adopting modified MMD in the p-dimensional embeddings:

$$\min_{V} \left\| \frac{1}{n_s^c}\sum_{k_i \in D_s^c} V^T k_i - \frac{1}{n_t^c}\sum_{k_j \in D_t^c} V^T k_j \right\|_{\mathcal{H}}^2 = \text{tr}(V^T KM_c K^T V) \tag{5}$$

where $D_s^c = \{k_i : k_i \in D_s \wedge y(k_i) = c\}$ is the set of samples belonging to class c in the source data, $y(k_i)$ is the given label of k_i, and n_s^c is the number of samples in $c - th$ class of the source domain. Similarly, $D_t^c = \{k_j : k_j \in D_t \wedge \hat{y}(k_j) = c\}$ is the set of samples belonging to class c in the target data, $\hat{y}(k_j)$ is the pseudo label of k_i, and n_t^c is the number of samples in $c - th$ class of the target domain. Thus here the Maximum Mean Discrepancy (MMD) matrix M_c which considers class labels can be computed as follows:

$$(M_c)_{ij} = \begin{cases} \frac{1}{n_s^c}\frac{1}{n_s^c} & k_i, k_j \in D_s^c \\ \frac{1}{n_t^c}\frac{1}{n_t^c} & k_i, k_j \in D_t^c \\ \frac{-1}{n_s^c}\frac{1}{n_t^c} & \begin{cases} k_i \in D_s^c,\, k_j \in D_t^c \\ k_j \in D_s^c,\, k_i \in D_t^c \end{cases} \\ 0 & \textbf{otherwise} \end{cases} \qquad (6)$$

Preserving Intrinsic Geometric Structure: Since there are no training samples available for the target domain, over-fitting may occur if we train the model only on data labeled with the source domain. One of the best ways to address the problem of over-fitting is to introduce a regularizer. Amongst the most popular regularizers is the Laplacian regularizer. This regularizer basically provides us the flexibility to include prior knowledge about the data. Incorporating prior means that if two data samples share the same label in the initial space, they are likely to share the same label (or similar embedding) in the learned space as well. Let us assume $\{k_i\}_{i=1}^n$ be the set of samples available in both domains, the relationship between nearby data samples can be modeled by using q-NN Graph. More specifically, we keep an edge between data samples i and j if k_i and k_j are "close", i.e., k_i and k_j are among q-nearest neighbors of each other. Then, the corresponding weight matrix W can be defined as follows:

$$W_{ij} = \begin{cases} 1, & \text{if } k_i \in N_q(k_j) \mid k_j \in N_q(k_i) \\ 0, & \text{otherwise} \end{cases} \qquad (7)$$

$N_q(k_j)$ and $N_q(k_i)$ are the set of data points of q-nearest neighbours k_j and k_i respectively. Thus, a natural regularizer is defined as follows:

$$\begin{aligned} L_r &= \sum_{i,j} \left(V^T k_i - V^T k_j\right)^2 W_{ij} \\ &= 2\sum_i V^T k_i \mathcal{D}_{ii} k_i^T V - 2\sum_{ij} V^T k_i W_{ij} k_j^T V \\ &= 2V^T K(\mathcal{D} - W)K^T V \\ &= 2V^T KLK^T V \end{aligned} \qquad (8)$$

where $\mathcal{D}_{ii} = \sum_j W_{ij}$ is the diagonal matrix and $L = \mathcal{D} - W$ is the Laplacian matrix.

3.3   Overall Objective Function

The overall objective function for KTFLM is determined by minimizing the difference of marginal and conditional distribution and preserving the geometrical similarity of both domains. By incorporating Eqs. (2), (3), (5) and (8), we obtain the KTFLM optimization problem:

$$\underset{V^T K H_c K^T V = I}{min} \text{tr}\left(V_T K (M + M_c + \eta L_r) K^T V\right) + \lambda \|V\|_F^2 \tag{9}$$

where η and λ are the trade-off parameters to regularize the Laplacian term and feature matching respectively.

3.4   Optimization of Overall Objective Function

The optimization of objective function is done by using the Lagrange multiplier σ, where $\sigma = diag(\sigma_1, \dots, \sigma_p) \in R^{p \times p}$. Apply Lagrange function in Eq. (9) and is modified as:

$$L_\sigma = \text{tr}(V^T(KMK^T + KM_cK^T + \eta KL_rK^T + \lambda I)V) + \text{tr}((I - V^T K H_c K^T V)\sigma) \tag{10}$$

$\frac{\partial L_\sigma}{\partial V} = 0$ will derive an optimum value for the transformation matrix V. Equation (10) is rewritten as

$$(K(M + M_c + \eta L_r)K^T + \lambda I)V = K H_c K^T V \sigma \tag{11}$$

Equation (11) derives an eigenvalue decomposition problem and compute $\sigma = diag(\sigma_1, \sigma_2, \dots, \sigma_p)$ are the leading p eigenvalues and the leading eigenvectors $V = (v_1, \dots, v_p)$.

Algorithm 1 gives the steps of KTFLM:

4   Experiments

The performance of KTFLM is obtained by simulating various real-world image recognition data sets such as PIE and Office-Caltech with Decaf features. Cross-domain pairs are created using these data sets and the accuracy of the results are compared with some of the existing non-DA and DA methods. From the observations, it is clear that our model excels in the majority of the tasks.

A. Experimented Image Recognition Data Sets

1. **PIE** (Pose, Illumination, and Expression) is one of the bench mark dataset used for face recognition in DA [12]. The database has 41,368 face images of 68 persons images are taken on 13 cameras under 21 lighting conditions to get different pose, illumination and expression. Based on poses it is classified into five groups: P1 (C05), P2 (C07), P3 (C09), P4 (C27), P5 (C29). Twenty cross-domain tasks are set up by considering one pose as the source domain and the other as the target domain.

Algorithm 1: KTFLM for Transfer Learning

Input : $\mathbb{X}_s$ and $\mathbb{X}_t$ are the source and target domain features. $\mathbb{Y}_s$ represent the labels of the source data samples. η, λ are the trade-off parameters and T is the number of iterations.

Output: Adaptation Matrix V_p and the accuracy of the model 'acc'.

1 Construct the kernel matrix K for both domains from feature matrix $\mathbb{X}$ which consists of $\mathbb{X}_s$ and $\mathbb{X}_t$. Find the laplacian matrix L as shown in Equation (8).

2 Initialize T=0. Using a k-NN classifier calculate the pseudo label $\hat{Y}$ from $\mathbb{X}_s$ and $\mathbb{X}_t$ and $\mathbb{Y}_s$.

3 *while (T < 10) do*

4 Construct M, M_c using the equation (4) and (6) and solve the objective function in equation (9) by substituting equation (3), (5), (8)

5 Solve the generalized eigenvalue decomposition and calculate the projection vector matrix V and select its leading p eigenvectors.

6 Find projected data for both domains using V_p. $X'_s = V_p[1:n_s]^T \mathbb{X}_s$ and $\mathbb{X}'_t = V_p[n_s + 1:n]^T \mathbb{X}_t$. Train k-NN classifier on $\{X'_s, X'_t, Y_s\}$ and generate new pseudo labels $Y_{tpseudo}$. Update matrices after every iteration.

7 *end*

8 Finally obtain the accuracy 'acc'.

---

2. **Office-Caltech** is one of the bench mark dataset used for object recognition. Office-31 has 3 object domains: DSLR(D), Amazon(A) and Webcam(W) and consists of 31 classes and 4652 images. Caltech-256 (C) has 256 classes and 30,607 images [13]. There are 10 common classes in both the data sets, so Office-Caltech is taken for conducting experiments. For this dataset DeCAF features are considered. Twelve tasks are selected from this 10 classes: $A \rightarrow C, \ldots, C \rightarrow D$.

Table 2. Estimation of accuracies (%) for PIE face recognition dataset.

Methods	P1 5->7	P25->9	P35->27	P45->29	P57->5	P57->27	P77->27	P77->29	P99->5	P109->7	P119->27	P129->29	P1327->5	P1427->7	P1527->9	P1627->29	P1729->5	P1829->7	P1929->9	P2029->27	Avg
NN	26.09	26.59	30.67	15.67	24.49	46.63	54.07	26.53	21.37	41.01	46.53	26.23	32.95	62.68	73.22	37.19	18.49	24.19	26.31	31.24	34.76
PCA	34.8	25.16	29.26	16.3	24.22	45.53	53.55	26.43	20.95	40.45	46.14	25.31	31.96	60.96	72.18	35.11	18.85	23.39	27.31	30.34	33.85
TCA	40.76	41.79	59.63	29.35	41.81	51.47	64.73	33.7	34.69	47.7	56.23	33.15	55.64	67.83	75.86	40.26	26.98	29.9	29.9	33.64	44.75
JDA	58.62	54.22	84.51	49.75	57.62	62.93	75.81	39.88	50.97	57.96	68.46	39.96	80.57	82.62	87.26	54.67	46.45	42.05	53.32	57.01	60.24
TJM	29.51	33.75	59.21	26.97	39.41	37.73	49.81	17.09	37.38	35.29	44.04	17.04	59.52	60.58	64.89	25.06	32.45	22.89	22.23	30.73	37.29
JGSA	68.06	67.53	82.86	46.5	25.22	54.77	56.97	35.42	22.82	44.19	56.87	41.36	72.13	88.28	86.09	74.53	17.53	41.06	49.22	34.76	53.39
MEDA	49.6	48.4	77.23	39.82	58.49	55.27	61.25	44.05	56.24	57.82	78.23	53.06	88.95	78.2	80.2	67.7	57.71	49.86	62.15	72.15	62.56
DGA-DA	65.32	62.81	83.54	56.07	63.69	61.27	82.37	45.63	56.72	61.26	77.83	44.24	61.84	85.27	90.95	53.8	57.44	53.84	55.37	61.82	65.09
KUFDA	67.67	70.34	94.32	49.02	72.62	74.34	87.86	61.7	73.91	72.56	96.96	69.85	90	88.4	84.62	75.24	54.05	67.46	70.77	76.78	74.42
KTFLM(LINEAR)	71.15	74.14	92.01	58.39	79.62	79.33	91.14	65.26	82.8	74.80	94.50	73.12	95.41	94.78	95.28	81.43	63.75	67.53	70.83	83.84	79.41
KTFLM(GAUSSIAN)	55.62	56.74	83.96	55.09	67.44	69.85	86.81	56.43	73.65	67.65	90.27	63.73	90.74	89.44	91.85	75.74	62.76	56.56	60.75	74.08	71.34
KTFLM(POLY)	55.8	61.03	86.26	55.92	69.69	71.51	67.8	53.43	73.23	70.35	90.48	65.99	92.11	91.53	92.59	76.23	67.59	54.63	62.93	75.13	72.61

B. Comparison with Primitive Machine Learning and Current DA Methods

Algorithms considered for the comparison of KTFLM with PIE data set are: Nearest Neighbour (NN) Classifier, PCA, Transfer Component analysis(TCA) [10], Joint Distribution Adaptation (JDA) [7], Transfer Joint Matching (TJM) [7], Joint Geometrical and Statistical Alignment (JGSA) [15], Manifold Embedded Distribution Alignment (MEDA) [14], Discriminative and Geometry Aware

Table 3. Estimation of accuracies (%) for Office-Caltech (DeCAF) data-set.

Methods	Tasks												
	C->A	C->W	C->D	A->C	A->W	A->D	W->C	W->A	W->D	D->C	D->A	D->W	Avg
NN	85.7	66.1	74.52	70.35	57.29	64.97	60.37	62.53	98.73	52.09	62.73	89.15	70.38
TCA	89.25	80	83.44	74.18	71.86	78.34	72.57	79.96	100	73.46	88.2	97.29	82.38
GFK	88.41	80.68	84.56	76.85	68.47	79.62	74.8	75.26	100	74.09	85.8	98.64	82.65
TJM	89.67	80.68	87.26	78.45	72.54	85.99	79.61	82.46	100	80.77	91.34	98.31	85.59
JDA	90.19	85.42	85.99	81.92	80.68	81.53	81.21	90.71	100	80.32	91.96	99.32	87.44
SCA	75.93	85.35	78.81	86.12	100	74.8	89.98	98.64	78.09	89.46	85.42	87.9	85.88
OTGL	92.15	84.17	87.25	85.51	83.05	85	81.45	90.62	96.25	84.11	92.31	96.29	88.18
JGSA	91.75	85.08	92.36	85.04	84.75	85.35	84.68	91.44	100	85.75	92.28	98.64	89.76
KTFLM (LINEAR)	**92.8**	**84.75**	**96.18**	**86.29**	84.75	**89.17**	**87.71**	**91.75**	100	**88.51**	**93.11**	**100**	**91.25**
KTFLM (GAUSSIAN)	**93.01**	**87.8**	95.54	**86.64**	**86.44**	**93.63**	**87.89**	91.23	100	88.33	93.11	100	**91.97**
KTFLM (POLY)	91.96	82.71	75.8	84.06	85.76	91.72	86.73	81.11	100	88.33	89.56	99.66	88.11

Unsupervised Domain Adaptation(DGA-DA) [8], Kernelized Unified Framework for Domain Adaptation(KUFDA) [12].

Algorithms considered for the comparison of KTFLM with Office-Caltech data set are: Nearest Neighbour (NN) Classifier, Transfer Component analysis(TCA) [10], Geodesic Flow Kernel (GFK) [3], Transfer Joint Matching (TJM) [7], Joint Distribution Adaptation (JDA) [7], Scatter Component Analysis(SCA) [2], Optimal Transport (OTGL) [1] and Joint Geometrical and Statistical Alignment (JGSA) [15].

5   Simulation Results and Discussion

KTFLM has given a good performance for the visual recognition tasks when compared to the existing DA algorithms. The accuracy of each task is evaluated for the two data sets and the simulated results are shown in Tables 2 and 3, respectively. The performance of classical machine learning methods such as NN and PCA is poor because of the large distribution difference between the source and target data. For the PIE dataset, JDA shows a good performance than TJM and TCA. JGSA results are comparatively better than JDA but less with MEDA. By preserving discriminative instances, DGA-DA shows an improvement in performance than MEDA and JGSA. KUFDA improves the above methods by considering statistical and geometrical alignment of data for both domains and its average accuracy is more compared to the other methods. KTFLM improves the above methods by considering three kernels such as linear, Gaussian and polynomial and Laplacian regularization. An average accuracy of 79.41% is obtained using linear kernel which is more than KUFDA for all the tasks. The polynomial and Gaussian kernel has given a mean accuracy of 72.81% and 71.34% which is more than DGA-DA but less than KUFDA. For Office-Caltech dataset TJM gives a good performance by considering instance re-weighting compared to TCA and GFK. JGSA performance is improved by taking the projection vector matrices for source and target domain and statistical alignment of data. KTFLM obtain an average accuracy of 91.97% and 91.25% by using Gaussian and linear kernel respectively. Its average accuracy is more than

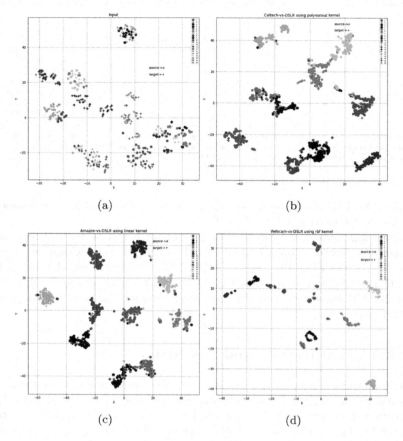

Fig. 2. t-SNE visualization of Input, $C \to D$, $A \to D$ and $W \to D$ for Office-Caltech data set.

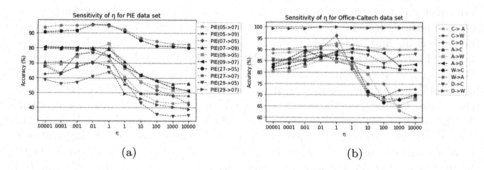

Fig. 3. Sensitivity of η for PIE and Office-Caltech data set.

the other methods but less than JGSA in some of the tasks. The polynomial kernel has a mean accuracy of 88.11% which is less compared to JGSA and OTGL.

To get satisfactory performance, the parameter values are tuned. For PIE and Office Caltech dataset the parameter values are fixed as $\eta = 0.1$, $\lambda = 0.1$, subspace dimension p = 120 (PIE) and p = 30 (Office-Caltech), $\gamma = 0.6$ for Gaussian kernel, degree (m) = 2 for the polynomial kernel, k-NN(q) = 7 and Cosine metric for Laplacian matrix. Using these values the accuracy of different tasks are shown in Table 2 and Table 3. We vary the η parameter value from 0.0001 to 10000 and the values of other parameters ($\lambda = 0.1$, p = 120 for PIE and p = 30 for Office-Caltech) kept as constant. The η parameter value is varied for all the tasks in PIE and Office-Caltech dataset. The η sensitivity analysis is shown in Fig. 3a and 3b and the other parameters such as $\lambda = 0.1$, subspace dimension (p) = 120 for PIE and p = 30 for Office-Caltech is kept constant. It is evident from the η sensitivity analysis, that KTFLM performs well for $\eta = 0.1$.

To have a better clarity on the classification of samples, we exploited t-SNE (t-Stochastic Neighbor Embedding) visualization [9]. t-SNE projects data into a low dimensional space to preserve data clustering of high dimensional space. Ten different colours such as ('black', 'lime', 'red', 'blue', 'orange', 'cyan', 'magenta', 'green', 'maroon' and 'chocolate') used to represent the source and target domain. 'Circles' indicate the source domain and 'pluses' for the target domain. Using t-SNE plots, we evaluated the results by identifying different clusters. When t-SNE is applied, the data points for the tasks whose accuracy is high, got assembled due to the high similarity score of the data samples. t-SNE visualizations for input and various tasks using poly, linear, and Gaussian kernels shown in Fig. 2(a), 2(b), (2c), and (2d) for Office- Caltech data set. Figure 2(a) shows the input, and from the visualization, it is evident that there is a large discrimination difference in the source and target domain. Figure 2(b) illustrates the $C \rightarrow D$ task (polynomial kernel), which has an accuracy of 75.8%, and the samples are not classified correctly. But in Fig. 2(c), the samples are much more classified because of the accuracy of 89.17% for the $A \rightarrow D$ task (linear kernel). It is obvious from the Fig. 2(d) that samples of similar classes are clustered for the task $W \rightarrow D$ (Gaussian kernel) task which has an accuracy of 100%. i.e. (the circles and pluses of same colour are grouped together).

6   Conclusion and Future Work

In this work, we put forward a novel framework Kernelized Transfer Feature Learning on Manifold (KTFLM) to address the limitations of existing primitive and domain adaptation approaches. The objective functions are defined on the basis of kernelization, KTFLM makes it possible to more efficiently utilize and preserve the intrinsic geometric information among the samples from both the source and the target domains with similar/ dissimilar labels. Extensive experiments show that KTFLM with different types of kernels such as (linear, poly and Gaussian) is robust and effective for different tasks of cross-domain adaptation problems, and can outperforms over state-of-the-art of existing non-DA and DA approaches.

In the future, we will extend the unsupervised DA by adapting the model to train on numerous source domains to a target domain. There is scope for improvement in accuracy by applying various deviation methods. Also, we will advance KTFLM by studying and using different kernels and regularizers.

References

1. Courty, N., Flamary, R., Tuia, D., Rakotomamonjy, A.: Optimal transport for domain adaptation. IEEE Trans. Pattern Anal. Mach. Intell. **39**(9), 1853–1865 (2016)
2. Ghifary, M., Balduzzi, D., Kleijn, W.B., Zhang, M.: Scatter component analysis: a unified framework for domain adaptation and domain generalization. IEEE Trans. Pattern Anal. Mach. Intell. **39**(7), 1414–1430 (2016)
3. Gong, B., Shi, Y., Sha, F., Grauman, K.: Geodesic flow kernel for unsupervised domain adaptation. In: 2012 IEEE Conference on Computer Vision and Pattern Recognition, pp. 2066–2073. IEEE (2012)
4. Gretton, A., Borgwardt, K., Rasch, M.J., Scholkopf, B., Smola, A.J.: A kernel method for the two-sample problem. arXiv preprint arXiv:0805.2368 (2008)
5. Long, M., Wang, J., Ding, G., Pan, S.J., Philip, S.Y.: Adaptation regularization: a general framework for transfer learning. IEEE Trans. Knowl. Data Eng. **26**(5), 1076–1089 (2013)
6. Long, M., Wang, J., Ding, G., Sun, J., Yu, P.S.: Transfer feature learning with joint distribution adaptation. In: Proceedings of the IEEE International Conference on Computer Vision, pp. 2200–2207 (2013)
7. Long, M., Wang, J., Ding, G., Sun, J., Yu, P.S.: Transfer joint matching for unsupervised domain adaptation. In: Proceedings of the IEEE Conference on Computer Vision and Pattern Recognition, pp. 1410–1417 (2014)
8. Luo, L., Chen, L., Hu, S., Lu, Y., Wang, X.: Discriminative and geometry-aware unsupervised domain adaptation. IEEE Trans. Cybern. **50**(9), 3914–3927 (2020)
9. Van der Maaten, L., Hinton, G.: Visualizing data using T-SNE. J. Mach. Learn. Res. **9**(11) (2008)
10. Pan, S.J., Tsang, I.W., Kwok, J.T., Yang, Q.: Domain adaptation via transfer component analysis. IEEE Trans. Neural Networks **22**(2), 199–210 (2010)
11. Pan, S.J., Yang, Q.: A survey on transfer learning. IEEE Trans. Knowl. Data Eng. **22**(10), 1345–1359 (2010)
12. Sanodiya, R.K., Mathew, J., Paul, B., Jose, B.A.: A kernelized unified framework for domain adaptation. IEEE Access **7**, 181381–181395 (2019)
13. Sanodiya, R.K., Yao, L.: A subspace based transfer joint matching with Laplacian regularization for visual domain adaptation. Sensors **20**(16), 4367 (2020)
14. Wang, J., Feng, W., Chen, Y., Yu, H., Huang, M., Yu, P.S.: Visual domain adaptation with manifold embedded distribution alignment. In: Proceedings of the 26th ACM International Conference on Multimedia, pp. 402–410 (2018)
15. Zhang, J., Li, W., Ogunbona, P.: Joint geometrical and statistical alignment for visual domain adaptation. In: Proceedings of the IEEE Conference on Computer Vision and Pattern Recognition, pp. 1859–1867 (2017)

Data-Free Knowledge Distillation with Positive-Unlabeled Learning

Jialiang Tang[1], Xiaoyan Yang[1], Xin Cheng[1], Ning Jiang[1(✉)], Wenxin Yu[1], and Peng Zhang[2]

[1] School of Computer Science and Technology, Southwest University of Science and Technology, Mianyang, China
jiangning@swust.edu.cn
[2] School of Science, Southwest University of Science and Technology, Mianyang, China

Abstract. In model compression, knowledge distillation is a popular algorithm, which trains a lightweight network (student) by learning the knowledge from a pre-trained complicated network (teacher). It is essential to acquire the training data that the teacher used since the knowledge is obtained by inputting training data to the teacher network. However, the data is often unavailable due to privacy problems or storage costs. Its lead exiting data-driven knowledge distillation methods is unable to apply to the real world. To solve these problems, in this paper, we propose a data-free knowledge distillation method called DFPU, which introduce positive-unlabeled (PU) learning. For training a compact neural network without data, a generator is introduced to generate pseudo data under the supervision of the teacher network. By feeding the generated data into the teacher network and student network, the attention features are extracted for knowledge transfer. The student network is promoted to produce more similar features to the teacher network by PU learning. Without any data, the efficient student network trained by DFPU contains only half parameters and calculations of the teacher network and achieves an accuracy similar to the teacher network.

Keywords: Model compression · Data-free knowledge distillation · Positive-unlabeled learning · Attention mechanism

1 Introduction

Convolutional Neural Networks (CNNs) have shown their outstanding capacity in many fields, for example image recognition [1–3], object detection [4–6], and semantic segmentation [7–9]. However, the high-ability CNNs often with huge parameters and calculations, which demands considerable memory and computing resources. It makes these advanced CNNs unable to apply to mobile artificial intelligence (AI) devices (such as smartphones and smart cameras) and hinders the application of CNN-based AI in real life. A series of works are proposed to compress the neural networks, such as knowledge distillation [10–12] and network

© Springer Nature Switzerland AG 2021
T. Mantoro et al. (Eds.): ICONIP 2021, LNCS 13109, pp. 309–320, 2021.
https://doi.org/10.1007/978-3-030-92270-2_27

pruning [13–15]. Knowledge distillation has achieved satisfactory results, which trains a high-ability student network with fewer parameters and calculations by the teacher network with more parameters and calculations. But the traditional knowledge distillation methods need to input the original data into the teacher network to extract the knowledge to train the student network. However, it is difficult to obtain datasets in actual situations due to privacy issues or the huge costs of dataset transmission and storage. To solve this problem, many data-free model compression methods [16–18] have been proposed. Still, the models compressed by these methods are generally less effective due to the lack of original data.

In this paper, we propose a new data-free knowledge distillation method named DFPU to deal with above problems. By utilizing the DFPU, the training of the student network only needs a trained teacher network. Figure 1 shows the overall structure of our proposed method. Similar to Generative Adversarial Networks [19,20], we introduce a generator to generate pseudo data and utilize the generated data to train the student network. Specifically, our DFPU composes of a teacher, generator, and student. And the training process of DFPU is divided into the generation stage and the student network training stage. During the generation stage, the teacher network is take as task of a discriminator. The generator generates fake images supervised by the teacher network. During the student network training stage, the generated data is fed into the teacher network and student network to get soft targets for knowledge transfer. To promote the student network to achieve considerable performance with the teacher network, we extract the attention features from the teacher network and student network for knowledge transfer. The features of CNN penultimate layer are discriminated as high-quality and low-quality based on the PU learning. The contributions of the paper can be summarized as follows:

1. We introduce positive-unlabeled (PU) learning into knowledge distillation to propose a data-free knowledge distillation method and successfully train a student network without data.
2. We utilize the PU learning to promote the student network to produce similar features to the teacher network and train the student network to achieve comparable performance with the teacher network without data.
3. Through extensive experiments and ablation studies, DFPU shows its effectiveness on popular datasets (MNIST [3], CIFAR [21], SVHN [22]) and achieves state-of-the-art performance compare to advance data-free knowledge distillation methods.

The follow parts of the paper is report as follows. Second 2 introduces the related works of this paper. Section 3 describes how to implement our DFPU. Section 4 demonstrates the effectiveness of DFPU through extensive experiments and ablation studies. Section 5 concludes the paper.

2   Related Work

Knowledge Distillation
Knowledge distillation is a widely studied model compression method. Ba et al. [10]

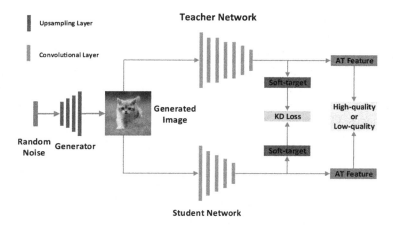

Fig. 1. The diagram of our proposed DFPU. The teacher network plays as a discriminator to supervise the generator to generate the images. Then, by training on the generated images, a high-capability student network can be trained through learning from the teacher network.

first propose to input the output of a neural network into a softmax function to get dark knowledge for knowledge transfer, which includes more information than the output of a neural network. Hinton et al. [11] systematic interpret the knowledge distillation, used a large trained neural network as the teacher network to train the small student network by the soft-target. BornNet [23] proposes to train a student network by generation and makes it perform well than the teacher network. PMAT [24] proposes that using attention feature maps as knowledge can train student networks more effectively. Although these data-driven knowledge distillation methods can effectively compress CNNs, these methods will fail when the data is unavailable. For the data-free knowledge distillation methods. Lopes et al. [16] reconstruct the data by the meta-data to train the student network. KegNet [25] extracts the knowledge from the teacher network to generate the artificial data based on the generator and discriminator. Data-free learning (DAFL) [18] is similar to us, which introduces the generative adversarial networks to implement data-free knowledge distillation, but the performance of the student network trained by DFAL unsatisfactory. Zero-Shot Knowledge Distillation (ZSKD) [26] obtains useful prior information of the underlying data distribution from the parameters of the teacher network to generate the train data for the student network. Wang [27] proposes to model the intermediate feature of the teacher by a multivariate normal distribution to gain pseudo samples.

Positive-Unlabeled Learning
PU learning explores a classifier to distinguish the positive data from the mixed data. Xu et al. [28] train a multi-class model to predict the label of the unlabeled data, which can determine the specific category of the unlabeled samples. Kiryo et al. [29] utilize the non-negative risk estimator to alleviate the overfitting problem of PU learning. Our DFPU also discriminate the features of the

teacher network and student network as the high-quality feature and low-quality feature, respectively.

Generative Adversarial Networks
Goodfellow et al. [30] propose the Generative Adversarial Networks (GANs) for the first time, which consists of a generator and a discriminator. InfoGAN [31] proposes the interpretable GANs and generated evenly distributed handwritten digital images. Guo et al. [32] propose to discriminate the images as high-quality samples and low-quality samples, instead of real samples and fake samples.

Attention Mechanism
Attention mechanism is used in computer vision broadly and can quickly extract important features of sparse data. Wang et al. [33] propose a residual attention network and add the attention modules to the neural network to promote the classification of the network. Hu et al. [34] propose SE-layer based on the attention mechanism, which can improve the ability of neural networks obviously.

3   Approach

In this section, we introduce our method in detail. First, we describe how the generator products the data. Then we show how to train the student network. In the end, we propose the objective function of DFPU. Figure 1 is the overall structure of our method, there are a teacher named $\mathcal{N}_T$, generator G, and student named $\mathcal{N}_S$:

Teacher network is a pre-trained advanced CNN, such as ResNet [1], VGGNet [35], which is used to supervise the training of the generator and student network.

Generator contains of many upsampled layers that are trained under the teacher network's supervision. The random noise vector first input into the generator, and the outputs are the pseudo images with a preset size.

Student network is a small CNN that is trained on the pseudo data and learns knowledge from the teacher network.

3.1   Generating Data by Generator

In our method, there is no true data, so the generator is supervised by the teacher. We hope that the generated samples can train the student to output features similar to the teacher. The teacher is a high-performance CNN. If the student can output features similar to the teacher, then the student can also achieve superior performance. Therefore, we introduce PU learning to distinguish the features of teachers and students as high-quality and low-quality. Specifically, the random noises z are first sampled and input into the generator to get the generated data $x = G(z)$. Then the generated data x are input into the teacher $\mathcal{N}_T$ and student $\mathcal{N}_S$ to get the penultimate layer's features as P_T and P_S, and the classification results of the teacher are obtained as y_T. Then, the outputs

y_T are used to calculate the pseudo label t through the argmax function: $t = \arg\max(y_T)$. The teacher network plays the role of a fixed discriminator, we only optimize the generator during the data generation. As the work [32], the optimization of generator can be implemented by follows:

$$\mathcal{L}_{pu} = \pi \mathcal{H}_{KL}\left(P_T \| t\right) + (1 - \pi)\mathcal{H}_{KL}\left(P_S \| t\right) \tag{1}$$

π is the class prior (gradually increase, up to 0.6). By minimizing the loss, we expect that the generated images are input into the student network to output high-quality features, which can get the same predictions with the teacher through the classification layer.

In the image classification dataset, for training the neural network better, each category has the roughly same number of images. However, the generating of the generator is uncontrollable, and the number of data generated by each category is largely different. To solve this problem, similar to DAFL [18], we use the information theory to adjust the generation of the generator. In detail, for a set of probability vectors p, the total amount of information contained in the vector p can be got by the Information-Entropy $\mathcal{H}_{info}(p) = -\frac{1}{n}\sum_i^n p_i \log(p_i)$, and when all elements in n are equal to $1/n$, p contains the maximum information. For the outputs y of the teacher network, the frequency distribution of data generated by each category is $\frac{1}{n}\sum_i^n y_T^i$. Base on the information theory, when the information entropy of the y is maximum, the frequency distribution of each category is similar to $1/n$. Based on the above analysis, the information entropy loss function of the generated data can be expressed as:

$$\mathcal{L}_{inf} = -\mathcal{H}_{info}\left(\frac{1}{n}\sum_i^n y_T^i\right) \tag{2}$$

The $\mathcal{H}_{info}$ is the information entropy function. By maximizing the $\mathcal{L}_{inf}$, the number of images in each category in the generated images is approximately the same. Therefore, the generated images can better train the student network.

3.2   Training the Student Network

Knowledge distillation induces student network training by the soft targets which are extracted from the teacher network. More specifically, the outputs of the neural network are the one-hot vectors $y = \{y^1, y^2, \cdots, y^n\}$, n is the vector number in the y. For a vector $y^i = [0, 1, ..., 0]$, there is only one element is '1' and the other elements are '0', which represents that y^i only contains the class information that value is one. To promote the student network to learn more knowledge from the teacher network, we input the y into the softmax function to get the soft targets $s = \{s^1, s^2, \cdots, s^n\}$, which $s^i = \frac{\exp(y^i/T)}{\sum_j^n \exp(y^i/T)}$, the T is a parameter to soften the y that is normally set as 4, a bigger T can result in more soften soft-targets that contain more non-zero values. For example, a vector among the soft-targets is $s^i = [0.1, 0.7, ..., 0.08]$, s^i contains many values that not zero, and it includes more knowledge than the outputs of neural network y^i. By using soft targets, we can calculate the loss function of knowledge distillation:

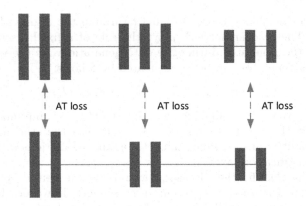

Fig. 2. The diagram of how DFPU extract attention feature maps from each convolution group and the red dotted arrow represents the loss function of attention maps. (Color figure online)

$$\mathcal{L}_{kd} = \frac{1}{n} \sum_{i=1}^{n} \mathcal{H}_{KL}\left(s_S^i, s_T^i\right). \tag{3}$$

The s_T and s_S are the soft targets of the teacher network and student network, respectively. The $\mathcal{H}_{KL}$ is the Kullback-Leibler Divergence function, which is widely used to calculate the degree of difference between two input features.

In the condition of without data, the student cannot improve the ability by training on the dataset. The teacher is a pre-trained model with superior performance and has learned massive knowledge from the real data. So, we improve the performance of the student network by learning more information from the teacher network. As shown in Fig. 2, we use the attention mechanism to extract more useful information from the teacher network. More specifically, the teacher network and student network contain multiple convolution groups. After each convolution group, the attention mechanism is used to extract attention maps. For example, there are the tensors A_j after the j-th convolution group, we calculate the sum of absolute values that are raised to the power 2 as the attention features of j-th convolution group F_j:

$$F_j = \sum_{i=1}^{n_j} \left|A_j^i\right|^2, \tag{4}$$

n_j is the channel number of j-th convolutional groups feature. By using the extracted attention features, we calculate the Euclidean distance as the loss function of attention features by the following formula:

$$\mathcal{L}_{at} = \frac{1}{l} \sum_{j=1}^{l} \frac{1}{n_j} \sum_{i=1}^{n_j} \left\|F_{Sj}^i - F_{Tj}^i\right\|_2^2 \tag{5}$$

Table 1. The results of LeNet training on the MNIST dataset.

Algorithm	Needed data	params	FLOPs	Accuracy
Teacher	Original data	~62K	~436K	98.84%
Standard BP	Original data	~16K	~144K	98.59%
Knowledge distillation [11]	Original data	~16K	~144K	98.74%
Meta data [16]	Meta data	~16K	~144K	92.47%
DAFL [18]	No data	~16K	~144K	98.20%
DFPU	No data	~16K	~144K	**98.38%**

The F_T and the F_S are the feature maps of the teacher network and student network respectively, the l is the number of the convolution group. The Euclidean distance can effectively calculate the similarity of two attention features. By minimizing the Euclidean distance between the attention features of the teacher and student, student network can get features similar to the teacher network, so achieves similar performance to the teacher network.

3.3 Objective Function

By combining these loss functions proposed above, the total objective function of our proposed DFPU can be defined as:

$$\mathcal{L}_{\text{Total}} = \mathcal{L}_{pu} + \alpha \cdot \mathcal{L}_{inf} + \beta \cdot \mathcal{L}_{kd} + \gamma \cdot \mathcal{L}_{at} \tag{6}$$

The α, β, γ are hyper-parameters to adjust each component in Eq. (6).

4 Experiments

In this section, there are extensive experiments conduct on the MNIST [3], CIFAR [21] and SVHN [22] datasets to verify the effectiveness of DFPU. DFPU is compared with data-free model compression methods DAFL [18], Meta data [16], ZSKD [26], DeepInversion [36], and the data-driven method that the student network trained by standard backpropagation (BP) and knowledge distillation (KD) [11]. And we prove the importance of each element in $\mathcal{L}_{Total}$ by ablation experiments. During the training of DFPU, the student network is trained by the generated data and only uses the testset of the dataset to validate its accuracy.

4.1 Experiments on MNIST Dataset

The MNIST dataset is composed of 60,000 training samples and 10,000 test samples. The samples in MNIST are the 28×28 pixels grayscale handwritten digital pictures in 10 categories from '0' to '9'. When conducting experiments on the MNIST dataset, the LeNet [3] is used as the teacher network and the LeNet

Table 2. The results of ResNet training on the CIFAR10 and CIFAR100, - means these algorithms did not report the accuracy on the CIFAR100.

Algorithm	Needed data	Params	FLOPs	CIFAR10	CIFAR100
Teacher	Original data	$\sim$ 21.28M	$\sim$1.16G	94.73%	76.74%
Standard BP	Original data	$\sim$11.17M	$\sim$0.55G	93.96%	75.79%
KD [11]	Original data	$\sim$11.17M	$\sim$0.55G	94.38%	76.63%
ZSKD [26]	No data	$\sim$11.17M	$\sim$0.55G	91.99%	–
DAFL [18]	No data	$\sim$11.17M	$\sim$0.55G	92.22%	73.52%
DeepInversion [36]	No data	$\sim$11.17M	$\sim$0.55G	93.26%	–
DFPU	No data	$\sim$11.17M	$\sim$0.55G	**94.07%**	**74.53%**

with half the number of channels of each layer is used as the student network. During the training process, Stochastic Gradient Descent (SGD) [37] is used as the optimizer, the learning rate is set to 0.1, and the hyperparameter α, β, and γ is set to 10, 0.1, and 1e-6, respectively.

Table 1 shows the experimental results on the MNIST dataset. It can be found that the DFPU achieves the best performance in data-free methods, reaching an accuracy of 98.38%. When compared with data-driven methods, the accuracy is only 0.21 % and 0.36% worse than standard backpropagation and knowledge distillation. Even compared with the accuracy of the teacher, the accuracy of the student is only 0.46 % lower. The experimental results show that our proposed DFPU can successfully generate data to simulate the MNIST dataset to train a convenient and efficient student network model.

4.2 Experimental Results on CIFAR Dataset

According the contained categories, CIFAR datasets are divided into CIFAR10 and CIFAR100, which include 60,000 RGB 32×32 pixels images, 50,000 images for training, and 10,000 images for testing. The CIFAR10 contains 10 categories, with 6,000 images in each category. And the CIFAR100 contains 100 categories, with 600 images in each category. When conducting experiments on the CIFAR dataset, we choose ResNet [1] as the neural network model.

When using the ResNet to experiment on CIFAR datasets, the 34-layer ResNet34 is used as the teacher network and the 18-layer ResNet18 is used as the student network. The amount of parameters (about 21 million, M) and calculation (about 1.16 giga, G) of the teacher network are about twice of the student network. During the training process, the training settings in CIFAR10 and CIFAR100 are the same, we choose the stochastic gradient descent (SGD) [37] as the optimizer. When training on the CIFAR10 the hyperparameters α, β and γ is set to 10, 0.25, and 1e-4, respectively. When training on the CIFAR100, the hyperparameters α, β and γ is set to 1, 0.2, and 1e-5.

Table 2 shows the experimental results of ResNet on the CIFAR dataset. When without data, our DFPU achieved the highest accuracy of 94.07% and

Table 3. The results of ablation experiments.

$\mathcal{L}_{kd}$		✓			✓	✓		✓
$\mathcal{L}_{at}$			✓		✓		✓	✓
$\mathcal{L}_{pu}$				✓		✓	✓	✓
Accuracy	89.01%	89.42%	90.21%	90.46%	91.05%	91.54%	91.86%	**92.05%**

Table 4. The results of extended experiments.

Dataset	Model		Accuracy	
	Teacher	Student	Teacher	Student
CIFAR10	VGGNet16	VGGNet13	92.86%	92.01%
CIFAR100	VGGNet16	VGGNet13	72.24%	70.27%
SHVN	VGGNet16	VGGNet13	94.51%	94.36%

74.53% on the CIFAR10 dataset and CIFAR100 dataset. The student network trained by DAFL gets 92.22% and 73.52% on the CIFAR10 and CIFAR100, respectively, which also uses a generator to generate the data. The student network trained by DFPU is significantly better than that trained by DAFL. It proves that our used positive-unlabeled learning is beneficial for data generation. The student network trained by ZSKD and DeepInversion achieve accuracies of 91.99% and 93.26% on CIFAR10. Our DFPU-trained student network is 2.08% and 0.81% higher than that of ZSKD and DeepInversion. Compared to the data-driven methods KD, the accuracy of the DFPU-trained student network is only lower 0.31% and lower 2.10% on CIFAR10 and CIFAR100. Even compared to the teacher network, the accuracy of the DFPU-trained student network is only lower 0.66% on CIFAR10 and lower 2.21% on CIFAR100. These experiments demonstrate that our DFPU can achieve the performance of state-of-the-art in the data-free knowledge distillation and can train the student network to achieve comparable capability with the teacher network.

4.3 Ablation Experiments

In this subsection, a series of ablation experiments are conducted to prove the effectiveness of each item in $\mathcal{L}_{Total}$. Because we find during the experiment, if there is no $\mathcal{L}_{inf}$, the student network and the generator are not convergence, so we use $\mathcal{L}_{inf}$ by default. We choose the 19-layers VGGNet19 as the teacher network and the 13-layers VGGNet13 as the student network to train on the CIFAR10 dataset. The student VGGNet13 only contains about 9.41 million (M) parameters and 0.23 giga (G) calculations, which is only half of that of the teacher. The training setups are the same as that the Subsect. 4.2.

Table 3 shows the results of the ablation experiments. When none of the three items are used, the accuracy is only 89.01%. When using $\mathcal{L}_{kd}$, $\mathcal{L}_{at}$, and $\mathcal{L}_{pu}$ alone, the student network achieved accuracy of 89.42%, 90.21%, and 90.46%, respectively. We can find that $\mathcal{L}_{pu}$ has the greatest improvement in student network performance. When using two-loss items at the same time, using $\mathcal{L}_{at}$ and $\mathcal{L}_{pu}$ together achieved the highest accuracy of 91.86%. When all three items are used, the highest accuracy is reached at 92.05%.

Ablation experiments show that each item in $\mathcal{L}_{Total}$ is meaningful, especially is $\mathcal{L}_{pu}$. By using $\mathcal{L}_{Total}$, the pseudo data required for the experiment can be successfully generated, and a student network with excellent ability is trained.

4.4 Extended Experiments

We also conduct experiments on several benchmarks to verify the performance of the DFPU. We select the VGGNet16 and VGGNet13 as the teacher network and student network to train on CIFAR10, CIFAR100 and SHVN datasets. The experimental setups are the same as Subsect. 4.2.

Table 4 reports the results of the extended experiments. For the teacher VGGNet16 trained on the CIFAR10 and CIFAR100 datasets, the student VGGNet13 without any truthful data achieves the accuracies of 92.01% and 70.27%, respectively, which are close to the teacher network (0.85% and 1.97% lower than the teacher). When the teacher is trained on the SVHN, the student obtains an accuracy of 94.36%, which is only 0.15% lower than that of the teacher. These results demonstrate that our proposed DFPU can effectively approximate the original training dataset that the teacher used and train the student well.

5   Conclusion

Existing CNNs are often unable to apply in practice due to model size and data unattainable. This paper proposes a data-free knowledge distillation method called DFPU to train a compact student network without data. We generate pseudo training data based on GANs and use the attention mechanism to extract effective information from the neural network for knowledge transfer. To generate the data that can well train the student and improve the student's capability, we utilize positive-unlabeled learning to discriminate the features of teacher and student as high-quality and low-quality. Extensive experimental results demonstrate that we propose DFPU can train a small compact student network without data to achieve a performance similar to the large high-ability neural network. And the ablation studies show that our DFPU gets SOTA performance among exiting data-free methods.

Acknowledgement. This work was supported by the Mianyang Science and Technology Program 2020YFZJ016, SWUST Doctoral Foundation under Grant 19zx7102, 21zx7114, Sichuan Science and Technology Program under Grant 2020YFS0307.

References

1. He, K., Zhang, X., Ren, S., Jian, S.: Deep residual learning for image recognition. In: 2016 IEEE Conference on Computer Vision and Pattern Recognition (CVPR) (2016)
2. Krizhevsky, A., Sutskever, I., Hinton, G.E.: ImageNet classification with deep convolutional neural networks. In: Advances in Neural Information Processing Systems, pp. 1097–1105 (2012)
3. LeCun, Y., Bottou, L., Bengio, Y., Haffner, P., et al.: Gradient-based learning applied to document recognition. Proc. IEEE **86**(11), 2278–2324 (1998)
4. Girshick, R.B., Donahue, J., Darrell, T., Malik, J.: Rich feature hierarchies for accurate object detection and semantic segmentation. In: CVPR, pp. 580–587. IEEE Computer Society (2014)
5. Girshick, R.: Fast R-CNN. In: Proceedings of the IEEE International Conference on Computer Vision, pp. 1440–1448 (2015)
6. Ren, S., He, K., Girshick, R., Sun, J.: Faster R-CNN: towards real-time object detection with region proposal networks. In: Advances in Neural Information Processing Systems, pp. 91–99 (2015)
7. Long, J., Shelhamer, E., Darrell, T.: Fully convolutional networks for semantic segmentation. In: Proceedings of the IEEE Conference on Computer Vision and Pattern Recognition, pp. 3431–3440 (2015)
8. Yu, F., Koltun, V.: Multi-scale context aggregation by dilated convolutions. arXiv preprint arXiv:1511.07122 (2015)
9. Liu, Y., Chen, K., Liu, C., Qin, Z., Luo, Z., Wang, J.: Structured knowledge distillation for semantic segmentation. In: Proceedings of the IEEE/CVF Conference on Computer Vision and Pattern Recognition, pp. 2604–2613 (2019)
10. Ba, L.J., Caruana, R.: Do deep nets really need to be deep? In: Advances in Neural Information Processing Systems, pp. 2654–2662 (2013)
11. Hinton, G., Vinyals, O., Dean, J.: Distilling the knowledge in a neural network. Comput. Sci. **14**(7), 38–39 (2015)
12. Romero, A., Ballas, N., Kahou, S.E., Chassang, A., Gatta, C., Bengio, Y.: FitNets: hints for thin deep nets. Comput. Sci. (2014)
13. Bolukbasi, T., Wang, J., Dekel, O., Saligrama, V.: Adaptive neural networks for efficient inference. In: ICML, Series Proceedings of Machine Learning Research, vol. 70. PMLR, pp. 527–536 (2017)
14. Huang, Z., Wang, N.: Data-driven sparse structure selection for deep neural networks. In: Ferrari, V., Hebert, M., Sminchisescu, C., Weiss, Y. (eds.) ECCV 2018. LNCS, vol. 11220, pp. 317–334. Springer, Cham (2018). https://doi.org/10.1007/978-3-030-01270-0_19
15. Xu, S., Ren, X., Ma, S., Wang, H.: meProp: Sparsified back propagation for accelerated deep learning with reduced overfitting. In: ICML 2017 (2017)
16. Lopes, R.G., Fenu, S., Starner, T.: Data-free knowledge distillation for deep neural networks (2017)
17. Liu, Z., et al.: MetaPruning: meta learning for automatic neural network channel pruning. arXiv preprint arXiv:1903.10258 (2019)
18. Chen, H.: Data-free learning of student networks. In: Proceedings of the IEEE/CVF International Conference on Computer Vision, pp. 3514–3522 (2019)
19. Goodfellow, I.: Nips 2016 tutorial: generative adversarial networks. arXiv preprint arXiv:1701.00160 (2016)

20. Odena, A.: Semi-supervised learning with generative adversarial networks. arXiv preprint arXiv:1606.01583 (2016)
21. Krizhevsky, A., Hinton, G., et al.: Learning multiple layers of features from tiny images (2009)
22. Netzer, Y., Wang, T., Coates, A., Bissacco, A., Wu, B., Ng, A.Y.: Reading digits in natural images with unsupervised feature learning (2011)
23. Furlanello, T., Lipton, Z.C., Tschannen, M., Itti, L., Anandkumar, A.: Born again neural networks (2018)
24. Zagoruyko, S., Komodakis, N.: Paying more attention to attention: improving the performance of convolutional neural networks via attention transfer. arXiv preprint arXiv:1612.03928 (2016)
25. Yoo, J., Cho, M., Kim, T., Kang, U.: Knowledge extraction with no observable data (2019)
26. Nayak, G.K., Mopuri, K.R., Shaj, V., Radhakrishnan, V.B., Chakraborty, A.: Zero-shot knowledge distillation in deep networks. In: International Conference on Machine Learning. PMLR, pp. 4743–4751 (2019)
27. Wang, Z.: Data-free knowledge distillation with soft targeted transfer set synthesis. arXiv preprint arXiv:2104.04868 (2021)
28. Xu, Y., Xu, C., Xu, C., Tao, D.: Multi-positive and unlabeled learning. In: IJCAI, pp. 3182–3188 (2017)
29. Kiryo, R., Niu, G., Plessis, M.C.D., Sugiyama, M.: Positive-unlabeled learning with non-negative risk estimator. arXiv preprint arXiv:1703.00593 (2017)
30. Goodfellow, I., et al.: Generative adversarial nets. In: Advances in Neural Information Processing Systems, vol. 27 (2014)
31. Xi, C., Yan, D., Houthooft, R., Schulman, J., Abbeel, P.: InfoGAN: interpretable representation learning by information maximizing generative adversarial nets. In: Neural Information Processing Systems (NIPS) (2016)
32. Guo, T.: On positive-unlabeled classification in GAN. In: Proceedings of the IEEE/CVF Conference on Computer Vision and Pattern Recognition, pp. 8385–8393 (2020)
33. Wang, F.: Residual attention network for image classification. In: Proceedings of the IEEE Conference on Computer Vision and Pattern Recognition, pp. 3156–3164 (2017)
34. Hu, J., Shen, L., Sun, G.: Squeeze-and-excitation networks. In: Proceedings of the IEEE Conference on Computer Vision and Pattern Recognition, pp. 7132–7141 (2018)
35. Simonyan, K., Zisserman, A.: Very deep convolutional networks for large-scale image recognition. Comput. Sci. (2014)
36. Yin, H.: Dreaming to distill: data-free knowledge transfer via DeepInversion. In: Proceedings of the IEEE/CVF Conference on Computer Vision and Pattern Recognition, pp. 8715–8724 (2020)
37. Bottou, L.: Stochastic gradient descent tricks (2012)

Manifold Discriminative Transfer Learning for Unsupervised Domain Adaptation

Xueliang Quan, Dongrui Wu$^{(\boxtimes)}$, Mengliang Zhu, Kun Xia, and Lingfei Deng

Huazhong University of Science and Technology, Wuhan, China
{quanxl,drwu,mlzhu,kxia,lfdeng}@hust.edu.cn

Abstract. This paper proposes manifold discriminative transfer learning (MDTL) for traditional unsupervised domain adaptation. It first utilizes manifold subspace learning to reconstruct the original data in both source and target domains, which can reduce the domains shifts, then performs simultaneously structural risk minimization, discriminative class level alignment, and manifold regularization for transfer learning. More specifically, it minimizes the intra-class distribution discrepancy to improve the domain transferability, maximizes the inter-class distribution discrepancy to improve the class discriminability, and also maintains geometrical structures of the data samples. Remarkably, MDTL is a traditional transfer learning approach and it has a closed-form solution, so the computational cost is low. Extensive experiments showed that MDTL outperforms several state-of-the-art traditional domain adaptation approaches.

Keywords: Domain adaptation · Discriminative class level adaptation · Manifold subspace learning

1 Introduction

A basic assumption in traditional machine learning is that the training and test data follow the same distribution, which rarely holds in real-world applications [15]. However, collecting and labeling more data consistent with the test data may be labor-intensive, time-consuming, or even unrealistic [12,20]. Transfer learning (TL), or domain adaptation (DA), is a promising solution to these challenges [15]. Recent years have witnessed its successful applications in many fields, including image recognition [12,22], text classification [14], brain-computer interfaces [21], etc.

Transfer learning applies data or knowledge in a well labeled source domain to a zero or sparsely labeled target domain [15]. It's important to reduce the domain distribution discrepancy to avoid negative transfer [23]. Many feature-based approaches [6,12,22] have been proposed, which reduce the domain distribution discrepancy through subspace learning or distribution adaptation. However, there are still two challenges: feature transformation degeneration and feature discriminability degradation.

Feature transformation degeneration means that both subspace learning and distribution adaptation can only reduce, but not completely eliminate, the distribution discrepancy. Specifically, a subspace transformation [6] is learned to obtain domain-invariant feature representations, but distribution discrepancy minimization is rarely

T. Mantoro et al. (Eds.): ICONIP 2021, LNCS 13109, pp. 321–334, 2021.
https://doi.org/10.1007/978-3-030-92270-2_28

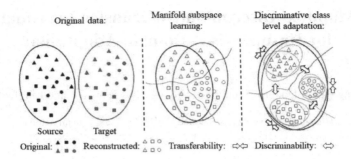

Fig. 1. Illustration of MDTL. Data reconstruction by manifold subspace learning makes the source and target domains more similar. The feature dimensionality is simultaneously reduced. Discriminative class level adaptation with structural risk minimization and manifold regularization is then performed to learn the final domain-invariant classifier.

considered afterwards. Most distribution adaptation approaches [12, 14] focus on the original feature space, where feature distortions may happen and lead to undesirable performance. Thus, it is necessary to combine subspace learning and distribution adaptation for better domain adaptation. A similar idea has been explored in Manifold Embedded Distribution Alignment (MEDA) [19], but it ignores feature discriminability in distribution adaptation.

Feature discriminability degradation means that existing feature distribution adaptation approaches [12, 18] only attempt to minimize the intra-class distribution discrepancy to enhance the feature transferability, which may hurt the feature discriminability [2]. It's important to maximize the inter-class distribution discrepancy to push feature representations of different classes away from the decision boundary for better discriminability.

This paper proposes manifold discriminative transfer learning (MDTL) to address the above issues. Figure 1 illustrates its main idea. Specifically, we use manifold subspace learning as a pre-processing step to reconstruct the original data, and then perform distribution adaptation on the reconstructed data to accommodate feature transformation degeneration. During distribution adaptation, to simultaneously consider feature transferability and discriminability, we not only minimize the intra-class distribution discrepancy but also maximize the inter-class distribution discrepancy. Manifold regularization is also used to preserve the local geometrical structure. Comprehensive experiments on six datasets demonstrated that MDTL outperforms several state-of-the-art traditional transfer learning approaches.

In summary, our contributions are:

– We consider feature transferability and discriminability simultaneously during distribution adaptation, whereas previous domain adaptation approaches usually consider only one of them.
– We use manifold subspace learning and distribution adaptation to get their respective virtues. Not only that, we also utilize manifold learning to preserve the local geometrical structure.

2   Related Work

Feature based DA can be divided into two categories: subspace learning and distribution adaptation.

Subspace Learning. Fernando *et al.* [4] aligned the bases of two domains' feature spaces to minimize the domain discrepancy. CORelation ALignment (CORAL) [17] further considers second-order statistics in the alignment. Both approaches align the distributions in the original feature space, which may not be optimal. Gopalan *et al.* [7] proposed to align the feature subspace projected onto a Grassmann manifold. Gong *et al.* [6] proposed a geodesic flow kernel (GFK) on the Grassmann manifold to make the source and target domains more similar. Our MDTL uses GFK to reconstruct the original source and target data.

Distribution Adaptation. Borgwardt *et al.* [1] proposed maximum mean discrepancy (MMD) to measure the distribution difference between two domains, which has been widely adopted in TL [12,14,24]. Pan *et al.* [14] proposed to match the marginal distribution of two domains, and Long *et al.* [12] matched simultaneously the marginal and conditional distributions. Balanced DA [18] further weights the marginal and conditional distribution discrepancies. Long *et al.* [11] proposed a framework to learn a domain-invariant classifier for DA, mainly based on joint distribution adaptation. Wang *et al.* [19] modified it by combining subspace learning and balanced DA. However, previous work has not consider the feature transferability and discriminability at the same time. Our MDTL simultaneously considers subspace learning and feature discriminability, which has never been done before.

3   Manifold Discriminative Transfer Learning

We consider unsupervised homogeneous transfer learning, i.e., there are labeled samples of the source domain $\mathcal{D}_s = \{(\boldsymbol{x}_{s,i}, y_{s,i})\}_{i=1}^{n_s}$ and unlabeled samples of the target domain $\mathcal{D}_t = \{\boldsymbol{x}_{t,i}\}_{i=1}^{n_t}$, where n_s and n_t are the number of samples in the source and target domains, respectively. $\mathcal{D}_s$ and $\mathcal{D}_t$ have the same feature space $\mathcal{X}_s = \mathcal{X}_t$, and also the same label space $\mathcal{Y}_s = \mathcal{Y}_t$.

Typically, there exist distribution shifts between $\mathcal{D}_s$ and $\mathcal{D}_t$, i.e., the marginal probability distributions $P_s(\boldsymbol{x}_s) \neq P_t(\boldsymbol{x}_t)$, and conditional probability distributions $Q_s(y_s|\boldsymbol{x}_s) \neq Q_t(y_t|\boldsymbol{x}_t)$. The task is to train a well performed classifier for the target data, by making use of the labeled source data and unlabeled target data.

3.1   Overview

MDTL consists of two steps:

1. Manifold subspace learning to accommodate degenerated feature transformation.
2. Classifier training by simultaneously considering structural risk minimization (SRM), discriminative class level adaptation, and manifold regularization.

Eventually, a domain-invariant classifier f^* is learned:

$$
f^* = \underset{f \in \mathcal{H}_K}{\arg\min} \sum_{i=1}^{n_s} \ell(f(\tilde{\boldsymbol{x}}_{s,i}), y_{s,i}) + \gamma \cdot \|f\|_K^2
$$
$$
+ \eta \cdot d(\tilde{\mathcal{D}}_s, \tilde{\mathcal{D}}_t) + \rho \cdot M_{f,K}(\tilde{\mathcal{D}}_s, \tilde{\mathcal{D}}_t), \tag{1}
$$

where $\tilde{\mathcal{D}}_s$ and $\tilde{\mathcal{D}}_t$ represent the reconstructed source and target domains, and γ, η and ρ are hyper-parameters. $\mathcal{H}_K$ is the Hilbert space induced by kernel function $k(\cdot, \cdot)$.

The first two terms in (1) implement SRM on $\tilde{\mathcal{D}}_s$, $d(\tilde{\mathcal{D}}_s, \tilde{\mathcal{D}}_t)$ implements discriminative class level adaptation, and $M_{f,K}(\tilde{\mathcal{D}}_s, \tilde{\mathcal{D}}_t)$ manifold regularization. More details are introduced next.

3.2 Manifold Subspace Learning

Subspace learning reduces the domain distribution discrepancy by projecting $\boldsymbol{x}_s$ and $\boldsymbol{x}_t$ into common or similar subspaces, which facilitates domain adaptation. GFK [6], a Grassmann manifold subspace learning approach, is used to overcome the feature distortions in MDTL. We briefly introduce the core ideas of GFK due to the space limitations and the details can be found in its original paper.

A Grassmann manifold $G(q, Q)$ is a collection of all q-dimensional linear subspaces of the Q-dimensional vector space V, where Q and q are the dimensionality of the original feature space and the selected manifold embedding subspace, respectively. GFK reduces the dimensionality of the source and target data from Q to q via principal component analysis, and views these two subspaces as two points on $G(q, Q)$, namely $\Phi(0)$ and $\Phi(1)$. Then, it finds a geodesic flow $\Phi(t)$, $t \in [0, 1]$, on $G(q, Q)$ to connect these two points.

The geodesic flow kernel is defined as:

$$
\langle \tilde{\boldsymbol{x}}_i, \tilde{\boldsymbol{x}}_j \rangle = \int_0^1 ((\Phi(t)^T \boldsymbol{x}_i)^T (\Phi(t)^T \boldsymbol{x}_j)) dt = \boldsymbol{x}_i^T G \boldsymbol{x}_j \tag{2}
$$

where $\tilde{\boldsymbol{x}}_i$ and $\tilde{\boldsymbol{x}}_j$ ($i, j = 1, ..., n_s + n_t$) are q-dimensional transformed feature vectors of $\boldsymbol{x}_i$ and $\boldsymbol{x}_j$, respectively. G can be computed by singular value decomposition and more details can be found in [6]. Next, the new feature representations $\tilde{\boldsymbol{x}}_s$ and $\tilde{\boldsymbol{x}}_t$ obtained from the geodesic flow kernel are used to learn the domain-invariant classifier f.

3.3 Structural Risk Minimization

Structural risk minimization is used and augmented to learn a robust classifier for target data in MDTL. SRM learns a classifier f^* by minimizing the following combined risk:

$$
\underset{f \in \mathcal{H}_K}{\min} \sum_{i=1}^{n_s} (y_i - f(\tilde{\boldsymbol{x}}_{s,i}))^2 + \gamma \|f\|_K^2. \tag{3}
$$

According to the Representer Theorem [16], the solution of (3) is

$$f^*(\tilde{x}) = \sum_{i=1}^{n_s} \alpha_i^* k(\tilde{\boldsymbol{x}}, \tilde{\boldsymbol{x}}_{s,i}),\tag{4}$$

where $k(\tilde{\boldsymbol{x}}, \tilde{\boldsymbol{x}}_{s,i})$ is a kernel function. Then, (3) can be re-expressed as

$$
\begin{aligned}
&\sum_{i=1}^{n_s} (y_i - f(\tilde{\boldsymbol{x}}_{s,i}))^2 + \gamma \|f\|_K^2 \\
=\ &\sum_{i=1}^{n} A_{ii} (y_i - f(\tilde{\boldsymbol{x}}_i))^2 + \gamma \|f\|_K^2 \\
=\ &\left\| (Y^T - \boldsymbol{\alpha}^T K) A \right\|_F^2 + \gamma \cdot \mathrm{tr}(\boldsymbol{\alpha}^T K \boldsymbol{\alpha}),
\end{aligned}\tag{5}
$$

where $n = n_s + n_t$, and $A \in R^{n \times n}$ is a diagonal domain indicator matrix with $A_{ii} = 1$ for $i = 1, ..., n_s$, and 0 otherwise. $Y = [y_{s,1}, ..., y_{s,n_s}, y_{t,1}, ..., y_{t,n_t}]^T \in R^{n \times 1}$ is the source label matrix. $K \in R^{n \times n}$ is a kernel matrix with $K_{ij} = k(\tilde{\boldsymbol{x}}_i, \tilde{\boldsymbol{x}}_j)$. $\boldsymbol{\alpha} = [\alpha_1, ..., \alpha_n]^T$ is an n-dimensional coefficient vector.

3.4   Discriminative Class Level Adaptation

In MDTL, to match the distributions of two domains, we introduced discriminative class level adaptation, which considers the domain transferability and class discriminability simultaneously [9,24]. It minimizes the intra-class distribution discrepancy d_{intra} to increase the feature transferability, and maximizes the inter-class distribution discrepancy d_{inter} to increases the feature discriminability.

In unsupervised domain adaptation, the target data do not have labels. We compute their pseudo labels $\hat{y}_t$, by applying a classifier trained on $\mathcal{D}_s$ to $\mathcal{D}_t$. Then, d_{intra} and d_{inter} are computed by MMD:

$$d_{intra} = \sum_{c=\hat{c}=1}^{C} \left\| P(y_s^c)\mathbb{E}\left[\tilde{\boldsymbol{x}}_s | y_s^c\right] - P(\hat{y}_t^{\hat{c}})\mathbb{E}\left[\tilde{\boldsymbol{x}}_t | \hat{y}_t^{\hat{c}}\right] \right\|^2 \tag{6}$$

$$d_{inter} = \sum_{c=1}^{C}\sum_{\hat{c} \neq c} \left\| P(y_s^c)\mathbb{E}\left[\tilde{\boldsymbol{x}}_s | y_s^c\right] - P(\hat{y}_t^{\hat{c}})\mathbb{E}\left[\tilde{\boldsymbol{x}}_t | \hat{y}_t^{\hat{c}}\right] \right\|^2 \tag{7}$$

where c and $\hat{c}$ are class labels, $\mathbb{E}[\cdot]$ denotes expectation of the subspace samples, and

$$P(y_s^c) = \frac{n_s^c}{n_s}, \quad P(\hat{y}_t^{\hat{c}}) = \frac{n_t^{\hat{c}}}{n_t}, \quad \mathbb{E}[\tilde{\boldsymbol{x}}_s | y_s^c] = \frac{1}{n_s^c}\sum_{i=1}^{n_s^c} \tilde{\boldsymbol{x}}_{s,i}^c, \quad \mathbb{E}[\tilde{\boldsymbol{x}}_t | \hat{y}_t^{\hat{c}}] = \frac{1}{n_t^{\hat{c}}}\sum_{i=1}^{n_t^{\hat{c}}} \tilde{\boldsymbol{x}}_{t,i}^{\hat{c}}$$

where $\tilde{\boldsymbol{x}}_{s,i}^c$ and $\tilde{\boldsymbol{x}}_{t,i}^{\hat{c}}$ are the samples in the c-th class of the source and target domain.

The distribution discrepancy of Class c and $\hat{c}$ between the two domains is:

$$\left\| P(y_s^c)\mathbb{E}[\tilde{\boldsymbol{x}}_s|y_s^c] - P(\hat{y}_t^{\hat{c}})\mathbb{E}[\tilde{\boldsymbol{x}}_t|\hat{y}_t^{\hat{c}}] \right\|^2 = \left\| \frac{1}{n_s}\sum_{i=1}^{n_s^c}\tilde{\boldsymbol{x}}_{s,i}^c - \frac{1}{n_t}\sum_{i=1}^{n_t^{\hat{c}}}\tilde{\boldsymbol{x}}_{t,i}^{\hat{c}} \right\|^2 \tag{8}$$

Let the source domain one-hot encoding label matrix be $Y_s = [y_{s,1}; ...; y_{s,n_s}] \in R^{n_s \times C}$, and the target domain one-hot encoding pseudo label matrix be $\hat{Y}_t = [\hat{y}_{t,1}; ...; \hat{y}_{t,n_t}] \in R^{n_t \times C}$. Then, (6) and (7) can be reformulated as:

$$d_{intra} = \sum_{c=\hat{c}=1}^{C} \left\| \frac{1}{n_s}\sum_{i=1}^{n_s^c}\tilde{\boldsymbol{x}}_{s,i}^c - \frac{1}{n_t}\sum_{j=1}^{n_t^{\hat{c}}}\tilde{\boldsymbol{x}}_{t,j}^{\hat{c}} \right\|^2 = \|\tilde{X}_s N_s - \tilde{X}_t N_t\|^2 \tag{9}$$

$$d_{inter} = \sum_{c=1}^{C}\sum_{\hat{c}\neq c} \left\| \frac{1}{n_s}\sum_{i=1}^{n_s^c}\tilde{\boldsymbol{x}}_{s,i}^c - \frac{1}{n_t}\sum_{i=j}^{n_t^{\hat{c}}}\tilde{\boldsymbol{x}}_{t,j}^{\hat{c}} \right\|^2 = \|\tilde{X}_s W_s - \tilde{X}_t W_t\|^2 \tag{10}$$

where

$$N_s = \frac{Y_s}{n_s}, \quad N_t = \frac{\hat{Y}_t}{n_t}, \quad W_s = \frac{F_s}{n_s}, \quad W_t = \frac{\hat{F}_t}{n_t}$$

$$F_s = [\overbrace{Y_s(:,1), ..., Y_s(:,1)}^{C-1}, \overbrace{Y_s(:,2), ..., Y_s(:,2)}^{C-1}, ..., \overbrace{Y_s(:,C), ..., Y_s(:,C)}^{C-1}]$$

$$\hat{F}_t = [\hat{Y}_t(:,2:C), \hat{Y}_t(:,[1\ 3:C]), ..., \hat{Y}_t(:,1:C-1)]$$

Discriminative class level adaptation is achieved by minimizing the following:

$$d(\tilde{\mathcal{D}}_s, \tilde{\mathcal{D}}_t) = d_{intra} - \mu d_{inter} = \mathrm{tr}(\tilde{X}M\tilde{X}^T), \tag{11}$$

where

$$M = \begin{bmatrix} N_s N_s^T & -N_s N_t^T \\ -N_t N_s^T & N_t N_t^T \end{bmatrix} - \mu \begin{bmatrix} W_s W_s^T & -W_s W_t^T \\ -W_t W_s^T & W_t W_t^T \end{bmatrix} \tag{12}$$

and μ is a trade-off hyper-parameter which is set to 0.1 in this paper.

To conveniently augment (11) to SRM in (5), it is re-expressed as

$$d(\tilde{\mathcal{D}}_s, \tilde{\mathcal{D}}_t) = \mathrm{tr}(\boldsymbol{\alpha}^T K M K^T \boldsymbol{\alpha}) \tag{13}$$

Algorithm 1. Manifold discriminative transfer learning (MDTL).

Inputs: $X_s/X_t \in R^{Q \times n_t}$: source/target data matrix; $Y_s \in R^{n_s \times 1}$: source label matrix

Parameters: q: manifold feature dimensionality; p: number of nearest neighbors in manifold
    regularization; T: maximum number of iterations; $\gamma, \eta,$ and ρ in (15)

Output: Classifier f^*

  1: GFK data reconstruction;
  2: Initialize the target domain pseudo labels;
  3: $iter = 0$;
  4: **repeat**
  5:    Construct kernel matrix K from the reconstructed data;
  6:    Compute matrix A in (5), M in (12), and L in (14);
  7:    Calculate $\boldsymbol{\alpha}^*$ by (16);
  8:    Obtain classifier f^* by (4);
  9:    Update the target domain pseudo labels;
10:    $iter = iter + 1$;
11: **until** $iter \geq T$
12: **return** Classifier f^*.

3.5 Manifold Regularization

Manifold regularization [16] is also used in MDTL to maintain the geometrical structures of the samples on manifold $G(q, Q)$:

$$M_{f,K}(\tilde{\mathcal{D}}_s, \tilde{\mathcal{D}}_t) = \sum_{i,j=1}^{n} (f(\tilde{\boldsymbol{x}}_i) - f(\tilde{\boldsymbol{x}}_j))^2 W_{ij}$$

$$= \sum_{i,j=1}^{n} f(\tilde{\boldsymbol{x}}_i) L_{ij} f(\tilde{\boldsymbol{x}}_j)$$

$$= \mathrm{tr}(\boldsymbol{\alpha}^T K L K \boldsymbol{\alpha}), \tag{14}$$

where W is the graph affinity matrix, built based on the cosine similarity function:

$$W_{ij} = \begin{cases} \cos(\tilde{\boldsymbol{x}}_i, \tilde{\boldsymbol{x}}_j), & \tilde{\boldsymbol{x}}_i \in \mathcal{N}_p(\tilde{\boldsymbol{x}}_j) \text{ or } \tilde{\boldsymbol{x}}_j \in \mathcal{N}_p(\tilde{\boldsymbol{x}}_i) \\ 0, & \text{otherwise} \end{cases}$$

in which $\mathcal{N}_p(\tilde{\boldsymbol{x}}_i)$ denotes the set of p-nearest neighbors of $\tilde{\boldsymbol{x}}_i$. Let $D \in R^{n \times n}$ be a diagonal matrix with $D_{ii} = \sum_{j=1}^{n} W_{ij}$. Then, the manifold matrix $L = I - D^{-1/2} W D^{-1/2}$.

3.6 The Closed-Form Solution

MDTL in (1) integrates SRM, discriminative class level adaptation and manifold regularization, i.e., (5), (13) and (14):

$$f^* = \underset{f \in \mathcal{H}_K}{\arg\min} \|(Y^T - \boldsymbol{\alpha}^T K)A\|_F^2 + \gamma \cdot \mathrm{tr}(\boldsymbol{\alpha}^T K \boldsymbol{\alpha})$$

$$+ \eta \cdot \mathrm{tr}(\boldsymbol{\alpha}^T K M K^T \boldsymbol{\alpha}) + \rho \cdot \mathrm{tr}(\boldsymbol{\alpha}^T K L K \boldsymbol{\alpha}). \tag{15}$$

By setting the partial derivative w.r.t. α to 0, we have the closed-form solution:

$$\alpha^* = [(A + \eta M + \rho L)K + \gamma I]^{-1} AY \tag{16}$$

The pseudo-code of MDTL is summarized in Algorithm 1.

4   Experiments

This section compares MDTL with several state-of-the-art TL approaches on six benchmark datasets. The code of MDTL is available at https://github.com/quanxl97/MDTL.

Table 1. Statistics of the six datasets.

Dataset	#Examples	#Features	#Classes	Domains
Offfice	1,410	4,096	10	A, W, D
Caltech	1,123	4,096	10	C
ImageCLEF	1,800	2,048	12	P, I, C
MNIST	2,000	256	10	M
USPS	800	256	10	U
COIL20	1,440	1,024	20	C1, C2

4.1   Datasets

The six datasets summarized in Table 1 were used in our experiments.

Office-Caltech10 [3] consists of four domains: Amazon (A), Webcam (W), DSLR (D) and Caltech (C), which are widely used in TL research. A → W denotes the transfer learning task that A is the source domain and W the target domain. There are 12 different transfer learning tasks. 4, 096 DeCaf6 [3] features were used in our experiments.

ImageCLEF-DA [25] consists of three domains: Pascal VOC 2012 (P), ImageNet ILSVRC 2012 (I) and Caltech-256 (C). Each domain has 12 categories, each with 50 images. There are six different transfer learning tasks. ResNet50 features were used in our experiments.

MNIST-USPS [8, 10] consists of two digit recognition datasets with different distributions. Each dataset contains 10 categories. We constructed two TL tasks from them: MNIST(M) → USPS(U) and USPS(U) → MNIST(M).

COIL20 [12] is a benchmark for object recognition, with 1, 440 images from 20 categories. The images of each object were taken from different angles (5 degrees increment). We partitioned the dataset into two equal subsets COIL1 (C1) and COIL2 (C2) for transfer learning.

4.2 Baselines

MDTL was compared with the following state-of-the-art statistical traditional transfer learning approaches: Transfer Component Analysis (TCA) [14], Geodesic Flow Kernel (GFK) [6], Joint Distribution Adaptation (JDA) [12], Transfer Joint Match (TJM) [13], Adaptation Regularization (ARTL) [11], CORelation ALignment (CORAL) [17], Scatter component analysis (SCA) [5], Joint Geometrical and Statistical Alignment (JGSA) [22], Manifold Embedded Distribution Alignment (MEDA) [19].

All parameters were set according to their original papers, except that the parameters of MEDA were identical to MDTL since they have similar components. For MDTL, the manifold feature dimension $q = 20$ for all datasets, the number of iterations $T = 10$, the number of nearest neighbors $p = 5$, and hyper-parameters $\eta = 10$, $\rho = 1$ and $\gamma = 0.01$. More discussions on the parameter sensitivity are given in Sect. 4.5.

4.3 Experimental Results

The classification accuracies are shown in Tables 2 and 3. MDTL had the best average performance on all datasets, and on most individual TL tasks. Its improvements over other approaches were particularly large on MNIST-USPS and COIL20.

MDTL outperformed both the two subspace learning approaches (GFK and CORAL) and the two distribution adaptation approaches (TCA and JDA). The two subspace learning approaches do not match the distribution between the two domains.

Table 2. Classification accuracies (%) on Office-Caltech10.

Task	C→A	C→W	C→D	A→C	A→W	A→D	W→C	W→A	W→D	D→C	D→A	D→W	Avg.
TCA	89.25	80.00	83.44	74.18	71.86	78.34	72.57	79.96	100.00	73.46	88.20	97.29	82.38
GFK	88.41	80.68	84.56	76.85	68.74	79.62	74.80	75.26	100.00	74.09	85.80	98.64	82.29
CORAL	89.98	78.64	85.99	83.88	74.58	80.25	74.98	77.14	100.00	79.25	85.80	99.66	84.18
TJM	89.67	80.68	87.26	78.45	72.54	85.99	79.61	82.46	100.00	80.77	91.34	98.31	85.59
JDA	90.08	82.03	88.54	83.44	76.27	78.98	79.25	84.34	100.00	82.01	88.10	98.98	86.00
ARTL	93.22	85.42	91.08	87.98	85.08	89.17	86.29	92.07	100.00	86.82	91.75	96.95	90.49
SCA	89.46	85.42	87.90	78.81	75.93	85.35	74.80	86.12	100.00	78.09	89.98	98.64	85.88
JGSA	91.44	86.78	**93.63**	84.86	81.02	88.54	84.95	90.71	100.00	86.20	91.65	**99.66**	89.95
MEDA	92.59	88.14	89.81	**89.23**	83.05	89.81	87.89	**92.59**	100.00	**87.62**	**92.69**	99.32	91.06
MDTL	**93.42**	**92.20**	88.54	88.78	**90.51**	**90.45**	**88.16**	92.38	**100.00**	87.27	92.48	98.98	**91.93**

Table 3. Classification accuracies (%) on ImageCLEF-DA, MNIST-USPS and COIL20.

Dataset	ImageCLEF-DA							MNIST-USPS			COIL20		
Task	P→I	P→C	I→P	I→C	C→P	C→I	Avg.	M→U	U→M	Avg.	C1→C2	C2→C1	Avg.
TCA	79.17	84.00	76.67	91.83	71.33	85.83	81.47	56.28	51.05	53.67	88.47	85.83	87.15
GFK	78.33	85.33	74.67	91.50	69.00	82.67	80.25	67.22	46.45	56.84	72.50	74.17	73.34
CORAL	79.17	82.17	75.33	90.83	68.50	84.00	80.00	66.44	44.80	55.62	83.61	82.92	83.27
TJM	80.33	86.17	77.00	93.83	**77.50**	91.33	84.36	63.28	52.25	57.77	94.37	93.33	93.85
JDA	78.83	81.50	77.67	92.67	76.50	91.00	83.03	67.28	59.65	63.47	89.31	88.47	88.89
ARTL	88.67	91.00	77.50	**94.50**	75.67	91.83	86.53	88.78	67.70	78.24	90.69	89.72	90.21
JGSA	81.17	85.00	77.67	**94.50**	77.33	**92.17**	84.64	80.44	68.15	77.81	92.64	90.83	91.74
MEDA	**91.67**	**93.17**	76.17	94.17	74.33	90.33	86.64	89.50	66.90	78.20	94.03	91.25	92.64
MDTL	**91.67**	**93.17**	**78.50**	94.17	76.33	91.00	**87.47**	**89.56**	**78.25**	**83.90**	**97.36**	**95.14**	**96.25**

The two distribution adaptation approaches ignore the feature discriminability when reducing the distribution discrepancy. MDTL uses simultaneously subspace learning to reconstruct the original data, and discriminative class level adaptation to reduce the distribution discrepancy. This represents the most significant contribution of our work. The differences between these approaches and MDTL are summarized in Table 4.

4.4   Ablation Analysis

Data Reconstruction (DR) via Manifold Learning. To investigate the effect of manifold subspace learning based DR, we ran MDTL on several tasks with and without DR. The results are shown in Table 5. Clearly, DR via manifold subspace learning improved the classification performance on all tasks, indicating that reconstructing the original data can indeed alleviate feature distortion in distribution adaptation.

Discriminative Class Level Adaptation (DCLA). To evaluate the effectiveness of DCLA, we ran DCLA and joint distribution (marginal and conditional distributions) adaptation on several tasks. The results are shown in Table 6. Clearly, aligning class-level distributions results in better performance than aligning the joint distributions. The reason is that, compared with joint distribution adaptation, DCLA can improve simultaneously the feature transferability and discriminability.

Effect of Each Component. MDTL includes three components in learning the final classifier: SRM, DCLA, and manifold regularization (MR). To investigate the importance and necessity of each component, we performed experiments on 4 randomly selected transfer learning tasks. The results are shown in Fig. 2. Clearly, the three components are complementary, and all of them are significant and necessary in MDTL.

Table 4. Differences between ARTL, SCA, JGSA, MEDA and MDTL.

	ARTL	SCA	JGSA	MEDA	MDTL
Data Reconstruction	–	✓	–	✓	✓
Structural Risk Minimization	✓	–	–	✓	✓
Distribution Adaptation	✓	✓	✓	✓	✓
Domain Transferability	✓	✓	✓	✓	✓
Class Discriminability	–	–	–	–	✓
Manifold Regularization	✓	–	–	✓	✓
Closed Form Solution	✓	✓	✓	✓	✓

Table 5. Classification accuracies (%) with and without DR.

Task	$M \rightarrow U$	$U \rightarrow M$	$A \rightarrow C$	$D \rightarrow C$	$C1 \rightarrow C2$	$C2 \rightarrow C1$
Without DR	87.00	67.90	88.42	86.55	90.56	92.78
With DR	89.56	78.25	88.78	87.27	97.36	95.14
Improvement	2.56	10.35	0.36	0.72	6.80	2.36

Table 6. Classification accuracies (%) of JDA and DCLA.

Task	M → U	U → M	C → A	C → W	C1 → C2	C2 → C1
JDA	67.28	59.65	90.08	82.03	89.31	88.47
DCLA	68.94	58.95	90.50	82.37	93.33	89.58
Improvement	1.66	−0.7	0.42	0.35	4.02	1.11

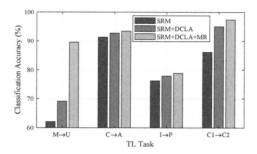

Fig. 2. Effect of each component in MDTL.

4.5   Parameter Sensitivity

This subsection analyzes the parameter sensitivity of MDTL. Due to page limit, we only report experimental results on C ↔ A, P ↔ I and C1 ↔ C2 tasks. Results on other tasks are similar.

Subspace Dimensionality q. The sensitivity of MDTL w.r.t. the manifold subspace dimensionality q, with $q \in \{10, 20, ..., 100\}$, is shown in Fig. 3(a). MDTL is robust w.r.t. q in $[30, 60]$.

Table 7. Running time (s) of JGSA, MEDA and MDTL.

Task	#samples × #features	JGSA	MEDA	MDTL
M → U	$2,800 \times 256$	**13.16**	21.15	33.99
C1 → C2	$1,440 \times 1,024$	12.50	**9.13**	9.20
P → I	$1,200 \times 2,048$	40.77	13.05	**10.81**
C → A	$2,081 \times 4,096$	164.15	59.74	**55.57**

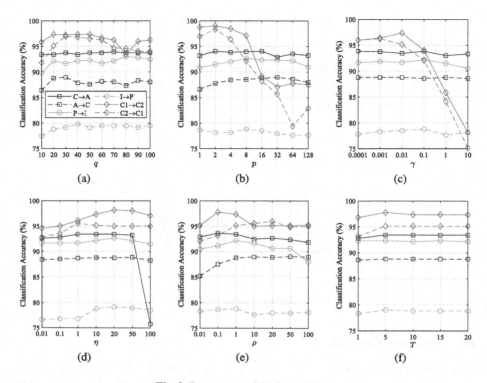

Fig. 3. Parameter sensitivity analysis.

Number of Nearest Neighbors p. The sensitivity of MDTL w.r.t. the number of nearest neighbors p, with $p \in \{1, 2, 4, ..., 128\}$, is shown in Fig. 3(b). MDTL is robust w.r.t. p in $[1, 8]$.

Regularization Parameters γ, η **and** ρ. We also investigated the sensitivity of MDTL w.r.t. regularization parameters γ, η and ρ. The results are shown in Fig. 3(c)–3(e). MDTL achieved good and stable performance when $\gamma \in [0.001, 0.1]$, $\eta \in [10, 50]$, and $\rho \in [0.1, 1]$.

Number of Iterations T. The number of iterations T determines the computational cost of MDTL. Figure 3(f) indicates that the classification accuracies increase as T increases, and converge when $T = 5$.

4.6 Computational Cost

We also compared the computational cost of MDTL with the top two baselines, JGSA and MEDA, on four random TL tasks. The platform was a laptop with Intel Core i7-8750H CPU and 8 G memory, running Windows 10×64 and Matlab 2019a. The results are shown in Table 7. The computational costs of JGSA, MEDA and MDTL are comparable. When the feature dimensionality is high, MDTL may even be slightly faster.

5   Conclusion

This paper proposes manifold discriminative transfer learning (MDTL), which first utilizes manifold subspace learning to reconstruct the original data in both domains, then simultaneously performs structural risk minimization, discriminative class level adaptation, and manifold regularization for transfer learning. More specifically, it minimizes the intra-class distribution discrepancy to improve the domain transferability, maximizes the inter-class distribution discrepancy to improve the class discriminability, and performs manifold regularization to maintain the geometrical structures of the data samples. MDTL is a traditional domain adaptation approach and it has a closed-form solution, so the computational cost is low. Extensive experiments showed that MDTL outperforms several state-of-the-art traditional transfer learning approaches.

Acknowledgments. This research was supported by the Hubei Province Funds for Distinguished Young Scholars under Grant 2020CFA050, the Key Laboratory of Brain Machine Collaborative Intelligence of Zhejiang Province under Grant 2020E10010-01, the Technology Innovation Project of Hubei Province of China under Grant 2019AEA171, the National Natural Science Foundation of China under Grants 61873321 and U1913207, and the International Science and Technology Cooperation Program of China under Grant 2017YFE0128300.

References

1. Borgwardt, K.M., Gretton, A., Rasch, M.J., Kriegel, H., Scholkopf, B., Smola, A.J.: Integrating structured biological data by kernel maximum mean discrepancy. In: Proceedings 14th International Conference on Intelligent Systems for Molecular Biology, Fortaleza, Brazil, pp. 49–57, August 2006
2. Chen, X., Wang, S., Long, M., Wang, J.: Transferability vs. discriminability: batch spectral penalization for adversarial domain adaptation. In: Proceedings 36th International Conference on Machine Learning, Long Beach, California, pp. 1081–1090, June 2019
3. Donahue, J., et al.: DeCAF: a deep convolutional activation feature for generic visual recognition. In: Proceedings International Conference on Machine Learning, Beijing, China, pp. 647–655, June 2014
4. Fernando, B., Habrard, A., Sebban, M., Tuytelaars, T.: Unsupervised visual domain adaptation using subspace alignment. In: Proceedings IEEE International Conference on Computer Vision, Sydney, Australia, pp. 2960–2967, December 2013
5. Ghifary, M., Balduzzi, D., Kleijn, W.B., Zhang, M.: Scatter component analysis: a unified framework for domain adaptation and domain generalization. IEEE Trans. Pattern Anal. Mach. Intell. **39**, 1414–1430 (2017)
6. Gong, B., Shi, Y., Sha, F., Grauman, K.: Geodesic flow kernel for unsupervised domain adaptation. In: Proceedings IEEE Conference on Computer Vision and Pattern Recognition, Providence, RI, pp. 2066–2073, June 2012
7. Gopalan, R., Li, R., Chellappa, R.: Domain adaptation for object recognition: an unsupervised approach. In: Proceedings IEEE International Conference on Computer Vision, Barcelona, Spain, pp. 999–1006, November 2011
8. Hull, J.J.: A database for handwritten text recognition research. IEEE Trans Pattern Anal. Mach. Intell. **16**(5), 550–554 (1994)
9. Kang, G., Jiang, L., Yang, Y., Hauptmann, A.G.: Contrastive adaptation network for unsupervised domain adaptation. In: Proceedings IEEE Conference on Computer Vision and Pattern Recognition, Long Beach, CA, pp. 4893–4902, June 2019

10. LeCun, Y., Bottou, L., Bengio, Y., Haffner, P.: Gradient-based learning applied to document recognition. Proc. IEEE **86**(11), 2278–2324 (1998)
11. Long, M., Wang, J., Ding, G., Pan, S.J., Yu, P.S.: Adaptation regularization: a general framework for transfer learning. IEEE Trans. Knowl. Data Eng. **26**(5), 1076–1089 (2014)
12. Long, M., Wang, J., Ding, G., Sun, J., Yu, P.S.: Transfer feature learning with joint distribution adaptation. In: Proceedings IEEE International Conference on Computer Vision, Sydney, Australia, pp. 2200–2207, December 2013
13. Long, M., Wang, J., Ding, G., Sun, J., Yu, P.S.: Transfer joint matching for unsupervised domain adaptation. In: Proceedings IEEE Conference on Computer Vision and Pattern Recognition, Columbus, OH, pp. 1410–1417, June 2014
14. Pan, S.J., Tsang, I.W., Kwok, J.T., Yang, Q.: Domain adaptation via transfer component analysis. IEEE Trans. Neural Netw. **22**(2), 199–210 (2011)
15. Pan, S.J., Yang, Q.: A survey on transfer learning. IEEE Trans. Knowl. Data Eng. **22**(10), 1345–1359 (2010)
16. Scholkopf, B., Smola, A.J.: Learning with Kernels: support vector machines, regularization, optimization, and beyond. Adaptive Computation and Machine Learning Series. MIT Press (2002)
17. Sun, B., Feng, J., Saenko, K.: Return of frustratingly easy domain adaptation. In: Proceedings 30th AAAI Conference on Artificial Intelligence, Phoenix, Arizona, pp. 2058–2065, February 2016
18. Wang, J., Chen, Y., Hao, S., Feng, W., Shen, Z.: Balanced distribution adaptation for transfer learning. In: Proceedings IEEE International Conference on Data Mining, New Orleans, LA, pp. 1129–1134, November 2017
19. Wang, J., Feng, W., Chen, Y., Yu, H., Huang, M., Yu, P.S.: Visual domain adaptation with manifold embedded distribution alignment. In: Proceedings ACM International Conference on Multimedia, Seoul, Republic of Korea, pp. 402–410, October 2018
20. Wu, D., Huang, J.: Affect estimation in 3D space using multi-task active learning for regression. IEEE Trans. Affect. Comput. (2021 in Press)
21. Wu, D., Xu, Y., Lu, B.L.: Transfer learning for EEG-based brain-computer interfaces: a review of progress made since 2016. IEEE Trans. Cogn. Dev. Syst. (2020, in Press)
22. Zhang, J., Li, W., Ogunbona, P.: Joint geometrical and statistical alignment for visual domain adaptation. In: Proceedings IEEE Conference on Computer Vision and Pattern Recognition, Honolulu, HI, pp. 5150–5158, July 2017
23. Zhang, W., Deng, L., Zhang, L., Wu, D.: Overcoming negative transfer: a survey (2020). https://arxiv.org/abs/2009.00909
24. Zhang, W., Wu, D.: Discriminative joint probability maximum mean discrepancy (DJP-MMD) for domain adaptation. In: Proceedings International Joint Conference on Neural Networks, Glasgow, UK, pp. 1–8, July 2020
25. Zhang, Y., Tang, H., Jia, K., Tan, M.: Domain-symmetric networks for adversarial domain adaptation. In: Proceedings IEEE Conference on Computer Vision and Pattern Recognition, Long Beach, CA, pp. 5031–5040, June 2019

Training-Free Multi-objective Evolutionary Neural Architecture Search via Neural Tangent Kernel and Number of Linear Regions

Tu Do[1,2] and Ngoc Hoang Luong[1,2(✉)]

[1] University of Information Technology, Ho Chi Minh City, Vietnam
18521578@gm.uit.edu.vn, hoangln@uit.edu.vn
[2] Vietnam National University, Ho Chi Minh City, Vietnam

Abstract. A newly introduced training-free neural architecture search (TE-NAS) framework suggests that candidate network architectures can be ranked via a combined metric of expressivity and trainability. Expressivity is measured by the number of linear regions in the input space that can be divided by a network. Trainability is assessed based on the condition number of the neural tangent kernel (NTK), which affects the convergence rate of training a network with gradient descent. These two measurements have been found to be correlated with network test accuracy. High-performance architectures can thus be searched for without incurring the intensive cost of network training as in a typical NAS run. In this paper, we suggest that TE-NAS can be incorporated with a multi-objective evolutionary algorithm (MOEA), in which expressivity and trainability are kept separate as two different objectives rather than being combined. We also add the minimization of floating-point operations (FLOPs) as the third objective to be optimized simultaneously. On NAS-Bench-101 and NAS-Bench-201 benchmarks, our approach achieves excellent efficiency in finding Pareto fronts of a wide range of architectures exhibiting optimal trade-offs among network expressivity, trainability, and complexity. Network architectures obtained by our approach on CIFAR-10 also show high transferability on CIFAR-100 and ImageNet.

Keywords: Neural architecture search · Multi-objective optimization · Evolutionary computation · Deep learning · Neural tangent kernels

1 Introduction

Neural Architecture Search (NAS) [32] problems involve automating the design process of high-performance neural network architecture. Most NAS algorithms, however, require some prohibitively high computational cost due to intensive network evaluations. For each candidate architecture encountered during an NAS run, we need to carry out many training epochs to obtain a set of proper network

© Springer Nature Switzerland AG 2021
T. Mantoro et al. (Eds.): ICONIP 2021, LNCS 13109, pp. 335–347, 2021.
https://doi.org/10.1007/978-3-030-92270-2_29

parameter values that can then be evaluated for validation/test error. Several earlier works on NAS reported thousands of GPU days for their search. There have been many efforts to speed up this training process, such as using surrogate models to predict network performance or employing a supernet to share its weights with candidate architectures (i.e., subnetworks sampled from the supernet). However, these methods might suffer from other problems that affect the search process, e.g., difficulty in training surrogate models and supernets, or the poor correlation between the performance of a supernet and its subnetworks [2].

The performance of a neural network is influenced by its *expressivity* and *trainability* [12, 28]. To achieve high accuracy, the network architecture needs to allow functions that are complex enough to be encoded. At the same time, the network architecture needs to enable its parameter to be trained efficiently by gradient descent. Chen et al. [2] introduced a training-free neural architecture search (TE-NAS) framework that evaluates candidate architectures in terms of expressivity (via the number of linear regions $R_\mathcal{N}$) and trainability (via the condition number of the NTK $\kappa_\mathcal{N}$). Both indicators $R_\mathcal{N}$ and $\kappa_\mathcal{N}$ can be computed using network parameter values at initialization and thus do not involve any training. Optimizing a combined function of $R_\mathcal{N}$ and $\kappa_\mathcal{N}$, instead of directly optimizing network accuracy, TE-NAS has been shown to obtain top-performance architectures within just four hours using a 1080Ti GPU [2].

In practice, accuracy is not the sole optimization objective when solving NAS problems. We need to consider multiple, possibly conflicting, objectives, such as predictive performance, model size, or network efficiency, where a single *utopian* architecture that achieves top accuracy while incurring little computational complexity does not exist. Instead, there is a Pareto set of different architectures that together represent the optimal trade-offs among competing objectives. We can then investigate these architectures and select the one exhibiting the suitable trade-off. It is more insightful to perform architecture selection based on a Pareto set of diverse alternatives than relying on a single solution that optimizes a single objective. Due to this inherent multi-objective nature, several recent works on NAS have been preferably handled by multi-objective evolutionary algorithms (MOEAs), which maintain and evolve a population of candidate solutions to approximate the set of Pareto-optimal architectures [6, 22]. Here, we propose a training-free multi-objective evolutionary neural architecture search (TF-MOENAS) approach, which integrates the two training-free network performance indicators $R_\mathcal{N}$ and $\kappa_\mathcal{N}$ with an MOEA. More specifically, we solve a tri-objective NAS problem formulation: optimizing network expressivity via the number of linear regions $R_\mathcal{N}$, network trainability via the condition number of the NTK $\kappa_\mathcal{N}$, and network complexity via the number of floating-point operations (FLOPs). In our approach, $R_\mathcal{N}$ and $\kappa_\mathcal{N}$ are kept as two separate objectives instead of being combined into a single indicator as in TE-NAS [2].

Architectures discovered with NAS using validation accuracy as network performance indicator tend to be overfitted to the specific dataset employed during the search [6]. In this paper, we also compare the transferability of our training-free MOENAS approach (which optimizes $R_\mathcal{N}$, $\kappa_\mathcal{N}$, and FLOPs) against a traditional MOENAS approach (which optimizes validation error and FLOPs). NAS runs are performed on CIFAR-10 and the resulting architectures are then transferred to CIFAR-100 and ImageNet16-120 datasets. Our codes are available at https://github.com/MinhTuDo/TF-MOENAS.

2 Backgrounds

2.1 Number of Linear Regions

Piecewise linear functions (e.g., ReLU) are often used as activation functions in deep neural networks. A ReLU network, which can be seen as a composition of piecewise linear functions, thus also represents a piecewise linear function. Such a network can divide its input space into *linear regions*, and within each region, the function encoded by the network is affine [29]. Figure 1 visualizes linear regions of a ReLU network.

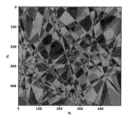

Fig. 1. 2D input space divided by a ReLU network into pieces of linear regions. (Generated using the code at https://colab.research.google.com/github/CIS-522/)

The expressivity of a neural network $\mathcal{N}$ can be indicated by the number of linear regions $R_{\mathcal{N}}$ it partitions the input space into [15]. Xiong et al. [29] defined a linear region $\boldsymbol{R}(\boldsymbol{P};\boldsymbol{\theta})$ as a set of input data points $\boldsymbol{x}$'s that, when being fed forward through a ReLU network $\mathcal{N}$ with parameter values $\boldsymbol{\theta}$, yield the same *activation pattern* $\boldsymbol{P}(z) \in \{-1, +1\}$ at neurons z's in $\mathcal{N}$. We have

$$\boldsymbol{R}(\boldsymbol{P};\boldsymbol{\theta}) = \{\boldsymbol{x} \in \texttt{Input Space} : z(\boldsymbol{x};\boldsymbol{\theta}) \cdot \boldsymbol{P}(z) > 0, \forall z \in \mathcal{N}\} \qquad (1)$$

where $z(\boldsymbol{x};\boldsymbol{\theta})$ denotes the pre-activation value of input $\boldsymbol{x}$ at a neuron z. A linear region associated with an activation pattern $\boldsymbol{P}$ exists if there is at least one input $\boldsymbol{x}$ satisfies $z(\boldsymbol{x};\boldsymbol{\theta}) \cdot \boldsymbol{P}(z) > 0, \forall z \in \mathcal{N}$. The number of linear regions $R_{\mathcal{N},\theta}$ at $\boldsymbol{\theta}$ can thus be computed by counting the number of unique activation patterns.

$$R_{\mathcal{N},\theta} = |\{\boldsymbol{R}(\boldsymbol{P};\boldsymbol{\theta}) : \boldsymbol{R}(\boldsymbol{P};\boldsymbol{\theta}) \neq \emptyset \text{ for some activation pattern } \boldsymbol{P}\}| \qquad (2)$$

While linear regions $\boldsymbol{R}(\boldsymbol{P};\boldsymbol{\theta})$'s depend on $\boldsymbol{\theta}$, Hanin et al. [15] showed that, during and after training, the number of linear regions $R_{\mathcal{N}}$ stays roughly similar to its value at initialization. Therefore, without gradient descent training to obtain the final $\boldsymbol{\theta}$, we can still approximate $R_{\mathcal{N}}$ using the initialized network parameter values. More precisely, $R_{\mathcal{N}} \approx \mathbb{E}_{\boldsymbol{\theta} \sim W}[R_{\mathcal{N}, \boldsymbol{\theta}}]$, where W is an initialization distribution such as the Kaiming He initialization [16].

Regarding NAS problems, Chen et al. [2] showed that $R_{\mathcal{N}}$ is positively correlated to the network test accuracy, with Kendall's τ correlation of 0.5. Therefore, maximizing $R_{\mathcal{N}}$ would encourage finding high-performance architectures.

2.2   Condition Number of Neural Tangent Kernel

While expressivity indicates how complex the set of functions a network architecture can represent theoretically, trainability (or learnability) involves how effectively the gradient descent algorithm can obtain a proper set of values for network parameters $\boldsymbol{\theta}$ such that a loss function $L(\boldsymbol{\theta}) = \sum_{i=1}^{n} e(f(\boldsymbol{\theta}, \boldsymbol{x}_i), y_i)$ is minimized [2]. $f(\boldsymbol{\theta}, \boldsymbol{x}_i)$ is the output of the neural network with parameters $\boldsymbol{\theta}$ regarding training input $\boldsymbol{x}_i$, y_i is the corresponding target label, and $e(,)$ is an error metric. At each iteration t of gradient descent, network parameters $\boldsymbol{\theta}$ are updated as $\boldsymbol{\theta}_{t+1} = \boldsymbol{\theta}_t - \eta \frac{\partial L(\boldsymbol{\theta}_t)}{\partial \boldsymbol{\theta}_t}$ where $\eta > 0$ is the learning rate.

The training dynamics of a neural network involves the sequence $\{L(\boldsymbol{\theta}_t)\}_{t=0}^{\infty}$. Recent works (e.g., Jacot et al. [17], Du et al. [10,11], Arora et al. [1], Hanin & Nica [14]) demonstrated that such training dynamics can be studied via a notion called the *neural tangent kernel* (NTK)

$$\hat{\Theta}(\boldsymbol{x}_i, \boldsymbol{x}_j) = \left\langle \frac{\partial f(\boldsymbol{\theta}, \boldsymbol{x}_i)}{\partial \boldsymbol{\theta}}, \frac{\partial f(\boldsymbol{\theta}, \boldsymbol{x}_j)}{\partial \boldsymbol{\theta}} \right\rangle \tag{3}$$

where $\partial f(\boldsymbol{\theta}, \boldsymbol{x})/\partial \boldsymbol{\theta}$ is the gradient of the network output of the training input $\boldsymbol{x}$ with respect to its parameters $\boldsymbol{\theta}$. Let $\hat{\Theta}_{\text{train}}$ be the matrix in which the (i, j) element represents the NTK between training inputs $\boldsymbol{x}_i, \boldsymbol{x}_j$. Several works (e.g., Du et al. [10,11], Lee et al. [19], Xiao et al. [28]) showed that when the neural network is over-parameterized enough (or the width of each layer is wide enough) the convergence rate of $\{L(\boldsymbol{\theta}_t)\}_{t=0}^{\infty}$ is governed by the spectrum of $\hat{\Theta}_{\text{train}}$ and the learning rate η:

$$\mathbb{E}_t[f(\boldsymbol{\theta}, \boldsymbol{x}_i)] = (\mathbf{I} - e^{-\eta \lambda_k t}) y_i \tag{4}$$

where λ_k is an eigenvalue of $\hat{\Theta}_{\text{train}}$. As t tends to infinity, $\mathbb{E}[f(\boldsymbol{\theta}, \boldsymbol{x}_i)] = y_i$. The condition number $\kappa_{\mathcal{N}}$ of the NTK $\hat{\Theta}_{\text{train}}$ is defined as

$$\kappa_{\mathcal{N}} = \frac{\lambda_{\max}}{\lambda_{\min}} \tag{5}$$

where $\lambda_{\max}, \lambda_{\min}$ are the largest and the smallest eigenvalues of the NTK $\hat{\Theta}_{\text{train}}$. Chen et al.[2] showed that as the maximum learning rate can be scaled up to $\eta \sim 2/\lambda_{\max}$, regarding Eq. 4, the convergence would occur at rate $1/\kappa_{\mathcal{N}}$. If $\kappa_{\mathcal{N}}$ diverges, the network $\mathcal{N}$ becomes untrainable and would have a poor predictive performance. The condition numbers of the NTK $\kappa_{\mathcal{N}}$ of architectures in NAS-Bench-201 [9] are shown to be negatively correlated (Kendall's τ correlation coefficient of -0.42) with their test accuracy values [2]. Minimizing $\kappa_{\mathcal{N}}$ during an NAS process would thus guide the search toward architectures that are efficiently trainable by gradient descent, and potentially have high performance.

3   Proposed Approach

NAS typically requires multiple conflicting objectives to be considered simultaneously. Desirable architectures are the ones that not only yield high accuracy performance but also incur acceptable computing costs (few floating-point operations (FLOPs)) or contain small numbers of trainable parameters. A single *utopian* architecture optimizing all these competing objectives does not exist. We aim to obtain a Pareto-optimal *front* of architectures that show the best possible *trade-offs* among the objectives. These architectures are all optimal in the sense that if we want to improve one objective (e.g., their accuracy), at least one other objective (e.g., their number of parameters) must be diminished, and vice versa. Figure 2 (left) shows an example bi-objective Pareto front of NAS involving the minimization of validation error and FLOPs. However, we here do not use validation error, but employ the number of linear regions $R_{\mathcal{N}}$ and the condition number of the NTK $\kappa_{\mathcal{N}}$ as network performance indicators.

Chen et al. [2] also employed $R_{\mathcal{N}}$ and $\kappa_{\mathcal{N}}$ in their TE-NAS framework. However, $R_{\mathcal{N}}$ and $\kappa_{\mathcal{N}}$ values are not used directly, but are first converted into relative rankings s_R and s_{κ} among sampled architectures each time, and then added together to obtain a combined measurement $s = s_{\kappa} + s_R$. It is challenging to combine $R_{\mathcal{N}}$ and $\kappa_{\mathcal{N}}$ values into a single performance indicator (to measure both expressivity and trainability) because their magnitudes are very different and their ranges are unknown *a priori*. In this paper, using a multi-objective optimization approach, we propose that $R_{\mathcal{N}}$ and $\kappa_{\mathcal{N}}$ values can be straightforwardly handled by keeping expressivity and trainability as two separate objectives to be optimized simultaneously with FLOPs as the third objective. Figure 2 (right) shows an example tri-objective Pareto front of NAS involving the maximization of $R_{\mathcal{N}}$ (network expressivity), the minimization of $\kappa_{\mathcal{N}}$ (network trainability) and the minimization of FLOPs (network complexity).

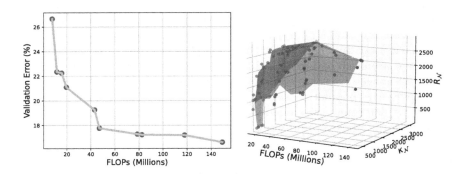

Fig. 2. Example Pareto fronts of Pareto-optimal network architectures.

The nondominated sorting genetic algorithm II (NSGA-II [5]) is often employed to solve multi-objective optimization problems. NSGA-II starts with an initial population of randomly generated individuals (i.e., each initial individual is a random candidate architecture). In every generation, a selection operator is used to form a selection set of promising individuals from the current population. Variation operators, such as crossover and mutation, are performed over the selection set to create offspring individuals (i.e., new candidate architectures). The current population and the offspring population are then merged into a pool, over which a Pareto non-dominated sorting procedure is run to partition the individuals into different non-domination ranks. Individuals from the rank 0, forming the so-called *non-dominated front*, are not dominated by any other individuals. An individual x^1 is said to Pareto dominate another individual x^2 if x^1 is not worse than x^2 in all objectives and x^1 is better than x^2 in at least one objective. Better-ranking individuals from the pool are then selected into the new population for the next generation. NSGA-II has been used to tackle multi-objective NAS problems in [6,22]. We here also employ NSGA-II as the optimization algorithm for our training-free multi-objective evolutionary neural architecture search (TF-MOENAS) approach.

4   Experiments and Results

We evaluate our proposed approach on two NAS benchmarks: NAS-Bench-101 [30] and NAS-Bench-201 [7,9]. Experiments are conducted on a Ubuntu 18.04 desktop workstation equipped with an NVIDIA GTX 1070 8GB GPU.

For the sake of computation time, when computing $R_\mathcal{N}$, we set the initial channel number as 1. When computing $\kappa_\mathcal{N}$ and FLOPs, the initial channel number is set as 16. Other network configurations remain unchanged. By definition, both $\kappa_\mathcal{N}$ and $R_\mathcal{N}$ are expected values [2], which require a large number of samples to be taken and would thus incur exorbitant computation time. Instead, every time $\kappa_\mathcal{N}$ and $R_\mathcal{N}$ of a candidate architecture are computed, we repeat the calculation three times and keep the worst values. Specifically, the lowest $R_\mathcal{N}$ value and the largest $\kappa_\mathcal{N}$ value are retained. In each time, 3000 images are sampled from the training set to obtain the activation patterns from all ReLU layers for estimating $R_\mathcal{N}$, and a mini-batch of 16 images are used to compute $\hat{\Theta}(x_i, x_j)$ for estimating $\kappa_\mathcal{N}$. In TE-NAS [2], the average values over three calculations of $R_\mathcal{N}$ and $\kappa_\mathcal{N}$ are used, and in each time, 5000 images are sampled for estimating $R_\mathcal{N}$ and 32 images are sampled for estimating $\kappa_\mathcal{N}$.

The NSGA-II population size is set as 50. We perform binary tournament selection to determine which individuals would survive each generation. We employ uniform crossover and integer mutation with probabilities 0.9 and 0.02, respectively. To preserve all non-dominated solutions obtained throughout a run, an elitist archive is implemented. Elitist members dominated by new solutions are removed from the archive. Solutions in the archive at the end of an NSGA-II run thus form an *approximation front* (i.e., a non-dominated front that approximates the Pareto-optimal front) and is regarded as the optimization result of that run.

We evaluate the approximation front $\mathcal{S}$ against the Pareto front $\mathcal{P}_\mathcal{F}$, i.e., the optimal trade-off front that minimizes two objectives: FLOPs and test error.

Because the front $\mathcal{S}$ has three objectives (i.e., FLOPs, $\kappa_{\mathcal{N}}$, and $R_{\mathcal{N}}$), $\mathcal{S}$ first needs to be projected into a bi-objective front of FLOPs and test error (which can be easily looked up in the corresponding benchmark). The quality of the converted front is then measured using the Inverted General Distance (IGD) as

$$IGD(\mathcal{S}, \mathcal{P}_{\mathcal{F}}) = \frac{1}{|\mathcal{P}_{\mathcal{F}}|} \sum_{p \in \mathcal{P}_{\mathcal{F}}} \min_{q \in \mathcal{S}} \{d(p, q)\} \qquad (6)$$

where $d(.,.)$ denotes the Euclidean distance between candidate architectures in the objective space. The IGD metric thus computes the average distance over each solution in the optimal front $\mathcal{P}_{\mathcal{F}}$ to the nearest solution in the approximation front $\mathcal{S}$. When comparing the resulting fronts $\mathcal{S}_1, \mathcal{S}_2$ of two MOEAs, the algorithm that has a smaller IGD value is considered to have better performance.

We compare our proposed approach against a baseline, which is the traditional MOENAS problem formulation that minimizes FLOPs and validation error. On each benchmark, for each approach, we perform 30 independent runs with different random seeds. During an NAS run, the optimization objectives are FLOPs, $\kappa_{\mathcal{N}}$, and $R_{\mathcal{N}}$ (for our approach) or FLOPs and validation error (for the baseline), but after the NAS run, the IGD score of the resulting approximation front is computed based on FLOPs and test error.

4.1   Results on NAS-Bench-101

NAS-Bench-101 [30] consists of approximately 423,000 unique architectures generated from a cell-structured search space. The benchmark contains a database of validation error values and test error values of each architecture evaluated on CIFAR-10 at four epoch lengths $\{4, 12, 36, 108\}$. When scoring IGD, we employ test error values at epoch 108, but during a baseline MOENAS run, we query the benchmark database for validation error values at epoch 36. This is to simulate NAS in real-world situations, where it would be too costly to accurately evaluate all candidate architectures by training them until convergence (which might require hundreds of epochs for each architecture). Instead, the performance of candidate architectures during an NAS run is typically estimated by training a few epochs, and only resulting architectures obtained at the end of the run are fully trained and evaluated. In each baseline run, the benchmark database is allowed to be queried 5,000 times for validation error values, i.e., equivalent to 5,000 architectures being considered. Our TF-MOENAS runs do not involve any training, and we only evaluate architectures based on FLOPs, $\kappa_{\mathcal{N}}$ and $R_{\mathcal{N}}$.

Figure 3 shows that while both approaches obtain good approximation fronts at the end of an NAS run (i.e., IGD values very close to 0), our TF-MOENAS exhibits a slightly better IGD convergence. Regarding the computation cost in terms of GPU hours, TF-MOENAS can reach the same IGD results with considerably fewer resources. The number of GPU hours of the baseline approach is taken from the benchmark with the reported hardware [30]. To evaluate the accuracy performance, at the end of each TF-MOENAS run, we extract the architecture with the best test error (by querying the benchmark database for

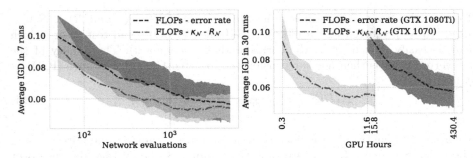

Fig. 3. Left: Average IGD score with respect to number of evaluations. Right: Average run-times on NAS-Bench-101.

test error values at epoch 108). Table 1 compares our best test accuracy averaged over 30 independent runs with other state-of-the-art NAS methods. It can be seen that both MOENAS approaches obtain the best (average) accuracy performance, and TF-MOENAS is just slightly worse than the baseline MOENAS.

Table 1. Test accuracy comparison with other NAS methods on NAS-Bench-101 [31]

Method	CIFAR-10	Search method	Multi-objective
ENAS [25]	91.83 ± 0.59	RL	
NAO [23]	92.59 ± 0.59	Gradient	
FBNET [27]	92.29 ± 1.25	Gradient	
DARTS [21]	92.21 ± 0.61	Gradient	
SPOS [13]	89.85 ± 3.80	Gradient	
FairNAS [4]	91.10 ± 1.84	Gradient	
NASWOT [24]	91.77 ± 0.05	Training-free	
REA [26]	$\mathbf{93.87 \pm 0.22}$	evolution	
MOENAS (baseline)	93.64 ± 0.16	Evolution	✓
TF-MOENAS (ours)	93.49 ± 0.20	Training-free, evolution	✓
Optimal test accuracy	94.31	(in the benchmark)	

4.2 Results on NAS-Bench-201

NAS-Bench-201 [9] consists of 16,625 candidate architectures constructed from 4 nodes and 5 operations, which are conv1x1, conv5x5, conv3x3, avgpool, zeroize. However, there are only 6,466 unique architectures; we set the computation budget as 3,000 network evaluations for each NAS run here. The baseline MOENAS would query the benchmark database for validation error values at epoch 25 during an NAS run. All IGD scores (for both baseline and training-free MOENAS) are computed based on test error values at epoch 200 and FLOPs. Note that

IGD scores are computed separately from all NAS runs *a posteriori* to assess the convergence of MOENAS algorithms in approaching Pareto-optimal fronts.

Figure 4 shows that our approach has much better IGD convergence performance than the baseline. While TF-MOENAS optimizes over FLOPs, $\kappa_{\mathcal{N}}$ and $R_{\mathcal{N}}$, the resulting architectures are found to form approximation fronts (FLOPs versus test error) that are better than those of the baseline MOENAS, which optimizes over FLOPs and validation error. In terms of GPU hours, Fig. 4 also indicates that our training-free approach is more efficient than the baseline approach, which requires training for estimating network performance via validation error. Table 2 compares our best test accuracy obtained from the final elitist archive at each NAS run and then averaged over 30 independent runs against the results reported by other state-of-the-art NAS methods. When optimizing network expressivity and trainability, TF-MOENAS is able to locate top-performing architectures in the NAS-Bench-201 search space more efficiently.

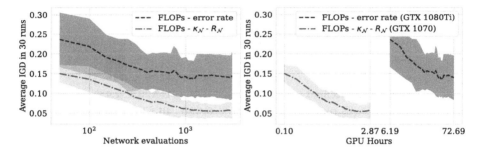

Fig. 4. Left: Average IGD score with respect to number of evaluations. Right: Average run-times on NAS-Bench-201.

Table 2. Test accuracy comparison with other NAS methods on NAS-Bench-201

Method	CIFAR-10	Search method	Multi-objective
RSPS [20]	87.66 ± 1.69	Random	
ENAS [25]	54.30	RL	
DARTS [21]	54.30	Gradient	
GDAS [8]	93.61 ± 0.09	Gradient	
NASWOT [24]	92.81 ± 0.99	Training-free	
TENAS [2]	93.9 ± 0.47	Training-free	
REA [26]	93.92 ± 0.3	evolution	
MOENAS (baseline)	93.611 ± 0.43	Evolution	✓
TF-MOENAS (ours)	**94.16 ± 0.22**	Training-free, evolution	✓
Optimal test accuracy	94.37	(in the benchmark)	

4.3   Results on Architecture Transfer

Network architectures obtained from an NAS run optimizing over validation accuracy on a dataset might be overfitted to that specific dataset and would suffer from worse performance when being employed on other different datasets [6]. Considering multiple datasets simultaneously during an NAS run is not a viable option because every candidate architecture would need separate training and evaluating for validation accuracy on each dataset, incurring prohibitively more computational cost. To verify the transferability of TF-MOENAS, we take the architectures obtained at the end of our NAS runs on CIFAR-10 (in the above NAS-Bench-201 experiments) and evaluate them on CIFAR-100 and ImageNet16-120. Table 3 shows that optimizing over $\kappa_{\mathcal{N}}$ and $\mathcal{R}_{\mathcal{N}}$ as in training-free MOENAS obtains architectures of higher test accuracy than optimizing over validation accuracy as in the baseline MOENAS when being transferred to different datasets.

Table 3. Test accuracy comparison of architecture transfer in NAS-Bench-201

Method	CIFAR-100 [18]	ImageNet16-120 [3]
REA [26]	71.84 ± 0.99	45.54 ± 1.03
NASWOT [24]	69.48 ± 1.70	43.10 ± 3.16
MOENAS (baseline)	70.517 ± 1.605	44.290 ± 1.485
TF-MOENAS (ours)	**72.749 ± 0.630**	**46.611 ± 0.458**
Optimal test accuracy	73.51	47.31

Table 4 compares the average IGD scores over 30 runs between the approximation fronts obtained by the two MOENAS methods on CIFAR-10 and the optimal fronts of CIFAR-100 and ImageNet16-120 in the NAS-Bench-201 benchmark. Note that, for computing IGD scores with the optimal fronts, all approximation fronts (optimizing FLOPS and validation error as in the baseline MOENAS, or optimizing FLOPS, $\kappa_{\mathcal{N}}$, and $\mathcal{R}_{\mathcal{N}}$ as in TF-MOENAS) are projected into FLOPs and test error. Table 4 and Fig. 5 indicate that the resulting fronts of TF-MOENAS are much closer to the optimal fronts than those of the baseline approach. The condition number of the NTK and the number of linear regions are thus effective tools for NAS frameworks to address overfitting problems when searching for top-performing architectures on only one specific dataset.

Table 4. Comparison of average IGD scores across 30 runs between approximation fronts found on CIFAR-10 and optimal fronts of other datasets in NAS-Bench-201

Method	CIFAR-100 [18]	ImageNet16-120 [3]
MOENAS (baseline)	0.165 ± 0.064	0.173 ± 0.054
TF-MOENAS (ours)	**0.073 ± 0.019**	**0.072 ± 0.026**

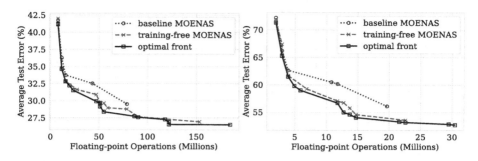

Fig. 5. Approximation fronts on CIFAR-100 (left) and ImageNet16-120 (right)

5 Conclusion

In this paper, we investigate the practicability of two neural network performance indicators, the number of linear regions $R_{\mathcal{N}}$ and the condition number of NTK $\kappa_{\mathcal{N}}$, in solving multi-objective NAS problems. Experiments on the NAS-Bench-101 and NAS-Bench-201 benchmarks show that, by optimizing these two indicators together with the number of floating-point operations (FLOPs), we could achieve competitive results compared to standard NAS methods, which involve the intensive computational cost of actual training candidate architectures to evaluate their validation accuracy. The vast reduction of computational cost has a tremendous meaning for real-world applications of the NAS, i.e., optimization algorithms could sample more candidate architectures to explore the search space better, thereby increasing the possibility of finding top-performing architectures. Furthermore, our architecture transferability experiment results indicate that optimizing $R_{\mathcal{N}}$ and $\kappa_{\mathcal{N}}$ help to address overfitting problems better than optimizing validation accuracy when carrying out NAS on a single dataset.

Acknowledgements. This research is funded by Vietnam National University HoChiMinh City (VNU-HCM) under grant number DSC2021-26-06.

References

1. Arora, S., Du, S.S., Hu, W., Li, Z., Salakhutdinov, R., Wang, R.: On exact computation with an infinitely wide neural net. In: NeurIPS, pp. 8139–8148 (2019)
2. Chen, W., Gong, X., Wang, Z.: Neural architecture search on ImageNet in four GPU hours: a theoretically inspired perspective. In: ICLR (2021)
3. Chrabaszcz, P., Loshchilov, I., Hutter, F.: A downsampled variant of ImageNet as an alternative to the CIFAR datasets. CoRR abs/1707.08819 (2017)
4. Chu, X., Zhang, B., Xu, R., Li, J.: FairNAS: rethinking evaluation fairness of weight sharing neural architecture search. CoRR abs/1907.01845 (2019)
5. Deb, K., Pratap, A., Agarwal, S., Meyarivan, T.: A fast and elitist multiobjective genetic algorithm: NSGA-II. IEEE Trans. Evol. Comput. **6**(2), 182–197 (2002)

6. Do, T., Luong, N.H.: Insightful and practical multi-objective convolutional neural network architecture search with evolutionary algorithms. In: Fujita, H., Selamat, A., Lin, J.C.-W., Ali, M. (eds.) IEA/AIE 2021. LNCS (LNAI), vol. 12798, pp. 473–479. Springer, Cham (2021). https://doi.org/10.1007/978-3-030-79457-6_40

7. Dong, X., Liu, L., Musial, K., Gabrys, B.: NATS-bench: benchmarking NAS algorithms for architecture topology and size. IEEE Trans. Pattern Anal. Mach. Intell., 1 (2021)

8. Dong, X., Yang, Y.: Searching for a robust neural architecture in four GPU hours. In: CVPR, pp. 1761–1770 (2019)

9. Dong, X., Yang, Y.: NAS-Bench-201: extending the scope of reproducible neural architecture search. In: ICLR (2020)

10. Du, S.S., Lee, J.D., Li, H., Wang, L., Zhai, X.: Gradient descent finds global minima of deep neural networks. In: ICML, vol. 97, pp. 1675–1685 (2019)

11. Du, S.S., Zhai, X., Póczos, B., Singh, A.: Gradient descent provably optimizes over-parameterized neural networks. In: ICLR (2019)

12. Giryes, R., Sapiro, G., Bronstein, A.M.: Deep neural networks with random gaussian weights: a universal classification strategy? IEEE Trans. Signal Process. **64**(13), 3444–3457 (2016)

13. Guo, Z., et al.: Single path one-shot neural architecture search with uniform sampling. In: Vedaldi, A., Bischof, H., Brox, T., Frahm, J.-M. (eds.) ECCV 2020. LNCS, vol. 12361, pp. 544–560. Springer, Cham (2020). https://doi.org/10.1007/978-3-030-58517-4_32

14. Hanin, B., Nica, M.: Finite depth and width corrections to the neural tangent kernel. In: ICLR (2020)

15. Hanin, B., Rolnick, D.: Complexity of linear regions in deep networks. In: ICML, vol. 97, pp. 2596–2604 (2019)

16. He, K., Zhang, X., Ren, S., Sun, J.: Deep residual learning for image recognition. In: CVPR, pp. 770–778 (2016)

17. Jacot, A., Hongler, C., Gabriel, F.: Neural tangent kernel: convergence and generalization in neural networks. In: NeurIPS, pp. 8580–8589 (2018)

18. Krizhevsky, A.: Learning multiple layers of features from tiny images. Technical report, University of Toronto, Toronto (2009)

19. Lee, J., et al.: Wide neural networks of any depth evolve as linear models under gradient descent. In: NeurIPS, pp. 8570–8581 (2019)

20. Li, L., Talwalkar, A.: Random search and reproducibility for neural architecture search. In: UAI, pp. 367–377 (2019)

21. Liu, H., Simonyan, K., Yang, Y.: DARTS: differentiable architecture search. In: ICLR (2019)

22. Lu, Z., et al.: NSGA-Net: neural architecture search using multi-objective genetic algorithm. In: GECCO, pp. 419–427 (2019)

23. Luo, R., Tian, F., Qin, T., Chen, E., Liu, T.: Neural architecture optimization. In: NeurIPS, pp. 7827–7838 (2018)

24. Mellor, J., Turner, J., Storkey, A.J., Crowley, E.J.: Neural architecture search without training. In: ICML, vol. 139, pp. 7588–7598 (2021)

25. Pham, H., Guan, M.Y., Zoph, B., Le, Q.V., Dean, J.: Efficient neural architecture search via parameter sharing. In: ICML, vol. 80, pp. 4092–4101 (2018)

26. Real, E., Aggarwal, A., Huang, Y., Le, Q.V.: Regularized evolution for image classifier architecture search. In: AAAI, pp. 4780–4789 (2019)

27. Wu, B., et al.: FBNet: hardware-aware efficient convnet design via differentiable neural architecture search. In: CVPR, pp. 10734–10742 (2019)

28. Xiao, L., Pennington, J., Schoenholz, S.: Disentangling trainability and generalization in deep neural networks. In: ICML, vol. 119, pp. 10462–10472 (2020)
29. Xiong, H., Huang, L., Yu, M., Liu, L., Zhu, F., Shao, L.: On the number of linear regions of convolutional neural networks. In: ICML, vol. 119, pp. 10514–10523 (2020)
30. Ying, C., Klein, A., Christiansen, E., Real, E., Murphy, K., Hutter, F.: NAS-Bench-101: towards reproducible neural architecture search. In: ICML, vol. 97, pp. 7105–7114 (2019)
31. Yu, K., Sciuto, C., Jaggi, M., Musat, C., Salzmann, M.: Evaluating the search phase of neural architecture search. In: ICLR (2020)
32. Zoph, B., Le, Q.V.: Neural architecture search with reinforcement learning. In: International Conference on Learning Representations (ICLR) (2017)

Neural Network Pruning via Genetic Wavelet Channel Search

Saijun Gong[1,2], Lin Chen[2(✉)], and Zhicheng Dong[1(✉)]

[1] School of Information Science and Technology, Tibet University,
Tibet 850011, China
[2] Chongqing Institute of Green and Intelligent Technology,
Chinese Academy of Sciences, Chongqing 400714, China
chenlin@cigit.ac.cn

Abstract. Neural network pruning has been commonly adopted for alleviating the computational cost of resource-limited devices. Neural architecture search (NAS) is an efficient approach to facilitate neural network compression. However, most existing NAS-based algorithms focus on searching the best sub-network architecture at the layer level, while ignoring the channels that contain richer information. Meanwhile, fixing one pruning rate could not be suitable for all the layers with various features. In this paper, we present a novel NAS-based network pruning strategy called genetic wavelet channel search (GWCS) model, which automatically prunes the network at the channel level with two-fold ideas: (a) Each channel in the whole network is gene coded and pruned adaptively in multiple stages with dynamic genetic selection, and (b) the fitness function of channels is carefully designed by calculating the similarity between the pre-trained network and pruned network using wavelet decomposition to discover the most representative and discriminative channels. Extensive experiments are conducted on CIFAR-10 and CIFAR-100 using ResNet series networks. The proposed GWCS method outperforms other existing pruning approaches in terms of accuracy with a higher compression rate. Our source code is freely available at https://github.com/MGongsj/Network-pruning.git.

Keywords: Network channel pruning · Neural architecture search · Wavelet transform · Genetic algorithm

1 Introduction

The emergence of deep Convolutional Neural Networks (CNNs) has aroused remarkable interest and achieved considerable success in a variety of image analysis. However, the number of parameters in deep CNNs (*e.g.*, Res-Net50 [3]) are over hundreds of megabytes, and thus it requires huge amounts of energy and computer power to processing the input data. This inevitably provides a vital challenge to the deployment of network models on some tiny devices having insufficient computational resources, hindering the extension of deep CNNs.

© Springer Nature Switzerland AG 2021
T. Mantoro et al. (Eds.): ICONIP 2021, LNCS 13109, pp. 348–358, 2021.
https://doi.org/10.1007/978-3-030-92270-2_30

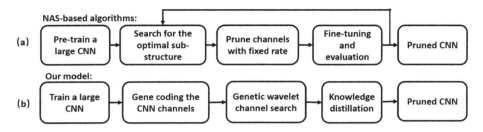

Fig. 1. Flow-charts of previous NAS-based pruning algorithms and the proposed model: (a) Previous NAS-based pruning methods consider the searching the optimal sub-network with the fixed pruning rate. (b) Our pruning method turns the network pruning into a GA-based combinatorial optimization procedure to obtain dynamical pruning results.

To address this issue, various works on neural network compression [13, 15, 18] are developed to slim the large deep CNNs for execution speedup without significantly decreasing the model performance. The most intuitive way to compress the CNNs is to slim unnecessary weights from a pre-trained model [12]. Recently, neural architecture search (NAS) approaches [1, 6, 8, 17, 19, 21], which can automate the compression with a structured pruning, becomes a very active research topic in this area.

A typical pipeline for NAS-based network pruning [2, 7, 10], as indicated in Fig. 1(a), is searching the potential sub-network. Then the final model can be achieved by fine-tuning the pruned CNNs. However, in the procedure of searching, prevailing NAS-based algorithms usually try to discover the optimal architecture of the network at coarser structures (i.e., find the best layer or block), while paying less attention to the specific channels. In addition, the candidate sub-networks are pruned according to various evaluation criteria with the predefined depth or width under a fixed pruning rate, e.g., pruning 50% of the network directly. As mentioned in [22], various features and functions exist in different CNN layers. Thus using one fixed pruning rate for the entire network could wrongly reduce the truly important (or discriminative) channels [10, 20], leading to a performance decrease of the compressed model.

To this end, we aim to address the drawbacks of both strategies, and proposed a novel NAS-based compression method based on the genetic algorithm (GA) named as genetic wavelet channel search (GWCS). As shown in Fig. 1(b), in contrast to current methods, we model the network compression at the channel level as a GA optimization problem. Specifically, GWCS prunes the channels in the pre-trained network adaptively and dramatically in multiple stages based on the carefully designed fitness function. To find the most representative and discriminative channels, we further propose a wavelet feature fusion method to aggregate the different candidates of channels based on the wavelet transform to guide the pruning process.

Our main contributions are in three aspects:

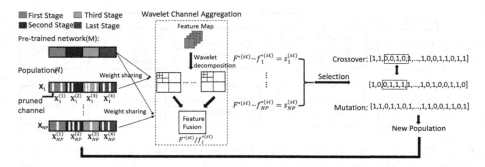

Fig. 2. Details of GWCS. We divide the network searching process into four stages with different number of iterations.

(1) We model the network pruning problem as a genetic search procedure and propose a GWCS strategy that automatically prunes the network at the channel level.
(2) We designed a fitness function based on the wavelet features aggregation method to select the best-pruned networks.
(3) We tested our pruning method by pruning ResNet series networks on CIFAR-10 and CIFAR-100 datasets. The experimental results indicate that the proposed GWCS method achieves more compressed CNNs models and performs better than tested network pruning methods in terms of pruning accuracy.

2   Methodology

2.1   Overview of the Pruning Architecture

We assume that we have a pre-trained network M with N channels. Our goal is to get a lightweight model O with guaranteed accuracy by reducing the number of channels in M. The proposed model contains three steps as follows: (1) Pre-training an unpruned, over-parameterized network (original network M). (2) Applying GWCS to remove the unimportant channels at each layer in M. (3) Fine-tuning the slimmed network to obtain the final pruned model O. The most important of the search process is effectively and adaptively removing the channels in M while maintaining accuracy. The reason for using GA algorithm is that the channel level coding of the neural network is very consistent with the characteristics of individuals in the algorithm. The codes of the network can be adaptively updated according to the fitness selection.

2.2   GWCS-Based Channel Pruning

Initialization. At the initial stage of pruning, the channels of pre-trained network M are encoded into a set of binary codes to produce the initial population

$\mathcal{A}$, and the pruned candidate $\mathbf{X}_i \in \mathcal{A}$, which denotes the ith instance defined as follows:

$$\mathbf{X}_i = \{c_i^1, c_i^2, ..., c_i^N\} \tag{1}$$

where $i \in \{1, 2, ..., NP\}$, NP is the total number of individuals, and c_i^j denotes the jth code in $\mathbf{X}_i$. $c_i^j = 0$ or 1 means that the channel is reduced or not. All the instances are assembled together to form the population set $\mathcal{A}$, which is given by

$$\mathcal{A} = \begin{cases} \mathbf{X}_1 = [1, 0, 1, 1, 0, 1, 0, ..., 0, 1, 1, 1] \\ \mathbf{X}_2 = [0, 1, 1, 0, 0, 1, 0, ..., 0, 1, 0, 1] \\ \vdots \\ \underbrace{\mathbf{X}_{NP} = [1, 0, 1, 0, 0, 1, 1, ..., 0, 0, 1, 0]}_{channels'\ code} \end{cases} \tag{2}$$

Hierarchical Search Space. Obviously, assessment of a huge number of channels in $\mathbf{X}_i$ is a non-trivial thing. This study proposed a novel Gradual Genetic Search (GGS) strategy to search the corresponding important channels in multiple stages. In fact, a great part of the success of the CNN is due to its hierarchical architecture from concrete to abstract: the lower-level of layers extract concrete features like color or edge, while the higher-level of layers can obtain the abstract semantic concept about the categories.

The proposed GGS is also consistent with this argument. We divide the entire CNNs model into multiple segments according to the down-sampling size, e.g., as illustrated in Fig. 2, the whole network M can be segmented into four sub-networks (i.e., each sub-network is a searching subspace) with multi-scales feature maps in sizes of $[4\times, 8\times, 16\times, 32\times]$. Thus the codes of $\mathbf{X}_i$ are divided into:

$$\mathbf{X}_i = [\mathbf{X}_i^{(1)}, \mathbf{X}_i^{(2)}, \mathbf{X}_i^{(3)}, \mathbf{X}_i^{(4)}] \tag{3}$$

where the sub-network $\mathbf{X}_i^{(st)} \in \mathbf{X}_i$ and $st \in [1, 4]$. Note that the number of maximum iterations $T^{(st)}$ is different as the size of searching space of $\mathbf{X}_i^{(st)}$ changes at each stage st .

Crossover. The crossover operator is applied to generate the new code of a possible network. It selects two chromosomes, called parents (e.g., random selected instances $\mathbf{X}_{r1}^{(st)}$ and $\mathbf{X}_{r2}^{(st)}$, from the mating pool and exchanges channel bits at specific points to generate a new generation offspring. We choose multi-point crossing strategy to create unique individuals owing to its popularity and robustness, which are defined as follows:

$$\mathbf{X}_{cr}^{(st)} = \mathbf{G} \circ (\mathbf{X}_{r1}^{(st)}) + |1 - \mathbf{G}| \circ (\mathbf{X}_{r2}^{(st)}) \tag{4}$$

where $\mathbf{G}$ contains the random logics (a set of 0 and 1) exchanging the binary codes between $\mathbf{X}_{r1}^{(st)}$ and $\mathbf{X}_{r2}^{(st)}$. After crossover, a new trial network $\mathbf{X}_{cr}^{(st)}$ can be produced with a certain percentage of codes exchanged between two individuals.

Mutation. To strengthen the diversity (various in channel codes of $\mathbf{X}$) of the individuals and the ability to avoid local optima. We choose the binary mutation strategy for randomly flipping the bits in $\mathbf{X}_{cr}^{(st)}$ to set the new individual $\mathbf{X}_m^{(st)}$, which is given by

$$\mathbf{X}_m^{(st)} = H(\mathbf{X}_{cr}^{(st)}) \tag{5}$$

where $H(\cdot)$ denote that $p_m\%$ of binary codes in $\mathbf{X}_{cr}^{(st)}$ are flipped to form various pruned CNN models with random removed channels.

Selection. Next, we evaluate the pruned networks at each stage $\mathbf{X}_i^{(st)}$, and then we employ the Roulette Wheel Algorithm to select top K individuals with the highest finesses for next iteration with the probability of P. Here, for the st_{th} stage, the $P_i^{(st)}$ of each $\mathbf{X}_i^{(st)}$ can be calculated as follows:

$$P_i^{(st)} = \frac{Ft(\mathbf{X}_i^{(st)})}{\sum_{i=1}^{NP} Ft(\mathbf{X}_i^{(st)})} \tag{6}$$

where the $Ft(\cdot)$ means the fitness function that chooses a potential pruned network for surviving. The detail of $Ft(\cdot)$ will be introduced below.

Fitness Selection via Wavelet Channel Aggregation. The fitness function $Ft(\cdot)$ for selecting the best individuals is based on the feature similarity between the unpruned network and pruned ones. More specifically, the feature maps of candidate networks with the various reduced channels can be firstly aggregated as the same vector using wavelet channel aggregation (WCA), then the cosine similarity of the wavelet fusion features can be generated to evaluate the pruned network to identify the most representative and discriminative channels.

Wavelet transform is able to obtain the different scale components of the input image. The mathematical function of the wavelet decomposition is given in Eq. (7).

$$F^{\star} = \frac{1}{\sqrt{a}} \int_{-\infty}^{+\infty} F * \psi(\frac{t-\tau}{a}) dt \tag{7}$$

where a controls the stretching, and τ controls the translation. F is the feature maps of the input network.

The Haar wavelet is chosen as the wavelet basis in our method as it is simple and effective. The formula for the fusion of transformed wavelet feature maps is described in Eq. (8).

$$F^* = \max(F_{HH}^{\star}) \oplus Avg(F_{LL}^{\star}) \tag{8}$$

where HH is high frequency information and LL contains low frequency information. The fusion rule is to add the maximum of HH and the total average of LL with element-wise addition $\oplus$. Different from common aggregate functions (i.e., global average-pooling (GAP) and global max-pooling (GMP)), various frequency components of the wavelet can capture more specific information [14], thus the fusion of high- and low-frequency features can improve the similarity estimation.

We aggregate the feature maps of both pre-trained sub-networks $F^{(st)}$ and the pruned sub-networks $f_i^{(st)}$ using Eq. (8), results are denoted as $F^{*(st)}$ and $f_i^{*(st)}$.

The calculation of cosine similarity $s_i^{(st)}$ is defined in Eq. (9).

$$s_i^{(st)} = \frac{F^{*(st)} \cdot f_i^{*(st)}}{\| F^{*(st)} \| \| f_i^{*(st)} \|} \tag{9}$$

After maximum iterations, the best-pruned network O with the maximal fitness value can be elected from the final population set $\mathcal{A}$.

2.3 Knowledge Distillation

The fine-tuning process is necessary for achieving robust pruning results [2]. This study employs the knowledge distillation (KD) [5] for accuracy recovery of the pruned network. The softmax output is approximately a one-hot vector for classification. There is no other relationship to learn except the final result. So the output can be softened by Eq. (10).

$$q_k = \frac{exp(z_k/T)}{\Sigma_j exp(z_j/T)} \tag{10}$$

where T is temperature. When the temperature T tends to zero, the softmax output will converge to a one-hot vector. Hard targets can only reflect a final result, while soft targets can reflect the relationship between various possible results. Since our compressed result O comes from the pre-trained model M, the pruned small network takes the soft targets output from M as a loss, and it will learn the feature distribution of M. Similar to previous work [2], the middle layer transfer of KD is adopted in our GWCS, defined in Eq. (11).

$$L_T = \rho_1 L_1 + ... + \rho_n L_n + (1 - \rho_1 - ... - \rho_n)L_{hard} \tag{11}$$

where L_T is total loss of KD, n is total number of the training stages and ρ_j is the proportional of loss L_j at jth stage, $j \in [1, n]$.

3 Experiment

3.1 Datasets and Settings

Datasets. We tested our model on CIFAR-10 and CIFAR-100 datasets. CIFAR-10 contains ten categories, per class has 5000 images and 1000 images for training and test, respectively. CIFAR-100 has 100 classes, with 500 training images and 100 verification images for each class. All the images in CIFAR-10 and CIFAR-100 are RGB pictures with the size of 32×32, including various objects in the real world.

Table 1. Network pruning results based on the ResNets. The values with bold and underlined are produced by the best and the second-best methods, respectively. "Prune Acc(Drop)" means pruning accuracy with the performance drop, "FLOPs (PR)" means FLOPs (pruning ratio).

Depth	Method	CIFAR-10		CIFAR-100	
		Prune Acc(Drop)	FLOPs(PR)	Prune Acc(Drop)	FLOPs(PR)
32	FPGM [4]	92.31%(0.32%)	4.03E7(41.5%)	68.52%(1.25%)	4.03E7(41.5%)
	ManiDP [16]	92.15%(0.48%)	2.56E7(63.2%)	–	–
	TAS [2]	92.92%(0.96%)	3.78E7(45.4%)	71.74%(-1.12%)	3.80E7(45.0%)
	Ours	92.97%(**0.11%**)	**2.23E7(73.6%)**	71.88%(-0.79%)	**2.29E7(73.1%)**
56	FPGM [4]	93.49%(**0.42%**)	5.94E7(52.6%)	69.66%(1.75%)	5.94E7((52.6%)
	HRank [9]	93.52%(0.94%)	6.58E7(37.9%)	–	–
	JST [11]	93.68%(0.73%)	6.32E7(49.7%)	70.63%(2.26%)	6.72E7(51.1%)
	TAS [2]	93.69%(0.77%)	5.95E7(52.7%)	72.25%(0.93%)	6.12E7(51.3%)
	Ours	93.75%(0.48%)	**5.05E7(60.3%)**	73.75%(-0.61%)	**5.12E7(59.7%)**
110	FPGM [4]	93.85%(-0.17%)	1.21E8(52.3%)	72.55%(1.59%)	1.21E8(52.3%)
	JST [11]	94.22%(0.61%)	**1.08E8(58.0%)**	72.26%(2.16%)	1.08E8(58.0%)
	TAS [2]	94.33%(0.64%)	1.19E8(53.0%)	73.16%(1.90%)	1.20E8(52.6%)
	Ours	94.78%(0.25%)	1.12E8(56.0%)	75.00%(**0.05%**)	**1.07E8(58.2%)**

Training Details. ResNets are trained using SGD for 150 and 200 epochs on CIFAR-10 and CIFAR-100, respectively. The learning rate initialized as 0.1 and will be gradually reduced during the training with the batch size of 128. The proposed and tested models are all implemented and tested with Pytorch on dual NVIDIA GTX1080ti. Finally, a total of 25 epochs of KD is applied to refine the pruned network.

In network pruning process, the pre-trained network is taken as the input for the GWCS algorithm. We randomly initialize 50 individuals derived from the original network. Each individual represents a diverse pruned network. We prune the ResNets in four search stages (the maximal number of iterations is 10, 10, 5, 5, respectively). The top 20 individuals having the higher finesses are selected to participate in mutation and crossover. Finally, a new population with randomly selected 30 individuals can be produced for the next round of searching.

3.2   Compare to the State-of-the-Art

Several existing state-of-the-art pruning methods are chosen to test and compare with our model, including kinds of filter or channel pruning approaches based on FPGM [4], TAS [2], HRank [9], JST [11] and ManiDP [16]. The pruning results are shown in Table 1.

Among all the pruned ResNet models, our method achieves the highest pruning rate so that it produces the minimum FLOps. Especially on ResNet-32, we can see that our model prunes 73.6% and 73.1% of the neurons on CIFAR-10 and CIFAR-100, which could be only half of the size of FPGM. For the aspect of pruning accuracy, our model performs best with such lower FLOPs among

Table 2. Comparison of GGS and OGS on CIFAR-10 using ResNet-32.

Method	Prune Acc(Drop)	FLOPs(PR)
OGS	92.19%(0.89%)	2.31E7(72.7%)
GGS	**92.97%(0.11%)**	**2.23E7(73.6%)**

Table 3. Comparison of channel fusion schemes on CIFAR-10.

Fitness function	Prune Acc(Drop)	FLOPs(PR)
GAP	91.41%(1.67%)	**2.15E7(75.4%)**
GMP	91.25%(1.83%)	2.21E7(74.0%)
GAP+GMP	91.84%(1.24%)	2.23E7(73.5%)
WCA	**92.97%(0.11%)**	2.23E7(73.6%)

the tested approaches. Turning to CIFAR-100, the improvement of our model is more obvious than that on CIFAR-10. For example, our model obtains 73.75% accuracy by pruning ResNet-56 on CIFAR-100, which outperforms TAS (the second-best pruning method) by 1.5% of the drop of prune accuracy.

We also need to point out that, on CIFAR-100, our pruned model is the only one among the tested models, whose prediction accuracy rates are even better than both pre-trained ResNet-32 and ReNet-56. All of these confirm that our model can elaborately search the high quality of the channels from the original networks.

3.3   Ablation Experiments

Ablation Study on GGS. The GGS algorithm is proposed to gradually obtain the pruned object network rather than inspecting all channels as a whole (i.e., Overall Genetic Search (OGS)). In this ablation experiment, ResNet-32 model is pruned to evaluate the different search strategies on CIFAR-10. Since the GGS divided the pre-trained network into 4 stages (the number of iterations for each stage is 10, 10, 5, and 5, respectively), the total iterations is 30 for the entire pruning process, which is also set as the maximum iterations of OGS method for a fair comparison. KD is applied to refine the pruned networks obtained by both search strategies.

As is shown in Table 2, GGS performs better than OGS in terms of prune accuracy with about 0.9% computation reduction. This demonstrates that the proposed GGS algorithm could be beneficial for pruning the large CNNs.

Ablation Study on WCA. In our GWCS algorithm, the selection of channels is based on the feature maps that are fused by wavelet transform in the fitness function. WCA is proposed to verify the importance of channels by using the wavelet transform to aggregate the feature maps with various channel sizes.

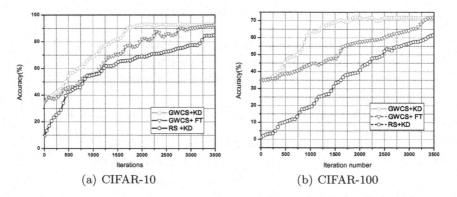

(a) CIFAR-10   (b) CIFAR-100

Fig. 3. Visualization of pruning results with different fine-tuning strategies.

In the ablation experiment, the fitness function based on the WCA is compared with traditional feature aggregation operations, including GAP, GMP, and their combination GAP+GMP. As we can observe from Table 3, the fitness function using WCA obtains the highest test accuracy values but with slightly increases FLOPs. As discussed in [14], both the high- and low-frequency information are important for feature representation of CNNs, thus, the fusion of all these frequencies components in our method could be more suitable for evaluating the valuable channels.

Ablation Study on KD. In our GWCS algorithm, the final step is to enhance the pruned network by KD. In this experiment, we tested our GWCS+KD with two other fine-tuning strategies using ResNet-32., including: (1) GWCS with the common fine-tune (FT) method by retraining the pruned network (called GWCS+FT). (2) A straightforward slimming method by randomly reducing the channels with fine-tuning using KD (called RS+KD).

As we can see in Fig. 3, GWCS+KD and GWCS+FT can reach almost the same test accuracy but with different speeds of convergence, as the model with KD can reach the highest accuracy in about 2000 iterations, while the model with FT method can obtain a close result after 3500 iterations. Meanwhile, the test accuracy of RS+KD is always lower than GWCS+KD and GWCS+FT at each iteration. These indicate that the GWCS+KD is more efficient than GWCS+FT and RS+KD in pruning and can reduce unnecessary channels.

4   Conclusion

We propose a GWCS method for automatic network compression at the channel level. The proposed GWCS method can adaptively search the most informative channels from the pre-trained network via multi-stage GA optimization. Furthermore, we also propose a WCA function to fuse the features maps with various channel sizes for evaluating the pruned network. On multiple benchmark datasets, our method can achieve a more compact CNNs model with less accuracy loss than existing pruning methods tested in experiments. We will employ our pruning method to compress the other CNNs applied in object localization or GAN in future work.

References

1. Chen, Y., et al.: Contrastive neural architecture search with neural architecture comparators. In: Proceedings of the IEEE/CVF Conference on Computer Vision and Pattern Recognition (CVPR), pp. 9502–9511, June 2021
2. Dong, X., Yang, Y.: Network pruning via transformable architecture search. In: The Conference on Neural Information Processing Systems (NeurIPS), pp. 760–771 (2019)
3. He, K., Zhang, X., Ren, S., Sun, J.: Deep residual learning for image recognition. In: Proceedings of the IEEE Conference on Computer Vision and Pattern Recognition (CVPR), pp. 770–778 (2016)
4. He, Y., Liu, P., Wang, Z., Hu, Z., Yang, Y.: Filter pruning via geometric median for deep convolutional neural networks acceleration. In: Proceedings of the IEEE/CVF Conference on Computer Vision and Pattern Recognition, pp. 4340–4349 (2019)
5. Hinton, G., Vinyals, O., Dean, J.: Distilling the knowledge in a neural network. arXiv preprint arXiv:1503.02531 (2015)
6. Jia, F., Wang, X., Guan, J., Li, H., Qiu, C., Qi, S.: Arank: toward specific model pruning via advantage rank for multiple salient objects detection. Image Vision Comput. **111**, 104192 (2021). https://doi.org/10.1016/j.imavis.2021.104192
7. Yu, J., Huang, T.: Network slimming by slimmable networks: towards one-shot architecture search for channel numbers. CoRR abs/1903.11728, http://arxiv.org/abs/1903.11728 (2019)
8. Liang, T., Wang, Y., Tang, Z., Hu, G., Ling, H.: Opanas: one-shot path aggregation network architecture search for object detection. In: Proceedings of the IEEE/CVF Conference on Computer Vision and Pattern Recognition (CVPR), pp. 10195–10203, June 2021
9. Lin, M., et al.: Hrank: filter pruning using high-rank feature map. In: 2020 IEEE/CVF Conference on Computer Vision and Pattern Recognition (CVPR), pp. 1529–1538 (2020)
10. Liu, Z., et al.: Metapruning: meta learning for automatic neural network channel pruning. In: Proceedings of the IEEE/CVF International Conference on Computer Vision (CVPR), pp. 3296–3305 (2019)
11. Lu, X., Huang, H., Dong, W., Li, X., Shi, G.: Beyond network pruning: a joint search-and-training approach. In: Twenty-Ninth International Joint Conference on Artificial Intelligence and Seventeenth Pacific Rim International Conference on Artificial Intelligence IJCAI-PRICAI-20, pp. 2583–2590 (2020)
12. Luo, J.H., Wu, J.: An entropy-based pruning method for cnn compression. arXiv preprint arXiv:1706.05791 (2017)
13. Park, S., Lee, J., Mo, S., Shin, J.: Lookahead: a far-sighted alternative of magnitude-based pruning. In: Proceedings of the International Conference on Learning Representations (ICLR) (2020). https://openreview.net/forum?id=ryl3ygHYDB

14. Qin, Z., Zhang, P., Wu, F., Li, X.: Fcanet: frequency channel attention networks. arXiv preprint arXiv:2012.11879 (2020)
15. Renda, A., Frankle, J., Carbin, M.: Comparing rewinding and fine-tuning in neural network pruning. In: Proceedings of the International Conference on Learning Representations (ICLR) (2020). https://openreview.net/forum?id=S1gSj0NKvB
16. Tang, Y., Wang, Y., Xu, Y., Deng, Y., Xu, C., Tao, D., Xu, C.: Manifold regularized dynamic network pruning. In: Proceedings of the IEEE/CVF Conference on Computer Vision and Pattern Recognition (CVPR), pp. 5018–5028, June 2021
17. Wang, D., Li, M., Gong, C., Chandra, V.: Attentivenas: improving neural architecture search via attentive sampling. In: Proceedings of the IEEE/CVF Conference on Computer Vision and Pattern Recognition (CVPR), pp. 6418–6427, June 2021
18. Xu, X., Feng, W., Jiang, Y., Xie, X., Sun, Z., Deng, Z.H.: Dynamically pruned message passing networks for large-scale knowledge graph reasoning. In: Proceedings of the International Conference on Learning Representations (ICLR) (2020). https://openreview.net/forum?id=rkeuAhVKvB
19. Xu, Y., et al.: Renas: relativistic evaluation of neural architecture search. In: Proceedings of the IEEE/CVF Conference on Computer Vision and Pattern Recognition (CVPR), pp. 4411–4420, June 2021
20. Yang, T.J., et al.: Netadapt: Pplatform-aware neural network adaptation for mobile applications. In: Proceedings of the European Conference on Computer Vision (ECCV), pp. 285–300, September 2018
21. Yang, Z., et al.: Hournas: extremely fast neural architecture search through an hourglass lens. In: Proceedings of the IEEE/CVF Conference on Computer Vision and Pattern Recognition (CVPR), pp. 10896–10906, June 2021
22. Yosinski, J., Clune, J., Fuchs, T., Lipson, H.: Understanding neural networks through deep visualization. In: In ICML Workshop on Deep Learning (2015)

Binary Label-Aware Transfer Learning for Cross-Domain Slot Filling

Gaoshuo Liu[1], Shenggen Ju[1(✉)], and Yu Chen[2]

[1] College of Computer Science, Sichuan University, Chengdu 610065, China
{gaoshuoliu,jsg}@scu.edu.cn
[2] College of Science and Technology, SichuanMinzu College, Kangding 626001, China

Abstract. Slot filling plays an important role in spoken language under-standing. Slot prediction need to use a lot of labeled data in a specific field for training, but in the real situation, there is often a lack of training data in a specific field, which is the biggest problem in cross-domain slot filling. In the previous works on cross-domain slot filling, many methods train their model through the sufficient source domain data, so that the model could predict the slot type in the unknown domain. However, previous approaches do not make good use of the small amount of labeled target domain data. In this paper, we proposed a cross-domain slot filling model with label-aware transfer learning. First, we classify words into three categories based on their BIO labels, calculate the MMD (maximum mean discrepancy) by computing hidden representations between two domains with the same ground truth label, which can participate in the loss function calculation, so that the model can better capture the overall characteristics of the target domain. Experimental results show that our proposed models significantly outperform other methods on average F1-score.

Keywords: Slot filling · Transfer learning · Cross domain

1 Introduction

SLU is a part of Natural Language Under-standing, including domain classification, slot filling and intention detection. Slot filling model is used to detect task-related slot types in utterances from users in a specific field, such as flight or movie ticket information. The traditional supervised slot filling models [1] have achieved great achievements. However, in the traditional supervised slot filling model, in order to achieve accurate prediction results, each field requires a lot of labeled samples, which makes it difficult to add new tasks, and it is expensive and time consuming at the same time. In order to deal with the lack of labeled data in specific domains in the real environment, domain adaptive methods [2] have achieved great success, which used the information learned from the source domain and as few labeled samples as possible from the target domain to make the model transferable. The biggest challenge of cross-domain slot filling is the

© Springer Nature Switzerland AG 2021
T. Mantoro et al. (Eds.): ICONIP 2021, LNCS 13109, pp. 359–368, 2021.
https://doi.org/10.1007/978-3-030-92270-2_31

lack of labeled samples in the target domain. The previous methods relied on the information of the slot description or example value, and led the user input to the new slot by capturing the semantic relationship between the user's input and the slot description. Based on the ways of slot filling, these methods are divided into two categories: one-stage slot filling and two-stage slot filling. In one-stage slot filling [3], the model fills the slots separately for each slot type, first, the model will generate a word-level representation, and then the representation will interact with the slot description information in the semantic space, get the final prediction based on the fusion feature, where the prediction results for each slot type are independent [4]. [2] proposed a framework for slot filling, which can achieve zero-shot adaptation. The framework first generates a word-level representation, then connects the word and the representation of the slot description, and makes predictions based on this connected feature. Since the slot entity will show differences in different fields, it is difficult for the framework to obtain a complete slot entity, at the same time, such forecasting method will also bring about the problem of multiple forecasts. For example, the label of the word "tune" may be recognized as "B" by the two slot types "music item" and "playlist" at the same time, which will cause errors in the final prediction. Aiming at the difficulty of entity recognition and multiple prediction problems caused by one-stage slot filling, [5] proposed a two-stage slot filling framework, which first used the BIO three-way classifier to determine whether the word is an entity, and then generates the entity representation, after that, the entity representation will be compared with the slot description for similarity, and finally the specific slot type of this entity could be obtained, this method effectively avoids the problem of multiple predictions, at the same time, the author also proposes a template regularization method, which can dephrenize the words of the slot entity in the sentence into labels of different slot types, and then generate correct and wrong templates to standardize the representation of the sentence, it makes the model to cluster the representations of semantically similar words into similar vector spaces to improve the transferability of the model. The previous work mainly worked on the direction of unsupervised domain adaptation. In fact, in the cross-domain sequence labeling task, labeled samples in target domains are very important. Inspired by [6], this paper proposes a Label-aware Transfer Learning module applied to cross-domain slot filling, which is applied to the first stage of a two-stage slot filling. We calculate the MMD [7] of the hidden state corresponding to the same BIO tag in the source domain and the target domain, and then add it to the loss function calculation to reduce the difference between domains and improve the generalization effect of the model.

2 Methodology

This section first introduces the overall architecture of the two-stage slot filling model, and then introduces the implementation of the Label-aware Transfer Learning module.

2.1   Neural Architectures

Aiming at the problem of not effectively using a small number of labeled samples in the target domain in cross-domain slot filling, this paper proposes a BLCS model that incorporates a label migration method. Our model is based on two-stage, the first stage is three-class classification of BIO tags with tag migration, which can identify the entities in the sentence, as shown in Fig. 1, in the second stage, the feature representation of the entity identified in the first stage is obtained through another Bi-LSTM network, and then the feature representation is compared with the slot description matrix to establish the final slot type.

Bi-LSTM Encoder. RNN [8] can get the context information in a whole sentence, so it can be widely used in natural language processing tasks. One of its variants is LSTM [9], which includes input gates and forget gates to capture long-term and short-term dependencies. The other is Bi-LSTM [10], which can process the input sequence separately from the forward and backward, and can better consider the context information, so it is very suitable for sequence labeling tasks.

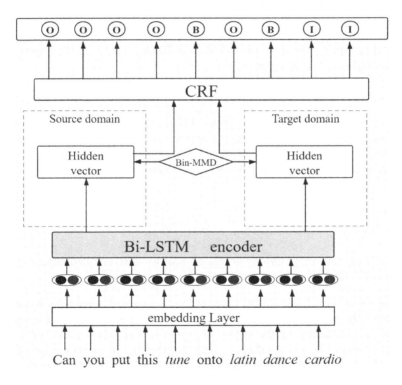

Fig. 1. BIO classifier.

Maximum Mean Discrepancy. MMD is often used to measure the distance between two distributions, and is a commonly used loss function in transfer learning. It is defined as follows:

$$MMD(F, p, q) = \sup_{||f||_{H \leq l}} E_{x \sim p}[f(x)] - E_{y \sim q}[f(y)] \tag{1}$$

"F" is defined as the unit ball in the Reproducing Kernel Hilbert Space (RKHS) [11], only when $p = q$, $MMD(F, p, q) = 0$. The given two sets of sample sets X and Y are independent and identically distributed in p and q on the data space X.

$$MMD(X, Y) = \left\| \frac{1}{m} \sum_{i=1}^{m} \phi(xi) - \frac{1}{n} \sum_{i=1}^{n} \phi(yi) \right\|_{\mathcal{H}} \tag{2}$$

Conditional Random Field. Conditional Random Field (CRF) [12] is a sequence marking algorithm, it accepts an input sequence X and outputs the target sequence Y, CRF is extensive Applied to the joint labeling task of terms in the sequence [8].

BIO Classification Module. The input of the model is n terms w=[w1,w2, ...,wn] in the user sentence, E represents the embedding layer of the input sentence [13], the source domain and the target domain share the embedding layer, by inputting the terms into Bi-LSTM, the hidden state of the input can be obtained, expressed as follows:

$$[h_1, h_2, \ldots, h_n] = BiLSTM(E(\boldsymbol{w})) \tag{3}$$

After getting the hidden state, for each label category $y \in y_v$, where y_v is the set of matching entity category labels in the source domain and the target domain, the model can calculate The sum of squares of the hidden states of the source domain and target domain samples with the same entity category label.

$$MMD^2 \left(R_y^s, R_y^t \right) = \frac{1}{(N_y^s)^2} \sum_{i,j=1}^{N_y^s} k \left(\mathbf{h}_i^s, \mathbf{h}_j^s \right) + \frac{1}{(N_y^t)^2} \sum_{i,j=1}^{N_y^t} k \left(\mathbf{h}_i^t, \mathbf{h}_j^t \right) - \frac{2}{N_y^s N_y^t} \sum_{i,j=1}^{N_y^s, N_y^t} k \left(\mathbf{h}_i^s, \mathbf{h}_j^t \right) \tag{4}$$

R_y^s and R_y^t are the set of hidden states $\mathbf{h}_i^t$ and $\mathbf{h}_j^t$ with corresponding numbers of N_y^s and N_y^t in the source domain and the target domain.

As shown in Fig. 2, the model divides the hidden states corresponding to the entries with the same BIO label in the source domain and the target domain into the same set according to the entity tag type of the entry. First, we calculate the MMD loss of the source domain and target domain of each type of hidden state set, and then add the MMD loss of all sets in the three categories of BIO to get the final total label migration loss, which is recorded as $L_{Bin-MMD}$:

$$L_{\text{Bin-MMD}} = \sum_{y \in \Sigma_v} \text{MMD}^2 \left(R_y^s, R_y^t \right) \tag{5}$$

Target domain

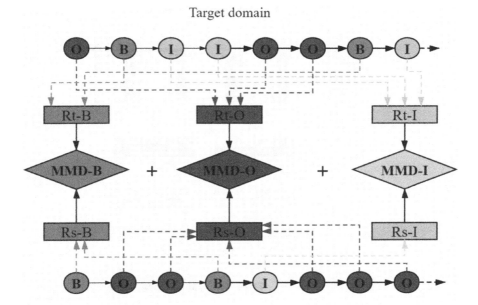

Source domain

Fig. 2. The framework of Bin-MMD.

When the label transfer loss is applied to Bi-LSTM learning, the distribution of instances with the same label in different domains will be closer, so that the model can better migrate from the source domain to the target domain. After that, the generated hidden state h_1, h_2, ..., h_n will be sent to the CRF layer to generate the final sequence, as shown below:

$$[p_1, p_2, \ldots, p_n] = \text{CRF}([h_1, h_2, \ldots, h_n]) \tag{6}$$

Slot Type Prediction Module. According to the classification results obtained by the first-stage slot filling of the model, as shown in Fig. 3, the second-stage slot filling first sends the corresponding hidden state sequence h_i, h_{i+1},..., h_j of the entity into Bi-LSTM in turn, and generate the hidden state representation information r_k of the entity sequence, where h_i, h_{i+1},..., h_j represents the hidden state of the k-th entity.

Whenever a vector r_k representing information is generated, it will be matched with the slot description matrix M_{desc} of shape $R^{n_s \times d_s}$ Similarity comparison. n_s represents the number of alternative slot types, and d_s represents the dimension of the slot description information. After that, the slot type prediction s_k for the k-th entity will be generated as:

$$s_k = M_{\text{desc}} \cdot r_k \tag{7}$$

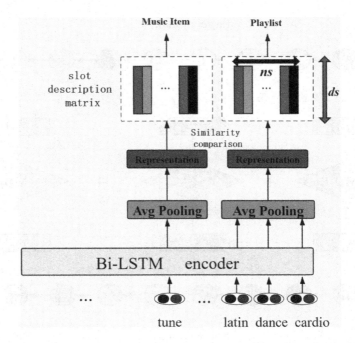

Fig. 3. The framework of Slot type prediction module.

In the slot description matrix, the representation vector r^{desc} of slot description information is obtained by adding the embeddings of N slot description entries:

$$r^{\text{desc}} = \sum_{i=l}^{N} \mathbf{E}\left(t_i\right) \tag{8}$$

$\mathbf{E}$ is the embedding layer shared with the input sentence.

2.2 Training

The BLCS model is the same as the two-stage slot filling model, and the loss function have two parts. In the first-stage slot filling, the training purpose of the model has two, one is to minimize the classification loss of the BIO label, and the other is to reduce the distribution difference between the source domain and the target domain, so the loss function is divided into two parts, one is CRF loss, the other is MMD loss, and is the coefficient of MMD loss.

$$L_{total} = L_{crf} + \alpha L_{Bin-MMD} + \beta L_r \tag{9}$$

In the slot type prediction module, we compare the representation information of each entity with the slot description matrix to generate a slot type prediction. The model will connect the BIO tags generated in a stage with the slot type information to get the final complete label prediction. In this module, the training goal is to minimize the classification loss of the slot type. In order to control model overfitting, we introduced a regularization term L_r:

$$L_r = \|\boldsymbol{\theta}_b\|_2^2 \tag{10}$$

θ_b is the same as the calculation method in [6].

3 Experiments

3.1 Data Set

We conducted experiments on the SNIPS [14] data set shown in Table 1. This is a personal voice assistant data set. According to different users' intentions, the sentences of SNIPS are classified into seven areas, and there are a total of 39 slot types in this data set.

Table 1. Detailed statistics of SNIPS dataset.

Domain	Samples	Slot types	Public types
AddToPlaylist	2042	5	3
BookRestaurant	2073	14	6
GetWeather	2100	9	5
PlayMusic	2100	10	5
RateBook	2056	7	2
SearchCreativeWork	2054	2	2
FindScreeningEvent	2059	7	3

3.2 Experimental Setup

The model will first train the remaining six samples of the source domain as the training set, and then the data in the target domain will be divided into several groups, 500 of which will be taken as the validation set, and then 50 samples will be taken to the training set. The remaining samples are used as the test set. During the training process, these 50 samples will be further calculated for the MMD difference with the entire training set data, and the extracted MMD difference will be added to the first-stage loss function calculation. We use a two-layer Bi-LSTM, the hidden state dimension is set to 200, and the dropout rate is set to 0.3. The parameters of the two layers of Bi-LSTM are not shared, and the hidden state dimension of the second layer of Bi-LSTM used to encode the hidden state is also set to 200, which is used to keep consistent with the embedding of words and characters. The optimizer used in this article is the Adam optimizer, the batch size of the training sample input is set to 32, the learning rate is set to 0.0005, and the patience of the early stop is set to 5.

3.3 Results and Analysis

- **CT (2017):** A one-stage slot filling model that proposes a slot filling framework that uses slot descriptions to deal with invisible types in the target domain.

- **RZT (2019):** A CT-based one-stage slot filling model that introduces slot instance values to improve the robustness of cross-domain adaptation.

- **Coach (2020):** A method proposed by Liu et al. For the first time divides the cross-domain slot filling task into two stages: firstly, the words are classified into three BIO tags in a coarse-grained manner, and then in a fine-grained manner. The words are classified into specific slot type labels, and the slot description information is used in the second stage to help identify unknown slot types.

- **Coach+TR (2020):** A variant of Coach, which further uses template regularization to improve the slot filling ability of similar or identical slot types, and achieves better results.

Table 2. Experimental results on SNIPS.

Model	ATP	BR	GW	PM	RB	SCW	FSE	AVG
CT	68.69	54.22	63.23	54.32	76.45	66.38	70.67	64.85
RZT	74.89	54.49	58.87	59.20	76.87	67.81	74.58	66.67
Coach	71.63	72.19	81.55	62.41	86.88	65.38	78.10	74.02
Coach+TR	74.68	74.82	79.64	66.38	84.62	64.56	83.85	75.51
BLCS	75.47	73.83	82.45	64.16	86.72	67.11	79.93	75.67
BLCS+TR	76.47	76.02	81.49	68.59	85.93	66.85*	84.87	77.17

As shown in Table 2, we compared existing cross-domain slot filling models, including one-stage and two-stage slot filling models. We observe that two-stage slot filling model with transfer learning performs better . The results show that adding Bin-MMD module between source domain and target domain can improve performance. At the same time, we noticed that compared with other two-stage slot filling models, our model does not perform well on some domains, such as RateBook. We think the reason is that the domain has a lot of domain-specific slot type but not many cross-domain shared slot type, so our transfer learning in step one has limited effect in this situation. Besides, we found that in SearchCreativeWork, two-stage slot filling model does not perform as well as one-stage slot filling, this is because there is no domain-specific slot type in this domain, so we think with enough cross-domain shared slot type, our model is more effective.

In the next experiments, we will study the hyper-parameters on model performance.

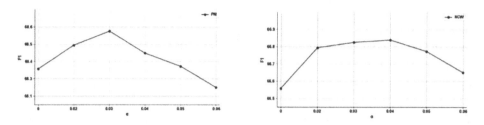

Fig. 4. Hyperparameter study for α on PlayMusic and SearchCreativeWork.

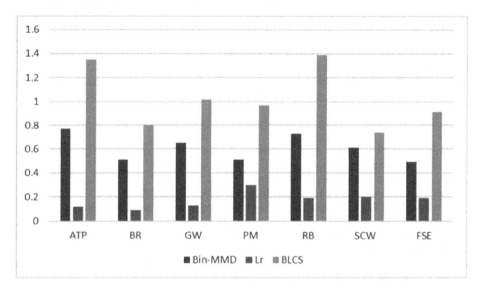

Fig. 5. Ablation experiments in various domains.

In order to verify the effectiveness of the Bin-MMD module, we try to change the value range of the parameter α. When the parameter is zero, it means that the tag migration module does not participate in model training. We show the experimental results in the Fig. 4, taking the PlayMusic domain and the SearchCreativeWork domain as examples. We study the influence of key hyperparameter α in BLCS on PlayMusic and SearchCreativeWork. The results in Fig. 4 indicate that setting $\alpha \in [0.02, 0.05]$ could better leverage BLST. Finally, we conduct an overall ablation experiment on the model. As shown in Fig. 5, the value corresponding to the histogram is the improvement brought by the module.

4   Conclusion

Experiments have proved that the Bin-MMD module can affect the overall generalization ability of the model by reducing the distribution difference, so that the model can be better identified in new domains. At the same time, we notice that the model performs better in the area with a small number of slot types than in the area with a large number of slot types.

References

1. Liu, B., Lane, I.: Recurrent neural network structured output prediction for spoken language understanding. In: NIPS Workshop on Machine Learning for Spoken Language Understanding and Interactions (2015)
2. Bapna, A., Tür, G., Hakkani-Tür, D., Heck, L.: Towards zero-shot frame semantic parsing for domain scaling. In: Interspeech 2017, August 2017
3. Shah, D., Gupta, R, Fayazi, A., Hakkani-Tür, D.: Robust zero-shot cross-domain slot filling with example values. In: Proceedings of the 57th Annual Meeting of the Association for Computational Linguistics, pp. 5484–5490 (2019)
4. Lee, S., Jha, R.: Zero-shot adaptive transfer for conversational language understanding. In: Proceedings of the AAAI Conference on Artificial Intelligence, vol. 33, pp. 6642–6649 (2019)
5. Liu, Z., Winata, G., Fung, P.: Zero-resource cross-domain named entity recognition. In: ACL, pp. 19–25 (2020)
6. Wang, Z., Qu, Y., Chen, L.: Label-aware double transfer learning for cross-specialty medi-cal named entity recognition. In: NAACL, pp. 1–15 (2018)
7. Gretton, A., Borgwardt, K.M., Rasch, M.J., Schölkopf, B., Smola, A.: A kernel two-sample test. J. Mach. Learn. Res. **13**, 723–773 (2012)
8. Elman, J.L.: Distributed representations, simple recurrent networks, and grammatical structure. Mach. Learn. **7**, 195–225 (1991)
9. Hochreiter, S., Schmidhuber, J.: Long short- term memory. Neural Comput. **9**(8), 1735–1780 (1997)
10. Liu, B., Lane, I.: Attention-based recurrent neural network models for joint intent detection and slot filling. In: Interspeech, pp. 685–689 (2016)
11. Yao, K.: Applications of reproducing kernel Hilbert spaces-bandlimited signal models. Inf. Control **11**(4), 429–444 (1967)
12. Raymond, C., Riccardi, G.: Generative and discriminative algorithms for spoken language understanding. In: Interspeech, pp. 1605–1608 (2007)
13. Goldberg, Y., Levy, O.: Word2vec explained: deriving Mikolov et al'.s negative sampling word-embedding method. arXiv preprint arXiv:1402.3722 (2014)
14. Coucke, A., et al.: Snips voice platform: an embedded spoken language understanding system for private-by-design voice interfaces. arXiv preprint arXiv:1805.10190 (2018)

Condition-Invariant Physical Adversarial Attacks via Pixel-Wise Adversarial Learning

Chenchen Zhao[1] and Hao Li[1,2]($\boxtimes$)

[1] Department of Automation, Shanghai Jiao Tong University, Shanghai, China
haoli@sjtu.edu.cn
[2] SPEIT, Shanghai Jiao Tong University, Shanghai, China

Abstract. Research has validated that deep neural networks are vulnerable to a series of methods named adversarial attacks. Such methods greatly impair performance of target networks by manipulating data with perturbations imperceptible to human. However, digital adversarial examples are vulnerable to various physical disturbances, while existing physical attack methods are strictly condition-variant and cannot generate adversarial data maintaining effectiveness with unforeseen disturbances. In this paper, we propose a physical adversarial attack method named Pixel-wise Adversarial Learning in Physical Attacks (P-ALPhA) to generate effective condition-invariant adversarial examples. Each adversarial example is decoupled into two separate clusters of pixels. Pixels in the first cluster are perturbed for adversarial purposes, while pixels in the second cluster are manipulated with a competitive optimization criterion against the adversarial aim. ALPhA enhances aggressiveness of the first cluster of pixels through intra-competition. Moreover, the original benign optimizations on the second cluster of pixels guide stochastic environmental disturbances to have adversarial impacts on the data. Experimental results show that P-ALPhA greatly enhances effectiveness of digital adversarial examples in establishments or simulations of multiple physical conditions. Success of P-ALPhA increases hazardness of adversarial attacks in physical applications of neural networks.

Keywords: Adversarial attack · Physical attack · Adversarial learning

1 Introduction

Deep neural networks are widely adopted in various kinds of vision processing tasks such as image classification and object detection with their performance validated. However, recent studies challenge neural networks and their performance by proposing a series of methods named adversarial attacks [1–6]. Adversarial attack methods make modifications (i.e. adversarial perturbations) on original data samples. Such adversarial perturbations are barely perceptible to human, yet they have the ability to completely change judgements of the

© Springer Nature Switzerland AG 2021
T. Mantoro et al. (Eds.): ICONIP 2021, LNCS 13109, pp. 369–380, 2021.
https://doi.org/10.1007/978-3-030-92270-2_32

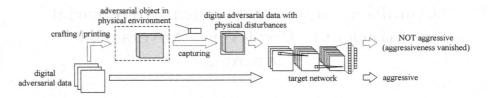

Fig. 1. The aggressiveness-vanishing problem in existing adversarial attack methods.

samples outputted by target neural networks. Therefore, success of adversarial attacks greatly impair performance of neural networks in digital scenarios. To handle the threats by adversarial attacks, another series of methods named adversarial defense are also proposed to enhance robustness of target neural networks against various attack methods. Multiple defense methods have been validated to have strong defense abilities against specific attacks.

However, in the real world, adversarial defenses may not be the strongest opponent of adversarial attacks. Imperceptibility of adversarial perturbations may result in attack failure in physical scenarios. In physical applications, adversarial attacks have to expose physical adversarial objects to perception sensors. In various physical conditions, the extremely-low-amplitude perturbations are rather likely to be 'submerged' by environmental disturbances, and aggressiveness of adversarial examples may be reduced and even vanishes. Two processes contribute most to the failure: conversion from digital data to physical adversarial objects; capturing of adversarial objects as input data to target networks. Disturbances in the former process include errors in object crafting or printing, while disturbances in the latter one include imbalanced illumination in the environment, large distances between the camera and the objects, errors in camera capturing, etc. The aggressiveness-vanishing problem above is shown in Fig. 1. Although the problem seriously restricts performance of adversarial attacks in practical use, few research works focus on proposing effective physical adversarial attacks. Among them, most works adopt general optimizations involving a variety of simulated physical conditions, and there are no condition-invariant methods maintaining effectiveness in complex or unforeseen environments.

In this paper, we propose a novel adversarial attack method named Pixel-wise Adversarial Learning in Physical Attacks (P-ALPhA) to generate condition-invariant physical adversarial examples. P-ALPhA decouples each adversarial example into two separate clusters of pixels. The adversarial optimizations are conducted on the first cluster of pixels. For pixels in the second cluster, P-ALPhA manipulates them with a competitive optimization criterion against the adversarial aim, which is the same as the original network training criterion in most cases. Such adversarial learning in adversarial example generation enhances aggressiveness of the first cluster of pixels through intra-competition. Moreover, the second cluster of pixels which are well optimized by the original criterion leave no room for environmental disturbances to have benign impacts on the data. Experimental results show that P-ALPhA greatly enhances effectiveness of digital adversarial examples in establishments or simulations of multiple physical conditions. Success of P-ALPhA increases hazardness of adversarial attacks in practical use, and poses threats to deep neural networks in their physical applications.

The paper is organized as follows: existing digital and physical adversarial attack methods are briefly discussed in Sect. 2; the proposed Pixel-wise Adversarial Learning in Physical Attacks is detailedly introduced in Sect. 3; experiments are conducted in Sect. 4 to validate performance of P-ALPhA; we conclude the paper in Sect. 5.

2   Related Work

2.1   Existing Adversarial Attack Methods

Digital adversarial attacks are well investigated. The first adversarial attack is conducted with L-BFGS optimizations [7]. In [1], the authors proposed the Fast Gradient Sign Method (FGSM) to determine the direction of perturbations according to gradient of data. FGSM acts as a fundamental method to a series of attack methods proposed afterwards [2–6,8,9,21].

For most methods, one of the attack criteria is to reduce the amplitude of adversarial perturbations, to ensure that adversarial examples are not distorted. The criterion is optimized in two possible ways: adding L2 loss in the iteration process, which is adopted by most attack methods; reducing the number of perturbed pixels, which is adopted by Jacobian Saliency Map Attack (JSMA) [3] and One Pixel Attack [10]. As stated in Sect. 1, these attack methods are not qualified for physical adversarial attacks partially because of the small amplitude of adversarial perturbations.

2.2   Physical Adversarial Attacks

Physical adversarial attacks are first proposed and experimented in [8]. To conduct physical attacks based on digital methods listed above, there are a few related research on improving the adversarial example generation process. Athalye et al. proposed the Expectation over Transformation (EoT) [11] method. EoT exposes adversarial examples to multiple simulated physical transformations and adds related criteria in adversarial iterations. In [17], total variation [18] and non-printability score are adopted to quantify errors in crafting or printing. EoT combined with total variation and non-printability score is the basis of most existing physical attack methods [12,13]. In addition, some of the other physical disturbances are also simulated and experimented in [14–16].

The physical attack methods listed above can be categorized as aposteriori methods (i.e. proposed after adversarial examples are influenced by physical conditions). Such methods are likely to suffer from conditions not simulated or foreseen, as their optimization criteria are strictly condition-variant. In addition, simulations of physical conditions in adversarial example generation require

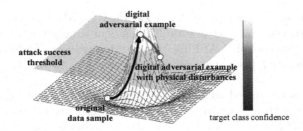

Fig. 2. Formulation of the aggressiveness-vanishing problem from the manifold perspective. The black and the red arrows respectively represent adversarial perturbations and physical disturbances. (Color figure online)

rather much computation resources. Instead, in this paper, we propose an apriori (i.e. condition-invariant) method to improve effectiveness of physical adversarial examples in varied foreseen or unforeseen conditions with similar or even fewer computation resources.

3   Pixel-Wise Adversarial Learning in Physical Attacks (P-ALPhA)

3.1   Problem Formulation

We make a definition in this paper that physical disturbances on a data sample refer to differences between the sample and the capturing result of its corresponding object. Factors that result in such disturbances are listed in Sect. 1.

Mathematically, in the high-dimensional decision manifold of the target network, direction of physical disturbances may be different from or opposite to the optimization direction of attacks. Therefore, adversarial examples with physical disturbances are not well optimized compared with their digital form. With the influences of physical disturbances becoming bigger, the aim of attacks cannot be guaranteed, and adversarial examples gradually lose their aggressiveness. An exemplary situation of the aggressiveness-vanishing problem is shown in Fig. 2.

3.2   P-ALPhA for Solving Aggressiveness-Vanishing

Instead of minimizing impacts of disturbances on adversarial examples, P-ALPhA attempts to 'guide' disturbances to have adversarial impacts via specially-designed adversarial data patterns, indicating that physical disturbances may instead be enhancements to physical adversarial attacks. Optimization aims of P-ALPhA are as follows:

$$F(\hat{x}) \neq F(x) \tag{1}$$

$$\forall \delta \in \Omega, \quad ||F(\hat{x} + \delta) - F(x)|| > ||F(\hat{x}) - F(x)|| \tag{2}$$

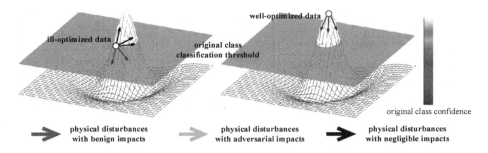

Fig. 3. The well-optimized benign data increases possibility of adversarial impacts by stochastic physical disturbances.

in which the adversarial data $\hat{x}$ is modified from x, F is the target neural network, and Ω is the set of all physical disturbances.

The proposed optimization approach of such specific patterns $\hat{x}$ is shown in Fig. 3. In ordinary cases, stochastic disturbances result in stochastic impacts. However, if the data is located on the local maxima of the decision manifold (with the largest confidence by the target network), chances that disturbances have benign impacts will be significantly reduced. Therefore, the pattern is set on the local maxima of the network decision manifold.

However, such patterns have the largest recognition confidence by the target network, meaning that they are not aggressive. Therefore, to obtain adversarial examples while maintaining characteristics of such patterns, P-ALPhA introduces adversarial learning in adversarial example generation. During adversarial iterations, the data is decoupled into two clusters of pixels according to a specific mask, which is determined by a proposed module named image saliency analyzer. Each disturbance can also be decoupled via this mask. One cluster is adversarially optimized, ensuring the adversarial purposes of P-ALPhA. Simultaneously, the other cluster is optimized towards the patterns stated above. Obviously, aims of the two clusters are exactly the opposite, forming competitive training in the process. Optimization aims of P-ALPhA involving adversarial training are shown as follows:

$$\hat{x}_2 = mask(\hat{x}), \quad \hat{x}_1 = \hat{x} - \hat{x}_2 \tag{3}$$

$$F(\hat{x}_1) = F(\hat{x}) \tag{4}$$

$$F(\hat{x}_2) = F(x) \neq F(\hat{x}) \tag{5}$$

$$\forall \delta \in \Omega, \quad ||F(\hat{x} + mask(\delta)) - F(x)|| > ||F(\hat{x}) - F(x)|| \tag{6}$$

P-ALPhA makes a joint optimization according to the aims to generate robust physical adversarial examples. Note that such decoupled optimizations are based on an assumption that:

$$\hat{x} = \hat{x}_1 + \hat{x}_2 \Rightarrow F(\hat{x}) = F(\hat{x}_1) \oplus F(\hat{x}_2) \tag{7}$$

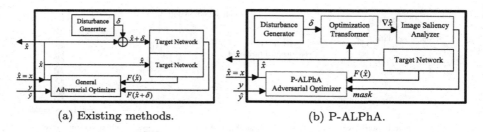

(a) Existing methods.   (b) P-ALPhA.

Fig. 4. Comparisons between P-ALPhA and existing physical attack methods in adversarial example generation. The red blocks are novel modules proposed in P-ALPhA. (Color figure online)

that the recognition result of the data is a non-linear combination of recognition results of both clusters, indicating that the results can be decoupled.

Comparisons of adversarial example generation by existing physical attacks and P-ALPhA are shown in Fig. 4, in which y and $\hat{y}$ are the original and the predefined adversarial labels. From the comparisons, advantages of the adversarial learning technique and the competitive criteria are listed as follows:

- P-ALPhA is an apriori condition-invariant method that generates more effective physical adversarial examples with similar or even fewer computation resources.
- The adversarial cluster has stronger attack ability via competitive optimizations.
- Through the competitive optimizations, $\hat{x}_2$ manages to convert stochastic physical disturbances to attack stabilizers.

In summary, with adversarial impacts of disturbances, the proposed P-ALPhA can generate more effective physical adversarial examples, and can also mitigate negative impacts of physical disturbances on them.

3.3 Method Structure

To avoid conditions that the benign patterns overwhelm aggressiveness of the adversarial cluster, the proportion of the two clusters requires precise control. To ensure enough impacts of disturbances on the benign patterns, the benign patterns should be more sensitive than the adversarial cluster. Both restrictions are satisfied by *mask* determined in the process. Inspired by [3], we determine *mask* according to the image saliency map and a predefined threshold. The adversarial cluster is less sensitive and is adversarially optimized; the other cluster is more sensitive and is optimized with the same criterion as network training. With *mask*, the optimization criterion is converted to be disturbance-independent, with details as follows:

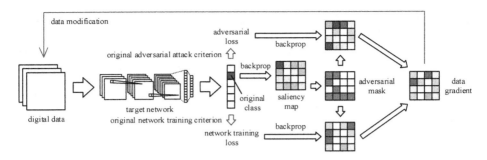

Fig. 5. Structure of P-ALPhA.

$$\hat{x}_1 = \hat{x}_1 - \overline{mask}(\frac{\partial CE(F(\hat{x}), \hat{y})}{\partial \hat{x}}) \times \lambda_1 \tag{8}$$

$$\hat{x}_2 = \hat{x}_2 - mask(\frac{\partial CE(F(\hat{x}), y)}{\partial \hat{x}}) \times \lambda_2 \tag{9}$$

in which CE is the cross entropy criterion, and λ_1 and λ_2 are the iteration step lengths of the two clusters.

The structure of P-ALPhA is shown in Fig. 5. From the structure, it is obvious that P-ALPhA is compatible with existing digital attack methods, and can be embedded into them to enhance their physical performance.

4   Experiments and Discussions

4.1   Experiment Setup

To estimate effectiveness of adversarial examples with physical disturbances, we propose a metric named survival rate (i.e. proportion of the condition-invariant adversarial examples over all examples). In addition, to validate that adversarial learning in P-ALPhA does not reduce digital attack performance, success rates of P-ALPhA are also recorded. Baseline comparisons on the two metrics are conducted to demonstrate the state-of-the-art attack performance of P-ALPhA. Performance with different as and εs is also studied. Note that in experiments, value of ε represents the proportion of the cluster with larger saliency values to the whole data instead of the saliency threshold. Note that since the adversarial mask varies during iterations, impacts of δ_1 and δ_2 cannot be precisely directed and quantified. Therefore, experiments involving $\hat{x}_1$, $\hat{x}_2$, δ_1 and δ_2 are not conducted.

All experiments are conducted on the ResNet18 [20] and VGG19 [22] network structures with the real-world KITTI dataset [19] for autonomous driving.

A series of physical conditions are established or simulated to verify physical performance of P-ALPhA, as listed in Sect. 4.2. Baseline comparisons involve targeted PGD [2], C&W [4] and EoT [11]. For all methods, data samples are perturbed to change the network's classification from Car/Pedestrian/Cyclist to Pedestrian/Cyclist/Car. All adversarial examples go through 500 iterations with step length 10^{-3} on ResNet18, and 1000 iterations with step length 10^{-5} on VGG19.

4.2 Physical Conditions Setup

A series of physical conditions are established or simulated:

- Camera internal errors, established by image capturing via a camera on a mobile phone. Adversarial images are displayed on a 1920×1080 laptop screen.
- Illumination, simulated by increasing pixel values with the criterion:

$$i_l = A_i(1 - \frac{l}{r}) \quad (l < \frac{r}{2}) \qquad i_l = A_i\frac{r}{4l} \quad (l >= \frac{r}{2}) \qquad (10)$$

in which i_l is the pixel increase; A_i is the illumination strength; l is the distance between the target pixel and the center of illumination; r is the predefined illumination radius. The center of illumination is randomly set at the top of the image; r is set $r_h = 0.3h$ and $r_w = 0.3w$ (h and w are the height and width of the image); A_i is randomly set in range $[100, 200]$.
- Size, simulated by downsampling images to smaller dimensions which are randomly set in range $[\frac{1}{4}, \frac{7}{8}]$ of the original size.
- Noise, simulated by adding a combination of gaussian and salt&pepper noise to images. The amplitude of the gaussian noise is set $0.03 \times 255 \approx 8$, and the frequency of the salt&pepper noise is set 10^{-4}.

None of the disturbances above heavily distort data samples, as shown in Fig. 6 that samples with different disturbances are very similar.

original   adversarial illuminated   resized   noisy

Fig. 6. An exemplary data sample, its corresponding adversarial example and the examples with effects of the stated physical conditions.

Table 1. Success rates and survival rates (against imbalanced illumination, size changes and noise) of P-ALPhA and the baselines.

Method (ResNet18)	Success rate	Illumination	Size	Noise	ε	a
PGD-P-ALPhA	**100.0**	98.0	**100.0**	**91.8**	0.50	0.05
PGD	100.0	6.07	4.49	3.34		
EoT	16.0	40.5	87.3	77.9		
C&W-P-ALPhA	100.0	**99.0**	79.6	59.2	0.10	0.20
C&W	100.0	92.9	75.5	30.6		
Method (VGG19)	Success rate	Illumination	Size	Noise	ε	a
PGD-P-ALPhA	99.0	**95.9**	**100.0**	**74.5**	0.50	0.02
PGD	100	8.08	6.06	12.1		
EoT	50.5	12.0	54.0	12.0		
C&W-P-ALPhA	**100.0**	39.4	53.5	11.1	0.02	0.20
C&W	100.0	33.3	53.5	13.1		

4.3 Horizontal Comparisons of P-ALPhA with Baselines

As stated above, P-ALPhA can be embedded into existing digital attack methods. In this section, P-ALPhA is embedded into PGD and C&W for portability validation and baseline comparisons.

Success rates and survival rates of P-ALPhA and the baselines are shown in Table 1. From the results, we can reach the conclusion that P-ALPhA is indeed capable of generating effective condition-invariant physical adversarial examples without sacrificing digital attack performance. Physical attack performance of P-ALPhA outperforms baselines by several times. Note that C&W-P-ALPhA has only slight performance increase compared with C&W. Attack criterion of C&W involves the class with the second-highest confidence by the target network, and optimizes it smaller. Since it is highly possible that the class with the second-highest confidence is exactly the original class, criterion of C&W is in conflict with the adversarial learning criteria in P-ALPhA. Therefore, adversarial learning cannot reach its optimal performance if combined with C&W.

Exemplary adversarial examples generated by P-ALPhA are shown in Fig. 7.

4.4 Ablation Study

PGD-P-ALPhA on VGG19 is adopted for ablation study. Attacks with $\varepsilon \in \{0, 0.05, 0.1, 0.2, 0.5, 0.75, 0.8, 0.9\}$, $a \in \{0, 1, 2, 5, 10, 20, 50\} \times 10^{-2}$ and numbers of iterations $n \in \{1, 5, 10, 20\} \times 10^2$ are experimented and compared.

With different εs and fixed $a = 2 \times 10^{-2}$, attack performance of PGD-P-ALPhA is shown in Fig. 8(a). From the results, we can reach the conclusion that adversarial learning is a significant factor in generation of effective condition-invariant adversarial examples. However, excessively large εs result in low success

Fig. 7. Exemplary adversarial examples generated by P-ALPhA. Each row contains several examples by P-ALPhA with one specific attack baseline on one specific model.

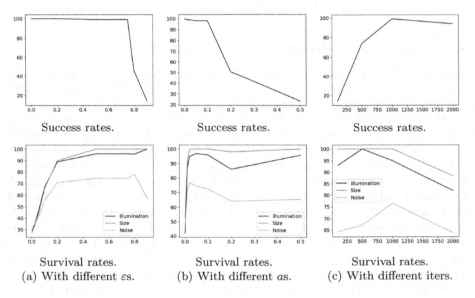

Fig. 8. Attack performance of PGD-P-ALPhA with different εs, as and numbers of iterations.

rates. In the specific experiments in this section, the optimal ε lies in range $[0.5, 0.75]$.

With different as and fixed $\varepsilon = 0.5$, attack performance of PGD-P-ALPhA is shown in Fig. 8(b). From the results, we can reach the conclusion that a is an important factor of digital attack performance. In the specific experiments in this section, the optimal a lies in range $[0.02, 0.05]$.

With fixed $\varepsilon = 0.5$ and $a = 0.02$, attack performance of PGD-P-ALPhA with different numbers of iterations is shown in Fig. 8(c). From the results, we can reach the conclusion that adversarial learning requires a certain number of iterations to reach its optimal performance. However, requirement of iterations by P-ALPhA is similar to other attack methods, and P-ALPhA still outperforms the selected baselines with similar or even fewer computation resources. In the specific experiments in this section, the optimal a lies in range $[500, 1000]$.

In summary, both ε and a have significant impacts on digital and physical attack performance of P-ALPhA. High general performance of P-ALPhA requires rational assignments of both factors.

5 Conclusions

In this paper, we propose a novel physical adversarial attack method named Pixel-wise Adversarial Learning in Physical Attacks (P-ALPhA) to solve the problem that digital adversarial examples are vulnerable to physical disturbances such as illumination, large distances and errors in camera capturing. P-ALPhA decouples each adversarial example into two clusters, and adopts competitive criteria to separately optimize them. With the competitive optimization criteria, aggressiveness of adversarial examples can be guaranteed and enhanced, while benign impacts of physical disturbances can also be effectively suppressed. Experimental results show that P-ALPhA greatly enhances effectiveness of digital adversarial examples in establishments or simulations of multiple physical conditions. The condition-invariant P-ALPhA makes existing digital adversarial attack methods more hazardous in physical applications of deep neural networks.

References

1. Goodfellow, I.J., Shlens, J., Szegedy, C.: Explaining and harnessing adversarial examples. arXiv preprint arXiv:1412.6572 (2014)
2. Madry, A., Makelov, A., Schmidt, L., Tsipras, D., Vladu, A.: Towards deep learning models resistant to adversarial attacks. arXiv preprint arXiv:1706.06083 (2017)
3. Papernot, N., McDaniel, P., Jha, S., Fredrikson, M., Celik, Z.B., Swami, A.: The limitations of deep learning in adversarial settings. In: 2016 IEEE European Symposium on Security and Privacy (EuroS&P), pp. 372–387. IEEE (2016)
4. Carlini, N., Wagner, D.: Towards evaluating the robustness of neural networks. In: 2017 IEEE Symposium on Security and Privacy (SP), pp. 39–57. IEEE (2017)
5. Inkawhich, N., Wen, W., Li, H.H., Chen, Y.: Feature space perturbations yield more transferable adversarial examples. In: Proceedings of the IEEE Conference on Computer Vision and Pattern Recognition, pp. 7066–7074 (2019)
6. Inkawhich, N., Liang, K.J., Carin, L., Chen, Y.: Transferable perturbations of deep feature distributions. arXiv preprint arXiv:2004.12519 (2020)
7. Szegedy, C., et al.: Intriguing properties of neural networks. arXiv preprint arXiv:1312.6199 (2013)
8. Kurakin, A., Goodfellow, I., Bengio, S.: Adversarial examples in the physical world. arXiv preprint arXiv:1607.02533 (2016)

9. Moosavi-Dezfooli, S.M., Fawzi, A., Frossard, P.: DeepFool: a simple and accurate method to fool deep neural networks. In: Proceedings of the IEEE Conference on Computer Vision and Pattern Recognition, pp. 2574–2582 (2016)

10. Su, J., Vargas, D.V., Sakurai, K.: One pixel attack for fooling deep neural networks. IEEE Trans. Evol. Comput. **23**(5), 828–841 (2019)

11. Athalye, A., Engstrom, L., Ilyas, A., Kwok, K.: Synthesizing robust adversarial examples. In: International Conference on Machine Learning, pp. 284–293. PMLR (2018)

12. Komkov, S., Petiushko, A.: AdvHat: real-world adversarial attack on ArcFace face ID system. arXiv preprint arXiv:1908.08705 (2019)

13. Brown, T.B., Mané, D., Roy, A., Abadi, M., Gilmer, J.: Adversarial patch. arXiv preprint arXiv:1712.09665 (2017)

14. Zhao, Y., Zhu, H., Liang, R., Shen, Q., Zhang, S., Chen, K.: Seeing isn't believing: practical adversarial attack against object detectors. arXiv preprint arXiv:1812.10217 (2018)

15. Thys, S., Van Ranst, W., Goedemé, T.: Fooling automated surveillance cameras: adversarial patches to attack person detection. In: Proceedings of the IEEE Conference on Computer Vision and Pattern Recognition Workshops (2019)

16. Sharif, M., Bhagavatula, S., Bauer, L., Reiter, M.K.: A general framework for adversarial examples with objectives. ACM Trans. Priv. Secur. (TOPS) **22**(3), 1–30 (2019)

17. Sharif, M., Bhagavatula, S., Bauer, L., Reiter, M.K.: Accessorize to a crime: real and stealthy attacks on state-of-the-art face recognition. In: Proceedings of the 2016 ACM SIGSAC Conference on Computer and Communications Security, pp. 1528–1540 (2016)

18. Mahendran, A., Vedaldi, A.: Understanding deep image representations by inverting them. In: Proceedings of the IEEE Conference on Computer Vision and Pattern Recognition, pp. 5188–5196 (2015)

19. Geiger, A., Lenz, P., Urtasun, R.: Are we ready for autonomous driving? The KITTI vision benchmark suite. In: 2012 IEEE Conference on Computer Vision and Pattern Recognition, pp. 3354–3361. IEEE (2012)

20. He, K., Zhang, X., Ren, S., Sun, J.: Deep residual learning for image recognition. In: Proceedings of the IEEE Conference on Computer Vision and Pattern Recognition, pp. 770–778 (2016)

21. Chen, S., He, Z., Sun, C., Yang, J., Huang, X.: Universal adversarial attack on attention and the resulting dataset damagenet. IEEE Trans. Pattern Anal. Mach. Intell. (2020)

22. Simonyan, K., Zisserman, A.: Very deep convolutional networks for large-scale image recognition. arXiv preprint arXiv:1409.1556 (2014)

Multiple Partitions Alignment with Adaptive Similarity Learning

Hao Dai$^{(\boxtimes)}$

College of Computer Science, Sichuan University, Chengdu 610065,
People's Republic of China
2018141461161@stu.scu.edu.cn

Abstract. Multi-view clustering is a hot topic which aims to utilize
the consistent and complementary information among different views
in unsupervised setting. Most existing methods integrate the information from different views in the data space. Thus the performance can
be deteriorated considering the unavoidable noises or inconsistency in
views. In this paper, we propose to construct the similarity matrix for
different views by utilizing adaptive similarity learning, and fuse multi-
view information in partition level instead of the original data space.
We further leverage an auto-weighted strategy to avoid bad performance
caused by certain views. Finally, we formulate the above principles into a
unified framework successfully and present the optimization algorithm to
optimize the corresponding objective function. Experimental results on
several commonly used datasets demonstrate the efficacy of our approach
against the state-of-the-art methods.

Keywords: Multi-view learning · Graph-based clustering · Adaptive
learning · Data fusion · Partition space

1 Introduction

Many classic methods in machine learning usually work in a single view manner.
However, single view learning is somehow not in conformity with real situations.
Thus there comes a concept named multi-view learning. The fact is that we
humans deal with the issue in a way of multiple views. It seems to be a more
comprehensive and objective way to approach the problem. Besides, many real
situations also provide with data generated from multiple views. For instance,
a text translated into several different languages; same image augmented by
crops, distort, blur, filtering, and so on. In this case, we can fully leverage the
information in each view to form a more explicit perspective of the problem.
Thereby, multi-view learning has raised a major concern recently due to the
complementary information among multiple views. In this field, mainly existing
studies have been summarized in [25, 29]. Moreover, with contrastive learning
popping up like mushrooms, which can benefit from multi-view data, particularly
in unsupervised way [4, 7, 8]. Clustering is a significant problem in unsupervised

© Springer Nature Switzerland AG 2021
T. Mantoro et al. (Eds.): ICONIP 2021, LNCS 13109, pp. 381–392, 2021.
https://doi.org/10.1007/978-3-030-92270-2_33

learning that groups data into different clusters without any label information. Therefore, well-performed multi-view clustering methods can draw increasing interest due to the high efficiency.

In the matter of multi-view clustering, the primary solution is to concatenating multi-view features straightforwardly. These analogous approaches expose the key problem how to reach an agreement of clustering among all views. In other word, we should take the view divergences into consideration rather than just scrabbling together. To tackle this challenging problem, increasing methods have been proposed lately [10,11], several representative categories of which exhibit a promising result. One is subspace clustering based method [5,9,13,14,20]. These methods follow a self-expressiveness property to learn the similarity graph matrix. In these methods, a data point can be represented as a linear combination of other points. Afterwards, [6,28] fuse the matrix of different view into a consensus graph. The graph matrix can be used to execute the subsequent clustering algorithm and get a final result. Plenty of multi-view subspace clustering methods have shown impressive success. However, these approaches share the same limitation that they can not make fully use of complementary information and are sensitive to initialization [16]. The main process of multi-view graph-based clustering methods is fusing varieties of graphs to a unified one. Such as [12,19,23,27], they learn a unified affinity graph and then execute additional clustering algorithm with the graph to produce the result.

However, the aforementioned methods are still limited in some cases. For example, they always fuse multi-view data information in early stage. That is to say, they need to rely on original data to share the information. But there comes to a problem that unavoidable existing noise and inconsistency among different views that can perturb the final output severely in data space. To conquer this challenge, [13,16] propose a new strategy which moves the fusion step to partition space. Broadly speaking, they put the data into a more suitable space for clustering and achieve a promising performance.

Orthogonal to previous methods, we propose a novel multi-view clustering method named Multiple Partitions Alignment with Adaptive Similarity Learning. Our method jointly performs the similarity graph construction, spectral clustering, and partitions fusion into a unified framework. Specifically, we adopt an adaptive learning strategy to construct the similarity graph for each view. Then multiple partitions can be obtained by using spectral clustering. Finally, we fuse the multiple partitions by providing each partition a weight coefficient to balance the impact on the consensus result. In summarize, our major contribution can be shown as follow:

- We propose to fuse the multi-view information in the partition level with the adaptive similarity graphs. In our model, similarity graphs can be adaptively constructed in each iteration, while the consensus partition can be calculated by reasonably integrating the multiple partitions. Hence the alternative updates can finally lead to a satisfied consensus result.
- We present a new objective function for our proposed model, which successfully combines the similarity graph construction, spectral clustering, and consensus fusion into a unified framework.

- We further design an exclusive alternative optimization algorithm to achieve the optimal result. Experiments showcase the superiority of our model, and outperforms state-of-the-art multi-view clustering methods.

2   Proposed Method

In this section, first we illustrate the process of assigning of adaptive neighbors for similarity graph construction. Then we formulate our model that combines the similarity graph construction, spectral clustering, and consensus fusion into a unified framework. The corresponding optimization algorithm is also described in detail.

As we mentioned before, the first step to process multi-view data is searching for the similarity between data points with the help of adaptive similarity learning. Many graph-based methods typically use the classical k-nearest neighbors (KNN) method. However, they always need to use Gaussian kernel to measure the similarity, which leaves a hyper-parameter to set manually. Moreover, KNN may destroy the local connectivity of data, giving rise to a suboptimal choice of neighbor number [23]. In that case, the study [19] explored a more reasonable way with probabilistic neighborhood assignment. Applying it to multi-view scenario, the core idea of the solution can be stated as

$$\min_{S^v} \sum_v \sum_{i,j} ||x_i^v - x_j^v||_2^2 \, s_{ij}^v + \alpha ||S^v||_F^2$$

$$s.t. \ \forall v, s_{ii}^v = 0, s_{ii}^v \geq 0, \mathbf{1}^T s_i^v = 1, \quad (1)$$

where x_i^v is the i-th data point of the v-th view x^v, s_{ij} denotes the similarity of x_i and x_j, and $\mathbf{1}^T s_i^v = 1$ is a constraint that guarantees the sparse of S^v. It has been proven that adaptive similarity learning strategy can offer a more explicit structure and is robust to noise and outliers [24]. Based on Eq. (1), we can obtain the clustering partition of each view by performing the spectral clustering with each similarity graph, which can be formulated as:

$$\min_{S^v, F_v} \sum_v \sum_{ij} ||x_i^v - x_j^v||_2^2 \, s_{ij}^v + \alpha ||S^v||_F^2 + \beta Tr(F_v^T L^v F_v)$$

$$s.t. \ s_{ii}^v = 0, s_{ii}^v \geq 0, \mathbf{1}^T s_i^v = 1, F_v^T F_v = I, \quad (2)$$

where F^v can be seen as an individual partition of each view. With these base partitions, it can simply fuse them into a consensus one by $||Y - F_v||_F^2$. However, directly measuring the differences may hurt consensus clustering Y badly because of the orientations of different partitions [13]. Thereby, the key to solve fusion problem can be described as

$$\min_Y \sum_v w_v ||YY^T - F_v F_v^T||_F^2 \quad s.t. \ Y^T Y = I, \quad (3)$$

where $Y \in \mathbb{R}^{N \times c}$ denotes the consensus result disturbed by each view's partition F^v, w_v denotes the weight of views which can be updated by

$$w_v = \frac{1}{2||YY^T - F_v F_v^T||_F}. \tag{4}$$

Note that Eq. (4) is nothing but the inverse distance weighting. It should be stressed that, with the involvement of w_v in Eq. (4), the consensus result can closely reside in some valuable partitions. Thus it shows strong robustness when a relatively error partition occurs. Let us put the substeps mentioned above together. The objective function of the proposed model can be finally formulated as

$$\min_{S^v, F_v, Y} \sum_v \sum_{ij} ||x_i^v - x_j^v||_2^2 \, s_{ij}^v + \alpha ||S^v||_F^2 + \beta Tr(F_v^T L^v F_v) + \gamma w_v ||YY^T - F_v F_v^T||_F^2 \tag{5}$$

$$s.t. \; s_{ii}^v = 0, s_{ii}^v \geq 0, \mathbf{1}^T s_i^v = 1, F_v^T F_v = I, Y^T Y = I.$$

3   Optimization

To handle the proposed objective function Eq. (5), we introduce an alternative optimization algorithm to solve it. As can be seen, the entire optimization problem can be divided into three subproblems in terms of the variables S^v, F_v, and Y.

3.1   Updating S^v

Fixing F_v and Y, the objective function Eq. (5) w.r.t. S^v is equivalent to

$$\min_{S^v} \sum_v \sum_{ij} ||x_i^v - x_j^v||_2^2 \, s_{ij}^v + \alpha ||S^v||_F^2 + \beta Tr(F_v L^v F_v) \tag{6}$$

$$s.t. \; s_{ii}^v = 0, s_{ii}^v \geq 0, \mathbf{1}^T s_i^v = 1.$$

Since $Tr(F_v L^v F_v) = \frac{1}{2} \sum_{ij} ||f_i^v - f_j^v||_2^2 s_{ij}^v$, and S^v of each view is independent, Eq. (6) can be rewritten as:

$$\min_{S^v} \sum_v \sum_{ij} ||x_i^v - x_j^v||_2^2 \, s_{ij}^v + \alpha ||S^v||_F^2 + \frac{\beta}{2} \sum_{ij} ||f_i^v - f_j^v||_2^2 \, s_{ij}^v \tag{7}$$

$$s.t. \; s_{ii}^v = 0, s_{ii}^v \geq 0, \mathbf{1}^T s_i^v = 1.$$

Now that Eq. (7) is independent from different i. Thus it can be solved with the following vector form

$$\min_{S_i^v} \sum_v \sum_j ||x_i^v - x_j^v||_2^2 \, s_{ij}^v + \alpha ||S^v||_F^2 + \frac{\beta}{2} \sum_{ij} ||f_i^v - f_j^v||_2^2 \, s_{ij}^v \tag{8}$$

$$s.t. \; s_{ii}^v = 0, s_{ii}^v \geq 0, \mathbf{1}^T s_i^v = 1.$$

Denote $d_{ij}^v = ||x_i^v - x_j^v||_2^2 + \frac{\beta}{2}||f_i^v - f_j^v||_2^2$, where $d_i^v \in \mathbb{R}^{n \times 1}$ is a vector with its j-th element being d_{ij}^v. Then Eq. (8) can be expressed as

$$\min_{s_i^v} ||s_i^v + \frac{d_i^v}{2\alpha}||_2^2$$
$$s.t.\ s_{ii}^v = 0, s_{ii}^v \geq 0, \mathbf{1}^T s_i^v = 1. \tag{9}$$

To calculate Eq. (9), we introduce the Lagrange multiplier ζ and η. Thus the corresponding Lagrange function is

$$L(s_i^v, \zeta, \eta) = \frac{1}{2}||s_i^v + \frac{d_i^v}{2\alpha}||_2^2 - \zeta((s_i^v)\mathbf{1} - 1) - \eta_i^T s_i^v. \tag{10}$$

Based on KKT condition [1], we have

$$s_{ij}^v = \left(\zeta - \frac{d_i^v}{2\alpha}\right)_+. \tag{11}$$

Reordering vector d_i^v from small to large allows us to assign a nonzero value to s_{ik}^v, where k is the number of neighborhood. Thus we get

$$\begin{cases} \zeta - \dfrac{d_{ik}^v}{2\alpha} > 0, \\ \zeta - \dfrac{d_{i,k+1}^v}{2\alpha} \leq 0. \end{cases} \tag{12}$$

Taking both Eq. (11) and constraint $(\mathbf{s}_i^v)^T \mathbf{1} = 1$ into account, we obtain

$$\zeta = \frac{1}{k} + \frac{1}{2k\alpha} \sum_h^k d_{ih}^v. \tag{13}$$

According to Eqs. (12) and (13), we have

$$\begin{cases} \alpha > \dfrac{kd_{ik}^v - \sum_h^k d_{ih}^v}{2}, \\ \alpha \leq \dfrac{kd_{i,k+1}^v - \sum_h^k d_{ih}^v}{2}. \end{cases} \tag{14}$$

To constrain the number of neighbors k in each s_i^v, α is set to

$$\alpha = \frac{kd_{i,k+1}^v - \sum_h^k d_{ih}^v}{2}. \tag{15}$$

In other words, parameter α is determined by k. According to Eqs. (12), (13) and (15), we draw the final solution of s_{ij}^v as

$$s_{ij}^v = \left(\frac{d_{i,k+1}^v - d_{ij}^v}{kd_{i,k+1}^v - \sum_h^k d_{ih}^v}\right)_+. \tag{16}$$

3.2   Updating F_v

Fixing S^v and Y, we extract an expression only with respect to F_v from Eq. (5), which is

$$\min_{F_v} \sum_v \beta Tr(F_v L^v F_v) + \gamma w_v ||YY^T - F_v F_v^T||_F^2 \quad s.t.\ F^T F = I. \tag{17}$$

It can further be transformed into a more simple form

$$\min_F Tr(F^T M F) \quad s.t.\ F^T F = I, \tag{18}$$

where $M = \beta L - 2\gamma w Y Y^T + \gamma w I$. Analogy with spectral clustering, we know that the optimal solution can be reached when F consists of the c eigenvector of M corresponding to the eigenvalues from small to large.

3.3   Updating Y

Same way as before, we can get the subproblem of Y when F_v and S^v are fixed, i.e.,

$$\min_Y Tr(Y^T P Y) \quad s.t.\ Y^T Y = I, \tag{19}$$

where $P = \sum_v (I - 2w_v F_v F_v^T)$. Treating F_v as a constant. It reaches the minimum value when Y consists of the c eigenvector of P corresponding to the eigenvalues from small to large.

In summary, our procedure to solve the objective function Eq. (5) is described in Algorithm 1. We empirically set the maximum iterations to 200 and run K-means on the convergence result Y to get the final indicator matrix.

Algorithm 1. Optimization algorithm for Eq. (5)

Input: Multiview matrix $X^1, ..., X^v$, number of clusters c, number of neighbors k, parameters β, γ.
Output: S^v, F_v, Y.
1: **Initialize:** Obtain similarity matrix S^v by Eq. (16), weight $w_v = \frac{1}{v}$, random matrix Y.
2: **repeat**
3:     **for** view 1 to v **do**
4:         Update S^v by using Eq. (16);
5:         Update F_v by using Eq. (18);
6:     **end for**
7:     Update Y by using Eq. (19);
8:     Update w_v by solving $\frac{1}{2||YY^T - F_v F_v^T||_F}$.
9: **until** Converges or the maximum iteration reached

---

4 Experiments

4.1 Experiment Datasets

Following [6,16,17,26], we adopt four benchmark datasets that are widely used in multi-view clustering. Table 1 describes the statistics of the dataset in detail.

- BBC is a dataset collected from BBC news website containing 4 views by splitting each document into four related segments. Segment consists of consecutive textual paragraphs and contains no less than 200 characters.
- Reuters contains feature characteristics of documents originally written in five different languages and their translations, over common 6 categories. In experiments, we use the subset which is the documents written or translated in English.
- HW consists of features of handwritten numerals 0–9 extracted from a collection of Dutch utility maps. 2000 instances in total have been digitized in binary images represented in terms of 6 feature sets.
- Caltech101 is a famous object recognition dataset containing pictures of objects belonging to 101 categories. Caltech20 is the subset of Caltech101, which contains only the widely used 20 classes respectively. We use the Caltech20 due to the unbalance of the number of data in each classes of Caltech101.

Table 1. Description of our dataset (number of features).

View	BBC	Reuters	HW	Caltech20
1	Segment1 (4659)	English (2000)	Profile correlations (216)	Gabor (48)
2	Segment2 (4633)	French (2000)	Fourier coefficients (76)	Wavelet moments (40)
3	Segment3 (4665)	German (2000)	Karhunen coefficients (64)	CENTRIST (254)
4	Segment4 (4684)	Spanish (2000)	Morphological (6)	HOG (1984)
5	–	Italian (2000)	Pixel averages (240)	GIST (512)
6	–	–	Zernike moments (47)	LBP (928)
Data points	145	1200	2000	2386
Clusters	2	6	10	20
Type	Text	Text	Image	Image

4.2 Experiment Setup

To evaluate the performance of the proposed method, we compare it with following competitive multi-view clustering methods: Co-train [17], Co-reg [18], MVKKM [22], RMKMC [2], MSPL [26], AMGL [19], MVEC [21], MVSC [6], DiMSC [3] and PMSC [16]. K-means is also included as a baseline by using concatenated features with equal weight. Besides, following [2], we normalize the raw data into range $[-1,1]$ for a fair comparison.

To further demonstrate how the clustering performance can be improved by our method, the popular evaluation metrics, accuracy (ACC), Purity and normalized mutual information (NMI) [15], are used in this paper.

4.3   Evaluation Results

In Table 2, 3, 4 and 5, we report the clustering result on all benchmark datasets. Note that the mean and standard deviation (std) values of each measure after 10 repetitions are also recorded. We can observe that our method achieves the best performance at most times. To be specific, when compared with the subspace clustering based method such as [16], our model always shows the superiority on the experimental datasets. That is to say, instead of traditional subspace clustering method, graph-based clustering method based on adaptive similarity learning may suit the partition level fusion better.

To prove that partition level fusion can work with the adaptive similarity graph construction, we plot the partition F_v of each view as well as our consensus partition Y. Take BBC dataset for an example, there are four different partitions with 145 instances and two classes. We visualize partitions F_v and consensus result Y over BBC dataset in Fig. 1. It turns out that partitions F_v and Y share a relatively similar pattern. We can also observe that the difference in orientations still exists. Therefore, we cannot directly measure the differences

Table 2. Evaluation of performance on BBC (%).

Method	ACC	Purity	NMI
KM	91.59 (0.31)	90.24 (0.24)	14.10 (1.30)
Co-train	91.27 (0.00)	87.57 (1.20)	3.50 (0.00)
Co-reg	90.90 (0.76)	90.78 (1.40)	6.80 (0.30)
MVKKM	84.00 (6.13)	89.01 (2.35)	8.30 (0.64)
RMKMC	91.31 (0.62)	89.67 (1.80)	8.00 (0.74)
MSPL	80.41 (13.24)	90.41 (0.00)	10.11 (9.48)
AMGL	89.66 (0.31)	91.00 (0.67)	11.20 (0.00)
DiMSC	93.79 (0.00)	94.62 (0.00)	13.71 (0.00)
MVEC	88.97 (0.00)	94.48 (0.00)	1.03 (0.00)
MVSC	91.03 (0.00)	95.62 (0.00)	0.41 (0.00)
PMSC	95.86 (0.00)	95.86 (0.00)	37.42 (0.00)
Ours	**96.55 (0.00)**	**96.55 (0.00)**	**44.70 (0.00)**

Table 3. Evaluation of performance on Reuters (%).

Method	ACC	Purity	NMI
KM	24.57 (4.52)	25.48 (4.37)	11.78 (5.01)
Co-train	17.00 (0.10)	17.15 (0.07)	9.40 (0.11)
Co-reg	20.62 (1.24)	20.95 (1.32)	2.33 (0.34)
MVKKM	20.48 (3.82)	20.65 (3.83)	5.77 (3.66)
RMKMC	22.42 (6.54)	22.55 (6.57)	7.21 (7.29)
MSPL	24.87 (5.98)	28.12 (4.97)	11.50 (4.28)
AMGL	18.35 (0.15)	20.08 (0.54)	6.38 (1.00)
DiMSC	39.60 (1.32)	46.28 (1.74)	18.17 (0.64)
MVEC	31.08 (0.00)	82.48 (0.09)	11.92 (0.01)
MVSC	25.08 (0.39)	80.11 (5.50)	6.60 (0.68)
PMSC	**40.18 (2.32)**	60.07 (3.56)	**21.83 (1.75)**
Ours	19.96 (0.00)	**92.24 (0.01)**	7.99 (0.01)

Table 4. Evaluation of performance on HW (%).

Method	ACC	Purity	NMI
KM	54.46 (5.60)	58.64 (2.92)	58.25 (0.85)
Co-train	71.42 (4.21)	74.86 (2.62)	71.06 (1.07)
C-reg	83.38 (7.35)	85.17 (4.98)	77.97 (2.92)
MVKKM	58.81 (3.50)	62.40 (3.40)	62.91 (2.60)
RMKMC	63.04 (3.36)	65.74 (2.16)	66.57 (1.18)
MSPL	68.00 (1.12)	68.99 (1.17)	70.42 (1.95)
AMGL	73.61 (10.29)	76.48 (8.54)	81.86 (4.53)
DiMSC	42.72 (1.94)	45.65 (0.97)	37.89 (0.87)
MVEC	66.93 (5.51)	79.95 (1.73)	70.69 (2.55)
MVSC	79.60 (2.54)	87.19 (1.48)	73.89 (1.93)
PMSC	83.81 (6.76)	87.34 (3.07)	82.05 (2.93)
Ours	**92.82 (0.05)**	**94.99 (0.01)**	**89.97 (0.02)**

Table 5. Evaluation of performance on Caltech20 (%).

Method	ACC	Purity	NMI
KM	31.40 (1.30)	60.06 (0.38)	37.05 (0.41)
Co-train	38.94 (2.10)	69.77 (1.42)	50.90 (1.12)
C-reg	34.38 (0.79)	65.59 (1.03)	46.42 (0.96)
MVKKM	44.87 (2.49)	72.84 (0.72)	54.06 (1.23)
RMKMC	33.35 (1.47)	64.22 (0.89)	42.44 (0.67)
MSPL	33.49 (0.00)	34.24 (0.00)	35.80 (0.00)
AMGL	52.28 (2.91)	67.60 (2.31)	56.61 (1.93)
DiMSC	33.89 (1.45)	37.78 (1.35)	39.33 (1.16)
MVEC	52.19 (4.25)	60.36 (3.21)	59.78 (1.10)
MVSC	44.96 (2.06)	50.87 (2.35)	45.36 (0.88)
PMSC	52.63 (0.89)	72.93 (2.57)	48.35 (2.82)
Ours	**61.71 (0.02)**	**75.05 (0.02)**	**63.16 (0.03)**

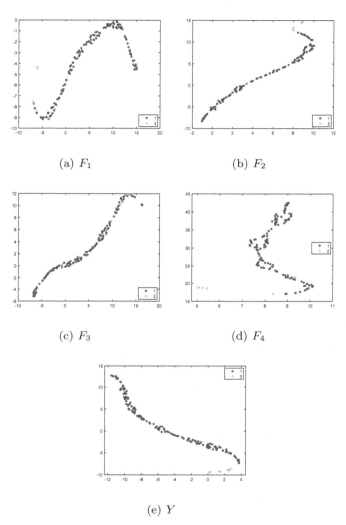

(a) F_1 (b) F_2

(c) F_3 (d) F_4

(e) Y

Fig. 1. Visualization of partitions F_v and consensus result Y, where red dots represent class 1 and blue dots represent class 2. (Color figure online)

among partitions based on the formulation $||Y - F_v||_F^2$. That's the reason why we choose to utilize Eq. (3). It is clear that our method can recognize a good clustering pattern as shown in Fig. 1(e), which validates the effectiveness of our proposed partition alignment method.

4.4 Convergence Study

In Fig. 2, we empirically show the convergence curve of each dataset. As we can see, our method always converges very fast. Especially in experimental dataset, our method only takes a few iterations to reach convergence.

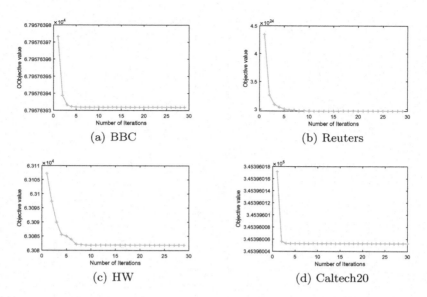

(a) BBC

(b) Reuters

(c) HW

(d) Caltech20

Fig. 2. The convergence curves on each dataset, where x-axis represents the number of iterations and y-axis represents the objective value.

4.5 Sensitivity Analysis

In this section, we illustrate the influence of different parameter settings to the proposed model. The accuracy results of Caltech-20 dataset with respect to different number of neighbors k, parameters β and γ are shown in Fig. 3. We can observe that the results are quite stable against a wide range of parameters, which demonstrate the effectiveness of the proposed method.

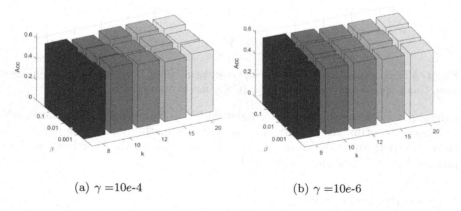

(a) $\gamma = 10e\text{-}4$

(b) $\gamma = 10e\text{-}6$

Fig. 3. Parameters sensitive analysis over Caltech-20 dataset. (We have selected 30 parameter combinations to discuss its stability, and divided them into two groups according to the value of γ.)

5   Conclusion

In this paper, we introduce a novel framework for multi-view clustering by adaptive similarity learning and multi-view information fusion in partition level. We further leverage an auto-weighted strategy to avoid bad performance caused by some views. Experimental results on four benchmark datasets demonstrate our method outperforms state-of-the-art multi-view method techniques in most case.

Acknowledgments. We thank the anonymous reviewers for their helpful comments and suggestions. This is part of an ongoing project with several collaborators, including Shudong Huang, and Qing Ye. This work was partially supported by the National Science Foundation of China under Grant 62106164, and the Fundamental Research Funds for the Central Universities under Grant 1082204112364.

References

1. Boyd, S., Boyd, S.P., Vandenberghe, L.: Convex Optimization. Cambridge University Press, Cambridge (2004)
2. Cai, X., Nie, F., Huang, H.: Multi-view k-means clustering on big data. In: Twenty-Third International Joint Conference on Artificial Intelligence (2013)
3. Cao, X., Zhang, C., Fu, H., Liu, S., Zhang, H.: Diversity-induced multi-view subspace clustering. In: Proceedings of the IEEE Conference on Computer Vision and Pattern Recognition, pp. 586–594 (2015)
4. Chen, T., Kornblith, S., Norouzi, M., Hinton, G.: A simple framework for contrastive learning of visual representations. In: International Conference on Machine Learning, pp. 1597–1607. PMLR (2020)
5. Chen, X., Ye, Y., Xu, X., Huang, J.Z.: A feature group weighting method for subspace clustering of high-dimensional data. Pattern Recogn. **45**(1), 434–446 (2012)
6. Gao, H., Nie, F., Li, X., Huang, H.: Multi-view subspace clustering. In: Proceedings of the IEEE International Conference on Computer Vision, pp. 4238–4246 (2015)
7. Huang, S., Kang, Z., Tsang, I.W., Xu, Z.: Auto-weighted multi-view clustering via kernelized graph learning. Pattern Recogn. **88**, 174–184 (2019)
8. Huang, S., Kang, Z., Xu, Z.: Self-weighted multi-view clustering with soft capped norm. Knowl.-Based Syst. **158**, 1–8 (2018)
9. Huang, S., Kang, Z., Xu, Z.: Auto-weighted multi-view clustering via deep matrix decomposition. Pattern Recogn. **97**, 107015 (2020)
10. Huang, S., Tsang, I., Xu, Z., Lv, J.C.: Measuring diversity in graph learning: a unified framework for structured multi-view clustering. IEEE Trans. Knowl. Data Eng. 1–15 (2021)
11. Huang, S., Tsang, I., Xu, Z., Lv, J.C.: Multiple partitions alignment via spectral rotation. Mach. Learn. 1–23 (2021)
12. Huang, S., Xu, Z., Tsang, I.W., Kang, Z.: Auto-weighted multi-view co-clustering with bipartite graphs. Inf. Sci. **512**, 18–30 (2020)
13. Kang, Z., et al.: Multiple partitions aligned clustering. arXiv preprint arXiv:1909.06008 (2019)
14. Kang, Z., Peng, C., Cheng, Q.: Kernel-driven similarity learning. Neurocomputing **267**, 210–219 (2017)
15. Kang, Z., Wen, L., Chen, W., Xu, Z.: Low-rank kernel learning for graph-based clustering. Knowl.-Based Syst. **163**, 510–517 (2019)

16. Kang, Z., et al.: Partition level multiview subspace clustering. Neural Netw. **122**, 279–288 (2020)

17. Kumar, A., Daumé, H.: A co-training approach for multi-view spectral clustering. In: Proceedings of the 28th International Conference on Machine Learning (ICML-2011), pp. 393–400 (2011)

18. Kumar, A., Rai, P., Daume, H.: Co-regularized multi-view spectral clustering. Adv. Neural. Inf. Process. Syst. **24**, 1413–1421 (2011)

19. Nie, F., Li, J., Li, X., et al.: Parameter-free auto-weighted multiple graph learning: a framework for multiview clustering and semi-supervised classification. In: IJCAI, pp. 1881–1887 (2016)

20. Peng, X., Feng, J., Xiao, S., Yau, W.Y., Zhou, J.T., Yang, S.: Structured autoencoders for subspace clustering. IEEE Trans. Image Process. **27**(10), 5076–5086 (2018)

21. Tao, Z., Liu, H., Li, S., Ding, Z., Fu, Y.: From ensemble clustering to multi-view clustering. In: IJCAI (2017)

22. Tzortzis, G., Likas, A.: Kernel-based weighted multi-view clustering. In: 2012 IEEE 12th International Conference on Data Mining, pp. 675–684. IEEE (2012)

23. Wang, H., Yang, Y., Liu, B.: GMC: graph-based multi-view clustering. IEEE Trans. Knowl. Data Eng. **32**(6), 1116–1129 (2019)

24. Wright, J., Yang, A.Y., Ganesh, A., Sastry, S.S., Ma, Y.: Robust face recognition via sparse representation. IEEE Trans. Pattern Anal. Mach. Intell. **31**(2), 210–227 (2008)

25. Xu, C., Tao, D., Xu, C.: A survey on multi-view learning. arXiv preprint arXiv:1304.5634 (2013)

26. Xu, C., Tao, D., Xu, C.: Multi-view self-paced learning for clustering. In: Twenty-Fourth International Joint Conference on Artificial Intelligence (2015)

27. Zhan, K., Zhang, C., Guan, J., Wang, J.: Graph learning for multiview clustering. IEEE Trans. Cybern. **48**(10), 2887–2895 (2017)

28. Zhang, C., Hu, Q., Fu, H., Zhu, P., Cao, X.: Latent multi-view subspace clustering. In: Proceedings of the IEEE Conference on Computer Vision and Pattern Recognition, pp. 4279–4287 (2017)

29. Zhao, J., Xie, X., Xu, X., Sun, S.: Multi-view learning overview: recent progress and new challenges. Inf. Fusion **38**, 43–54 (2017)

Recommending Best Course of Treatment Based on Similarities of Prognostic Markers

Sudhanshu, Narinder Singh Punn$^{(\boxtimes)}$, Sanjay Kumar Sonbhadra, and Sonali Agarwal

Indian Institute of Information Technology Allahabad, Prayagraj, India
{ism2016004,pse2017002,rsi2017502,sonali}@iiita.ac.in

Abstract. With the advancement in the technology sector spanning over every field, a huge influx of information is inevitable. Among all the opportunities that the advancements in the technology have brought, one of them is to propose efficient solutions for data retrieval. This means that from an enormous pile of data, the retrieval methods should allow the users to fetch the relevant and recent data over time. In the field of entertainment and e-commerce, recommender systems have been functioning to provide the aforementioned. Employing the same systems in the medical domain could definitely prove to be useful in variety of ways. Following this context, the goal of this paper is to propose collaborative filtering based recommender system in the healthcare sector to recommend remedies based on the symptoms experienced by the patients. Furthermore, a new dataset is developed consisting of remedies concerning various diseases to address the limited availability of the data. The proposed recommender system accepts the prognostic markers of a patient as the input and generates the best remedy course. With several experimental trials, the proposed model achieved promising results in recommending the possible remedy for given prognostic markers.

Keywords: Health recommender system · Prognostic markers · Collaborative filtering · Machine learning

1 Introduction

Recently, recommender systems have become an important part of many different sectors. Major e-commerce platforms employ the use of recommender systems to display the filtered results for every customer. These recommendations keep updating over time with the aim to improve the user's experience of the platform. The other factors that a recommendation system [11,24] brings to a platform

---

Sudhanshu, N. S. Punn, S. K. Sonbhadra and S. Agarwal—All authors contributed equally.

T. Mantoro et al. (Eds.): ICONIP 2021, LNCS 13109, pp. 393–404, 2021.
https://doi.org/10.1007/978-3-030-92270-2_34

are: sales boost, enhanced customer engagement, transform shoppers to clients, increase average order value, lower manual work and overhead, and bring more traffic on the e-commerce platform. Recommender systems are also utilized in the following areas: entertainment and media (movie/song/book/news), economy (stocks), banking, telecommunications, etc.

The recommender systems are also termed as SaaS (Software as a Service) and hence could be employed in various other fields given the right dataset. Following this, many recommender system based approaches have been utilized in the healthcare domain to aim for better healthcare services. One of the applications is to recommend remedies to a patient that would be most effective in the treatment process. The system proposed in the paper analyzes the prognostic markers for the patient in question by using the collaborative filtering to match the profiles of other patients that had similar prognostic markers. The top remedies will be listed as the output which have proven to be effective on the other patients' health status when their prognostic markers were on similar levels.

1.1   Motivation and Contribution

Recommender systems are being widely used to recommend the relevant items in the context of e-commerce and infotainment. Recommender systems have helped the businesses because they [27] improve the inventory value, user experience etc.

The crux of what is happening behind the scenes is, users are being delivered the relevant content from a huge stack of information. Consequently, when the same approach was applied in the healthcare domain, it gave rise to health recommender systems (HRS). Following this, there was an undeniable boost in automated healthcare and tele-medicine. Health recommender systems (HRS) have been put to use in the following forms:

1. Enterprise Resource Planning (ERP system) [4]
2. A Doctor recommendation algorithm [3]
3. Web-based RS suggesting online health resources [23]
4. A diet recommendation system to a patient [1]
5. Chronic Disease Diagnosis Prediction and Recommendation System [2]

According to Park et al. [10], the research in the field of health recommender systems has increased immensely but the practical implementations of such systems still requires more research. The major takeaway is that, even with the adequate knowledge in the field of health recommender systems, they are not being put in practical use on a large scale. In this paper, we propose a system in the field of diagnosing and treating diseases in the essence of automating healthcare. The major contributions of the present article are as follows:

1. Developing a recommender system that when given a set of symptoms, will perform a diagnosis and then recommend the next best course of treatment.
2. Creating a data set that contains 'course of treatment' corresponding to their diseases. And these diseases range from mild and acute to chronic states.

1.2 Organization of the Paper

The rest of the paper is divided into several sections. Second section mentions the related works in the field of health recommender systems and the associated challenges. Third section presents the methodology behind building the proposed system. It also mentions the pre-processing of the dataset, and the inclusion of a newly created data file in the dataset. Fourth section presents the results obtained by the proposed approach. Last section concludes the paper with the outlines of the future scope of recommender systems in the healthcare domain.

2 Related Work

The massive growth and advancements in deep learning algorithms across vivid domains such as healthcare, image processing, etc. [7,15,22,26] has resulted in immense applications by developing real-world applications. In the earlier survey of recommender systems by Park et al. [10] it was observed that scholastic investigation on recommender systems have expanded fundamentally over the last ten years, but more insights are needed to develop real-world applications. The research field on health recommender systems is potentially wide, however there are less developments in the practical scenarios. In like manner, the current articles on recommender systems should also be reviewed up to the coming age of health recommender systems. Hors-Fraile et al. [13] also discovered the need of increasing and improving the research in the area of HRS that covers the proposed multidisciplinary taxonomy. This includes the features like integration with electronic health records, incorporation of health promotion theoretical factors and behavior change theories. Kamran and Javed [16] presented a survey of RS in healthcare and also proposed a hybrid recommender system that takes into account the features like hospital quality (measured objectively based on doctors communication, nurses communication, staff behavior, pain control procedures, medicine explanation, guidance during recovery at home, surrounding cleanliness, quietness in patient's surrounding) and patients similarity. This recommender system suggests the best hospitals for a patient based on the above factors.

Pincay et al. [5] presented a state-of-the-art review providing insights about methods and techniques used in the design and development of HRS(s), focusing on the areas or types of the recommendations these systems provide and the data representations that are employed to build a knowledge base. Sezgin et al. [4] outlined the major approaches of HRS which included current developments in the market, challenges, and opportunities regarding HRS and emerging approaches. Huang et al. [3] proposed an algorithm which improves the performance of the medical appointment procedure. This algorithm creates a 'doctor performance model' based on the reception and appointment status. It also creates a 'patient preference model' based on the current and historical reservation choices which help in the accurate recommendation. It prevents the situation where a doctor is under-appointed or over-appointed and the patients are not being treated even if doctors are available. Peito [21] proposed a HRS for patient-doctor match-making based on patients' individual health profiles and consultation history. Another utility HRS was proposed by Kim et al. [1] that personalized diet recommended service through considering the real-time vital sign, family history, food preference, and intake of users to solve the limitations in the existing diet recommendation services.

Hussein et al. [2] proposed a HRS with the hypothesis that, if a patient's chronic disease diagnosis and set of medical advice are predicted and recommended with high accuracy, it is expected to reflect the improvement of patients' health conditions and lifestyle adjustments along with reducing the healthcare services costs. This can be considered as a 'core health recommendation system', as it directly focuses on the disease and the preventive side of the healthcare field, whereas the other HRS usually help a medical institution function better in other aspects. In another similar work, Kuanr et al. [6] proposed a HRS to help women by providing information on the features responsible for prognosis of cervical cancer in women. Cheung et al. [8] presented another review which outlines that incorporating multiple filtering, i.e. making a hybrid system could potentially add value to traditional tailoring with regard to enhancing the user experience. This study illustrates how recommender systems, especially hybrid programs, may have the potential to bring tailored digital health forward.

Considering the nature of recommender systems, it's not easy to confine them to some specific sectors. Traditional recommender systems are either collaborative or content-based (broadly speaking). In HRS, which type of recommender system should be used depends on the application. For instance, collaborative filtering might be used in an educational context, whereas content-based filtering would prove to have more impact in creating a doctor recommendation algorithm that takes into account the performance of doctors as well. In collaborative filtering, only the objective information regarding the items are stored, whereas, in content-based filtering, more comprehensive information is stored which gives rise to the following major privacy issue [4]: 'Combining data from multiple users (probably from different geographical locations) can be seen as an intrusion to the individual private data. It may even uncover some confidential data of healthcare institutions'. This poses a major challenge that violates the delicate topic about privacy which must be confidential in a healthcare system.

Apart from this major healthcare sector confined flaw, in general there are some other basic challenges that a recommender system faces (which are also applicable to HRS):

1. *Data sparsity*: If the recommender system is employed in very few places then the performance of the system may not be very promising. As it will only infer suggestions based on the data samples considered within its limited range, it will not follow into the standard footsteps of a recommender system which usually works on large data samples.
2. *Scalability*: If the users in the system scale to a very large number, say in millions, then the collaborative filtering algorithm will fail. The linear run time

algorithm (i.e. $\mathcal{O}(n)$ time-complexity) is undesirable in healthcare scenarios because the results should be generated in real-time.

3. *Diversity and the long tail*: The system won't be able to recommend remedies with a limited amount of historical data. If a remedy is not recommended to a set of patients with similar prognostic markers as the current user, then that remedy is unlikely to be recommended to the current user, even though that specific remedy could prove more beneficial than the rest.

In the healthcare sector, every peripheral context is as important as working on ailments and their remedies. All five health recommender systems [1,3,4,20] mentioned in Sect. 1.1, are aimed at providing recommendations in the healthcare context. Among these HRS(s), Hussein et al. [2] approach works on predicting diseases or recommending treatment. The CDD recommender system (as they coined it), acts as a core HRS since it predicts the disease and recommends medical advice. This type of HRS acts as an extra tool, by assisting the physicians and patients in controlling and managing the diseases. They have employed 'decision tree' algorithm in 'random forest' manner for disease prediction and used a 'unified collaborative filtering' approach for advice recommendation. This complete model seems a breakthrough in the HRS sector, however the model is built to predict and diagnose only the 'chronic diseases'. Following this context, the present research work aims to develop an HRS that:

1. Acts as a core HRS (acting as a tool to help recommend medical remedies).
2. Provides medical remedies to a wide range of diseases, not just the chronic diseases.

3   Methodology

Collaborative filtering and content-based filtering are the most common approaches to build a recommendation system. The proposed model of HRS is built using the collaborative filtering technique.

3.1   Collaborative Filtering

Collaborative filtering [29] as the name suggests, employs the use of collaboration. The underlying presumption of this approach is that if an individual A has a similar assessment as an individual B on a context, A is probable to have B's assessment on a different issue in comparison to that of a randomly picked individual. For instance, a collaborative filtering recommendation system for shopping preferences could make predictions about which outfits and accessories a client would like, given an partial list of that client's preferences. These preferences may include likes or dislikes, frequency of buying from a particular brand, the average spending amount, etc. It should be noted that, even though these predictions use data gathered from numerous clients, ultimately provides tailored predictions to individual client(s). This contrasts from the simpler methodology of giving a normalized rating for every item of interest.

The analogy to the system proposed in this paper holds as: If a person A was cured by the same treatment as person B given the same set of symptoms, A and B are more likely to be cured by the same treatment for a new set of common symptoms.

3.2 Dataset Synthesis

Acquiring the dataset for this system was one of the biggest challenges. There are plenty of datasets publicly available for the healthcare domain but most of them conform only to some specific category of illness (like heart diseases only, nervous system disorder only, etc.). This system requires a dataset that contains a list of diseases spread over various domains. The base dataset[1] hence chosen is taken from the profile of P. Larmuseau, Pharmacist at Gaver Apotheek (Harelbeke, Flanders, Belgium). The dataset consists of 8 files in .csv format. The primary data from the dataset contained the information arranged in tuples. Following list shows the labeling (of the tuples) as found in the files of the dataset in the format as *(file name: (tuple labels) = (corresponding alias))*:

1. sym_t.csv: (syd, symptom) = (Symptom identifier, Symptom name)
2. dia_t.csv: (did, diagnose) = (Disease identifier, Disease name)
3. diffsydiw.csv: (syd, did, wei) = (Symptom identifier, Disease identifier, Weight of the symptom on the disease)
4. prec_t.csv: (did, diagnose, pid) = (Disease identifier, Disease name, treatment course)

There was no good data sources available for all diseases (from generic to chronic) and their treatment courses. Hence the data file *prec_t.csv*[2] (name as in the code repository) is created by exploring several medical websites and resources [9, 12, 14, 17–19, 28].

3.3 Dataset Pre-processing

In order to develop the complete dataset for the proposed system, the base dataset (See Footnote 1) was cleaned using the following steps:

1. Dropping the rows if any of the attributes was *NULL*.
2. Removing unrecognizable delimiters and replacing them with commas (,).

3.4 Building a Sparse Matrix

The pre-processed data from the data-files is used to create a new matrix which is sparse in nature. For instance, the columns from files: sym_t.csv - (syd, symptom), dia_t.csv - (did, diagnose) and diffsydiw.csv - (syd, did, wei) are transformed into a sparse matrix $Data(i, j)$ such that $Data[i][j] \geq 0$ represents the

---

[1] https://www.kaggle.com/plarmuseau/sdsort.
[2] https://github.com/sud0lancer/Diagonosis-Precaution-dataset.

weight of the jth symptom on i^{th} disease, where higher value represents larger weight of a symptom for a disease and 0 represents that the symptom doesn't give rise to the corresponding disease. This matrix is considered as the source from where the system will generate the recommendation.

3.5 Normalization Using BM25 Weighting

Normalization was done to calculate the average weight (or importance) of a symptom for a disease in presence of other symptoms. BM25 weighting [25] scheme is used for this purpose. BM25 is considered to be a better version of the TF-IDF ranking functions. Main motive of these functions is to estimate the relevance (score, ranking) of a term in a huge text corpus. These functions employ the frequency and rarity of a term to compute their importance to the corpus. TF-IDF uses the Eq. 1 to compute the relevance score:

$$R_{score}(D,T) = termFrequency(D,T) * log\left(\frac{N}{docFrequency(T)}\right) \quad (1)$$

Here:

1. $R_{score}(D,T)$ = score of a term T in a document D,
2. $termFrequency(D,T)$ = how many times does the term T occur in document D.
3. $docFrequency(T)$ = in how many documents does the term T occur.
4. N = size of the search index or corpus.

Scores for all such documents in the corpus are added to get the final score (ranking) of a term. In contrast, BM25 adds modifications to compute the ranking score as follows:

1. A document length variable is added, where larger documents are penalised having the same term frequency as those of smaller documents.
2. After the term's frequency has reached saturation, further occurrences don't affect its score.

In BM25 [25], the ranking score can be computed using Eq. 2.

$$R_{score}(D,Q) = \sum_{t \in Q} \frac{f_{t,D} \cdot (k_1 + 1)}{f_{t,D} + k_1 \cdot (1 - b + b \cdot \frac{|D|}{avg(dl)})} \cdot log\frac{N - n_t + 0.5}{n_t + 0.5} \quad (2)$$

Here:

1. $\sum_{t \in Q}$ = sum the scores of each query term,
2. $\frac{f_{t,D}}{f_{t,D} + k_1}$ = term frequency saturation trick,
3. $\frac{1}{(1 - b + b \cdot \frac{|D|}{avg(dl)})}$ = adjust saturation curve based on document length,

4. $\frac{N - n_t + 0.5}{n_t + 0.5}$ = probabilistic flavour of IDF.

In this case, the analogy holds as: 'symptom' is 'term', 'list of diseases' is 'the huge text corpus'. After the normalization, the sparse matrix will now contain updated values. For example, $Data[i][j] = 3$ will have changed to either 2.8736 or 3.1252 (exact values may vary) depending on the effect of the other symptoms on the corresponding disease.

3.6 SVD and Cosine Similarity

Single value decomposition (SVD) in the context of recommendation systems is used as a collaborative filtering (CF) algorithm. It is used as a tool to factorize the sparse matrix to get better recommendations.

Let $R \in \mathbb{R}^{m \times n}$ be the original data matrix. Then after applying SVD, R breaks into the following 3 matrices as shown in Eq. 3:

1. U is a $m \times r$ orthogonal left singular matrix,
2. V is a $r \times n$ orthogonal right singular matrix,
3. S is a $r \times r$ diagonal matrix, such that

$$R = USV^T \tag{3}$$

The SVD decreases the dimensions of the original matrix R from $m \times n$ to $m \times r$ and $r \times n$ by extracting its latent factors. In our case, $R \in \mathbb{R}^{1145 \times 272}$ is reduced as: $U \in \mathbb{R}^{1145 \times 50}$, $S \in \mathbb{R}^{50 \times 50}$ and $V \in \mathbb{R}^{50 \times 272}$. The matrices U and V are used to find the recommendations.

Cosine similarity is the measure of similarity between two vectors. This similarity is calculated by measuring the cosine of the angle between the two vectors which may be projected into multidimensional space as shown in Eq. 4. In the proposed system, cosine similarity finds the n rows (where each row is represented by a disease) from the decomposed matrices, that have the maximum sum of the symptoms weight (which in turn means the diseases having the maximum matching symptoms). Here n is the number (a manual threshold) of diseases that we wish to generate for the given inputs.

$$\cos \theta = \frac{A \cdot B}{||A|| \cdot ||B||} = \frac{\sum_{i=1}^{n} Ai \cdot Bi}{\sqrt{\sum_{i=1}^{n} (Ai)^2} \cdot \sqrt{\sum_{i=1}^{n} (Bi)^2}} \tag{4}$$

4   Results and Discussion

The proposed system was tested in two phases in order to determine the working and the performance of the system respectively. In the first phase, the experiments are conducted keeping in mind the related and unrelated symptoms experienced by the patients that signify the real life scenarios. Sometimes a patient might be experiencing multiple symptoms but most of them hint towards a common disease and in the other cases, those symptoms may be completely unrelated to each other and hence the patient might be having multiple diseases. In the case of related symptoms, the HRS is expected to recommend the remedy for the disease that is most likely to happen because of the given multiple symptoms that are related to each other. And in the case of unrelated symptoms, the HRS must recommend the remedies for all the different possible diseases. Table 1 shows the two cases as mentioned above. 'Case1: Unrelated-symptoms' has an array of 'symptom_id' as input. It predicts the probable disease(s) and then recommends the best treatment. Likewise for 'Case2: Related-symptoms'. In both the cases, the system takes in an array of symptom IDs, then it predicts the most probable diseases as a result of the symptoms and then recommends the course of treatment.

Table 1. The remedy recommendation results of the proposed HRS for the given symptoms.

Case	Symptom ID	Symptoms	Most probable disease(s)	Best treatment(s)
Related symptoms	1	Upper abdominal pain	1: Ventral hernia: bulging of the abdominal wall	1: Eating smaller meals may help prevent bloating and swelling
	2	Lower abdominal pain	2: Diverticulosis: weakening of the large intestine wall	
Unrelated symptoms	2	Lower abdominal pain	1: Ventral hernia: bulging of the abdominal wall 2: Vitiligo: loss of skin pigment	1: Laparoscopic surgery 2: Photodynamic therapy, Medications: Steroid and
	81	Rash		Immunosuppresive drug

The second phase consists of further evaluations which incorporates the analysis of the quality of the predictions based on various types of testing. The first level of testing is the 'sanity testing'. In sanity testing, the dataset is divided into two halves while preserving the uniformity of the symptoms subgroups in the dataset. Both the halves are fed as the input to the system, and the corresponding similarity matrices generated must show minimal difference in the values (along the diagonal). Euclidean distance is used to find the similarity between the matrices. Henceforth it can be said that the proposed system is un-biased towards data. Matrices M_1 and M_2 show that the euclidean distance matrix has all diagonal values near 0, implying that the two similarity matrices are similar. M_1 is composed of two similarity matrices belonging to full dataset and one of the halves of the dataset, respectively. Similar results hold for the similarity of the other half of the dataset with the full dataset. M_2 is composed of two similarity matrices belonging to both the halves of the dataset.

$$M_1 = \begin{bmatrix} \mathbf{0.3610} & 2.8490 & 3.3920 & ... \\ 2.9624 & \mathbf{0.3843} & 3.2234 & ... \\ 3.3868 & 3.1321 & \mathbf{0.3062} & ... \\ ... & ... & ... & ... \end{bmatrix} \quad M_2 = \begin{bmatrix} \mathbf{0.3031} & 2.8507 & 3.3858 & ... \\ 2.8626 & \mathbf{0.1469} & 3.1463 & ... \\ 3.4013 & 3.1316 & \mathbf{0.4758} & ... \\ ... & ... & ... & ... \end{bmatrix}$$

```
A =    test_input = [1, 47, 67, 91]
```

```
       Most probable disease(s)
B =    1115 ['Ventral hernia : bulging of the abdominal wall']
       825 ['Vocal cord paralysis : voice box weakness']
       760 ['Thyroiditis : thyroid inflammation']
       1503 ['Scar']
```

```
       Most probable expected disease(s))
C =    1115 ['Ventral hernia : bulging of the abdominal wall']
       825 ['Vocal cord paralysis : voice box weakness']
       771 ['Torticollis, spasmodic : neck muscle spasm']
       1503 ['Scar']
```

Fig. 1. Given a set of symptoms (A), the predicted (B) and expected outputs (C) are shown respectively.

The second level of testing is the regression testing, in which we use a subset of the training data to generate the output and match with the training set, which proves that the model is correctly created and has proper similarity matrices. Figure 1 shows that when given a set of symptoms, system predicts 3 out of 4 diseases which resemble the ground truth (since they had the maximum weight in the training dataset). The remaining 4th prediction is also correct but it had a lower weight in the training dataset. Hence it can be said that the system predicts the most probable disease(s) for a given set of symptoms together, not the most probable disease for each individual symptom(s).

5   Conclusion and Future Scope

Indeed recommender systems play a major role in everybody's daily life covering online shopping, movie streaming, etc. With state-of-the-art potential of recommender system, this can be extended to healthcare department to aid in a variety of tasks such as managing resources of a healthcare institution, replacement suggestions for equipment(s), recommending medical advice and suggestions, etc. In this paper, the proposed model recommends remedies for the predicted disease(s) followed from the given symptoms by using the generated dataset[3] consisting of a list of most favourable remedies corresponding to a wide range of disease(s). The future scope for this system includes improving the dataset by categorising the mentioned remedies under labels such as: self-care, medication, surgical

---

[3] https://github.com/sud0lancer/Diagonosis-Precaution-dataset.

procedures, non-surgical procedures, therapies, etc., incorporating more filtering algorithms for better results, creating a web based UI for better interaction with the proposed system.

Acknowledgment. This research is supported by "ASEAN- India Science & Technology Development Fund (AISTDF)", SERB, Sanction letter no. - IMRC/AISTDF/R&D/P-6/2017. Authors are also thankful to the authorities of "Indian Institute of Information Technology, Allahabad at Prayagraj", for providing us with the infrastructure and necessary support.

References

1. Kim, J.-H., Lee, J.-H., Park, J.-S., Lee, Y.-H., Rim, K.-W.: Design of diet recommendation system for healthcare service based on user information. In: 2009 Fourth International Conference on Computer Sciences and Convergence Information Technology, November 2009
2. Hussein, A.S., Omar, W.M., Li, X., Ati, M.: Efficient chronic disease diagnosis prediction and recommendation system. In: 2012 IEEE-EMBS Conference on Biomedical Engineering and Sciences, December 2012
3. Huang, Y.-F., Liu, P., Pan, Q., Lin, J.-S.: A doctor recommendation algorithm based on doctor performances and patient preferences. In: 2012 International Conference on Wavelet Active Media Technology and Information Processing (ICWAMTIP), December 2012
4. Sezgin, E., Ozkan, S.: A systematic literature review on health recommender systems. In: 2013 E-Health and Bioengineering Conference (EHB), November 2013
5. Pincay, J., Terán, L., Portmann, E.: Health recommender systems: a state-of-the-art review. In: 2019 Sixth International Conference on eDemocracy & eGovernment (ICEDEG), April 2019
6. Kuanr, M., Mohapatra, P., Piri, J.: Health recommender system for cervical cancer prognosis in women. In: 2021 6th International Conference on Inventive Computation Technologies (ICICT), January 2021
7. Agarwal, S., et al.: Unleashing the power of disruptive and emerging technologies amid COVID-19: a detailed review. arXiv preprint arXiv:2005.11507 (2020)
8. Cheung, K.L., Durusu, D., Sui, X., de Vries, H.: How recommender systems could support and enhance computer-tailored digital health programs: a scoping review. Digit. Health **5** (2019). https://doi.org/10.1177/2055207618824727
9. Cleavelandclinic: my.cleavelandclinic.org (2021). https://my.cleavelandclinic.org. Accessed 1 May 2021
10. Park, D.-H., Kim, H.-K., Choi, I.-Y., Kim, J.K.: A literature review and classification of recommender systems on academic journals. J. Intell. Inf. Syst. **17**(1), 139–152 (2011)
11. GEOVIZ: Advantages of a recommendation system (2021). https://geo-viz.com/blog/advantages-of-a-recommendation-system. Accessed 4 June 2021
12. Healthline: healthline.com (2021). https://healthline.com. Accessed 2 May 2021
13. Hors-Fraile, S., et al.: Analyzing recommender systems for health promotion using a multidisciplinary taxonomy: a scoping review. Int. J. Med. Inform. **114**, 143–155 (2018)
14. Albert Einstein Hospital, Brazil (2021). https://einstein.br. Accessed 3 May 2021

15. Lu, J., Wu, D., Mao, M., Wang, W., Zhang, G.: Recommender system application developments: a survey. Decis. Support Syst. **74**, 12–32 (2015)
16. Kamran, M., Javed, A.: A survey of recommender systems and their application in healthcare. Tech. J. **20**, 111 (2015). University of Engineering and Technology (UET) Taxila, Pakistan
17. Mayoclinic: mayoclinic.org (2021). https://mayoclinic.org. Accessed 2 May 2021
18. Medicalnewstoday: medicalnewstoday.com (2021). https://medicalnewstoday.com. Accessed 3 May 2021
19. Medlineplus: Trusted health information for you (2021). https://medlineplus.gov. Accessed 2 May 2021
20. Mohammadi, N., Babaei, M.H.: Recommending an appropriate doctor to a patient based on fuzzy logic. Int. J. Curr. Life Sci. **4**(2), 403–407 (2014)
21. Peito, J.: Incorporating complex domain knowledge into a recommender system in the healthcare sector. Master's thesis, NSBE: NOVA - School of Business and Economics (2020)
22. Punn, N.S., Agarwal, S.: Multi-modality encoded fusion with 3D inception U-net and decoder model for brain tumor segmentation. Multimed. Tools Appl. **80**(20), 30305–30320 (2021). https://doi.org/10.1007/s11042-020-09271-0
23. Schäfer, H., et al.: Towards health (aware) recommender systems. In: Proceedings of the 2017 International Conference on Digital Health, pp. 157–161 (2017)
24. Towards Data Science: Recommendation systems - models and evaluation (2021). https://towardsdatascience.com/recommendation-systems-models-and-evaluation-84944a84fb8e. Accessed 4 June 2021
25. Seitz, R.: Understanding TF-IDF and BM25 (2020). https://kmwllc.com. Accessed 28 Oct 2020
26. Sumanth, U., Punn, N.S., Sonbhadra, S.K., Agarwal, S.: Enhanced behavioral cloning based self-driving car using transfer learning. arXiv preprint arXiv:2007.05740 (2020)
27. Underwood, C.: Use cases of recommendation systems in business - current applications and methods (2020). https://emerj.com/ai-sector-overviews/use-cases-recommendation-systems/. Accessed 8 Oct 2020
28. Webmd: Better information, better health (2021). https://webmd.com. Accessed 2 May 2021
29. Wikipedia: Collaborative filtering - Wikipedia, the free encyclopedia (2021). https://en.wikipedia.org/wiki/Recommender_system. Accessed 15 Dec 2020

Generative Adversarial Negative Imitation Learning from Noisy Demonstrations

Xin Cao[1,2] and Xiu Li[1(✉)]

[1] Tsinghua Shenzhen International Graduate School, Tsinghua University,
Shenzhen 518055, People's Republic of China
caox19@mails.tsinghua.edu.cn, li.xiu@sz.tsinghua.edu.cn
[2] Department of Automation, Tsinghua University, Beijing 100084,
People's Republic of China

Abstract. In imitation learning, we aim to learn a good policy from demonstrations directly. A limitation of previous imitation learning algorithms is that the demonstrations are required to be optimal, but collecting expert demonstrations is costly. Existing few robust imitation learning algorithms need additional annotations or strict noise distribution assumptions, which is usually impractical. In this work, we propose a novel generative adversarial negative imitation learning method to learn from noise demonstrations by training a classifier simultaneously exploiting the difference in the intrinsic attributes of the trajectories. Meanwhile, we derive a stopping criterion for the classifier for more stable performance. In particular, our method can be easily integrated into other adversarial imitation learning algorithms to improve their robustness. Experimental results in several continuous-control tasks and noise-rate demonstrations show that our method achieves state-of-the-art performance with only unlabeled trajectories.

Keywords: Imitation learning · Noisy demonstrations ·
Positive-unlabeled learning

1 Introduction

While Reinforcement Learning has shown to be a powerful tool for sequential decision-making [11,19,25], hand-crafted reward functions are usually extremely difficult to obtain, especially in complex settings [4,5]. Imitation learning methods have the potential to close this gap by learning a good policy directly from expert demonstrations. Common imitation learning methods include behavioral cloning and apprenticeship learning. Behavioral cloning does not require environment interactions, which learns from sufficient expert data through supervised learning, but is vulnerable to compounding error and covariate shift [15]. Adversarial imitation learning [9] and inverse reinforcement learning [1,13,26], two effective and relatively novel approaches to apprenticeship learning, which

T. Mantoro et al. (Eds.): ICONIP 2021, LNCS 13109, pp. 405–416, 2021.
https://doi.org/10.1007/978-3-030-92270-2_35

assume that several trajectories generated by an expert are available, can learn a reward function that is used to score trajectories. They all assume the optimal demonstrations are available. However, it is difficult to get high-quality demonstrations in many non-trivial applications.

There have been several works that try to learn good policies from suboptimal demonstrations. Valko et al. [23] assume that some expert demonstrations are available, and use a semi-supervised learning method to label the noisy demonstrations. Method of Wu et al. [24] needs a confidence score related to each trajectory determining the probability that unlabeled demonstrations come from experts. Brown et al. [3] require real demonstrations to be ranked according to their relative performance. Meanwhile, the method of Tangkaratt et al. [21] assumes that the distribution of noise is known in advance. These methods all require additional annotations or strict noise distribution assumptions, implying that their application scenarios are limited.

Recent studies focus on learning directly from unlabeled demonstrations. Tangkaratt et al. [20] proposed a robust imitation learning with a symmetric loss, but is unable to converge to the optimal performance without knowing the noise rate. Additionally, Sasaki et al. [16] put forward a behavioral cloning method, but experimental results show that it is difficult to deal with high noise rate. In conclusion, prior works do not perform well under high noise-rate conditions.

To address the limitations mentioned above, we propose a new generative adversarial negative imitation learning method for learning from noise demonstrations. A novel classifier is designed in this paper, reducing the noise rate in mixture demonstrations by utilizing the similarity between the trajectories generated in the imitation learning and the non-expert trajectories. It is trained simultaneously with the general imitation learning algorithm, which is actually a negative learning process. For the compounding error that may occur during training, we remit its impact by improving the generator and classifier structure. Lastly, we build on positive unlabeled learning to derive a stopping criterion for our method, making learning more stable. We call our method, combing Generative Adversarial Negative imitation Learning and Positive Unlabeled learning, GANL-PU. We summarize the main contributions of this paper as follows:

- We study the problem of learning from noisy demonstrations robustly. To address this problem, we propose a model-agnostic algorithm that can easily integrated into other imitation learning algorithms. It does not need additional annotations or strict noise distribution assumptions.
- We propose a novel classifier that can obtain competitive classification accuracy compared to other unsupervised clustering algorithms.
- We conduct experiments on different tasks and different noise-rate demonstrations, empirical results show that our method surpasses all baselines in the final performance.

2 Background

2.1 Preliminaries

We consider an infinite-horizon Markov Decision Process (MDP), in which the environment is specified by a tuple $(\mathcal{S}, \mathcal{A}, \mathcal{P}, \rho_0, r, \gamma)$ with state space $\mathcal{S}$, action space $\mathcal{A}$, dynamics $\mathcal{P} : \mathcal{S} \times \mathcal{A} \times \mathcal{S} \rightarrow \mathbb{R}_+$, initial state distribution ρ_0, reward function $r : \mathcal{S} \rightarrow \mathbb{R}$, and discount factor $\gamma \in (0,1)$. The expected discounted return of policy π is given by $\eta(\pi) = \mathbb{E}_\tau \left[\sum_{t=0} \gamma^t r_t \right]$, where $\tau = (s_0, a_0, \cdots, a_{T-1}, s_T)$ denotes the trajectory, $s_0 \sim \mathcal{P}$, $a_t \sim \pi(a_t \mid s_t)$, and $s_{t+1} \sim \mathcal{P}(s_{t+1} \mid s_t, a_t)$.

2.2 Generative Adversarial Imitation Learning

Generative Adversarial Imitation Learning (GAIL) [9] is an imitation learning method inspired by Generative Adversarial Networks (GANs) [6]. In GAIL, the generator imitates the expert policy by matching state-action (s, a) distribution with the expert's distribution. The discriminator serves as a surrogate reward to measure the similarity between the real and generated data. The formal GAIL objective is denoted as follows,

$$\min_\pi \max_{D \in (0,1)} \mathbb{E}_\pi[\log D(s,a)] + \mathbb{E}_{\pi_E}[\log(1 - D(s,a))], \tag{1}$$

where π is the agent policy we want to imitate π_E with, and D is the discriminator to guide π to improve. In practice, we often use Trust Region Policy Optimization (TRPO) [17] or Proximal Policy Optimization (PPO) [18] to update π with the surrogate reward function: $r = -\log D(s,a)$.

2.3 Learning from Non-optimal Trajectories

In this paper, we are concerned with learning from non-optimal trajectories, which consist of both expert trajectories $\{\tau_i^E\}_{i=1}^m$ and non-expert trajectories $\{\tau_i^N\}_{i=1}^n$. Expert trajectories have higher expected discounted returns than non-expert trajectories. However, we do not have access to the reward function of the MDP, accordingly, we can't label these trajectories for training. The main challenge is to learn a good policy from this mixed data set. Particularly, we assume that expert trajectories account for more than half of the collected data: $m > n$, which is a typical assumption in previous works [2,12].

3 Methodology

We propose a generative adversarial negative imitation learning method to separate non-expert demonstrations from expert demonstrations, consisting of a discriminator, a generator and two classifiers. Demonstrations screening and generator training are executed simultaneously, illustrated in the following sections.

3.1 Generative Adversarial Negative Imitation Learning

Given a sequence of k demonstrations with different qualities, we want to obtain a classifier that can select expert demonstrations without prior information about their labels. To settle this, we consider suboptimal demonstrations as anomalies whose inherent properties have a significant difference from optimal ones.

Figure 1 depicts a visualization of trajectories obtained by different levels of demonstrators. We collect these demonstrations using five policy snapshots trained by PPO in MuJoCo PointMaze [22]. As shown in Fig. 1, *the suboptimal demonstrations are not as concentrated as the optimal ones*, which is consistent with previous researches [21] and as the basic assumption of this paper.

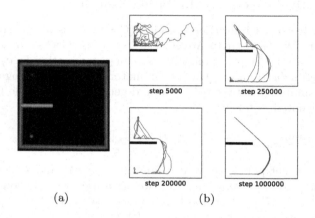

step 5000   step 250000

step 200000   step 1000000

(a)   (b)

Fig. 1. (a) The Point-Maze environment, (b) Trajectories of different time steps generated by PPO. Each subgraph has 5 trajectories, and the length of each demonstration is 100

Based on the above phenomenon and perception, we characterize the concentration of τ by the relative density $\rho(\tau)$ and respectively label expert or non-expert demonstrations as 1 or 0. If $\rho(\tau)$ is less than a certain threshold β, τ is classified as a suboptimal trajectory, and vice versa. Hence, we can construct a classifier $\mathcal{C}$ as

$$\mathcal{C}(\tau \mid \rho(\tau) \geq \beta) \rightarrow 1 \text{ and } \mathcal{C}(\tau \mid \rho(\tau) \leq \beta) \rightarrow 0. \tag{2}$$

However, since the length of the demonstrations is unequal, the calculation of $\rho(\tau)$ is complicated, precisely when the dimension of states increases. We randomly generate k trajectories as the potential suboptimal trajectories with a lower density in comparison to overcome this issue. In practice, we replace Eq. (2) with Eq. (3) to omit the calculation of $\rho(\tau)$:

$$\mathcal{L}_{\mathcal{C}} = -\frac{1}{2k} \sum_{i=1}^{2k} \left(y_i \log\left(\mathcal{C}\left(\tau_i\right)\right) + (1 - y_i) \log\left(1 - \mathcal{C}\left(\tau_i\right)\right) \right), \tag{3}$$

$$\mathcal{C}\left(\tau_i\right) = \mathbb{E}_{(s,a)\sim\tau_i}\left(\mathcal{C}(s,a)\right).   \tag{4}$$

where y_i is labeled as 1 or 0 when τ_i is drawn from real data or generated ones. By solving the above optimization problem, the classifier will give a higher value to expert demonstrations with a higher relative density $\rho(\tau)$. However, due to the lack of the actual distribution of non-expert trajectories and expert trajectories, those randomly generated trajectories far from the real data can't provide sufficient information for the classifier. Thus, we need to develop more informative negative samples to carve out the division boundary.

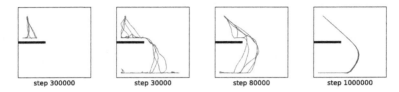

Fig. 2. Trajectories of different time steps generated by GAIL. Each subgraph has 5 trajectories, and the length of each demonstration is 100

Figure 2 shows the trajectories generated by the agent of GAIL during the training process. We observe that these trajectories are relatively close to the suboptimal samples in Fig. 1. Specifically, the agent can generate diverse-quality trajectories near the expert levels during the training process corresponding to experts of different levels. Therefore, the generated trajectories are very suitable as negative samples of the classifier. In practice, we associate agent trajectories with label 0, which is actually a negative learning method. We call this method Generative Adversarial Negative imitation Learning (GANL). We emphasize that we do not need to get the distribution of non-expert trajectories accurately but to access the decision boundary around the expert samples. The ultimate goal is to reduce suboptimal trajectories in the mixture data used for imitation learning. Our method's significant advantage over traditional binary classification algorithms is that the generator can capture deep representations of complex real data to generate more valuable negative samples to address the sample imbalance problem in classification.

3.2   Stopping Criterion

In addition, there are still two issues that need to be discussed. *The first issue is what the stopping criterion for the classifier is?* In the early stage, the agent behaves randomly, the classifiers can easily distinguish expert demonstrations from generated demonstrations. However, as policy improves, the distribution of generated data will be very close to the distribution of expert data. Then it will be difficult for the classifier to separate them. If we still label generated trajectories as negative, the classifiers will overfit to select some non-expert trajectories as positive, causing the unstable training. Inspired by the recent theoretical results of positive-unlabeled learning in GAN [8], we instead consider the agent trajectories τ_g as a mixture of positive outcomes τ_{gp} and negative outcomes τ_{gn}. Let $p_g(\tau)$ be the marginal density of τ_g, we have

$$p_g(\tau) = \mu p_{gp}(\tau) + (1 - \mu)p_{gn}(\tau), \tag{5}$$

where μ is the unknown class prior. We define a new classifier $\hat{\mathcal{C}}$ about τ_{gp} and τ_{gn}, its missclassification risks $R(\hat{\mathcal{C}})$ is calculated by

$$\min_{\hat{\mathcal{C}}} R(\hat{\mathcal{C}}) = \mu \mathbb{E}_{\tau \sim p_{gp}(\tau)}[\ell(\hat{\mathcal{C}}(\tau), 1)] + (1 - \mu)\mathbb{E}_{\tau \sim p_{gn}(\tau)}[\ell(\hat{\mathcal{C}}(\tau), 0)]. \tag{6}$$

However, we know very little about τ_{gp}. Based on the assumption in Sect. 2.3, half of the trajectories with the highest value τ_h are similar to expert trajectories at the end of the learning process, thus we can approximate τ_{gp} to τ_h. Additionally, the negative risk can be instead of positive and unlabeled generated data according to Eq. (5):

$$(1 - \mu)\mathbb{E}_{p_{gn}}[\ell(\hat{\mathcal{C}}(\tau), 0)] = \mathbb{E}_{p_g}[\ell(\hat{\mathcal{C}}(\tau), 0)] - \mu\mathbb{E}_{p_{gp}}[\ell(\hat{\mathcal{C}}(\tau), 0)]. \tag{7}$$

Then $R(\hat{\mathcal{C}})$ can be rewrited as follows,

$$\min_{\hat{\mathcal{C}}} R(\hat{\mathcal{C}}) = \mu\mathbb{E}_{p_h}[\ell(\hat{\mathcal{C}}(\tau), 1)] + \mathbb{E}_{p_g}[\ell(\hat{\mathcal{C}}(\tau), 0)] - \mu\mathbb{E}_{p_h}[\ell(\hat{\mathcal{C}}(\tau), 0)]. \tag{8}$$

When $\hat{\mathcal{C}}$ is complex, Eq. (8) has an overfitting problem as well. We find the original loss $(1 - \mu)\mathbb{E}_{p_{gn}}[\ell(\hat{\mathcal{C}}(\tau), 0)]$ is non-negative, but $\mathbb{E}_{p_g}[\ell(\hat{\mathcal{C}}(\tau), 0)] - \mu\mathbb{E}_{p_h}[\ell(\hat{\mathcal{C}}(\tau), 0)]$ may be less than 0. Through constructing a non-negative empirical estimator:

$$\min_{\hat{\mathcal{C}}} R(\hat{\mathcal{C}}) = \mu\mathbb{E}_{p_h}[\ell(\hat{\mathcal{C}}(\tau), 1)]$$
$$+ \max\left\{0, \mathbb{E}_{p_g}[\ell(\hat{\mathcal{C}}(\tau), 0)] - \mu\mathbb{E}_{p_h}[\ell(\hat{\mathcal{C}}(\tau), 0)]\right\}, \tag{9}$$

when τ_h is composed entirely of expert trajectories, the estimation error of Eq. (9) can be bounded [10]. Meanwhile, this method enjoys the same global equilibrium point as general GANs, proved by Guo et al. [8].

The task of the above classifier is to separate high-quality trajectories from low-quality trajectories, so it can be used not only as a classifier for τ_{gp} and τ_{gn}, but also as a classifier for expert trajectories τ^E and non-expert trajectories τ^N. It is implied that $\hat{\mathcal{C}}$ and $\mathcal{C}$ are equivalent, expert trajectories will be selected by minimizing Eq. (9).

3.3   Compounding Error

The second issue is how to alleviate the compounding error in the training process? We change the network structure to handle this problem and reduce the risks of misclassification. For the generator, we adopt the network structure of $d * d * \ldots * d$ with random orthogonal initial weights to generate data in the entire sample space. For the classifier, we train two classifiers denoted by $\hat{\mathcal{C}}_1$ and $\hat{\mathcal{C}}_2$ by minimizing the following empirical risks simultaneously:

$$
\begin{aligned}
\min_{\hat{\mathcal{C}}_1} R\left(\hat{\mathcal{C}}_1\right) &= \mu \mathbb{E}_{p_{h_1}}\left[\ell\left(\hat{\mathcal{C}}_1(\tau), 1\right)\right] \\
&+ \max\left\{0, \mathbb{E}_{p_g}\left[\ell\left(\hat{\mathcal{C}}_1(\tau), 0\right)\right] - \mu \mathbb{E}_{p_{h_1}}\left[\ell\left(\hat{\mathcal{C}}_1(\tau), 0\right)\right]\right\},
\end{aligned}
\tag{10}
$$

$$
\begin{aligned}
\min_{\hat{\mathcal{C}}_2} R\left(\hat{\mathcal{C}}_2\right) &= \mu \mathbb{E}_{p_{h_2}}\left[\ell\left(\hat{\mathcal{C}}_2(\tau), 1\right)\right] \\
&+ \max\left\{0, \mathbb{E}_{p_g}\left[\ell\left(\hat{\mathcal{C}}_2(\tau), 0\right)\right] - \mu \mathbb{E}_{p_{h_2}}\left[\ell\left(\hat{\mathcal{C}}_2(\tau), 0\right)\right]\right\},
\end{aligned}
\tag{11}
$$

where p_{h_1} and p_{h_2} respectively correspond to two disjoint subsets T_1 and T_2 of real data T: $\{\tau_i\}_{i=1}^k$. Meanwhile, T_{h1} from T_1 are selected according to the loss $\ell\left(\hat{\mathcal{C}}_2(\tau), 1\right)$, which are the half trajectories with the smallest loss in every sample, and T_{h2} are selected according to $\ell\left(\hat{\mathcal{C}}_1(\tau), 1\right)$. We use this method to reduce the influence of over-confident of single classifier [20]. We summarize the training procedure in Algorithm 1.

Algorithm 1. Generative Adversarial Negative Imitation Learning with PU (GANL-PU)

1: **Inputs:** policy π, discriminator D, classifiers $\hat{\mathcal{C}}_1$ and $\hat{\mathcal{C}}_2$, real data T, replay buffer $\mathcal{B}$, hyperparameter $N = 50$, $k_1 = 16$, $k_2 = 16$
2: Divide T equally into two subsets T_1 and T_2
3: **for** i = 1, 2, 3, ... **do**,
4:     sample k_1 trajectories from T, and choose $k_1/2$ samples with highest value based on $\hat{\mathcal{C}}_1$ or $\hat{\mathcal{C}}_2$ as $\mathcal{P}$
5:     **for** i = 1, 2, 3, ..., N **do**,
6:         run π to sample and save trajectories into $\mathcal{B}$
7:         sample k_2 trajectories from T_2, and choose $k_2/2$ samples with highest value based on $\hat{\mathcal{C}}_2$ as T_{h1}
8:         sample k_2 trajectories from T_1, and choose $k_2/2$ samples with highest value based on $\hat{\mathcal{C}}_1$ as T_{h2}
9:         update the parameters of $\hat{\mathcal{C}}_1$ by minimizing Eq. (10) using T_{h1} and $\mathcal{B}$
10:        update the parameters of $\hat{\mathcal{C}}_2$ by minimizing Eq. (11) using T_{h2} and $\mathcal{B}$
11:        update the parameters of D by minimizing Eq. (1) using $\mathcal{P}$ as expert data and $\mathcal{B}$
12:        update the parameters of policy π via a RL method using (s, a) from $\mathcal{B}$ and D with rewards: $r = -\log D(s, a)$
13:     **end for**
14: **end for**

---

4   Experiments and Results

We use three continuous control tasks from the Mujoco simulator within OpenAI Gym, namely HalfCheetah, Walker2D and Ant, to evaluate our method. Expert trajectories are obtained by running an agent on each task using the rewards defined in the simulator, the agent is well trained by PPO. One hundred trajectories with a length of 1000 state-action pairs are generated in each simulator by the optimal agent. In addition, we collect demonstrations of different levels by adding uniform noise $P \sim U(-\xi, \xi)$ to action samples drawn from the expert policy. As for each task, we choose five different ξ to obtain five sets of trajectories with the same amount, their average reward is shown in Table 1.

Table 1. Average cumulative rewards of diversity-quality demonstrations on benchmarks

Task	Expert	ξ_1	ξ_2	ξ_3	ξ_4	ξ_5
HalfCheetah	2500	−900	0	600	900	1300
Walker2D	2700	400	700	1100	1300	1600
Ant	3500	0	700	1000	1300	1900

We evaluate the performance of GANL-PU by comparing it against four baselines: GAIL, FAIRL [7], VILD [21] and RIL-Co [20]. Since our algorithm can be combined with any adversarial imitation learning method, we choose GAIL, the most widely used adversarial imitation learning algorithm, as our primary model. These methods have the same generators and discriminators. Generators with two hidden-layers of 64 units are orthogonally initialized and optimized using TRPO. The network structure of classifiers in GANL-PU is similiar to discriminators adopted, which have 2 hidden layers of 100 units. As for VILD and RIL-Co, we employ the same implementation in [20]. We choose 0.5 as the class prior μ in Eq. (9) for all the experiments.

4.1   Performance on Different Tasks

In this experiment, we generate a high noise-rate dataset including 100 optimal demonstrations and 80 suboptimal demonstrations, each ξ has 16 demonstrations, which corresponds to the noise rate of 0.4. We note that all experiments are tried over five same random seeds $\{0, 1, 2, 3, 4\}$. Empirical results in Fig. 3 show that our method can improve GAIL and outperform the SOTA method in HalfCheetah and Walker2D. More specially, our approach is able to take advantage of VILD and RIL-Co, and perform better and more efficiently than the original algorithm.

GAIL and FAIRL with logistic function have poor robustness, easily be affected by trajectory quality. Although they learn quickly at the beginning, their performance is terrible in the end. VILD assumes that the distribution of noise is known, so it has difficulty coping with unknown noise. In RIL-Co, discriminator regard the mixed data as expert samples, policy is hard to converge to the optimal without knowing noise rate. GANL in the experiment refers to GANL-PU in Algorithm 1 without stopping criterion. We can see that GANL's classification accuracy decreases in the later stage of Ant, but GANL-PU alleviates this issue by drawing inspiration from positive unlabeled learning. To sum up, Our method achieves the best performance with high-noise demonstrations in most tasks. Nevertheless, the flaw of GANL-PU is insufficient data-efficient, we attribute this to the fact that the expert data used for imitation learning changes in the early stage, which can be resolved by increasing the update frequency of expert data to improve the early learning speed.

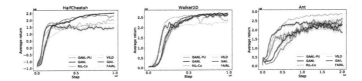

Fig. 3. Comparison on three evaluation tasks. Each curve is the average of 5 trials with a confidence interval of 0.95

4.2   Performance on Different Noise Rates

We further test our algorithm under different noise-rate demonstrations. The way of generating noise demonstrations is the same as the above section, but the number is reduced accordingly. We compare with baselines under four different noise rates in HalfCheetah. Figure 4 shows our method has the best final results under various noise rates, and RIL-Co comes second. However, RIL-Co destroys the integrity of the trajectory, so it will never reach the optimal under noise rate 0.

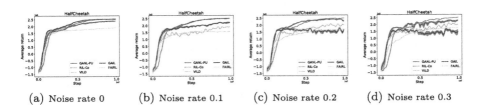

(a) Noise rate 0   (b) Noise rate 0.1   (c) Noise rate 0.2   (d) Noise rate 0.3

Fig. 4. Comparison on HalfCheetah under different noise rates. Each curve is the average of 5 trials with a confidence interval of 0.95

4.3   Evaluating the Impact of Class Prior μ

In this section, we study the impact of μ in Eq. (9). We test two additional choices 0.2 and 0.8. The other experiment settings are consistent with the previous sections. Empirical results shown in Fig. 5 indicate that our algorithm is not sensitive to μ. In contrast, 0.5 is the best choice, this agrees with the actual situation, agent trajectories gradually improve from entirely random to optimal.

Fig. 5. Evaluation result about the impact of class prior μ

When μ is 0.8, the learning curve will drop sharply in the later stage. The reason is that if μ is set too high, the stopping criterion is reached shortly after the start of the training, which results in the classifier not learning the mode of suboptimal trajectories well. With the continuous improvement of policy, when the threshold is reached again, the training of the part about generated trajectories in the loss function is restarted. At this time, misclassification is easy to happen, the high-quality trajectory is judged as negative, and the same problem as the original GANL occurs.

4.4   Evaluating the Accuracy of Classifier

To illustrate the advantage of our classifier, we naturally compare it with other unsupervised clustering algorithms. We experiment with four common algorithms: K-means, GMM, DBSCAN and Mean-shift, implemented on a common framework [14]. In practice, the length of the trajectories is uncertain, the trajectories cannot be clustered directly, so we use state-action pair (s, a) instead, the label of each trajectory is determined by the sum of the outcomes about (s, a). Meanwhile, we utilize HalfCheetah as the emulator and the noise rate is 0.4. Table 2 shows the experimental results, everyone in the table is the average value of five trials. It can be seen that density-based algorithms, DBSCAN and Mean-shift, achieve good classification results on our data set, which verifies our hypothesis: expert trajectories have higher relative density. And our method produces the best classification results on two data sets, which reveals why GAIL can be improved in our work.

Table 2. Comparison results of classification accuracy on high noise-rate demonstrations

Task	GANL-PU	K-means	GMM	DBSCAN	Mean-shift
HalfCheetah	1	0.636	0.988	0.971	0.943
Walker2D	1	0.767	0.722	0.733	0.600
Ant	0.868	0.630	0.741	0.945	**0.999**

5   Conclusion

In this paper, we introduce a novel generative adversarial negative learning method for learning from noise unlabeled demonstrations. This is an unsupervised and model-agnostic technique, which can be combined with most generative adversarial imitation learning algorithm. Especially, our method does not require any annotation nor any assumptions about the noise distribution. We empirically evaluate our approach on a variety of tasks, the results indicate it outperforms state-of-the-art robust imitation learning techniques. The main disadvantage of our algorithm is low sample efficiency. We defer a more effective method for addressing this problem to future works.

Acknowledgments. The paper is supported by project of the Guoqiang Research Institute of Tsinghua University (No. 2020GQG1001).

References

1. Abbeel, P., Ng, A.Y.: Apprenticeship learning via inverse reinforcement learning. In: Proceedings of the Twenty-first International Conference on Machine Learning, p. 1 (2004)
2. Angluin, D., Laird, P.: Learning from noisy examples. Mach. Learn. **2**(4), 343–370 (1988)
3. Brown, D.S., Goo, W., Nagarajan, P., Niekum, S.: Extrapolating beyond suboptimal demonstrations via inverse reinforcement learning from observations. arXiv e-prints pp. arXiv-1904 (2019)
4. Chen, S., Chen, Z., Wang, D.: Adaptive adversarial training for meta reinforcement learning. arXiv preprint arXiv:2104.13302 (2021)
5. Chen, Z., Wang, D.: Multi-initialization meta-learning with domain adaptation. In: ICASSP 2021–2021 IEEE International Conference on Acoustics, Speech and Signal Processing (ICASSP), pp. 1390–1394. IEEE (2021)
6. Creswell, A., White, T., Dumoulin, V., Arulkumaran, K., Sengupta, B., Bharath, A.A.: Generative adversarial networks: An overview. IEEE Sig. Process. Mag. **35**(1), 53–65 (2018)
7. Ghasemipour, S.K.S., Zemel, R., Gu, S.: A divergence minimization perspective on imitation learning methods. In: Conference on Robot Learning, pp. 1259–1277. PMLR (2020)
8. Guo, T., et al.: On positive-unlabeled classification in GAN . In: Proceedings of the IEEE/CVF Conference on Computer Vision and Pattern Recognition, pp. 8385–8393 (2020)

9. Ho, J., Ermon, S.: Generative adversarial imitation learning. In: Advances in Neural Information Processing Systems 29 (NIPS 2016), pp. 4565–4573 (2016)
10. Kiryo, R., Niu, G., Plessis, M.C.d., Sugiyama, M.: Positive-unlabeled learning with non-negative risk estimator. arXiv preprint arXiv:1703.00593 (2017)
11. Lyu, J., Ma, X., Yan, J., Li, X.: Efficient continuous control with double actors and regularized critics. arXiv preprint arXiv:2106.03050 (2021)
12. Natarajan, N., Dhillon, I.S., Ravikumar, P., Tewari, A.: Learning with noisy labels. In: NIPS (2013)
13. Ng, A.Y., Russell, S.J., et al.: Algorithms for inverse reinforcement learning. In: ICML. vol. 1, p. 2 (2000)
14. Pedregosa, F., et al.: Scikit-learn: machine learning in Python. J. Mach. Learn. Res. **12**, 2825–2830 (2011)
15. Ross, S., Bagnell, D.: Efficient reductions for imitation learning. In: Proceedings of the Thirteenth International Conference on Artificial Intelligence and Statistics, pp. 661–668. JMLR Workshop and Conference Proceedings (2010)
16. Sasaki, F., Yamashina, R.: Behavioral cloning from noisy demonstrations. In: International Conference on Learning Representations (2020)
17. Schulman, J., Levine, S., Abbeel, P., Jordan, M., Moritz, P.: Trust region policy optimization. In: International Conference on Machine Learning, pp. 1889–1897. PMLR (2015)
18. Schulman, J., Wolski, F., Dhariwal, P., Radford, A., Klimov, O.: Proximal policy optimization algorithms. arXiv preprint arXiv:1707.06347 (2017)
19. Singh, S., Okun, A., Jackson, A.: Learning to play go from scratch. Nature **550**(7676), 336–337 (2017)
20. Tangkaratt, V., Charoenphakdee, N., Sugiyama, M.: Robust imitation learning from noisy demonstrations. arXiv preprint arXiv:2010.10181 (2020)
21. Tangkaratt, V., Han, B., Khan, M.E., Sugiyama, M.: Variational imitation learning with diverse-quality demonstrations. In: International Conference on Machine Learning, pp. 9407–9417. PMLR (2020)
22. Todorov, E., Erez, T., Tassa, Y.: Mujoco: A physics engine for model-based control. In: 2012 IEEE/RSJ International Conference on Intelligent Robots and Systems, pp. 5026–5033. IEEE (2012)
23. Valko, M., Ghavamzadeh, M., Lazaric, A.: Semi-supervised apprenticeship learning. In: European Workshop on Reinforcement Learning, pp. 131–142. PMLR (2013)
24. Wu, Y.H., Charoenphakdee, N., Bao, H., Tangkaratt, V., Sugiyama, M.: Imitation learning from imperfect demonstration. In: International Conference on Machine Learning, pp. 6818–6827. PMLR (2019)
25. Yang, R., Yan, J., Li, X.: Survey of sparse reward algorithms in reinforcement learning - theory and experiment. CAAI Trans. Intell. Syst. **15**(5), 888–899 (2020)
26. Ziebart, B.D., Maas, A.L., Bagnell, J.A., Dey, A.K., et al.: Maximum entropy inverse reinforcement learning. In: AAAI, vol. 8, pp. 1433–1438. Chicago, IL, USA (2008)

Detecting Helmets on Motorcyclists by Deep Neural Networks with a Dual-Detection Scheme

Chun-Hong Li[1,2] and Dong Huang[1,2(✉)]

[1] College of Mathematics and Informatics, South China Agricultural University,
Guangzhou, China
lch666@stu.scau.edu.cn, huangdonghere@gmail.com
[2] Guangzhou Key Laboratory of Intelligent Agriculture, Guangzhou, China

Abstract. This paper deals with the problem of helmet detection on motorcyclists in single images. Some previous attempts have been made to detect helmets on motorcyclists, most of which are designed for videos and not suitable for single-image helmet detection. In this paper, we propose a single-image motorcyclist helmet detection method via deep neural networks with a dual-detection scheme. Two types of detectors are first trained, namely, the rider detection and the head-shoulder detection. A post-processing strategy is devised to remove duplicate detections and pedestrians (without motorbikes). Furthermore, we present a new benchmark dataset named SCAU helmet detection on motorcyclists (SCAU-HDM). Extensive experimental results demonstrate the effectiveness of the proposed method.

1 Introduction

Wearing a helmet is important for the safety of the rider of motorbike (i.e., the motorcyclist). In fact, it is a mandatory measure in many countries for the motorcyclists to wear a helmet. And there is a need to automatically detect whether a motorcyclist is wearing a helmet for public management.

In the field of computer vision, there have been a variety of previous works aiming to deal with the task of helmet detection on motorcyclists, where the techniques of the histograms of oriented gradient (HOG) [3], the scale-invariant feature transform [11], the local binary pattern (LBP) [5], and the deep convolutional neural network (CNN) [19] are often employed. Dahiya et al. [2] proposed an automatic motorcyclist helmet detection method for surveillance videos. It first detects the motorcyclist through background subtraction and object segmentation by Gaussian mixture model (GMM) [4]. Then it locates the head of the rider and decides whether the helmet is worn in the upper area of the motorcyclist [4]. Silva et al. [15] exploited a different feature extraction method after the background subtraction, where the HOG, the LBP, and the circular Hough transform are used for motorcyclist helmet detection. Instead of using hand-crafted features, Vishnu et al. [19] utilized CNN to detect motorcyclists

T. Mantoro et al. (Eds.): ICONIP 2021, LNCS 13109, pp. 417–427, 2021.
https://doi.org/10.1007/978-3-030-92270-2_36

Fig. 1. Some examples of false detection. *up*: Duplicate detections. *down*: Interference of head-shoulder detections caused by pedestrians.

without helmets in videos. They also use adaptive background subtraction on video frames to discover moving objects. Similar to Silva et al. [15], they cropped the upper part of an image as the potential head region of the motorcyclist [19]. On the one hand, these methods rely on the GMM for background subtraction and are only applicable to the videos with static background, which cannot be used for motorcyclist helmet detection in the dynamic background or a single image. On the other hand, these methods tend to locate the potential region of the motorcyclist's head coarsely and cannot provide accurate location of head and helmet.

To address the above issues, in this paper, we propose a novel approach for single-image helmet detection on motorcyclists with a dual-detection scheme by deep neural networks. Particularly, the rider detector and the head-shoulder detector [6] are trained, respectively. Considering the difference between bikes and motorbikes, bicyclists and motorcyclists are also treated separately. In addition, for better helmet detection, we utilize head-shoulder detection instead of head detection to precisely detect head region with/without a helmet. An efficient and accurate one-stage object detector by EfficientDet [18] is exploited for helmet detection on motorcyclists. Through the analysis of typical false detections (as shown in Fig. 1), we present a post-processing strategy to remove duplicate detections and eliminate the head-shoulder bounding box of pedestrians. Furthermore, a new benchmark dataset termed SCAU Helmet Detection on Motorcyclists (SCAU-HDM) is constructed for the motorcyclist helmet detection task, which includes 2900 images in the training set and 300 images in the test set. Experimental results on the SCAU-HDM dataset have demonstrated the effectiveness of the proposed motorcyclist helmet detection approach.

2 Proposed Approach

This section describes the proposed approach for detecting helmets on motorcyclists. We first present the dual-detection scheme of our approach in Sect. 2.1, then introduce the network architecture in Sect. 2.2, and finally describe the post-processing strategy in Sect. 2.3.

2.1 Dual-Detection Scheme

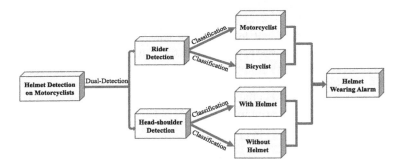

Fig. 2. The architecture of the dual-detection scheme.

For detecting helmets on motorcyclists, many attempts have been made in the literature. Lin et al. [7] utilized deep learning to directly detect motorbikes and helmets by one detector, which locates the motorcyclist and classifies the helmet-wearing status in the same bounding box, which, however, fails to distinguish motorcyclists and bicyclists and cannot locate the head and helmet of riders. In fact, bicyclists are very similar to motorcyclists, and bicyclists are generally not required to wear helmets, so it is necessary to distinguish bicyclists from motorcyclists. Another category of methods [2,14,15,19] exploit two types of classifiers. First, they discover motorcyclists by background subtraction and feature extraction. After the region of interest (ROI) extraction about the head of the motorcyclist, they further judge whether the rider is wearing a helmet or not through another binary classifier. These methods [2,14,15,19] treat the upper region of the image or the motorcyclist as the ROI of the motorcyclist's head, which may not be suitable for more complicated scenarios.

Different from previous works on motorcyclist helmet detection, in this paper, we divide the task of helmet detection on motorcyclists into two sub-tasks of object detection, namely, rider detection and head-shoulder detection. The overall process of the dual-detection scheme is shown in Fig. 2. Considering that bicyclists and motorcyclists have very similar characteristics, we first treat the bicyclist detection as a branch of rider detection. Notably, bicyclist detection also acts like the negative sample to motorcyclist detection, which leads to better distinctions between bicyclists and motorcyclists. Besides, instead of detecting the head, we exploit the head-shoulder detection in this paper. On the one hand, the head-shoulder detection not only locates precisely the rider's head, but also makes helmet-wearing detection independent of motorcyclist detection. On the other hand, compared with head detection, the head-shoulder detection can achieve more robust performance by taking advantage of more structure information (of head and shoulder).

2.2   Network Architecture

In the proposed approach, we adopt the EfiicientDet [18] for detecting helmets on motorcyclists. Typically, EfficientDet exploits EfficientNet [17] as the backbone network, the bi-directional feature pyramid network (BiFPN) [18] as the feature network, and the focal loss to solve the extreme imbalance of positive and negative samples in one-stage object detection.

EfficientNet achieves adaptive network optimization based on neural structure search technique with a compound scaling method. The compound scaling method uses a compound coefficient ϕ to uniformly scale the network width, depth, and resolution in a principled way:

$$depth : d = \alpha^{\phi} \tag{1}$$
$$width : w = \beta^{\phi}$$
$$resolution : r = \gamma^{\phi}$$
$$s.t. \quad \alpha \cdot \beta^2 \cdot \gamma^2 \approx 2$$
$$\alpha \geq 1, \beta \geq 1, \gamma \geq 1$$

where α, β, and γ respectively manage the extra resources assignment to network depth, width, and resolution. In Eq. (1), we first fix $\phi = 1$, through a grid search of α, β, and γ, and obtain the best values of α, β, and γ for EfficientNet-B0. To scale up the baseline network and obtain EfficientNet-B1 to B7, we can fix α, β, and γ as constants and adjust different ϕ.

EfficientDet first introduces BiFPN as the feature network, which employs cross-scale feature fusion and introduces learnable weights to reflect the importance of different input features. Note that the layers of BiFPN can be increased linearly by using the compound coefficient ϕ.

With the different compound coefficient ϕ, we can build different models with EfficientDet-D0 ($\phi = 0$) to D7 ($\phi = 7$). As shown in Tables 2 and 3, EfficientDet-D5 has achieved the best performance for our task. More experimental details are provided in Sect. 3.

2.3 Post-Processing Module

Many detection models adopt non-maximum suppression (NMS) as the post-processing method, which is an essential component for removing duplicate detections. However, NMS fails to eliminate duplicate detections with cross-class. As shown in Fig. 1, the common post-processing can not remove the duplicate detections with different classes or well eliminate the interference of head-shoulder detections caused by pedestrians.

Fig. 3. Detections after the proposed post-process. Duplicate detections and interference of head-shoulder detections caused by pedestrians are suppressed (compare to Fig. 1).

Based on the observation, this paper introduces a new post-processing strategy based on NMS to refine the detection outputs. To pursue dispelling tedious duplicate detections, our intuition is simple: in the first place, given a threshold, we think of two classes as the mutex in the same object detection. For example, the motorcyclist and the bicyclist are treated as mutex in rider detection. Finally, Once the value of intersection over union (IoU) of two different classes which are treated as the mutex is higher than the threshold, we obtain the higher score of the bounding box and eliminate the lower one. Moreover, for more accurate detection, we present a heuristic judgment to remove pedestrians detections. We remove the bounding box of head-shoulder detection which is out of any rider boxes.

Figure 3 illustrates the results after using our post-processing strategy, which shows that the duplicate detections and pedestrians detections are successfully removed.

3   Experiments

3.1   Dataset and Evaluation Metrics

In this paper, we constructed a new helmet detection on motorcyclists dataset named SCAU-HDM in the wild to facilitate the research. Table 1 shows the statistics of the collected images and their annotations. Our dataset contains 300 test images and 2900 training images, which include more than 700 images from the Kaggle dataset[1] and the others from the Internet. Besides, all reported results follow the standard COCO-style Average Precision (AP) metric, which averages mAP of IoUs from 0.5 to 0.95 with an interval of 0.05.

Table 1. Statistics of annotations in SCAU-HDM.

Dataset	# image	# With helmet	# Without helmet	# Bicyclist	# Motorcyclist
Trian	2900	2139	2743	902	3250
Test	300	242	343	65	437
Total	3200	2381	3086	967	3687

3.2   Implementation Details

This paper uses the open-source implementation of EfficientDet for experiments. By default, the model of EfficientDet-D5 is trained on 2 GPUs of 1080Ti with a total of 8 images per batch (4 images per GPU). Meanwhile, training every 1000 batches, we set up a checkpoint. Moreover, we use AdamW optimizer with $\alpha = 1e-4$ and employ commonly-used focal loss with $\alpha = 0.25$ and $\gamma = 2.0$. In addition, We only use simple flipping augmentation. Notably, no auto-enhancement [20] is employed in the EfficientDet model. Finally, for model inference, we employ NMS with a threshold of 0.2.

---

[1] https://www.kaggle.com/brendan45774/bike-helmets-detection.

Table 2. Results in EfficientDet D0-D6 on SCAU-HDM. Best result is **EfficientDet-D5**.

Model	AP	AP_{50}	AP_{75}	AP_S	AP_M	AP_L
EfficientDet-D0	46.4	64.4	53.6	8.4	41.4	47.7
EfficientDet-D1	57.6	76.5	65.2	1.9	45.9	59.4
EfficientDet-D2	61.4	80.2	70.3	16.8	47.8	63.3
EfficientDet-D3	64.6	82.5	73.6	16.3	56	66.3
EfficientDet-D4	68.5	**86.4**	79.7	**26.2**	**60.3**	70.3
EfficientDet-D5	**69.7**	**86.4**	**81.3**	22.4	58.9	**70.9**
EfficientDet-D6	53.1	70.7	61.5	6.7	44.6	54.9

3.3 Results and Analysis

In this section, we evaluate EfficientDet on SCAU-HDM test set and compare it with other state-of-the-art one-stage and classic two-stage detectors. And we will discuss some of the reasons for the results.

The Accuracy of EfficientDet. All results about EfficientDet are shown in Table 2. Obviously, As the value of ϕ increases, the network of the model becomes more complex. In the other words, different EfficientDet models are obtained through compound scaling. At the same, During the EfficientDet-D0 ($\phi = 0$) to D5 ($\phi = 5$), the accuracy and efficiency increase rapidly with the augment of . However, when $\phi = 6$, EfficientDet-D6 achieves 53.1 AP, which is 16.6 points lower than EfficientDet-D5. We finally find EfficientDet-D5 is the most suitable model for our task.

Table 3. The results of different models on SCAU-HDM.

Methods	Backbone	AP	AP_{50}	AP_{75}
One-stage detectors				
EfficientDet-D5	EfficientNet-B5	**69.7**	86.4	**81.3**
YOLOv3	Darknet	66.7	**89.9**	80.8
YOLOv4	CSPdarknet	67.2	89.1	78.6
RetinaNet	ResNet-50	66.8	87.2	78.3
RetinaNet	ResNet-101	64.2	86.5	74.9
SSD	VGG-16	66.1	87.5	75.4
Two-stage detectors				
Faster R-CNN on FPN	VGG-16	58.5	87.5	64.2
Faster R-CNN on FPN	ResNet-50	54.6	83.2	57.4

Comparison with Other Models. The results of comparing detectors are shown in Table 3. As for the classic two-stage detector, using feature pyramid

Table 4. Ablation study for post-process module. The symbol 'P' means our post-process.

	AP	AP_{50}	AP_{75}
EfficientDet-D5	69.7	86.4	81.3
EfficientDet-D5 with P	**69.8**	**86.5**	81.3

network (FPN) [8] as the feature network, Faster R-CNN [13] using VGG-16 [16] as backbone achieves 58.5 AP, which is 3.9 points higher than Faster R-CNN based on ResNet50. Besides, As state-of-the-art one-stage detectors, SSD [10], YOLOv3 [12], and YOLOv4 [1] skip the region proposal step and conduct classification and location directly through from anchor. RetinaNet [9] constructs a backbone network similar to FPN by using the residual network principle and uses focal loss to solve the problem of imbalance of positive and negative samples. YOLOv3, YOLOv4, RetinaNet-50, and SSD achieve accuracy and efficiency closely. Furthermore, EfficientDet outperforms YOLOv4 by 2.5 AP, which achieves the highest AP of these comparing detectors. Noticeably, RetinaNet using ResNet-101 as backbone declines using ResNet-50 by 4.6 AP. It further proves that the particular task needs a specific network instead of the more complex network. Finally, there is no doubt that EfficientDet-D5 is the most appropriate model for helmet detection on motorcyclists.

3.4   Ablation Analysis

In this section, we conduct two ablation experiments to verify the effects of the proposed post-process in the task on SCAU-HDM.

We use the EfficientDet-D5 model for detecting helmets on motorcyclists. As shown in Table 4, utilizing the proposed post-process contributes higher to the performance of helmet detection on motorcyclists.

3.5   Visualization of Inference Results

Through using the method which is proposed by this paper, we achieve the task of helmets in motorcyclists detection with high accuracy and efficiency. Meanwhile, the duplicate detections and pedestrians detections are suppressed effectively. The examples of test images on SCAU-HDM are shown in Fig. 4.

(a)   (b)   (c)

(d)   (e)   (f)

Fig. 4. Examples of test images of the helmet detection on motorcyclists.

4   Conclusion

In this paper, we propose a novel method for the task of single-image motorcyclist helmet detection with a dual-detection scheme via deep neural networks. First, a dual-detection scheme is proposed for independent motorcyclist and helmet detections. EfficientDet is used with compound scaling. In addition, considering the potential duplicate detections and interference detections caused by pedestrians, we introduce a post-processing strategy to refine the detection results. Moreover, in this paper, we present a new benchmark dataset with object detection annotations for the task of single-image helmet detection on motorcyclists, which is termed SCAU-HDM. Finally, experimental results on the benchmark dataset have shown the effectiveness of the proposed method.

Acknowledgement. This work was supported by the Project of Guangzhou Key Laboratory of Intelligent Agriculture (201902010081), the NSFC (61976097), and the Natural Science Foundation of Guangdong Province (2021A1515012203).

References

1. Bochkovskiy, A., Wang, C.Y., Liao, H.Y.M.: YOLOv4: optimal speed and accuracy of object detection. arXiv preprint arXiv:2004.10934 (2020)
2. Dahiya, K., Singh, D., Mohan, C.K.: Automatic Detection of Bike-riders without Helmet using surveillance videos in real-time. In: Proceedings of the International Joint Conference on Neural Networks, pp. 3046–3051 (2016)
3. Dalal, N., Triggs, B.: Histograms of oriented gradients for human detection. In: Proceedings of the IEEE Conference on Computer Vision and Pattern Recognition, pp. 886–893 (2005)
4. Friedman, N., Russell, S.: Image segmentation in video sequences: a probabilistic approach. arxiv preprint arXiv:1302.1539 (2013)
5. Guo, Z., Zhang, L., Zhang, D.: A Completed modeling of local binary pattern operator for texture classification. IEEE Trans. Image Process. **19**, 1657–1663 (2010)
6. Li, M., Zhang, Z., Huang, K., Tan, T.: Estimating the number of people in crowded scenes by MID based foreground segmentation and head-shoulder detection. In: Proceedings of the International Conference on Pattern Recognition, pp. 1–4 (2008)
7. Lin, H., Deng, J.D., Albers, D., Siebert, F.W.: Helmet use detection of tracked motorcycles using cnn-based multi-task learning. IEEE Access **8**, 162073–162084 (2020)
8. Lin, T.Y., Dollár, P., Girshick, R., He, K., Hariharan, B., Belongie, S.: Feature pyramid networks for object detection. In: Proceedings of the IEEE Conference on Computer Vision and Pattern Recognition, pp. 2117–2125 (2017)
9. Lin, T.Y., Goyal, P., Girshick, R., He, K., Dollár, P.: Focal Loss for Dense Object Detection. In: Proceedings of the IEEE International Conference on Computer Vision. pp. 2980–2988 (2017)
10. Liu, W., et al.: SSD: single Shot multibox detector. In: Proceedings of the European Conference on Computer Vision, pp. 21–37 (2016)
11. Lowe, D.G.: Distinctive image features from scale-invariant keypoints. Int. J. Comput. Vis.**60**, 91–110 (2004)
12. Redmon, J., Farhadi, A.: YOLOv3: an incremental improvement. arXiv preprint arXiv:1804.02767 (2018)
13. Ren, S., He, K., Girshick, R., Sun, J.: Faster R-CNN: towards real-time object detection with region proposal networks. arXiv preprint arXiv:1506.01497 (2015)
14. Shine, L., Jiji, C.V.: Automated detection of helmet on motorcyclists from traffic surveillance videos: a comparative analysis using hand-crafted features and CNN. Multimedia Tools Appl, **79**, 14179–14199 (2020)
15. e Silva, R.R., Aires, K.R., de MS Veras, R.: Detection of helmets on motorcyclists. Multimedia Tools Appl. **77**, 5659–5683 (2018)
16. Simonyan, K., Zisserman, A.: Very deep convolutional networks for large-scale image recognition. arXiv preprint arXiv:1409.1556 (2014)
17. Tan, M., Le, Q.: EfficientNet: rethinking model scaling for convolutional neural networks. In: Proceedings of the International Conference on Machine Learning, pp. 6105–6114 (2019)

18. Tan, M., Pang, R., Le, Q.V.: EfficientDet: scalable and efficient object detection. In: Proceedings of the IEEE Conference on Computer Vision and Pattern Recognition, pp. 10781–10790 (2020)
19. Vishnu, C., Singh, D., Mohan, C.K., Babu, S.: Detection of motorcyclists without helmet in videos using convolutional neural network. In: Proceedings of the International Joint Conference on Neural Networks, pp. 3036–3041 (2017)
20. Zoph, B., Cubuk, E.D., Ghiasi, G., Lin, T.Y., Shlens, J., Le, Q.V.: Learning data augmentation strategies for object detection. In: Proceedings of the European Conference on Computer Vision, pp. 566–583 (2020)

Short-Long Correlation Based Graph Neural Networks for Residential Load Forecasting

Yiran Deng[1], Yingjie Zhou[1(✉)], and Zhiyong Zhang[2]

[1] College of Computer Science, Sichuan University, Chengdu 610065, China
yjzhou@scu.edu.cn
[2] Cyberspace Security Key Laboratory of Sichuan Province, China Electronic Technology Cyber Security Co. Ltd., Chengdu 610041, China

Abstract. Accurate residential load forecasting is crucial to the future smart grid since its fundamental role in efficient energy distribution and dispatch. Compared with aggregated electricity consumption forecasting, predicting residential load of an individual user is more challenging due to the stochastic and dynamic characteristics of electricity consumption behaviors. Existing methods did not fully explore the intrinsic correlations among different types of electricity consumption behaviors, which restricts the performance of these methods. To fill this gap, this paper proposes a residential load forecasting method employing graph neural networks (GNN) to make full use of the intrinsic dependencies among various types of electricity consumption behaviors. Specifically, two kinds of graphs are constructed to leverage the dependence information, *i.e.*, short-term dynamic graphs are constructed for describing correlations among different appliances' electricity consumption behaviors only a short time ago, while long-term static graphs are built to profile a more general pattern for the internal structure of individual electricity consumption. Both short-term and long-term correlations are restricted mutually through the fusion of these two graphs. GNN is then employed to learn the implied dependencies from both the fused graphs and time-series data for load forecasting. Experiment results on a real-world dataset demonstrate the advantages of the proposed model.

Keywords: Time-series forecasting · Graph neural networks · Smart grid · Residential electricity consumption

1 Introduction

Residential load forecasting aims at obtaining an estimation of the residents' future electricity consumption. Accurate load forecasting could be used to guide the process of electricity transmission, which is able to reduce the transmission loss due to unscheduled distribution and dispatch. In smart grids, real-time load forecasting is also the fundamental step of demand response to realize peak load

© Springer Nature Switzerland AG 2021
T. Mantoro et al. (Eds.): ICONIP 2021, LNCS 13109, pp. 428–438, 2021.
https://doi.org/10.1007/978-3-030-92270-2_37

shaving [1]. The peak load snowballs with an increasing number of end users and devices connected to the power grids. Peak load shaving could decrease the risk of power system failure triggered by high peak load and maintain the stable operation of power grids [2]. Therefore, forecasting residential electricity consumption is of vital importance to the efficient operation of power grids [3].

Residential load forecasting could be categorized into two kinds: aggregated residential load forecasting and individual residential load forecasting. The former predicts the total load of multiple users in a region (*e.g.*, a neighborhood), while the latter estimates a single user's electricity consumption. Aggregated residential load forecasting [4–7] has been well studied. However, individual residential load forecasting is still a challenging task due to the stochasticity of various electricity consumption behaviors and the volatility of appliances' power consumption. Recently, a number of deep learning-based approaches [1,8,9] have been developed to address the problem and try to obtain satisfying performance. Nevertheless, they only utilize the overall load of a resident, neglecting the dependencies among different types of electricity consumption behaviors. Among the few methods that employ appliance-level load data to make use of the intrinsic correlations, they simply treat the correlations equally, which restricts the performance.

To overcome the limitations in existing methods, we intend to use graphs to represent the inherent correlations among electricity consumption behaviors of different appliances. Along with the historical load data reflecting the temporal correlations, they constitute a relationship that contains the aforementioned correlations, which could also be regarded as a special type of spatio-temporal correlation. We note that neither short-term nor long-term correlations could completely represent the internal correlations. On the one hand, short-term correlations contain more detailed information from recent time-series data yet introducing some unnecessary interference. On the other hand, long-term correlations reflect a general pattern of individual users' electricity consumption habits, but it lacks time-variant information. Thus, we expect to take the strengths of both short-term and long-term correlations to benefit the load forecasting.

In this paper, we propose a method based on graph neural networks (GNN) for individual residential load forecasting that fully leverages the internal correlations among various electricity consumption behaviors. To this end, two kinds of graphs—short-term dynamic graphs and long-term static graphs—are constructed to describe the short-term and long-term correlations respectively. Then, these two kinds of graphs are fused, which compensate for each other to get a more precise graph structure. Last but not least, spatio-temporal relationship learning is performed to capture spatial and temporal correlations among the fused graphs and the load data for load forecasting. Experimental results on real-world datasets show that our proposed model's forecasting performance is enhanced due to the graph construction and graph fusion strategies.

2   Related Work

In this section, we briefly introduce the related work regarding load forecasting problem. In recent years, extensive researches about aggregated residential short-term load forecasting (STLF) have been studied. Traditional aggregated residential STLF methods contain autoregressive moving average (ARMA) [4], autoregressive integrated moving average (ARIMA) [5], support vector regression (SVR) [6], *etc.* They have been replaced by deep learning-based methods gradually, such as deep neural networks (DNN) [10], recurrent neural networks (RNN) [11] and residual neural networks (ResNet) [7].

Besides aggregated residential STLF, there are also numbers of research works conducting residential STLF for individual user [8,9,12]. Most of them leverage the historical information to enhance the prediction performance, while only a few works explore using the relationships among different appliances to improve the accuracy of load forecasting. Some methods considering from the aspect of appliance level achieve superior results. Razghandi *et al.* [13] forecasted the appliance-level electricity consumption of an individual house through LSTM. Dinesh *et al.* [14] proposed a method that predicts each appliance's power consumption separately using graph spectral clustering and then aggregates them to obtain the total household load forecasting. Wu *et al.* [15] put forward a multi-variate time series forecasting model based on GNN which could also be applied to individual load forecasting problems.

3   Proposed Model

In this section, we introduce the proposed residential load forecasting model *ShortLong* for individual user. An overview of the model is illustrated in Fig. 1. In general, this model is made up of three layers, 1) a graph structure extraction layer, 2) a spatio-temporal relationship learning layer and 3) an output layer. The input of the model, $X = [l_{t_1}, l_{t_2}, \ldots, l_{t_P}]$, is a sequence of n appliances' load data in P time steps, where l_{t_i} represents the values of n appliances' load at time step t_i. $l_{t_i}^j$ is the load of the jth appliance at time step t_i. The output of the proposed model, $Y = [l_{t_P+1}]$, is n appliances' load at time step $P + 1$. Details are shown as follows.

Firstly, fused graphs are obtained through the graph structure extraction layer by fusing the constructed short-term dynamic graphs and long-term static graph. Next, the model borrows the idea from the spatio-temporal graph neural network to design the spatio-temporal relationship learning layer, where interleaved temporal convolution (TC) modules and graph convolution (GC) modules capture the temporal and spatial dependencies respectively. The fused graphs are utilized as the input of the GC module. Finally, the output layer aims to project the hidden features back to one dimension to get the load forecasting of n appliances at the next time step. Residual connections are introduced to avoid gradient vanishing.

3.1   Graph Construction

Short-Term Dynamic Graph Construction. Short-term dynamic graph reflects correlations among different types of electricity consumption behaviors

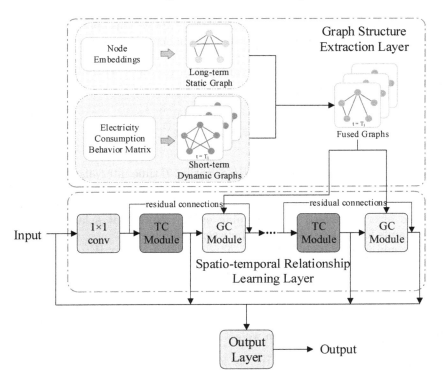

Fig. 1. The architecture of *ShortLong* model

in a short time period. The first step to construct this kind of graph is to obtain an electricity consumption behavior matrix E regarding whether each appliance performs electricity consumption behaviors in T time intervals (a relatively short period of time). Suppose that N is the number of appliances, the size of E is $T \times N$. When an appliance is connected to a power supply at a switch-off mode or standby mode, it always yields standby power, which is small but could not be ignored. Each appliance's standby power threshold is manually chosen based on the electricity consumption behavior profile we draw. We consider that an electricity consumption behavior occurs when the actual power of an appliance is greater than the threshold, which is labeled by 1. On the contrary, 0 means there is no electricity consumption behavior happened. In this way, the original time-series data is transformed into the electricity consumption behavior matrix, which is made up of 0 and 1. E is updated every T time intervals.

Next, the short-term dynamic graph is constructed based on the electricity consumption behavior matrix E by a probability-based method. We observed that whether appliance A_i performs an electricity consumption behavior depends differentially on appliance A_j and A_k. Meanwhile, the effect of A_i consuming energy on A_j consuming energy is different from the effect of A_j consuming energy on A_i consuming energy. Consequently, the constructed dynamic graph is a directed graph with edge weights, where nodes are various appliances, and edge weights denote the correlations between appliances with respect to electricity consumption behaviors.

The short-term correlation between appliance A_i and A_j could be calculated by the following probability formula:

$$P(A_i \to A_j | t = T) = \begin{cases} \frac{N_{A_i}(T)}{N_{A_j}^{A_i}(T)} & i \neq j \\ 0 & i = j \end{cases} \tag{1}$$

where $N_{A_i}(T)$ is the number of times within T time intervals that electricity consumption behaviors of appliance A_i occurred, and $N_{A_j}^{A_i}(T)$ is the number of times within T time intervals that electricity consumption behaviors of A_j occurred after appliance A_i consuming energy. Note that a large probability value stands for a close correlation between appliances.

We could further obtain the $N \times N$ edge weight matrix W:

$$W_{ij} = \sigma(P(A_i \to A_j | t = T)) \tag{2}$$

where σ represents the sigmoid activation function.

To eliminate the accidental error, we manually choose the threshold θ to cut off some redundant edges. Then we get the optimized edge weight matrix $\widehat{W}$:

$$\widehat{W}_{ij} = \rho(W_{ij} - \theta) \tag{3}$$

where ρ represents the ReLU activation function. In this way, only essential edge relations are preserved.

Hence, the adjacent matrix A_{short} of the constructed short-term dynamic graph is defined as:

$$A_{ij}^{short} = \begin{cases} 0 & \widehat{W}_{ij} \leq 0 \\ 1 & \widehat{W}_{ij} > 0 \end{cases} \tag{4}$$

The short-term spatial correlations among appliances are represented by A_{short}, which is updated every time T intervals with the short-term dynamic graph.

Long-Term Static Graph Construction. Unlike the probability-based approach to construct the short-term dynamic graph, we are supposed to use a data-driven method to build the long-term static graph, which is also a directed graph concerning the correlation among appliances' usage patterns from a long time period, e.g., the whole training set.

The adjacent matrix of the long-term static graph A_{long} is obtained similar to [15]. The process to extract the long-term correlations among appliances' electricity consumption behaviors is shown as follows:

$$H_1 = \tau(\alpha \mathcal{L}(\mathbb{E}_1, \Theta_1)) \tag{5}$$

$$H_2 = \tau(\alpha \mathcal{L}(\mathbb{E}_2, \Theta_2)) \tag{6}$$

$$A_{long} = \rho(\tau(\alpha(H_1 H_2{}^T - H_2 H_1{}^T)) \tag{7}$$

where τ is the tanh activation function, α determines the saturation rate of tanh function, $\mathcal{L}(\cdot, \cdot)$ is a linear layer, of which Θ_1, Θ_2 are learnable parameters, $\mathbb{E}_1$, $\mathbb{E}_2$ are node embeddings learned from the long-period data, and ρ represents the ReLU activation function. Both the subtraction term and ReLU function in Eq. 7 make A_{long} asymmetric.

3.2 Graph Fusion

Due to the dynamic characteristic of individual users' electricity consumption behaviors, the short-term dynamic graph may contain some redundant edge relations compared with the actual situation. Meanwhile, the long-term static graph suffers from the imprecise representation of the correlations for it only reflects a general usage pattern. We investigate that fusing these two graphs could make up their disadvantages and get a more comprehensive view of the spatial correlations.

The fused graph's adjacent matrix A is calculated as follows:

$$A = A_{short} \circ A_{long} \tag{8}$$

where $\circ$ represents the fusion operation. In this paper, we simply choose Hadamard product as the fusion operation.

In this way, the short-term dynamic graphs and long-term static graph are restricted mutually for the reason that graph fusion not only eliminates the redundant edge relations in short-term dynamic graphs but also preserves the necessary graph structure from a long period.

3.3 Spatio-Temporal Relationship Learning for Load Forecasting

The spatio-temporal relationship learning layer consists of several interleaved GC modules and TC modules to handle the spatio-temporal dependencies of the multi-appliance time-series data. We introduce these two kinds of modules from [15]. An output layer is followed to get the final load forecasting. More details are expounded below.

Graph Convolution Module. A GC module is made up of two mix-hop propagation layers, where the adjacency matrix A of the fused graph and its transpose matrix are utilized as the input respectively. The final output of GC module is the sum of the two mix-hop propagation layers' outputs.

The mix-hop propagation layer works in a two-step process, *i.e.*, the information propagation step and the information selection step. M multi-layer perceptions are applied to the second step. The output of mix-hop propagation layer P_{out} is defined as follows:

$$P_{out} = \sum_{i=0}^{M} P_i w_i \tag{9}$$

where P_i is the output of the information propagation step and w_i is the weight matrix learned by each MLP. $P_i = \beta P_{in} + (1 - \beta)\widehat{A}P_{i-1}$, where β is a hyper-parameter determining the proportion of the original node states to be reserved, P_{in} is the hidden states of the previous layer, $P_0 = P_{in}$, $\widehat{A}$ is the transformed adjacency matrix of the fused graph. The calculation of $\widehat{A}$ is the same as [15].

Temporal Convolution Module. A TC module contains two dilated inception layers which are inspired by dilated convolution [16]. This module aims to capture not only the potential temporal relationship but also the sequential patterns of the multi-appliance time series. The two dilated inception layers are respectively followed by a tanh activation function which is regarded as a filter as well as a sigmoid activation function which is regarded as a gate, and are gathered together in the end.

Output Layer. The output layer is made up of two 2D convolution layers to project the features learned from spatio-temporal relationship learning layer to one dimension getting the final output Y:

$$Y = \mathcal{F}_1(\rho(\mathcal{F}_2(X'))) \tag{10}$$

where $\mathcal{F}_1(\cdot)$ and $\mathcal{F}_2(\cdot)$ denote two 1×1 convolution layers with different input channel size and output channel size, ρ is the ReLU activation function and X' represents the output of the spatio-temporal relationship learning layer.

4 Experiments

4.1 Experimental Setup

Dataset. In this paper, we use the version 2 of the almanac of minutely power dataset (AMPds[1]), which is commonly utilized in energy disaggregation problems. It contains 727-complete-day appliance-level load data of an individual house in Canada collected every minute ranging from April 1th, 2012 to March 29th, 2014. Of all the 19 appliances, we preserved 14 appliances whose electricity consumption takes up more than 1% of the total electricity usage. Besides, we convert AMPds into datasets with three kinds of time intervals, i.e., 15 min, 30 min and 1 h.

Performance Metrics. To evaluate the performance of the proposed model, we choose two widely used metrics for forecasting problems, i.e., mean absolute error (MAE) and mean absolute percentage error (MAPE). The definitions of them are listed below:

$$MAE = \frac{1}{n}\sum_{i=1}^{n}|\hat{y}_i - y_i| \tag{11}$$

[1] https://dataverse.harvard.edu/dataset.xhtml?persistentId=doi:10.7910/DVN/FIE0S4.

$$MAPE = \frac{100\%}{n} \sum_{i=1}^{n} |\frac{\hat{y}_i - y_i}{y_i}| \tag{12}$$

where y_i is the actual load value and $\hat{y}_i$ is the predicted load value.

4.2   Competing Methods

ShortLong is compared with five methods, containing ARMA [4], DNN [10], LSTM [1], ResNetPlus [7] and MTGNN [15]. The chosen competing methods cover traditional statistical methods, deep-learning based methods and GNN-based methods. ARMA only use the overall load as input. MTGNN only use the appliance-level load, while the remaining models use both appliance-level load and overall load as models' inputs.

4.3   Results and Analysis

Each experiment conducted in this section is repeated for 10 times. And all the results shown in the tables and figures are the average values. The best results are highlighted. The datasets are split into a training set of 60%, a validation set of 20% and a test set of 20% in chronological order.

Main Results. To evaluate the effectiveness of our proposed model, we compare *ShortLong* with five load forecasting models. The competing methods could be classified into two types, *i.e.*, methods not using graphs and GNN-based method. The first four methods of Table 1 belong to the first type, while MTGNN is a GNN-based method.

We conduct experiments on AMPds of three different time intervals. Since ResNetPlus is a model for load forecasting of one hour, we only employ ResNet-Plus model on one-hour interval dataset. For our proposed model, the hyper-parameter settings of GC module and TC module are the same as those described in [15]. And the hyper-parameter T is set to 16, which will be illustrated in detail later in the parameter study section.

Table 1 shows the main results. First, we find that the results of GNN-based methods surpass those of the first type of methods on all the datasets of three time intervals, which demonstrates that utilizing graphs to represent correlations among different types of electricity consumption behaviors is useful to improve the performance. Second, *ShortLong* achieves the best results among all the competing methods. Compared with MTGNN, *ShortLong* brings a 2.90% reduction of MAPE on the 15-min interval dataset, a 1.61% reduction of MAPE on the 30-min interval dataset and a 4.23% reduction of MAPE on the one-hour interval dataset. The reason why *ShortLong* performs better than MTGNN is that the fusion of short-term dynamic graphs and long-term static graph provides the model with a comprehensive graph with more useful information.

Table 1. Main results on AMPds dataset

	Interval = 15 min		Interval = 30 min		Interval = 1 h	
	MAE	MAPE	MAE	MAPE	MAE	MAPE
ARMA	4.1925	40.3747	10.5700	50.3961	21.4767	55.0448
DNN	3.7095	33.9846	9.6932	40.6433	21.1873	50.7147
LSTM	3.3873	30.4601	8.9172	37.3026	18.3652	40.8831
ResNetPlus	–	–	–	–	19.4258	34.9819
MTGNN	3.0994	23.1981	8.8840	26.6461	17.9697	30.1992
ShortLong	**3.0520**	**22.5258**	**8.8437**	**26.2177**	**17.6697**	**28.9231**

Ablation Study. To verify that the fused graphs we construct in this paper contribute to the increase of our proposed method, we conduct the ablation study by constructing six various graphs, *i.e.*, three kinds of basic graphs and three kinds of fused graphs. Since the performance of pure data-driven approaches is not satisfying when the amount of data is relatively small, the data-driven method is not suitable to construct the short-term dynamic graphs. Thus, we construct three kinds of basic graphs, *i.e.*, long-term graph using probability-based method ($long_P$), long-term graph using data-driven method ($long_{DD}$) and short-term graph using probability-based method ($short_P$). Three fused graphs, which are respectively represented by $long_P + long_{DD}$, $long_P + short_P$ and $long_{DD} + short_P$, are constructed based on the combination of the three basic graphs. The last kind of fused graph is the one proposed in this paper. Experiments are conducted on one-hour interval AMPds dataset.

The results are shown in Table 2. In general, we observe that the performance of models with fused graphs are better than those with basic graphs. It proves that fused graphs contain more information than basic graphs in describing the internal correlations among different types of electricity consumption behaviors. Among the three models with fused graphs, the best results of MAE and MAPE are achieved when we leverage the ones fusing long-term graph using data-driven method and short-term graphs using probability-based method (*ShortLong*). In this way, we extract the graph structures from two time scales using different methods, which helps to obtain precise graph structures. Thus, the way we construct the fused graphs of *ShortLong* model is proved to be effective.

Table 2. Ablation study on AMPds dataset with 1-h intervals

	MAE	MAPE
$long_P$	17.7413	29.8134
$long_{DD}$	17.9895	30.5641
$short_P$	17.7411	30.4895
$long_P + long_{DD}$	17.8778	29.6328
$long_P + short_P$	17.6968	29.3640
$long_{DD} + short_P$ (ShortLong)	**17.6697**	**28.9231**

Parameter Study. T is the hyper-parameter determining the number of time intervals among which a short-term dynamic graph is constructed. Experiments are only conducted on AMPds dataset of one-hour time interval. We set T to 3, 12, 16, 24, 32, 64 and 84, while the other parameters remain the same as the proposed method. The line of MAPE values is shown in Fig. 2. We find that T is not sensitive when it is relatively small. Increasing T from 3 to 32 makes slight change to the MAPE value of the proposed method. However, when T becomes larger, the performance drops drastically. Among this set of experiments, the best result (MAPE = 28.9231%) is achieved when $T = 16$.

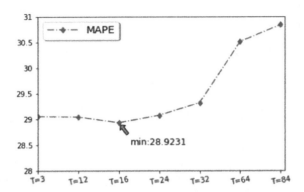

Fig. 2. Parameter sensitivity study on AMPds of 1-h interval

5 Conclusions

In this paper, we propose an effective GNN-based method *ShortLong* to address the individual residential load forecasting problem. Compared with other time series forecasting models, we fully leverage the short-term and long-term correlations among different types of electricity consumption behaviors. To be more specific, our proposed method constructs two kinds of graphs, *i.e.*, short-term dynamic graphs and long-term static graphs, from two time scales. Fused graphs with more useful information are obtained through the fusion of these two kinds of graphs. Spatio-temporal relationships for load forecasting are captured via spatio-temporal graph neural networks from the fused graphs and the load data. The performance of *ShortLong* is proved to be superior in MAE and MAPE on AMPds dataset of three time intervals compared with state-of-art methods. We consider introducing extra knowledge, such as weather conditions and public holiday information, to the graph neural networks in future works.

Acknowledgement. This work is partly supported by National Natural Science Foundation of China (NSFC) with grant number 61801315 and 62171302.

References

1. Kong, W., Dong, Z.Y., Jia, Y., Hill, D.J., Xu, Y., Zhang, Y.: Short-term residential load forecasting based on LSTM recurrent neural network. IEEE Trans. Smart Grid **10**(1), 841–851 (2017)
2. Uddin, M., Romlie, M.F., Abdullah, M.F., Abd Halim, S., Kwang, T.C., et al.: Areview on peak load shaving strategies. Renew. Sustain. Energy Rev. **82**, 3323–3332 (2018)
3. Zhang, W., Quan, H., Srinivasan, D.: An improved quantile regression neural network for probabilistic load forecasting. IEEE Trans. Smart Grid **10**(4), 4425–4434 (2019)
4. Huang, S.J., Shih, K.R.: Short-term load forecasting via ARMA model identification including non-Gaussian process considerations. IEEE Trans. Power Syst. **18**(2), 673–679 (2003)
5. Li, W., Zhang, Z.: Based on time sequence of ARIMA model in the application of short-term electricity load forecasting. In: 2009 International Conference on Research Challenges in Computer Science, pp. 11–14. IEEE (2009). https://doi.org/10.1109/ICRCCS.2009.12
6. Ghelardoni, L., Ghio, A., Anguita, D.: Energy load forecasting using empirical mode decomposition and support vector regression. IEEE Trans. Smart Grid **4**(1), 549–556 (2013)
7. Chen, K., Chen, K., Wang, Q., He, Z., Hu, J., He, J.: Short-term load forecasting with deep residual networks. IEEE Trans. Smart Grid **10**(4), 3943–3952 (2018)
8. Shi, H., Xu, M., Li, R.: Deep learning for household load forecasting-a novel pooling deep RNN. IEEE Trans. Smart Grid **9**(5), 5271–5280 (2017)
9. Hossen, T., Nair, A.S., Chinnathambi, R.A., Ranganathan, P.: Residential load forecasting using deep neural networks (DNN). In: 2018 North American Power Symposium (NAPS), pp. 1–5 (2018). https://doi.org/10.1109/NAPS.2018.8600549
10. Sze, V., Chen, Y.H., Yang, T.J., Emer, J.S.: Efficient processing of deep neural networks: a tutorial and survey. Proc. IEEE **105**(12), 2295–2329 (2017)
11. Rahman, A., Srikumar, V., Smith, A.D.: Predicting electricity consumption for commercial and residential buildings using deep recurrent neural networks. Appl. Energy **212**, 372–385 (2018)
12. Chaouch, M.: Clustering-based improvement of nonparametric functional time series forecasting: application to intra-day household-level load curves. IEEE Trans. Smart Grid **5**(1), 411–419 (2013)
13. Razghandi, M., Turgut, D.: Residential appliance-level load forecasting with deep learning. In: GLOBECOM 2020–2020 IEEE Global Communications Conference, pp. 1–6 (2020). https://doi.org/10.1109/GLOBECOM42002.2020.9348197
14. Dinesh, C., Makonin, S., Bajić, I.V.: Residential power forecasting using load identification and graph spectral clustering. IEEE Trans. Circuits Syst. II Express Briefs **66**(11), 1900–1904 (2019)
15. Wu, Z., Pan, S., Long, G., Jiang, J., Chang, X., Zhang, C.: Connecting the dots: multivariate time series forecasting with graph neural networks. In: Proceedings of the 26th ACM SIGKDD International Conference on Knowledge Discovery & Data Mining, pp. 753–763 (2020). https://doi.org/10.1145/3394486.3403118
16. Yu, F., Koltun, V.: Multi-scale context aggregation by dilated convolutions. arXiv preprint arXiv:1511.07122 (2015)

Disentangled Feature Network for Fine-Grained Recognition

Shuyu Miao[1], Shuaicheng Li[2], Lin Zheng[1], Wei Yu[3], Jingjing Liu[1],
Mingming Gong[1], and Rui Feng[2(✉)]

[1] Ant Group, Hangzhou, China
miaoshuyu.msy@antgroup.com

[2] School of Computer Science, Shanghai Key Laboratory of Intelligent Information
Processing, Fudan University, Shanghai, China
fengrui@fudan.edu.cn

[3] College of Computer, National University of Defense Technology, Changsha, China

Abstract. Most of fine-grained recognition researches are implemented
based on generic classification models as the backbone. However, it is
a sub-optimal choice because the differences between similar categories
in this task are so small that the models must capture discriminative
fine-grained subtle variances. In this paper, we design a dedicated backbone network for fine-grained recognition. To this end, we propose a
novel *Disentangled Feature Network (DFN)* that gradually disentangles
and incorporates coarse- and fine-grained features to explicitly capture
multi-grained features. Thus, it promotes the models to learn more representative features that potentially determine the classification results
via easily replacing the original inappropriate backbone. Moreover, we
further present an optional *error correction loss* to adaptively penalize
misclassification between extremely similar categories and guide to capture fine-grained feature diversity. Extensive experiments fully demonstrate that when adopting our DFN as the backbone, like freebies, the
baseline models boost the performance by about 2% with negligible extra
parameters on widely used CUB, AirCraft, and Stanford Car dataset.

Keywords: Fine-grained recognition · Disentangled feature learning ·
Backbone models · Error correction loss

1 Introduction

Fine-grained recognition aims to identify fine granularity, which is a significant
but challenging task. Compared to the generic classification, fine-grained recognition suffers from high intra-class variances and low inter-class variances. Thus,
it requires to pay more attention to fine-grained features. However, almost all the
existing backbones of fine-grained recognition models are based on the generic
classification models (e.g., VGG [17], ResNet [8], DensenNet [11], etc.). It raises a
dilemma that coarse-grained features captured by these backbone models cannot

© Springer Nature Switzerland AG 2021
T. Mantoro et al. (Eds.): ICONIP 2021, LNCS 13109, pp. 439–450, 2021.
https://doi.org/10.1007/978-3-030-92270-2_38

effectively identify similar categories. As shown in Fig. 1, the generic classification model can identify a dog by focusing on the coarse-grained features of the head region, while more fine-grained feature relationships need to be further considered to identify the specific bird categories. This means that the general classification network cannot meet the backbone requirements of fine-grained tasks. Thus, designing the suitable backbones with the ability of capturing coarse- and fine-grained features sheds new light on the development of fine-grained recognition.

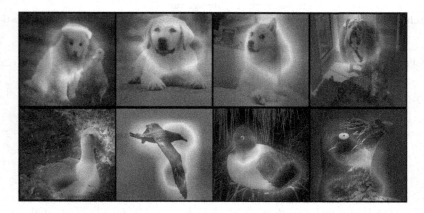

Fig. 1. Comparison visualization between coarse-grained and fine-grained recognition.

Recently, some efforts have been made for optimizing the feature representation ability of the generic backbone models. Some methods [10,20] introduced attention mechanisms to make models focus on interested areas. Some works [6,7] designed dedicated connected modules to capture multi-scale features. However, these approaches are not optimal as the backbone model for the fine-grained recognition task. The backbone models of this task need to meet the following **three requirements** at the same time. (1) Seeing more regions is the basis for obtaining multi-grained features. The limited local field of view limits the perceptual ability of features, thus more multi-grained features should be covered by expanding the visual regions. (2) Capturing the coarse- and fine-grained particle features plays an important role in distinguishing subtle variances. The coarse-grained features are helpful to capture the general classification information, and the fine-grained features facilitate extracting the differences of subtle details. (3) Comprehensive use of the feature relations of various regions promotes to determine classification results. It is helpful to enlarge the local differences between similar categories via facilitating the relationship of regional features.

In this paper, we propose a novel *Disentangled Feature Network*, called **DFN**, to gradually disentangle and incorporate coarse- and fine-grained features to explicitly capture multi-grained features. Firstly, we imagine the ghost backbone with the shared parameters with the main backbone model to interrelate the high-level fine-grained feature with the low-level coarse-grained feature to see more regions (*meet the requirements. (1)*). Secondly, we disentangle coarse- and fine-grained features via the main backbone and the ghost backbone, which are effectively constrained to learn different granularity features (*meet the requirements. (2)*). Thirdly, we distill the correlations between regional features to enhance the relation interaction between different regions (*meet the requirements. (3)*). Moreover, as an optional choice, the *error correction loss* (**Ecloss**) is further proposed to adaptively penalize misclassification between extremely similar categories and guide to capture fine-grained feature diversit.

The proposed DFN can easily replace existing generic classification backbones with negligible additional parameters for fine-grained recognition. We conduct quantitative and qualitative experiments on nearly all generic backbone models and recent sota models in fine-grained recognition task as the baselines, and the performances of baselines are improved by about 2% on widely used datasets. Meanwhile, compared with the best backbone models in recent years, (e.g., ResNeSt [20], Res2Net [7], SCNet [15], etc.), the experimental results show that our methods outperform them.

2 Related Work

Fine-Grained Recognition. Fine-grained recognition aims to distinguish similar categories, and many researches have focused on it recently. Region localization subnetworks were introduced to concentrate on the regions with objects [5,12] by means of employing detection and segmentation usages, adopting deep activations, or using attention mechanisms. Instead of leveraging localization subnetworks, some works directly encoded feature via an end-to-end pipeline. Specific loss functions [1,18] were proposed to promote models to distinguish similar features. However, almost all of these works are based on generic classification models as the backbone, which is not the fittest for fine-grained recognition task. Exploiting fittest backbone models will have a positive impact on the development of fine-grained recognition task.

Backbone Models. A great deal of previous researches on backbone models have explored to effectively boost the performance in computer vision task. Res2Net [7] constructed hierarchical residual-like connections within one single residual block to represent the multi-scale features. ResNeSt [20] proposed a modular Split-Attention block that enabled attention across feature-map groups to aid downstream tasks. SCNet [15] adaptively built inter-channel and long-range spatial dependencies around each spatial location via a novel self-calibration process to help models generate more discriminative representations by incorporating richer information. However, these backbones are not well qualified for fine-grained recognition, because they cannot guarantee to explicitly capture required multiple grained features for distinguishing similar categories.

3   Our Method

3.1   Overview

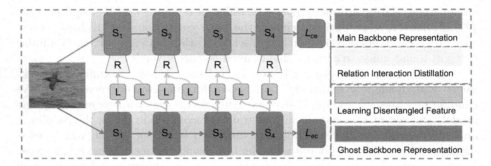

Fig. 2. The pipeline of our proposed DFN. DFN consists of Ghost Backbone Representation (GBR) module, Learning Disentangled Feature (LDF) module, Relation Interaction Distillation (RID) module, and optional Error Correction Loss (Ecloss). Each proposed module corresponds to one color like the right colors of figure. 'L' denotes the proposed LDF, 'R' indicates the proposed RID, and 'S_i' represents the i-th stage in the model.

The pipeline of our method is illustrated in Fig. 2. The biggest challenge of fine-grained recognition lies in leveraging extreme subtle features to distinguish similar categories. The top pipe in the figure, that almost entirely adopts a generic classification model as the backbone, is not enough to capture discriminative subtle variances. To break this restriction, We propose a Disentangled Feature Network (DFN) to make the backbone models more appropriate for collecting requisite relation interaction between coarse- and fine-grained features. Proposed DFN consists of the following subtasks: 1) Ghost Backbone Representation (GBR) is employed to interrelate high-level fine-grained features and low-level coarse-grained features to see more regions; 2) Learning Disentangled Feature (LDF) decouples coarse- and fine-grained features to distinguish subtle variances; 3) Relation Interaction Distillation (RID) distills correlation between regional grained features; and (4) Error Correction Loss (Ecloss) is further introduced to amplify the discrimination ability of extremely similar categories.

3.2   Ghost Backbone Representation

For fine-grained recognition task, merely focusing on local regions may not be enough to distinguish the similar categories. For example, sometimes it can not obtain correct category results just by the mouth region of two birds since both are too similar. Therefore, it is progressive to see more regions to cover more discriminative features.

The generic classification backbones (e.g., VGG, ResNet, DensenNet) consist of several CNN stages. They can be denoted by $\mathcal{S} := \langle S_1, S_2, \cdots, S_n \rangle$, where n is the number of the model stage and S_n is n-th stage. The seen region or called receptive field of the k_{th} layer of the CNN model can be expressed as [16]

$$l_k = l_{k-1} + ((f_k - 1) * \prod_{i=1}^{k-1} s_i), \tag{1}$$

where l_{k-1} is the receptive field of $k - 1^{th}$ layer, f_k is the filter size, and s_i is the stride of layer. Because $(f_k - 1) * \prod_{i=1}^{k-1} s_i > 0$, naturally, we can obtain

$$l_{S_n} > \cdots > l_{S_2} > l_{S_1}. \tag{2}$$

It is a natural idea to fuse high-level feature F_h with the larger seen region directly to low-level feature F_l with smaller seen region, like Feature Pyramid Networks (FPN) [13]. However, native FPN-style design is not suitable to be directly applied as fine-grained backbone model. Because (1) such design is not competent for capturing coarse- and fine-grained features, and (2) single level output causes the model trapped in loop optimization and multi-level output cannot match the design pattern of the backbone.

Inspired by siamese network that can accomplish feature extraction and feature interaction via two parallel networks, we imagine the ghost backbone having the shared parameters with the main backbone. The features of main backbone and ghost backbone are defined by $\langle F_{S_m(1)}, F_{S_m(2)}, \cdots, F_{S_m(n)} \rangle$ and $\langle F_{S_g(1)}, F_{S_g(2)}, \cdots, F_{S_g(n)} \rangle$, respectively. To make the main backbone model see more to obtain multi-grained features, we firstly fuse the high-level features and low-level features of ghost backbone as

$$F_{S_{g'}(i)} = \alpha * f_i(F_{S_g(i)}) + \beta * f_{i+1}(F_{S_g(i+1)} \uparrow^2), \tag{3}$$

where α and β are super-parameters with $\alpha + \beta = 1$, and $\uparrow^2$ means the upsampling operation by two times, $f_i(\cdot)$ and $f_{i+1}(\cdot)$ are the restrictive conditions to drive ghost backbone to capture fine-grained features via **LDF**. $F_{S_{g'}(i)}$ are interrelated with $F_{S_m(i)}$ via **RID**, that is denoted as $\Delta(\cdot)$, to guide the main backbone learning fine-grained features as:

$$F_{S_{m'}(i)} = \Delta(F_{S_m(i)}, F_{S_{g'}(i)}). \tag{4}$$

According to Eq. (2), it is easy to obtain that $l_{S'_m(i)} > l_{S_g(i+1)} > l_{S_g(i)} = l_{S_m(i)}$. Features of main backbone sees more regions that contributes to capturing more detailed features and learns the relations between regional features.

3.3   Learning Disentangled Feature

Because of the characteristics of high intra-class variances and low inter-class variances of fine-grained recognition task, it is very critical effectively leverage the limited subtle variances to distinguish categories. Thus, we decouple the fine-grained feature from the coarse-grained feature, and obtain the coarse- and fine-grained features explicitly via main backbone and ghost backbone, respectively. The main backbone and $\mathcal{L}_{ce}$ in the top pipe, in Fig. 2, is the generic classification pipeline. Such native models tend to extract coarse-grained features. Like Fig. 1, the dog can be classified correctly by coarse-grained feature of dog's head, since the heads of different animals have obvious discrepancies. However, it is not enough for fine-grained task since different birds may have extreme similar heads.

It pointed out that dual skipping networks could simultaneously deal with both coarse- and fine-grained recognition tasks inspired by recent biological experiments [4]. In the light of this work, disentangled feature learning enables that the main backbone model focuses on coarse-grained features and the ghost backbone model focuses on fine-grained features. However, how to effectively limit the ghost backbone to learn the fine-grained features is pending. In the Eq. (3), $f_i(\cdot)$ is the restrictive conditions to guide the ghost backbone to learn more fine-grained features. In one feature map abbreviated as $F \in \mathcal{R}^{H \times W \times C}$, the spatial regions containing extremely distinguishing subtle features tend to have high response weights. We obtain this responded feature by the $f_i(\cdot)$ by

$$F = f_i(F), \quad F^{(i,j)}_{\{i \in H, j \in W\}} = \sum_{k=1}^{C} \gamma_k * F^{(i,j,k)}. \tag{5}$$

where γ is a learnable weight parameter. This process carries out learnable feature response for each unit value of feature map in the spatial dimension, which ensures the learned features are fine-grained based on unit process.

By means of above processes, the fine-grained features that benefit to subtle variances are then mapped to the coarse-grained features and blended to conduct fine-grained recognition. Moreover, responded features will be further refined for facilitating the optimization of subtle features by learning regional relation interaction distillation (RID) and error correction loss (Ecloss).

3.4   Relation Interaction Distillation

It has analyzed that strengthening the connections among regions is beneficial to fine-grained recognition [21]. Go a step further, making full use of the connections of regional features in the backbone will benefit to obtain more profits. Inspired by knowledge distillation (KD) [9] that drives the model to leverage the "dark" knowledge, we propose Relation Interaction Distillation (**RID**) to explicitly learn the "dark" relationship interaction among regional features.

KD is firstly revisited. Neural models typically produce class probabilities by using a *softmax* output layer. To drive the model to learn more knowledge, the output layer is reimplemented as

$$q_i = \frac{exp(z_i/T)}{\sum_j exp(z_j/T)}. \tag{6}$$

It converts the logit z_i that is computed for each class into a probability to q_i by comparing z_i with the other logits. T is a temperature value that is normally set as 1. T with a higher value produces a softer probability distribution over classes. The model learns the "dark" relations between potential classes. In our implement, the relations among regional features in the feature map is calculated in the spatial dimension. Therefore, we expand the Eq. (6) to the following Eq. (7).

$$F_{i,j} = \frac{exp(F_{i,j}/T)}{\sum_a \sum_b exp(F_{a,b}/T)}. \tag{7}$$

In the Eq. (4), the feature map $F_{S_{g'}(i)}$ is fed into the Eq. (7) to learn the relation interaction among regional features. To blend the coarse- and fine-grained features in the main backbone, $F_{S_m(i)}$ and $F_{S_{g'}(i)}$ are then fused by

$$F_{S_{m'}(i)} = (1 + F_{S_{g'}(i)}) \otimes F_{S_m(i)}. \tag{8}$$

By relation interaction distillation, the features in the main backbone cover coarse- and fine-grained features with relations among fine-grained features, which promotes the backbone to distinguish the discriminative subtle variances.

3.5 Error Correction Loss

With the above processes, the model enables to learn coarse- and fine-grained features separately, which has solved the coarse- and fine-grained feature fusion learning of the backbone model well. Like a dessert and a snack, we find that if the model could give the ghost backbone branch an extra level of supervision, it is further conducive to the fine-grained feature extraction. As an optional choice, we move forward a step to introduce Error Correction Loss (**Ecloss**) to adaptively penalize misclassification between extremely similar categories and guide to capture fine-grained feature diversity.

Let's recall the traditional cross-entropy loss widely used in fine-grained recognition task, which is defined as

$$\mathcal{L}(x, class) = -log(\frac{exp(x[class])}{\sum_j exp(x[j])}), \tag{9}$$

where $class$ is the truth label and x is the predicted scores. For the fine-grained recognition, the models are easily misclassified into the most similar categories since they have such small discriminative subtle variances. Naturally, when this happens, the predicted value of the misclassification result will be larger than that of the corresponding truth label. Instead of ignoring the implications of these misclassification results in the model training loss, we consider it in loss to force the ghost backbone to improve the fine-grained feature extraction ability to increase the immunity to interference. Thus, we propose Ecloss based on cross-entropy loss function, which is formulated as:

$$\mathcal{L}_{mis}(x, misclass) = -\sum_{\{C\}} log(\frac{exp(x[misclass])}{\sum_j exp(x[j])}), \tag{10}$$

$$\mathcal{L}_{ec} = (1 - \eta)\mathcal{L}(x, class) + \eta\mathcal{L}_{mis}(x, misclass), \tag{11}$$

where C meets the condition with $x[misclass] > x[class]$ and η is set as 0.01 empirically.

As an alternative scheme, Ecloss is used to restrict the fine-grained features of ghost backbone learning to distinguish extremely similar categories. It drives the backbone model to become more robust during the training phase and to some extent reduces the risk of overfitting the model learning.

4   Experiments

4.1   Datasets and Implementation Details

Datasets. All our experiments are conducted on the following widely used datasets.

CUB-200-2011. CUB consists of 11,788 images from 200 bird classes, where 5,994 is used for training and 5,794 for testing.

FGVC Aircraft. Aircraft contains 10,000 images from 100 aircraft classes, where 6,667 is conducted for training and 3,333 for testing.

Stanford Car. Stanford Car covers 16,185 images from 196 cars classes, where 8,144 is adopted for training and 8,041 for testing.

For all the experiments, we use the same strategy to validate our proposed DFN. For the sake of fairness, we conduct the experiments with the same image preprocessing that just resizes the image size to 448 × 448 instead of complex image augmentation tricks to show the essential performance of the models. SGD with the 0.9 momentum and 0.0001 weight decay is utilized in all our experiments. All the experimental results are tested without Ecloss except for additional instructions. Note that all our source codes are implemented with Pytorch.

Table 1. Overall performance of DFN with other generic classification models as the backbone on CUB, AirCraft, and Standford Car.

Backbone	CUB	AirCraft	Stanford Car
ResNet50 [8]	83.1	90.3	92.5
ResNet50 [8]/*DFN*	85.4	91.9	94.3
ResNet152 [8]	85.7	91.5	94.6
ResNet152 [8]/*DFN*	86.8	92.9	96.0
VGG16 [17]	79.3	87.4	88.2
VGG16 [17]/*DFN*	80.7	88.6	88.7
DenseNet121 [11]	80.8	90.9	90.0
DenseNet121 [11]/*DFN*	84.5	92.0	92.2
DenseNet169 [11]	82.1	91.8	92.6
DenseNet169 [11]/*DFN*	84.7	92.7	93.7
DCL [3]	87.8%	94.5%	93.0%
DCL [3]/*DFN*	89.1%	95.6%	94.2%
APINet [22]	90.1%	93.9%	95.3%
APINet [22]/*DFN*	**91.1%**	**95.1%**	**96.3%**

4.2 Overall Performance

We perform experiments to compare with generic classification models, like [8,11,17] on fine-grained task, and apply our DFN as the backbone to famous models to show the performance intuitively in Table 1. All experiments are implemented on the same public datasets.

As can be seen from the table, the following results can be concluded. 1) DFN makes available for various backbone models. Closer inspection of the table shows that DFN improves the performance of VGG, ResNet, and DenseNet. 2) DFN adapts to different model depth. No matter $<ResNet50, ResNet152>$ or $<DenseNet121, DenseNet169>$, DFN outperforms the generic classification models. 3) DFN boosts the performance of sota models and achieve more excellent results. 4) DFN works well on different datasets. For the datasets of CUB, AirCraft, and Stanford Car, our DFN owns a stable positive income. All the experimental results show the effectiveness of our methods by about 2% performance improvement with negligible parameter of several 1×1 convolutional layers. DFN decouples the coarse- and fine-grained features and learns the relations among regional features of birds, aircrafts, and cars to capture discriminative subtle variances.

Table 2. Comparison results for DFN with other state-of-the-art backbone models on CUB, AirCraft, and Standford Car.

Models	CUB	AirCraft	Stanford Car
ResNet50 (baseline)	83.1	90.3	92.5
SENet50 [10]	83.7	89.8	92.7
SCNet50 [15]	84.5	91.3	93.2
ResNeSt50 [20]	84.8	91.2	93.6
Res2net50 [7]	83.9	90.7	93.4
DFN50 (ours)	**85.4**	**91.9**	**94.3**

4.3 Comparison with Other Backbone Models

To demonstrate the effectiveness of our proposed DFN, we perform the related experiments with other state-of-the-art backbone models like SENet [10], SCNet [15], ResNeSt [20] and Res2Net [7]. All the backbone models build on similar settings corresponding to ResNet50. The related experimental results are shown in Table 2.

It can be found from the table that DFN50 outperforms all other backbone models, which shows our DFN is fitter for fine-grained recognition. Closer inspection of the table shows the following findings. 1) DFN outperforms SENet by 1.7%, 2.1%, and 1.5% on CUB, AirCraft, and Stanford Car separately, suggesting that simple attentional mechanisms fail to focus on the areas where need the most attention. 2) DFN performs better than SCNet, demonstrating that it is not well suited to the fine-grained task by adaptively establishing remote spatial and channel dependencies around each spatial location. 3) DFN outperforms ResNeSt, indicating that capturing relationships across channels is not sufficient for fine-grained tasks. 4) DFN performs better than Res2Net, showing that simply increasing the receptive field of each layer is not sufficient to distinguish the most effective subtle features. Although they excel at generic classification tasks, they are not most effective against fine-grained tasks because they do not meet the three requirements required for fine-grained recognition. In conclusion, compared to these backbone models, DFN is more competent as the backbone for fine-grained recognition.

Table 3. Ablation study results for the modules of DFN on CUB, AirCraft, and Standford Car. 'W/o' means proposed DFN without corresponding module, and 'W' means proposed DFN with corresponding module.

Modules	CUB	AirCraft	Stanford Car
ResNet50 (baseline)	83.1	90.3	92.5
W/o GBR	85.0	91.4	93.8
W/o LDF	84.8	91.2	93.8
W/o RID	84.5	91.2	93.6
DFN50	85.4	91.9	94.3
W Ecloss	86.0	92.2	95.0

Table 4. Ablation study results for the parameter $[\alpha, \beta]$ in GBR module with various values on CUB.

$[\alpha, \beta]$	$[1.0, 0]$	$[0.7, 0.3]$	$[0.5, 0.5]$	$[0.3, 0.7]$	$[0, 1.0]$
CUB	85.1	85.2	85.3	**85.4**	**85.4**

4.4   Ablation Study

Experiments on the Modules of DFN. To verify the reasonability and reliability of proposed DFN, we perform the related ablation studies on each module of DFN, shown as Table 3. We validate the performance of DFN under different conditions like without GBR module, without LDF module, without RID module, and with Ecloss. The table shows that, 1) Without GBR module, the performance of DFN is reduced by 0.4%, 0.5%, and 0.5% on CUB, Aircraft, and Standford Car separately, which illustrates that interrelating high-level fine-grained and low-level coarse-grained features leverages seeing more regions to meet the *requirements. (a)*; 2) Without LDF module, the performance of DFN is reduced by 0.6%, 0.7%, and 0.5% separately, which mentions that disentangling coarse- and fine-grained features promotes the model to extract more subtle features to meet the *requirements. (b)*; 3) Without RID module, the performance of

DFN is reduced by 0.9%, 0.7%, and 0.7% separately, which reveals that enhancing the relation interaction between different regions drives the model to learn more regions associations for aiding recognizing to meet the *requirements. (c)*. As an optimal module, when the ghost backbone with the supervision of Ecloss, the performance of DFN is further increased by 0.6%, 0.3%, and 0.7% separately., manifesting that Ecloss is beneficial to extracting fine-grained features. All the experimental results can really verify the reasonability and reliability of our DFN.

Experiments on the Setting of Super-Parameters. In the module of GBR, there are two parameters α and β, which corresponds to the fusion scale. To choose the best parameters and validate the effectiveness of seeing more regions, detailed experiments are conducted on CUB based on DFN50. The experimental results are shown in Table 4. What we can find from the table is that the model performance is positively related to the value of β. This also demonstrates that obtaining information with the larger receptive field to seen more contributes to maximizing model performance. The performance is same when β is set as 0.7 and 1.0. Thus, α is set as 0 and β as 1.0 for simplicity and efficiency in our setting. All the ablation study experiments can prove that for fine-grained recognition task the three requirements discussed earlier are essential.

5   Conclusions

Fine-grained recognition meets a dilemma, where most of the related models are based on generic classification models as the backbone those are not enough to capture discriminative subtle variances. To break this dilemma, we introduce a Disentangled Feature Network (DFN) to promote the backbone to gradually disentangle and incorporate coarse- and fine-grained features. DFN consists of ghost backbone representation, learning disentangled feature, relation interaction distillation, and error correction loss, meeting the three requirements in such task. Like freebies, our DFN can easily replace generic classification backbone to promote the performance with negligible additional parameter. We hope that our DFN can shed new light on the development of the specialized backbone of various downstream task.

References

1. Chang, D., et al.: The devil is in the channels: mutual-channel loss for fine-grained image classification. In: TIP, pp. 4683–4695 (2020)
2. Chen, Y., Dai, X., Liu, M., Chen, D., Yuan, L., Liu, Z.: Dynamic convolution: attention over convolution kernels. In: CVPR, June 2020
3. Chen, Y., Bai, Y., Zhang, W., Mei, T.: Destruction and construction learning for fine-grained image recognition. In: CVPR, pp. 5157–5166 (2019)
4. Cheng, C., et al.: Dual skipping networks. In: CVPR (2018)
5. Ding, Y., Zhou, Y., Zhu, Y., Ye, Q., Jiao, J.: Selective sparse sampling for fine-grained image recognition. In: ICCV, October 2019

6. Duta, I.C., Liu, L., Zhu, F., Shao, L.: Pyramidal convolution: rethinking convolutional neural networks for visual recognition (2020)
7. Gao, S., Cheng, M., Zhao, K., Zhang, X., Yang, M., Torr, P.H.S.: Res2Net: a new multi-scale backbone architecture. In: TPAMI, p. 1 (2019)
8. He, K., Zhang, X., Ren, S., Sun, J.: Deep residual learning for image recognition. In: CVPR, June 2016
9. Hinton, G.E., Vinyals, O., Dean, J.: Distilling the knowledge in a neural network. ArXiv (2015)
10. Hu, J., Shen, L., Sun, G.: Squeeze-and-excitation networks. In: CVPR, June 2018
11. Huang, G., Liu, Z., van der Maaten, L., Weinberger, K.Q.: Densely connected convolutional networks. In: CVPR, July 2017
12. Ji, R., et al.: Attention convolutional binary neural tree for fine-grained visual categorization. In: CVPR, June 2020
13. Lin, T.Y., Dollar, P., Girshick, R., He, K., Hariharan, B., Belongie, S.: Feature pyramid networks for object detection. In: CVPR, July 2017
14. Lin, T.Y., RoyChowdhury, A., Maji, S.: Bilinear CNN models for fine-grained visual recognition. In: ICCV, December 2015
15. Liu, J.J., Hou, Q., Cheng, M.M., Feng, J., Wang, C.: Improving convolutional networks with self-calibrated convolutions. In: CVPR (2020)
16. shawnleezx: calculating receptive field of CNN (2017). http://shawnleezx.github.io/blog/2017/02/11/calculating-receptive-field-of-cnn
17. Simonyan, K., Zisserman, A.: Very deep convolutional networks for large-scale image recognition. In: CVPR (2014)
18. Sun, M., Yuan, Y., Zhou, F., Ding, E.: Multi-attention multi-class constraint for fine-grained image recognition. In: Ferrari, V., Hebert, M., Sminchisescu, C., Weiss, Y. (eds.) ECCV 2018. LNCS, vol. 11220, pp. 834–850. Springer, Cham (2018). https://doi.org/10.1007/978-3-030-01270-0_49
19. Yu, C., Zhao, X., Zheng, Q., Zhang, P., You, X.: Hierarchical bilinear pooling for fine-grained visual recognition. In: Ferrari, V., Hebert, M., Sminchisescu, C., Weiss, Y. (eds.) ECCV 2018. LNCS, vol. 11220, pp. 595–610. Springer, Cham (2018). https://doi.org/10.1007/978-3-030-01270-0_35
20. Zhang, H., et al.: ResNeSt: split-attention networks. ArXiv (2020)
21. Zhou, M., Bai, Y., Zhang, W., Zhao, T., Mei, T.: Look-into-object: self-supervised structure modeling for object recognition. In: CVPR, June 2020
22. Zhuang, P., Wang, Y., Qiao, Y.: Learning attentive pairwise interaction for fine-grained classification. In: AAAI, vol. 34, pp. 13130–13137 (2020)

Large-Scale Topological Radar Localization Using Learned Descriptors

Jacek Komorowski$^{(\boxtimes)}$, Monika Wysoczanska , and Tomasz Trzcinski

Warsaw University of Technology, Warsaw, Poland
{jacek.Komorowski,monika.wysoczanska,tomasz.trzcinski}@pw.edu.pl

Abstract. In this work, we propose a method for large-scale topological localization based on radar scan images using learned descriptors. We present a simple yet efficient deep network architecture to compute a rotationally invariant discriminative global descriptor from a radar scan image. The performance and generalization ability of the proposed method is experimentally evaluated on two large scale driving datasets: MulRan and Oxford Radar RobotCar. Additionally, we present a comparative evaluation of radar-based and LiDAR-based localization using learned global descriptors. Our code and trained models are publicly available on the project website (https://github.com/jac99/RadarLoc).

Keywords: Topological localization · Radar-based place recognition · LiDAR-based place recognition · Global descriptors

1 Introduction

Place recognition is an important problem in robotics and autonomous driving community. It aims at recognizing previously visited places based on an input from a sensor, such an RGB camera or a LiDAR scanner, installed on the moving vehicle. Place recognition plays an important part in mapping and localization methods. It allows detecting and closing loops during a map creation. It can improve localization accuracy in areas with restricted or limited GPS coverage [24].

Sensing capabilities of modern robotic and autonomous driving solutions constantly improve as technology advances and becomes more affordable [23]. Therefore, methods for visual place recognition span from classical approaches using RGB camera images to 360° range-based measuring systems, such as LiDARs and radars. Appearance-based methods [2,5] leverage fine details of observed scenes, such as a texture of visible buildings. However, they fail under light-condition variance and seasonal changes. Structure information-based place recognition methods address these limitations. Modern 3D LiDARs, such as Velodyne HDL-64E, can capture up to 100 meters range providing a rich geometrical information about the observed scene. LiDAR-based topological localization

The project was funded by POB Research Centre for Artificial Intelligence and Robotics of Warsaw University of Technology within the Excellence Initiative Program - Research University (ID-UB).

T. Mantoro et al. (Eds.): ICONIP 2021, LNCS 13109, pp. 451–462, 2021.
https://doi.org/10.1007/978-3-030-92270-2_39

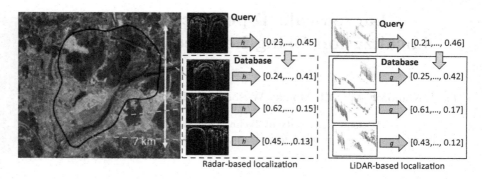

Fig. 1. Radar-based and LiDAR-based topological localization. Trained neural network (blue arrow) is used to compute a global descriptor from a query reading. Localization is performed by searching the database for a geo-tagged sensor readings with closest descriptors. (Color figure online)

using learned descriptors is currently an active field of research, with a larger number of published methods [1,7,11,13,14,16,21,22,26]. Nevertheless, high-end LiDARs are too expensive to be widely applicable, with the price as high as 70k USD per unit. Additionally, LiDAR readings are adversely affected by extreme environmental conditions such as fog, heavy rain or snow [12]. In such challenging conditions, radars show a great potential as they are more robust against atmospheric phenomena. Moreover, modern frequency-modulated continuous wave (FMCW) scanning radars cover broader area, having an operating range up to 200 m. However, radar-based localization is relatively little exploited [12,19]. This can be attributed to limited availability of sufficiently large and diverse datasets. The situation has improved recently with the release of large-scale MulRan [12] and Radar RobotCar [3] datasets.

In this work, we propose a method for topological localization using learned descriptors based on Frequency-Modulated Continuous Wave (FMCW) radar scan images. The idea is illustrated in Fig. 1. A trained neural network computes low-dimensional descriptors from sensor readings. Localization is performed by searching the database for geotagged elements with descriptors closests, in Euclidean distance sense, to the descriptor of the query scan. We present a simple and efficient deep network architecture to compute rotationally invariant discriminative global descriptor from a radar scan image. Rotational invariance is an important property for place recognition tasks, as the same place can be revisited from different directions.

An interesting research question is the comparison of performance of radar-based and LiDAR-based localization methods. In this work we perform such analysis by comparing our radar-based descriptor with LiDAR-based global descriptor. Contributions of this work can be summarized as follows. First, we propose a simple yet efficient deep neural network architecture to compute a discriminative global descriptor from a radar scan image. Second, present a comparative evaluation of radar-based and LiDAR-based localization using learned descriptors.

2   Related Work

Radar-Based Place Recognition. Radar-based place recognition using learned descriptors is relatively unexplored area. In [12] authors adapt a handcrafted ScanContext [11] descriptor, originally used for point cloud-based place recognition. [19] presents a learned a global descriptor computed using convolutional neural network with NetVLAD [2] aggregation layer, commonly used in visual domain. In order to achieve rotational invariance authors introduce couple of modifications, such as cylindrical convolutions, anti-aliasing blurring, and azimuth-wise max-pooling. [6] extends this method by developing a two-stage approach which integrates global descriptor learning with precise pose estimation using spectral landmark-based techniques.

LiDAR-Based Place Recognition. Place recognition methods based on LiDAR scans can be split into two categories: handcrafted and learned descriptors. Among the first group, one of the most effective methods is ScanContext [11], which represents a 3D point cloud as a 2D image. 3D points above the ground plane level are converted to egocentric polar coordinates and projected to a 2D plane. This idea was extended in a couple of later works [4,8].

The second group of methods leverages deep neural networks in order to compute a discriminative global descriptor in a learned manner. One of the first is PointNetVLAD [1], which uses PointNet architecture to extract local features, followed by NetVLAD pooling producing a global descriptor. It was followed by a number of later works [13,16,21,26] using the same principle: local features extracted from the 3D point cloud are pooled to yield a discriminative global descriptor. However all these works operate on relatively small point clouds constructed by accumulating and downsampling a few consecutive scans from a 2D LiDAR. Thus, they do not scale well to larger point clouds generated by a single 360° sweep from a 3D LiDAR with an order of magnitude more points. To mitigate this limitation another line of methods uses an intermediary representation of an input point cloud before feeding it to a neural network. DiSCO [22] first converts a point cloud into a multi-layered representation, then uses a convolutional neural network to extract features in a polar domain and produce a global descriptor.

3   Topological Localization Using Learned Descriptors

The idea behind a topological localization using learned descriptors is illustrated in Fig. 1. A trained neural network is used to compute a discriminative global descriptor from a sensor reading (radar scan image or 3D point cloud from LiDAR). Place recognition is performed by searching the database of geotagged sensor readings for descriptors closests, in Euclidean distance sense, to the descriptor of the query reading. The geo-position of the reading with a closest descriptor found in the database approximates query location. This coarse localization may be followed by a re-ranking step and a precise 6DoF pose estimation based on local descriptors. But in this work we focus only on coarse-level localization using learned global descriptors.

3.1 Radar Scan-Based Global Descriptor

This section describes the architecture of our network to compute a discriminative global descriptor of an input radar scan image. The high-level network architecture is shown in Fig. 2. It consists of two parts: local feature extraction network followed by the generalized-mean (GeM) [18] pooling layer. Our initial experiments proved that GeM yields better results compared to commonly used NetVLAD [2] pooling. One of the reasons is much smaller number of learnable parameters that reduces the risk of overfitting to a moderately-sized training set. The input to the network is a single-channel radar scan image in polar coordinates. The scan image is processed using a 2D convolutional network modelled after FPN [15] design pattern. Upper part of the network, with left-to-right data flow, contains five convolutional blocks producing 2D feature maps with decreasing spatial resolution and increasing receptive field. The bottom part, with right-to-left data flow, contains a transposed convolution generating an upsampled feature map. Upsampled feature map is concatenated with the skipped features from the corresponding block in the upper pass using a lateral connection. Such design is intended to produce a feature map with higher spatial resolution, having a large receptive field. Our experiments proved its advantage over a simple convolutional architecture with one-directional data flow. Table 1 shows details of each network block. The first convolutional block ($2dConv_0$) has a bigger 5×5 kernel, in order to aggregate information from a larger neighbourhood. Subsequent blocks ($2dConv_1 \ldots 2dConv_4$) are made of a stride two convolution, which decreases spatial resolution by two, followed by residual block consisting of two convolutional layers with 3×3 kernel and ECA [20] channel attention layer. All convolutional layers are followed by batch normalization [10] layer and ReLU non-linearity. Two 1xConv blocks have the same structure, both contain a single convolutional layer with 1×1 kernel. The aim of these blocks is to unify the number of channels in feature maps produced by the blocks in the left-to-right, before they are merged with feature maps from the right-to-left pass through the network. The bottom part of the network (left-to-right pass) consists of a single transposed convolution layer ($2dTConv_4$) with 2×2 kernel and stride 2. The feature map $\mathcal{F}_h$ computed by the feature extraction network is pooled with generalized-mean (GeM) [18] pooling to produce a radar scan descriptor $\mathcal{H}$.

Important property for loop closure applications is rotational invariance, as the same place may be revisited from different directions. Invariance to the viewpoint rotation translates into shift invariance along the angular direction of the scan image in polar coordinates. To ensure this invariance we use a circular padding along the angular direction in all convolutions in the radar scan descriptor extraction network. The left boundary of the scan image (corresponding to

0° angular coordinate) is padded with values on the right boundary of the image (corresponding to 360° angle) and vice verse. This only partially solves the problem, as convolutions and pooling with stride 2 are not translational invariant due to the aliasing effect. To mitigate this problem we augment the training data, by randomly rolling the image over axis corresponding to the angular direction, which gives the same effect as rotating an image in Cartesian coordinates. Alternative approach is to apply anti-aliasing blurring [25], where stride 2 max-pooling and convolutions are replaced with stride 1 operations, followed by a stride 2 Gaussian blur. But we found it giving worse results in practice.

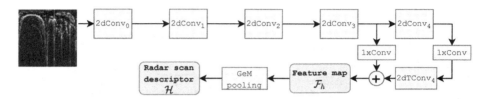

Fig. 2. Architecture of a radar scan descriptor extraction network. The input is processed by a 2D convolutional network with FPN [15] architecture to produce a feature map $\mathcal{F}_h$. The feature map is pooled with generalized-mean (GeM) [18] pooling to produce a radar scan descriptor $\mathcal{H}$.

Table 1. Details of the descriptor extractor network for radar scan images.

Block	Layers
2dConv$_0$	2d convolutions with 32 filters 5×5 - BN - ReLU
2dConv$_k$	2d convolution with c_k filters 2×2 stride 2 - BN - ReLU
	2d convolution with c_k filters 3×3 stride 1 - BN - ReLU
	2d convolution with c_k filters 3×3 stride 1 - BN - ReLU
	Efficient Channel Attention (ECA) layer [20]
	where $c_1 = 32, c_{2,3} = 64, c_{4,5} = 128$
2dTConv$_6$	2d transposed conv. with 128 filters 2×2 stride 2
1xConv	2d convolution with 128 filters 1×1 stride 1
GeM pooling	Generalized-mean Pooling layer [18]

3.2 LiDAR-Based Global Descriptor

This section describes the architecture of the network used to compute a discriminative global descriptor of a 3D point cloud. We choose a 3D convolutional architecture using sparse volumetric representation that produced state of the art results in our previous MinkLoc3D [13] work. However, MinkLoc3D is architectured to process relatively small point clouds, constructed by concatenating multiple 2D LiDAR scans. In this work we use point clouds build from a single

360° scans from 3D LiDAR, covering much larger area, app. 160 m in diameter, and containing an order of magnitude more points. To extract informative features from such larger point clouds we enhanced the network architecture to increase the receptive field. The number of blocks in upper and lower part of the network is increased compared to MinkLoc3D design. Figure 2 shows high-level architecture and details of each network block are given in Table 1. For more information we refer the reader to our MinkLoc3D [13] paper. To ensure rotational invariance of the resultant global descriptor, necessary for loop closure applications, we resort to data augmentation. Point clouds are randomly rotated around the z-axis before they are fed to the network during the training.

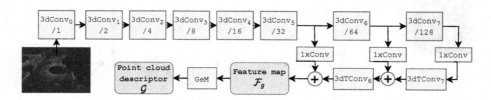

Fig. 3. Architecture of a LiDAR point cloud descriptor extraction network. The input is quantized into a sparse voxelized representation and processed by a 3D convolutional network with FPN [15] architecture. The resultant sparse 3D feature map is pooled with generalized-mean (GeM) pooling to produce a global point cloud descriptor $\mathcal{G}$.

Table 2. Details of the descriptor extractor network for LiDAR point clouds.

Block	Layers
3dConv$_0$	32 filters $5 \times 5 \times 5$ - BN - ReLU
3dConv$_k$	3d convolution with c_k filters $2 \times 2 \times 2$ stride 2 - BN - ReLU
	3d convolution with c_k filters $3 \times 3 \times 3$ stride 1 - BN - ReLU
	3d convolution with c_k filters $3 \times 3 \times 3$ stride 1 - BN - ReLU
	Efficient Channel Attention (ECA) layer [20]
	where $c_1 = 32, c_2 = c_3 = 64, c_{4...7} = 128$
3dTConv$_k$, $k = 6, 7$	3d transposed conv. with 128 filters $2 \times 2 \times 2$ stride 2
1xConv	3d convolution with 128 filters $1 \times 1 \times 1$ stride 1
GeM pooling	Generalized-mean Pooling layer [18]

3.3   Network Training

To train both networks to generate discriminative global descriptors we use a deep metric learning approach [17] with a triplet margin loss [9] defined as:

$$L(a_i, p_i, n_i) = \max \{d(a_i, p_i) - d(a_i, n_i) + m, 0\},$$

where $d(x, y) = ||x - y||_2$ is an Euclidean distance between embeddings x and y; a_i, p_i, n_i are embeddings of an anchor, a positive and a negative elements in i-th training triplet and m is a margin hyperparameter. The loss is formulated to make embeddings of structurally dissimilar sensor readings (representing different places) further away, in the descriptor space, than embeddings of structurally similar readings (representing the same place). The loss is minimized using a stochastic gradient descent with Adam optimizer.

To improve effectiveness of the training process we use batch hard negative mining [9] strategy to construct informative triplets. Each triplet is build using the hardest negative example found within a batch. The hardest negative example is a structurally dissimilar element that has the closest embedding, computed using current network weights, to the anchor.

To increase variability of the training data, reduce overfitting and ensure rotational invariance of global descriptors, we use on-the-fly data augmentation. For radar scan images, we use random erasing augmentation [27] and random cyclical shift over x-axis (angular dimension). For points clouds, it includes random jitter and random rotation around z-axis. We also adapted random erasing augmentation to remove 3D points within the randomly selected frontoparallel cuboid.

4   Experimental Results

4.1   Datasets and Evaluation Methodology

To train and evaluate our models we use two recently published large-scale datasets: MulRan [12] and Oxford Radar RobotCar [3]. MuRan dataset is gathered using a vehicle equipped with Ouster OS1-64 3D LiDAR with 120 m. range and Navtech CIR204-H FMCW scanning radar with 200 operating range. Radar RobotCar data is acquired using Velodyne HDL-32E LiDAR with 100 m. range and Navtech CTS350-X scanning radar with 160 m. operating range.

In both datasets each trajectory is traversed multiple times, at different times of day and year, allowing a realistic evaluation of place recognition methods. Radar scans are provided in similar format, as 360° polar images with 400 (angular) by 3360 (radial dimension) pixel resolution. To decrease computational resources requirements we downsample them to 384 (angular) by 128 (radial) resolution. LiDAR scans are given as unordered set of points, containing between 40–60 thousand points. To speed up the processing, we remove uninformative ground plane points with z-coordinate below the ground plane level. Both datasets contain ground truth positions for each traversal.

Table 3. Length and number of scans in training and evaluation sets.

Trajectory	Split	Length	Number of scans (map/query)	
			Radar	LiDAR
Sejong	Train	19 km	13 611	33 698
Sejong	Test	4 km	1 366/1 337	3 358/3 315
KAIST	Test	6 km	3 209/3 420	7 975/8 521
Riverside	Test	7 km	2 117/2 043	5 267/5 084
Radar RobotCar	test	10 km	5 904/5 683	26 719/25 944

The longest and most diverse trajectory from MulRan dataset, Sejong, is split into disjoint training and evaluation parts. We evaluate our models using disjoint evaluation split from Sejong trajectory and two other trajectories: KAIST and Riverside, acquired at different geographic locations. Each evaluation trajectory contains two traversals gathered at different times of a day and year. The first traversal (Sejong01, KAIST01, Riverside01) forms a query set and the second one (Sejong02, KAIST02, Riverside02) is used as a map. To test generalization ability of our model, we test it using two traversals from Radar RobotCar dataset acquired at different days: traversal 019-01-15-13-06-37-radar-oxford-10k as a map split and 2019-01-18-14-14-42-radar-oxford-10k as a query split.

To avoid processing multiple scans of the same place when the car doesn't move, we ignore consecutive readings with less than 0.1 m displacement in the ground truth position. We also ignore readings for which a ground truth pose in not given with 1 s. tolerance. See Fig. 4 for visualization of training and evaluation trajectories. Details of the training and evaluation sets are given in Table 3.

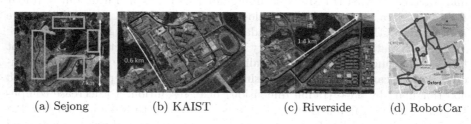

(a) Sejong   (b) KAIST   (c) Riverside   (d) RobotCar

Fig. 4. Visualization of three trajectories in MulRan dataset and one in Radar Robot-Car. Red rectangles in Sejong trajectory delimit the training split. (Color figure online)

Training triplets are generated using the ground truth coordinates provided in each dataset. Positive examples is chosen from structurally similar readings, that is readings that are at most 5 m apart. Negative examples are sampled from dissimilar readings, that is readings that are at least 20 m apart.

Evaluation Metrics. To evaluate the global descriptor performance we follow similar evaluation protocol as in other point cloud-based or radar-based place recognition works [1]. Each evaluation set is split into two parts: a query and a database set, covering the same geographic area. A query is formed from sensor readings acquired during one traversal and the database is build from data gathered during a different traversal, on a different day. For each query element we find database element with closests, in Euclidean distance sense, global descriptors. Localization is successful if at least one of the top N retrieved database elements is within d meters from the query ground truth position. *Recall@N* is defined as the percentage of correctly localized queries.

4.2 Results and Discussion

We compare the performance of our radar-based global descriptor with a hand-crafted ScanContext [11] method and our re-implementation of the VGG-16/NetVLAD architecture used as a baseline in [19]. Evaluation results are shown in Table 4. Our method (RadarLoc) consistently outperforms other approaches on all evaluation sets at both 5 m and 10 m threshold. Recall@1 with 5 m threshold is between 2–14 p.p. higher than the runner-up ScanContext. At larger 10 m threshold, learning-based VGG-16/NetVLAD architecture scores higher than ScanContext, but it's still about 5–10 p.p. lower than RadarLoc. Our method generalizes well to a different dataset. The model trained on a subset of Sejong traversal, from MulRan dataset, has a top performance when evaluated on Radar RobotCar. It must be noted that these two datasets are acquired using a different suite of sensors, although having similar operational characteristics. Figure 5 shows Recall@N plots, for N ranging from 1 to 10, of all evaluated methods on different evaluation sets from MulRan dataset. For Sejong and KAIST trajectory, the performance of our method increases with N, quickly reaching near-100% accuracy. However characteristics of the environment in which data was gathered impacts all evaluated methods. The results for all traversals acquired in the city centre areas (Sejong, KAIST, Radar RobotCar) are relatively better, for all evaluated methods. Riverside traversal is acquired outside the city, where significantly fewer structural elements, such as buildings or lamp posts, are present. Hence it gives considerably worse results for all evaluated methods.

Table 4. Evaluation results (Recall@1) of a radar-based descriptor.

Method	Sejong		KAIST		Riverside		RadarRobotCar	
	5 m	10 m	5 m	10 m	5 m	10 m	5 m	10 m
Ring key [11]	0.503	0.594	0.805	0.838	0.497	0.595	0.747	0.786
ScanContext [11]	0.868	0.879	0.935	0.946	0.671	0.772	0.906	0.933
VGG-16/NetVLAD	0.789	0.938	0.8885	0.937	0.613	0.834	0.883	0.939
RadarLoc (ours)	**0.929**	**0.988**	**0.959**	**0.988**	**0.744**	**0.923**	**0.949**	**0.981**

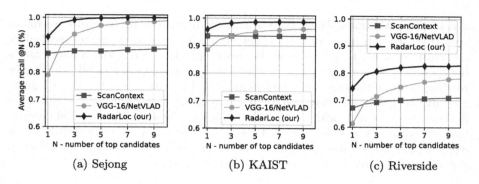

(a) Sejong   (b) KAIST   (c) Riverside

Fig. 5. Average Recall@N with 5 m threshold of radar-based descriptor.

Table 5 compares performance of radar-based and LiDAR-based localization on three different evaluation sets from MulRan dataset. LiDAR has 2–3 times higher scanning frequency than radar and our datasets contain 2–3 more LiDAR 3D point clouds than radar scan images. To cater for this difference, we compare the performance of radar versus LiDAR-based topological localization in two scenarios. First, we use all available data gathered during vehicle traversals. In this scenario, LiDAR scans cover the mapped area denser than radar. Second, we evaluate the performance, using a smaller, subsampled LiDAR dataset with the same number of point clouds as radar scans. In this scenario, LiDAR and radar scans cover the mapped area equally dense. It can be seen, that better map coverage caused by higher LiDAR scanning frequency has little impact. Performance of the LiDAR-based method on full evaluation sets and on downsampled evaluation sets is very close.

Intuitively, LiDAR provides richer 3D information about the structure of the observed scene, whereas radar captures only 2D range data. We would expect LiDAR-based approach to produce more discriminative global descriptors. Interestingly, this not happen and localization using radar scan images generally yields better results, especially with larger 10 m threshold. We hypothesize that the reason is bigger complexity and larger susceptibility to overfitting of a LiDAR-based model. The performance of both methods on MulRan evaluation split, having scenes with similar characteristic as in the training split, is very close. When

Table 5. Comparison (Recall@1) of radar-based and LiDAR-based descriptors.

Method	Sejong		KAIST		Riverside	
	5 m	10 m	5 m	10 m	5 m	10 m
RadarLoc (our)	0.929	**0.988**	**0.959**	**0.988**	0.744	**0.923**
LiDAR-based	**0.950**	0.986	0.901	0.930	**0.748**	0.881
LiDAR-based (subsampled dataset)	0.941	0.986	0.897	0.929	0.740	0.881

evaluated on different traversals, covering locations with different characteristic (e.g. suburbs versus city center), LiDAR-based model generalizes worse.

5   Conclusion

Topological localization using radar scan images is relatively unexplored area. In this work we demonstrate that it has a high potential and is competitive to LiDAR-based approaches. Our proposed radar-based localization method, having a simple and efficient architecture, outperforms baseline methods and has better generalization abilities than LiDAR-based approach. LiDAR, however, provides richer 3D information about the structure of the observed scene, whereas radar captures only 2D range data. An interesting research question is how to fully exploit structural information available in 3D LiDAR scans to increase discriminative power of resultant global descriptors and outperform radar-based methods.

References

1. Angelina Uy, M., Hee Lee, G.: PointNetVlad: deep point cloud based retrieval for large-scale place recognition. In: Proceedings of the IEEE Conference on Computer Vision and Pattern Recognition, pp. 4470–4479 (2018)
2. Arandjelovic, R., Gronat, P., Torii, A., Pajdla, T., Sivic, J.: NetVlad: CNN architecture for weakly supervised place recognition. In: Proceedings of the IEEE Conference on Computer Vision and Pattern Recognition, pp. 5297–5307 (2016)
3. Barnes, D., Gadd, M., Murcutt, P., Newman, P., Posner, I.: The Oxford radar RobotCar dataset: a radar extension to the Oxford RobotCar dataset. In: Proceedings of the IEEE International Conference on Robotics and Automation (ICRA), Paris (2020)
4. Cai, X., Yin, W.: Weighted Scan Context: global descriptor with sparse height feature for loop closure detection. In: 2021 International Conference on Computer, Control and Robotics (ICCCR), pp. 214–219. IEEE (2021)
5. Cummins, M., Newman, P.: FAB-MAP: probabilistic localization and mapping in the space of appearance. Int. J. Robot. Res. **27**(6), 647–665 (2008)
6. De Martini, D., Gadd, M., Newman, P.: kRadar++: coarse-to-fine FMCW scanning radar localisation. Sensors **20**(21), 6002 (2020)
7. Du, J., Wang, R., Cremers, D.: DH3D: deep hierarchical 3d descriptors for robust large-scale 6DoF relocalization. In: Vedaldi, A., Bischof, H., Brox, T., Frahm, J.-M. (eds.) ECCV 2020. LNCS, vol. 12349, pp. 744–762. Springer, Cham (2020). https://doi.org/10.1007/978-3-030-58548-8_43
8. Fan, Y., He, Y., Tan, U.X.: Seed: a segmentation-based egocentric 3D point cloud descriptor for loop closure detection. In: 2020 IEEE/RSJ International Conference on Intelligent Robots and Systems (IROS), pp. 5158–5163. IEEE (2020)
9. Hermans, A., Beyer, L., Leibe, B.: In defense of the triplet loss for person re-identification. arXiv preprint arXiv:1703.07737 (2017)
10. Ioffe, S., Szegedy, C.: Batch normalization: accelerating deep network training by reducing internal covariate shift. In: Proceedings of Machine Learning Research, vol. 37, pp. 448–456. PMLR, Lille (2015)

11. Kim, G., Kim, A.: Scan Context: egocentric spatial descriptor for place recognition within 3D point cloud map. In: 2018 IEEE/RSJ International Conference on Intelligent Robots and Systems (IROS), pp. 4802–4809. IEEE (2018)
12. Kim, G., Park, Y.S., Cho, Y., Jeong, J., Kim, A.: MulRan: multimodal range dataset for urban place recognition. In: Proceedings of the IEEE International Conference on Robotics and Automation (ICRA), Paris, pp. 6246–6253 (2020)
13. Komorowski, J.: MinkLoc3D: point cloud based large-scale place recognition. In: Proceedings of the IEEE/CVF Winter Conference on Applications of Computer Vision (WACV), pp. 1790–1799 (2021)
14. Komorowski, J., Wysoczanska, M., Trzcinski, T.: MinkLoc++: lidar and monocular image fusion for place recognition. In: 2021 International Joint Conference on Neural Networks (IJCNN) (2021)
15. Lin, T.Y., Dollar, P., Girshick, R., He, K., Hariharan, B., Belongie, S.: Feature pyramid networks for object detection. In: Proceedings of the IEEE Conference on Computer Vision and Pattern Recognition, pp. 2117–2125 (2017)
16. Liu, Z., et al.: LPD-Net: 3D point cloud learning for large-scale place recognition and environment analysis. In: Proceedings of the IEEE International Conference on Computer Vision, pp. 2831–2840 (2019)
17. Lu, J., Hu, J., Zhou, J.: Deep metric learning for visual understanding: an overview of recent advances. IEEE Signal Process. Mag. **34**(6), 76–84 (2017)
18. Radenovic, F., Tolias, G., Chum, O.: Fine-tuning CNN image retrieval with no human annotation. IEEE Trans. Pattern Anal. Mach. Intell. **41**(7), 1655–1668 (2018)
19. Saftescu, S., Gadd, M., De Martini, D., Barnes, D., Newman, P.: Kidnapped radar: topological radar localisation using rotationally-invariant metric learning. In: 2020 IEEE International Conference on Robotics and Automation (ICRA), pp. 4358–4364. IEEE (2020)
20. Wang, Q., Wu, B., Zhu, P., Li, P., Zuo, W., Hu, Q.: ECA-Net: efficient channel attention for deep convolutional neural networks. In: Proceedings of the IEEE/CVF Conference on Computer Vision and Pattern Recognition (CVPR), June 2020
21. Xia, Y., et al.: SOE-Net: a self-attention and orientation encoding network for point cloud based place recognition. arXiv preprint arXiv:2011.12430 (2020)
22. Xu, X., Yin, H., Chen, Z., Li, Y., Wang, Y., Xiong, R.: DISCO: differentiable scan context with orientation. IEEE Robot. Autom. Lett. **6**(2), 2791–2798 (2021)
23. Yeong, D.J., Velasco-Hernandez, G., Barry, J., Walsh, J.: Sensor and sensor fusion technology in autonomous vehicles: a review. Sensors **21**(6), 2140 (2021)
24. Yin, H., Xu, X., Wang, Y., Xiong, R.: Radar-to-lidar: heterogeneous place recognition via joint learning. Front. Robot. AI **8**, 101 (2021)
25. Zhang, R.: Making convolutional networks shift-invariant again. In: International Conference on Machine Learning, pp. 7324–7334. PMLR (2019)
26. Zhang, W., Xiao, C.: PCAN: 3d attention map learning using contextual information for point cloud based retrieval. In: Proceedings of the IEEE Conference on Computer Vision and Pattern Recognition, pp. 12436–12445 (2019)
27. Zhong, Z., Zheng, L., Kang, G., Li, S., Yang, Y.: Random erasing data augmentation. arXiv preprint arXiv:1708.04896 (2017)

Rethinking Binary Hyperparameters for Deep Transfer Learning

Jo Plested[1]([✉]), Xuyang Shen[2], and Tom Gedeon[2]

[1] Department of Computer Science, University of New South Wales,
Kensington, Australia
j.plested@unsw.edu.au
[2] School of Computer Science, Australian National University, Canberra, Australia
{xuyang.shen,tom.gedeon}@anu.edu.au

Abstract. The current standard for a variety of computer vision tasks using smaller numbers of labelled training examples is to fine-tune from weights pre-trained on a large image classification dataset such as ImageNet. The application of transfer learning and transfer learning methods tends to be rigidly binary. A model is either pre-trained or not pre-trained. Pre-training a model either increases performance or decreases it, the latter being defined as negative transfer. Application of L2-SP regularisation that decays the weights towards their pre-trained values is either applied or all weights are decayed towards 0. This paper re-examines these assumptions. Our recommendations are based on extensive empirical evaluation that demonstrate the application of a non-binary approach to achieve optimal results. (1) Achieving best performance on each individual dataset requires careful adjustment of various transfer learning hyperparameters not usually considered, including number of layers to transfer, different learning rates for different layers and different combinations of L2SP and L2 regularization. (2) Best practice can be achieved using a number of measures of how well the pre-trained weights fit the target dataset to guide optimal hyperparameters. We present methods for non-binary transfer learning including combining L2SP and L2 regularization and performing non-traditional fine-tuning hyperparameter searches. Finally we suggest heuristics for determining the optimal transfer learning hyperparameters. The benefits of using a non-binary approach are supported by final results that come close to or exceed state of the art performance on a variety of tasks that have traditionally been more difficult for transfer learning.

Keywords: Transfer learning · Image classification · Computer vision

1 Introduction

Convolutional neural networks (CNNs) have achieved many successes in image classification in recent years [12, 14, 17, 22]. It has been consistently demonstrated that CNNs work best when there is abundant labelled data available for the task

© Springer Nature Switzerland AG 2021
T. Mantoro et al. (Eds.): ICONIP 2021, LNCS 13109, pp. 463–475, 2021.
https://doi.org/10.1007/978-3-030-92270-2_40

and very deep models can be trained [13,20,24]. However, there are many real world scenarios where the large amounts of training data required to obtain the best performance cannot be met or are prohibitively expensive. Transfer learning has been shown to improve performance in a wide variety of computer vision tasks, particularly when the source and target tasks are closely related and the target task is small [6,18–20,23,24,29]. It has become standard practice to pre-train on Imagenet 1K for many different tasks where the available labeled datasets are orders of magnitude smaller than Imagenet 1K [17–19,22,23,29]. Several papers published in recent years have questioned this established paradigm. They have shown that when the target dataset is very different from Imagenet 1K and a reasonable amount of data is available, training from scratch can match or even out perform pre-trained and fine-tuned models [11,20].

The standard transfer learning strategy is to transfer all layers apart from the final classification layer, and either use a single initial learning rate and other hyperparameters for fine-tuning all layers, or freeze some layers. Given that lower layers in a deep neural network are known to be more general and higher layers more specialised [31], we argue that the binary way of approaching transfer learning and fine-tuning is counter intuitive. There is no intuitive reason to think that all layers should be treated the same when fine-tuning, or that pre-trained weights from all layers will be applicable to the new task. If transferring all layers results in negative transfer, could transferring some number of lower more general layers improve performance? If using an L2SP weight decay on all transferred layers for regularisation decreases performance over decaying towards 0, might applying the L2SP regularisation to some number of lower layers that are more applicable to the target dataset result in improved performance?

We performed extensive experiments across four different datasets to:

- re-examine the assumptions that transfer learning hyperparameters should be binary, and
- find the optimal settings for number of layers to transfer, initial learning rates for different layers, and number of layers to apply L2SP regularisation vs decaying towards 0.

We developed methods for non-binary transfer learning including combining L2SP and L2 regularization and performing non-traditional fine-tuning hyperparameter searches. We show that the optimal settings result in significantly better performance than binary settings for all datasets except the most closely related. Finally we suggest heuristics for determining the optimal transfer learning hyperparameters.

2   Related Work

The standard transfer learning strategy is to transfer all layers apart from the final classification layer, then use a search strategy to find the best single initial learning rate and other hyperparameters. Several studies include extensive hyperparameter searches over learning rate and weight decay [14,20], momentum [16], and L2SP [19]. This commonly used strategy originates from various works showing that performance on the target task increases as the number of layers transferred increases [1,2,4,31]. All these works were completed prior to advances in residual networks [12] and other very deep models, and searches for optimal combinations of learning rates and number of layers to transfer were not performed. Additionally, in two of the works layers that were not transferred were discarded completely rather than reinitialising and training them from scratch. This resulted in smaller models with less layers transferred and a strong bias towards better results when more layers were transferred [1,2]. Further studies have shown the combination of a lower learning rate and fewer layers transferred may be optimal [25]. However, again modern very deep networks were not used and only very similar source and target datasets were selected.

It has been shown that the similarity between source and target datasets has a strong impact on performance for transfer learning:

1. More closely related datasets can be better than more source data for pre-training [6,20,24].
2. A multi-step pre-training process where the interim dataset is smaller and more closely related to the target dataset can outperform a single step pre-training process when originating from a very different, large source dataset [23].
3. Self-supervised pre-training on a closely related source dataset can be better than supervised training on a less closely related dataset [33].
4. L2SP regularization, where the weights are decayed towards their pre-trained values rather than 0 during fine-tuning, improves performance when the source and target dataset are closely related, but hinders it. when they are less related [18,19,29]
5. Momentum should be lower for more closely related source and target datasets [16].

These five factors demonstrate the importance of the relationship between the source and target dataset in transfer learning.

3 Methodology

3.1 Datasets

We performed evaluations on four different small target datasets, each being less than 10,000 training images. We chose one that is very similar to Imagenet 1K that transfer learning and sub methods have traditionally performed best on. This is used as a baseline for comparison to show that the default methods that perform badly on the other datasets chosen do perform well on this one. For each of the other target datasets it has been shown that traditional transfer learning strategies:

- do not perform well on them and/or
- they are very different to the source dataset used for pre-training.

We used the standard available train, validation and test splits for the three datasets for which they were available. For Caltech256-30 we used the first 30 items from each class for the train split, the next 20 for the validation split and the remainder for the test split.

Source Dataset. We used Imagenet 1K [7] as the source dataset as it is most commonly used source dataset and therefore the most suitable to demonstrate our approach.

Target Datasets. These are the final datasets that we transferred the models to and measured performance on. We used a standard 299 × 299 image size for all datasets.

Most Similar to Imagenet 1K. We chose Caltech 256-30 (Caltech) [9] as the most similar to Imagenet. It contains 256 different general subordinate and superordinate classes. As our focus is on small target datasets, we chose the smallest commonly used split with 30 examples per class.

Fine-Grained. Fine-grained object classification datasets contain subordinate classes from one particular superordinate class. We chose two where standard transfer learning has performed badly [10,14].

- Stanford Cars (Cars): Contains 196 different makes and models of cars with 8,144 training examples and 8,041 test examples [15].
- FGVC Aircraft (Aircraft): Contains 100 different makes and models of aircraft with 6,667 training examples and 3,333 test examples [21].

Very Different to Imagenet 1K. We evaluated performance on another dataset that is intuitively very different to Imagenet 1K as it consists of images depicting a djectives instead of nouns.

- Describable Textures (DTD) [5]: consists of 3,760 training examples of texture images jointly annotated with 47 attributes. The texture images are collected "in the wild" by searching for texture adjectives on the web.

3.2   Model

Inception v4 [28] was selected for evaluation as it has been shown to have state of the art performance for transferring from Imagenet 1K to a broad range of target tasks [14]. Our code is adapted from [30] and we used the publicly available pre-trained Inception v4 model. Our code is available at https://github.com/XuyangSHEN/Non-binary-deep-transfer-learning-for-image-classification. We did not experiment with pre-training settings, for example removing regularization settings as suggested by [14], as it was beyond the scope of this work and the capacity of our compute resources.

3.3 Evaluation Metric

We used top-1 accuracy for all results for easiest comparison with existing results on the chosen datasets. Graphs showing ablation studies with different hyperparameters show results on one run. Final reported results are averages and standard deviations over four runs. For all experiments we used single crop and for our final comparison to state of the art for each dataset we used an ensemble of all four runs and 10-crop.

4 Results

4.1 Assessing the Suitability of the Fixed Features

We first examined performance using the pre-trained Inception v4 model as a fixed feature extractor for comparison and insight into the suitability of the pre-trained features. We trained a simple fully connected neural network classification layer for each dataset and compared it to training the full model from random initialization. The comparisons for all datasets are shown in Table 1.

The class normalized Fisher score on target datasets using the pre-trained but not fine-tuned weights, was used as a measure of how well the pre-trained weights separate the classes in the target dataset [8]. A larger normalized Fisher score shows more separation between classes in the feature space.

We also calculated domain similarity between the fixed features of the source and target domain using the earth mover distance (EMD) [27] and applying the procedure defined in [6].

The domain similarity calculated using the EMD has been shown to correlate well with the improvement in performance on the target task from pre-training with a particular source dataset [6].

Table 1. Domain measures

	Trained from random initialization	Fixed features classification	EMD domain similarity	Normalised Fisher score	State of the art
Caltech	67.2	83.4	0.568	1.35	84.9 [18]
Cars	92.7	64.2	0.536	1.18	96.0 (@448) [26]
Aircraft	88.8	59.9	0.557	0.83	93.9 (@448) [32]
DTD	66.8	74.6	0.540	3.47	78.9 [3]

4.2 Default Settings

For the first experiment we transferred all but the final classification layer and trained all layers at the same learning rate as per standard transfer learning practice. We performed a search using the validation set to find the optimal single fine-tuning learning rate for each dataset. Final results on the test set for each dataset are shown in Table 2. The optimal single learning rate for Caltech, the most similar dataset to ImageNet, is an order of magnitude lower than the optimal learning rate for the fine-grained classification tasks Stanford Cars and Aircraft. This shows that the optimal weights for Caltech are much closer to the pre-trained values. The surprising result was that the optimal learning rate for DTD was very similar to Caltech and also an order of magnitude lower than Stanford Cars and Aircraft.

Table 2. Final results default settings

	Caltech	Stanford cars	Aircraft	DTD
Learning rate	0.0025	0.025	0.025	0.002
Accuracy	83.69 ± 0.0784	94.59 ± 0.110	93.78 ± 0.137	77.30 ± 0.726

4.3   Optimal Learning Rate Decreases as More Layers Are Reinitialized and Transferring all Possible Layers Often Reduces Performance on the Target Task

To the best of our knowledge the combination of learning rate and number of layers reinitialized when performing transfer learning has not been examined in modern deep models. To examine the relationship between learning rate and number of layers reinitialized, we searched for the optimal learning rate for each of 1–3 blocks of layers reinitialized across each of the four datasets. One layer is the default with the final classification layer only being reinitialized. Two and three involve reinitializing an additional one or two Inception C blocks of layers respectively as well as the final layer. For consistency we refer to both the final classification layer and the Inception C blocks as layers. Table 3 shows the final results for the optimal learning rate and accuracy for each number of layers reinitialized. The optimal learning rate when reinitializing more than one layer is always lower than when reinitializing only the final classification layer. Also the optimal number of layers to reinitialize is more than one for all datasets except Cars.

Reinitializing More Layers has a More Significant Effect When the Optimal Learning Rate for Reinitializing Just One Layer Is Higher. Caltech and DTD have an order of magnitude lower optimal learning rates than Cars and Aircraft, but reinitializing more layers for the former results in a significant increase in accuracy whereas there is none for the latter. This result initially seems counter intuitive as Cars and Aircraft are less similar to Imagenet than Caltech and DTD. However, a lower learning rate tempers the ability of the upper layers of the model to specialise to the new domain and even very closely related source and target domains have different final classes and thus optimal feature spaces. The combination of a lower learning rate and reinitializing more layers likely allows the models to keep the closely related lower and middle layers and create new specialist upper layers.

Table 3. Final optimal learning rate for each number of layers reinitialized

	Caltech	Cars	Aircraft	DTD
1 layer lr	0.0025	0.025	0.025	0.002
Accuracy	83.69 ± 0.078	94.59 ± 0.110	93.78 ± 0.137	77.30 ± 0.726
2 layers lr	0.015	0.015	0.02	0.001
Accuracy	84.31 ± 0.114	94.59 ± 0.205	93.56 ± 0.942	78.92 ± 0.191
3 layers lr	0.015	0.015	0.02	0.0015
Accuracy	83.57 ± 0.104	94.46 ± 0.348	92.96 ± 1.588	78.90 ± 0.243

4.4 Lower Learning Rates for Lower Layers Works Better When Optimal Learning Rate Is Higher

Conventionally learning rates for different layers are set so that either all layers are fine-tuned with the same learning rate or some are frozen. The thinking is that as lower layers are more general and higher layers more task specific [31] the lower layers do not need to be fine-tuned. Recent work has shown that fine-tuning tends to work better in most cases [25]. However, setting different learning rates for different layers is not generally considered. We examined the effects of applying lower learning rates to lower layers that are likely to generalise better to the target task. Table 4 shows that for the Stanford Cars and Aircraft where the optimal initial learning rate is higher, setting lower learning rates for lower layers significantly improves performance whereas for Caltech and DTD it does not.

Table 4. Optimal learning rates for each number of layers reinitialized with different learning rates for lower layers. Learning rates are in the format (high layers learning rate, low layers learning rate, number of low layers)

	Caltech	Cars	Aircraft	DTD
Default accuracy	83.69 ± 0.0784	94.59 ± 0.110	93.78 ± 0.137	77.30 ± 0.726
1 layer lrs	0.002 0.001 12	0.025 0.01 8	0.025 0.015 10	0.0025 0.0015 10
Accuracy	83.95 ± 0.196	94.86 ± 0.0460	94.32 ± 0.144	77.753 ± 0.456
2 layers lrs	0.002 0.001 12	0.025 0.01 8	0.025 0.015 10	0.0015 0.001 14
Accuracy	84.15 ± 0.553	94.78 ± 0.132	94.17 ± 0.311	78.59 ± 0.127
3 layers lrs	0.002 0.001 12	0.025 0.01 10	0.01 0.025 0.01 8	0.0015 0.001 12
Accuracy	83.46 ± 0.143	94.83 ± 0.0832	94.50 ± 0.192	78.84 ± 0.464

4.5 L2SP

The L2SP regularizer decays pre-trained weights towards their pre-trained values rather than towards zero during fine-tuning. In the standard transfer learning paradigm with only the final classification reinitialized, when the L2SP regularizer is applied the final classification layer only is decayed towards 0.

The values α and β are tuneable hyperparameters to control the amount of regularization applied to the pre-trained and randomly initialized layers respectively.

The original experiments showing the effectiveness of the L2SP regularizer [19] were done on target datasets that are extremely similar to the source datasets used. They showed that a high level of regularization decaying towards the pre-trained weights is beneficial on these datasets. It has since been shown that the L2SP regularizer can result in minimal improvement or even worse performance when the source and target datasets are less related [16,29].

Our results shown in Fig. 1 align with the original paper [19] for the dataset Caltech showing that standard L2SP regularization does result in an improvement in performance over L2 regularization. Our results also align with [16,29] in showing that for the datasets we have chosen to be different from Imagenet, and known to be more difficult to transfer to, L2SP regularization performs worse than L2 regularization. In general the lower the setting for alpha (the L2SP regularization hyperparameter) the better the performance.

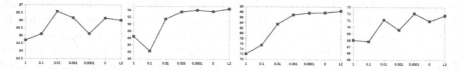

Fig. 1. L2SP with default settings. Left to right Caltech, Cars, Aircraft, DTD

We relaxed the binary assumption that L2SP regularization must be applied to all pre-trained layers. We used L2SP regularization for lower layers that we expected to be more similar to the source dataset and L2 regularization for upper layers to allow them to specialise to the target dataset. We searched for the optimal combinations of L2SP and L2 weight regularization along with the α and β hyperparameters for each dataset. We show the best settings for number of layers and amount of L2SP regularization in Table 5.

We make the following observations:

1. A combination of L2SP and L2 regularization is optimal for most settings of the L2SP regularization hyperparameter (α) for all datasets except Caltech.
2. When more layers are trained with L2 rather than L2SP regularization the optimal L2 regularization hyperparameter (β) is lower as the squared sum of the weights in these layers will be larger for the same model.

Further experiments were performed to search for the combination of optimal L2SP vs L2 regularization settings with optimal number of layers transferred and different learning rates for different layers. These results are also shown in Table 5.

Table 5. Best default and optimal L2SP settings and results

	Caltech	Cars	Aircraft	DTD
L2	83.694	94.59	93.78	77.30
Default L2SP 1 new layer α, β	0.01 0.01	0.0001 0.01	0.0001 0.01	0.0001 0.01
Result	84.52	94.42	93.29	78.32
Optimal L2SP 1 new layer α, β	0.01 0.01 L2SP layers all	0.01 0.001 L2SP layers 10	0.0001 0.001 L2SP layers 10	0.001 0.001 L2SP layers 10
Result	84.52	94.57	93.86	78.32
Optimal L2SP overall α, β	0.01 0.01 L2SP layers all	0.01 0.001 L2SP layers 10	0.0001 0.001 L2SP layers 10	0.001 0.001 1.0 L2SP layers 14
Result	84.52	94.65	94.22	78.90

4.6   Final Optimal Settings and How to Predict Them

The final results and optimal settings are shown in Table 6 and Fig. 2. Table 7 shows a clear distinction between target datasets that are more similar to the source dataset and for which the pre-trained weights are better able to separate the classes in feature space and those that are less similar with pre-trained weights that fit the target task poorly. The best measure for determining whether the pre-trained weights are well suited is a comparison of the performance achieved with fixed pre-trained features and that achieved through training from random initialization. As this method is computationally intensive a reasonable alternative may be the normalized Fisher Ratio, but as the differences are not as pronounced in all cases it should be further investigated on more datasets to see how reliable it is as a heuristic. The EMD measure of domain similarity is a poor predictor of the suitability of the pre-trained weights.

Targets datasets for which the pre-trained weights are well suited need:

– a much lower learning rate for fine-tuning,
– more than one layer to be reinitialized from random weights, and
– some or all layers trained with L2SP regularization.

Target datasets where the pre-trained weights are not well suited need:

– a much higher learning rate for fine-tuning with a lower learning rate for lower layers,
– only the final classification layer reinitialized, and
– L2 rather than L2SP regularization.

Using the above best practice, non-binary transfer learning procedures we achieved state of the art or close to, on three out of the four datasets. We used publicly available pre-trained weights and no additional methods for either pre-training or fine-tuning.

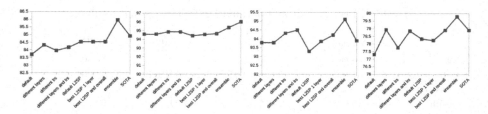

Fig. 2. Optimal settings versus best default settings. Left to right Caltech, Cars, Aircraft, DTD

Table 6. Optimal settings versus best default settings

	Caltech	Cars	Aircraft	DTD
Default settings	0.0025	0.025	0.025	0.002
Default result	83.69	94.59	93.78	77.30
Default L2SP	84.52	94.42	93.29	78.32
Optimal settings	1 new layer lr 0.0025 L2SP 0.01 0.01	1 new layer high lr 0.025 low lr 0.01 low layers 8 no L2SP	3 new layers FC lr 0.01 high lr 0.025 low lr 0.01 low layers 8 no L2SP	new layers 2 lr 0.002 L2SP 0.001 0.001 L2SP layers 14
Optimal result	84.52	94.86	94.50	78.90
Ensemble	85.94	95.35	95.11	79.79
State of the art	84.9 [18]	96.0 (@448) [26]	93.9 (@448) [32]	78.9 [3]

Table 7. Predicting optimal settings based on pre-trained features

	Caltech	DTD	Cars	Aircraft
Fisher ratio	1.35	3.47	1.18	0.83
EMD similarity	0.568	0.540	0.536	0.557
Random initialisation minus fixed features	−16.2	−7.8	28.5	28.9
Optimal learning rate	Low	Low	High	High
L2SP	Yes	Yes	No	No
More layers reinitiailized	Yes	Yes	No	No
Low layers at low learning rate	No	No	Yes	Yes

5   Discussion

Traditional binary assumptions about transfer learning hyperparameters should be discarded in favour of a tailored approach to each individual dataset. These assumptions include transferring all possible layers or none, training all layers at the same learning rate or freezing some, and using L2SP regularization or L2 regularization for all layers. Our work demonstrates that optimal transfer learning non-binary hyperparameters are dataset dependent and strongly influenced by how well the pre-trained weights fit the target task. For a particular dataset, optimal non-binary transfer learning hyperparameters can be determined based on the difference between model performance when fixed features are used and when the full model is trained from random initialization as shown in Table 7. We recommend using the settings shown on the left in this table for target datasets where the difference is negative and settings shown on the right for positive differences. Target datasets for which the pre-trained weights are well suited and target datasets for which they are not result in large differences in this value. These differences should still be pronounced even if suboptimal learning hyperparameters are used for this initial test due to limited resources for hyperparameter search. This heuristic for determining optimal hyperparameters should be useful in most transfer learning for image classification cases. The normalized Fisher Ratio may be useful in some cases, however, care should be taken because the differences are not as pronounced. The EMD domain similarity measure should not be used to determine transfer learning hyperparameters.

References

1. Agrawal, P., Girshick, R., Malik, J.: Analyzing the performance of multilayer neural networks for object recognition. In: Fleet, D., Pajdla, T., Schiele, B., Tuytelaars, T. (eds.) ECCV 2014. LNCS, vol. 8695, pp. 329–344. Springer, Cham (2014). https://doi.org/10.1007/978-3-319-10584-0_22
2. Azizpour, H., Razavian, A.S., Sullivan, J., Maki, A., Carlsson, S.: Factors of transferability for a generic convnet representation. IEEE Trans. Pattern Anal. Mach. Intell. **38**(9), 1790–1802 (2015)
3. Chen, T., Kornblith, S., Norouzi, M., Hinton, G.: A simple framework for contrastive learning of visual representations. In: International Conference on Machine Learning, pp. 1597–1607. PMLR (2020)
4. Chu, B., Madhavan, V., Beijbom, O., Hoffman, J., Darrell, T.: Best practices for fine-tuning visual classifiers to new domains. In: Hua, G., Jégou, H. (eds.) ECCV 2016. LNCS, vol. 9915, pp. 435–442. Springer, Cham (2016). https://doi.org/10.1007/978-3-319-49409-8_34
5. Cimpoi, M., Maji, S., Kokkinos, I., Mohamed, S., Vedaldi, A.: Describing textures in the wild. In: Proceedings of the IEEE Conference on Computer Vision and Pattern Recognition, pp. 3606–3613 (2014)
6. Cui, Y., Song, Y., Sun, C., Howard, A., Belongie, S.: Large scale fine-grained categorization and domain-specific transfer learning. In: Proceedings of the IEEE Conference on Computer Vision and Pattern Recognition, pp. 4109–4118 (2018)
7. Deng, J., Dong, W., Socher, R., Li, L.J., Li, K., Fei-Fei, L.: ImageNet: a large-scale hierarchical image database. In: CVPR 2009 (2009)
8. Fukenaga, K.: Introduction to Statistical Pattern Recognition, 2nd edn. Academic Press, San Diego (1990)
9. Griffin, G., Holub, A., Perona, P.: Caltech256 object category dataset. California Institute of Technology (2007)
10. Guo, Y., Shi, H., Kumar, A., Grauman, K., Rosing, T., Feris, R.: SpotTune: transfer learning through adaptive fine-tuning. In: Proceedings of the IEEE Conference on Computer Vision and Pattern Recognition, pp. 4805–4814 (2019)

11. He, K., Girshick, R., Dollár, P.: Rethinking ImageNet pre-training. arXiv preprint arXiv:1811.08883 (2018)
12. He, K., Zhang, X., Ren, S., Sun, J.: Deep residual learning for image recognition. In: Proceedings of the IEEE Conference on Computer Vision and Pattern Recognition, pp. 770–778 (2016)
13. Kolesnikov, A., et al.: Big transfer (bit): general visual representation learning. arXiv preprint arXiv:1912.11370 (2019)
14. Kornblith, S., Shlens, J., Le, Q.V.: Do better ImageNet models transfer better? In: Proceedings of the IEEE Conference on Computer Vision and Pattern Recognition, pp. 2661–2671 (2019)
15. Krause, J., Stark, M., Deng, J., Fei-Fei, L.: 3D object representations for fine-grained categorization. In: 4th International IEEE Workshop on 3D Representation and Recognition (3dRR 2013), Sydney, Australia (2013)
16. Li, H., et al.: Rethinking the hyperparameters for fine-tuning. arXiv preprint arXiv:2002.11770 (2020)
17. Li, S., Deng, W.: Deep facial expression recognition: a survey. IEEE Trans. Affect. Comput. (2020)
18. Li, X., Xiong, H., Wang, H., Rao, Y., Liu, L., Huan, J.: Delta: deep learning transfer using feature map with attention for convolutional networks. arXiv preprint arXiv:1901.09229 (2019)
19. Li, X., Grandvalet, Y., Davoine, F.: Explicit inductive bias for transfer learning with convolutional networks. arXiv preprint arXiv:1802.01483 (2018)
20. Mahajan, D., et al.: Exploring the limits of weakly supervised pretraining. In: Ferrari, V., Hebert, M., Sminchisescu, C., Weiss, Y. (eds.) ECCV 2018. LNCS, vol. 11206, pp. 185–201. Springer, Cham (2018). https://doi.org/10.1007/978-3-030-01216-8_12
21. Maji, S., Kannala, J., Rahtu, E., Blaschko, M., Vedaldi, A.: Fine-grained visual classification of aircraft. Technical report, Toyota Technological Institute at Chicago (2013)
22. Masi, I., Wu, Y., Hassner, T., Natarajan, P.: Deep face recognition: a survey. In: 2018 31st SIBGRAPI Conference on Graphics, Patterns and Images (SIBGRAPI), pp. 471–478. IEEE (2018)
23. Ng, H.W., Nguyen, V.D., Vonikakis, V., Winkler, S.: Deep learning for emotion recognition on small datasets using transfer learning. In: Proceedings of the 2015 ACM on International Conference on Multimodal Interaction, pp. 443–449 (2015)
24. Ngiam, J., Peng, D., Vasudevan, V., Kornblith, S., Le, Q.V., Pang, R.: Domain adaptive transfer learning with specialist models. arXiv preprint arXiv:1811.07056 (2018)
25. Plested, J., Gedeon, T.: An analysis of the interaction between transfer learning protocols in deep neural networks. In: Gedeon, T., Wong, K.W., Lee, M. (eds.) ICONIP 2019. LNCS, vol. 11953, pp. 312–323. Springer, Cham (2019). https://doi.org/10.1007/978-3-030-36708-4_26
26. Ridnik, T., Lawen, H., Noy, A., Friedman, I.: TResNet: high performance GPU-dedicated architecture. arXiv preprint arXiv:2003.13630 (2020)
27. Rubner, Y., Tomasi, C., Guibas, L.J.: The earth mover's distance as a metric for image retrieval. Int. J. Comput. Vision $40(2)$, 99–121 (2000)
28. Szegedy, C., Ioffe, S., Vanhoucke, V., Alemi, A.A.: Inception-v4, inception-resNet and the impact of residual connections on learning. In: Thirty-First AAAI Conference on Artificial Intelligence (2017)

29. Wan, R., Xiong, H., Li, X., Zhu, Z., Huan, J.: Towards making deep transfer learning never hurt. In: 2019 IEEE International Conference on Data Mining (ICDM), pp. 578–587. IEEE (2019)
30. Wightman, R.: PyTorch image models. https://github.com/rwightman/pytorch-image-models (2019). https://doi.org/10.5281/zenodo.4414861
31. Yosinski, J., Clune, J., Bengio, Y., Lipson, H.: How transferable are features in deep neural networks? In: Advances in Neural Information Processing Systems, pp. 3320–3328 (2014)
32. Zhuang, P., Wang, Y., Qiao, Y.: Learning attentive pairwise interaction for fine-grained classification. In: Proceedings of the AAAI Conference on Artificial Intelligence, pp. 13130–13137, No. 07 in 34 (2020)
33. Zoph, B., et al.: Rethinking pre-training and self-training. arXiv preprint arXiv:2006.06882 (2020)

Human Centred Computing

Hierarchical Features Integration and Attention Iteration Network for Juvenile Refractive Power Prediction

Yang Zhang[1,2], Risa Higashita[1,3], Guodong Long[2], Rong Li[4], Daisuke Santo[3], and Jiang Liu[1,5,6,7](✉)

[1] Department of Computer Science and Engineering, Southern University of Science and Technology, Shenzhen, China
zhangy2018@mail.sustech.edu.cn, liuj@sustech.edu.cn
[2] University of Technology Sydney, Sydney, Australia
[3] Tomey Corporation, Nagoya, Japan
[4] Xiao Ai Eye Clinic, Chengdu, China
[5] Cixi Institute of Biomedical Engineering, Chinese Academy of Sciences, Beijing, China
[6] Research Institute of Trustworthy Autonomous Systems, Southern University of Science and Technology, Shenzhen, China
[7] Guangdong Provincial Key Laboratory of Brain-Inspired Intelligent Computation, Department of Computer Science and Engineering, Southern University of Science and Technology, Shenzhen, China

Abstract. Refraction power has been accredited as one of the significant indicators for the myopia detection in clinical medical practice. Standard refraction power acquirement technique based on cycloplegic autorefraction needs to induce with specific medicine lotions, which may cause side-effects and sequelae for juvenile students. Besides, several fundus lesions and ocular disorders will degenerate the performance of the objective measurement of the refraction power due to equipment limitations. To tackle these problems, we firstly propose a novel hierarchical features integration method and an attention iteration network to automatically obtain the refractive power by reasoning from relevant biomarkers. In our method, hierarchical features integration is used to generate ensembled features of different levels. Then, an end-to-end deep neural network is designed to encode the feature map in parallel and exploit an inter-scale attentive parallel module to enhance the representation through an up-bottom fusion path. The experiment results have demonstrated that the proposed approach is superior to other baselines in the refraction power prediction task, which could further be clinically deployed to assist the ophthalmologists and optometric physicians to infer the related ocular disease progression.

Keywords: Refractive power prediction · Non-cycloplegic refraction records · Attention iteration

T. Mantoro et al. (Eds.): ICONIP 2021, LNCS 13109, pp. 479–490, 2021.
https://doi.org/10.1007/978-3-030-92270-2_41

1   Introduction

In recent years, the high incidence rate of myopia has been identified as the leading cause of correctable vision impairment among juvenile students, and dealing with this problem is a priority of WHO Vision 2020 [1]. The essential characteristics of myopia are summarised as the unconformity between the elongation of ocular axial length and the integrated optical power derived from the combination of the corneal and crystalline lens, which is also known as ametropia. The illustration of the ocular structure and light pathway of myopia and the normal eye is revealed in the Fig. 1. In clinical, the refractive error has been defined as a significant indicator of the commencement of myopia. Because of children's stronger accommodation, in medical practice, refraction power measurement under cycloplegic autorefraction by dropping 1.0% cyclopentolate or 0.5% − 1.0% atropine eye ointment is regarded as golden standard [2,3]. However, these medicine lotions may lead to side effects and sequelae for juvenile students. Besides, the existence of some specific fundus lesions, such as amblyopia, strong nystagmus, and lens opacity, will be notable obstructions in detecting the accurate refractive power.

As seen from previous medicine literature, classical statistical and meta-analysis principles are exploited to identify and evaluate the myopia distribution characteristics of particular participated groups [4–7]. However, these methods may lead to over-fitting conditions with spurious diagnoses. In recent years, several kinds of neural network architectures have revealed great success in pattern recognition and representation learning tasks [8–10]. However, most of the superiority of canonical deep methods decreased when dealing with the structural spreadsheet form database, i.e. tabular data. Ke et al. regard the feature groups assignment task as an NP-hard problem that could be solved by a heuristic algorithm [11] and Arik et al. use sequential attention to select meaningful features to reason at each forward step empowering the model interpretability [12]. In contrast, the ensembled tree-based methods [13], such as GBDT [14–18] and random forests [19], have been widely used in structural data problems, as the decision manifolds of the tree-based model are neat hyperplanes, which leads to outstanding performance. To alleviate the non-differentiable problem of the tree-based model, the soft decision tree and neural decision tree have been proposed by introducing differentiable functions into the tree models [20–25]. Furthermore, the neural oblivious decision ensembles (NODE) tree model [26] generalizes the CatBoost to make the splitting node feature selection and tree model construction differentiable, which take advantage of the end-to-end neural network optimization process and the hierarchical representation learning ability.

In order to estimate the refractive power in a more reliable and innocuous way, inspired by previous work [11,26], we firstly investigate an automated juvenile refractive power prediction approach, based on related biomedical parameters, such as axial length and corneal curvature, etc. The advantage of the method could be divided into two points: 1) *Hierarchical Features Integration*. Firstly, we calculated the feature significance score of all original biomedical

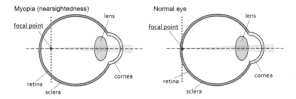

Fig. 1. The simplified illustration of the ocular structure and light pathway of myopia and normal eye. The combined refractive power of the cornea and crystal lens could project the image to the appropriate location onto the retina. Extreme high or low refractive error will guide the light focusing onto the in front of or behind the retina. The axial length of ocular and cornea curvature are significant indicators of the refractive power prediction in the clinical setting.

variables. Then, more instructive combination items are proposed through low-level categorical variables interaction. Finally, the top-k created features are collected into the candidature predictor sets according to the significance score order for the subsequent steps. 2) *Attention Iteration Network*. Similar to the recurrent neural network, the corresponding predictors will be sent into the model in batch, based on their significance score. To emphasize the vital predictors in the network computing process, the most valuable predictors are successively stressed on each concatenation node. With the high-level features extraction, an inter-scale iterative attention module implicitly delivering vital features from top to down is embedded to establish a hierarchical relationship. Thus the information concentration could be realized in horizontal and vertical concurrently. Finally, the loss function value calculated from the iterative multi-branches decoder structure would be back-propagated through the whole architecture.

To evaluate the approach, we conduct experiments on 24,970 non-cycloplegic clinical electronic medical records of juveniles recruited from 22 primary and junior high schools. The cogent evidence could empirically prove the predominant performance of the proposed approach compared with other baseline rivals with a lightweight framework. Furthermore, an ablative study is conducted to assess how much each modification of the network architecture contributes to performance improvement.

2   Methodology

2.1   Deep Refraction Power Prediction Framework

To further leverage obscure information of the records, we design an effective approach utilizing a hierarchical feature integration method and an attention iteration structure to predict the refractive power. In particular, the final biomedical feature sets used in the model prediction are named as *predictors* in our research, which is denoted as P. The major components contain two phases: (1) Hierarchical Features Engineering: to create high-level *predictors* and (2) Attention Iteration Network: to encode the features in sequential to jointly predict the refractive power. In the rest of this section, we will detail the components implementation of the approach.

2.2 Hierarchical Features Engineering

As mentioned previously, selecting significant features to be the carved variables is one of the superiority of the tree-based model. It can pay more attention to crucial attributes and extract distinguished knowledge from the original search space.

Algorithm 1. Hierarchical Feature Engineering

Input: $\mathcal{D}$: Dataset, Var_{cat}: categorical variables, $\mathcal{P}$: predictors, $\mathcal{G}$: created features, k;
1: Initial: $\mathcal{P} \leftarrow \varnothing$, $\mathcal{H}_i \leftarrow \varnothing$, $\mathcal{G} \leftarrow \varnothing$;
2: $\mathcal{P} \leftarrow$ Calculate the significance score for all original features;
3: // Start to create 2-level features;
4: **for** $i = 0$ **to** $(len(Var_{cat}) - 1)$ **do**
5:      **for** $j = 1$ **to** $(len(Var_{cat}))$ **do**
6:          // Each cardinal value after one-hot-encode
7:          Interaction of $Var_{cat}[i]$ and $Var_{cat}[j]$;
8:          **for** each carved point **do**
9:              $\mathcal{G} \leftarrow$ Features with the calculated score;
10:         **end for**
11:         $\mathcal{H}_2 \leftarrow$ Select the top-k of $\mathcal{G}$;
12:         $\mathcal{P} \leftarrow$ Select the top-k of $\mathcal{G}$;
13:     **end for**
14: **end for**
15: Create 3-level features based on $\mathcal{H}_2$ by adding each first-level feature in series;

16: $\vdots$
17: Create n-level features based on $\mathcal{H}_{n-1}$;
Output: Dictionary of $\mathcal{P}$ and the corresponding significance scores.

However, the combination of original features sometimes possesses the capability to capture crucial features of the tasks, such as AutoCross [27]. Thus, hierarchical feature engineering aims to create effective categorical feature combination items to be explicitly delivered to the model. The initial *predictors* contains all original features, including normalized numerical features and one-hot-encoded categorical features. The feature combination process begins with calculating the significance score of all the original features, which is designed under the guidance of the carved node selection principle of the regression tree. The carved node is determined according to the information gain difference between the before and after the splitting. For regression problem, when the specific carved variable j and carved point s are determined, the dataset will be divided into two separate districts $R_1(j,s) = \{x \mid x^{(j)} \leq s\}$ and $R_2(j,s) = \{x \mid x^{(j)} > s\}$. Specifically, the average value of the target variable y_i of the examples x_i in the split district R_m could be denoted as $C_m = ave(y_i \mid x_i \in R_m)$.

To be specific, the significance score could be calculated as the mean square error variation before the split as the Eq. (1) and after the split as the Eq. (2) to calculate the information gain S, through $S = S1 - S2$, contributed by the selected carved variable. Similar, the average value of the target variable y_i of the examples x_i in the primitive district R could be denoted as $C = ave(y_i \mid x_i \in R)$.

$$S1 = \sum_{x_i \in R(j,s)} (y_i - c)^2 \tag{1}$$

$$S2 = \sum_{x_i \in R_1(j,s)} (y_i - c_1)^2 + \sum_{x_i \in R_2(j,s)} (y_i - c_2)^2 \tag{2}$$

Nevertheless, when measuring the significance score of the numerical variables, there is a challenge that the large scale of the cardinal of the numerical variables makes the best-carved point search to be a huge burden. Concerning this issue, we introduce the α interval search, which means the best-carved point search will be proceeding with each α percentage of the whole range, from the minimum value to the maximum value. With the increasing of α, the search space will be reduced leading to a decrease in the computation complexity.

In our method, to keep the search high efficiency, we exploited the beam search to locate the best-carved variables and the carved points. Due to the categorical features have been one-hot-encoded, the selected features could be derived from the same feature with different values. Based on the first-level candidature *predictors*, we could calculate the crossing feature of them and regard the created feature as second-level features. Also, based on the beam search algorithm, we just select the top-k significance score categorical features of the low-level candidature *predictors* to generate the high-level *predictors* in the following steps. For example, after the $2nd$ features are available, the significance score calculation is conducted again and top-k score $2nd$ features are joined into the *predictors* for $3rd$ features calculation. The process will be conducted recursively until the $k - th$ highest-level features are generated. When the original features contain m numerical features and n categorical features, more formally, the number of the final *predictors* is denoted as p:

$$p = m + O(n) + k * (n - 1) \tag{3}$$

in which $O(n)$ represents the whole cardinal value number of all of the categorical *predictors*.

2.3 Attention Iteration Network

Multi-channel Input Gate. After the hierarchical feature engineering, a dictionary of *predictors* and the corresponding significance score could be provided for the network. To exploit key features, we designed a novel feature set entry mode according to the previously calculated significance score as shown in Fig. 2, in which the circle's number indicate the scale of the input *predictors* through

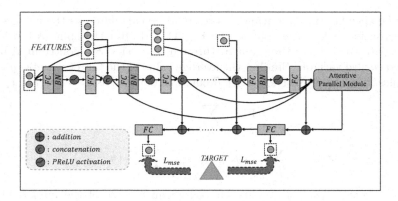

Fig. 2. The overview of our proposed automated refractive power prediction framework. In the scheme, the *predictors* are divided into 5 batches and sent into the model in series. The key *predictors* are emphasized repeatedly during the forward encoding route and concatenated with encoding features and other input *predictors* in batch.

each gate. Conformed with the subsequent description, the input *predictors* scale is decreasing from entrance to the end.

We assume the element number of the final *predictors* is p and the final *predictors* set as P. For the reason that the key *predictors* should be the most crucial ones in a controllable scale, we define the main *predictors* number of feature groups through the main entrance gate as $\lceil \log_2 p \rceil$. After the *predictors* of the main entrance gate are determined, the remaining *predictors* are sorted descend in significance score order to be assigned to other feature groups.

To regularize the scale of the input gates, we reduce the *predictors'* number in half iteratively. Thus, the *predictors* number of $i - th$ input gate should be $\lceil \frac{p - \lceil \log_2 p \rceil}{2^{i-1}} \rceil$ with $2 \leq i \leq \lceil \log_2 (p - \lceil \log_2 p \rceil + 1) \rceil$. The corresponding *predictors* set P_i could be the bottom half significance score *predictors* of the $P - \{P_1 \bigcup P_2 \cdots \bigcup P_{i-1}\}$. To make the vital *predictors* provide explicit contribution to the model, moreover, we arrange them at a closer location to the decoder output structure. As a result of the particular input structure, each front layer l_i before the embedded parallel module is concatenated by P_1, P_{i+1}, and fc_{2i}, in which the fc_{2i} is the output of the previous fully connected layer. The concatenation layers will be further processed in the successive module.

Attentive Parallel Module. Despite the model has emphasized the vital *predictors* in the forward encoding route and delivered the *predictors* in batch according to the significance scores, the extracted meaningful features of the neural network could vanish during the serial training. To deal with this problem, we utilized the attentive parallel module to aggregate the features $(f_1, f_2, \ldots, f_5)$ derived from different levels with a parallel connection and inter-scale interaction through a top-down fusion path to obtain the refined intermediate feature f^* and the aggregation feature f_{agg}, *i.e.*,

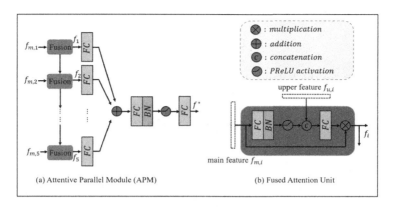

Fig. 3. The visual illustration of the (a) attentive parallel module (APM) and the (b) fused attention unit. In the single input condition of the fused attention unit, the refined feature is only produced by the main feature without concatenation with the upper feature.

$$f^* = F\left(\sigma\left(b\left(F_{agg}\right)\right)\right) \tag{4}$$

$$f_{agg} = F_1 \oplus F_2 \cdots \oplus F_5 \tag{5}$$

where $F(\cdot)$ denotes the fully connected layer, i.e. $F_i = F(f_i)$, followed by the batch normalization $b(\cdot)$, and $\oplus$ denotes the pixel-wise addition operation. Different from traditional selection, we choose the parametric rectified linear unit (PReLU) as our non-linear activation function σ. Similar, the f_i is defined as:

$$f_i = f_{m,i} \otimes F_c \tag{6}$$

$$f_c = C\left(f_{u,i}, \sigma\left(b\left(F_{m,i}\right)\right)\right) \tag{7}$$

where $f_{m,i}$ and $f_{u,i}$ denotes the main feature and upper feature respectively, $\otimes$ denotes the multiplication operation, and C denotes the concatenation operation to obtain the intermediate feature f_c. The specific scheme of the attentive parallel module and the fused attention unit is exhibited in Fig. 3.

Iterative Decoder Structure. The rough result following the attentive parallel module is not accurate enough. To fully exploit the forward encoded features, we design an iterative decoder structure to output the results through multi-branches, in which each previous output will combine with an encoded feature of different granularity in the series training route. Furthermore, the refractive power prediction loss function $\mathcal{L}_{ref}$ is defined as the combination of the mean square error (MSE) $\mathcal{L}_i$ of each output branches:

$$\mathcal{L}_{ref} = \sum_{i=1}^{n} \gamma_i \cdot \mathcal{L}_i \tag{8}$$

where γ_i indicates the balance factor of each branch and n is the branch's number. Especially, the synthesized loss value is calculated as the weighted average loss function values in each output branch and the final result would be exported in the last layer of the structure as shown in Fig. 2.

3 Experiments

In this section, we conducted comprehensive experiments to evaluate the performance of our proposed approach. The experimental results are compared with several outstanding baselines and provide cogent evidence that our method could outperform or approach the existing baselines. Besides, we conducted an ablative study to evaluate the contributions of each component of the proposed method.

3.1 Implementation Details

The neurons of the frontal layer and the hidden layer are defined as 20 and 50. The training is implemented by Python v3.7.5 and Pytorch v1.4.0 framework on TITAN V GPUs with 12 GB memory. We adopt the standard Adam optimizer and keep the learning rate at 0.001. The batch size is defined as 20. We use the default setting ($\alpha = 5$ and $\gamma = 1$) in our research, which means the carved point is enumerated with 5% of the value range interval and the weighted loss value is the average of each output branch.

The original clinical records are split into the training set, valid set, and test set according to the proportion of 5:2:3, i.e. $12.5K$, $5.0K$, and $7.5K$ examples, and the valid set is used to fine-tune the model's hyper-parameters each 100 training rounds. Considering the particular learning requirements of the neural network, we normalize the numerical features and conduct one-hot-encode on categorical features. Eventually, the final model is generated when there is no further improvement after 5000 training rounds.

3.2 Refractive Power Prediction

At first, we evaluate the effectiveness of the clinical refraction power prediction of participants from several primary and junior high schools. The model is constructed based on specific clinical biomedical parameters, such as axial length, Mean Keratometry, and ASTigmatism. In medical practice, myopia could be determined as having a refractive power error of Spherical Equivalent (SE) of less than -0.50 diopter (D), which is calculated by the Sphere degree (S) plus half of the Cylinder degree (C), i.e. $SE = S + \frac{1}{2}C$. For clinical requirements, the model performance is evaluated by the hierarchical statistics according to the absolute value of the refraction error in a specific range, like $\pm 0.5D$ and $\pm 1D$,

Table 1. The sphere equivalent prediction performance comparison of MSE and hierarchical statistics.

	$MSE(sd)$	Under $0.5D$	Under $1D$	Under $1.5D$	Under $2D$
FCNN†	0.5242(2e−4)	55.78%	86.13%	96.41%	99.98%
Catboost†	0.5172(2e−4)	56.12%	86.27%	96.56%	99.98%
XGBoost	0.5131(2e−4)	56.45%	87.24%	96.89%	99.98%
XGBoost†	0.4996(1e−4)	57.79%	87.15%	96.95%	99.98%
NODE	0.5104(2e−4)	56.71%	86.52%	96.83%	99.98%
NODE†	0.4887(2e−4)	57.80%	87.37%	97.14%	99.98%
Ours	0.5149(3e−4)	55.57%	86.35%	97.11%	99.98%
Ours†	0.4893(2e−4)	57.90%	87.34%	96.94%	99.98%
Ours†+$\mathcal{A}$	0.4734(2e−4)	57.92%	87.37%	**97.23%**	99.98%
Ours†+$\mathcal{A}$+$\mathcal{B}$	**0.4646(2e−4)**	**58.33%**	**87.45%**	97.15%	99.98%

and most of the absolute value of the refraction power prediction error is less than $2D$.

In the comprehensive experiments, we compare the proposed method with the following baselines:

- FCNN: General deep neural network with several fully connected layers and PReLU non-linear activation function.
- XGBoost [16]: The classical XGBoost defined a customized loss function by the feat of the second-order of Taylor series expansion and combined the regularization item of the leaf node to the loss function.
- Catboost [18]: A universal implementation of the GBDT model integrated the oblivious decision tree into gradient boosting mode.
- NODE (Neural Oblivious Decision Ensemble) [26]: An efficient deep method generalized the CatBoost by utilizing the oblivious decision tree to make the carved node feature selection differentiable.

The experiments begin with all baselines with the hierarchical feature engineering (†) to utilize the generated high-level *predictors*. To distinguish the contribution of the feature enrichment, we select the XGBoost and the NODE as the representative methods, respectively. Expressly, iterative decoder structure and the multi-channel input gate of the *predictors'* are marked as $\mathcal{A}$ and $\mathcal{B}$.

The final results of the Spherical Equivalent (SE) and the Sphere degree (S) prediction are recorded as the mean square error (MSE) and the hierarchical statistics of the prediction results. The hierarchical statistics exhibit the examples proportion in different levels of the absolute value of the refraction power prediction error from under $0.5D$ to under $2D$, separately. The quantitative results comparison of MSE and hierarchical statistics of SE and S are distinctively present in Table 1 and Table 2, which are obtained from 10 times test average with 10 different random seeds.

Table 2. The sphere degree prediction performance comparison of MSE and hierarchical statistics.

	$MSE(sd)$	Under 0.5D	Under 1D	Under 1.5D	Under 2D
FCNN†	0.5289(2e−4)	56.41%	86.14%	96.21%	99.98%
Catboost†	0.5153(2e−4)	56.78%	86.49%	96.34%	99.98%
XGBoost	0.5119(2e−4)	57.23%	87.21%	96.89%	99.98%
XGBoost†	0.4923(1e−4)	57.91%	87.33%	96.94%	99.98%
NODE	0.5138(2e−4)	56.51%	86.67%	96.89%	99.98%
NODE†	0.4943(2e−4)	57.40%	87.26%	**97.34%**	99.98%
Ours	0.5129(3e−4)	55.66%	86.50%	96.96%	99.98%
Ours†	0.4967(2e−4)	57.48%	87.14%	96.95%	99.98%
Ours†+$\mathcal{A}$	0.4954(2e−4)	57.72%	87.23%	97.33%	99.98%
Ours†+$\mathcal{A}$+$\mathcal{B}$	**0.4802(2e−4)**	**58.30%**	**87.42%**	97.09%	99.98%

As an assistant model for clinical medical practice, our approach is expected to provide crucial information regarding the refractive power. The quantitative assessment of the experiment results has revealed that our method is in the leading position on most of the evaluation indicators compared with other baselines. For example, in terms of the accuracy of the refractive error under 0.5D, the proposed pipeline outperforms the NODE and XGBoost by 0.4% and 0.6% on sphere equivalent and sphere degree prediction, respectively. Considering the high volume of the enrolled participants' size, the advances of our approach could benefit more than 4,000 patients in the real clinical environment of our study.

Through the contrast of whether to employ the integration of the hierarchical features (†), XGBoost, NODE, and our method have all substantiated the handled features generation could exploit the existing features' information sufficiently. The hierarchical features engineering creates a few high-level *predictors* based on their significance score and selects the top-k score ones recursively. The selected *predictors* carries the most information gains through the guided equation calculation in each level to provide essential information related to the refractive power,i.e. sphere equivalent/sphere degree, which is more effective and pertinent than automatic feature crossing of deep neural network.

Additionally, subsequent to the feature engineering, the collected *predictors* are delivered into the network in batch to lay stress on the key *predictors* according to their significance score order. After the original input *predictors* and created high-level *predictors* are encoded in sequential and parallel mode serially, the key *predictors* is more closer to the output gate to guide the prediction and the representative features of them jointly predict the refractive power value. Furthermore, the attentive parallel module could leverage the stratified relations between different levels via a downward path to complete inter-scale interaction to deliver the crucial information across different scales. Also, the fused unit could use an attention mechanism to blend the upper feature from the previous level and the main feature to enrich the course intermediate feature in each scale. In a final, the refined features would be incorporated to yield the relatively accurate initial results.

Relying on the experiment's result, we conclude the network benefits from both of the modifications ($\mathcal{A}$ and $\mathcal{B}$) more or less. Especially for the absolute value of the refraction power prediction error under $0.5D$ and $1D$, the proposed model could acquire the best clinical prediction performance on both tasks, which can provide the optometrists and ophthalmologists more reliable guidance and reference for future refractive treatment. As most of the refractive error is under $2D$ except for anomalous patients, the prediction error under $2D$ could achieve 99.98% among the whole test set, as shown in the last column of the Table 1 and Table 2.

4 Conclusion

To the best of our knowledge, this is the most leading research to explore the deep learning-based method for non-cycloplegic juvenile clinical records refractive power prediction. In this paper, we present a novel approach for refractive power prediction, including the hierarchical features generation and an end-to-end deep neural network. In the experiments, we evaluate the performance of the proposed method on a high volume of collected clinical health records, by comparing it to other baseline rivals with a lightweight framework. Moreover, our method has achieved state-of-the-art performance in the clinical refractive power prediction area, which could be deployed in further medical practice to predict the refractive power and assist the ophthalmologists and optometric physicians for related ocular disease progression diagnosis.

References

1. Harrington, S.C., et al.: Refractive error and visual impairment in Ireland schoolchildren. Br. J. Ophthalmol. **103**, 1112–1118 (2019)
2. Zadnik, K., et al.: Ocular predictors of the onset of juvenile myopia. Invest. Ophthalmol. Vis. Sci. **40**(9), 1936–1943 (1999)
3. Mutti, D.O., et al.: Refractive error, axial length, and relative peripheral refractive error before and after the onset of myopia. Invest. Ophthalmol. Vis. Sci. **47**(3), 2510–2519 (2007)
4. Varma, R., et al.: Visual impairment and blindness in adults in the United States: demographic and geographic variations from 2015 to 2050. JAMA Ophthalmol. **134**, 802–809 (2016)
5. Tideman, J.W.L., et al.: Axial length growth and the risk of developing myopia in European children. Acta Ophthalmol. **96**(3), 301–309 (2018)
6. Zhang, M., et al.: Validating the accuracy of a model to predict the onset of myopia in children. Invest. Ophthalmol. Vis. Sci. **52**(8), 5836–5841 (2011)
7. Lin, H., et al.: Prediction of myopia development among Chinese school-aged children using refraction data from electronic medical records: a retrospective, multi-centre machine learning study. PLoS Med. **15**(11), 1–17 (2018)

8. Ni, Z.-L., et al.: RAUNet: residual attention U-Net for semantic segmentation of cataract surgical instruments. In: Gedeon, T., Wong, K.W., Lee, M. (eds.) ICONIP 2019. LNCS, vol. 11954, pp. 139–149. Springer, Cham (2019). https://doi.org/10.1007/978-3-030-36711-4_13
9. Shavitt I., Segal, E.: Regularization learning networks: deep learning for tabular datasets. In: Advances in Neural Information Processing Systems. LNCS, pp. 1379–1389 (2018)
10. Tian, Z., et al.: Prior guided feature enrichment network for few-shot segmentation. IEEE Trans. Pattern Anal. Mach. Intell. arXiv preprint arXiv:2008.01449 (2020)
11. Ke, G., et al.: TabNN: a universal neural network solution for tabular data (2018). https://openreview.net/pdf?id=r1eJssCqY7
12. Arik, S.O., Pfister, T.: TabNet: attentive interpretable tabular learning. arXiv preprint arXiv:1908.07442 (2019)
13. Xu, Y., et al.: FREEtree: a tree-based approach for high dimensional longitudinal data with correlated features. arXiv preprint arXiv:2006.09693 (2020)
14. Friedman, J.H., et al.: Greedy function approximation: a gradient boosting machine. Ann. Stat. **29**(5), 1189–1232 (2001)
15. De'ath, G.: Boosted trees for ecological modeling and prediction. Ecology **88**(1), 243–251 (2007)
16. Chen, T., Guestrin, C.: XGBoost: a scalable tree boosting system. In: Proceedings of the ACM SIGKDD International Conference on Knowledge Discovery and Data Mining, pp. 785–794 (2016)
17. Ke, G., et al.: LightGBM: a highly efficient gradient boosting decision tree. Adv. Neural. Inf. Process. Syst. **30**, 3149–3157 (2017)
18. Prokhorenkova, L., et al.: CatBoost: unbiased boosting with categorical features. In: Advances in Neural Information Processing Systems, pp. 6638–6648 (2018)
19. Breiman, L.: Random forests. Mach. Learn. **45**(1), 5–32 (2001)
20. Ke, G., et al.: DeepGBM: a deep learning framework distilled by GBDT for online prediction tasks. In: Proceedings of the 25th ACM SIGKDD International Conference on Knowledge Discovery & Data Mining, pp. 384–394 (2019)
21. İrsoy, O., et al.: Soft decision trees. In: Proceedings of the 21st International Conference on Pattern Recognition, pp. 1819–1822 (2012)
22. Peter, K., et al.: Deep neural decision forests. In: Proceedings of the IEEE International Conference on Computer Vision, pp. 1467–1475 (2015)
23. Frosst, N., Hinton, G.: Distilling a neural network into a soft decision tree. arXiv preprint arXiv:1711.09784, April 2017
24. Wang, S., et al.: Using a random forest to inspire a neural network and improving on it. In: Proceedings of the 2017 SIAM International Conference on Data Mining, pp. 1–9 (2017)
25. Yang, Y., et al.: Deep neural decision trees. arXiv preprint arXiv:1806.06988 (2018)
26. Popov, S., et al.: Neural oblivious decision ensembles for deep learning on tabular data. arXiv preprint arXiv:1909.06312 (2020)
27. Luo, Y., et al.: AutoCross: automatic feature crossing for tabular data in real-world applications. In: Proceedings of the ACM SIGKDD International Conference on Knowledge Discovery and Data Mining, pp. 1936–1945 (2019)

Stress Recognition in Thermal Videos Using Bi-directional Long-Term Recurrent Convolutional Neural Networks

Siyuan Yan$^{(\boxtimes)}$ and Abhijit Adhikary

School of Computing, The Australian National University,
Canberra, ACT 2600, Australia
{Siyuan.Yan,Abhijit.Adhikary}@anu.edu.au

Abstract. Stress is a serious day-to-day concern capable of affecting our health, and for this, the development of stress recognition models is of great importance. Contact-free stress recognition methods, *i.e.* RGB or thermal image-based methods have been widely explored. However, the use of visual perceptual features in static images is limited. In this paper, we first propose a Bidirectional Neural Network (BDNN) for stress recognition and show that Bi-directional training (BDT) achieves a better generalization performance by using a compressed dataset. Then, we propose a Bi-directional Long-term Recurrent Convolutional Network (BD-LRCN) that can automatically recognize stress from thermal videos by jointly learning visual perceptual and time-sequential representations, and achieves a classification accuracy of 68.4% on the ANUStressDB dataset. Furthermore, to facilitate our model to learn both forward and backward information from video clips, we use the BDT scheme to better learn temporal information (Our code and data is publicly available at: https://github.com/redlessme/Stress-Recognition-in-Thermal-Videos-using-BDLRCN).

Keywords: Stress recognition · Bimodal · Thermal video · Long-term recurrent convolutional network · Bi-directional training

1 Introduction

Stress is a common factor in our daily life capable of causing negative emotions, *i.e.* anger, anxiety, and fear. The persistence of stress can cause cardiovascular diseases [12] and cancer [19]. To overcome this, researchers have used physiological signals, *i.e.* heart rate [18] and skin temperature [20] for stress recognition, which can be done with touch-sensitive sensors but are time-consuming. A better way is to use facial data, *i.e.* thermal and RGB data, which can be simply captured by cameras. Sharma et al. [16] used a feature-level fusion of RGB and thermal images while Irani et al. [5] proposed a method to extract features from the super-pixels of facial images, and Kumar et al. [10] developed a Spatio-temporal network to predict ISTI signals from thermal videos and then

© Springer Nature Switzerland AG 2021
T. Mantoro et al. (Eds.): ICONIP 2021, LNCS 13109, pp. 491–501, 2021.
https://doi.org/10.1007/978-3-030-92270-2_42

fed them into a classifier to exploit both spatial and temporal features for stress recognition. The latter achieves better performance than the static image-based method, but is a complex and non-end-to-end model.

Traditional Recurrent Neural Networks (RNN) try to simulate the human brain and can learn information through time via back-propagation [14]. One problem is when we use the RNN to learn temporal information from thermal videos, it only relies on past and current information, which fails to model global information. Using a bidirectional learning scheme, RNNs can learn information from both past and future sequences, which generalizes better for classification. Inspired by the human brain, bidirectional associative memories [7,8] have been developed for bidirectional learning. But both of them have limited capacities and showcase inconsistent learning in different layers. BDNN [13] avoids these problems and can perform back-propagation in both the forward and reverse direction and can remember input patterns as well as output classes, given either of them. Inspired by BDNN, we adapt its Bi-directional training scheme into the Long Short-Term Memory (LSTM) unit. To get the spatial feature of each time step of the LSTM, we use a Convolutional Neural Network (CNN) [9] as our feature extractor.

In this paper, we first train a neural network on the compressed ANUS-tressDB dataset as our baseline and compare it with the performance of the BDNN. Then, we implement a Bi-directional Long-term Recurrent Convolutional Network (BD-LRCN) consisting of stacked CNNs to extract spatial features for each time step and Bi-LSTMs to learn temporal information and train on the ANUStressDB dataset containing video sequences. Our contribution can be summarized as (1) We use a BDNN to learn the bidirectional mapping between PCA features and classes and show that Bi-directional training has a better generalization. (2) We propose an end-to-end trainable BD-LRCN for classifying thermal videos. (3) A BD-LSTM module to learn both the forward information and backward information from time sequences. (4) Extensive experiments demonstrate that our proposed Bi-LRCN can significantly perform better than traditional sequence models on video datasets.

2   Method

2.1   Dataset Description

Dataset. We use the ANUStressDB dataset [5], which contains 31 subjects where each subject watches 20 different films. Their facial data are then collected by an RGB and a thermal camera and contains a total of 620 samples. We use two versions of the dataset: i) data v1: a compressed version where each pattern consists of 10 principal component analysis (PCA) features extracted from raw videos with labels "Stressful" and "Calm". For further analysis, we use PCA to visualize the data (shown in Fig. 1) and find that there are overlapped samples and also find that they are on different scales. ii) data v2: the complete video dataset containing 31 videos for 31 subjects, and each video is about 32 min long and can be divided into 20 "film reactions" each 1.5 mins long. There is a

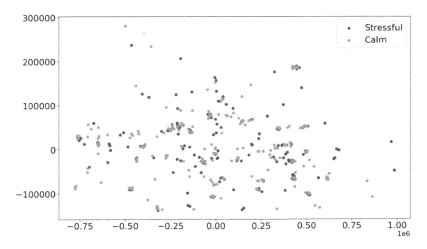

Fig. 1. Visualization of dataset v1 using PCA.

5–8 s set up at the beginning and about a 6 s gap between every 1.5 mins "film reaction". For the remaining part of the paper, we will use "video" to denote 32 videos and "film" to denote 1.5 min long "film reaction".

Data Pre-processing. For data v1, we normalize each of the 10 PCA features using the z-score in Eq. 1 to keep them on the same scale.

$$X' = \frac{X - \mu}{\sigma} \tag{1}$$

For data v2, we perform video processing and frame extraction (described in Fig. 2) and divide each video into 20 equal-length films. As the change of face in videos is slow, we only sample 36 frames for each film. To avoid the wrong segmentation between adjacent films and neutralize the subjects' emotions before playing the next film, we discard the first 4 frames and the last 4 frames of each film. We finally get 28 frames for each film. Also, we standardize all frames.

2.2   Baseline Method

We use a fully connected neural network with only one hidden layer for the baseline model as it is sufficient to obtain high accuracy in the training dataset. We use binary cross-entropy as our loss function (Eq. 2) with a Sigmoid function at the end of the last layer for the binary classification problem. In Sect. 3, we will investigate the performance with different numbers of hidden neurons, optimizers, activation functions, and modalities.

$$\mathcal{L}_{bce} = -\frac{1}{N} \sum_{i=1}^{N} t_i log(y_i) + (1 - t_i log(1 - y_i)) \tag{2}$$

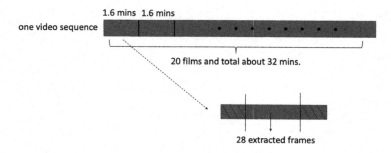

Fig. 2. Video processing and frame extraction.

where N is the sample size, t_i and y_i represent the i-th ground-truth predicted labels respectively.

2.3 Bidirectional Neural Network

Architecture. In contrast to the baseline model, the BDNN (Fig. 3) recognizes both stress and produces input patterns by classes. So, the relation between input and output should be invertible, and the BDNN should be a one-to-one mapping function. However, stress recognition is a many-to-one problem, which means BDNN cannot directly be used on it. Inspired by [13], we add an extra node on the output of BDNN. For each sample, we add the mean value to the ground truth for each pattern to perform the one-to-one mapping. Again, in contrast to the baseline model, the BDNN has two output neurons, one for classification and another for regression with the mean value of the input pattern. We also removed all the biases to simplify the training process and found no change in performance.

Training BDNN. We applied back-propagation in both forward and reverse directions. For the first 50 epochs, we trained in the forward direction and used two loss functions, Binary Cross-Entropy (BCE) for stress classification and Mean Squared Error (MSE) for regression. We reverse the training direction if the maximum epochs are reached or the BCE becomes lower than the threshold. For reverse training, we combine the two output values as the input vector, and the model predicts the input patterns using the MSE loss. The complete loss for the forward direction can be summarized in Eq. 3.

$$\mathcal{L}_{bce} = -\frac{1}{N} \sum_{i=1}^{N} t_i log(y_i) + (1 - t_i log(1 - y_i)),$$

$$\mathcal{L}_{mse} = \frac{1}{N} \sum_{i=1}^{N} (E_i - \hat{E}_i)^2,$$

$$\mathcal{L}_{total} = \mathcal{L}_{bce} + \mathcal{L}_{mse}.$$

(3)

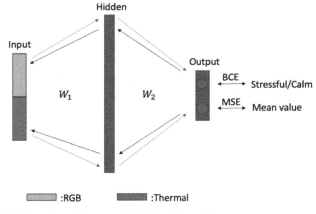

Fig. 3. Illustration of our BDNN model. We follow a similar architecture as the baseline, and the difference is the output of our BDNN contains an extra neuron, performing regression with an extra value (mean value).

where E_i and $\hat{E}_i$ are the i-th extra and the mean value respectively. The loss in the reverse direction is shown in Eq. 4.

$$\mathcal{L}_{reversed} = \frac{1}{N} \sum_{i=1}^{N} (X_i - \hat{X}_i)^2 \qquad (4)$$

where X_i and $\hat{X}_i$ are the i-th input and predicted patterns respectively.

2.4   Bi-directional Long-Term Recurrent Convolutional Network

Inspired by the LSTM [2] for Action recognition, we propose the Bi-LRCN combining a spatial feature extractor with a sequential model for a long-range temporal feature for stress recognition in videos (Fig. 4). Let us first define each video: $D_v = \{x_i\}_i^T \to y$, where x_i is the input frame in the i-th time step and y is the label for the video D_v. We feed frames $x_1, ..., x_T$ into parallel spatial CNNs $F_1, ..., F_T$ to extract fixed-length spatial features $F_1(x_1), ..., F_T(x_T)$, then we feed them into the time-sequential modeling module (Bi-LSTM), the results are $y_i, ..., y_T$. Finally, we use two fully connected layers to get the final classification results of the video. Notice that our spatial CNNs can be trained parallel over time steps and thus save a lot of computation time. Also, our Bi-LCRN model is end-end trainable, which is much simpler than previous work [10].

Spatial CNN. Residual Network (ResNet) [3] overcomes the vanishing and exploding gradient problem in deep neural networks via skip-connections between layers. Our spatial CNN starts with a pre-trained ResNet followed by Batch Normalization, Relu, and fully connected layers.

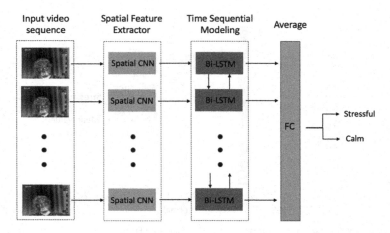

Fig. 4. Illustration of our BD-LRCN model. The model uses parallel spatial CNNs to extract fix-length visual features and then feed them into stacked sequence models Bi-LSTMs. Finally, the results from stacked sequence models are fused and get the final classification result.

Bi-LSTM Module. LSTMs can capture long-range dependencies of sequence and avoid the gradient vanishing problem. A simple example of the LSTM is shown in Fig. 5a. As stress is related to both previous and future video frames, to further consider the forward and backward direction information in video sequences, we extend the LSTM into Bi-LSTM, Fig. 5b, where the blue and green arrows denote forward and backward passes respectively and f_i and b_i denote forward and backward hidden states respectively.

Training Bi-LRCN. We use the conventional back-propagation to train the spatial CNNs. For Bi-directional training in the Bi-LSTMs, we follow Bi-directional RNN training [15]. It can be summarized as:

1. *FORWARD PASS:*
    (a) Do forward pass for forward hidden states $f_1, ... f_T$
    (b) Do forward pass for backward hidden states $b_T, ..., b_1$.
    (c) Do forward pass for all outputs $\hat{y}_1, ... \hat{y}_T$.
2. *BACKWARD PASS*
    (a) Do backward pass for all outputs $\hat{y}_1, ... \hat{y}_T$.
    (b) Do backward pass for forward hidden states $f_T, ... f_1$
    (c) Do backward pass for all backward hidden states $b_1, ... b_T$.
3. *UPDATE WEIGHTS*

Fig. 5. (a) LSTM (left) (b) Bi-LSTM (right) (Color figure online)

3   Results and Discussion

3.1   Experiment Setup

For data v1, we use 80%, 10%, and 10% of data as training, validation, and testing and fuse both RGB and thermal features. We use the Adam optimizer with a learning rate of $1e^{-2}$ and train for 1000 epochs with 1000 neurons. We also use the same split proportion for data v2. However, we only use the first 6 "films" for each video to stay consistent with previous work [16]. We use ResNet152 as the backbone network and train our Bi-LRCN for 40 epochs. For a fair comparison, we also train all completing models on data v2 for 40 epochs and fine-tune them. For the optimizer, we use Adam with an initial learning rate of $1e^{-3}$. The training takes about 3 h with 3 NVIDIA GeForce RTX 2080Ti GPUs with a batch size of 64. As we have a balanced distribution of labels, we only use accuracy as our evaluation metric.

3.2   Baseline Method

Model Tuning. We trained our baseline model with different settings and found that the number of hidden neurons and activation functions (ReLU and Sigmoid) can significantly affect the performance. We also trained with varying numbers of hidden neurons (10, 50, 200, and 500) and found the optimal to be 200. The results are shown in Table 1. We found that our model performs better using the Sigmoid rather than the ReLU. One cause might be the "dead ReLU" effect.

Table 1. Accuracy (%) for different activation functions and number of hidden neurons

n	Train with Relu	Test with Relu	Train with Sigmoid	Test with Sigmoid
10	73.99	50.00	73.99	41.94
50	85.48	45.97	94.96	50.81
200	88.31	48.39	**95.77**	**51.61**
500	91.33	48.39	92.54	49.19

The Model with Different Modalities. We also explore the performance when using different modalities and optimizers. For this experiment, we follow the setting in Sect. 3.1. We train the model for 1000 epochs using the Sigmoid function and 200 neurons. For optimizer, we sue Adam and SGD with a learning rate of 0.01. For the input data, we use RGB, thermal, and bimodal inputs. The RGB input contains the top 5 PCA features from RGB images, the thermal input contains the top 5 PCA features from thermal images, and bimodal input combines the RGB input and thermal input using feature-level fusion. Our result is shown in Table 2. We find the model with Adam performs better than the model with SGD. Also, the model with bimodal input achieves the best performance. So, we can conclude the complementary information between RGB features and thermal features help to improve performance. Finally, the best setting we choose for the baseline is the bimodal input with Adam optimizer and Relu function with 200 neurons.

Table 2. Accuracy (%) for different modalities and optimizers

Modal	Train with Adam	Test with Adam	Train with SGD	Test with SGD
RGB	81.65	51.61	52.82	50.81
Thermal	81.05	45.97	51.81	48.39
RGB+thermal	**95.77**	**51.63**	53.63	51.61

3.3 Experiments on Bidirectional Neural Network

Performance with the Bidirectional Neural Network. We compare the performance between our baseline and our BDNN. The result is shown in Table 3. We found that the BDNN does not converge easily. We also found that the BDNN has a better generalization ability due to the advantage of Bi-directional training. However, the performance is still not satisfactory. Although the model has a high training accuracy, the features in our dataset are not enough to predict stress in the test set. For further analysis, we move to the video dataset in the following sections.

Table 3. Comparison between the baseline and BDNN

Model	Train accuracy	Test accuracy
Baseline	95.77	51.61
BDNN	85.69	**55.61**

3.4   Experiments on Bi-BRCN Using Data v2

Model Evaluation. We compare our model with traditional 2DCNN, 3DCNN [6], and LRCN [2]. All models are trained for 40 epochs and are fine-tuned. The result is shown in Table 4. By observing the table, it can be seen that the 2DCNN has the worst performance as CNN only learns spatial features. The 3DCNN and the LRCN further improve the performance compared to 2DCNN as they both additionally consider the temporal information. We find that the 3DCNN costs much more memory, and we can only choose a small batch size. Finally, we find that our model outperforms all models on the test set, which proves the effectiveness of our model on the video dataset. In the following sections, we perform some ablation studies to better demonstrate the effectiveness of various components in our model.

Table 4. Comparison between different models

Model	Dataset	Test accuracy
2DCNN	v2	58.4
3DCNN [6]	v2	60.84
LRCN [2]	v2	61.4
Ours	v2	**68.4**

Impact of Different Backbones. We report our performance using different backbones in Table 5. The first model's backbone is a manually designed CNN, while the other two are VGG [17] and ResNet [3]. We can find ResNet outperforms all other models as the skip connections in ResNet can avoid the vanishing gradient problem and is useful to train deep networks and also demonstrates the importance of the ResNet backbone in our model.

Table 5. Comparison with different backbones

Model	Test accuracy
CNN+Bi-LSTM	58.3
VGG+Bi-LSTM	65.4
ResNet+Bi-LSTM	**68.4**

Table 6. Comparison with different sequence models

Model	Test accuracy
ResNet+RNN	53.2
ResNet+GRU	61.1
ResNet+LSTM	61.4
ResNet+Bi-LSTM	**68.4**

Impact of Different Sequence Models. We also experiment with different sequence models, which are RNN [11], GRU [1], LSTM [4], and our Bi-LSTM. The results are shown in Table 6. By comparing the first three models, we can find that the LSTM performs best as it has more gates to memorize the long-range sequence and can avoid the vanishing gradient problem. But by further experiments, we find that the GRU can achieve similar performance as the LSTM while using less memory and is faster. So, LSTM and GRU both have their advantages, and we can choose them according to our needs. Finally, we observe that the performance of the Bi-LSTM improves by 7% compared to the LSTM, which proves the effectiveness of our Bi-directional training scheme.

4   Conclusion and Future Work

In this paper, we studied stress recognition problems using BDNNs on the compressed ANUStressDB dataset and using BD-LSTM on the video version dataset. We explored different modalities for stress recognition and showed that the bimodal data is helpful to improve performance. We also found that Bi-directional training is harder to converge but has a better generalization ability. Moreover, we extended the Bi-directional training scheme into LSTM and found it is helpful for modeling video sequence data. Extensive experiments showed that our proposed Bi-LRCN can significantly improve performance than traditional sequence modeling-based models on video datasets.

Although our model achieves the best results on the ANUStressDB dataset compared to other models, the performance is still not satisfactory. We can summarize problems as (1) We simply treated the input video as a flat image sequence while ignoring the different granularity information in video content. (2) The fusion method in our model is just two fully connected layers at the end of the Bi-LSTMs module, which might have not fully exploited the feature representations.

For future work, our architecture can be extended by combining both CNN, 3DCNN, and Bi-LSTM in a hierarchical way such that each part can focus on different granularity information. Also, a more reasonable fusion method can be explored.

References

1. Chung, J., Gulcehre, C., Cho, K., Bengio, Y.: Empirical evaluation of gated recurrent neural networks on sequence modeling. In: NIPS 2014 Workshop on Deep Learning, December 2014
2. Donahue, J., et al.: Long-term recurrent convolutional networks for visual recognition and description. In: 2015 IEEE Conference on Computer Vision and Pattern Recognition (CVPR), pp. 2625–2634 (2015)
3. He, K., Zhang, X., Ren, S., Sun, J.: Deep residual learning for image recognition. In: 2016 IEEE Conference on Computer Vision and Pattern Recognition (CVPR), pp. 770–778 (2016)
4. Hochreiter, S., Schmidhuber, J.: Long short-term memory. Neural Comput. **9**, 1735–1780 (1997)
5. Irani, R., Nasrollahi, K., Dhall, A., Moeslund, T.B., Gedeon, T.: Thermal superpixels for bimodal stress recognition. In: 2016 Sixth International Conference on Image Processing Theory, Tools and Applications (IPTA), pp. 1–6 (2016)

6. Ji, S., Xu, W., Yang, M., Yu, K.: 3D convolutional neural networks for human action recognition. IEEE Trans. Pattern Anal. Mach. Intell. **35**, 221–231 (2013)
7. Kosko, B.: Bidirectional associative memories. IEEE Trans. Syst. Man Cybern. **18**, 49–60 (1988)
8. Kosko, B.: Neural Networks and Fuzzy Systems: A Dynamical Systems Approach to Machine Intelligence. Prentice-Hall Inc., Englewood Cliffs (1991)
9. Krizhevsky, A., Sutskever, I., Hinton, G.E.: ImageNet classification with deep convolutional neural networks. Commun. ACM **60**, 84–90 (2012)
10. Kumar, S., et al.: StressNet: detecting stress in thermal videos. In: 2021 IEEE Winter Conference on Applications of Computer Vision (WACV), pp. 998–1008 (2021)
11. Mcculloch, W., Pitts, W.: A logical calculus of the ideas immanent in nervous activity. Bull. Math. Biophys. **5**, 127–147 (1943). https://doi.org/10.1007/BF02478259
12. Miller, G.E., Cohen, S., Ritchey, A.K.: Chronic psychological stress and the regulation of pro-inflammatory cytokines: a glucocorticoid-resistance model. Health Psychol. Off. J. Div. Health Psychol. **21**(6), 531–41 (2002)
13. Nejad, A.F., Gedeon, T.: Bidirectional neural networks and class prototypes. In: Proceedings of ICNN 1995 - International Conference on Neural Networks, vol. 3, pp. 1322–1327 (1995)
14. Rumelhart, D., Hinton, G., Williams, R.: Leaning internal representations by back-propagating errors. Nature **323**(99), 533–536 (1986)
15. Schuster, M., Paliwal, K.K.: Bidirectional recurrent neural networks. IEEE Trans. Sig. Process. **45**, 2673–2681 (1997)
16. Sharma, N., Dhall, A., Gedeon, T., Göcke, R.: Modeling stress using thermal facial patterns: a spatio-temporal approach. In: 2013 Humaine Association Conference on Affective Computing and Intelligent Interaction, pp. 387–392 (2013)
17. Simonyan, K., Zisserman, A.: Very deep convolutional networks for large-scale image recognition. In: International Conference on Learning Representations (2015)
18. Ushiyama, T., et al.: Analysis of heart rate variability as an index of noncardiac surgical stress. Heart Vessels **23**, 53–59 (2007). https://doi.org/10.1007/s00380-007-0997-6
19. Vitetta, L., Anton, B., Cortizo, F.G., Sali, A.: Mind-body medicine: stress and its impact on overall health and longevity. Ann. N. Y. Acad. Sci. **1057**(1), 492–505 (2005)
20. Zhai, J., Barreto, A.B.: Stress recognition using non-invasive technology. In: FLAIRS Conference (2006)

StressNet: A Deep Neural Network Based on Dynamic Dropout Layers for Stress Recognition

Hao Wang[✉] and Abhijit Adhikary

School of Computing, The Australian National University,
Canberra, ACT 2600, Australia
{Hao.Wang,Abhijit.Adhikary}@anu.edu.au

Abstract. Stress is a body response to the changing of environmental conditions, such as facing time pressure, threats, or scary things. Being in a stressful state for a long time affects our physical and mental health. Therefore, we need to regularly monitor our stress. In this paper, we propose a deep neural network with novel dynamic dropout layers to address the stress recognition task through thermal images. Dropout regularization has been widely used in various deep neural networks for combating overfitting. In the task of stress recognition, overfitting is a common phenomenon. Our experiments show that our proposed dynamic dropout layers speed up both the training process and alleviate overfitting, but also make the network focus on the important features while ignoring unimportant features at the same time. The proposed approach was evaluated in comparison with the baseline models [5,10] over the ANUStressDB dataset. The experimental results show that our model achieves 95.8% classification accuracy on the test set. The code publicly is available at https://github.com/onehotwh/StressNet.

Keywords: Stress recognition · Deep neural network · Dropout · Dynamic masks

1 Introduction

Stress is a regular phenomenon of our life, whether slight or intense, it always there and affects our health conditions. It can affect our central nervous system (CNS) [6] and has destructive effects on our memory, cognition, and learning systems [14]. Traditional technologies for stress detection are mainly contact-based such as face-to-face stress consulting by a psychologist. Self-reporting systems are the alternative way but may not be able to detect stress situations in a short period of time [5]. Nowadays, the world has come to a standstill due to the COVID-19 pandemic [9]. It is imperative to need of reliable contact-less monitoring systems that could detect the changing of stressful situations in our daily life. Stress is highly correlated with our physiological response that can be collected by some sensors such as an RGB or a thermal camera. If we are stressed,

© Springer Nature Switzerland AG 2021
T. Mantoro et al. (Eds.): ICONIP 2021, LNCS 13109, pp. 502–512, 2021.
https://doi.org/10.1007/978-3-030-92270-2_43

the features extracted from images produced by these cameras will be different from the features when we are calm. Researchers were attempting to detect and classify stress states from calm states by measuring physiological signal features extracted from the RGB or thermal images. Pavlidis et al. [7,8] measured the stress by using a thermal sensor to detect the increase of blood flow in one's forehead region caused by stress conditions. Sharma et al. [10] apply the support vector machines on the Spatio-temporal features extracted from thermal images and achieved 86% accuracy. Shastri et al. [11] proposed a physiological function to measure the transient perspiration captured by thermal images. Irani et al. [5] extracted features from super-pixels by fusing the features extracted by RGB and thermal images and achieved 89% classification accuracy on the ANUstressDB dataset.

Dropout is a stochastic regularization technique [3,12] that has a good effect on combating overfitting, so it is widely used in deep neural networks (DNN). Conventionally, it works mainly in fully connected (FC) layers by randomly "dropping" out the activation of a neuron with a certain probability p for each training case. The essence of the dropout method is to randomly prune the neural network. The process has the effect of model averaging by simulating a large number of networks with different network structures, which, in turn, making node activations in the network more robust to the inputs.

1.1 Motivations

Inspired by the mechanism of dropout, other stochastic methods were proposed to create the dynamic sparsity within the network. For example, spatial dropout [13] extended the dropout approach to the convolutional layer to remove the feature map activations in the object localization task, accounting for a strong spatial correlation of nearby pixels in natural images. The standard dropout and spatial drop at the output of 2D convolutions are shown in Fig. 1. The standard dropout is a pixel-wise zeroed approach while the spatial dropout is a channel-wise zeroed approach on the feature maps. The sparsity generated by these two approaches can induce more randomness to the network during the training phase, which can improve the robustness of the model.

Overfitting is a common phenomenon on the stress recognition task since stress is a subjective response, which means that different people will respond to the same situation differently. On the other hand, stress has a strong coherence, and this coherence will make the body's response to changes in the external environment lagging behind. These problems make the detection of pressure very difficult. Overcoming the overfitting problem seems to be the top priority of the stress recognition task. The biggest problem in Dropout is that we need to set all fixed probability values for all dropout layers. These probability values are the hyperparameters of the model that need to tune. Although we can use cross-validation or a genetic algorithm (GA) to obtain the best hyperparameters combination on the validation set, this will greatly increase the training time.

On the other hand, cross-validation wastes some training data, which is undoubtedly worse for stress detection tasks that are prone to overfitting. Therefore, we propose a dynamic dropout layer that could generate masks to eliminate some elements in the feature maps. These masks are generated by some independent convolutional layers whose parameters will update during training via backpropagation. Unlike the traditional dropout layer, the dynamic dropout layer does not retain the values that have not been zeroed but scales them according to different parameters. The scaled parameters can still be updated during training.

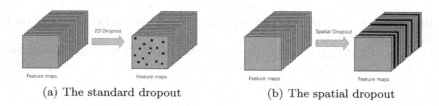

(a) The standard dropout   (b) The spatial dropout

Fig. 1. Two different dropout approaches. (a) Standard dropout will zero each element of its input by a fixed probability p. (b) The spatial dropout will zero each channel of its input by a fixed probability p.

1.2 Technical Contributions

- A Deep Neural Network StressNet that achieved high accuracy on the ANUStressDB dataset.
- Dynamic dropout layers could speed up training and reduce overfitting.
- An exploration of model parameter settings that is different from traditional strategies such as cross-validation and genetic algorithms, but let the model learn the parameters during training.

The rest of this paper is organized as follows: Sect. 2 explains the task we designed and the data pre-processing method we applied, then we introduce the architecture of the proposed model. Section 3 shows the results of the experiments. Section 4 provides a discussion and ablation study and Sect. 5 includes the future work and concludes this paper.

2 Methodology

2.1 Dataset and Data Pre-processing

Dataset: ANUStressDB. The dataset we used in this paper is the ANUStressDB dataset [5] containing thermal videos of 35 subjects watching stressed and not-stressed film clips validated by the subjects. Each video has 20 video clips that half of which belong to the label "Stressful" and the other half of the videos belongs to "Calm". Therefore, it is a well-balanced dataset. A schematic diagram of the experiment setup of ANUStressDB and the example image frame extracted from one of these videos are shown in Fig. 2 [10].

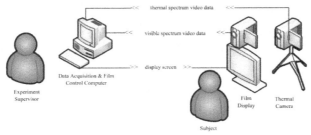

(a) Setup for the film experiment of ANUStressDB [10]

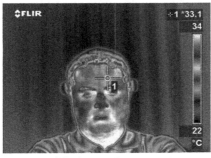

(b) A sample image

Fig. 2. The setup process of ANUStressDB and one sample images extracted from one of the videos. The image has the size of 640 × 480.

Data Pre-processing. The data pre-processing in this paper is shown as Fig. 3. From the given information about the experiment generating ANUStressDB, there exist about 30 s for preparing at the beginning of each video and there exist about 6 s between each pair of clips. We use FFMPEG to segment each video into 20 video segments by the order of labels. Then for each video segment, we extract 30 frames evenly to form the image dataset. We randomly split the image dataset into the training set and test set by the ratio of 80:20. This is a big dataset, so we also extract a smaller dataset extracting images only from the first 6 labels of each video for evaluating our proposed approaches. For each image in the training set and test set, we apply center crop on it to get the image of size 480 × 480 and then resize it to 240 × 240.

2.2   Network Architecture

The architecture of the proposed model is shown in Fig. 4. The model is mainly composed of a Shallow Convolutional Layer (SCL) and a Deep Convolutional

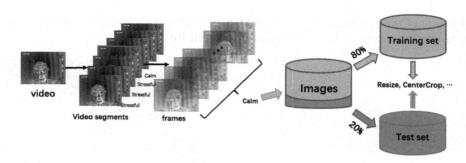

Fig. 3. Data preprocessing

Layer (DCL). There also exists a standard dropout Layer before the final fully connected layer. We used ReLU [1] as the activation function in the Dynamic Dropout Layer to generate the mask and the sigmoid activation function in the last fully connected layer. Other activation functions are all GELUs [2] which could generate a more robust model and speed up training considerably. Our experiments show that GELUs could reduce the training time by about 40% compared with ReLU and 45% PReLU respectively.

$$\text{GELU}(x) = xP(X < x) = x\Phi(x) = \frac{1}{2}x[1 + \text{erf}(\frac{1}{\sqrt{2}})] \qquad (1)$$

where $X \sim N(0, 1)$ and $\Phi(x)$ its cumulative distribution function.

Shallow Convolutional Layer (SCL). The SCL contains three convolutional layers. Each convolutional layer contains a 2D convolution with kernel size of 3 followed by a Batch Normalization [4] and a GELU activation function. We have the equation of the CL as

$$\mathbf{f_o} = \text{CL}(\mathbf{f_i}) = \text{GELU}(\text{BatchNorm}(\text{Conv}(\mathbf{f_i}))) \qquad (2)$$

The reason why we do not use the Dynamic Dropout Layer (DDL) in SCL is that the network needs to fully extract the features of the image in the shallow layer. If dropout is added to the shallow network, it will cause serious interference to the network, so that the deep network cannot extract useful feature information, which will makes the training phase slow and even underfitting.

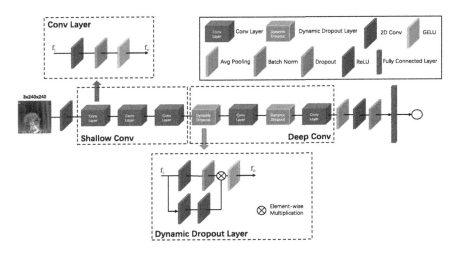

Fig. 4. Model architecture.

Deep Convolutional Layer (DCL). The DCL contains two convolutional layers and two Dynamic Dropout Layers (DDL). A DDL contains the same architecture as the convolutional layer as the main trunk. And it also has a side branch network including convolution and a ReLU activation function. The side branch network can generate feature maps with the same shape as the output of the main trunk network. Then the feature maps generated by the backbone network and the masks are correspondingly multiplied, and the result goes through the GELU function to generate the output of DDL. Note that the convolutions in DDL have kernels with a stride of 1 while the convolutional layer in DCL has kernels with a stride of 2. The process of DDL can be written as:

$$\mathbf{f_o} = \mathrm{DDL}(\mathbf{f_i}) = \mathrm{GELU}(\mathrm{BatchNorm}(\mathbf{Conv_1}(\mathbf{f_i})) \star \mathrm{ReLU}(\mathbf{Conv_2}(\mathbf{f_i}))) \quad (3)$$

Deep layers in the network will receive the out feature maps from previous layers. However, we do not expect the neural network to learn all the feature information, because this can easily lead to overfitting. In the experiment section, we will show that if dropout is not added to the deeper layers, although the loss of the network in the training set continues to decline and the accuracy rate continues to increase, it will struggle in the test set, and the accuracy rate will stay in a certain range for a long time and even drop. At the end of the two nets, there exists an average pooling layer and a standard dropout layer. And the result will pass through a fully connected layer for classification. We use the Binary Cross Entropy loss:

$$\mathcal{L} = -\frac{1}{N} \sum_{n=1}^{N} \{t_n \log(y_n) + (1 - t_n) \log(1 - y_n)\} \quad (4)$$

3 Experiments

We perform the experiment section by comparing the results of different model settings and we used the model of [5] and [10] as baselines.

3.1 Experimental Setup

As we discussed in Sect. 2.1, we use the smaller image dataset containing the first 6 labels of each video to evaluate our proposed model and the whole image dataset to compare with the baseline. We use PyTorch 1.6.0 to construct our models. We use a NVIDIA RTX 2070 SUPER GPU to training our models on the Windows 10 platform. The hyperparameters used in this paper is shown in Table 1.

Table 1. Hyperparameters

Dataset	Epochs	Learning rate	Batch size	Optimizer
Small (6 labels)	30	5e−4	16	Adam
Whole (20 labels)	100	1e−3	16	Adam

3.2 Results and Discussion

The results of our experiments are shown in Table 2, Fig. 5 and Fig. 6. Obviously, the Dynamic Dropout Layer could speed up training and alleviate the overfitting phenomenon which is quite common in the stress recognition task. However, Batch Normalization is generally considered to have the same series of advantages as above, but why in our experiments, the model with batch normalization layers has a serious overfitting phenomenon? Let us try to analyze the mechanisms of Batch Normalization and dropout against overfitting. The mechanism for Batch Normalization to achieve good generalization is to try to find the solution of the problem in a smoother solution subspace. The emphasis is on the smoothness of the process of processing the problem. The implicit idea is that the smoother solution has better generalization ability, For dropout, robustness is the goal, that is, the solution is required to be insensitive to the disturbance of the network configuration, and the implicit idea is that the more robust generalization ability is better. From this mechanism, Dropout's control mechanism for over-fitting is actually more direct, superficial, and simpler, while Batch Normalization is more indirect, but at a lower level and more essential. However, the mechanisms of the two are different and overlapped. The smoothness and robustness are often the same, but they are not completely consistent. It cannot be simply said which effect is better, but from the mechanical point of view, the Batch Normalization idea seems to be statistical It is better, but because Batch Normalization's method for constraining smoothness is not complete, it just chooses a control mode, so it cannot fully reflect its control smoothness idea, so Batch Normalization is not had much effect on the generalization performance compared with the dropout.

Table 2. Performances of our models on the test set.

	Baseline1 [5]	Baseline2 [10]	StressNet (proposed)	StressNet (no dropout)
Small dataset	/	/	**89.7%**	73.3%
Whole dataset	89.0%	86.0%	**95.8%**	71.4%

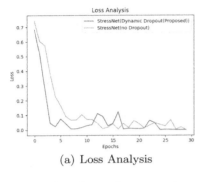 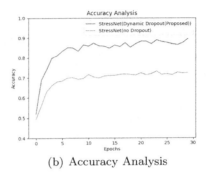

(a) Loss Analysis   (b) Accuracy Analysis

Fig. 5. Results of experiment on the smaller dataset of the StressNet with Dynamic Dropout Layers and without any Dropout layer.

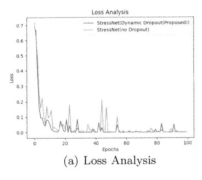 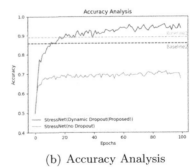

(a) Loss Analysis   (b) Accuracy Analysis

Fig. 6. Results of experiment on the whole dataset of the StressNet with Dynamic Dropout Layers and without any Dropout layers.

4   Discussion

4.1   Visualization of Learned Masks

The masks generated by the first and second Dynamic Dropout Layers are shown in Appendix Fig. 8. We randomly select 64 of these masks to visualize. As the figure shows, the Dynamic Dropout Layer could generate sparse masks. These masks will drop out the background part of the image that has less information about whether the person is stressed or calm and makes the network focus on the facial part in the image. In addition, the deep DDL will produce more sparse masks, and the non-zero part of these masks contains the key features to distinguish the participants from stressful or calm.

4.2   Ablation Study

We evaluate performances of different kinds of dropout layers in our StressNet.

Standard Dropout v.s. Spatial Dropout v.s. DDL. The performances of StressNet with different kinds of dropout layers on the smaller dataset are shown in Fig. 7. It can be seen from the loss curves that the performance of spatial dropout is the worst because his zero-setting strategy is too aggressive, which will cause many useful feature maps to be discarded during the forward propagation. This will greatly reduce the training speed of the network. The result of standard dropout is acceptable. Compared with the previous strategy without dropout, the introduction of dropout improves the generalization of the model. However, as we discussed above, the setting of hyper-parameters is a critical factor, and during the training process, the hyper-parameters are fixed, which brings many restrictions, and it may require multiple training processes to find the best combination of hyperparameters. The Dynamic Dropout Layer proposed in this paper has obvious advantages in speeding up the training process and improving the generalization ability of the model, and it also avoids the hyperparameter turning process.

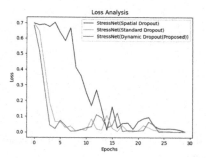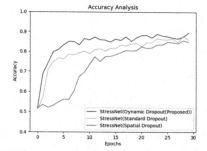

Fig. 7. Ablation study.

5   Conclusions and Future Work

In this paper, we proposed an A Dynamic Dropout Layer-based Neural Network for Stress Recognition named StressNet to solve the stress recognition task. Our experiments prove that in the task of stress recognition, we need the dropout approach to combat the overfitting problem, especially in the deep part of the neural network. In addition, different from the traditional over-fitting method applied to the fully connected layer, this paper explored the impact of different types of dropout methods in the convolutional layer on the generalization of classification tasks. This paper proposes a dynamic dropout method to generate masks. Its parameters can be updated during the training process using backpropagation. Compared with the traditional dropout method, this method improves the training speed and the generalization ability of the model. Due to limited time, we did not explore the limitations of the model and optimize the model structure, or test the effect of the model on other data sets. Therefore, in the future, I may explore other conventional RGB data sets, such as CIFAR-10, ImageNet, etc.

6   Appendix

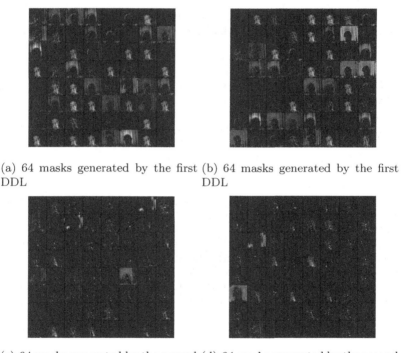

(a) 64 masks generated by the first DDL

(b) 64 masks generated by the first DDL

(c) 64 masks generated by the second DDL

(d) 64 masks generated by the second DDL

Fig. 8. Visualization of 64 masks generated by two Dynamic Dropout Layers.

References

1. Glorot, X., Bordes, A., Bengio, Y.: Deep sparse rectifier neural networks. In: Proceedings of the Fourteenth International Conference on Artificial Intelligence and Statistics, pp. 315–323. JMLR Workshop and Conference Proceedings (2011)
2. Hendrycks, D., Gimpel, K.: Gaussian error linear units (GELUs). arXiv preprint arXiv:1606.08415 (2016)
3. Hinton, G.E., Srivastava, N., Krizhevsky, A., Sutskever, I., Salakhutdinov, R.R.: Improving neural networks by preventing co-adaptation of feature detectors. arXiv preprint arXiv:1207.0580 (2012)
4. Ioffe, S., Szegedy, C.: Batch normalization: accelerating deep network training by reducing internal covariate shift. In: International Conference on Machine Learning, pp. 448–456. PMLR (2015)
5. Irani, R., Nasrollahi, K., Dhall, A., Moeslund, T.B., Gedeon, T.: Thermal superpixels for bimodal stress recognition. In: 2016 Sixth International Conference on Image Processing Theory, Tools and Applications (IPTA), pp. 1–6. IEEE (2016)
6. Lupien, S.J., Lepage, M.: Stress, memory, and the hippocampus: can't live with it, can't live without it. Behav. Brain Res. **127**(1–2), 137–158 (2001)
7. Pavlidis, I., Eberhardt, N.L., Levine, J.A.: Seeing through the face of deception. Nature **415**(6867), 35–35 (2002)
8. Pavlidis, I., Levine, J.: Thermal image analysis for polygraph testing. IEEE Eng. Med. Biol. Mag. **21**(6), 56–64 (2002)
9. Roser, M., Ritchie, H., Ortiz-Ospina, E., Hasell, J.: Coronavirus pandemic (COVID-19). Our World in Data (2020)
10. Sharma, N., Dhall, A., Gedeon, T., Goecke, R.: Thermal spatio-temporal data for stress recognition. EURASIP J. Image Video Process. **2014**(1), 1–12 (2014). https://doi.org/10.1186/1687-5281-2014-28
11. Shastri, D., Papadakis, M., Tsiamyrtzis, P., Bass, B., Pavlidis, I.: Perinasal imaging of physiological stress and its affective potential. IEEE Trans. Affect. Comput. **3**(3), 366–378 (2012)
12. Srivastava, N., Hinton, G., Krizhevsky, A., Sutskever, I., Salakhutdinov, R.: Dropout: a simple way to prevent neural networks from overfitting. J. Mach. Learn. Res. **15**(1), 1929–1958 (2014)
13. Tompson, J., Goroshin, R., Jain, A., LeCun, Y., Bregler, C.: Efficient object localization using convolutional networks. In: Proceedings of the IEEE Conference on Computer Vision and Pattern Recognition, pp. 648–656 (2015)
14. Yaribeygi, H., Panahi, Y., Sahraei, H., Johnston, T.P., Sahebkar, A.: The impact of stress on body function: a review. EXCLI J. **16**, 1057 (2017)

Analyzing Vietnamese Legal Questions Using Deep Neural Networks with Biaffine Classifiers

Nguyen Anh Tu[1,2], Hoang Thi Thu Uyen[1], Tu Minh Phuong[1], and Ngo Xuan Bach[1(✉)]

[1] Department of Computer Science, Posts and Telecommunications Institute of Technology, Hanoi, Vietnam
{anhtunguyen446,thuuyenptit}@gmail.com,
{phuongtm,bachnx}@ptit.edu.vn
[2] FPT Technology Research Institute, FPT University, Hanoi, Vietnam

Abstract. In this paper, we propose using deep neural networks to extract important information from Vietnamese legal questions, a fundamental task towards building a question answering system in the legal domain. Given a legal question in natural language, the goal is to extract all the segments that contain the needed information to answer the question. We introduce a deep model that solves the task in three stages. First, our model leverages recent advanced autoencoding language models to produce contextual word embeddings, which are then combined with character-level and POS-tag information to form word representations. Next, bidirectional long short-term memory networks are employed to capture the relations among words and generate sentence-level representations. At the third stage, borrowing ideas from graph-based dependency parsing methods which provide a global view on the input sentence, we use biaffine classifiers to estimate the probability of each pair of start-end words to be an important segment. Experimental results on a public Vietnamese legal dataset show that our model outperforms the previous work by a large margin, achieving 94.79% in the F_1 score. The results also prove the effectiveness of using contextual features extracted from pre-trained language models combined with other types of features such as character-level and POS-tag features when training on a limited dataset.

Keywords: Question answering · Legal domain · Deep neural network · Biaffine classifier · BERT · BiLSTM

1 Introduction

Question answering (QA) [7,12,22,28–30], a sub-field of natural language processing (NLP) and information retrieval (IR), aims to build computer systems that can automatically answer questions in natural languages. There are two main approaches for building QA systems: IR-based and knowledge-based. The former approach finds and extracts answers from a collection of unstructured text, while the latter utilizes structured text or knowledge bases. Although two

© Springer Nature Switzerland AG 2021
T. Mantoro et al. (Eds.): ICONIP 2021, LNCS 13109, pp. 513–525, 2021.
https://doi.org/10.1007/978-3-030-92270-2_44

Question 1:

<TL>Đèn xanh</TL> thì <QT>được đi chưa</QT> ?

(<QT>Is it possible to go</QT> when <TL>the light is green</TL> ?)

Question 2:

Người đi <TV>xe máy</TV> có <AC>nồng độ cồn</AC> là <V>0,3 miligam/1 lít khí thở</V> <QT>có bị phạt không</QT> ?

(Are <TV>motorcyclists</TV> with an <AC>alcohol level</AC> of <V>0.3 milligrams per liter of breathing</V> <QT>fined</QT> ?)

Fig. 1. Examples of two legal questions and their important information.

approaches exploit different types of resources, they share the first step, question analysis, which extracts the needed information to answer the input question. Such information is then used to form a query in various ways which serves as the input for the next step. Question analysis is therefore a crucial task for both IR-based and knowledge-based question answering.

In this paper, we target at the question analysis task in the legal domain, which is undoubtedly important but has received less attention. Legal and regulatory documents are ubiquitous and have a great impact on our life. Figure 1 shows two examples of Vietnamese legal questions annotated with key information. The goal is to correctly extract two types of information in the first question (Traffic Light-TL and Question Type-QT), and four types of information in the second question (Type of Vehicle-TV, Alcohol Concentration-AC, Value-V, and Question Type-QT). We call the segments that contain important information important segments.

Traditional methods often frame the task as a sequence labeling problem and exploit probabilistic graphical models like conditional random fields (CRFs) [14] to solve it. The main advantage of those methods is that they can build an accurate model using a relatively small annotated corpus with a handcrafted feature set. Recently, however, deep neural networks have made tremendous breakthroughs in various NLP tasks and applications, including sentence classification [13,26], sequence labeling [2,9,33], syntactic and dependency parsing [21,34], natural language inference [5,10], machine translation [27,35], as well as question answering [10,12]. Furthermore, the well-known limitation of deep models, i.e. data hungry, could be mitigated by using advanced pre-trained models and fine-tuning [5,10]. All of these make deep neural networks the preferred choice for NLP tasks in general and QA in particular.

Here, we propose to use deep neural networks for Vietnamese question analysis in the legal domain. Our method combines several recent advanced techniques in the NLP and deep learning research communities: pre-trained language models [5] for contextual word embeddings, convolutional neural networks (CNNs) [16] for extracting character-level features, and bidirectional long-short term memory networks (BiLSTM) [8] for sentence-level representations. Furthermore, instead of formulating it as a sequence labeling problem, we employ biaffine classifiers to estimate directly the possibility that a pair of words becomes an important segment. The main advantage of biaffine classifiers is that they provide a global view on the input sentence, which has been shown to be effective in dependency paring [6,17], and named entity and relation extraction [23,36]. Experimental results on a Vietnamese corpus consisting of 1678 legal questions show that our model outperforms a SOTA method by a large margin, showing the F_1 score of 94.79%. The effectiveness of these components of the model is also validated by an ablation study.

The remainder is organized as follows. Section 2 reviews related work. Section 3 presents our model for extracting important information from Vietnamese legal questions. The model architecture is presented first, and its key components are then described in more detail. Experimental results and error analysis are introduced in Sect. 4. Finally, Sect. 5 concludes the paper and shows some future work.

2   Related Work

Several studies have been performed on Vietnamese QA in various domains, including travel, education, as well as legal. Tran et al. [28] introduce a Vietnamese QA system, which can answer simple questions in the travel domain by mining information from the Web. Bach et al. [3] focus on the task of analyzing Vietnamese question in education. Using deep neural networks (BiLSTM-CRFs) with a rich set of features, their model can accurately extract 14 types of vital information from education utterances. In the legal domain, Duong and Ho [7] develop a QA system to answer Vietnamese questions about provisions, procedures, processes, etc. in enterprise laws. Kien et al. [12] introduce a retrieval-based method for answering Vietnamese legal questions by learning text representation. Their model leverages CNNs and attention mechanisms to extract and align important information between a question and a legal article. Other works on Vietnamese question answering include Nguyen et al. [22], Tran et al. [29], and Le-Hong and Bui [15].

Perhaps the most closely work to ours is the one of Bach et al. [1], which also focuses on analyzing Vietnamese legal questions. Our method, however, is distinguished from theirs in two aspects. First, we formulate the task as a multi-class classification problem instead of sequence labeling. Second, we utilize deep neural networks instead of using traditional models like CRFs.

3   Method

3.1   Model Overview

Our goal is to extract all important segments from an input legal question, where each segment is a triple of start/end positions and a label for the information type. Figure 2 illustrates our proposed architecture. First, we create word representations by concatenating different types of features: contextual word

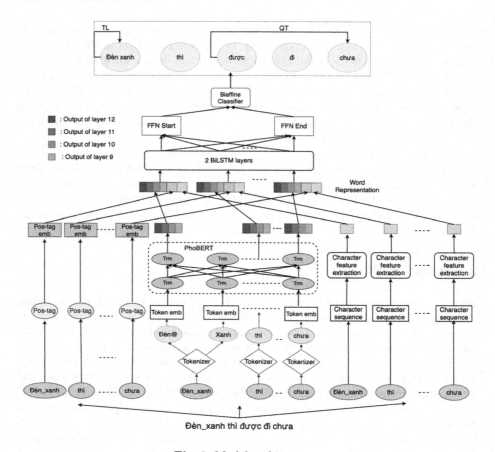

Fig. 2. Model architecture.

embeddings, character-level features, and part-of-speech (POS) tag embeddings. The outputs of the word representation layer are then fed into two stacked BiLSTMs to obtain the sentence-level representations. After that, we use two feed forward neural networks (FFN) to generate different representations for the start/end of segments. Finally, we employ a biaffine classifier to create a $n \times n \times c$ scoring tensor R, where n denotes the length of the input question, and c is the number of labels (including Null for non important information). Tensor R provides scores for all possible segments that could contain key information.

In the up-coming sections, we show the model components in detail. For notation, we denote vectors, matrices, and scalars with bold lower-case (e.g., $\mathbf{x}_t$, $\mathbf{h}_t$, $\mathbf{b}$), bold upper-case (e.g., $\mathbf{H}$, $\mathbf{W}_i$, $\mathbf{V}_i$), and italic lower-case (e.g., n, c), respectively.

3.2 Word Representations

Because words are basic elements to form written languages, a good word representation method is the first and crucial step to build successful NLP systems. In this work, we create rich information word representations by integrating multiple information sources, including contextual word embeddings, character-level features, and POS-tag embeddings.

Contextual Word Embeddings. Traditional NLP approaches usually represent words by one-hot vectors with one value 1 and the rest 0. These high-dimensional sparse vectors are memory consuming and cannot capture the word semantics. Distributed word representation methods, which use low-dimensional continuous vectors, have been introduced to handle these issues. Word2vec [20], Fasttext [4], and Glove [25] are successful examples of such methods that represent similar words with similar vectors. Although these methods have made many breakthroughs in NLP research, they represent words by fix vectors which are context independent. Static word embedding methods like word2vec are therefore limited in representing polysemous words.

Recently, contextual word embedding methods have been shown to be the key component of many SOTA NLP systems [5,18]. The main advantage of these methods it that they can learn different representations for polysemous words by considering the sequence of all words in sentences/documents. Perhaps BERT proposed by Devlin et al. [5] is the most famous and popular contextual word embedding method. The key technical innovation of BERT is applying the bidirectional training of Transformer [31] to language modeling with two strategies: masked-language modeling and next sentence prediction.

In this work, we use PhoBERT [24], a monolingual variant of RoBERTa [18] pre-trained on a 20GB word-level Vietnamese dataset. Like BERT and RoBERTa, PhoBERT segments the input sentence into sub-words, which brings the balance between character- and word-level hybrid representations and enables the encoding of rare words with appropriate sub-words. We represent each sub-word by concatenating embeddings of the last four encoding layers (9 to 12) of PhoBERT-base, and the contextual embedding of a word is the embedding of its first sub-word.

Character-Level Features. Beside contextual word embeddings, we also utilize morphological information from characters. Additional character embeddings are derived from character-level convolutional (charCNN) networks. As shown in Fig. 3, charCNN consists of 1D operations: convolution and max pooling. Feature maps are then concatenated to produce a character-level representation for the word.

POS-Tag Embeddings. We suppose POS tags are another useful source of information. Therefore, we also use POS tag embeddings to represent words. These embedding vectors are initialized randomly in range $(-\sqrt{3/dim}, \sqrt{3/dim})$, where dim denotes their dimension. We use VnCoreNLP [32] for word segmentation and POS tagging for input questions.

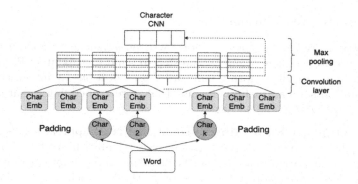

Fig. 3. CNN-based character-level features.

Finally, all feature vectors are concatenated into a single embedding for representing a word. Word vectors are then fed into BiLSTM networks to create sentence-level representations.

3.3 BiLSTM

Long short-term memory (LSTM) networks [11] are designed for sequence data modeling problem. Let $\mathbf{X} = (\mathbf{x}_1, \mathbf{x}_2, \ldots, \mathbf{x}_n)$ denote the input question where $\mathbf{x}_i$ is the embedding of the ith word. At each position t, the LSTM computes an intermediate representation using a hidden state $\mathbf{h}$:

$$\mathbf{h}_t = f(\mathbf{h}_{t-1}, \mathbf{x}_t)$$

where f includes an input gate, a forget gate, an output gate, and a memory cell (denoted by $\mathbf{i}_t, \mathbf{f}_t, \mathbf{o}_t, \mathbf{c}_t$, respectively) to update $\mathbf{h}_t$:

$$\mathbf{i}_t = \sigma(\mathbf{W}_i \mathbf{x}_t + \mathbf{V}_i \mathbf{h}_{t-1} + \mathbf{b}_i),$$
$$\mathbf{f}_t = \sigma(\mathbf{W}_f \mathbf{x}_t + \mathbf{V}_f \mathbf{h}_{t-1} + \mathbf{b}_f),$$
$$\mathbf{o}_t = \sigma(\mathbf{W}_o \mathbf{x}_t + \mathbf{V}_o \mathbf{h}_{t-1} + \mathbf{b}_o),$$
$$\mathbf{c}_t = \mathbf{f}_t \odot \mathbf{c}_{t-1} + \mathbf{i}_t \odot \tanh(\mathbf{W}_c \mathbf{x}_t + \mathbf{V}_c \mathbf{h}_{t-1} + \mathbf{b}_c),$$
$$\mathbf{h}_t = \mathbf{o}_t \odot \tanh(\mathbf{c}_t),$$

and the output $\mathbf{y}_t$ can be produced based on $\mathbf{h}_t$:

$$\mathbf{y}_t = \sigma(\mathbf{W}_y \mathbf{h}_t + \mathbf{b}_y),$$

where $\odot$ indicates the multiplication operator function, σ is the element-wise softmax, and $\mathbf{W}_*, \mathbf{V}_*$, and $\mathbf{b}_*$ ($*$ denotes i, f, o, c, y) are weight matrices and vectors to be learned during the training process.

Bidirectional long short-term memory (BiLSTM) [8] combine two LSTMs: one network moves from the left to the right and the other network moves from the right to the left of the sequence. Two BiLSTMs are exploited to learn a higher level of semantics within the sentence.

Table 1. Types of important information and their occurrence numbers

Label	Meaning	#	Label	Meaning	#
A	Action of vehicle	1087	L	Location	426
AC	Alcohol concentration	44	QT	Question type	1678
ANO	Annotation	75	SP	Speed	115
DL	Driving license	119	TI	Traffic instructor	93
IF1	Add. info. about vehicle	196	TL	Traffic light	31
IF2	Add. info. about traffic light	12	TP	Traffic participant	20
IF3	Add. info. about traffic participant	287	TV	Type of vehicle	1245
IF4	Add. info.	227	V	Value	231

3.4 Biaffine Layer

We employ two feed forward neural networks (FFNs) to create different representations for start/end positions. The outputs of two FFNs at position t are denoted by $\mathbf{g}_t^{start}$ and $\mathbf{g}_t^{end}$:

$$\mathbf{g}_t^{start} = \mathrm{FFN}^{start}(\mathbf{y}_t)$$

$$\mathbf{g}_t^{end} = \mathrm{FFN}^{end}(\mathbf{y}_t)$$

For each start-end candidate pair (i, j), $1 \leq i \leq j \leq n$, we apply the biaffine classifier:

$$\mathbf{r}_{i,j} = \mathrm{Biaffine}(\mathbf{g}_i^{start}, \mathbf{g}_j^{end}) = (\mathbf{g}_i^{start})^\top \mathbf{U}\mathbf{g}_j^{end} + \mathbf{W}(\mathbf{g}_i^{start} \oplus \mathbf{g}_j^{end}) + \mathbf{b},$$

where $\mathbf{U}$, $\mathbf{W}$, $\mathbf{b}$ are a $d \times c \times d$ tensor, a $c \times 2d$ matrix, and a bias vector, respectively, and d is the size of the output layers of both FFN^{start} and FFN^{end}.

Vector $\mathbf{r}_{i,j}$ is then fed into a softmax layer to produce probability scores $\mathbf{s}_{i,j}$:

$$\mathbf{s}_{i,j}(k) = \frac{exp(\mathbf{r}_{i,j}(k))}{\sum_{k'=1}^{c} exp(\mathbf{r}_{i,j}(k'))}$$

The label of segment (i, j) can be determined as: $\widehat{l} = \arg\max_k \mathbf{s}_{i,j}(k)$.

The question analysis task now becomes a multi-class classification problem and model parameters are learned to minimize the cross-entropy loss function.

4 Experiments

4.1 Data and Evaluation Method

In our experiments we used the Vietnamese dataset of legal questions introduced by Bach et al. [1]. This dataset consists of 1678 legal questions about the traffic law in Vietnam. Questions were annotated with 16 labels reflecting different aspects of the domain listed in Table 1.

We performed cross-validation tests with the same training/test data splits of Bach et al. [1]. For each fold, we used 10% of the training set as a validation set. The performance of extraction models was measured using popular metrics such as precision, recall and the F_1 score.

4.2   Network Training

Our models were implemented in PyTorch[1] using Huggingface's Transformers[2]. In all experiments, we set the batch size to 64. The max character length was set to 15, and the max sequence length was tuned in $[50, 60, 80, 100]$ for all models, and the best value was 60. We set the dimensions of character-level features and POS-tag embeddings to 256 and 100, respectively. We used dimension of 300 for FFNs, and kernel sizes of 3 and 4 for charCNN. To mitigate overfitting, we used a dropout rate of 0.3 for each hidden layer. Our models were trained using the AdamW optimizer [19]. We set the epsilon and weight decay to default values in PyTorch, i.e. $1e-8$. The learning rate was tuned in $[3e-5, 4e-5, 5e-5]$ and the best learning rate value was $5e-5$. For each model, we trained for 30 epochs and selected the version that obtained the highest F_1 score on the validation set to apply to the test set.

4.3   Experimental Results

Our Model vs. Baseline We first conducted experiments to compare our model with the previous work of Bach et al. [1] (our baseline). Table 2 shows experimental results on each type of information and the overall scores of two models. Our model outperformed the baseline for 15 out of 16 types of information. Types with the biggest improvements in the F_1 score include TL (Traffic light: 19.79%), ANO (Annotation: 9.61%), TP (Traffic participant: 7.48%), A (Action: 4.47%), and SP (Speed: 4.33%). Overall, our model achieved a micro F_1 score of 94.79%, which improved 1.85% (26.20% error rate reduction) in comparison with the baseline. Experimental results demonstrated the effectiveness of deep neural networks compared to traditional methods like CRFs.

Ablation Study. Next we evaluated the contribution of individual components of our model by performing an ablation study as follows:

- **Biaffine classifier:** We replaced the biaffine classifier with a CRF/softmax layer and reformulated the task as a sequence labeling problem
- **Character-level features, POS-tag embeddings, BiLSTMs:** We removed each component in turn.
- **PhoBERT Embeddings:** We replaced contextual word embeddings with static word embeddings. In our experiments, we used Fasttext [4], a variant of Word2Vec [20] which deals with unknown words and sparsity in languages by using sub-word models.

Experimental results in Table 3 proved that all the components contributed to the success of our model. Our first observation is that the performance of the system degraded when we modified or removed a component. While the full model got 94.79%, the F_1 score reduced to 94.62% and 93.59% when we

---

[1] https://pytorch.org/.

[2] https://huggingface.co/transformers/.

Table 2. Experimental results of our model compared with the baseline (the improvements are indicated in bold)

Type	Baseline (Bach et al. [1])			Our model		
	Prec. (%)	Rec. (%)	F_1 (%)	Prec. (%)	Rec. (%)	F_1 (%)
A	88.20	89.14	88.66	92.91	93.35	93.13 (**4.47**↑)
AC	95.78	95.78	95.78	96.12	97.28	96.70 (**0.92**↑)
ANO	85.58	60.57	68.82	73.17	84.51	78.43 (**9.61**↑)
DL	97.97	99.20	98.54	100.00	98.26	99.12 (**0.58**↑)
IF1	94.64	87.35	90.67	90.34	91.48	90.91 (**0.24**↑)
IF2	100.00	73.33	82.67	86.72	70.44	77.74 (4.93↓)
IF3	88.08	75.06	80.91	85.77	83.45	84.59 (**3.68**↑)
IF4	85.77	74.34	79.51	80.14	82.97	81.53 (**2.02**↑)
L	92.23	92.71	92.44	93.18	96.31	94.72 (**2.28**↑)
QT	96.03	94.84	95.42	95.89	96.07	95.98 (**0.56**↑)
SP	95.61	91.44	93.23	100.00	95.24	97.56 (**4.33**↑)
TI	99.05	95.53	97.24	100.00	100.00	100.00 (**2.76**↑)
TL	86.95	60.35	68.59	87.33	89.46	88.38 (**19.79**↑)
TP	80.00	60.67	67.21	76.35	73.11	74.69 (**7.48**↑)
TV	97.32	98.78	98.04	99.57	97.99	98.77 (**0.73**↑)
V	98.18	99.26	98.71	100.00	98.23	99.11 (**0.40**↑)
Overall	93.84	92.05	92.94	94.30	95.28	94.79 (**1.85**↑)

replaced the biaffine classifier with CRF/softmax function and reformulated the task as a sequence labeling problem. The score was only 94.46%, 94.33%, and 93.77%, when we removed POS-tag features, character-level features, and the BiLSTM layers, respectively. The results showed that the BiLSTM layers have a high impact on the performance of our model. The second observation is that replacing contextual word embeddings by static word embeddings leads to the biggest decrease of 1.98%. This indicated that contextual word embeddings from pre-trained language models like PhoBERT played a critical role in our model.

4.4   Error Analysis

This section discusses the cases in which our model failed to extract important information. By analyzing the model's predictions on the test sets, we found that most errors belong to one of two following types:

- **Type I: Incorrect segments.** Our model identified segments that are shorter or longer than the gold segments.
- **Type II: Incorrect information types.** Our model recognized segments (start and end positions) correctly but assigned wrong information types (labels) to the segments.

Table 3. Ablation study (the decrease in the F_1 score of the modified models is indicated in bold)

Component	Modification	Prec. (%)	Rec. (%)	F_1 (%)
Full	None	94.30	95.28	94.79
Biaffine classifier	Seq. labeling, CRF	93.92	95.33	94.62 (**0.17↓**)
	Seq. labeling, Softmax	94.02	93.17	93.59 (**1.20↓**)
POS-tag embeddings	Removal	94.54	94.39	94.46 (**0.33↓**)
Character features	Removal	93.51	95.16	94.33 (**0.46↓**)
BiLSTM	Removal	92.95	94.60	93.77 (**1.02↓**)
PhoBERT	Fasttext	92.00	93.63	92.81 (**1.98↓**)

Figure 4 shows four examples of error cases. Among them, the first two examples belong to Type I and the others belong to Type II. While our model identified a longer segment in the first case, it detected a shorter segment in the second case. In the third and fourth examples, our model made a mistake on labels (TP instead of ANO, and IF4 instead of A).

#	Error examples
1	**Gold**: người đủ \<QT>bao nhiêu tuổi trở lên\</QT> thì được điều khiển xe \<TV>ô tô\</TV> chở người đến \<IF1>**9 chỗ ngồi**\</IF1> ? **Predicted**: người đủ \<QT>bao nhiêu tuổi trở lên\</QT> thì được điều khiển xe \<TV>ô tô\</TV> chở người \<IF1>**đến 9 chỗ ngồi**\</IF1> ? (How old do people have to be to operate a passenger car with up to 9 seats?)
2	**Gold**: khi điều khiển xe chạy trên đường biết \<IF4>**có xe sau xin vượt nếu đủ điều kiện an toàn**\</IF4> người lái xe \<QT>phải làm gì\</QT> ? **Predicted**: khi điều khiển xe chạy trên đường biết \<IF4>**có xe sau xin vượt**\</IF4> nếu đủ điều kiện an toàn người lái xe \<QT>phải làm gì\</QT> ? (While driving the car on the road, what should the driver do if he knows that there is a car behind him to overtake and meet all safety conditions?)
3	**Gold**: \<ANO>Người tham gia giao thông\</ANO> \<QT>gồm những đối tượng nào\</QT> ? **Predicted**: \<TP>Người tham gia giao thông\</TP> \<QT>gồm những đối tượng nào\</QT> ? (Who are the traffic participants?)
4	**Gold**: khi điều khiển \<TV>ô tô\</TV> \<A>lên dốc cao\</A>, người lái xe cần \<QT>thực hiện các thao tác nào\</QT> ? **Predicted**: khi điều khiển \<TV>ô tô\</TV> \<IF4>lên dốc cao\</IF4>, người lái xe cần \<QT>thực hiện các thao tác nào\</QT> ? (When driving a car up a steep slope, what actions should the driver take?)

Fig. 4. Examples of error cases.

5 Conclusion

We have introduced a deep neural network model for analyzing Vietnamese legal questions, a key step towards building an automatically question answering system in the legal domain. By utilizing recent advanced techniques in the NLP and deep learning research communities, our model can correctly extract 16 types of important information. For future work, we plan to develop a question answering system for Vietnamese legal questions. Studying deep neural network models for other NLP tasks in Vietnamese language is another direction for future work.

Acknowledgements. We would like to thank FPT Technology Research Institute, FPT University for financial support which made this work possible.

References

1. Bach, N.X., Cham, L.T.N., Thien, T.H.N., Phuong, T.M.: Question analysis for vietnamese legal question answering. In: Proceedings of KSE, pp. 154–159 (2017)
2. Xuan Bach, N., Khuong Duy, T., Minh Phuong, T.: A POS tagging model for Vietnamese social media text using BiLSTM-CRF with rich features. In: Nayak, A.C., Sharma, A. (eds.) PRICAI 2019. LNCS (LNAI), vol. 11672, pp. 206–219. Springer, Cham (2019). https://doi.org/10.1007/978-3-030-29894-4_16
3. Bach, N.X., Thanh, P.D., Oanh, T.T.: Question analysis towards a Vietnamese question answering system in the education domain. Cybern. Inf. Technol. **20**(1), 112–128 (2020)
4. Bojanowski, P., Grave, E., Joulin, A., Mikolov, T.: Enriching word vectors with subword information. TACL **5**, 135–146 (2017)
5. Devlin, J., Chang, M.W., Lee, K., Toutanova, K.: BERT: pre-training of deep bidirectional transformers for language understanding. In: Proceedings of NAACL-HLT, pp. 4171–4186 (2019)
6. Dozat, T., Manning, C.D.: Deep biaffine attention for neural dependency parsing. In: Proceedings of ICLR (2017)
7. Duong, H.-T., Ho, B.-Q.: A Vietnamese question answering system in Vietnam's legal documents. In: Saeed, K., Snášel, V. (eds.) CISIM 2014. LNCS, vol. 8838, pp. 186–197. Springer, Heidelberg (2014). https://doi.org/10.1007/978-3-662-45237-0_19
8. Graves, A., Schmidhuber, J.: Framewise phoneme classification with bidirectional LSTM and other neural network architectures. Neural Netw. **18**(5–6), 602–610 (2005)
9. He, Z., Wang, X., Wei, W., Feng, S., Mao, X., Jiang, S.: A survey on recent advances in sequence labeling from deep learning models. arXiv preprint arXiv:2011.06727v1 (2020)
10. He, P., Liu, X., Gao, J., Chen, W.: DeBERTa: decoding-enhanced BERT with disentangled attention. In: Proceedings of ICLR (2021)
11. Hochreiter, S., Schmidhuber, J.: Long short-term memory. Neural Comput. **9**(8), 1735–1780 (1997)
12. Kien, P.M., Nguyen, H.T., Bach, N.X., Tran, V., Nguyen, M.L., Phuong, T.M.: Answering legal questions by learning neural attentive text representation. In: Proceedings of COLING, pp. 988–998 (2020)

13. Kim, Y.: Convolutional neural networks for sentence classification. In: Proceedings of EMNLP, pp. 1746–1751 (2014)
14. Lafferty, J., McCallum, A., Pereira, F.: Conditional random fields: probabilistic models for segmenting and labeling sequence data. In: Proceedings of ICML, pp. 282–289 (2001)
15. Le-Hong, P., Bui, D.T.: A factoid question answering system for Vietnamese. In: Proceedings of Web Conference Companion, Workshop Track, pp. 1049–1055 (2018)
16. LeCun, Y., Bottou, L., Bengio, Y., Haffner, P.: Gradient-based learning applied to document recognition. Proc. IEEE **86**(110), 2278–2324 (1998)
17. Li, Y., Li, Z., Zhang, M., Wang, R., Li, S., Si, L.: Self-attentive biaffine dependency parsing. In: Proceedings of IJCAI, pp. 5067–5073 (2019)
18. Liu, Y., et al.: RoBERTa: a robustly optimized BERT pretraining approach. arXiv preprint arXiv:1907.11692v1 (2019)
19. Loshchilov, I., Hutter, F.: Decoupled weight decay regularization. In: Proceedings of ICLR (2019)
20. Mikolov, T., Chen, K., Corrado, G.S., Dean, J.: Efficient estimation of word representations in vector space. In: Proceedings of ICLR (2013)
21. Mrini, K., Dernoncourt, F., Tran, Q.H., Bui, T., Chang, W., Nakashole, N.: Rethinking self-attention: towards interpretability in neural parsing. In: Proceedings of EMNLP Findings, pp. 731–742 (2020)
22. Nguyen, D.Q., Nguyen, D.Q., Pham., S.Q.: A Vietnamese question answering system. In: Proceedings of KSE, pp. 26–32 (2009)
23. Nguyen, D.Q., Verspoor, K.: End-to-end neural relation extraction using deep biaffine attention. In: Azzopardi, L., Stein, B., Fuhr, N., Mayr, P., Hauff, C., Hiemstra, D. (eds.) ECIR 2019. LNCS, vol. 11437, pp. 729–738. Springer, Cham (2019). https://doi.org/10.1007/978-3-030-15712-8_47
24. Nguyen, D.Q., Nguyen, A.T.: PhoBERT: pre-trained language models for Vietnamese. In: Proceedings of EMNLP, pp. 1037–1042 (2020)
25. Pennington, J., Socher, R., Manning, C.: GloVe: global vectors for word representation. In: Proceedings of EMNLP, pp. 1532–1543 (2014)
26. Song, X., Petrak, J., Roberts, A.: A deep neural network sentence level classification method with context information. In: Proceedings of EMNLP, pp. 900–904 (2018)
27. Sutskever, I., Vinyals, O., Le, Q.V.: Sequence to sequence learning with neural networks. In: Proceedings of NIPS (2014)
28. Tran, V.M., Nguyen, V.D., Tran, O.T., Pham, U.T.T., Ha, T.Q.: An experimental study of Vietnamese question answering system. In: Proceedings of IALP, pp. 152–155 (2009)
29. Tran, V.M., Le, D.T., Tran, X.T., Nguyen, T.T.: A model of Vietnamese person named entity question answering system. In: Proceedings of PACLIC, pp. 325–332 (2012)
30. Tran, O.T., Ngo, B.X., Le Nguyen, M., Shimazu, A.: Answering legal questions by mining reference information. In: Nakano, Y., Satoh, K., Bekki, D. (eds.) JSAI-isAI 2013. LNCS (LNAI), vol. 8417, pp. 214–229. Springer, Cham (2014). https://doi.org/10.1007/978-3-319-10061-6_15
31. Vaswani, A., et al.: Attention is all you need. In: Proceedings of NIPS, pp. 6000–6010 (2017)
32. Vu, T., Nguyen, D.Q., Nguyen, D.Q., Dras, M., Johnson, M.: VnCoreNLP: a Vietnamese natural language processing toolkit. In: Proceedings NAACL Demonstrations, pp. 56–60 (2018)

33. Yadav, V., Bethard, S.: A survey on recent advances in named entity recognition from deep learning models. In: Proceedings of COLING, pp. 2145–2158 (2018)
34. Yang, K., Deng, J.: Strongly incremental constituency parsing with graph neural networks. In: Proceedings of NeurIPS (2020)
35. Yang, S., Wang, Y., Chu, X.: A survey of deep learning techniques for neural machine translation. arXiv preprint arXiv:2002.07526v1 (2020)
36. Yu, J., Bohnet, B., Poesio, M.: Named entity recognition as dependency parsing. In: Proceedings of ACL, pp. 6470–6476 (2020)

BenAV: a Bengali Audio-Visual Corpus for Visual Speech Recognition

Ashish Pondit[1], Muhammad Eshaque Ali Rukon[1], Anik Das[2,3(✉)], and Muhammad Ashad Kabir[4]

[1] Department of Computer Science and Engineering, Chittagong University of Engineering and Technology (CUET), Chattogram 4349, Bangladesh
{u1604071,u1604023}@student.cuet.ac.bd
[2] Department of Computer Science, St. Francis Xavier University, Nova Scotia B2G 2W5, Canada
x2021gmg@stfx.ca
[3] Department of Computer Science and Engineering, Bangladesh University, Dhaka 1207, Bangladesh
[4] School of Computing, Mathematics and Engineering, Charles Sturt University, NSW 2795, Australia
akabir@csu.edu.au

Abstract. Visual speech recognition (VSR) is a very challenging task. It has many applications such as facilitating speech recognition when the acoustic data is noisy or missing, assisting hearing impaired people, etc. Modern VSR systems require a large amount of data to achieve a good performance. Popular VSR datasets are mostly available for the English language and none in Bengali. In this paper, we have introduced a large-scale Bengali audio-visual dataset, named "BenAV". To the best of our knowledge, BenAV is the first publicly available large-scale dataset in the Bengali language. BenAV contains a lexicon of 50 words from 128 speakers with a total number of 26,300 utterances. We have also applied three existing deep learning based VSR models to provide a baseline performance of our BenAV dataset. We run extensive experiments in two different configurations of the dataset to study the robustness of those models and achieved 98.70% and 82.5% accuracy, respectively. We believe that this research provides a basis to develop Bengali lip reading systems and opens the doors to conduct further research on this topic.

Keywords: Visual speech recognition · Audio-visual dataset · Lip reading · Corpus · Bengali · Deep learning

1 Introduction

Visual speech recognition, also known as lip reading, is the process of recognising what is being said from visual information alone, typically the movement of the lips while ignoring the audio signal [14]. It is useful under scenarios where the acoustic data is unavailable. It has many applications, e.g., silent speech control systems [19], improving hearing aids and biomedical authentication [2],

T. Mantoro et al. (Eds.): ICONIP 2021, LNCS 13109, pp. 526–535, 2021.
https://doi.org/10.1007/978-3-030-92270-2_45

Table 1. A summary of recent audio-visual datasets

Name	Classes	Speakers	Task	Utterences	Duration	Language
CUAVE [12]	10	36	Digits	7,000	14 min	English
Grid [5]	51	34	Sentences	34,000	28 h	English
OuluVs2 [1]	550	53	Digits, Phrases, Sent	1,350	15 h	English
LRW [4]	500	1000+	Words	400,000	100+ h	English
LRW-1000 [21]	1000	2000+	Words	718,018	57+ h	Mandarin
BenAV (our)	50	128	Words	26,300	7.3 h	Bengali

and assist people with speech impairment [15]. In addition, lip reading systems can complement acoustic speech recognition in noisy environments as the visual data is not susceptible to noise [4].

Though lip reading has a lot of applications, the automatic recognition of it is still a very difficult task. Humans have outperformed different recognition tasks (e.g., action, sound, speech, etc.) but the performance of lip reading recognition still lacks preciseness because lip reading task requires fine-grained analysis which is beyond human capabilities [7]. In recent times, however, the availability of datasets and state-of-art machine learning techniques have improved the performance of lip reading significantly [4, 21].

Visual speech recognition approaches can be divided into two categories [18]: (a) modeling words, i.e., recognizing the whole word from a series of lip motions [4], and (b) modeling visemes, i.e., the basic visual unit of speech that represents a gesture of the mouth [2]. The former category is suitable for isolated word recognition, classification and detection, whereas the latter is preferable for sentence-level classification and large vocabulary continuous speech recognition (LVCSR). Our work falls under the first category. In this category, given a video clip of fixed length that contains a sequence of lip movement data, the model predicts the whole word.

Research on lip reading recognition requires a large amount of data. The early lip reading dataset was very small in speakers and data instances, e.g., CUAVE [12]. Grid [5] was the first sentence level lip reading dataset. OuluVS2 [1] was developed for digits, phrases and sentences. In 2016, a large scale audio-visual dataset called LRW [4] was published. This dataset consists of 400,000 data and contains 500 classes. In 2019, a naturally distributed mandarin dataset, called LRW-1000 [21], was published that comprises 1000 classes in the Mandarin language. LRW and LRW-1000 are both word-level lip reading datasets, used for English and Mandarin lip reading, respectively. They are currently considered as the benchmark datasets in visual speech recognition. However, to the best of our knowledge, there is no audio-visual dataset for the Bengali language. A summary of these datasets is presented in Table 1.

Lip reading research improved a lot from early traditional approaches to recent end-to-end deep learning based models [6]. In the early stages of lip reading, due to lack of dataset, HMMs [15] were widely used. Convolutional Neural Networks (CNN) with HMMs approach [11] improved the accuracy of lip reading because CNN models are good at extracting important features. Later HMM was replaced by RNN which achieves significantly higher accuracy [6].

The BenAV audio-visual dataset has a lot of potentials. The BenAV dataset is not only limited to Bengali lip reading or Bengali visual speech recognition. Isolated word recognition task or word level lip reading is not strongly language dependent. It can be used for general lip reading tasks via transfer learning [8]. Also, BenAV can be used for silent speech control systems [19], face recognition [3], and so on. Thus, in this paper, we have focused on developing the BenAV dataset and evaluated this dataset by applying recent deep learning based models on our BenAV dataset for visual speech recognition.

The main contributions of this paper are two-fold:

(i) We have built a new Bengali Audio-Visual (BenAV) dataset that contains a lexicon of 50 words from 128 speakers with a total number of 26,300 utterances. This is the first Bengali audio-visual dataset that can be used for various research including acoustic speech recognition [16] and audio-visual speech recognition [13].
(ii) We have applied the state-of-the-art deep learning based visual speech recognition models on our BenAV dataset and evaluated their accuracy in two different configurations to provide a baseline performance of existing visual speech recognition models on our BenAV dataset.

2 Methodology

In this section, we describe the steps of building our BenAV dataset and discuss the existing machine learning models that we have applied to demonstrate a baseline performance of our BenAV dataset.

2.1 Building the BenAV Dataset

We have applied a multi-step procedure for collecting and processing the BenAV dataset. The data collection pipeline (see Fig. 1) comprises six steps as described below.

Step 1: Word Selection. First, we have selected 50 Bengali words for recording those that appear more often in our day to day life. Although the words were chosen randomly, there were two criteria while choosing the words. Each instance of our dataset is a video of one second in duration, so the words that can be uttered within one second were selected. We only choose words containing three to four Bengali characters. Among the selected words, 34 words are three Bengali characters and 16 words are four Bengali characters. Table 2 contains a list of words that were chosen to be recorded for our dataset. These 50 words were then grouped into five batches where each batch contains ten words. Then each batch of words are assigned to an individual participant for recording.

Step 2: Participant Details and Recording Process. Participation in our data collection process was voluntary. All participants were students of

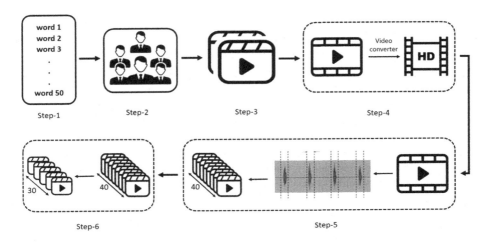

Fig. 1. Dataset development pipeline

Table 2. List of words that were chosen to be recorded for our dataset

আগুন (agun),	মানুষ (manush),	নির্বাচন (nirbachon),	মিছিল (michil),	আক্রান্ত (akranto),
নাগরিক (nagorik),	আশ্রয় (asroy),	নিকট (nikot),	বানিজ্য (banijjo),	আজব (ajob),
ব্যবসা (bebsha)	, নিরপেক্ষ (niropekkho),	বিখ্যাত (bikkhyato),	নৌকা (nouka),	বক্তব্য (boktobb),
অভিনয় (obhinoy),	বন্দর (bondor),	অবরোধ (oborodh),	বন্দুক (bonduk),	অজ্ঞাত (oggato)
বয়স (boyosh),	অনুমতি (onumoti),	চাকরী (chakri),	অতিথি (otithi),	করোনা (corona),
পায়েস (payesh),	দূরবীন (durbin),	পরিবহণ (poribohon),	ফসল (foshol),	পশ্চিম (poshchim),
ফুচকা (fuchka),	পৃথিবী (prithibi),	গবেষণা (gobeshona),	প্রকাশ (prokash),	ইতিহাস (itihash)
প্রকৃতি (prokriti)	, জনসভা (jonoshobha),	প্রশাসন (proshashon)	, কাগজ (kagoj),	প্রয়োজন (proyojon)
খবর (khobor),	সামরিক (shamorik),	কবিতা (kobita),	ঠিকানা (thikana),	কলম (kolom),
উদ্যান (uddan)	, কল্পনা (kolpona),	উপকার (upokar),	কৃষক (krishok),	উত্তর (uttor)

Bangladesh University. We had informed them of the purpose of the study and their consent were taken accordingly. Ethical approval of this study was obtained from the ethics committee of the Department of Computer Science and Engineering, Bangladesh University (ref id: BUCSEResearch_1017).

Initially, 144 speakers (116 male and 28 female) participated in our data collection. The speakers' ages were between 21 and 35 years, with a mean of 24 years. Each speaker was assigned to a batch of 10 words (described in Step 1) randomly to record utterances. Each batch of words was distributed among the speakers in such a way that we get an equal distribution of data for all of the words. Speakers are asked to record each of their assigned words by consecutively uttering it 40 times. They were asked to complete the utterance of a word within one second. The speakers were not constrained to maintain a fixed time interval between consecutive utterances, but keeping a minimum of five seconds time interval was advised. So, each speaker records ten video clips of ten words for a single batch where each recorded video contains 40 utterances of a word.

All speakers recorded their videos in their home environment. The speakers were advised to place the camera at eye level, face the light source, keep the head motionless and maintain neutral facial expression while recording.

Step 3: Video Collection and Selection. After collecting the videos from the speakers, we manually inspected all the videos and discarded those that did not meet our prerequisite, i.e., facing the light source, keeping the camera still while recording, having the face clearly visible while uttering, the subject is not too far from the camera, etc. Videos with resolutions lower than 1280×720 were also excluded. In the case of the audio, if it was extremely noisy or the speaker's voice was not clearly audible, then it was also excluded. Finally, out of 144 participants, video data of 128 speakers (male 107 and female 21) were accepted.

Step 4: Video Processing. We converted all the videos to the same format in this step. All the recorded videos' resolutions were 1280×720 or higher. Using a video converter software named FormatFactory[1], we converted all the videos into 1280×720 resolution, 16,000 video bitrate, and 30 frame per second (like other benchmark datasets such as LRW [4]).

Step 5: Video Clipping. In this step, we segment the utterances of the videos. Each video contains 40 utterances. We employed a semi-automatic screening procedure to segment these utterances. Using the pydub module[2], we analyzed the acoustic part and located the utterances in a video, and then cropped them into segments, each containing a single utterance of a word. Each video segment was one-second duration. At the end of this step, we got 40 video segments from each recorded video of a word.

Step 6: Video Clips Selection. All segmented video clips obtained in step 5 were manually checked to verify whether the video segment is contained the actual utterance without any noise. Among the group of 40 segmented video clips, the best 30 clips were selected for our dataset.

The final BenAV dataset[3] contains a lexicon of 50 words from 128 speakers (107 male and 21 female) with a total number of 26,300 utterances. The average number of speakers for each word is 18 (max 20, min 12, and standard deviation 1.826). The total duration of the dataset is 7.3 h. The number of utterances for each word is shown in Fig. 2.

2.2   Deep Learning Models for Lip Reading

A number of models [13,17,18,20–22] have been proposed in the literature for lip reading. Though lip reading research mainly focused on visual recognition, researchers found that the multimodal system which considers both audio and video data can improve the system further [13]. Stafylakis et al. [17] introduced

---

[1] http://www.pcfreetime.com/formatfactory.
[2] https://github.com/jiaaro/pydub.
[3] https://github.com/AnikNicks/BenAV-A-New-Bengali-Audio-Visual-Corpus.

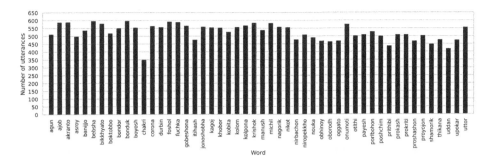

Fig. 2. Number of utterances in each word

an audio-visual speech recognition approach where they combined residual network with bidirectional LSTMs. Xu et al. [20] proposed an audio-visual model that works in two steps. In the first step, the voice is separated from the noise, and in the second step, audio is combined with visual modality to get a better performance. Yang et al. [21] experimented with different front-end and back-end for their LRW-1000 dataset. In this paper, we have only focused on visual speech recognition. Thus, in this paper, we have selected the below three models of visual speech recognition.

Stafylakis et al. [18] proposed model combines three sub-networks: (i) the front-end, which applies spatiotemporal convolution to the preprocessed frame stream, (ii) a 34-layer ResNet that is applied to each time frame, and (iii) the back-end, which is a Bidirectional LSTM network.

Zhao et al. [22] proposed a deep learning model where they introduced mutual information maximization technique on both the local feature's level and global sequence level. Firstly, they choose a baseline model. Their baseline model uses the 3D CNN layer, a deep 2D ResNet-18 and a gap layer. They used 3 layers BiGRU as the back-end. The training process involves three steps. First, they used the baseline model to train on the dataset. Second, the local information maximization was applied and the weight of the baseline model was used here. Finally, the global mutual information maximization was applied and the previous weight was used.

Petridis et al. [14] argued that their model is suitable for small-scale datasets (i.e., datasets that contain fewer classes and few data instances) such as CUAVE [12], AVLetters [10], and OuluVS2 [1]. The Model consists of two streams: one which extracts features from raw mouth images and the other one extracts features from different images (i.e., the difference between two consecutive video frames). The model is based on fully connected layers and Bidirectional Long Short Term Memory (BiLSTM) networks. CNNs were not used as encoders in this model because the author claimed that CNN models do not reach their full potential if the dataset is not large enough and, in such cases, the fully connected layers can be better suited.

Table 3. Dataset settings. Config. 1 – same speaker in training, validation and testing for each word, Config. 2 – different speaker in training and testing for each word

Config. No.	Train set	Validation set	Test set
1	60% data of each speaker	20% data of each speaker	20% data of each speaker
2	80% data of 80% speakers	20% data of 80% speakers	100% data of 20% speakers

We have applied these models to our dataset to provide a baseline accuracy. A comparative analysis of these models has been presented in Sect. 3. Among these three models, the implementation was available only for Zhao et al. [22]. We have implemented other two models [14,18] and selected the model hyperparameters (e.g., learning rate, batch size, optimizer, etc.) as described in those papers.

3   Experimental Evaluation

3.1   Dataset Settings

We conducted experiments in two different configurations of the dataset to evaluate the performance of the selected VSR models in different levels of difficulty.

In configuration 1, video data from each speaker is split into 60%, 20%, and 20% among train, test, and validation sets respectively. The purpose of this configuration is to evaluate the model performance in recognizing the same speaker's data that the model was trained for.

Configuration 2 is more challenging for VSR models as it evaluates their ability to recognize a word uttered by an unknown speaker. In this configuration, for each word, 20% speakers' data are kept in the test set and the remaining speakers' data are split into 80% and 20% among train and validation sets respectively. For example, if a word has 15 speakers, and there are 30 videos for each speaker (i.e., a total of 450 data instances) then 20% of 15 speakers is 3 speakers. So we separate these 3 speakers' data (i.e., a total of 90 data instances) for testing. In this dataset configuration, train and test speakers are totally different for each word. The purpose of this configuration is to train the model on some speakers' data and test the performance of the model on new speakers' data. Dataset configuration 1 and 2 are presented in Table 3.

3.2   Dataset Preprocessing

Using OpenCV[4], each video clip is first converted to a sequence of grayscale images. After performing face detection on each frame of the sequence, the Dlib facial point tracker [9] is used to track sixty-eight points on the face. Using the tracked mouth points, the mouth ROI (region of interest) is cropped which is then downscaled according to the requirements of the VSR models. Zhao et al. [22] and Petridis et al. [14] models require input of 96 × 96, and Stafylakis et al. [18] model requires 112 × 112 input shape. Figure 3 shows a mouth sequence collected from a data instance.

---

[4] https://github.com/itseez/opencv.

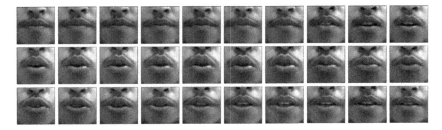

Fig. 3. A sample mouth sequence of lip movement for the word " অনুমতি (onumoti)"

3.3 Results and Discussion

Table 4 reports the accuracy of the two configurations for different models. The results show that Zhao et al. [22] model outperforms all other models in both configurations. It uses the mutual information maximization (MIM) technique which helps the model to capture effective spatial features. Because of this fine-grained analysis, this model achieves the best results.

For configuration 2 which is a difficult setting, the accuracy of Stafylakis et al. [18] model is significantly less than the other two models. Stafylakis et al. [18] model uses 3D CNN, ResNet-34, and LSTM based architecture that fails to capture discriminative features. The accuracy of Petridis et al. [14] model is 20.06%. The model used raw images and the difference between two images for more robust classification. Although the basic Zhao et al. [22] model obtained the best accuracy (31.76%) compared to others, it is quite low to accept. Thus, we used pre-trained weight from the LRW dataset to further improve its accuracy. It results a significant improvement in accuracy, from 31.76% to 82.5%.

Overall, we can observe a difference between the performance of configurations 1 and 2. The reason behind the high accuracy of configuration 1 for all models is that the train, test, and validation sets contain the same speakers. But it was totally different for configuration 2 as the unseen speakers' were in the train set.

Table 4. Comparison of accuracy of the two configurations for different models

Configuration	Model	Accuracy (%)
Configuration 1	Stafylakis et al. [18]	93.00
	Petridis et al. [14]	95.31
	Zhao et al. [22]	**98.70**
Configuration 2	Stafylakis et al. [18]	2.40
	Petridis et al. [14]	20.06
	Zhao et al. [22]	31.76
	Zhao et al. [22]+pretrained on LRW	**82.50**

Table 5. Most frequent similar word pairs. Here, (word1 ↔ word2) means model wrongly predicts word2 while predicting word1, and vice versa.

(asroy ↔ ajob)	(krishok ↔ nikot)	(bonduk ↔ bondor)	(boyosh ↔ bondor)
(otithi ↔ uddan)	(nikot ↔ nagorik)	(corona ↔ kolpona)	(oborodh ↔ bondor)
(bikkhyato ↔ banijjo)	(boktobbo ↔ bondor)		

One of the main challenges in lip reading is the ambiguity that arises in word-level lip reading due to homophemes (i.e., characters that have similar lip movement). Among the 50 words that we used for our classification, there are some pair of words which have similar lip movement. We observed the confusion matrix of all the models and focused on incorrectly identified word pairs (i.e., off-diagonal words in the confusion matrix). A list of frequently confused word pairs is presented in Table 5. This similarity among words was common among all the models and configurations but the frequency was different.

4   Conclusion

We have developed a new Bengali audio-visual dataset, named BenAV, for automatic lip reading. To the best of our knowledge, this is the first audio-visual dataset for the Bengali language. Our dataset contains 26,300 utterances of 50 words from 128 speakers. We have used a semi-automatic pipeline for creating a standard dataset. We have experimented with three lip reading models in two different dataset configurations to evaluate their performance on our dataset. The comparative analysis of these models shows the potentiality of the BenAV dataset for Bengali lip reading systems.

References

1. Anina, I., Zhou, Z., Zhao, G., Pietikäinen, M.: Ouluvs2: a multi-view audiovisual database for non-rigid mouth motion analysis. In: 11th International Conference and Workshops on Automatic Face and Gesture Recognition, vol. 1, pp. 1–5. IEEE (2015)
2. Assael, Y.M., Shillingford, B., Whiteson, S., de Freitas, N.: Lipnet: End-to-end sentence-level lipreading. arXiv (2016)
3. Barr, J.R., Bowyer, K.W., Flynn, P.J., Biswas, S.: Face recognition from video: a review. Int. J. Patt. Recogon. Artif. Intell. **26**(05), 1266002 (2012)
4. Chung, J.S., Zisserman, A.: Lip reading in the wild. In: Lai, S.-H., Lepetit, V., Nishino, K., Sato, Y. (eds.) ACCV 2016. LNCS, vol. 10112, pp. 87–103. Springer, Cham (2017). https://doi.org/10.1007/978-3-319-54184-6_6
5. Cooke, M., Barker, J., Cunningham, S., Shao, X.: An audio-visual corpus for speech perception and automatic speech recognition. J. Acoust. Soc. Am. **120**(5), 2421–2424 (2006)
6. Fernandez-Lopez, A., Sukno, F.M.: Survey on automatic lip-reading in the era of deep learning. Image Vis. Comput. **78**, 53–72 (2018)

7. Hilder, S., Harvey, R.W., Theobald, B.J.: Comparison of human and machine-based lip-reading. In: International Conference on Auditory-Visual Speech Processing (AVSP), pp. 86–89. ISCA (2009)
8. Jitaru, A.C., Abdulamit, Ş., Ionescu, B.: LRRO: a lip reading data set for the under-resourced Romanian language. In: Proceedings of the 11th ACM Multimedia Systems Conference, pp. 267–272 (2020)
9. Kazemi, V., Sullivan, J.: One millisecond face alignment with an ensemble of regression trees. In: IEEE Conference on Computer Vision and Pattern Recognition, pp. 1867–1874. IEEE (2014)
10. Matthews, I., Cootes, T.F., Bangham, J.A., Cox, S., Harvey, R.: Extraction of visual features for lipreading. IEEE Trans. Patt. Anal. Mach. Intell. **24**(2), 198–213 (2002)
11. Noda, K., Yamaguchi, Y., Nakadai, K., Okuno, H.G., Ogata, T.: Lipreading using convolutional neural network. In: 15th Annual Conference of the International Speech Communication Association, pp. 1149–1153. ISCA (2014)
12. Patterson, E.K., Gurbuz, S., Tufekci, Z., Gowdy, J.N.: Cuave: A new audio-visual database for multimodal human-computer interface research. In: International Conference on Acoustics, Speech, and Signal Processing (ICASSP), vol. 2, pp. II-2017-II-2020. IEEE (2002)
13. Petridis, S., Stafylakis, T., Ma, P., Cai, F., Tzimiropoulos, G., Pantic, M.: End-to-end audiovisual speech recognition. In: IEEE International Conference on Acoustics, Speech and Signal Processing (ICASSP), pp. 6548–6552. IEEE (2018)
14. Petridis, S., Wang, Y., Ma, P., Li, Z., Pantic, M.: End-to-end visual speech recognition for small-scale datasets. Patt. Recogn. Lett. **131**, 421–427 (2020)
15. Puviarasan, N., Palanivel, S.: Lip reading of hearing impaired persons using HMM. Exp. Syst. Appl. **38**(4), 4477–4481 (2011)
16. Sak, H., Senior, A., Rao, K., Beaufays, F.: Fast and accurate recurrent neural network acoustic models for speech recognition. In: 16th Annual Conference of the International Speech Communication Association, pp. 1468–1472. ISCA (2015)
17. Stafylakis, T., Khan, M.H., Tzimiropoulos, G.: Pushing the boundaries of audio-visual word recognition using residual networks and LSTMS. Comput. Vis. Image Underst. **176**, 22–32 (2018)
18. Stafylakis, T., Tzimiropoulos, G.: Combining residual networks with LSTMs for lipreading. In: 18th Annual Conference of the International Speech Communication Association (INTERSPEECH), pp. 3652–3656. ISCA (2017)
19. Sun, K., Yu, C., Shi, W., Liu, L., Shi, Y.: Lip-interact: improving mobile device interaction with silent speech commands. In: Proceedings of the 31st Annual ACM Symposium on User Interface Software and Technology, pp. 581–593. ACM (2018)
20. Xu, B., Lu, C., Guo, Y., Wang, J.: Discriminative multi-modality speech recognition. In: Proceedings of the IEEE/CVF Conference on Computer Vision and Pattern Recognition (CVPR), pp. 14433–14442. IEEE (2020)
21. Yang, S., et al.: LRW-1000: a naturally-distributed large-scale benchmark for lip reading in the wild. In: 14th IEEE International Conference on Automatic Face and Gesture Recognition (FG), pp. 1–8. IEEE (2019)
22. Zhao, X., Yang, S., Shan, S., Chen, X.: Mutual information maximization for effective lip reading. In: 15th International Conference on Automatic Face and Gesture Recognition (FG), pp. 420–427. IEEE (2020)

Investigation of Different G2P Schemes for Speech Recognition in Sanskrit

C. S. Anoop[(✉)] [iD] and A. G. Ramakrishnan[iD]

Indian Institute of Science, Bengaluru, Karnataka 560012, India
{anoopcs,agr}@iisc.ac.in

Abstract. In this work, we explore the impact of different grapheme to phoneme (G2P) conversion schemes for the task of automatic speech recognition (ASR) in Sanskrit. The performance of four different G2P conversion schemes is evaluated on the ASR task in Sanskrit using a speech corpus of around 15.5 h duration. We also benchmark the traditional and neural network based Kaldi ASR systems on our corpus using these G2P schemes. Modified Sanskrit library phonetic (SLP1-M) encoding scheme performs the best in all Kaldi models except for the recent end-to-end (E2E) models trained with flat-start LF-MMI objective. We achieve the best results with factorized time-delay neural networks (TDNN-F) trained on lattice-free maximum mutual information (LF-MMI) objective when SLP1-M is employed. In this case, SLP1 achieves a word error rate (WER) of 8.4% on the test set with a relative improvement of 7.7% over SLP1. The best E2E models have a WER of 13.3% with the basic SLP1 scheme. The use of G2P schemes employing schwa deletion (as in Hindi, which uses the same Devanagari script as Sanskrit) degrades the performance of GMM-HMM models considerably.

Keywords: G2P · ASR · Sanskrit · LF-MMI · E2E · Kaldi

1 Introduction and Motivation for the Study

Sanskrit is one of the oldest languages known to the human race. It is one of the 22 scheduled languages listed in the eighth schedule to the Indian constitution and has official status in the states of Himachal Pradesh and Uttarakhand. It is the only scheduled language in India with less than 1 million native speakers. Though not used in active communication currently, Sanskrit has a lasting impact on the languages of South Asia, Southeast Asia, and East Asia, especially in their formal and learned vocabularies. Many of the modern-day Indo-Aryan languages directly borrow the grammar and vocabulary from Sanskrit. In addition, it encompasses a vast body of literature in various areas spanning from mathematics, astronomy, science, linguistics, mythology, history, and mysticisms.

Though Sanskrit studies bear importance due to historical and cultural reasons, very little focus has been put on developing necessary computational tools to digitize the vast body of literature available in the language. Specifically,

© Springer Nature Switzerland AG 2021
T. Mantoro et al. (Eds.): ICONIP 2021, LNCS 13109, pp. 536–547, 2021.
https://doi.org/10.1007/978-3-030-92270-2_46

building a large vocabulary automatic speech recognition (ASR) system can expedite the digitization of a large volume of literature available across various written forms in Sanskrit. However, a major challenge in building such a system for any Indian language is the lack of readily available training data. Furthermore, an ASR system for Sanskrit poses additional challenges like the high rate of inflection, free word order, highly intonated speech, and large word lengths due to sandhi. These issues make speech recognition in Sanskrit both unique and challenging compared to other spoken languages.

Despite the above-mentioned issues, there have been a few attempts to build ASR systems for Sanskrit in recent times. [3] employ Gaussian mixture model (GMM) - hidden Markov model (HMM) based approaches using HMM toolkit (HTK) [28]. Recently [1] has attempted Sanskrit speech recognition using Kaldi [23] time-delay neural network (TDNN) models [19]. End-to-end speech recognition with connectionist temporal classification (CTC) objective [10] has been performed on Sanskrit with limited amount of data using spectrogram augmentation [18] techniques in [4]. [2] employs domain adaptation approaches to learn the acoustic models for Sanskrit from the annotated data in Hindi.

Pronunciation modeling is one of the major components of all traditional ASR systems. Sanskrit possesses alphabets that are largely designed to avoid multiple pronunciations for the same alphabet. Thus, there is a one-to-one correspondence between the alphabets and the corresponding sound units in most cases. However, there are a few instances where the context modifies the pronunciation of the alphabet. The presence of *chandra bindu* (̐), *anusvāra* (ं), *visarga*(ः), *nukta*(़) or *virama*(्) in the alphabet alters the pronunciation.

The pronunciation rules for handling them are described in [3]. Sanskrit is normally written in Devanagari script, just like Hindi. Hence the pronunciation variations due to *virama* are handled in most grapheme to phoneme (G2P) converters for Devanagari, as they also occur in Hindi. However, most of the G2P schemes consider *anusvara* and *visarga* as individual phonemes without considering their context. They also appear quite frequently in Sanskrit text. How the performance of the ASR for Sanskrit will be impacted when these contextual pronunciation rules are incorporated into the G2P converter is unclear. Also, being a more popular Indian language, Hindi has many off-the-shelf G2P conversion tools [5,17]. One might be tempted to use the same tools for Sanskrit as they both share the same orthography. However, Hindi G2P conversion tools are designed to handle schwa deletion, a phenomenon where the inherent vowel in the orthography is not pronounced in some contexts. The impact of using such G2P tools for speech recognition in Sanskrit has not been studied so far. In this work, we address the above issues and investigate the performance of four different G2P schemes on Sanskrit ASR. We evaluate their performance on multiple Kaldi-based ASR systems on a Sanskrit dataset of 15.5 h duration that we have collected.

The remaining part of the paper is organized as follows: Sect. 2 introduces the Sanskrit dataset used in this work. Section 3 gives a theoretical overview of different speech recognition systems in Kaldi, which we explore for the evaluation of G2Ps. Section 4 describes the setup used for our experiments. This section also details the four G2P schemes and the Kaldi systems used. Results of our experiments are discussed in Sect. 5 followed by conclusions in Sect. 6.

2   Sanskrit Speech Dataset Collected for the Study

The Sanskrit dataset we use in this work has 15.5 h of speech data consisting of 7900 utterances from 41 speakers. The speech data is mainly from 3 domains: (a) news recordings, (b) recordings of short stories, and (c) video lectures. Speech data is in raw wav file format with 16 kHz sampling frequency and 16 bits per sample. There are 41 speakers: 21 male and 20 female. Each audio file contains recordings from a single speaker. The corpus contains around 15100 unique words. The dataset is split into train and test subsets with approximately 12 and 3.5 h of data, respectively, with no overlap between speakers in the subsets.

The details of the dataset are given in Table 1.

Table 1. Details of the Sanskrit speech dataset collected by us from the recordings of news, short stories, and lectures, for our ASR experiments. The "Words" column gives the number of unique words in the subsets.

	Details		Speakers			Duration		
Data	Utterances	Words	Total	Male	Female	Total	Male	Female
Train	6067	11537	16	8	8	11:54:25	5:22:16	6:32:09
Test	1813	5145	25	13	12	3:28:52	1:44:12	1:44:39

3   Summary of Different Kaldi-Based ASR Systems

ASR can be treated as a sequence detection problem which maps a sequence of T-length input acoustic features $O = \{\mathbf{o}_t \in \mathcal{R}^m, t = 1, 2, \ldots, T\}$ to an L-length word sequence $W = \{\mathbf{w}_l \in V, l = 1, 2, \ldots, L\}$. $\mathbf{o}_t$ denote the m-dimensional acoustic feature vector at time t, $\mathbf{w}_l$ denote word at position l and V denote the vocabulary in that language. The ASR estimates the most probable word sequence $\hat{W}$ given O.

For ASR systems employing HMMs [24],

$$\hat{W} = \arg\max_{W} P(W|O) \tag{1}$$

$$= \arg\max_{W} \sum_{S} P(O|S, W) P(S|W) P(W) \tag{2}$$

$$\approx \arg\max_{W} \sum_{S} P(O|S) P(S|W) P(W) \tag{3}$$

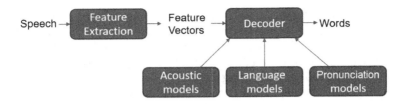

Fig. 1. Block diagram of an ASR.

where S denote the underlying HMM state sequence $S = \{s_t \in \mathcal{S}, t = 1, 2, \ldots, T\}$, $\mathcal{S}$ being the set of all possible HMM states. The factors $P(O|S)$, $P(S|W)$, and $P(W)$ in (3) are called the acoustic models, pronunciation models, and language models (LM), respectively [26]. A block diagram of an ASR depicting each of these components are shown in Fig. 1.

The acoustic models can be further factorised as:

$$P(O|S) = \prod_{t=1}^{T} P(\mathbf{o}_t|\mathbf{o}_1, \mathbf{o}_2, \ldots, \mathbf{o}_{t-1}, S) \approx \prod_{t=1}^{T} P(\mathbf{o}_t|s_t) \qquad (4)$$

The inner term $P(\mathbf{o}_t|s_t)$ represents the probability of observing the feature vector $\mathbf{o}_t$ given that the HMM state is $\mathbf{s}_t$ at time t. Traditional GMM-HMM based systems make use of GMM to model this probability. With the arrival of deep neural networks (DNN) for acoustic modeling in speech [12] GMMs made way for multi-layer perceptron (MLP) classifiers which computed the frame-wise posterior $P(s_t|\mathbf{o}_t)$. Frame-wise likelihood $P(\mathbf{o}_t|s_t)$ is then approximated as $\frac{P(s_t|\mathbf{o}_t)}{P(s_t)}$ using the MLP output. However, to obtain the HMM state (senone) labels needed to train MLP, they still depended on the alignments from a GMM-HMM system. These alignments are also used to compute the priors on the senones.

The pronunciation model can be factorized as:

$$P(S|W) = \prod_{t=1}^{T} P(s_t|s_1, s_2, \ldots, s_{t-1}, W) \approx \prod_{t=1}^{T} P(s_t|s_{t-1}, W) \qquad (5)$$

HMM state transition probabilities model this term. HMMs are usually prepared at the sub-word level (phonemes/triphones).

n-gram models are used for language modeling.

$$P(W) \approx \prod_{l=1}^{L} P(w_l|w_{l-n+1}, \ldots, w_{l-1}) \qquad (6)$$

Decoding typically employs weighted finite state transducers (WFST) [16]. A decoding graph, normally referred to as HCLG, is created as:

$$S \equiv HCLG = min(det(H \ o \ C \ o \ L \ o \ G)) \qquad (7)$$

where o, min and det are the FST operations; composition, minimisation and determinization and S is the search graph. Here, H, C, L, and G denote HMM structure, context-dependency of phonemes, lexicon, and grammar. In this graph, each arc has input label as an identifier for the HMM state and output label as a word or ϵ (indicating no label). The weight typically represents the negative log probability of the state. Decoding finds the best path through this graph. Usually, the search space is huge, and beam pruning [13] is employed instead of an exact search, where all the paths not within the beam are discarded.

The assumption in HMM that the observations are independent of past/future phone states, given the current state does not hold in practice. Due to the incorrectness in the model maximum likelihood may not be an optimal training criteria. A lattice-free version of maximum mutual information (MMI) objective [6] for training neural network acoustic models was proposed in [22]. MMI is a sequence-level discriminative criteria given by:

$$\mathcal{F}_{MMI}(\theta) = \sum_{u=1}^{U} log \frac{P_\theta(O_u|W_u)P(W_u)}{\sum_W P_\theta(O_u|W)P(W)} \tag{8}$$

Here U and θ represent the total number of utterances and the model's parameters, respectively. The objective tries to maximize the probability of the reference transcription while minimizing the probability of all other transcriptions. Both numerator and denominator state sequences are represented as HCLG FSTs. The denominator forward-backward computation is parallelised on a graphical processing unit (GPU). They replace word-level LM with a 4-gram phone LM for efficiency. They also use 2-state skip HMMs for phones, where the skip connections allowed for sub-sampling of frames at the DNN output.

Regular LF-MMI still depends on the alignments from GMM-HMM models. The output of the network is tied triphone/biphone states. The tying is done using a decision-tree built from GMM-HMM alignments. Flat-start LF-MMI [11] removes this dependency and makes the training completely end-to-end (E2E). It trains the DNN in a single stage without going through the usual GMM-HMM training and tree-building pipeline. Unlike regular LF-MMI, there is no prior alignment information in the numerator graph here. HMM transitions and state priors are fixed and uniform as in the regular case. Context dependency modeling is achieved by using full left-biphones, where separate HMMs are assumed for each possible pair of phonemes. The phoneme language models for the denominator graph are estimated using the training transcriptions and pronunciation dictionary.

4   Setup for the ASR Experiments

In this work, we investigate the performance of different G2P conversion schemes in the ASR task in Sanskrit. We also benchmark different Kaldi-based systems on our Sanskrit dataset described in Sect. 2. We explore the GMM-HMM, MLP, and factored time-delay neural network (TDNN-F) [21] architectures. MLP architectures are trained with frame-level cross-entropy objective. Training of TDNN-F models is explored with both regular and flat-start LF-MMI objectives. All the experiments are performed using Kaldi toolkit [23].

4.1 Extraction of the MFCC-Based Features

In our experiments, we use Mel-frequency cepstral coefficients (MFCC) as features. They are extracted from frames with a window size of 25 ms and frameshifts of 10 ms. For the training of the GMM-HMM and MLP models, low-resolution MFCC features with 13 coefficients are used. Splicing with left and right contexts of 9 gives a 247-dimensional vector at the input of the MLP. High-resolution features with 40 coefficients are used for training the TDNN-F models with the LF-MMI objective. They also use 100-dimensional i-vectors [7] computed from chunks of 150 consecutive frames. In the E2E scheme, only high-resolution MFCC features are used as features. With three consecutive MFCC vectors, the input dimensions to the regular and E2E TDNN-F networks are 220 and 120, respectively.

4.2 The Different Pronunciation Models Studied

A pronunciation lexicon maps the words to the corresponding phoneme sequence with the help of G2P converters. We investigate four G2P schemes for Sanskrit pronunciation modeling and evaluate their relative performance in speech recognition. They are:

1. Epitran [17]: Epitran is a massive multilingual G2P system supporting 61 languages, including Hindi. Though this system supports the Devanagari script, it is primarily designed to handle the schwa deletions in Hindi. Performance evaluation of the Sanskrit ASR using this scheme can give an idea about the negative impact of using Hindi G2P converters for Sanskrit.
2. Indian language speech sound label set (ILSL12) [14]: ILSL12 provides a standard set of labels (in Roman script) for speech sounds commonly used in Indian languages. Similar sounds in different languages are given a single label. It is commonly used in multilingual speech recognition systems in Indian languages.
3. Sanskrit library phonetic (SLP1) basic encoding scheme [25]: SLP1 can represent phonetic segments, phonetic features, and punctuation in addition to the basic Devanagari characters. SLP1 also describes how to encode classical and Vedic Sanskrit.
4. SLP1 modified with specific pronunciation rules for *anusvāra* and *visarga* (SLP1-M): In SLP1-M, we incorporate the pronunciation modification rules in Sanskrit to the basic SLP1 scheme hoping for a better ASR performance. A few examples for indicating the differences in the encoding schemes are shown in Table 2. In row 1, Epitran removes the final schwa, whereas the other three schemes retain it. Also note that the *anusvāra* is followed by a velar consonant, क (ka) and hence it is modified to "N" denoting the velar resonant ङ (nga) in SLP1-M. In row 2, SLP1-M modifies the final *visarga*

Table 2. G2P schemes evaluated for speech recognition in Sanskrit.

Sl. No.	Devanagari	Epitran	ILSL12	SLP1	SLP1-M
1	शंकर	ʃ ə ŋ k ə r	sh a q k a r a	S a M k a r a	S a N k a r a
2	गुरुः	g u r u ə h	g u r u hq	g u r u H	g u r u h u

(ೲ:), by adding a voiced echo of the previous vowel. ILSL12 and SLP1 use separate phonemes for the *visarga*. Epitran maps it to "h", which denotes the ह (h) sound.

4.3 Training of the Language Models

We use bigram language models trained with a large text corpus using IRSTLM toolkit [8]. The transcriptions of the speech data are extended with the text data from Sanskrit data dump [27] in wiki and from several websites. The collected text is filtered such that the sentences contain only Devanagari Unicode. There were 244652 sentences with 413828 distinct words in the final text corpus used for language modeling.

4.4 Training of the Different Acoustic Models

GMM-HMM Models: The development of GMM-HMM systems involves step-wise refinement of models. At each step, the complexity of the model is increased to refine the alignments of the training data with the model and passed to next step. Initially, context-independent monophone HMMs are built. There are 218 position-dependent phones (i.e., phones marked with their word-internal positions - beginning/ending/standalone/internal). An equal-alignment scheme is used for initial alignments, and the parameters are learned using maximum likelihood estimation.

Next, context-dependent phone (triphone) models are constructed using the left and right contexts of phones. To avoid the explosion in the number of HMM states due to the triphone context, a decision tree is trained to cluster the states that are acoustically similar. This way, the same acoustic model is shared across multiple HMM states. The application of the following transforms further refines these models:

- Linear discriminant analysis (LDA) + maximum likelihood linear transform (MLLT) [9]: Here MFCC frames are spliced with left and right contexts of 3 each, and LDA is applied to reduce the dimension to 40. Now the diagonalizing transform MLLT is estimated over multiple iterations.
- Speaker adaptive training (SAT): This trains a feature-based maximum likelihood linear regression (fMLLR) transform to normalize the features such that they better fit the speaker.

Subspace Gaussian mixture models (SGMM) [20] can compactly represent a large collection of GMMs. Here model parameters are derived from a set of state-specific parameters, and from a set of globally-shared parameters which can capture phonetic and speaker variations.

Hybrid DNN-HMM Models with MLP Architecture: We use a multilayer perceptron (MLP) with seven hidden layers and 1024 hidden nodes per layer. The output layer has 2576 nodes corresponding to the senones in the final GMM-HMM model. We randomly select 10% of the training data as the validation set. The network is trained with frame-level cross-entropy (CE) objective.

TDNN-F Models with LF-MMI Objective: We use TDNN-F models for training with both regular and flat-start LF-MMI objectives. They have similar structure as TDNN, but are trained with the constraint that one of the two factors of each weight matrix should be semi-orthogonal. TDNN-F blocks have a bottleneck architecture with a linear layer followed by an affine layer. The linear layer transforms the hidden-layer dimension to a lower bottleneck dimension, and the affine layer transforms it back to the original dimension. In regular LF-MMI training, the hidden and bottleneck dimensions are 768 and 96, respectively. In flat-start LF-MMI (E2E) scheme, we use hidden and bottleneck dimensions of 1024 and 128, respectively. Linear and affine layers are followed by ReLU activation and batch normalization. They also have a kind of residual connection where the output of the current block is added to the down-scaled (by a factor of 0.66) output of the previous block. There are 12 such blocks. The final output layer has dimensions of 1424 and 108, respectively, for the regular and flat-start LF-MMI networks. Speed and volume perturbations [15] are used in regular LF-MMI training. In the E2E scheme, all the training utterances are modified to be around 30 distinct lengths. Utterances with the same lengths are placed in the same mini-batch during training. Speed perturbation is used to modify the length of each utterance to the nearest of the distinct lengths.

5   Experimental Results

We use word error rate (WER) as the metric to assess the performance of the Kaldi-based systems. WER is defined as:

$$WER = (I + D + S)/N \qquad (9)$$

where I, D, S, and N denote the number of insertions, deletions, substitutions, and total words in the test set. $N = 14135$ in our case.

The results of speech recognition experiments for different G2P conversion schemes with the conventional GMM-HMM models are listed in Table 3. The well-curated SLP1-M scheme provides the best results across all the GMM-HMM models. The performance of ILSL12 and SLP1 are almost similar as they are just direct mappings between the orthography and the phonology, except for the

Table 3. The performance (WER in %) of GMM-HMM models on the test set using different G2P schemes for Sanskrit.

No.	Model	Epitran	ILSL12	SLP1	SLP1-M
M1	Mono	27.6	24.4	23.9	**23.0**
M2	Tri	21.1	18.2	18.1	**17.3**
M3	Tri+LDA+MLLT	22.6	20.8	20.0	**18.7**
M4	Tri+LDA+MLLT+SAT	23.3	20.7	20.6	**19.8**
M5	SGMM	17.0	15.7	15.8	**15.4**

differences in handling the pitch accents of Vedic Sanskrit. There is considerable deterioration in the performance of Epitran over the other G2P converters. The minimal degradation ranges from around 1.2% in SGMM models to around 3.2% in monophone HMM models. Though Epitran supports Devanagari graphemes, it is primarily designed to support Hindi, where the schwa deletion is prominent. However, schwa deletion is not present in Sanskrit, and hence its performance is slightly worse than the other G2P conversion schemes. Among the GMM-HMM models, SGMM provides the best results across all the G2P converters. They provide an improvement of at least 1.9% over the best triphone models. All the triphone models are better than the monophone models. The application of LDA+MLLT and SAT does not improve the results. We have assumed that each utterance in the training set belongs to a different speaker to have the effect of more number of speakers in the i-vector training for TDNN-F models. This could be the reason for the poor performance of SAT.

Table 4. The performance (WER in %) of different neural network models on the test set using different G2P schemes for Sanskrit.

No.	Model	Objective function	Model size	Epitran	ILSL12	SLP1	SLP1-M
M6	MLP	CE	9.0 M*	22.6	22.3	21.9	**20.0**
M7	TDNN-F	Regular LF-MMI	6.8 M	9.4	9.4	9.1	**8.4**
M8	E2E	Flat-start LF-MMI	4.9 M	14.9	14.6	**13.3**	14.7

*M stands for million.

The results of neural network models are listed in Table 4. SLP1-M gives the best results for the MLP and TDNN-F models. The best results for E2E models are achieved using the basic SLP1 scheme. E2E models employ full left-biphones for context dependency modeling. Retaining separate phonemes for *visarga* (◌ः) and *anusvāra* (◌ं) seems to help the basic SLP1 scheme in better modeling the biphone HMMs having *visarga/anusvara* as one of the phonemes. Surprisingly the Epitran G2P scheme is not as bad as in the case of GMM-HMM models, when it comes to the neural network models. They provide comparable results with ILSL12. TDNN-F architectures (rows 2 and 3 of Table 4)

Table 5. Performance (WER in %) of the SLP1-M scheme when a separate schwa phoneme is used for the vowel inherent in the consonant characters.

G2P scheme	M1	M2	M3	M4	M5	M6	M7	M8
SLP1-M	23.0	17.3	18.7	19.8	15.4	20.0	8.4	14.7
SLP1-M+schwa phoneme	23.1	18.3	20.6	20.7	15.8	21.4	8.7	14.2

have better performances and smaller model sizes compared to the normal MLP architectures. TDNN-F model trained with regular LF-MMI objective gives the best performance on the corpus with the SLP1-M G2P scheme. However, they depend on GMM-HMM models for the initial alignments and tree building for tying states. On the other hand, the E2E scheme eliminates such dependencies on GMM-HMM models and makes the LF-MMI training flat-start. Also, the i-vector extraction process is not required in the E2E scheme. Thus they can be trained easily. They provide around 5.3–8.6% improvement over the MLP with all the G2P schemes on the Sanskrit dataset. However, when compared to the regular LF-MMI training, their performance is worse by at least 4.2%.

We further extend the SLP1-M with a separate phoneme label for schwa, the vowel inherent in the consonant characters; i.e., we use a different label "ə" instead of mapping them to "a", the label for the vowel अ . The results of this are shown in Table 5. The first row of this Table is replicated from Tables 3 and 4. WER degrades for all the models in this scheme except for the E2E models.

6 Conclusions

This work explores the significance of G2P conversion schemes on the speech recognition task in Sanskrit. We evaluate the performance of four different G2P conversion schemes, viz. Epitran, ILSL12, SLP1, and SLP1-M (SLP1 modified to include some contextual pronunciation rules) using traditional and neural network-based Kaldi models. SLP1-M performs the best among all the models except E2E. For E2E models, basic SLP1 performs the best. SLP1-M brings about some improvement in MLP and TDNN-F models. The relative improvements in WER over SLP1 are 8.7% and 7.7% for MLP and the state-of-the-art TDNN-F models trained with LF-MMI objective, respectively. Epitran scheme, which employs schwa deletion, deteriorates the performance of GMM-HMM models. TDNN-F models trained with LF-MMI objective perform the best among all the Kaldi models. They provide a WER of 8.4% on the Sanskrit test set with SLP1-M. E2E models with flat-start LF-MMI objective achieve a WER of 13.3% with the basic SLP1. However, the WER performance of E2E models is worse by at least 4.2% than the TDNN-F models trained with the regular LF-MMI objective.

Acknowledgments. We thank Science and Engineering Research Board, Government of India for partially funding this research through the IMPRINT2 project, IMP/2018/000504.

References

1. Adiga, D., Kumar, R.A.K., Jyothi, P., Ramakrishnan, G., Goyal, P.: Automatic speech recognition in Sanskrit: a new speech corpus and modelling insights. In: 59th Annual Meeting of the Association for Computational Linguistics, ACLFindings (2021), pp. 5039–5050. https://doi.org/10.18653/v1/2021.findings-acl.447
2. Anoop, C.S., Prathosh, A.P., Ramakrishnan, A.G.: Unsupervised domain adaptation schemes for building ASR in low-resource languages. In: Proceedings of Workshop on Automatic Speech Recognition and Understanding, ASRU (2021)
3. Anoop, C.S., Ramakrishnan, A.G.: Automatic speech recognition for Sanskrit. In: 2nd International Conference on Intelligent Computing, Instrumentation and Control Technologies, ICICICT, vol. 1, pp. 1146–1151 (2019). https://doi.org/10.1109/ICICICT46008.2019.8993283
4. Anoop, C.S., Ramakrishnan, A.G.: CTC-based end-to-end ASR for the low resource Sanskrit language with spectrogram augmentation. 27th National Conference on Communications, NCC (2021). pp. 1–6. https://doi.org/10.1109/NCC52529.2021.9530162
5. Arora, A., Gessler, L., Schneider, N.: Supervised grapheme-to-phoneme conversion of orthographic schwas in Hindi and Punjabi. In: Proceedings of the 58th Annual Meeting of the Association for Computational Linguistics, pp. 7791–7795. Association for Computational Linguistics, July 2020. https://doi.org/10.18653/v1/2020.acl-main.696
6. Bahl, L.R., Brown, P.F., de Souza, P.V., Mercer, R.L.: Maximum mutual information estimation of hidden Markov model parameters for speech recognition. In: ICASSP, IEEE International Conference on Acoustics, Speech and Signal Processing - Proceedings, pp. 49–52 (1986)
7. Dehak, N., Kenny, P., Dehak, R., Dumouchel, P., Ouellet, P.: Front-end factor analysis for speaker verification. IEEE Trans. Audio Speech Lang. Process. **19**, 788–798 (2011)
8. Federico, M., Bertoldi, N., Cettolo, M.: IRSTLM: an open source toolkit for handling large scale language models. In: Proceedings of the Annual Conference of the International Speech Communication Association, INTERSPEECH (2008)
9. Gales, M.: Semi-tied covariance matrices for hidden Markov models. IEEE Trans. Speech Audio Proces. **7**(3), 272–281 (1999). https://doi.org/10.1109/89.759034
10. Graves, A., Fernández, S., Gomez, F.: Connectionist temporal classification: labelling unsegmented sequence data with recurrent neural networks. In: Proceedings of the International Conference on Machine Learning, ICML, pp. 369–376 (2006)
11. Hadian, H., Sameti, H., Povey, D., Khudanpur, S.: End-to-end speech recognition using lattice-free MMI. In: Proceedings of the Annual Conference of the International Speech Communication Association, INTERSPEECH, pp. 12–16 (2018). https://doi.org/10.21437/Interspeech
12. Hinton, G., et al.: Deep neural networks for acoustic modeling in speech recognition: the shared views of four research groups. IEEE Sig. Process. Mag. **29**(6), 82–97 (2012). https://doi.org/10.1109/MSP.2012.2205597
13. Hugo Van, H., Filip Van, A.: An adaptive-beam pruning technique for continuous speech recognition. In: Proceeding of Fourth International Conference on Spoken Language Processing, ICSLP, vol. 4, pp. 2083–2086. IEEE (1996)
14. Samudravijaya, K., Murthy H.A.: Indian language speech sound label set (ILSL12). In: Indian Language TTS Consortium & ASR Consortium (2012) https://www.iitm.ac.in/donlab/tts/downloads/cls/cls_v2.1.6.pdf

15. Ko, T., Peddinti, V., Povey, D., Khudanpur, S.: Audio augmentation for speech recognition. In: Proceedings of the Annual Conference of the International Speech Communication Association, INTERSPEECH. vol. January, pp. 3586–3589 (2015)
16. Mohri, M., Pereira, F., Riley, M.: Weighted finite-state transducers in speech recognition. Comput. Speech Lang. **16**(1), 69–88 (2002). https://doi.org/10.1006/csla.2001.0184
17. Mortensen, D., Dalmia, S., Littell, P.: Epitran: Precision G2P for many languages. In: 11th International Conference on Language Resources and Evaluation, LREC 2018. pp. 2710–2714 (2019)
18. Park, D., Chan, W., Zhang, Y., Chiu, C.C., Zoph, B., Cubuk, E., Le, Q.: Specaugment: A simple data augmentation method for automatic speech recognition. In: Proceedings of the Annual Conference of the International Speech Communication Association, INTERSPEECH. vol. September, pp. 2613–2617 (2019)
19. Peddinti, V., Povey, D., Khudanpur, S.: A time delay neural network architecture for efficient modeling of long temporal contexts. In: Proceedings of the Annual Conference of the International Speech Communication Association, INTERSPEECH. vol. 2015-January, pp. 3214–3218 (2015)
20. Povey, D., et al.: The subspace Gaussian mixture model - a structured model for speech recognition. Comput. Speech Lang. **25**(2), 404–439 (2011). https://doi.org/10.1016/j.csl.2010.06.003
21. Povey, D., Cheng, G., Wang, Y., Li, K., Xu, H., Yarmohamadi, M., Khudanpur, S.: Semi-orthogonal low-rank matrix factorization for deep neural networks. In: Proceedings of the Annual Conference of the International Speech Communication Association, INTERSPEECH, vol. 2018-September, pp. 3743–3747 (2018). https://doi.org/10.21437/Interspeech
22. Povey, D., Peddinti, V., Galvez, D., Ghahremani, P., Manohar, V., Na, X., Wang, Y., Khudanpur, S.: Purely sequence-trained neural networks for ASR based on lattice-free MMI. In: Proceedings of the Annual Conference of the International Speech Communication Association, INTERSPEECH, pp. 2751–2755 (2016). https://doi.org/10.21437/Interspeech
23. Povey, D., et al.: The Kaldi speech recognition toolkit. In: IEEE Workshop on Automatic Speech Recognition and Understanding, December 2011
24. Rabiner, L.: A tutorial on hidden Markov models and selected applications in speech recognition. Proc. IEEE **77**(2), 257–286 (1989). https://doi.org/10.1109/5.18626
25. Scharf, P.M., Hyman, M.D.: Linguistic Issues in Encoding Sanskrit. The Sanskrit Library, Providence(2012)
26. Watanabe, S., Hori, T., Kim, S., Hershey, J., Hayashi, T.: Hybrid CTC/attention architecture for end-to-end speech recognition. IEEE J. Select. Topics Sig. Process. **11**(8), 1240–1253 (2017). https://doi.org/10.1109/JSTSP.2017.2763455
27. Wikimedia: Wiki Sanskrit data dump. https://dumps.wikimedia.org/sawiki/
28. Young, S.J., Kershaw, D., Odell, J., Ollason, D., Valtchev, V., Woodland, P.: The HTK Book Version 3.4. Cambridge University Press, Cambridge (2006)

GRU with Level-Aware Attention for Rumor Early Detection in Social Networks

Yu Wang, Wei Zhou$^{(\boxtimes)}$, Junhao Wen, Jun Zeng, Haoran He, and Lin Liu

School of Big Data and Software Engineering, Chongqing University,
Chongqing, China
{wang_y,zhouwei,jhwen,zengjun,hehaoran,linliu}@cqu.edu.cn

Abstract. Social networks have developed rapidly in recent years, which gives people the opportunity to access more information. Nevertheless, social networks are illegally used to spread rumors. To clean up the network environment, rumor early detection is urgent. But the detection accuracy of existing rumor early detection methods is still not high enough in the early stage of news spread. They neglected to focus on more critical tweets by measuring the contribution of each retweet node to news events based on the propagation structure feature. In this paper, a novel method based on Gate Recurrent Unit with Level-Aware Attention Mechanism is proposed to improve the accuracy of rumor early detection. In this method, text features and user features are extracted from tweets about a given news event to generate a unified node representation of each tweet. Meanwhile, the process of news propagation is simulated to encode the node's forwarding level according to time span and forwarding hops between source tweet and retweet. In order to pay different attention to tweets according to the forwarding level of nodes, a new method based on attention mechanism is proposed to update node representation. Finally, the news event feature is learned from related tweet nodes representation in time sequence via a GRU-based classifier to predict the label (rumor or non-rumor). Extensive experimental results on two real-world public datasets show that the performance of the proposed model is higher than the baseline model in the early stage of rumors spread in the social network.

Keywords: Social networks · Rumor detection · Attention mechanism · Neural network models

1 Introduction

In recent years, social media's rapid development and popularization have made it more convenient for low-cost mass production and dissemination of rumors. Rumors not only seriously mislead people's thinking and damages people's spiritual civilization but also pose a further threat to public safety. In social networks, the delay time of detection is approximately linear with the life cycle

T. Mantoro et al. (Eds.): ICONIP 2021, LNCS 13109, pp. 548–559, 2021.
https://doi.org/10.1007/978-3-030-92270-2_47

of rumors [15]. Rumors will also be more likely to be found by users with the time of rumor spread increasing. But when the rumors spread on a large scale and even caused a lot of damage, it's too late to discover it. Consequently, it is crucial and necessary to detect rumors in the early stage of rumors spread (i.e., Rumor early detection).

There are many studies focusing on rumor detection on social networks. Most proposed approaches [16,18] detect rumors that rely on text or images. However, some rumors are carefully created by imitating the actual news to make people more convinced. These approaches detecting based on the features extracted only from text or images have limitations. Recently, some researchers detect rumors based on social content. Liu et al. [9] learn user characteristics and Song et al. [13] mine user credibility for detecting rumor. But they all ignore the topology of the spread of rumors. Rumors in social networks have formed specific propagation structures due to users interaction (e.g., forward, follow, comment). Ma et al. [11] and Kumar et al. [6] organized the source tweet and its response tweets into a tree structure, but they all ignore the time sequence of news spread. Khoo et al. [5], and Lao et al. [7] paid attention to both the linear structure and the nonlinear structure of rumor spread and systematically integrated the features of most aspects.

Although these methods achieved high accuracy in detection after the rumor was forwarded a large number of times, they ignored that tweet Nodes with different time span and forwarding hops have different contributions to news events. Specifically, firstly, the content of a tweet that directly forwards the source tweet is often more related to the news topic, while the content of a tweet that forwards a retweet may deviate from the news topic; secondly, based on interest recommendations in social networks, users who can participate in a topic in a short period are usually interested in or understand these topics, so they tend to have more say. Tweets need to be paid different attention according to the number of intermediate retweets (forwarding hops) and the time span between retweets and source tweets.

In this paper, a novel model based on Gate Recurrent Unit (GRU) neural network and attention mechanism for rumor detection (GLA4RD) is proposed to deal with these research problems. First, GLA4RD extracts the forwarding hops and the time span between the propagation structure, then combines these two features to measure the forwarding level of the tweet nodes. Meanwhile, GLA4RD extracts the text feature vector and user feature vector from text content and user profile, then combines them as the tweet feature vector. In order to promote early rumor detection, a novel Level-aware attention mechanism is proposed to update the feature vector of tweet nodes. Then tweet features in time sequence are used as the input of each unit of the GRU classifier, and the final output is used to classify the news to rumor or non-rumor.

The contributions of this paper are as follows:

- GLA4RD is proposed, a novel method based on GRU neural network and attention mechanism. It integrates the text features and user features, measure tweet nodes and give them different attention so that it can facilitate early rumor detection.

- A level-aware attention mechanism is proposed to mine the hidden feature of propagation structure from time span and forwarding hops and to update the tweet representation.
- Extensive experiments on two real-world public datasets were conducted to evaluate the proposed method. Experimental results showed the performance of GLA4RD more excellent than baseline models, especially within forwarding ten times after news starts to spread, and it maintains high accuracy when news is widespread.

2   Related Work

In recent years, rumor detection on social media has triggered heated discussions, and research on automatic detection based on artificial intelligence has become a hot topic. In general, the existing rumor detection methods based on artificial intelligence can be divided into the following three categories:

Some studies have adopted the method of inferring the label of the rumor by extracting linguistic features from the text content. For example, Wawer et al. [18] have predicted the reliability of the website by Utilizing the capabilities of text and language based on the vector space and psycholinguistic dimensions of the word bag. Castillo et al. [3] have used a set of language features to detect rumors such as specific characters, emotional words, emoticons, etc. Khattar et al. [4] have used both text content and image content to extract multimodal features by a classifier with a bimodal variational auto encoder to detect rumors.

And some studies have considered user profiles extracted from the social context feature of the news in social networks. Liu et al. [9] have treated the text content and corresponding user information as a crowded response arranged in chronological order, using a convolutional neural network to learn response characteristics. Wang et al. [17] have proposed a reinforced weakly-supervised framework, which enlarges the amount of available training data by leveraging users' reports as weak supervision.

And some studies have adopted the spread-based method that uses information related to rumor spread, i.e., how it propagates. Ma et al. [11] have constructed the propagation and diffusion processes into two tree structures, top-down and bottom-up, respectively, and aggregated the news features through a recurrent neural network along the process path. Lao et al. [7] have proposed a novel model with the field of linear and non-linear propagation (RDLNP), which automatically detects rumors by taking advantage of tweet-related information, including social content and temporal information.

3   Approach

Rumor detection in this paper is to classify news events according the learned the feature representation vector of news events. The proposed model GLA4RD consists of four main modules: basic feature extraction, forwarding level measure, tweet feature representation, and rumor detection (see Fig. 1). This paper's early rumor detection task is based on the first k retweets and uses the number of posts

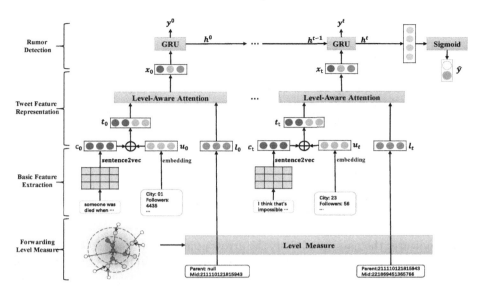

Fig. 1. The architecture of rumor detection model GLA4RD.

related to the event as the threshold for the detection period. The reason is that many hot events will have an explosive increase in the number of tweets in a short period of time, causing great social impact. Rumor early detection model strives to be efficient when the propagation range is still small.

3.1 Problem Formulation

Let $\mathcal{E} = \{E_1, E_2, ..., E_{|\mathcal{E}|}\}$ be a set of news events, each of which is associated with a label $y_i \in \{0, 1\}$, where $y_i = 0$ when E_i is none rumor and $y_i = 1$ when it is rumor. $E_i = \{e_i^0, e_i^1, ..., e_i^{|e_i|-1}, G_i\}$, $e_i^j(c_i^j, u_i^j)$ is j-th retweet arranged in time sequence, where c_i^j, u_i^j, respectively represents the text content and user profile (e.g.screen name, location, followers and so on) of tweet. e_i^0 refers to the first tweet (i.e. source tweet) of news event i. And G_i refers to the propagation structure.

Given a dataset, the goal of rumor detection is to learn a classifier

$$f : (\mathcal{E}, k) \to Y, \tag{1}$$

where Y refers to a set of labels of news events, k is a threshold of maximum number of detection that can be set.

3.2 Forwarding Level Measure

The spread of news event in social networks is promoted by forwarding, which can be simulated as a propagation graph (see Fig. 2). The depth of the background

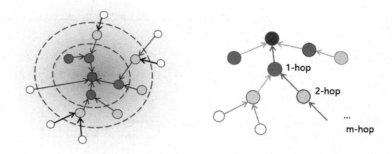

Fig. 2. The news spread process simulation graph.

color indicates the passage of time, the arrow points to the node to be responded from the response node, and the depth of the node color indicates the forwarding level of the tweet node. The task of this stage is to generate the forwarding level of nodes.

Denote the propagation graph as $G_i =< V_i, E_i >$, V_i is a set of tweet nodes and E_i is a set of edges between interacting nodes. $v_i^j = (t_i^j, h_i^j)$, $v_i^j \in V_i$, where t_i^j refers to the time span of tweet and h_i^j refers to the forwarding hops.

The forwarding level of each tweet denotes as:

$$l_i^j = \text{Relu}\left(W_{l_i}\left(\left(\max\{t_i^0, t_i^1, ..., t_i^k\} - t_i^j\right) \oplus \frac{h_i^j}{\max\{h_i^0, h_i^1, ..., h_i^k\}}\right) + b_{l_i}\right), \quad (2)$$

where $l_i^j \in \mathbb{L}_i^{d_l}$ refers to the forwarding level of tweet node v_i^j, where d_l refer to the dimension of forwarding level feature vectors, $\oplus$ is concatenate operation, W_{l_i} is space transformation matrix to transforms the fusion feature into a higher dimensional vector space and b_{l_i} is bias.

3.3   Basic Feature Extraction

Every tweet in social network contains text content and is associated with the user profile of the user who posted it.

Text Feature. Word vectors $w \in \mathbb{W}^{d_w}$ are extracted from sentence of text via word2vec [8], where d_w refer to the dimension of vector. In previous rumor detection models, neural network models such as CNN are often used for text feature learning. But most word embedding methods seek to capture word cooccurrence probabilities using inner vector product, end up giving unnecessarily large inner products to word pairs. The sentence vectors in this paper are extracted via a simple but very effective method Sentence2Vec [1] which emphasize the characteristics of low-frequency words but also retain the content of high-frequency words:

$$s_i^j = \frac{1}{|S_i|} \sum_{w \in s_i^j} \frac{a}{a + p(w)} w, \tag{3}$$

$$c_i^j = s_i^j - uu^T s_i^j, \tag{4}$$

where $c_i^j \in \mathbb{C}^{d_c}$ refers to the text feature vector, where d_c refer to the dimension of vector, S_i refers to the sentence from tweet node v_i^j, a is a parameter, $p(w)$ refers to the estimated probability of the word and u refers to the first singular vector of matrix whose columns are s_i^j.

User Feature. The way of embedding the user feature vector is the method mentioned in FNED [9]. All user features are normalized using the Z-score into the range of [0,1] to represent a user feature vector u_i^j of the user who posted the tweet.

3.4 Tweet Feature Representation

text feature and user feature are combined to represent the basic feature as follows:

$$x_i^j = p_i^j \oplus u_i^j, \tag{5}$$

where $x_i^j \in \mathbb{T}^{d_t}$ refers to the basic feature vector, where d_x refer to the dimension of vectors, $\oplus$ is concatenates operation.

Level-Aware Attention. The model should know the importance of each tweet node then give them different weights according to the forwarding level of tweets. Level-aware attention mechanism is proposed to solve this problem. It is an extension of the basic attention mechanism [2].

For each basic feature vector x_i^j from Eq. (5), its attention weight and transformed vector is calculated as follows:

$$F_w(x_i^j) = \text{ReLu}(W_{a_i^j}^T(x_i^j \oplus l_i^j) + b_{a_i^j}) \tag{6}$$

$$\tilde{x}_i^j = \frac{\exp(F_w(x_i^j))}{\sum_n \exp(F_w(x_i^j))} x_i^j \tag{7}$$

where $\oplus$ concatenates basic feature vector and the forwarding level of the tweet. F_w is an attention score function with weights $W_{a_i^j}$, bias $b_{a_i^j}$.

3.5 Rumor Detection

GRU-Based Neural Network. In order to obtain an enhanced feature vector, Gate Recurrent Unit (GRU) is used to learn the feature of the news event. In a tweets sequence, indicative features cluster along with the propagation history. The GRU is extended to simulate the long-distance interaction of nodes, for it reduces the required parameters while becoming more efficient. The forward propagation process of node j is as follows:

$$r_i^j = \sigma\left(W_{r_i}\tilde{x}_i^j + U_{r_i}h_i^{j-1}\right),$$
$$z_i^j = \sigma\left(W_{z_i}\tilde{x}_i^j + U_{z_i}h_i^{j-1}\right),$$
$$\tilde{h}_i^j = tanh\left(W_{h_i}\tilde{x}_i^j + U_{h_i}(h_i^{j-1} \odot r_i^j)\right),$$
$$h_i^j = (1 - z_i^j) \odot \tilde{h}_i^j + z_i^j \odot h_i^{j-1}$$

(8)

Where $\tilde{x}_i^j$ is the transformed representation of tweets by the Level-aware attention mechanism, $[W_*, U_*]$ are the weight connection inside GRU, and h_i^j refer to the hidden state of node v_i^j, $\odot$ denotes element-wise multiplication. r_i^j and z_i^j represents reset gate and update gate, respectively, where the reset gate determines how to combine the input of the current unit with the memory of the previous unit, and the update gate determines how to save the memory of the previous unit to the current unit.

After the node features are aggregated in time sequence, the final output is the overall feature representation of the news event. Dropout [14] is applied on GRU Layer to avoid over-fitting.

Classification. Finally, a multi-layer perceptron (MLP) block is adopted to predict a class label for the news event, simply denoted as:

$$\hat{y}_i = \text{Sigmoid}(W_i h_i^k + b_i),$$

(9)

where h_i^k is the learned hidden vector of threshold k, W_i and b_i are the weights and bias in output layer.

Optimization. Let $\mathcal{N}(\cdot;\theta)$ be the rumors classifier, where θ denotes all of the included parameters. Cross-entropy function measure the detection loss:

$$L(\theta,k) = -\mathbb{E}_{(c_i,y_i)\sim(C,Y)}\left[y_i \log \mathcal{N}(c_i,k) + (1-y_i)\log(1 - \mathcal{N}(c_i,k))\right].$$

(10)

Where Y be the set of labels (rumor or non-rumor). Given the detection threshold k, the optimization goal is to find the optimal θ that minimizes the detection loss:

$$\hat{\theta} = \arg\min_\theta L(\theta,k).$$

(11)

The optimization solved by stochastic gradient descent–based optimization approaches.

4   Experiments

Experiments on two real-world datasets were performed to evaluate our model, and the ablation studies on each proposed component were also conducted. We implemented the proposed model by using Tensorflow. When preprocessing the text, the jieba toolkit is used to segment and regularize the characters. The model is verified by five-fold cross-training and testing. In each round of cross-validation, the entire data set is randomly divided into five folds of the same size. We train the model for 1,000 epochs to minimize its loss. We use stochastic gradient descent and Adadelta [19] update to update the weight and bias rules.

4.1 Datasets

We conducted experiments on two real-world datasets: Twitter [9] and Weibo [10]. Weibo contains 4664 news events, and Twitter contains 680 news events. Weibo and Twitter all provide user profiles that include name length, followers number, friends number, city number, verified states, et al. The Twitter dataset does not provide the text content of retweets, so we treat the content of the source tweets as the text content of each retweet.

We constructed a propagation graph based on the message-id and the source id of the tweet and designed a simple program to obtain the forwarding hops of each tweet node according to traversing all nodes in the path of the graph. Table 1 shows statistics of these two datasets:

Table 1. Dataset statistics.

Statistic	Weibo	Twitter
News events	4664	680
Rumors	2313	325
None-rumors	2351	328
Users	2,746,818	215,691
Tweets	3,805,656	258,047

4.2 Baselines

We compared GLA4RD to a series of following baseline models: CSI [12]: A deep hybrid model based on three characteristics: text content, social content, and time sequence, which identifies rumor via article and user characteristics. RvNN [11]: A model based on tree RNN as the framework for the rumor detection, it includes a bottom-up model and a top-down model. FNED [9]: A deep learning framework that proposes a novel attention mechanism and a pooling method based on CNN and uses the user characteristics of news communicators to improve the effectiveness of rumor detection. RDLNP [7]: A rumor detection method that considers both the linear structure and the non-linear structure of news propagation, it comprehensively integrates various features of news.

4.3   Comparison with Baselines

Table 2 lists the hyper-parameters of the model. We use standard effectiveness indicators, including accuracy, precision, recall, and F1-score, all of which are used to evaluate all models. We use the number of retweets as the detection threshold.

Table 2. Hyper-parameters setting.

Hyper-parameter	Value	Test range
Parameter a of senten2vec	1e$-$3	1e$-$3,1e$-$4
Dimension of word feature d_w	2^8	2^5–2^8
Dimension of text feature d_c	2^8	2^5–2^8
Dimension of tweet feature d_t	2^7	2^5–2^8
Dropout rate	0.15	0–0.6

Table 3. Comparison of overall performance when k = 10.

Approach	Twitter				Weibo			
	Acc.	Pre.	Rec.	F1	Acc.	Pre.	Rec.	F1
CSI	0.786	0.805	0.781	0.795	0.825	0.797	0.838	0.823
RvNN	0.789	0.779	0.801	0.798	0.807	0.791	0.827	0.814
RDLNP	0.844	0.843	0.844	0.842	0.876	0.865	0.874	0.867
FNED	0.906	0.902	0.906	0.899	0.910	0.901	0.918	0.914
GLA4RD	**0.933**	**0.931**	**0.915**	**0.923**	**0.921**	**0.924**	**0.922**	**0.915**

Table 3 shows the performance comparison between GLA4RD and baselines when numbers of retweet arrive 10, i.e., k = 10, GLA4RD outperforms the baseline models in all evaluation metric. Then We compared GLA4RD with the baselines when the number of retweets is the less than 80, i.e., $k \in [10, 80]$, to get an overall effectiveness (See Fig. 3). We make the following observations through these experiments: GLA4RD has great performance at rumor detection, especially in the stage of rumor early detection. More precisely, GLA4RD improves over the strongest baselines with Accuracy by 2.75% and 1.21% in Twitter and Weibo, respectively. For the precision, GLA4RD outperforms the strongest baselines by 3.21% and 2.55%, respectively. This means it is effective to predict the rumor by paying different attention to the tweet nodes based on the forwarding level measured according to time span and forwarding hops.

4.4   Ablation Analysis

To investigate the impact of each key module on performance, we researched several simplified variants of GLA4RD on Weibo dataset. The following is a list of simplified models:

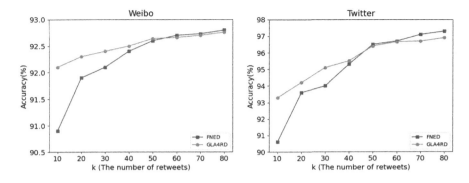

Fig. 3. Comparison with the optimal baseline on two datasets.

Table 4. Comparison of overall performance when k = 10.

Approach	Weibo			
	Acc.	Pre.	Rec.	F1
GLA4RD-TS	0.901	0.914	0.916	0.902
GLA4RD-RH	0.896	0.871	0.865	0.894
GLA4RD-LA	0.885	0.893	0.904	0.898
GLA4RD	**0.921**	**0.924**	**0.922**	**0.915**

- GLA4RD-LA: Level-Aware Attention Layer were not included. Table 4 shows the comparison of the reduced models and the full GLA4RD model when k = 10.
- GLA4RD-TS: Only forwarding hops is considered, ignore the time span.
- GLA4RD-RH: Only time span is considered, ignore the forwarding hops.

In addition, we also compared GLA4RD-LA and GLA4RD with different settings of k (See Fig. 4).

The ablation experiments results show:

- When one key component was removed, our proposed model's performance would drop.
- Among the two key factors considered by GLA4RD, time span and forwarding hops, forwarding hops affected the detection accuracy more significantly.
- GLA4RD-LA can achieve good results when the detection threshold is large, while the effect is significantly reduced when the detection threshold is small. This further indicates that it is effective to measure tweet nodes according to time span and forwarding hops for rumor early detection.

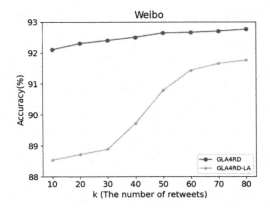

Fig. 4. Ablation analysis with the GLA4RD-LA on Weibo dataset.

5   Conclusion

In this paper, a novel rumor early detection model based on GRU with Level-aware attention mechanism that aims to measure each tweet node by propagation structure is proposed for rumor early detection in social networks. The idea is to reveal the hidden characteristics of the propagation structure. GLA4RD integrates propagation structure, text content, and user profile, and a level-aware attention mechanism is proposed to give different attention to tweet nodes. This allows the model to maintain the same detection performance at different stages of propagation, also perform well in the early stage of propagation. Finally, the labels are predicted by a GRU-based classifier. Extensive experiments on two real-world datasets also demonstrated the effectiveness of our approach, and an ablation study quantitatively verified that each component made an important contribution. In the future, optimizing the measuring method of tweet nodes may be a promising way to improve early detection performance further.

Acknowledgements. This work is supported by National Natural Science Foundation of China (Grant No. 72074036, 62072060), Special Funds for the Central Government to Guide Local Scientific and Technological Development YDZX20195000004725.

References

1. Arora, S., Liang, Y., Ma, T.: A simple but tough-to-beat baseline for sentence embeddings (2016)
2. Bahdanau, D., Cho, K., Bengio, Y.: Neural machine translation by jointly learning to align and translate. arXiv preprint arXiv:1409.0473 (2014)
3. Castillo, C., Mendoza, M., Poblete, B.: Information credibility on twitter. In: Proceedings of the 20th International Conference on World Wide Web, pp. 675–684 (2011)

4. Khattar, D., Goud, J.S., Gupta, M., Varma, V.: Mvae: Multimodal variational autoencoder for fake news detection. In: The World Wide Web Conference, pp. 2915–2921 (2019)
5. Khoo, L.M.S., Chieu, H.L., Qian, Z., Jiang, J.: Interpretable rumor detection in microblogs by attending to user interactions. In: Proceedings of the AAAI Conference on Artificial Intelligence, vol. 34, pp. 8783–8790 (2020)
6. Kumar, S., Carley, K.M.: Tree lstms with convolution units to predict stance and rumor veracity in social media conversations. In: Proceedings of the 57th Annual Meeting of the Association for Computational Linguistics, pp. 5047–5058 (2019)
7. Lao, A., Shi, C., Yang, Y.: Rumor detection with field of linear and non-linear propagation. In: Proceedings of the Web Conference 2021, pp. 3178–3187 (2021)
8. Le, Q., Mikolov, T.: Distributed representations of sentences and documents. In: International Conference on Machine Learning, pp. 1188–1196. PMLR (2014)
9. Liu, Y., Wu, Y.F.B.: Fned: a deep network for fake news early detection on social media. ACM Trans. Inf. Syst. (TOIS) 38, 1–33 (2020)
10. Ma, J., et al.: Detecting rumors from microblogs with recurrent neural networks (2016)
11. Ma, J., Gao, W., Wong, K.F.: Rumor detection on twitter with tree-structured recursive neural networks. In: Proceedings of the 56th Annual Meeting of the Association for Computational Linguistics, ACL 2018, Melbourne, Australia, July 15–20, 2018, Vol. 1 Long Papers, Association for Computational Linguistics (2018)
12. Ruchansky, N., Seo, S., Liu, Y.: Csi: a hybrid deep model for fake news detection. In: Proceedings of the 2017 ACM on Conference on Information and Knowledge Management, pp. 797–806 (2017)
13. Song, C., Yang, C., Chen, H., Tu, C., Liu, Z., Sun, M.: CED: credible early detection of social media rumors. IEEE Trans. Knowl. Data Eng. 33(8), 3035–3047 (2019)
14. Srivastava, N., Hinton, G., Krizhevsky, A., Sutskever, I., Salakhutdinov, R.: Dropout: a simple way to prevent neural networks from overfitting. J. Mach. Learn. Res. 15(1), 1929–1958 (2014)
15. Tripathy, R.M., Bagchi, A., Mehta, S.: A study of rumor control strategies on social networks. In: Proceedings of the 19th ACM International Conference on Information and Knowledge Management, pp. 1817–1820 (2010)
16. Vishwakarma, D.K., Varshney, D., Yadav, A.: Detection and veracity analysis of fake news via scrapping and authenticating the web search. Cogn. Syst. Res. 58, 217–229 (2019)
17. Wang, Y., et al.: Weak supervision for fake news detection via reinforcement learning. In: Proceedings of the AAAI Conference on Artificial Intelligence. vol. 34, pp. 516–523 (2020)
18. Wawer, A., Nielek, R., Wierzbicki, A.: Predicting webpage credibility using linguistic features. In: Proceedings of the 23rd International Conference on World Wide Web, pp. 1135–1140 (2014)
19. Zeiler, M.D.: Adadelta: an adaptive learning rate method. arXiv preprint arXiv:1212.5701 (2012)

Convolutional Feature-Interacted Factorization Machines for Sparse Contextual Prediction

Ruoran Huang[1,2(✉)], Chuanqi Han[1,2], and Li Cui[1]

[1] Institute of Computing Technology, Chinese Academy of Sciences, Beijing, China
[2] University of Chinese Academy of Sciences, Beijing, China
{huangruoran,hanchuanqi18b,lcui}@ict.ac.cn

Abstract. Factorization Machines (FMs) are extensively used for sparse contextual prediction tasks by modeling feature interactions. Despite successful application of FM and its abundant deep learning variants, these improved FMs mainly focus on capturing feature interaction at the vector-wise level while ignoring the more sophisticated bit-wise information. In this paper, we propose a novel Convolutional Feature-interacted Factorization Machine (CFFM), which learns crucial interactive patterns from enhanced feature-interacted maps with non-linearity. Specifically, in the high-order feature interactions part of CFFM, we propose a special Convolutional Max Pooling (Conv-MP) block to adequately learn interaction patterns from both vector-wise and bit-wise perspectives. Besides, we improve linear regression in FMs by incorporating a linear attention mechanism. Extensive experiments on two public datasets demonstrate that CFFM outperforms several state-of-the-art approaches.

Keywords: Factorization machines · Convolutional neural networks · Attention mechanism · Bit-wise perspectives

1 Introduction

In personalized recommender systems, prediction plays a significant role in web search, online advertising as well as shopping guidance [1]. The main goal of prediction is to infer the function that maps predictor variables to one target. This function is often represented as a predicted probability in a given context, which indicates the user's interest on the specific item such as a commercial advertising, a news article or a commodity item [2]. Technically speaking, to better represent these predictor variables which are mostly in multi-field categorical forms, a universal solution is to convert them into high-dimensional binary features via one-hot encoding [3], resulting in discreteness and sparsity of input vectors when building predictive models. For instance, *[City = New*

Supported by the National Natural Science Foundation of China (NSFC) under Grant No. 61672498.

York, Daytime = Wednesday, Gender = Female] are three categorical variables in restaurant recommendations, which can be normally transformed into high-dimensional sparse binary features as follows:

$$\underbrace{[0,0,0,1,\dots,0,0]}_{\text{City=New York}} \quad \underbrace{[0,0,1,0,0,0,0]}_{\text{Daytime=Wednesday}} \quad \underbrace{[1,0]}_{\text{Gender=Female}}$$

Despite that existing approaches (e.g., LR or SVM) can work on these sparse vectors, they suffer from intensive labor and abundant time, which highly depends on manual feature engineering and may degrade performances, especially in a sparse dataset. To address these problems, factorization machines (FMs) [4], which learn feature interactions by constructing cross features, are proposed to factorize coefficients into a product of two latent vectors. However, FMs only capture low-order interaction information and suffer from unreasonable weight assignments by treating each interaction fairly, leading to inferior performances.

Recently, some improved methods are successively proposed to address these limitations. For instance, He et al. [5] improve FM by performing a Bi-Interaction operation, based on the neural network model to extract more informative feature interactions at the vectorial level in the latent space of the second-order feature interactions. PNN [6] is proposed to learn expressive feature factors by embedding high-dimensional sparse features into a low-dimensional latent space. To obtain complex representations, DeepCross [7] introduces a cross network in learning bounded-degree feature interactions. Wide&Deep [3] memorizes sparse feature interactions and generalizes unseen feature interactions by using linear wide part and non-linear deep part. Subsequently, DeepFM [8] improves Wide&Deep [3] by replacing the original wide part with a second-order FM component while allowing FM and DNN to share the same embedding. Lian et al. [9] propose to incorporate a compressed interaction network to further learn certain bounded-degree feature interactions. Xiao et al. [11] propose AFM to discriminate the importance among interactions with utilizing the neural attention mechanism. However, these efforts weight factorized interactions in a shallow vector-wise way and belong to coarse-grained in relational models, which lose the fine-grained semantic information of the bit-wise perspective, reducing the performance of sparse predictions.

To tackle aforementioned drawbacks of existing solutions, in this paper, we present a novel model named Convolutional Feature-interacted Factorization Machines (CFFM) for sparse prediction tasks in recommender systems, which comprehensively learns interactions from feature-interacted patterns. More specifically, we first construct inner product feature interactions, and enhance the learning ability above these interactions by devising a feature interaction enhancement layer. Then, we propose a novel Convolutional Max Pooling (Conv-MP) block composed of three components: 1D kernel CNN, 1D max pooling and element-wise addition, and further capturing crucial interaction factors at both the vector-wise level and the bit-wise level. Inspired by [11], we also improve the linear regression in FMs by incorporating the linear attention mechanism, promoting the performance with limited parameter growth.

The main contributions of this paper are summarized as follows: 1) We propose a novel convolutional feature-interacted framework named CFFM to learn complicated interaction signals, which fuses complicated feature-interacted patterns. 2) We design an innovative convolutional max pooling block, effectively boosting the representation learning at both the vector-wise and bit-wise perspectives. 3) We redistribute weights of linear regression factors in original FMs by incorporating the linear feature attention mechanism. 4) Extensive experiments conducted on two publicly accessible datasets comparatively demonstrate the superiority of CFFM over existing state-of-the-art methods.

2   METHODS

2.1   Preliminaries

Factorization Machines [4] attempt to study relationship of feature interactions on sparse data, which combines the advantages of support vector machines with the flexibility of feature engineering. We assume that each instance has attributions $\mathcal{X} = \{x_1, x_2, \ldots, x_m\}$ from a target y, where m is the number of features and x_i is the real valued feature in i-th feature. Let y' represent the value predicted by FM model. Generally, FM with degree-2 can be defined as follows:

$$y' = w_0 + \underbrace{\sum_{i=1}^{m} w_i x_i}_{\text{linear regression}} + \underbrace{\sum_{i=1}^{m} \sum_{j=i+1}^{m} \hat{w}_{i,j} x_i x_j}_{\text{pair-wise feature interactions}} , \qquad (1)$$

where w_0 denotes the global bias, w_i denotes the model coefficient for the i-th feature. The pairwise feature interaction $\hat{w}_{i,j}$ denotes the weight of the cross feature $x_i x_j$, which is factorized as: $\langle \boldsymbol{v}_i, \boldsymbol{v}_j \rangle = \sum_{f=1}^{d} v_{i,f} v_{j,f}$, where $\boldsymbol{v}_i \in \mathbb{R}^d$ is an embedding vector for non-zero feature i and d is the embedding size. $\langle \cdot, \cdot \rangle$ represents the inner product of two vectors and $v_{i,f}$ denotes the f-th element of vector $\boldsymbol{v}_i$. Therefore, the time complexity of Equation (1) can drop from $O(dm^2)$ to linear time complexity $O(dm)$ by reformulating it as follows:

$$\sum_{i=1}^{m} \sum_{j=i+1}^{m} \langle \boldsymbol{v}_i, \boldsymbol{v}_j \rangle x_i x_j = \frac{1}{2} \sum_{f=1}^{d} \left[\left(\sum_{i=1}^{m} v_{i,f} x_i \right)^2 - \sum_{i=1}^{m} v_{i,f}^2 x_i^2 \right]. \qquad (2)$$

2.2   The Structure of CFFM

Equation (1) only models second-order feature interactions in a linear way, which fails to learn complex signals. Different from original FM, our model considers the high-order interaction signals at both vector-wise and bit-wise perspectives, which is reformulated as:

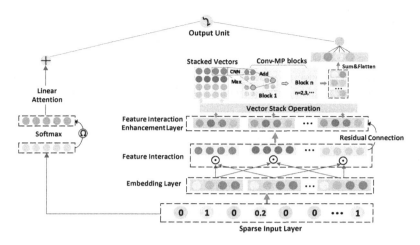

Fig. 1. The overall structure of CFFM. We first construct feature interactions among embedding vectors. Then, we enhance the learning ability by feeding embedding vectors into a feature interaction enhancement layer. After stacking these vectors together, we leverage Convolutional Max Pooling (Conv-MP) blocks to capture high-order interaction signals at both the vector-wise and bit-wise levels. Meanwhile, we improve the low-order linear part by incorporating the linear attention mechanism. Finally, we make the prediction by explicitly adding low- and high-order interaction signals together.

$$\hat{y}\left(\mathbf{x}\right) = w_0 + \sum_{i=1}^{m} \langle \boldsymbol{a}_i, \boldsymbol{w}_i \rangle \, x_i + f_\theta(\mathbf{x}), \tag{3}$$

where $\boldsymbol{a}_i$ is a linear weight factor discriminating the importance of different features and $f_\theta(\mathbf{x})$ denotes the core component to model feature interactions. In the following parts, we will elaborate how to learn $f_\theta(\mathbf{x})$ with the stacked inner products and convolutional neural networks, which explicitly captures high-order and finer-grained interaction signals. Figure 1 illustrates the whole network architecture of CFFM model and we will elaborate each component in following part.

Embedding Layer. As in FMs, each non-zero x_i of input can be converted into a learnable vector that is randomly initialized from the looking-up table. Formally, let $\boldsymbol{v}_i \in \mathbb{R}^d$ be the embedding vector which is randomly initialized with $\mathcal{N}(\mu, \sigma^2)$ for the i-th feature. Owing to sparse representation of $\mathcal{X}$, only non-zero features p are included (i.e., $x_i \neq 0$). We define embedded vectors as $\mathcal{V}_{\mathbf{x}} = \{\boldsymbol{v}_1, \boldsymbol{v}_2 \cdots, \boldsymbol{v}_m\}$, where each element is represented by a d-dimensional vector.

Feature Interaction Enhancement Layer. In this part, we explore re-weighting interaction factors among inner products to improve the robustness and generalization of the model. Let $\boldsymbol{v}_i$ and $\boldsymbol{v}_j$ be a pair of features from embedding vectors, and we can produce an interaction map by employing an element-wise operation on these two features via:

$$e_{ij} = x_i \boldsymbol{v}_i \odot x_j \boldsymbol{v}_j = x_i x_j [(v_{i,1} v_{j,1}), \dots, (v_{i,d} v_{j,d})], \tag{4}$$

where $e_{ij} \in \mathbb{R}^d$ denotes the feature interaction among embedding vectors and $\odot$ indicates element-wise product. Suppose the contextual input contains p non-zero features, the total number of generated interaction elements is $(p-1)p/2$. Thus, we can obtain $\mathbf{E} = [e_{12}, e_{13}, \cdots, e_{(p-1)p}]$, where $\mathbf{E} \in \mathbb{R}^{\frac{(p-1)p}{2} \times d}$. For further enhancing the learning ability of vectors at the vector-wise level, we apply a nonlinear transformation and the residual connection to enhance representation of feature interactions by:

$$\mathcal{P} = \sigma(\mathbf{E} H^T + b_e) + \mathbf{E}, \tag{5}$$

where $H \in \mathbb{R}^{d \times d}$, $b_e \in \mathbb{R}^d$ are the learnable weight matrix and the bias vector respectively, and σ denotes the $ReLU$ activation function. Clearly, $\mathcal{P} \in \mathbb{R}^{\frac{(p-1)p}{2} \times d}$ is a specific matrix that enhances the second-order interactions between features in the embedding space. It is noteworthy that compared with Bi-Interaction operation in NFM [5], we further devise a fully connected layer and its residual connection to capture more semantic interaction of high order.

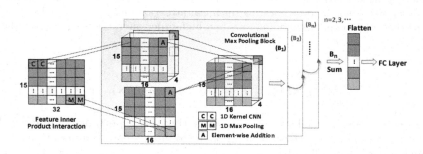

Fig. 2. The major process of Conv-MP blocks.

Convolutional Max Pooling Block. Afterwards, we stack $\mathcal{P}_{ij}$ together as a mapping matrix: $\boldsymbol{\xi} = [\mathcal{P}_{12}, \mathcal{P}_{13}, \cdots, \mathcal{P}_{(p-1)p}]$. For convenience, assume that p is 6, e.g., Book-Crossing dataset contains 6 non-zero features, while the embedding size d is 32. By performing Eq. 4 for each pair of interacted features, we then obtain vector representations as rows of $\boldsymbol{\xi} \in \mathbb{R}^{15 \times 32}$, which serves as the next input. Distinct from the conventional operation at the vector-wise [13], such as sum pooling operation and average pooling operation above feature interactions, we attempt to exploit more semantic information at both the vector-wise and bit-wise levels. As convolutional neural networks (CNN) take full advantage of the hierarchical pattern and local spatial coherence, it is well suited to distill relevant information at a low computational cost [13].

If we regard enhanced interactions $\mathcal{P}$ as "images" and directly leverage 2D

CNN to extract correlations above these interactions without considering the interaction order, it may disrupt the semantic integrity, especially when interpretable semantics only appear in one direction. Therefore, we introduce 1D kernel CNN to extract potential impact factors among interacted features. Innovatively, we design a special component named convolutional max pooling (Conv-MP) block above enhanced feature interactions, which can dynamically learn crucial factors from feature interactions at the bit-wise level. As shown in Fig. 2, the Conv-MP blocks can be divided into several blocks, and each block is comprised of three basic and efficient components: 1D kernel CNN, 1D max pooling and element-wise addition.

Firstly, stacked interaction matrix $\boldsymbol{\xi}$, which can be symbolized as ξ^0, is fed into the first Conv-MP block (abbreviated as B_1). For 1D kernel CNN, we customize channel size C as 4 and set the 1D kernel size to 2 in block n ($n = 1, 2, 3, \cdots$). After 1D kernel CNN, we can obtain matrix ξ_c^n, and the column size of matrix ξ_c^n is half of its previous ξ_c^{n-1} while the row size stays in step with the interaction size, e.g., $\xi^0 \in \mathbb{R}^{15 \times 32}$ in block n-1 and $\xi_c^1 \in \mathbb{R}^{15 \times 16}$ in block n. For 1D max pooling component, we set the stride to 2 for each row to obtain the matrix ξ_{max}^n after the operation of 1D max pooling, while keeping the same shape as ξ_c^n. Then, the element-wise addition is applied to these two matrices by:

$$\xi_{i,j,c}^n = \underbrace{\max\big\{\xi_{(i,2j,c)}, \xi_{(i,2j+1,c)}\big\}}_{\text{1D max pooling}} + \Delta \sum_{q=0}^{1} \underbrace{\mathcal{F}_{(q,c)} \cdot \xi_{(i,2j+q,c)}^{n-1}}_{\text{1D kernel CNN}}, \qquad (6)$$

where i, j and c denote the row, the column and the channel of matrix ξ^n, respectively. Δ denotes the activation function of $ReLU$, which improves the expression ability of CNN by adding nonlinear factors. After obtaining ξ^n, we input it into next Conv-MP block where number of channels could be customized as needed. Repeat above blocks several times, we not only distill vital weighting factors, but also re-weight importance for every interaction at the bit-wise level. In that follows, we overlay the column elements and channel elements with sum pooling, then flatten the remaining elements to reach a block output vector via:

$$\boldsymbol{\xi}' = \sum_{c=0}^{c_l} \sum_{j=0}^{\frac{d}{2^n}} \xi_{i,j,c}^n. \qquad (7)$$

Finally, $\boldsymbol{\xi}'$ is flattened to expediently tune each hidden factor, which can be implemented at the vector-wise level:

$$f_\theta(\mathbf{x}) = W_1 \cdot \text{Flatten}(\boldsymbol{\xi}') + b_1, \qquad (8)$$

where W_1 and b_1 denote weight matrix and bias term, respectively.

Linear Feature Attention Mechanism. In the linear terms $\sum_{i=1}^{m} w_i x_i$ of Eq. 1, we note that all features are modeled with the same weight, since treating each feature interaction fairly may result in suboptimal prediction and adversely degrading the performance. To alleviate this problem, we propose to leverage attention network to advance the fitting ability of linear regression by distinguishing the different contribution among interactions. Theoretically, we compute the attention vector with the following equation:

$$\hat{a}_i = ReLU\left(W_2\left(w_i x_i\right) + b_2\right), \ \boldsymbol{a}_i = \frac{\exp\left(\hat{a}_i/\eta\right)}{\sum_{i\in\mathcal{X}} \exp\left(\hat{a}_i/\eta\right)}, \tag{9}$$

where $W_2 \in \mathbb{R}^{m\times m}$ and $b_2 \in \mathbb{R}^m$ are the weight matrix and bias vector that project the input into a hidden space. η is a hyperparameter aiming to control the randomness of predictions by scaling logits. $\boldsymbol{a}_i$ is the corresponding attention score modeling the feature aspect, and the output is formulated as:

$$\Omega\left(\boldsymbol{x}\right) = \sum_{x_i \in \mathcal{X}} \langle \boldsymbol{a}_i, w_i \rangle x_i. \tag{10}$$

In fact, the improved linear feature attention can be viewed as linear calculation with $O(1)$ computing overhead and extra constant parameters.

2.3 Training and Analysis

Model Learning. As an extension of FMs, the goal of CFFM includes a variety of prediction tasks, such as classification, regression and ranking. In this paper, we primarily discuss the regression task and optimize the squared loss, and the analogous optimization strategies can be raised for classification and ranking tasks. To optimize the objective function, we employ adaptive gradient descent (AdaGrad) which is an adaptive gradient algorithm computing individual learning rates for each parameter. Meanwhile, L_2 regularization is also employed to control the overfitting of CFFM, then final objective function is defined:

$$\mathcal{L}_{loss} = \sum_{x\in\mathcal{X}} (\hat{y}(x) - y(x))^2 + \omega \left\|\Phi_{f(x)}\right\|_2^2, \tag{11}$$

where $\Phi_{f(x)}$ denote weight parameters and ω controls the regularization strength.

Time Complexity. The cost of computing $e_{i,j}$ (Eq. 4) and $\mathcal{P}$ (Eq. 5) are $O(dm^2)$ and $O(d^2)$. Due to $d \ll m$, this process can be summarized to $O(dm^2)$. Meanwhile, performing Eq. 6 and 7 consumes $O(c_l p^2 d + c_l d)$, where c_l is a small value indicating the number of channels in Conv-MP block, which is usually a fixed constant and can be ignored normally. So the time cost could be summarized as $O(p^2 d)$. For linear feature attention, due to its linear complexity, all calculations could be completed in $O(1)$. Stated thus, the total time complexity of CFFM is approximately represented as $O(dm^2 + p^2 d)$. And it is moderate compared with some advanced models such as CFM [13]. Remarkably, CFFM is generally smaller than tower structure models, e.g., Wide&Deep [3], PNN [6] and DeepFM [8], with more than 3 standard hidden layers.

Table 1. Statistics of datasets

Dataset	Features	Fields	Records	Sparsity
Book-crossing	226,336	6	1,213,367	99.97%
Frappe	5,382	10	288,606	99.81%

3   Experimental Setup

Dataset. Our proposed model is evaluated on two publicly accessible datasets: Book-Crossing[1] and Frappe[2]. Since two datasets only contain positive records, we adopt the same strategy of NFM [5], IFM [12], and sample negative records randomly by pairing two negative samples with each positive instance. Datasets are divided into a training set (70%), a validation set (20%), and a test set (10%). The statistics are summarized in Table 1.

Evaluation Metrics. The performance is evaluated by the Root Mean Square Error (RMSE) and coefficient of determination (R2-score). RMSE can well reflect the precision of prediction where a lower score indicates better performance, while R2-score is a statistical measure of how well regression predictions approximate the real data points.

Baselines. 1) **LibFM** [4] is the original factorization machine trained by squared loss. 2) **FFM** [10] assumes that features should have different latent factors when faced with features of different fields. 3) **Wide&Deep** [3] models low- and high-order feature interactions by leveraging "Wide" and "Deep" components, where the former is responsible for linear calculations, and the latter is an MLP layer. 4) **DeepCross** [7] introduces a deep residual network to explicitly learn cross features. 5) **PNN** [6] designs a product layer to learn high-order latent patterns. 6) **NFM** [5] combines the linearity of FM and the non-linearity of the neural network. 7) **IFM** [12] merges field aspects and feature aspects to distinguish the strength of interactions. 8) **CFM** [13] applies 3D convolution above self-attention interactions to extract higher-order signals. 9) **AFM** [11] learns one coefficient for every feature interaction to enable feature interactions that contribute differently to the prediction. 10) **DeepFM** [8] trains a deep component and FM component jointly for feature learning. 11) **XDeepFM** [9] takes out product at the vector-wise level and applies the convolution operations to obtain the high-order explicit and implicit features.

Parameter Settings. For a fair comparison, all methods are learned by optimizing the squared loss with the Adagrad optimizer. The learning rate is searched from $\{0.001, 0.005, 0.01, 0.05, 0.1, 0.5\}$, and the regularization coefficient is tuned amongst $\{0.001, 0.005, 0.01, 0.05\}$, where the best record is selected in the validation set for all models. For most datasets and baselines, the batch size for

---

[1] http://www2.informatik.uni-freiburg.de/%7ecziegler/BX.
[2] http://baltrunas.info/research-menu/frappe.

Book-Crossing and Frappe is set to 512 and 256, respectively, while the embedding size is fixed to 64 without special mention. Specially, 1) For AFM and IFM, we commonly set attention factor to the same value as the embedding, which is an empirical strategy reported in their papers. 2) For DeepFM, xDeepFM, DeepCross and PNN, we set the MLP according to original works. 3) As for multiply versions of models, such as IFM and PNN, we report the best result among these variants. Specially, the block size and the number of channel for each block is set to 2 and 4 respectively.

4   Results and Analysis

4.1   Performance Comparison

We report the parameters (abbreviated as #Param) of the mainstream algorithms in training and list the detailed comparison results with regard to metrics of RMSE and R2-score on two test sets in Table 2, from which the following observations are made:

Table 2. The symbol $*$ denotes our proposed method and the boldface indicates the best performance with a statistical significance $p < 0.05$. "M" stands for million and the downarrow (uparrow) expresses the lower (higher) score brings the better performance.

Dataset	Book-crossing			Frappe		
Metrics	RMSE↓	R2↑	#Param	RMSE↓	R2↑	#Param
LibFM	0.7323	0.4333	12.95M	0.3519	0.6860	0.32M
FFM	0.7167	0.4452	77.69M	0.3488	0.6935	3.18M
Wide&Deep	0.6935	0.4680	15.54M	0.3483	0.6947	0.49M
DeepCross	0.6917	0.4693	16.78M	0.3495	0.6924	0.53M
PNN	0.7271	0.4364	14.74M	0.3501	0.6905	0.39M
NFM	0.6883	0.4716	14.71M	0.3218	0.7467	0.36M
IFM	0.6860	0.4670	14.73M	0.3205	0.7503	0.37M
CFM	0.7088	0.4579	14.83M	0.3231	0.7433	0.36M
AFM	0.6876	0.4721	14.71M	0.3225	0.7450	0.36M
DeepFM	0.6933	0.4683	15.68M	0.3329	0.7249	0.44M
XDeepFM	0.6941	0.4678	15.82M	0.3228	0.7421	0.46M
$CFFM^*$	0.6798*	0.4797*	14.72M*	0.3174*	0.7561*	0.35M*

1) We can see that CFFM consistently achieves the best performance on Book-Crossing and Frappe datasets w.r.t. metrics of RMSE and R2-score, demonstrating the overall effectiveness of convolutional feature-interacted representation for prediction with sparse data. 2) We also note that improved FM models based on neural networks, such as NFM, AFM and CFM, are consistently better than original FM, which is attributed to the feature extraction by DNN

for capturing high-order interaction signals. However, these methods are usually parameters-intensive works due to fully connected structure, making it harder to train compared with the local connected CNN. 3) FFM consumes the most number of parameters, but it can not achieve the desired improvements and the results are limited compared with other DNN variants. This may be attributed to that the field-aware mechanism of FFM is still constrained by second-order interactions and losses the high-order feature interaction signals. Due to fact that the time complexity of FFM cannot be reduced to linear time by Eq. 2, the cost of its calculation is expensive [12]. 4) Some exquisite-simple structures could achieve better results than the stacked-complex structures, e.g., AFM are superior than DeepCross and PNN on two metrics. 5) The superiority among different approaches is not fixed, on the contrary, it varies with different datasets. For instance, NFM shows marginally improvement over IFM on Book-Crossing, but it is inferior to IFM on Frappe. This phenomenon can be explained that NFM uses Bi-interaction to horizontally connect features while IFM leverage the field of features to vertically estimate feature vectors, leading to more benefits for IFM when the number of fields is greater on Frappe. 6) Instead of increasing the depth and width of the structure, our work focuses on extracting critical factors by using 1D kernel CNN, effectively reducing redundant and useless parameters. Compared with other competitive models such as IFM, AFM and NFM, our CFFM can use fewer parameters to achieve better performance.

4.2   Efficacy of Hyper-parameters

Analysis of Embedding Dimensionality. Figure 3 shows R2-score with the embedding dimension d varying from 16 to 256. We find that the performance of each model tends to converge as the dimensionality increases. But the overall improvement rate drops a lot when the embedding size increases from 64 to 256. This is probably caused by overfitting of the model, and the increased burden of hyperparameters is much heavier and it does not bring corresponding desired benefits. Our model consistently outperforms all other baselines on two datasets even with a relatively small embedding dimensionality, which indicates that CFFM can have strong ability of feature interaction extraction and morphological-affluent representation.

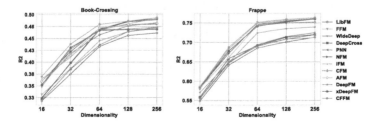

Fig. 3. The effect of embedding dimensionality d on R2-score for various models.

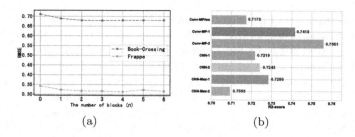

(a)   (b)

Fig. 4. (a) represents the study of the number of blocks n w.r.t. RMSE on two datasets. (b) means that the impact of Conv-MP component w.r.t. R2-score on Frappe dataset.

Study of the Block Size. The result is represented in Fig. 4(a). Notably, due to the default dimension is 64, only a maximum of 6 blocks that can be stacked. We observe that with the help of Conv-MP block, CFFM achieves 4.41% and 7.26% improvements than the variant of CFFM without Conv-MP block ($n = 0$) on Book-Crossing and Frappe respectively, revealing that the proposed Conv-MP block can promote the performance by capturing sophisticated feature interactions at both the vector-wise level and bit-wise level. With the block size increasing, the model achieves a limited gain since the potential deep information of neural units may have been exploited by the first two blocks. Considering the different characteristics of the datasets, we need to reasonably allocate the block size according to the dimensionality and dataset attributes, e.g., applying 5 blocks is not suitable on Frappe with the size is increasing.

4.3 Ablation Study

Impact of Conv-MP Component. To verify the contribution of proposed Conv-MP blocks, we develop several variants of CFFM: 1) **Conv-MP/wo** is a variant of CFFM which discards the whole Conv-MP blocks. 2) **Conv-MP-n** denotes that CFFM composed of **n** Conv-MP blocks. 3) **CNN-n** is a version of CFFM by replacing Conv-MP blocks with **n** CNN layers whose filters are 2×2 while the step size remains 2. 4) **CNN-Max-n** is a derivant of **CNN-n** by integrating max pooling after convolution operation. As shown in Fig. 4(b), we find that the performance of **Conv-MP/wo** is inferior to **Conv-MP-n** and **CNN-n**, which supports that the application of CNN to FM can indeed extract key information. In fact, **Conv-MP/wo** only contains the basic interaction signals, which can degenerate into vanilla-FM by discarding the feature interaction enhancement layer. Compared with the **CNN-n** and **CNN-Max-n** variants, the Conv-MP component demonstrates the more robust feature extraction capability, revealing that the validity of our schemes. Notably, **Conv-MP-2** achieves the worst results among all versions, this may be attributed to that the limited signals of the prediction is preserved when the size sharply reduces after the second max pooling operation.

Investigation of Feature Attention. To show the effect of linear feature attention mechanism (**Linear-A**), we perform comparative experiments for the linear part over two dataset. The results are listed in Table 3. Compared with

Table 3. Effect of feature attention mechanism. + denotes the additional parameters.

Dataset	Indicator	Linear	DNN-1	DNN-2	Linear-A	$FM_{Linear-A}$
Book	R2-score	0.1747	0.1563	0.1547	0.2364	0.4407
	RMSE	1.4153	1.5725	1.5890	1.0833	0.7205
	Param#	226,337	+42	+162	+42	#$\mathbf{P_{FM}}$+42
Frappe	R2-score	0.2501	0.2631	0.2653	0.2830	0.6910
	RMSE	0.9653	0.9133	0.9105	0.8654	0.3498
	Param#	5,383	+110	+430	+110	#$\mathbf{P_{FM}}$+110

original linear regression (**Linear**) in FM, **Linear-A** brings the substantial relative improvements with 32.95% and 12.34% when only 42 and 110 additional parameters are consumed, respectively. To eliminate the improvements caused by introducing extra parameters, we conducted other two comparative experiments by stacking DNNs on the original linear part. The unit size for depth-1 is the same as the number of the total fields, while the unit size for depth-2 is twice as many as depth-1 layer. we clearly see that simply stacking DNNs with increasing parameters does not help a lot in improving the experimental results, on the contrary, it sometimes adversely degrades the performance to some extent, e.g. the performances of DNN-1 and DNN-2 are even inferior to the original linear result on Book-Crossing. Meanwhile, we apply **Linear-A** to FM for investigating its improvement (**FM$_{Linear-A}$**), and the results from two datasets verify that the proposed mechanism can substantially improve the low-order fitting ability of vanilla-FM by reassigning weights of linear features.

5   Conclusion

In this paper, we proposed a novel convolutional feature-interacted framework named CFFM for sparse contextual prediction. To acquire semantic-rich interactions, we enhance representation of feature interactions by applying a nonlinear transformation and the residual connection, and design a novel convolutional max pooling block to extract more sophisticated signals from both vector-wise and bit-wise perspectives. Besides, we improve the linear regression in FM by introducing a linear attention network, which boosts the representation ability with the lower cost. Comprehensive experiments over two datasets demonstrate that CFFM outperforms state-of-the-art approaches.

References

1. He, X., Zhang, H., Kan, M.Y., Chua, T.S.: Fast matrix factorization for online recommendation with implicit feedback. In: SIGIR, pp. 549–558. ACM (2016)
2. Zhou, G., et al.: Deep interest network for click-through rate prediction. In: KDD, pp. 1059–1068. ACM (2018)

3. Cheng, H.T., et al.: Wide & deep learning for recommender systems. In: DLRS, pp. 7–10. ACM (2016)
4. Rendle, S.: Factorization machines with libfm. In: TIST, vol. 3, no. 3, pp. 1–22. ACM (2012)
5. He, X., Chua, T.S.: Neural factorization machines for sparse predictive analytics. In: SIGIR, pp. 355–364. ACM (2017)
6. Qu, Y., et al.: Product-based neural networks for user response prediction. In: ICDM, pp. 1149–1154. IEEE (2016)
7. Wang, R., Fu, B., Fu, G., Wang, M.: Deep & cross network for ad click predictions. In: ADKDD, pp. 1–7. ACM (2017)
8. Guo, H., Tang, R., Ye, Y., Li, Z., He, X.: DeepFM: a factorization-machine based neural network for CTR prediction. In: IJCAI, pp. 1725–1731. Morgan Kaufmann (2017)
9. Lian, J., Zhou, X., Zhang, F., Chen, Z., Xie, X., Sun, G.: XDeepFM: combining explicit and implicit feature interactions for recommender systems. In: KDD, pp. 1754–1763. ACM (2018)
10. Juan, Y., Zhuang, Y., Chin, W.S., Lin, C.J: Field-aware factorization machines for CTR prediction. In: RecSys, pp. 43–50. ACM (2016)
11. Xiao, J., Ye, H., He, X., Zhang, H., Wu, F., Chua, T.S: Attentional factorization machines: learning the weight of feature interactions via attention networks. In: IJCAI, pp. 3119–3125. Morgan Kaufmann (2017)
12. Hong, F., Huang, D., Chen, G.: Interaction-aware factorization machines for recommender systems. In: AAAI, vol. 33, no. 1, pp. 3804–3811. AAAI (2019)
13. Xin, X., Chen, B., He, X., Wang, D., Ding, Y., Jose, J.: CFM: convolutional factorization machines for context-aware recommendation. In: IJCAI, pp. 3926–3932. Morgan Kaufmann (2019)

A Lightweight Multidimensional Self-attention Network for Fine-Grained Action Recognition

Hao Liu[1], Shenglan Liu[2(✉)], Lin Feng[2(✉)], Lianyu Hu[1], Xiang Li[1], and Heyu Fu[3]

[1] Faculty of Electronic Information and Electrical Engineering,
Dalian University of Technology, Dalian 116024, Liaoning, China
[2] School of Innovation and Entrepreneurship, Dalian University of Technology,
Dalian 116024, Liaoning, China
{liusl,fenglin}@dlut.edu.cn
[3] School of Kinesiology and Health Promotion, Dalian University of Technology,
Dalian 116024, Liaoning, China

Abstract. Fine-grained action recognition is a challengeable task due to its background independence and complex semantics over time and space. Multidimensional attention is essential for this task to capture discriminative spatial details, temporal and channel features. However, multidimensional attention has challenge in keeping the balance between adaptive feature perception and computational overhead. To address this issue, this paper proposes a Lightweight Multidimensional Self-Attention Network (LMSA-Net) which can adaptively capture the discriminative features over multiple dimensions in an efficient manner. It is worth remarking that the contextual relationship between time and channel is established in temporal stream, which is complementary for spatial attention in spatial stream. Compared with the RGB based models, LMSA-Net achieves state-of-the-art performance in two fine-grained action recognition datasets, i.e. FSD-10 and Diving48-V2. In addition, it can be found that the streams of LMSA-Net are end-to-end trainable to reduce the overhead of computation and storage, and the recognition accuracy can reach the level of two-stage two-stream models.

Keywords: Fine-grained action recognition · Multidimensional attention · Adaptive feature perception · End-to-end trainable model

1 Introduction

In action recognition, a video tends to contain abundant spatial or temporal features which may distract models to focus on the discriminative ones [1,2].

This study was funded by National Natural Science Foundation of Peoples Republic of China (61672130, 61972064), the Fundamental Research Fund for Dalian Youth Star of Science and Technology (No. 2019RQ035), LiaoNing Revitalization Talents Program (XLYC1806006) and CCF-Baidu Open Fund (No. 2021PP15002000).

T. Mantoro et al. (Eds.): ICONIP 2021, LNCS 13109, pp. 573–584, 2021.
https://doi.org/10.1007/978-3-030-92270-2_49

Therefore, attention mechanism has been adopted to focus on the important features on space or time [3,4]. Spatial attention is effective to capture the scenes, objects or tools for recognizing background-dependent actions [3,5]. Temporal attention performs well on temporal modeling for time-related action recognition [4]. However, these methods based on the single dimensional attention are limited in addressing fine-grained action recognition.

Distinguishing from other actions, fine-grained actions are difficult to recognize in three aspects: 1) Spatial details. The discrimination of fine-grained actions does not depend on the background factors, including scenes, objects and tools (e.g. ice rink for skating, soccer for playing soccer), yet the important spatial details of actions. 2) Abundant action units. A series of action units are collected in an action sequence (e.g. a jump action contains preparing, taking off, turns and landing), and only specific ones can determine the category of an action. 3) Fast movement. The quality of motion features extracted is relatively low because some swift action units may blur the human body in RGB images (e.g. turn and twist). These issues may distract models to capture the discriminative features over space, time and channel, therefore multidimensional attention mechanism is essential for fine-grained action recognition.

Motion features of actions cannot be effectively represented in RGB images sampled from videos. To this end, [6] proposes flow modality to describe motion, which is complementary with RGB modality by fusion. This approach is advanced at coarse-grained (background-dependent) action recognition since background factors in RGB modality can contribute to the fusion. By contrast, the effect of RGB modality for fusion is limited in fine-grained action recognition with the single background (see details in Sect. 4). Although flow modality can improve the motion representation, there exists notable drawbacks in flow based models, including the two-stage pipeline, high computational cost and the burden of storage space. In further work, efficient motion cues have been explored [4,7,8]. For example, Persistent Appearance Network (PAN) [4] highlights the motion boundary and suppresses static features by adopting the Persistence of Appearance (PA) modality which is highly relevant to flow in effect. PA modality is beneficial to supplement motion features for RGB images because it can speed up computation by over 1000 times than flow and free up extra storage space. However, spatial and channel attention are ignored in PAN. The advantage of multidimensional attention on action recognition is demonstrated by Non-local Networks [9], W3 [10], MS-AAGCN [11], etc. The Non-local Networks adopts 3D convolution operators with expensive computation to compute the response at a position as a weighted sum of the features at all positions [9]. In order to reduce the computational cost, pooling operations are utilized in W3 and MS-AAGCN, while the pooling based approaches result in homogeneous context information being perceived by all features [12].

In order to solve the problems above, this paper proposes a Lightweight Multidimensional Self-Attention Network (LMSA-Net). Spatial stream and temporal stream of the model are trained cooperatively to achieve multidimensional self-attention in an end-to-end manner. The proposed model takes RGB images as input, and averages the class prediction scores from two streams to obtain the final result. In the data preprocessing, RGB images are extracted along the time sequence by the sparse sampling scheme. In temporal stream, four adjacent frames are processed into a PA image, which describes the motion of a RGB image in the spatial stream. Inspired by Non-local Networks [9] and cross self-attention [12], contextual relationship between time and channel is established adaptively and efficiently in temporal stream. Moreover, the temporal-channel self-attention weight can be utilized to improve the motion perception of RGB modality by connecting two streams. Taken together, this work contributes to fine-grained action recognition in three folds:

- This paper proposes the temporal-channel self-attention module, which can not only adaptively establish the contextual relationship between channel and time, but also promote the capability of motion perception for RGB modality.
- This paper proposes a Lightweight Multidimensional Self-Attention Network (LMSA-Net) which can realize the integration of multidimensional self-attention in an end-to-end manner. For attention modules, adaptive feature perception and computational overhead are balanced by adopting 2D CNNs based cross self-attention mechanism. The advantages make LMSA-Net an alternative approach for the two-stage two-stream model.
- Compared with RGB-based action recognition models, LMSA-Net achieves state-of-the-art performance on two fine-grained action datasets, including FSD-10 [13] and Diving48 [14].

2   Related Works

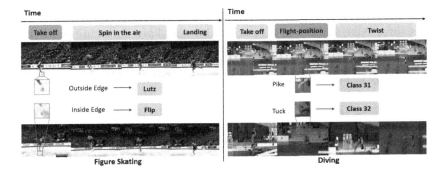

Fig. 1. The spatial-temporal properties of fine-grained actions. **Orange boxes** represent the discriminative action units along the timeline, and **orange lines** highlight the discriminative spatial details. **Green boxes** represent the indiscriminative action-units. **Red boxes** represent categories of actions. (Color figure online)

In the field of action recognition, attention mechanism is widely applied to focus on discriminative features. The methods are based on either temporal attention [4,15], spatial attention [5] or spatial-temporal attention [16]. However, the methods cannot perform well on fine-grained action recognition due to its rich yet redundant features over space, temporal and channel dimensions, as represented in Fig. 1. The multidimentional attention mechanism is therefore of great significance for this task. Generally, multidimentional attention methods can be summarized into two categories: (1) Pooling based attention methods. For example, W3 [10] and MS-AAGCN [11] stack pooling based attention modules together to achieve multidimensional attention. However, similar contextual information is perceived by all features, and contextual information can not be collected adaptively [12]. (2) 3D CNNs based self-attention methods. Multidimensional contextual relationship can be collected adaptively, while the models based on 3D CNNs are too expensive to deploy [12], such as Non-local Networks [9]. By contrast, LMSA-Net proposed in this paper can adaptively establish the contextual relationship of features by 2D convolution operators. Moreover, multidimensional self-attention is achieved by the connection of spatial stream and temporal stream, which are end-to-end trainable.

3   Model Formulation

3.1   Structure of LMSA-Net

In order to efficiently capture the discriminative spatiotemporal features in fine-grained actions (see Fig. 1), this paper proposes a multidimensional self-attention model based on 2D convolution operators. It can be seen in Fig. 2 that the architecture consists of temporal stream and spatial stream which are end-to-end trainable. In data preparation, sparsely sampled RGB images serve as the input of spatial stream, and PA images processed from consecutive RGB images [6] serve as the input of temporal stream. In order to capture crucial static details, such as inner edge and outer edge in skating and knee curvature in diving (see Fig. 1), spatial attention module is adopted in spatial stream (see Sect. 3.3). To further perceive motion features, relationship between time and channel is established by temporal-channel self-attention module in temporal stream (see Sect. 3.3). It can be observed in Sect. 4.4 that the capability of RGB modality on perceiving motion features can be improved by temporal-channel weights from temporal stream. Moreover, multidimensional self-attention is realized by connection and fusion of two streams, as shown in Fig. 2.

3.2   Data Preparation

Fine-grained action recognition mainly depend on motion features (see Fig. 1), which are difficult to be represented in RGB modality because swift action units may blur the human body. To this end, motion cues are proposed to represent the motion features of human body, such as flow modality [6] and skeleton modality [17]. However, it is inefficient for them to interact with RGB modality in the process of training due to the expensive computation and additional storage of

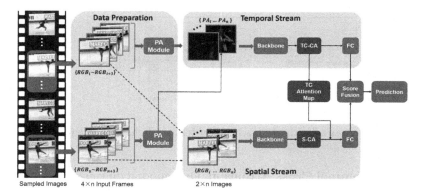

Fig. 2. The proposed network architecture. S-CA is spatial cross self-attention module, and TC-CA is temporal-channel cross self-attention module. The streams are connected by the temporal-channel attention map to achieve the multidimensional attention.

flow and skeleton modality. Another line of works make efforts to find efficient motion cues in an end-to-end manner, such as [4,7]. It is worth remarking that PA [4] can be processed over 1000 times faster than flow without additional storage. In effect, the PA_i image can represent the motion boundaries of $RGB_i \sim RGB_{i+3}$.

In addition to motion features, static human details are also discriminative for fine-grained actions (see Fig. 1). Given this, the complementary effect between static details (RGB modality) and motion features (PA modality) in training stage is considered in LMSA-Net. For the input of two streams, n RGB images are sparsely sampled along time sequence, and then each RGB image and its adjacent three frames are processed into a PA image through the PA module. To be specific, $4 \times n$ RGB images are processed to obtain n RGB images for spatial stream and n PA images for temporal stream. Consequently, it is possible for PA_i supplement RGB_i with motion features from $RGB_i \sim RGB_{i+3}$ in training stage.

3.3 Multidimensional Attention

As discussed in Sect. 2, the multidimentional attention is essential for fine-grained action recognition, while it is challengable to balance the relationship between adaptive feature perception and computional cost. In the proposed model, computing overhead is reduced by cross self-attention module based on 2D convolution, and multidimensional self-attention is realized by training two streams cooperatively.

Spatial Cross Attention. In order to balance adaptive feature perception and computational cost better, the cross self-attention mechanism is adopted in CCNet [12], reducing the computational complexity from $O((H \times W) \times (H \times W))$ to $O((H \times W) \times (H + W - 1))$. Given this, spatial stream of LMSA-Net utilizes the spatial cross self-attention module to capture the static details of human body.

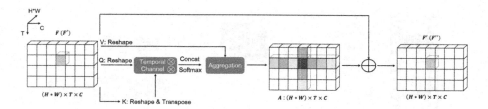

Fig. 3. The cross self-attention scheme of temporal-channel attention module.

Temporal-Channel Cross Attention. On temporal dimension, a fine-grained action contains abundant action units, whereas only specific units are discriminative. On channel dimension, SE-Net [18] can learn the importance of each channel to selectively enhance significant features and suppress unnecessary ones. Inspired by SE-Net [18] and CCNet [12], the contextual relationship between time and channel is constructed by extending the concept of cross self-attention on spatial dimension to temporal and channel dimensions (see Fig. 3). The module is adopted to the temporal stream because PA modality is more advanced at representing motion features (see Sect. 4.3 for more details).

Temporal-channel self-attention module obtains attention weight from horizontal and vertical context information (see Fig. 3). Firstly, the attention module takes the feature $F \in R^{(H*W) \times T \times C}$ as the input, which is transposed from the feature $X \in R^{H \times W \times T \times C}$. Different from the spatial self-attention in CCNet [12], convolutional operation is not employed in the proposed module to embed features before computing relationships between temporal and channel dimensions. In order to maintain the relationship between channels and time sequences, the input of the proposed module is reshaped into three feature maps (Q, K and V). Then Q and K are transposed to $\{Q_t \in R^{C \times T \times (H*W)}, Q_c \in R^{T \times (H*W) \times C}\}$ and $\{K_t \in R^{C \times (H*W) \times T}, K_c \in R^{T \times C \times (H*W)}\}$. Based on the features above, similarity matrixes $\{\alpha_t \in R^{T \times C \times T}, \alpha_c \in R^{T \times C \times C}\}$ are obtained by matrix multiplication and dimensional transposition conducted on $\{Q_t, K_t\}$ and $\{Q_c, K_c\}$ which can be denoted as $\alpha_t = Q_t \otimes K_t$ and $\alpha_c = Q_c \otimes K_c$, respectively. On the last dimension of α_t and α_c, concat and softmax operations are conducted to produce temporal-channel similarity matrix $\alpha_{tc} \in R^{T \times C \times (T+C-1)}$ which can be represented as $\alpha_{tc} = SoftMax([\alpha_t, \alpha_c])$.

Then α_{tc} is split in the last dimension and transposed into $\{\alpha_t' \in R^{C \times T \times T}, \alpha_c' \in R^{T \times C \times C}\}$. Attention weight A is obtained by the sum of two transposed matrix products of $\{\alpha_t', V_t\}$ and $\{\alpha_c', V_c\}$, which can be denoted as:

$$A = transpose(V_t \otimes \alpha_t') + transpose(V_c \otimes \alpha_c') \tag{1}$$

where $\{V_t \in R^{C \times HW \times T}, V_c \in R^{T \times C \times HW}\}$ are transposed from feature map V. Each element in A contains the contextual information along the axial (horizontal and vertical) direction. To be specific, attention weight over channel can be perceived in each frame x_t, and attention weight over time can be perceived in each channel x_c, which can be represented as:

$$x_t = \beta \sum\nolimits_{c=1}^{C} (A_{t,c}, x_{t,c}) + x_{t,c} \qquad (2)$$

$$x_c = \beta \sum\nolimits_{t=1}^{T} (A_{t,c}, x_{t,c}) + x_{t,c} \qquad (3)$$

where β denotes a learnable parameter.

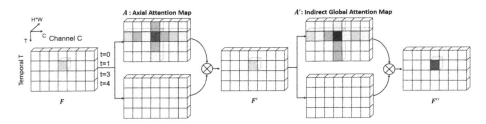

Fig. 4. Perception of global contextual features with temporal-channel self-attention module. The weights (**blue boxes**) are proportional to the intensity of color. For one position (**green box**), it achieves the weights along the axial direction after the first axial attention (**yellow box**). After the repeated operation, it (**red box**) indirectly collects the weights from all positions. (Color figure online)

With the repeated cross attention operation, each element can indirectly obtain the global attention weight A', as shown in Fig. 4. At the first input, $F \in R^{(H*W) \times T \times C}$ serves as the input of the cross attention module, producing the axial attention weight A and the axial feature enhanced F'. At the second input, F' is input into the cross attention module, and the global weight of attention A' is obtained indirectly. For example, although the elements $a(a_c, a_t)$ and $b(b_c, b_t)$ are not in the same row or column, the contextual features of (a_c, a_t) and (b_c, b_t) are perceptible for each other because (a_c, a_t) and (b_c, b_t) in F' contain the contextual features along the axial direction. For element $a(a_c, a_t)$, the perception of global contextual features can be formulated as $F''_a = [f(A, a_c, b_t, b_c, b_t) \cdot f(A', a_c, a_t, a_c, b_t) + f(A, b_c, a_t, b_c, b_t) \cdot f(A', a_c, a_t, b_c, a_t)] \cdot F_b$ where the mapping function from the element (c', t') to attention weight $A_{i,c,t}$ can be defined as $A_{i,c,t} = f(A, c, t, c', t')$. In the process of training, the global attention weight A' is transformed to the spatial stream to promote the capability of motion perception for RGB modality (see Fig. 2).

4 Experiments

4.1 Datasets

Experiments are conducted on FSD-10 and Diving48. FSD-10 is composed of 1,484 videos and 10 categories collected from the World Figure Skating Championships, and Diving48 consists of more than 18,000 videos and 48 categories from competitive diving. Note that we use the Diving48-V2 released in Oct 2020, which removed poorly segmented videos and cleaned up annotations. Actions in FSD-10 and Diving48 are challengeable to be recognized accurately due to the background independence and complex semantics over space and time (see Fig. 1).

4.2   Training Details

The stochastic gradient descent (SGD) with momentum (0.9) is selected as the optimization strategy and the cross-entropy is selected as the loss function of the back propagation. For TSM based models, they are trained for 80 epochs and 16 batches, starting at an initial learning rate of 0.01. The method of fusion is the average of the class prediction scores from the two streams. Backbone of LMSA-Net is ResNet-50 [19] based on TSM [20]. For other models, the training protocols follow the original papers. Note that Top-1 accuracy is calculated for evaluating the performance.

4.3   Ablation Study

Collocation Between Modalities and Attention Modules. The effects of attention modules differ from various modalities. Spatial attention (S-CA) is suitable for RGB modality due to its static spatial details. While PA modality is more suitable to adopt temporal-channel attention (TC-CA) rather than spatial attention, because PA modality highlights the boundary of motion and inhibits the expression of static feature. The performace of attention modules on various modalities are listed in Table 1.

Table 1. The impact of attention modules on RGB and PA modalities. S-CA is spatial cross self-attention module. TC-CA is temporal-channel cross self-attention module.

Methods	Acc. on PA (%)	Acc. on RGB (%)
TSM	74.1	63.3
TSM + S-CA	72.2	**69.4**
TSM + TC-CA	**75.2**	67.6

Two Streams or Stacking Modules? The attention modules of LMSA-Net are based on 2D convolution operators, which are limited to capture features over space, time and channel for a single module. In order to achieve multidimensional attention, we validate two potential architectures respectively, including stacking attention modules and two-stream interaction. As shown in Table 2, we take the input of 4×8 frames as an example to test the two architectures. The LMSA-Net achieves the advanced performance, because PA_i is benefical for RGB_i to perceive the motion features of the adjacent three frames at the moment i.

How the Two Streams Interact? Accroding to the discussion above, the effect of two-stream architecture is significant. Then we compare the different designs of two-stream architecture. As shown in Table 4, we take input 4×16 frames as an example to test three interaction methods. The advanced performance is achieved by the LMSA-Net which transmits the attention weight generated by temporal stream (PA) to spatial stream (RGB). This is because the ability of RGB modality on perceiving motion features can be improved by temporal-channel weights from temporal stream (PA). Interaction-1 and Interaction-2 are less effective, possibly because spatial attention weights focus on static spatial details in RGB images, while static features are difficult to be expressed in PA modality.

Table 2. The comparison between stacking modules and two-stream architecture.

Methods	Acc.(%)
TSM	63.3
TSM + S-CA	69.4
TSM + TC-CA	67.6
TSM + S-CA + TC-CA	65.1
LMSA-Net	**76.1**

Table 3. Comparison of computational cost between non-local and ours. Architectures of these methods are two-stream, and input of each stream is 8 frames.

Methods	GFLOPs(G)
TSM	65.0
TSM + Nonlocal	100.9 (+35.9)
LMSA-Net	77.2(+12.2)

Table 4. The comparison between different interaction strategies. SAt. denotes the spatial attention map. TCAt. denotes the temporal-channel attention map.

Methods	SAt. to RGB	TCAt. to PA	SAt. to PA	TCAt. to RGB	Acc.(%)
No interaction	✓	✓			83.3
Interaction-1			✓	✓	80.5
Interaction-2	✓	✓	✓	✓	80.1
LMSA-Net	✓	✓		✓	**87.9**

4.4 Comparison with the State of the Art

This section compares LMSA-Net with the state-of-the-art methods on FSD-10 and Diving48-V2. LMSA-Net can not only improve the performance of RGB modality, but also obtain the effect of the two-stage two-stream models based on flow and RGB.

In FSD-10, it is difficult to capture discriminative features over time and space. Spatially, with the assistance of skeleton modality, HF-Net [21] outlines the human body in RGB images to perceive significant features over space. However, HF-Net is a two-stage method since skeleton modality needs to be extracted

582   H. Liu et al.

before training the model [17]. Temporally, extensive motion features are collected in I3D [22] and ST-GCN [23], while they fail to capture the important spatial-temporal features due to the lack of attention mechanism. Although Non-local Networks [9] combined with TSM [20] achieves the effect of multidimensional attention, 3D convolution operators bring higher overhead (see Table 3). Without multidimensional attention, the improvement of other attention modules is limited, such as VAP (from PAN [4]) and CBAM [24]. With lightweight self-attention modules and the end-to-end trainable architecture, multidimensional self-attention and computational cost are balanced in LMSA-Net (see Table 3). Key static details in RGB modality are captured by spatial attention. Moreover, the ability of RGB modality to perceive discriminative motion features is significantly improved with the assistance of temporal-channel attention weight from temporal stream (PA). As shown in Table 5, LMSA-Net model achieves the best performance in RGB modality. With less computational cost, LMSA-Net can reach the level of the flow-based two-stream model in an end-to-end manner.

In Diving48-V2, we compares LMSA-Net with the state-of-the-art methods. I3D is difficult to capture motion features in the case of sparse sampling. Dense

Table 5. Comparison with state-of-the-art methods on FSD-10.

Methods	Frames	Modality	Acc.(%)
I3D	64	RGB+Flow	63.3
ST-GCN	350	SK^2	84.2
HF-Net	64	RGB+SK	85.5
TSM	8	RGB	63.3
TSM+NL[1]	8	RGB	69.1
TSM+cbam	8	RGB	66.2
PAN	8	RGB	65.4
TSM	16	RGB	82.9
		Flow	88.1
		Fusion	88.3
TSM+NL[1]	16	RGB	82.5
		Flow	88.3
		Fusion	88.4
LMSA-Net	8	RGB	**76.1**
		PA	75.2
		Fusion	77.4
LMSA-Net	16	RGB	**86.6**
		PA	85.8
		Fusion	87.9

Table 6. Comparison with state-of-the-art methods on Diving48-V2.

Methods	Frames	Modality	Acc.(%)
TimeSformer	96	RGB	81.0
SlowFast	16×8	RGB	77.6
TQN	dense	RGB	74.5
GST	8	RGB	69.5
I3D	8	RGB	33.2
	8	RGB	77.2
		Flow	78.7
		Fusion	80.4
TSM+NL[1]	8	RGB	78.5
		Flow	79.4
		Fusion	81.8
	8	RGB	77.9
		PA	78.2
		Fusion	80.7
LMSA-Net	8	RGB	**82.3**
		PA	79.9
		Fusion	**83.0**

NL[1] denotes the Non-local module.
SK^2 denotes the skeleton modality.

sampling methods perform well, such as Timesformer [16] and SlowFast [25], because more motion features are provided for temporal modeling. Timesformer-L achieves great advances by the spatial-temporal transformer module, yet it neglects the attention on channel dimension. SlowFast combines the slow and fast motion features for complementary effect, yet the achieved performance is unsatisfactory due to the lack of attention mechanism. Based on multidimensional self-attention and cooperative two streams, the significant spatial and temporal-channel features complement each other in LMSA-Net. As shown in Table 6, this model achieves the state-of-the-art performance in both RGB modality and two-stream fusion.

5   Conclusion

In this paper, we propose a Lightweight Multidimensional Self-Attention Network (LMSA-Net), which can balance the relationship between adaptive feature perception and computational overhead of multidimensional attention. With the cross self-attention mechanism based on 2D convolution operators, the contextual relationship between time and channel can be constructed in temporal stream. Combined with spatial self-attention in spatial stream, multidimensional self-attention is achieved by training two streams cooperatively. The experiments on FSD-10 and Diving48-V2 validate the effect of LMSA-Net in fine-grained action recognition. In RGB modality, LMSA-Net achieves the state-of-the-art performance. Compared with the two-stage two-stream models based on RGB and flow, LMSA-Net provides an alternative manner to make RGB modality and motion cues complement each other in an end-to-end method.

References

1. Shao, D., Zhao, Y., Dai, B., Lin, D.: Finegym: a hierarchical video dataset for fine-grained action understanding. In: Proceedings of the IEEE/CVF Conference on Computer Vision and Pattern Recognition, pp. 2616–2625 (2020)
2. Soomro, K., Zamir, A.R., Shah, M.: Ucf101: a dataset of 101 human actions classes from videos in the wild. arXiv preprint arXiv:1212.0402 (2012)
3. Wang, Y., Wang, S., Tang, J., O'Hare, N., Chang, Y., Li, B.: Hierarchical attention network for action recognition in videos. arXiv preprint arXiv:1607.06416 (2016)
4. Zhang, C., Zou, Y., Chen, G., Gan, L.: Pan: towards fast action recognition via learning persistence of appearance. arXiv preprint arXiv:2008.03462 (2020)
5. Sharma, S., Kiros, R., Salakhutdinov, R.: Action recognition using visual attention. arXiv preprint arXiv:1511.04119 (2015)
6. Simonyan, K., Zisserman, A.: Two-stream convolutional networks for action recognition in videos. arXiv preprint arXiv:1406.2199 (2014)
7. Fan, L., Huang, W., Gan, C., Ermon, S., Gong, B., Huang, J.: End-to-end learning of motion representation for video understanding. In: Proceedings of the IEEE Conference on Computer Vision and Pattern Recognition, pp. 6016–6025 (2018)
8. Huang, G., Bors, A.G.: Video classification with finecoarse networks. arXiv preprint arXiv:2103.15584 (2021)

9. Wang, X., Girshick, R., Gupta, A., He, K.: Non-local neural network. In: Proceedings of the IEEE Conference on Computer Vision and Pattern Recognition, pp. 7794–7803 (2018)
10. Perez-Rua, J.-M., Martinez, B., Zhu, X., Toisoul, A., Escorcia, V., Xiang, T.: Knowing what, where and when to look: efficient video action modeling with attention. arXiv preprint arXiv:2004.01278 (2020)
11. Shi, L., Zhang, Y., Cheng, J., Lu, H.: Skeleton-based action recognition with multi-stream adaptive graph convolutional networks. IEEE Trans. Image Process. **29**, 9532–9545 (2020)
12. Huang, Z., Wang, X., Huang, L., Huang, C., Wei, Y., Liu, W.: Ccnet: criss-cross attention for semantic segmentation. In: Proceedings of the IEEE/CVF International Conference on Computer Vision, pp. 603–612 (2019)
13. Liu, S., et al.: FSD-10: a fine-grained classification dataset for figure skating. Neurocomputing **413**, 360–367 (2020)
14. Li, Y., Li, Y., Vasconcelos, N.: Resound: towards action recognition without representation bias. In: Proceedings of the European Conference on Computer Vision (ECCV), pp. 513–528 (2018)
15. Zhang, H., Xin, M., Wang, S., Yang, Y., Zhang, L., Wang, H.: End-to-end temporal attention extraction and human action recognition. Mach. Vis. Appl. **29**(7), 1127–1142 (2018). https://doi.org/10.1007/s00138-018-0956-5
16. Bertasius, G., Wang, H., Torresani, L.: Is space-time attention all you need for video understanding? arXiv preprint arXiv:2102.05095 (2021)
17. Cao, Z., Simon, T., Wei, S.-E., Sheikh, Y.: Realtime multi-person 2d pose estimation using part affinity fields. In: Proceedings of the IEEE Conference on Computer Vision and Pattern Recognition, pp. 7291–7299 (2017)
18. Hu, J., Shen, L., Sun, G.: Squeeze-and-excitation networks. In: Proceedings of the IEEE Conference on Computer Vision and Pattern Recognition, pp. 7132–7141 (2018)
19. He, K., Zhang, X., Ren, S., Sun, J.: Deep residual learning for image recognition. In: Proceedings of the IEEE Conference on Computer Vision and Pattern Recognition, pp. 770–778 (2016)
20. Lin, J., Gan, C., Han, S.: TSM: temporal shift module for efficient video understanding. In: Proceedings of the IEEE/CVF International Conference on Computer Vision, pp. 7083–7093 (2019)
21. Hu, L., Feng, L., Liu, S.: Hfnet: a novel model for human focused sports action recognition. In: Proceedings of the 3rd International Workshop on Multimedia Content Analysis in Sports, pp. 35–43 (2020)
22. Carreira, J., Zisserman, J.: Quo vadis, action recognition? A new model and the kinetics dataset. In: Proceedings of the IEEE Conference on Computer Vision and Pattern Recognition, pp. 6299–6308 (2017)
23. Yu, B., Yin, H., Zhu, Z.: Spatio-temporal graph convolutional networks: a deep learning framework for traffic forecasting. arXiv preprint arXiv:1709.04875 (2017)
24. Woo, S., Park, J., Lee, J.-Y., Kweon, I.S. : Cbam: convolutional block attention module. In: Proceedings of the European Conference on Computer Vision (ECCV), pp. 3–19 (2018)
25. Feichtenhofer, C., Fan, H., Malik, H., He, K.: Slowfast networks for video recognition. In: Proceedings of the IEEE/CVF International Conference on Computer Vision, pp. 6202–6211 (2019)

Unsupervised Domain Adaptation with Self-selected Active Learning for Cross-domain OCT Image Segmentation

Xiaohui Li[1], Sijie Niu[1(✉)], Xizhan Gao[1], Tingting Liu[2], and Jiwen Dong[1]

[1] Shandong Provincial Key Laboratory of Network Based Intelligent Computing, School of Information Science and Engineering, University of Jinan, Jinan, China
sjniu@hotmail.com

[2] Shandong Eye Hospital, State Key Laboratory Cultivation Base, Shandong Provincial Key Laboratory of Ophthalmology, Shandong Eye Institute, Shandong First Medical University and Shandong Academy of Medical Sciences, Jinan, China

Abstract. Segmentation of optical coherence tomography (OCT) images of retinal tissue has become an important task for the diagnosis and management of eye diseases. Deep convolutional neural networks have shown great success in retinal image segmentation. However, a well-trained deep learning model on OCT images from one device often fail when it is deployed on images from a different device since these images have different data distributions. Unsupervised domain adaptation (UDA) can solve the above problem by aligning the data distribution between labeled data (source domain) and unlabeled data (target domain). In this paper, we propose an UDA adversarial learning framework with self-selected active learning. The framework consists of two parts: domain adaptation module (DAM) and self-selected active learning module (SALM). The DAM learns domain-invariant features (i.e., common features) gradually to narrow the distribution discrepancy between two domains. The SALM introduces the target data into source domain through discrepancy method and similarity method, which promotes the DAM to learn unique features of target domain. Extensive experiments show the effectiveness of our method. Compared with the state-of-the-art UDA methods, our method has achieved better performance on two medical cross-domain datasets.

Keywords: Unsupervised domain adaptation · Optical coherence tomography · Images segmentation · Adversarial learning · Active learning

1 Introduction

Macular edema is a swelling of the central retina caused by excess fluid, which leads to an increase in the retina thickness and a decrease in vision. The optical coherence tomography (OCT) imaging technology can clearly present the various cell layers of retina. Accurate segmentation of retinopathy regions in OCT images

T. Mantoro et al. (Eds.): ICONIP 2021, LNCS 13109, pp. 585–596, 2021.
https://doi.org/10.1007/978-3-030-92270-2_50

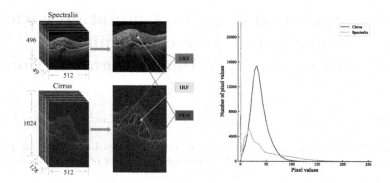

Fig. 1. Illustration of domain shift in OCT images. The OCT images acquired by two imaging devices (Spectralis and Cirrus) are shown on the left, which include three types of lesions (IRF: IntraRetinal Fluid, SRF: SubRetinal Fluid, PED: Pigment Epithelial Detachment). The right side shows gray histogram of two kinds of images.

is very important for disease diagnosis. With the rise of neural networks, deep learning has achieved great success in retinal image segmentation tasks [1]. However, for OCT images, as shown in Fig. 1, the data obtained by different imaging devices have different resolution, image appearance and data distribution. And the individual discrepancy makes the same lesion present different shapes and positions. Due to the above domain shift problems, a well-trained deep learning model on OCT images from one device often fail when it is deployed on images from a different device.

Unsupervised domain adaptation (UDA) aims to address the problem of domain shift and improve the segmentation performance of target domain only when the annotation of source domain is available. In the previous domain adaptation (DA) works, the distance measurement between source features and target features, such as Maximum Mean Discrepancy distance [2,3], Correlation Alignment distance [4] and Wasserstein distance [5], is directly introduced into the objective function to reduce the distribution discrepancy. Although these DA works have achieved a series of good performances, the distance measurements require artificial design, and it is difficult to find the most suitable measurement for each task. To solve the above problems, the idea of generative adversarial network (GAN) [6] is introduced into the DA work. It can learn a measurement from the data spontaneously to reduce domain discrepancy. The domain discriminator is used to distinguish the features of two domains, and the feature extractor tries to confuse the domain discriminator by learning domain-invariant features. Some works use CycleGAN [7,8] to reduce the distribution discrepancy of OCT image by synthesizing the target image into an image similar to the source image. Its disadvantage is the strong dependence on the quality of synthetic image. Some other works [9–12] apply the extracted knowledge from OCT image, fundus image or nature image to train the domain discriminator, such as appearance, feature map, entropy map or boundary. The above work is trying to find the similar (common) features between source domain and target domain, while ignoring the unique information of target domain. Our motivation is to introduce the data distribution of target domain into source domain, which can promote the network to learn the unique features of target domain.

In this paper, we propose an UDA adversarial learning framework with self-selected active learning, where the self-selected active learning does not need human participation. Our method consists of two parts: domain adaptation module (DAM) and self-selected active learning module (SALM). Specifically, the feature extractor and classifier in DAM is used to extract image features and predict segmentation map, while the discriminator in DAM is use to reduce the distribution discrepancy of two domains at feature-level. After that, the DAM can learn the domain-invariant (common) features. Then the SALM selects target data with better segmentation results by discrepancy method and similarity method and introduces them into source domain, which can promote the DAM to learn the common and unique features of the target domain simultaneously. Compared with the results without active learning, the introduction of self-selected active learning could significantly improve the segmentation results of two cross-domain datasets.

Our contributions are: a) We propose an UDA adversarial learning framework for cross-domain semantic segmentation of retinal OCT images. b) We propose a self-selected active learning module, which can promote the framework to learn the common and unique features of target domain simultaneously. c) Compared with the state-of-the-art methods, our method has achieved better performance on two medical cross-domain datasets.

2 Method

Overview. Suppose we have two datasets with different sources. We define them as source domain (s) and target domain (t). The source domain is labeled manually, denoted by $X^s = \{(x_i^s, y_i^s)\}_{i=1}^N$; the target domain is unlabeled, denoted by $X^t = \{(x_i^t)\}_{i=1}^M$, where x and y represent images and ground-truth; N and M are the number of images of source domain and target domain. As shown in the Fig. 2, our DAM consists of two parts: segmenter S_E and discriminator D_F. S_E contains feature extractor F_E and classifier C_F, and the parameters of S_E are shared by source domain and target domain. It takes the original image as input and outputs a probability map: $P(x) \in R^{H \times W \times C}$, where H, W, C represent height, width of image and class number of segmentation.

I. Single Domain Training. The data in source domain is trained in a single domain by using the segmenter S_E, and the supervised loss L_{seg}^{ss} is used to optimize S_E. The loss L_{seg}^{ss} is generalized dice loss, which can alleviate the problem of unstable training caused by small targets. L_{seg}^{ss} is defined as:

$$L_{seg}^{ss} = 1 - \frac{1}{C} \frac{2 \sum_{j=1}^C \omega_j \sum_{i=1}^N y_{ij}^s \cdot \hat{y}_{ij}^s}{\sum_{j=1}^C \omega_j \sum_{i=1}^N (y_{ij}^s + \hat{y}_{ij}^s)} \tag{1}$$

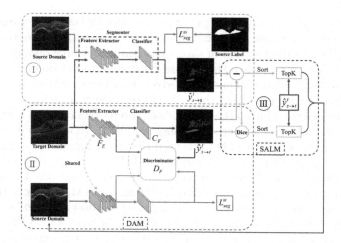

Fig. 2. Overview of our method. Blue arrows represent the flow of source data; red, pink and orange arrows represent the flow of target data. (I) Train source domain, and then directly test target domain training set. (II) Train target domain using the DAM. (III) Comparing the segmentation results of target domain obtained by (I) and (II), the data with better segmentation results were selected for active learning. (Color figure online)

where C is class number of segmentation; $\hat{y}_{ij}^s$ refers to the probability map of the j-th category of the i-th image; ω_j represents the weight of each class, the formula is $\omega_j = \frac{1}{(\sum_{i=1}^{N} y_{ij}^s)^2}$.

After training, we can get a model $M_{s \to s}$ that performs well on source domain. Using this model to directly test the training set of target domain, we can obtain the segmentation result $\hat{y}_{s \to s}^t = M_{s \to s}(s^t)$ without DA. Obviously, most of the results $\hat{y}_{s \to s}^t$ are unsegmentation or undersegmentation.

II. Domain Adaptation. The training sets of source domain and target domain are fed into F_E, and the features $F_E^s = F_E(x^s)$ and $F_E^t = F_E(x^t)$ of two domains are extracted. We can obtain source domain segmentation results through inputting F_E^s into C_F, and calculate the loss L_{seg}^{st} with its ground-truth to optimize the segmenter S_E. L_{seg}^{st} is the weighted sum of BinaryCrossEntropy Loss and Generalized Dice Loss, and defined as:

$$L_{seg}^{st} = -\frac{1}{N} \sum_{i=1}^{N} \sum_{j=1}^{C} (y_{ij}^s \log \hat{y}_{ij}^s + (1 - \hat{y}_{ij}^s) \log(1 - \hat{y}_{ij}^s)) + \lambda_G L_{seg}^{ss} \qquad (2)$$

Meanwhile, we feed F_E^s and F_E^t into D_F to obtain feature maps $D_F^s = D_F(F_E^s)$ and $D_F^t = D_F(F_E^t)$. The D_F can not only distinguish different domains, but also align class-level information. It is derived from the FADA [14], and the implementation details are shown in Fig. 3. The discrimination loss L_D is defined as:

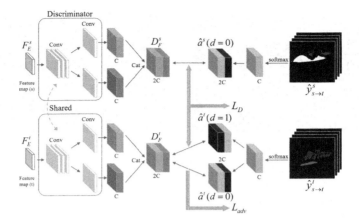

Fig. 3. Details of discriminator. The discriminator is a fully convolutional network. Cat means that outputs are concatenated in channel dimension. The black tensor is **0** with the same shape as $\hat{y}^t_{s \to s}$ and $\hat{y}^t_{s \to t}$.

$$L_D = -\lambda_D \sum_{i=1}^{N} \sum_{j=1}^{2C} \hat{a}^s_{ij}(d=0) \log(D^s_F)_{ij} - \lambda_D \sum_{k=1}^{M} \sum_{j=1}^{2C} \hat{a}^t_{kj}(d=1) \log(D^t_F)_{kj} \quad (3)$$

where d refers to the domain variable and 0 refers to source domain, 1 refers to target domain; $\hat{a}^s(d=0) = [\hat{y}^s, \mathbf{0}]$, $\hat{a}^t(d=1) = [\mathbf{0}, \hat{y}^t]$, and $\mathbf{0}$ is a tensor with the same shape as $\hat{y}^s$ and $\hat{y}^t$.

The adversarial loss L_{adv} is used to confuse D_F and promotes F_E to learn domain-invariant features:

$$L_{adv} = -\lambda_{adv} \sum_{k=1}^{M} \sum_{j=1}^{2C} \hat{a}^t_{kj}(d=0) \log(D^t_F)_{kj} \quad (4)$$

where $\hat{a}^t(d=0) = [\hat{y}^t, \mathbf{0}]$.

After DA training, we can get a model $M_{s \to t}$ that performs well on target domain. Using this model to test the training set of target domain, we can also obtain the segmentation result $\hat{y}^t_{s \to t} = M_{s \to t}(s^t)$.

III. Self-selected Active Learning. In step **I** and **II**, two kinds of segmentation results $\hat{y}^t_{s \to s}$ and $\hat{y}^t_{s \to t}$ are obtained. We introduce the target data with better segmentation into source domain by our proposed SALM to encourage the DAM to learn unique features of target domain. The SALM uses two methods to select images with accurate segmentation: discrepancy method and similarity method.

Discrepancy Method. In the experiment, we found that most of $\hat{y}^t_{s \to s}$ are unsegmented, while most of $\hat{y}^t_{s \to t}$ are under-segmented, so we conclude that the

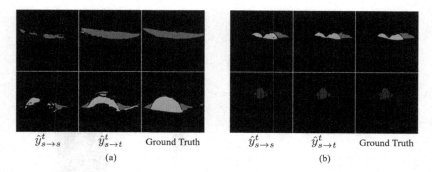

$\hat{y}^t_{s \to s}$ $\hat{y}^t_{s \to t}$ Ground Truth $\hat{y}^t_{s \to s}$ $\hat{y}^t_{s \to t}$ Ground Truth

(a)   (b)

Fig. 4. Segmentation results selected by SALM. (a) shows the segmentation results selected by discrepancy method. (b) shows the segmentation results selected by similarity method.

greater discrepancy between $\hat{y}^t_{s \to s}$ and $\hat{y}^t_{s \to t}$, the closer $\hat{y}^t_{s \to t}$ is to ground-truth. The discrepancy is defined as:

$$\sum_{k=1}^{M}\sum_{j=1}^{C}((\hat{y}^t_{s \to t})_{kj} - (\hat{y}^t_{s \to s})_{kj})^2 \tag{5}$$

We sort the discrepancy from high to low and then obtain the top K target images with segmentation results $\{x^t, \hat{y}^t_{s \to t}\}$.

Similarity Method. The discrepancy method only considers that $\hat{y}^t_{s \to s}$ is worse and $\hat{y}^t_{s \to t}$ is better, but ignores the data that are better in $\hat{y}^t_{s \to s}$ and $\hat{y}^t_{s \to t}$, so we introduce the similarity method to consider the results that are both better. We utilize dice coefficient to calculate similarity, defined as:

$$\frac{1}{C}\sum_{j=1}^{C}\frac{2\sum_{k=1}^{M}(\hat{y}^t_{s \to t})_{kj} \cdot (\hat{y}^t_{s \to s})_{kj}}{\sum_{k=1}^{M}(\hat{y}^t_{s \to t})_{kj} + (\hat{y}^t_{s \to s})_{kj}} \tag{6}$$

We also sort the similarity like discrepancy and then obtain the top K target images with segmentation results $\{x^t, \hat{y}^t_{s \to t}\}$. Because if two segmentation results $\hat{y}^t_{s \to s}$ and $\hat{y}^t_{s \to t}$ are very similar, we conclude that the result is satisfactory. The visualization is shown in Fig. 4. It can be seen that the data selected by our method ($\hat{y}^t_{s \to t}$ column) is close to the ground truth.

3   Experiments and Results

The performance of our method was evaluated in retinal OCT cross-domain dataset from Retinal OCT Fluid Challenge [16]. For all the following experiments, we used Dice Coefficients as evaluation metrics.

Dataset and Preprocessing. The datasets were collected by Cirrus and Spectralis imaging devices. For both devices, we had 24 volumes for training and 14 volumes for testing. For Cirrus, each volume contained 128 slices, so we collected 3072 training slices and 1792 testing slices. For Spectralis, each volume contained 49 slices, so we collected 1176 training slices and 686 testing slices. We resized the data into 512×512 pixel grayscale images uniformly, and conducted two types of experiments with the Cirrus data as source domain and the Spectralis data as target domain (Cirrus→Spectralis); the Spectralis as source domain and the Cirrus data as target domain (Spectralis→Cirrus).

Table 1. Details of our network. ConvBn is the convolution plus batch normalization layer, and the parameter of Conv/ConvBn[∗] represents [kernel_size, output_channels, stride, dilation, padding]; []×n means that the resnet block runs n times in serial, {∗} represents that the operations run in parallel, and then the outputs are concatenated in channel dimension.

Output size	Feature Extractor	Classifier	Discriminator
128×128	ConvBn$[7 \times 7, 32, s=2, d=1, p=3]$ ReLU MaxPool$[s=2]$		
128×128	$\begin{bmatrix}\text{ConvBn}[1 \times 1, 32, s=1, d=1, p=0]\\ \text{ConvBn}[3 \times 3, 32, s=1, d=1, p=1]\\ \text{ConvBn}[1 \times 1, 128, s=1, d=1, p=0]\\ \text{ReLU}\end{bmatrix} \times 3$		
64×64	$\begin{bmatrix}\text{ConvBn}[1 \times 1, 64, s=1, d=1, p=0]\\ \text{ConvBn}[3 \times 3, 64, s=\bar{s}, d=1, p=1]\\ \text{ConvBn}[1 \times 1, 256, s=1, d=1, p=0]\\ \text{ReLU}\end{bmatrix} \times 4$ $\bar{s}=2,1,1,1$	AdaptiveAvgPool$[\text{output\_size}=(1,1)]$ Conv$[1 \times 1, 256, s=1, d=1, p=0]$ Bilinear Interpolate	
64×64	$\begin{bmatrix}\text{ConvBn}[1 \times 1, 128, s=1, d=1, p=0]\\ \text{ConvBn}[3 \times 3, 128, s=1, d=2, p=2]\\ \text{ConvBn}[1 \times 1, 512, s=1, d=1, p=0]\\ \text{ReLU}\end{bmatrix} \times 6$	$\begin{cases}\text{Identity}\\ \text{Conv}[1 \times 1, 256, s=1, d=1, p=0]\\ \text{Conv}[3 \times 3, 256, s=1, d=6, p=6]\\ \text{Conv}[3 \times 3, 256, s=1, d=12, p=12]\\ \text{Conv}[3 \times 3, 256, s=1, d=18, p=18]\end{cases}$	Conv$[1 \times 1, 128, s=1, d=1, p=0]$ LeakyReLU[negative_slope=0.2] Conv$[3 \times 3, 256, s=1, d=1, p=1]$ LeakyReLU[negative_slope=0.2] Conv$[3 \times 3, 512, s=1, d=1, p=1]$
64×64	$\begin{bmatrix}\text{ConvBn}[1 \times 1, 256, s=1, d=1, p=0]\\ \text{ConvBn}[3 \times 3, 256, s=1, d=\bar{d}, p=\bar{p}]\\ \text{ConvBn}[1 \times 1, 1024, s=1, d=1, p=0]\\ \text{ReLU}\end{bmatrix} \times 3$ $\bar{d}=\bar{p}=1,2,4$	Conv$[1 \times 1, 256, s=1, d=1, p=0]$ ConvBn$[3 \times 3, 64, s=1, d=1, p=1]$ ReLU Conv$[1 \times 1, C, s=1, d=1, p=0]$	$\begin{cases}\text{Conv}[1 \times 1, C, s=1, d=1, p=0]\\ \text{Conv}[1 \times 1, C, s=1, d=1, p=0]\end{cases}$
512×512		Bilinear Interpolate	Bilinear Interpolate

Implementation Details. The proposed method was implemented in python using the deep learning framework Pytorch. The network architecture of our framework is shown in Table 1. Each experiment was trained from scratch, without using pre-trained model. The input image size was 512×512, batch-size was 16 and model parameters were optimized by Adam algorithm. For S_E and D_F, the learning rate was set to 1e−4 and 2e−5. And λ_G, λ_D and λ_{adv} were set to 0.001, 0.5 and 0.001 respectively.

592   X. Li et al.

Table 2. Quantitative comparison of segmentation results by different methods. No adaptation means training in source domain and testing in target domain without DA. Target model means to train and test in target domain. $K = 0$ means the experimental result without active learning. $K = 75$ indicates that 75 images are selected by discrepancy method and similar method for active learning, respectively.

Methods	Cirrus→Spectralis				Spectralis→Cirrus			
	IRF	SRF	PED	Avg	IRF	SRF	PED	Avg
No adaptation	44.2	55.0	33.7	44.3	24.7	8.5	1.0	11.4
ADVENT [11]	28.3	29.3	29.1	29.0	42.7	50.7	18.4	37.3
FADA [14]	44.0	55.1	45.0	48.0	56.6	60.9	33.4	50.3
Ours ($K = 0$)	49.5	63.1	47.0	53.2	62.3	76.2	48.0	62.1
Ours ($K = 75$)	53.6	66.7	56.8	59.0	65.3	77.8	51.6	64.9
Target model	65.1	71.5	57.8	64.8	67.5	80.5	52.3	66.8

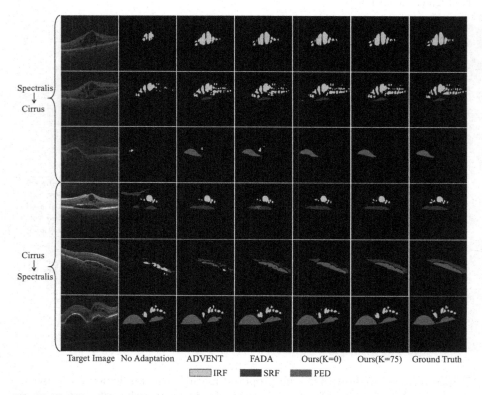

Fig. 5. Qualitative comparison of segmentation results by different methods. The first three rows are the segmentation results of Cirrus data in Spectralis→Cirrus, while the last three rows are the segmentation results of Spectralis data in Cirrus→Spectralis. The first and last columns are the input image and ground-truth. The five columns in middle are the results obtained by our method and the state-of-the-art UDA methods.

Experimental Results. We compared the proposed method with two state-of-the-art methods: ADVENT[1] [11] and FADA[2] [14]. For ADVENT, an entropy discriminator was introduced to narrow the entropy map discrepancy of source domain and target domain. For FADA, a fine-grained discriminator was used for aligning two domains at both image-level and class-level. The quantitative experiment results are shown in Table 2. Both in two experiments, our results are better than ADVENT and FADA, reporting an average Dice of 59.0% and 64.9%, respectively. The qualitative experimental results are shown in Fig. 5.

Table 3. Quantitative comparison of different K-values of our method.

Top K	Cirrus→Spectralis				Spectralis→Cirrus			
	IRF	SRF	PED	Avg	IRF	SRF	PED	Avg
0	49.5	63.1	47.0	53.2	62.3	76.2	48.0	62.1
50	52.5	**67.7**	54.3	58.2	64.7	77.2	**52.0**	64.6
75	53.6	66.7	**56.8**	**59.0**	65.3	**77.7**	51.6	**64.9**
100	**54.1**	67.1	54.4	58.5	64.8	77.5	51.5	64.6

Ablation Study. We conducted an ablation study to investigate the influence of K-value. The value of K was set to 0, 50, 75 and 100. The quantitative experiment results are shown in Table 3. Compared with the baseline DA model without active learning ($K = 0$), the other three K-values had achieved better performance. With the increase of K, the Dice coefficient showed a trend of rising first and then falling, and when K was 75, the best segmentation result could be obtained. This was because a larger K value will make target data selected by SLAM have more wrong information. These experiments also demonstrated the effectiveness of self-selected active learning in our method.

MRI-CT Experiment. We also applied the proposed method to the cross-modality segmentation experiment of cardiac structure in MRI-CT. The dataset came from UCMDA [15], and consisted of unpaired 20 MRI and 20 CT volumes collected at different clinical sites. The ground truth masks included myocardium of the left ventricle (LV-m), left atrium blood cavity (LA-b), left ventricle blood cavity (LV-b), and ascending aorta (AA). The quantitative experiment results are shown in Table 4. It can be seen from the data that our method ($K = 0$) has a slightly lower performance than UCMDA, but it is higher than ADVENT and FADA. After adding SLAM ($K = 75$), compared with the original method ($K = 0$), there is a significant increase of about 6% in both experiments, and it is higher than UCMDA, which demonstrates the robustness of our proposed method. The qualitative experimental results are shown in Fig. 6.

---

[1] https://github.com/valeoai/ADVENT.
[2] https://github.com/JDAI-CV/FADA.

Table 4. Quantitative comparison of MRI/CT image segmentation results.

Methods	MRI→CT					CT→MRI				
	LV-m	LA-b	LV-b	AA	Avg	LV-m	LA-b	LV-b	AA	Avg
No adaptation	0.8	9.1	0.0	11.5	5.4	1.5	6.4	3.1	1.9	3.2
ADVENT [11]	13.9	19.6	42.2	43.3	29.8	20.3	33.0	50.0	17.9	30.3
FADA [14]	36.1	55.3	52.7	43.0	46.8	33.7	47.4	55.6	37.7	43.6
UCMDA [15]	47.8	51.1	57.2	**74.8**	57.7	-	-	-	-	-
Ours ($K = 0$)	43.4	**68.1**	64.5	44.6	55.2	44.1	36.2	70.8	36.0	46.8
Ours ($K = 75$)	**52.0**	67.4	**73.2**	52.7	**61.3**	**45.4**	**50.9**	**76.4**	**38.2**	**52.8**
Target model	87.9	86.9	90.1	88.9	88.4	76.4	76.5	90.6	81.9	81.4

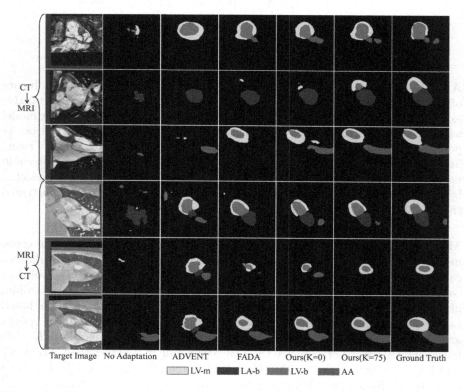

Fig. 6. Qualitative comparison of segmentation results by different methods. The first three rows are the segmentation results of MRI data in CT→MRI, while the last three rows are the segmentation results of CT data in MRI→CT.

4   Conclusion

In summary, we present an UDA adversarial learning framework with self-selected active learning, where the self-selected active learning process does not need human participation. Our method consists of a domain adaptation module and a self-selected active learning module, which can learn the domain-invariant features of two domains and unique features of target domain simultaneously. Our proposed method is validated on the retina OCT and MRI-CT cross-domain datasets, showing that our method can achieve better performance than the state-of-the-art UDA methods.

Acknowledgments. This work is supported by the National Natural Science Foundation of China under Grant No. 61701192, No. 61872419, No. 61873324, the Natural Science Foundation of Shandong Province, China, under Grant No. ZR2020QF107, No. ZR2020MF137, No. ZR2019MF040, No. ZR2019MH106, No. ZR2018BF023, the China Postdoctoral Science Foundation under Grants No. 2017M612178. University Innovation Team Project of Jinan (2019GXRC015), Key Science & Technology Innovation Project of Shandong Province (2019JZZY010324, 2019JZZY010448), and the Higher Educational Science and Technology Program of Jinan City under Grant with No. 2020GXRC057. The National Key Research and Development Program of China (No. 2016YFC13055004).

References

1. Mahapatra, D., Bozorgtabar, B., et al.: Pathological retinal region segmentation from oct images using geometric relation based augmentation. In: Proceedings of the IEEE Conference on Computer Vision and Pattern Recognition, pp. 9611–9620 (2020)
2. Tzeng, E., Hoffman, J., et al.: Deep domain confusion: maximizing for domain invariance. CoRR abs/1412.3474 (2014)
3. Long, M., Cao, Y., et al.: Learning transferable features with deep adaptation networks. In: Proceedings of the 32nd International Conference on Machine Learning. JMLR, Lille, France, pp. 97–105 (2015)
4. Sun, B., Saenko, K.: Deep CORAL: correlation alignment for deep domain adaptation. In: Hua, G., Jégou, H. (eds.) ECCV 2016. LNCS, vol. 9915, pp. 443–450. Springer, Cham (2016). https://doi.org/10.1007/978-3-319-49409-8_35
5. Shen, J., Qu, Y., et al.: Wasserstein distance guided representation learning for domain adaptation. In: Proceedings of the Thirty-Second Conference on Artificial Intelligence. AAAI, New Orleans, Louisiana, USA, pp. 4058–4065 (2018)
6. Goodfellow, I.J., Pouget, A.J., et al.: Generative adversarial nets. In: Proceedings of Annual Conference on Neural Information Processing Systems. ACM, Montreal, Quebec, Canada, pp. 2672–2680 (2014)
7. Philipp, S., David, R., et al.: Using CycleGANs for effectively reducing image variability across OCT devices and improving retinal fluid segmentation. In: Proceedings of the IEEE International Symposium on Biomedical Imaging, pp. 605–609 (2019)
8. David, R., Philipp, S., et al.: Reducing image variability across OCT devices with unsupervised unpaired learning for improved segmentation of retina. Biomed. Opt. Express **11**(1), 346–363 (2020)

9. Wang, J., Chen, Y.W., et al.: Domain adaptation model for retinopathy detection from cross-domain OCT images. In: Proceedings of the Third Conference on Medical Imaging with Deep Learning, in Proceedings of Machine Learning Research, vol. 121, pp. 795–810 (2020)

10. Wang, S., Yu, L., Li, K., Yang, X., Fu, C.-W., Heng, P.-A.: Boundary and entropy-driven adversarial learning for fundus image segmentation. In: Shen, D., et al. (eds.) MICCAI 2019. LNCS, vol. 11764, pp. 102–110. Springer, Cham (2019). https://doi.org/10.1007/978-3-030-32239-7_12

11. Vu, T.H., Jain, H., et al.: ADVENT: adversarial entropy minimization for domain adaptation in semantic segmentation. In: Proceedings of the IEEE Conference on Computer Vision and Pattern Recognition (2019)

12. Chen C., Dou Q., et al.: Synergistic image and feature adaptation: towards cross-modality domain adaptation for medical image segmentation. In: Proceedings of the AAAI Conference on Artificial Intelligence, vol. 33, no. 01, pp. 865–872 (2019)

13. Wang, S., Yu, L., et al.: Patch-Based output space adversarial learning for joint optic disc and cup segmentation. IEEE Trans. Med. Imag. 38(11), 2485–2495 (2019)

14. Wang, H., Shen, T., Zhang, W., Duan, L.-Y., Mei, T.: Classes matter: a fine-grained adversarial approach to cross-domain semantic segmentation. In: Vedaldi, A., Bischof, H., Brox, T., Frahm, J.-M. (eds.) ECCV 2020. LNCS, vol. 12359, pp. 642–659. Springer, Cham (2020). https://doi.org/10.1007/978-3-030-58568-6_38

15. Dou, Q., Chen, O.Y., et al.: Unsupervised cross-modality domain adaptation of ConvNets for biomedical image segmentations with adversarial loss. In: International Joint Conference on Artificial Intelligence (2018)

16. Hrvoje, B., Freerk, V., et al.: RETOUCH-the retinal OCT fluid detection and segmentation benchmark and challenge. IEEE Trans. Med. Imag. 38(8), 1858–1874 (2019)

Adaptive Graph Convolutional Network with Prior Knowledge for Action Recognition

Guihong Lao[1], Lianyu Hu[1], Shenglan Liu[2(✉)], Zhuben Dong[1], and Wujun Wen[1]

[1] Faculty of Electronic Information and Electrical Engineering, Dalian University of Technology, Dalian 116024, Liaoning, China
[2] School of Innovation and Entrepreneurship, Dalian University of Technology, Dalian 116024, Liaoning, China
`liusl@dlut.edu.cn`

Abstract. Skeleton-based action recognition has been paid more and more attention in recent years. Previous researches mainly depend on CNNs or RNNs to capture dependencies among sequences. Recently, graph convolution networks are widely used due to its extraordinary ability to exploit node relationships. We propose a new GCN-based model named PK-GCN which utilizes prior knowledge to design learnable node connections. It can be proved that models can learn adaptive connections by itself, because the node connections can be learned with random initialization. The prior knowledge can be used to design node connections by selecting prominent pairs of joints in actions. By combining the proposed methods above, PK-GCN achieves the best performance in ablation study. Compared with other single-stream GCN-based models, PK-GCN on two large-scale datasets NTU-RGB+D and Kinetics achieves state-of-the-art results.

Keywords: Graph convolutional netowrks · Prior knowledge · Skeleton-based action recognition

1 Introduction

With the evolution of Kinect Series equipments for locating body points and the development of pose estimation algorithms, skeleton-based action recognition becomes a hot topic in recent years. Skeleton data is widely used due to its characteristics: 1) The interference of background scenes has been filtered. 2) The appearance of actors has been ignored. 3) It is easy to store and merge into one file. Besides, skeleton data is also efficient to compute in the real world because of its small size and potential way to perform real-time action recognition.

This study was funded by CCF- Baidu Open Fund (NO. 2021PP15002000), The Fundamental Research Funds for the Central Universities (DUT19RC(3)01) and LiaoNing Revitalization Talents Program (XLYC1806006).

T. Mantoro et al. (Eds.): ICONIP 2021, LNCS 13109, pp. 597–607, 2021.
https://doi.org/10.1007/978-3-030-92270-2_51

The early attempts of using skeleton data encode all body points in a frame into a representation vector to capture their intrinsic relations. Though it is straight, they largely miss the internal relationships among body points thus they are unable to fully excavate the potential of skeleton data. To fully utilize relationships among body joints, graph convolution networks (GCNs) have been widely used recently by constructing spatial-temporal graphs and representing body points as nodes. As human body is a natural instantiation of graph, it's effective to represent relationships among body joints as a graph and further extract high-dimensional features for nodes. According to the predefined graph, each node in the graph updates features with its neighbors and features flow. Each node can connect with distant ones and has a large receptive field through iterations.

Spatial temporal graph convolutional networks [21] (ST-GCN) first introduces GCN into skeleton-based action recognition. It constructs a spatial-temporal graph according to the natural structure of human body and proposes three strategies to divide nodes' neighbors into groups. How to construct node connections, e.g. the adjacent matrix is one of the most important factors in GCNs. However, ST-GCN only adopts the predefined graph to represent node connections which is fixed, and it's unable to capture the dynamic and latent relationships of nodes in complex circumstances. Due to the varieties of actions and variation of poses at different moments, a flexible and learnable adjacent matrix is needed to handle them and improve the flexibility of model.

We attempt to build a flexible adjacent matrix to allow model dealing with complex circumstances by itself and excavate latent node connections with prior knowledge. Firstly, it is crucial to explore the weight of adjacent matrix in GCN and investigate whether the model can learn useful connections by itself. Therefore, the adjacent matrix is initialized randomly to train the model from scratch (the adjacent matrix is learnable). In this case, the model could learn beneficial information from a random distribution, and it achieves similar results with predefined adjacent matrix. Taking together, PK-GCN employs a self-learning strategy for the model to learn node connections and introduce the prior knowledge from the prominent connections of joints to excavate the latent connections of model. The prior knowledge has to be carefully designed by exploring intrinsic relationships during actions and is not limited to natural connections of joints. Thus our method is proposed to select prominent pairs of joints that matter in actions, which is able to promote final results. To obtain intrinsic latent connections of joints, the prominent pair of joints are selected by calculating in each frame and weighting them on histograms. Combining the above methods makes our model adaptive and powerful.

To verify the effectiveness of PK-GCN, we conduct extensive experiments on two large-scale datasets NTU-RGB+D [13] and Kinetics [2]. The experiment results prove that our method is effective in learning adaptive relationships and constructing latent node connections. Besides, our method achieves state-of-the-art performance on these two datasets.

The contributions of this paper are summarized as follows:

1. PK-GCN is able to learn beneficial information from scratch, which can be proved by conducting experiments that set the adjacent matrix randomly initialized and learnable.
2. PK-GCN utilizes a method to design the adjacent matrix with prior knowledge. Besides, extensive experiments are conducted on two large-scale datasets (NTU-RGB+D and Kinetics) and achieves state-of-the-art results, compared with other single-stream GCN-based models.

2   Related Work

2.1   Skeleton-Based Action Recognition

In the early stage of skeleton-based action recognition, methods rely on hand-crafted features for recognition. For example, [4, 20] employ rotation and translation to encode skeleton data for further features extracting. However, these methods are constricted to manual design and unable to utilize the intrinsic relationships of human body structure. With the application of deep learning in action recognition, recurrent neural networks(RNNs) and convolutional neural networks (CNNs) show their great ability in processing skeleton data as sequences of images to establish its internal relationships and extracting high-dimensional features. For example, [3] employs a bi-directional RNN to model skeleton data and send different parts of joints into sub-networks. [22] introduces view-adaptive transformations for skeleton data to make it robust against view changes. [8] encodes skeleton data as an image and employs multi-scale residual networks to process them. However, both RNN and CNN are not able to fully utilize relationships among joints.

2.2   Graph Convolution Networks

Models based on graphs have natural advantages over models which construct intrinsic relationships among joints by representing joints as graph nodes. Although early methods like conditional random field [7, 17, 23] (CRF) and hidden markov model [11, 12] (HMM) are widely used, they can not extract high-dimensional features while GCNs make up this drawback. GCNs fall into two categories: spatial perspective and spectral perspective. The methods in spatial perspective define convolution operations within neighbors while those in spectral perspective utilize graph Fourier transform and Laplacian operator to process data. Methods in spatial perspective are widely used in action recognition due to the fact that it fits the natural graph structure of human body. Since ST-GCN [21] introduces GCN into action recognition, a lot of GCN-based methods have emerged and achieved great performances. Part-based graph convolutional network [19] (PB-GCN) divides human body into small parts and conducts GCN in them separately. Then PB-GCN aggregates their results for classification. Actional-structural graph convolutional networks [10] (AS-GCN) makes use of A-Links and S-Links to capture latent dependencies among body joints and explore structural information. Global relation reasoning graph convolutional networks [5] (GR-GCN) divides relationships among joints into strong parts and weak parts for capturing them separately, and concatenating local frames to capture local information.

Among recent works, two-stream adaptive graph convolutional networks [14] (2S-AGCN) shares some similarities with our work. It adds a learnable matrix and a data-dependent matrix to predefined adjacent matrix for latent relationships capturing. Although both our methods and 2S-AGCN utilize a learnable matrix, we share different intuitions. We mask the adjacent matrix to emphasize prominent nodes which are only a part of all nodes, while 2S-AGCN updates matrix weights for all nodes adaptively. Besides, prior knowledge is introduced to excavate new latent prominent pairs, while 2S-AGCN calculates data-dependent matrix by attention mechanism which does not introduce new connections.

3   Training Models Without Prior Knowledge

In this section, an experiment without any prior knowledge is set up to prove GCN-based model could learn beneficial information from data, even though the adjacent matrix is random initialization. That means the initial adjacent matrix doesn't contain any relationships among joints, and it needs to learn nodes connection from scratch.

For GCN models, the calculation quantity is proportional to the square of node number. Therefore, A threshold value α is set to limit the selection of those prominent nodes and avoid too many computations. Our strategy can be described that those values of the adjacent matrix higher than α will be selected and subtract α. By changing the threshold value α, we are able to control the number of prominent nodes and their connection weights, as well as computation complexities.

Table 1. Experiment results with different α

α	−0.2	−0.1	0	0.1	0.2
Accuracy (frames = 128)	80.4%	80.0%	80.3%	78.9%	2.13%
Accuracy (frames = 256)	81.5%	80.2%	81.7%	80.1%	1.67%

Setup: ST-GCN was employed as our backbone to conduct experiments on NTU-RGB+D. Besides, the average of initialized adjacent matrix is 0 and the variance is 0.001. And other settings are the same as what ST-GCN adopts.

Experiment Results: Our experiment results are listed in Table 1 and the histogram of activated nodes with diverse α in Fig. 1. As Table 1 shows, when α is 0.2, there are only 4 activated nodes so that the network won't converge. Furthermore, too few activated nodes are inadequate for the model to update features. But the value of α between -0.2 and 0.1 can converge naturally for model and the peak performance can be observed when α is 0. Being a threshold value, a negative α means that most nodes are encouraged to be activated in the model. Although it brings large amounts of computations, the performance of model is similar to the situation that α is 0. Thus it's a better choice to set α as 0 due to its balance between accuracy and computations. Our results demonstrate that models are able to learn node connections from data without prior knowledge and achieve similar accuracy with ST-GCN (81.5%)

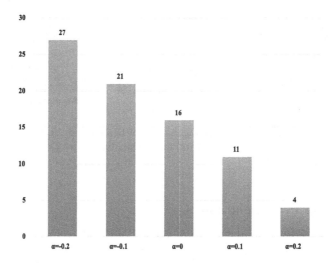

Fig. 1. The number of activated nodes. When α is 0.2, the network won't converge due to the lack of node connections. When α adopts other values, the model can achieve similar results with ST-GCN.

4   PK-GCN

In this section, The prior knowledge is utilized to design the adjacent matrix and help model learn intrinsic relationships among joints. The prominent pairs of joints are selected as prior knowledge by extracting other features manually. This is a complementary experiment for the learnable adjacent matrix explored in the previous section.

4.1   Human Body Structure

Human body structure, which can be expressed to nodes by body joints, is a natural instantiation of graph. Such a characteristic is widely used in ST-GCN and other GCN models. The number of nodes is defined according to the setting of depth sensors or pose estimation algorithms, and the edges are defined according to the natural connections of body joints. It is worth noting that the adjacent matrix is fixed and won't be updated during training. Hence the adjacent matrix only models the physical structure of human body and is unable to capture the dynamic and latent relationships among joints.

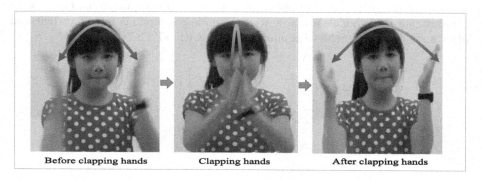

Fig. 2. An example of actions: clapping hands. When human clap hands, hands are shaking violently while others joints stand still. Thus hands are the prominent pairs in this action.

4.2   Selecting Prominent Joints

The predefined edges can be determined by physical connections from body structure, which are called apparent connections. However, in many cases, the latent connections which can not be captured easily, are crucial for recognizing actions. For example, as shown in Fig. 2, the connection between left hand and right hand is crucial when training the model to recognize the action clapping hands. To introduce latent connections as prior knowledge into the GCN model, the key is to excavate latent connections of joints and model physical connections to further express the intrinsic relationships of joints in actions. The prominent joint histogram is essential to select those prominent joints and calculate their weights. More details about our method can be described as follows:

Empirically, for a certain action, some joints are almost immobile as the fulcrum of human body while some joints change around them largely. The moving joints are crucial to recognize actions, and are more prominent than other joints in most cases. Therefore, the prominent connections will be saved and used to recognize actions. The weights of the prominent connections are assigned by normalization. And the latent prominent connections among joints are established successfully.

To obtain the prominent pairs of joints, two joints whose coordinates change most in consecutive two frames will be selected. The two joints are marked as prominent pairs because they carry prominent features in actions. For example, as shown in Fig. 2, when clapping hands, our hand movements are most intense while other joints move little. Thus hands are the prominent pairs of joints that can reflect intrinsic features of clapping hands. For all videos in training set, all frames are processed in the same way. Then the two joints which change most dramatically as prominent pairs are selected and plot the corresponding frequencies on the histogram. Finally, the top K pairs are selected and normalized with their sum equal to M to control their weights. These prominent pairs are designed to introduce prior knowledge into model.

Figure 3 shows the top 10 pairs in all actions as an example. From the figure, we can see that the most intense of pairs are distributed in the limbs as latent connections.

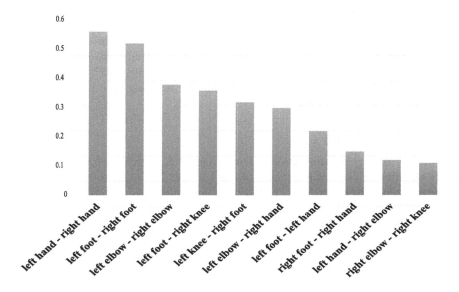

Fig. 3. Distribution of prominent pairs.

5   Experiments

5.1   Datasets

NTU-RGB+D. NTU-RGB+D [13] is a large-scale RGB+D dataset containing 56880 skeleton action sequences and 60 classes. The dataset's annotations include 3D coordinates of 25 joints. It can be evaluated by two protocols: Cross-Subject and Cross-View. The Cross-Subject part provides 40320 sequences performed by 20 subjects for training set and others for test set. The Cross-View part divides data by camera views into 37920 training sequences and 18960 test sequences, respectively.

Kinetics. Kinetics [2] is a large-scale dataset for action recognition containing about 240000 clips and 400 classes. The skeleton data is extracted by the public available toolkit OpenPose [1]. The skeleton data provides 3D coordinates including X, Y coordinates and 1D confidence score for actions with 18 joints. In our experiments, the paper only uses the 2D coordinates and discard the confidence score provided by OpenPose.

5.2   Experiment Setup

Our model adopts the same backbone as ST-GCN for fair comparison, which contains 9 units and follows a global average pooling layer and a SoftMax classifier for recognition. Instead of window sampling strategy, uniform sampling strategy is a better strategy to Normalize the data size and the sample numbers can be set to 128 or 256. Input data should be centralized before fed into model. About training details, the learning rate is 0.1 and the batch size is 64. The model is trained for 300 epochs totally and the learning rate is updated to 10 times after epoch 150 and 240.

5.3   Ablation Study

Numbers of Prominent Pairs. The values of K including 9, 12, 15, 18, generate the different experimental results in Table 2. The table shows that the best accuracy comes in when K is 15, and too many pairs lead to more computations and similar accuracy.

Table 2. Experiment results with different K

K	9	12	15	18
Accuracy (%)	77.4	82.1	82.5	82.3

Total Weights of Prominent Pairs. To control the weights of prominent pairs, their total weights can be calculated as a certain number M and normalize the weights with M. The experiment with M in 3, 5, 7 lists the results in Table 3. From the table, the model achieves the best result when setting M as 5. And the accuracy degrades if prominent pairs make too much effect.

Table 3. Experiment results with different M

M	3	5	7
Accuracy (%)	78.3	82.5	82.0

Setting the Adjacent Matrix Learnable with Prior Knowledge. The adjacent matrix of our model can be divided into three parts, including A_k,

B_k and C_k. A_k, which is predefined and fixed, is the physical structure of human body adopted by common GCN-based models. B_k, which represents latent prominent connections, is the adjacent matrix calculated by prominent pairs. C_k, which is a learnable matrix and can be updated during training, is able to learn complementary connections from data and the initialization is set to zeros. According to the results in Sect. 3, the mask value α is set to 0.

The ablation study depends on these three factors (A_k, B_k and C_k) and more experiments can be listed by combining them together. From the Table 4, Employing separate matrix leads to similar results with ST-GCN, while Combining them together brings the best accuracy up to 85.5% which get a promotion of 4.0% compared to ST-GCN.

Table 4. Experiment results with different type of adjacent matrixs

Type of adjacent matrixs	A_k	B_k	C_k	A_k+B_k+C_k
Accuracy (%)	81.5	82.5	81.7	85.5

6   Comparisons with the State-of-the-Art

It is an indispensable work to compare our method with other state-of-the-art methods on NTU-RGB+D and Kinetics. On NTU-RGB+D, our method is trained with two benchmarks: Cross-Subject and Cross-View. The results of these two datasets are listed in Table 5 and Table 6, respectively. On the two datasets, PK-GCN surpasses CNN-based methods largely. Generally speaking, our method surpasses most methods except for 2s-AGCN On NTU-RGB+D.

Table 5. Experiment results on NTU-RGB+D

Methods	Cross-Subject (%)	Cross-View (%)
Temporal Conv [6]	74.3	83.1
ST-GCN [21]	81.5	88.3
DPRL [18]	83.5	89.8
SR-TSL [15]	84.8	92.4
HCN [9]	86.5	91.1
AS-GCN [10]	86.8	94.2
2s-AGCN [14]	88.3	95.4
PB-GCN [19]	87.5	93.2
RA-GCN [16]	85.9	93.5
GR-GCN [5]	87.5	94.3
PK-GCN (128 frames)	**85.5**	**92.9**
PK-GCN (256 frames)	**87.8**	**95.2**

Table 6. Experiment results on Kinetics

Methods	Accuracy (%)
TCN [6]	20.3
ST-GCN [21]	31.5
2s-AGCN [14]	34.5
AS-GCN [10]	34.8
PK-GCN (128 frames)	**33.7**
PK-GCN (256 frames)	**35.6**

However, our PK-GCN only uses one stream while 2s-AGCN uses one extra stream named bone stream. What's more, our method establishes a state-of-the-art result on Kinetics using skeleton.

7   Conclusion

In this paper, we propose a novel GCN-based model named PK-GCN which introduces prior knowledge into model design and allow the model to learn its adjacent matrix from data. It can be proved that without prior knowledge, the model can learn intrinsic connections among joints from data from scratch. Then PK-GCN is come up to introduce prior knowledge into model design by selecting prominent pairs of joints. By combining the methods above, PK-GCN achieves state-of-the-art results on two large-scale datasets. We hope our method can provide some insights to excavate connections among joints for future researchers.

References

1. Cao, Z., Simon, T., Wei, S.E., Sheikh, Y.: Realtime multi-person 2D pose estimation using part affinity fields. In: CVPR (2017)
2. Carreira, J., Zisserman, A.: Quo vadis, action recognition? a new model and the kinetics dataset. In: proceedings of the IEEE Conference on Computer Vision and Pattern Recognition, pp. 6299–6308 (2017)
3. Du, Y., Wang, W., Wang, L.: Hierarchical recurrent neural network for skeleton based action recognition. In: Proceedings of the IEEE Conference on Computer Vision and Pattern Recognition, pp. 1110–1118 (2015)
4. Fernando, B., Gavves, E., Oramas, J.M., Ghodrati, A., Tuytelaars, T.: Modeling video evolution for action recognition. In: Proceedings of the IEEE Conference on Computer Vision and Pattern Recognition, pp. 5378–5387 (2015)
5. Gao, X., Hu, W., Tang, J., Liu, J., Guo, Z.: Optimized skeleton-based action recognition via sparsified graph regression. In: Proceedings of the 27th ACM International Conference on Multimedia, pp. 601–610 (2019)
6. Kim, T.S., Reiter, A.: Interpretable 3D human action analysis with temporal convolutional networks. In: 2017 IEEE Conference on Computer Vision and Pattern Recognition Workshops (CVPRW), pp. 1623–1631. IEEE (2017)

7. Lafferty, J., McCallum, A., Pereira, F.C.: Conditional random fields: probabilistic models for segmenting and labeling sequence data (2001)
8. Li, B., Dai, Y., Cheng, X., Chen, H., Lin, Y., He, M.: Skeleton based action recognition using translation-scale invariant image mapping and multi-scale deep CNN. In: 2017 IEEE International Conference on Multimedia & Expo Workshops (ICMEW), pp. 601–604. IEEE (2017)
9. Li, C., Zhong, Q., Xie, D., Pu, S.: Co-occurrence feature learning from skeleton data for action recognition and detection with hierarchical aggregation. arXiv preprint arXiv:1804.06055 (2018)
10. Li, M., Chen, S., Chen, X., Zhang, Y., Wang, Y., Tian, Q.: Actional-structural graph convolutional networks for skeleton-based action recognition. In: Proceedings of the IEEE Conference on Computer Vision and Pattern Recognition, pp. 3595–3603 (2019)
11. Rabiner, L., Juang, B.: An introduction to hidden Markov models. IEEE ASSP Mag. **3**(1), 4–16 (1986)
12. Rabiner, L.R.: A tutorial on hidden Markov models and selected applications in speech recognition. Proc. IEEE **77**(2), 257–286 (1989)
13. Shahroudy, A., Liu, J., Ng, T.T., Wang, G.: NTU RGB+D: a large scale dataset for 3D human activity analysis. In: Proceedings of the IEEE Conference on Computer Vision and Pattern Recognition, pp. 1010–1019 (2016)
14. Shi, L., Zhang, Y., Cheng, J., Lu, H.: Two-stream adaptive graph convolutional networks for skeleton-based action recognition. In: Proceedings of the IEEE Conference on Computer Vision and Pattern Recognition, pp. 12026–12035 (2019)
15. Si, C., Jing, Y., Wang, W., Wang, L., Tan, T.: Skeleton-based action recognition with spatial reasoning and temporal stack learning. In: Proceedings of the European Conference on Computer Vision (ECCV), pp. 103–118 (2018)
16. Song, Y.F., Zhang, Z., Wang, L.: Richly activated graph convolutional network for action recognition with incomplete skeletons. In: 2019 IEEE International Conference on Image Processing (ICIP), pp. 1–5. IEEE (2019)
17. Sutton, C., McCallum, A., et al.: An introduction to conditional random fields. Found. Trends® Mach. Learn. **4**(4), 267–373 (2012)
18. Tang, Y., Tian, Y., Lu, J., Li, P., Zhou, J.: Deep progressive reinforcement learning for skeleton-based action recognition. In: Proceedings of the IEEE Conference on Computer Vision and Pattern Recognition, pp. 5323–5332 (2018)
19. Thakkar, K., Narayanan, P.: Part-based graph convolutional network for action recognition. arXiv preprint arXiv:1809.04983 (2018)
20. Vemulapalli, R., Arrate, F., Chellappa, R.: Human action recognition by representing 3D skeletons as points in a lie group. In: Proceedings of the IEEE Conference on Computer Vision and Pattern Recognition, pp. 588–595 (2014)
21. Yan, S., Xiong, Y., Lin, D.: Spatial temporal graph convolutional networks for skeleton-based action recognition. In: Thirty-Second AAAI Conference on Artificial Intelligence (2018)
22. Zhang, P., Lan, C., Xing, J., Zeng, W., Xue, J., Zheng, N.: View adaptive recurrent neural networks for high performance human action recognition from skeleton data. In: Proceedings of the IEEE International Conference on Computer Vision, pp. 2117–2126 (2017)
23. Zheng, S., et al.: Conditional random fields as recurrent neural networks. In: Proceedings of the IEEE International Conference on Computer Vision, pp. 1529–1537 (2015)

Self-adaptive Graph Neural Networks for Personalized Sequential Recommendation

Yansen Zhang[1], Chenhao Hu[1], Genan Dai[1], Weiyang Kong[1],
and Yubao Liu[1,2(✉)]

[1] Sun Yat-Sen University, Guangzhou, China
{zhangys7,huchh8,daign,kongwy3}@mail2.sysu.edu.cn
[2] Guangdong Key Laboratory of Big Data Analysis and Processing,
Guangzhou, China
liuyubao@mail.sysu.edu.cn

Abstract. Sequential recommendation systems have attracted much attention for the practical applications, and various methods have been proposed. Existing methods based on graph neural networks (GNNs) mostly capture the sequential dependencies on an item graph by the historical interactions. However, due to the pre-defined item graph, there are some unsuitable edges connecting the items that may be weakly relevant or even irrelevant to each other, which will limit the ability of hidden representation in GNNs and reduce the recommendation performance. To address this limitation, we design a new method called Self-Adaptive Graph Neural Networks (SA-GNN). In particular, we employ a self-adaptive adjacency matrix to improve the flexibility of learning by adjusting the weights of the edges in the item graph, so as to weaken the effect of unsuitable connections. Empirical studies on three real-world datasets demonstrate the effectiveness of our proposed method.

Keywords: Sequential recommendation · Graph neural networks · Self-adaptive adjacency matrix

1 Introduction

Sequential recommendation predicts the item(s) that a user may interact with in the future given the user's historical interactions. It has attracted much attention, since the sequential manner in user historical interactions can be used for future behavior prediction and recommendation in many real-world applications. Numerous methods based on user's past interactions have been proposed for sequential recommendation, such as the Markov Chain based methods [3,15], the recurrent neural networks (RNNs) [5,6,17,18] and the convolutional neural networks (CNNs) [19,28,29], etc. Due to the ability of aggregating neighborhood information and learning local structure [16], GNNs are a good match to model the user's

Supported by the National Nature Science Foundation of China (61572537, U1501252).

T. Mantoro et al. (Eds.): ICONIP 2021, LNCS 13109, pp. 608–619, 2021.
https://doi.org/10.1007/978-3-030-92270-2_52

personalized intents over time. Recently, a surge of works have employed GNNs for sequential recommendation and obtained promising results [12, 23–26]. In this paper, we also pay close attention to the GNN-based methods.

Specifically, existing methods need to construct an item graph for all sub-sequences generated in a pre-defined manner based on the given user-item interactions. Due to the uncertainty of user behaviors or external factors [22], some items in the sequence may be weakly relevant or even irrelevant. These items are connected by some unsuitable edges, which are assigned the fixed weights in the item graph. These weights will not change during the learning process of GNN. In this way, the items will aggregate inappropriate information from neighbors and result in bad effects on the recommendation performance. For example, given a sub-sequence of a user, s = (MacBook, iPhone, Bread, iPad, Apple Pencil). The connections among the 'Bread' and the other items will negatively affect the learning of items representations and cannot effectively capture the true interests of the user. To address this issue, in our model, we design a self-adaptive adjacency matrix to weaken the effect of unsuitable connections by adjusting the weights of the edges in the item graph. The self-adaptive method can learn the weights among item 'Bread' and others based on user-item interactions. In particular, it can learn the different weights on any pair of connections in the sub-sequence, while existing methods will not. By doing so, the true relation among items can be learned in our method, and more accurate item embedding representations can be modeled for next prediction tasks.

In general, our contributions are listed below. First, we propose a GNN-based method named SA-GNN for personalized sequential recommendation. In SA-GNN, we automatically learn the weights among the items in the sequences by a self-adaptive adjacency matrix. In addition, we try to capture both local interest and global interest of user to enhance the prediction performance. For instance, the user's local interest hidden in the given sub-sequence s is the electronic product. Accordingly, the global interest of the user is often hidden in long-range sequence. Second, detailed experiments conducted on three representative datasets from real-world demonstrate the effectiveness of our method compared with the state-of-art approachs.

2   Related Work

The sequential recommendation refers to recommend which items the user will visit in the future. It usually converts a user's interactions into a sequence as input. Naturally the most conventional popular Markov Chain based methods [3, 15] capture the sequential patterns by item-item transition matrices. Recently, some deep neural network-based methods of learning sequential dynamics have been proposed due to the success of sequence learning in the field of natural language processing. On one hand, some session-based recommendation methods [5, 6, 17, 18] adopted RNNs to learn the sequential patterns. The general thought of these models is to represent user's interest into a vector with different recurrent architectures and loss functions given the historical records. Apart from RNNs, some CNN-based models [19,27–29] are also proposed for recommendation. For instance, pioneering work [19] proposes to apply the CNNs on item embedding sequence, and capture the local contexts by the convolutional operations. On the other hand, attention mechanisms [8,10,17] is utilized in sequential recommendation and achieves promising performance. They adopt self-attention methods to consider interacted items adaptively. In addition, memory networks [1,12] are utilized to memorize the important items in predicting the user's future behaviors. GNN can not only process graph structure data, but also have superiority in capturing richer information in sequence data [25]. For example, in session-based recommendation, [24] first uses GNN to capture complex relationships among items in a session sequence. GC-SAN [26] uses GNN with self-attention mechanism to process session data. MA-GNN [12] applies GNN and external memories to model the user interests. However, different from these studies, we learn the weights in the item graph by a self-adaptive adjacency matrix to achieve better learning.

3   Problem Statement

The task of sequential recommendation uses the historical user-item interaction sequences as input and predicts the next-item that the user will interact with. Let $\mathcal{U} = \{u_1, u_2, \ldots, u_{|\mathcal{U}|}\}$ denotes a set of users, and $\mathcal{I} = \{i_1, i_2, \ldots, i_{|\mathcal{I}|}\}$ denotes a set of items, where $|\mathcal{U}|$ and $|\mathcal{I}|$ are the number of users and items, respectively. The user-item interaction sequences can be represented by a sequence $\mathcal{S}^u = (i_1^u, i_2^u, \ldots, i_{|\mathcal{S}^u|}^u)$ in chronological order, where $i_*^u \in \mathcal{I}$ represents a clicked item of the user u from $\mathcal{U}$ within the sequence $\mathcal{S}^u$. We define the sequential recommendation task by the above notations as follows. Given the historical sub-sequence $\mathcal{S}_{1:t}^u (t < |\mathcal{S}^u|)$ of M users, the goal is to recommend top K items from all of $|\mathcal{I}|$ items $(K < |\mathcal{I}|)$ for every user and assess whether the items in $\mathcal{S}_{t+1:|\mathcal{S}^u|}^u$ appear in the recommended list.

4   Proposed Methodology

In this section, we present the personalized sequential recommendation model SA-GNN proposed. It consists of three components, i.e., embedding layer, local interest modeling layer and prediction layer, demonstrated in Fig. 1. We first introduce the input data and how to generate the training data with future contexts. Next, we illustrate the self-adaptive adjacency matrix and attention module to get the more accurate items representations and select representative items that can reflect user's preference, respectively. Then we could generate the user's local interest. Moreover, we model unified interests representation of the user by combining the global interest with the local interest to capture the personalized user's interests. Lastly, we present the prediction layer and how to train the SA-GNN.

4.1   Embedding Layer

To obtain local interest of the user explicitly, we generate fine-grained subsequences by splitting the historical sequence and adopting a way of sliding

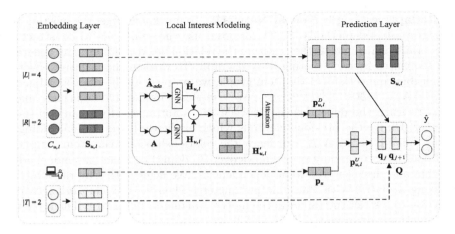

Fig. 1. The architecture of SA-GNN. In the embedding layer, the top four items of the $C_{u,l}$ represent the past contexts, and the two items below represent the future contexts. The $\odot$ represents the element-wise product.

window. Meanwhile, to better exploit the past and future contexts information, we extract every left $|L|$ successive items and right $|R|$ successive items as input and the middle $|T|$ items as the targets to be predicted for each user u, where $C_{u,l} = \{i_l, \ldots, i_{l+|L|-1}, i_{l+|L|+|T|}, \ldots, i_{l+|L|+|R|+|T|-1}\}$ is the l-th sub-sequence of the user u. And items in $T_{u,l} = \{i_{l+|L|}, \ldots, i_{l+|L|+|T|-1}\}$ represent the corresponding targets. The input is a sequence of $|L| + |R|$ items. We embed every item $i \in \mathcal{I}$ by an item embedding matrix $\mathbf{E} \in \mathbb{R}^{d \times |\mathcal{I}|}$, where d and $|\mathcal{I}|$ are the item embedding dimension size and the number of items, respectively. Then the item sub-sequence $C_{u,l}$ embeddings for user u are represented as follows:

$$\mathbf{S}_{u,l} = [\mathbf{e}_l, \cdots, \mathbf{e}_{l+|L|-1}, \mathbf{e}_{l+|L|+|T|}, \cdots, \mathbf{e}_{l+|L|+|R|+|T|-1}] \tag{1}$$

To distinguish the input item embeddings, we use the $\mathbf{Q} \in \mathbb{R}^{d \times |\mathcal{I}|}$ to represent the output item embeddings. Similar to the item embeddings, for a user u, we also have an user embedding $\mathbf{p}_u \in \mathbb{R}^{d'}$, and d' is the user embedding dimension size.

4.2 Local Interest Modeling

A user's local interest is based on several recently visited items in a short-term period and reflects the user's current intent [12]. Here we will further introduce how to apply the self-adaptive adjacency matrix with GNN and attention mechanism for modeling user's local interest. First, we illustrate that how to get better item representations by the self-adaptive adjacency matrix in detail. Second, with an attention module, we model the user's local interest by considering the different important score of items in the sub-sequence.

Self-adaptive Graph Neural Network. To our knowledge, the existing popular graph-based methods [12,24,26] often use the same rule to construct the item graph, which means they capture the relation between items in all generated sub-sequences by constructing a fixed graph. To weaken the negative effects of unsuitable connections in the item graph, we propose a self-adaptive adjacency matrix $\hat{\mathbf{A}}_{ada} \in \mathbb{R}^{(|L|+|R|) \times (|L|+|R|)}$. We use this self-adaptive adjacency matrix $\hat{\mathbf{A}}_{ada}$ to learn the dependencies between items automatically in an end-to-end manner. Unlike the previous methods which explicitly learn the embeddings of items, our proposed $\hat{\mathbf{A}}_{ada}$ can learn the relationship between items implicitly and does not require any prior knowledge. By doing so, we let the model adjust weights and discover true item dependencies. We achieve this by randomly initializing a matrix with learnable parameters. Then we use the *tanh* activation function to restrict the values in the matrix ranging from -1 to 1. The layer-wise propagation rule in matrix form as follows.

$$\hat{\mathbf{H}}_{u,l}^{(r)} = tanh(\mathbf{W}_{h_1} \mathbf{S}_{u,l}^{(r)} \hat{\mathbf{A}}_{ada}) \tag{2}$$

where $\mathbf{W}_{h_1} \in \mathbb{R}^{d \times d}$ controls the weights in the GNN. $\mathbf{S}_{u,l}^{(r)} \in \mathbb{R}^{d \times (|L|+|R|)}$ is the final hidden state of sub-sequence $C_{u,l}$ after propagation step r and $\mathbf{S}_{u,l}^{(0)} = \mathbf{S}_{u,l}$.

To better model the accurate item embeddings, we also consider the existing GNN methods. The item graph construction process as follows. For every item in interacted sequences, we extract two subsequent items and build edges between them. According to the process of generating sub-sequences mentioned before, we will not add edges for items in the middle of the sequence unless the length of target $|T|$ is less than the number of subsequent items to be added edges. We apply this for every user, and for all users with the item pairs extracted, we count the total number of edges. Then we perform row regularization on the adjacency matrix. We use $\mathbf{A} \in \mathbb{R}^{(|L|+|R|) \times (|L|+|R|)}$ to represent the adjacency matrix. Similarly, we use the $\mathbf{A}$ to aggregate the neighboring information as follows:

$$\mathbf{H}_{u,l}^{(r)} = tanh(\mathbf{W}_{h_2} \mathbf{S}_{u,l}^{(r)} \mathbf{A}) \tag{3}$$

where $\mathbf{W}_{h_2} \in \mathbb{R}^{d \times d}$ is the weights in the GNN. Then we incorporate these two embeddings by the element-wise product.

$$\mathbf{H}_{u,l}^{\prime(r)} = \hat{\mathbf{H}}_{u,l}^{(r)} \odot \mathbf{H}_{u,l}^{(r)} \tag{4}$$

After the propagation of R layers, we can get the final latent vector $\mathbf{H}_{u,l}^{\prime(\mathrm{R})}$ of each sub-sequence $C_{u,l}$. For simplicity, we mark $\mathbf{H}_{u,l}^{\prime}$ rather than $\mathbf{H}_{u,l}^{\prime(\mathrm{R})}$. The final latent state of each sub-sequence not only mines its own characteristics, but also gathers its R-order neighbors' information.

Attention Module. As we described before, to infer the user's local preference better, we capture the different important scores for each item embeddings generated by GNN. Inspired by [12,13,20], we assign each item embedding with an attention weight vector by an importance score matrix.

$$\mathbf{S}'_{u,l} = softmax(\mathbf{W}_{a_2}(tanh(\mathbf{W}_{a_1}\mathbf{H}'_{u,l} + \mathbf{b}_{a_1}) + \mathbf{b}_{a_2})) \qquad (5)$$

where $\mathbf{S}'_{u,l} \in \mathbb{R}^{d_a \times (|L|+|R|)}$ is the matrix of attention weight, $\mathbf{W}_{a_1} \in \mathbb{R}^{d \times d}$, $\mathbf{W}_{a_2} \in \mathbb{R}^{d_a \times d}$, $\mathbf{b}_{a_1} \in \mathbb{R}^d$, and $\mathbf{b}_{a_2} \in \mathbb{R}^{d_a}$ are the parameters. The d_a represents that we want to assign the important scores from the embeddings with d_a aspects in $\mathbf{H}'_{u,l}$.

By multiplying sub-sequence embeddings with the attention weight matrix, we can represent the matrix form of a sub-sequence:

$$\mathbf{Z}_{u,l} = \mathbf{S}'_{u,l}\mathbf{H}'^{\mathrm{T}}_{u,l} \qquad (6)$$

where $\mathbf{Z}_{u,l} \in \mathbb{R}^{d_a \times d}$ is the matrix representation of the sub-sequence. Then we have a *Avg* (average) function to aggregate the sub-sequence matrix representation into a vector representation. So we can obtain the user's local interest representation as follows:

$$\mathbf{p}^D_{u,l} = Avg(tanh(\mathbf{Z}_{u,l})) \qquad (7)$$

4.3 Unified Interest Modeling

The user's global interest reflects the intrinsic characteristics of a user and can be represented as the embedding $\mathbf{p}_u$. So we concatenate the local and global interest of the user, and then adopt the linear transformation operation to obtain the unified personalized representation of the user:

$$\mathbf{p}^U_{u,l} = [\mathbf{p}^D_{u,l}; \mathbf{p}_u] \cdot \mathbf{W}_u \qquad (8)$$

where $[\cdot; \cdot] \in \mathbb{R}^{d+d'}$ denotes vertical concatenation, and matrix $\mathbf{W}_u \in \mathbb{R}^{(d+d') \times d}$ is the weight of linear transformation.

4.4 Prediction and Training

As shown in [7,11,12], it is important for the sequential recommendation to consider the effect of item co-occurrence. To capture co-occurrence patterns of the item, following [11], we also adopt the inner product to model the item relations between the input item embeddings and the output item embeddings:

$$\sum_{\mathbf{e}_k \in \mathbf{S}_{u,l}} \mathbf{e}_k^{\mathrm{T}} \cdot \mathbf{q}_j \qquad (9)$$

where $\mathbf{q}_j$ is the j-th column of the embedding matrix $\mathbf{Q}$.

We combine the aforementioned factors together to infer user preference. The prediction value of user u on item j of the l-th sub-sequence is:

$$\hat{y}_{u,j} = \mathbf{p}^{U\mathrm{T}}_{u,l} \cdot \mathbf{q}_j + \sum_{\mathbf{e}_k \in \mathbf{S}_{u,l}} \mathbf{e}_k^{\mathrm{T}} \cdot \mathbf{q}_j \qquad (10)$$

We adopt the Bayesian Personalized Ranking objective [14] via gradient descent to optimize the proposed model. That is, given the positive (observed) and negative (non-observed) items, we optimize the pairwise ranking:

$$\underset{\mathbf{P},\mathbf{Q},\mathbf{E},\Theta}{\arg\min} \sum_{(u,S_u,j_+,j_-)\in\mathcal{D}} -\log\sigma(\hat{y}_{u,j_+} - \hat{y}_{u,j_-})$$

$$+ \lambda(||\mathbf{P}||^2 + ||\mathbf{Q}||^2 + ||\mathbf{E}||^2 + ||\Theta||^2) \quad (11)$$

where $(u, S_u, j_+, j_-) \in \mathcal{D}$ denotes the generated set of pairwise preference order, S_u represents one of the $|L| + |R|$ successive items of target user u, j_+ and j_- represent the positive items in $T_{u,l}$ and randomly sample negative items, respectively. σ is the sigmoid function, λ is the regularization term. Θ denotes other learnable parameters in the GNN, $\mathbf{p}_*$, $\mathbf{q}_*$ and $\mathbf{e}_*$ are column vectors of $\mathbf{P}$, $\mathbf{Q}$ and $\mathbf{E}$, respectively.

5   Experiments and Analysis

5.1   Experiment Setup

Datasets. We use three real-world datasets with various sparsities in different domains to assess the proposed model SA-GNN: *Amazon-CDs* [4], *Goodreads-Children* and *Goodreads-Comics* [8]. We also follow the general rules in [11,12,17, 19] to process datasets. For all datasets, we convert all the presence of a review or numeric ratings into implicit feedback of 1 and keep those with ratings out of five as positive feedback. To guarantee the quality of the dataset, following [11,12], we remove the items with less than five ratings and the users with less than ten ratings. The statistics of the processed datasets are summarized in Table 1.

Table 1. The statistics of datasets.

Dataset	#Users	#Items	#Interactions	Avg. length	Density
CDs	17,052	35,118	472,265	27.696	0.079%
Children	48,296	32,871	2,784,423	57.653	0.175%
Comics	34,445	33,121	2,411,314	70.005	0.211%

Evaluation Metrics. We use both Recall@K and NDCG@K to evaluate all the methods. For each user, Recall@K (R@K) indicates the proportion of correctly recommended items that emerge in the top K recommended items. NDCG@K (N@K) is the normalized discounted cumulative gain at K, which considers the location of correctly recommended items. Here, we employ Top-K (K = 5, 10) for recommendation. For all the models, we perform five times and report the average results.

Baselines. We compare our model with the following representative baselines to verify the effectiveness: The non graph-based methods include: the pairwise learning based on matrix factorization **BPRMF** [14], a GRU-based method to capture sequential dependencies **GRU4Rec** [6], a CNN-based method **Caser** [19] from both horizontal and vertical aspects to model high-order Markov Chains, bidirectional self-attention mechanism **BERT4Rec** [17], an encoder-decoder framework **GRec** [28] and a hierarchical gating network method **HGN** [11]. The graph-based methods: GNN with self-attention mechanism **GC-SAN** [26] and GNN with memory network **MA-GNN** [12].

Parameter Settings. To ensure the comparison is as fair as possible, we fix the length of sub-sequence to 8, the length of input to 6 and the length of targets to 2 in all methods. The latent dimension of all baseline models is set to 50 in all experiments. For GRU4Rec, we set the batch size to 50 and the learning rate to 0.001. Adopting Top1 loss can obtain better results. For Caser, we set the number of vertical filters and horizontal filters to 4 and 16, respectively. For BERT4Rec, we set the number of transformer layers and attention heads to 4 and 2, respectively. For HGN, we use the default hyperparameters following the settings in the author-provided code[1]. For GRec, we set the dilated levels to 1 and 4, and set the kernel size to 3. For GC-SAN, we set the number of self-attention blocks k and the weight factor ω to 4 and 0.6, respectively. For MA-GNN, we implement it with PyTorch and set both the number of attention dimensions and memory units to 20.

For SA-GNN, we set the same step R = 2 as MA-GNN and dimension of item $d = 100$, and set user embedding dimension $d' = 50$ for all datasets. The source code of SA-GNN is available online[2]. The value of d_a is selected from {5, 10, 15, 20}. Both λ and the learning rate are set to 0.001. We set the batch size to 4096 and adopt Adam [9] optimizer to train the model. Hyper-parameters are tuned by grid search on the validation set.

5.2 Comparisons of Performance

Table 2 illustrates comparison among the experiment results. We have the following observations:

Observations About Our Model. First, on all three datasets with all evaluation metrics, the proposed model SA-GNN achieves the best performance, which illustrates the superiority of our model. Second, SA-GNN obtains better results than MA-GNN and HGN. Despite MA-GNN and HGN capture both local and global user intents and consider item co-occurrence patterns, they all neglect the effect of future contexts information in modeling the user's interests. Third, SA-GNN obtains better results than GC-SAN. One possible main reason is that GC-SAN fails to capture the global item dependencies but only models the user's local

---

[1] https://github.com/allenjack/HGN.
[2] https://github.com/Forrest-Stone/SA-GNN.

Table 2. Comparisons among the performance of all methods according to R@5, R@10, N@5 and N@10. In each row, the best scores are bold and the second best are underlined.

Method		BPRMF	GRU4Rec	Caser	BERT4Rec	GRec	HGN	GC-SAN	MA-GNN	SA-GNN	Improv
CDs	R@5	1.104	1.137	1.163	2.064	1.684	_2.737_	2.477	2.502	**3.162**	15.53%
	R@10	2.018	2.054	2.081	3.732	2.473	_4.459_	4.042	4.202	**4.946**	10.92%
	N@5	1.408	1.425	1.441	2.278	2.012	_3.233_	2.609	2.853	**3.655**	13.05%
	N@10	1.709	1.749	1.780	2.705	2.428	_3.817_	3.378	3.504	**4.271**	11.89%
Children	R@5	5.178	5.773	6.689	7.622	7.506	_7.825_	7.801	7.664	**8.514**	8.81%
	R@10	8.674	9.182	10.706	12.016	11.863	_12.218_	12.183	12.123	**12.832**	5.03%
	N@5	8.164	8.697	10.993	12.961	12.248	_13.001_	12.988	12.511	**14.026**	7.88%
	N@10	9.004	9.588	11.633	13.518	13.025	_13.538_	13.526	13.190	**14.351**	6.01%
Comics	R@5	5.312	6.017	8.954	11.910	11.597	11.859	_11.984_	10.204	**13.273**	10.76%
	R@10	8.421	9.222	13.276	16.902	16.453	16.894	_16.922_	15.187	**18.458**	9.08%
	N@5	8.687	9.435	17.504	22.784	22.438	23.464	_23.467_	19.567	**25.952**	10.59%
	N@10	9.219	10.128	16.874	21.960	21.555	21.951	_21.968_	18.985	**24.180**	10.07%

interest in a short-term window. Moreover, GC-SAN does not consider the item co-occurrence patterns. Fourth, SA-GNN gains better results than GRec and BERT4Rec. Although GRec and BERT4Rec adopt the bidirectional model to improve the learning ability, they neglect the item co-occurrence patterns between two closely related items, which can be learned by our item-item inter product. Moreover, they use the RNN or CNN to learn while we use the GNN. Fifth, SA-GNN obtains better results than Caser and GRU4Rec. One possible main reason is that Caser and GRU4Rec not consider the item attention score for various users, but only apply CNN or RNN to model the group-level representation of several successive items. Last, SA-GNN outperforms BPRMF, since BPRMF does not model the sequential patterns of user-item interactions, but only captures the global interest of the user.

Other Observations. First, all the results reported on dataset *CDs* is worse than the results on dataset *Children* and *Comics*. The possible main reason is that the data sparsity can reduce the performance of recommendation and dataset *CDs* is more sparse than others. Second, on all the datasets, GC-SAN, MA-GNN, HGN, and BERT4Rec achieve better results than Caser. The main possible reason is that these methods may capture more personalized user representation because they can learn different importance scores of items in the item sequence. Third, GC-SAN and HGN obtain better results than MA-GNN. This shows that considering the importance of different items is necessary for local interest modeling. Fourth, by comparing GC-SAN, HGN and BERT4Rec. We can see that the BERT4Rec and GC-SAN are worse than HGN on the *CDs* and *Children* datasets, while they are better on the *Comics* dataset. One possible reason may be that HGN with item co-occurrence patterns performs better on the more sparse datasets. On the other hand, GC-SAN outperforms BERT4Rec in all datasets. This shows the superiority of GNNs in sequential recommendation. Fifth, BERT4Rec is much better than GRec. The main reason may be that BERT4Rec distinguishes the different importance of items by the attention mechanism. Sixth, Caser achieves better results than GRU4Rec. One main reason is that Caser can learn user's global interest in prediction layer by explicitly feeding the user embeddings. Last, all the methods achieves better results than BPRMF. This shows that it is not enough to model user's global interest without capturing the user's sequential behaviors.

5.3   Ablation Analysis

To evaluate the contribution of each component for final user intent representations, we conducted experiments of different variants of SA-GNN. (1) **SA-GNN (-U)** is a kind of SA-GNN without the global interest but only with the local interest. (2) **SA-GNN(-U-Att)** is a sort of SA-GNN without the global interest or attention module. (3) **SA-GNN(-U-Att-A)** and (4) **SA-GNN(-U-Att-SA)** denote we model user's local interest by self-adaptive adjacency matrix A_{ada} and vanilla matrix A, respectively. (5) **SA-GNN(-U-Att-SA+GAT)**: we replace the self-adaptive GNN in SA-GNN(-U-Att) with graph attention network (GAT) [21].

Table 3. The ablation analysis.

Method	CDs		Children		Comics	
	R@10	N@10	R@10	N@10	R@10	N@10
(1) SA-GNN(-U-Att-SA)	4.434	3.848	12.344	13.821	17.892	23.598
(2) SA-GNN(-U-Att-SA+GAT)	4.488	3.902	12.472	13.948	17.923	23.666
(3) SA-GNN(-U-Att-A)	4.635	4.029	12.697	14.189	18.316	24.111
(4) SA-GNN(-U-Att)	4.837	4.210	12.711	14.257	18.332	24.219
(5) SA-GNN(-U)	4.870	4.266	12.728	14.277	18.363	**24.294**
(6) SA-GNN	**4.946**	**4.271**	**12.832**	**14.351**	**18.458**	24.180

According to the results shown in Table 2 and Table 3, we make the following observations. First, comparing (1) with baseline models, we can observe that better results can be achieved by utilizing future contexts only based on GNN in *Children* and *Comics* datasets. Second, we compare (1)–(4). On one hand, compared with (1), (2) and (3) can learn different weights in the item graph, which helps improve performance. Moreover, (3) achieves better performance than (2). One possible reason may be that (2) will be limited by unsuitable connections of the pre-defined matrix, while (3) will not because of the self-adaptive adjacency matrix. On the other hand, (4) achieves significant improvement on all indicators and all datasets. This demonstrates that element-wise product can improve the ability of model representation and boost the performance for recommendation. Third, from (4) and (5), we observe that by considering the different important scores for items in the sequence, the performance can be improved. Last, The performance of (6) is better than (5), which illustrates that the user's global interest makes our model more capable of capturing personalized interests.

6 Conclusion

In this paper, we propose a self-adaptive graph neural network (SA-GNN) for sequential recommendation. SA-GNN applies GNN to model user's local interest, and utilizes a self-adaptive adjacency matrix to weaken the negative effect of unsuitable connections in the item graph. The experiments on three real-world datasets verify the effectiveness of SA-GNN. In the future, we will extend the SA-GNN to capture the user's long-term interest more efficiently.

References

1. Chen, X., et al.: Sequential recommendation with user memory networks. In: WSDM 2018, pp. 108–116. ACM, Marina Del Rey, CA, USA (2018)
2. He, R., Kang, W., McAuley, J.J.: Translation-based recommendation. In: RecSys 2017, pp. 161–169. ACM, Como, Italy (2017)
3. He, R., McAuley, J.J.: Fusing similarity models with Markov chains for sparse sequential recommendation. In: ICDM 2016, pp. 191–200. IEEE Computer Society, Barcelona, Spain (2016)
4. He, R., McAuley, J.J.: Ups and downs: Modeling the visual evolution of fashion trends with one-class collaborative filtering. In: WWW 2016, pp. 507–517. ACM, Montreal, Canada (2016)
5. Hidasi, B., Karatzoglou, A.: Recurrent neural networks with top-k gains for session-based recommendations. In: CIKM 2018, pp. 843–852. ACM, Torino, Italy (2018)
6. Hidasi, B., Karatzoglou, A., Baltrunas, L., Tikk, D.: Session-based recommendations with recurrent neural networks. In: ICLR 2016, San Juan, Puerto Rico (2016)
7. Kabbur, S., Ning, X., Karypis, G.: FISM: factored item similarity models for top-n recommender systems. In: SIGKDD 2013, pp. 659–667. ACM, Chicago, IL, USA (2013)
8. Kang, W., McAuley, J.J.: Self-attentive sequential recommendation. In: ICDM 2018, pp. 197–206. IEEE Computer Society, Singapore (2018)
9. Kingma, D.P., Ba, J.: Adam: a method for stochastic optimization. In: ICLR 2015, San Diego, CA, USA (2015)
10. Liu, Q., Zeng, Y., Mokhosi, R., Zhang, H.: STAMP: short-term attention/memory priority model for session-based recommendation. In: SIGKDD 2014, pp. 1831–1839. ACM, London, UK (2018)
11. Ma, C., Kang, P., Liu, X.: Hierarchical gating networks for sequential recommendation. In: SIGKDD 2019, pp. 825–833. ACM, Anchorage, AK, USA (2019)
12. Ma, C., Ma, L., Zhang, Y., Sun, J., Liu, X., Coates, M.: Memory augmented graph neural networks for sequential recommendation. In: AAAI 2020, pp. 5045–5052. AAAI Press, New York, NY, USA (2020)
13. Ma, C., Zhang, Y., Wang, Q., Liu, X.: Point-of-interest recommendation: exploiting self-attentive autoencoders with neighbor-aware influence. In: CIKM 2018, pp. 697–706. ACM, Torino, Italy (2018)
14. Rendle, S., Freudenthaler, C., Gantner, Z., Schmidt-Thieme, L.: BPR: bayesian personalized ranking from implicit feedback. In: UAI 2009, pp. 452–461. AUAI Press, Montreal, QC, Canada (2009)
15. Rendle, S., Freudenthaler, C., Schmidt-Thieme, L.: Factorizing personalized Markov chains for next-basket recommendation. In: WWW 2010, pp. 811–820. ACM, Raleigh, North Carolina, USA (2010)

16. Scarselli, F., Gori, M., Tsoi, A.C., Hagenbuchner, M., Monfardini, G.: The graph neural network model. IEEE Trans. Neural Networks **20**(1), 61–80 (2009)
17. Sun, F., et al.: BERT4Rec: sequential recommendation with bidirectional encoder representations from transformer. In: CIKM 2019, pp. 1441–1450. ACM, Beijing, China (2019)
18. Tan, Y.K., Xu, X., Liu, Y.: Improved recurrent neural networks for session-based recommendations. In: DLRS@RecSys 2016, pp. 17–22. ACM, Boston, MA, USA (2016)
19. Tang, J., Wang, K.: Personalized top-n sequential recommendation via convolutional sequence embedding. In: WSDM 2018, pp. 565–573. ACM, Marina Del Rey, CA, USA (2018)
20. Vaswani, A., et al.: Attention is all you need. In: NIPS 2017, pp. 5998–6008. Long Beach, CA, USA (2017)
21. Velickovic, P., Cucurull, G., Casanova, A., Romero, A., Lió, P., Bengio, Y.: Graph attention networks. CoRR abs/1710.10903 (2017)
22. Wang, S., Hu, L., Wang, Y., Cao, L., Sheng, Q.Z., Orgun, M.A.: Sequential recommender systems: challenges, progress and prospects. In: IJCAI 2019, pp. 6332–6338. ijcai.org, Macao, China (2019)
23. Wu, S., Zhang, W., Sun, F., Cui, B.: Graph neural networks in recommender systems: a survey. CoRR abs/2011.02260 (2020)
24. Wu, S., Tang, Y., Zhu, Y., Wang, L., Xie, X., Tan, T.: Session-based recommendation with graph neural networks. In: AAAI 2019, pp. 346–353. AAAI Press, Honolulu, Hawaii, USA (2019)
25. Wu, S., Zhang, M., Jiang, X., Xu, K., Wang, L.: Personalizing graph neural networks with attention mechanism for session-based recommendation. CoRR abs/1910.08887 (2019)
26. Xu, C., et al.: Graph contextualized self-attention network for session-based recommendation. In: IJCAI 2019, pp. 3940–3946. ijcai.org, Macao, China (2019)
27. Xu, C., et al.: Recurrent convolutional neural network for sequential recommendation. In: WWW 2019, pp. 3398–3404. ACM, San Francisco, CA, USA (2019)
28. Yuan, F., et al.: Future data helps training: Modeling future contexts for session-based recommendation. In: WWW 2020, pp. 303–313. ACM/IW3C2, Taipei, Taiwan (2020)
29. Yuan, F., Karatzoglou, A., Arapakis, I., Jose, J.M., He, X.: A simple convolutional generative network for next item recommendation. In: WSDM 2019, pp. 582–590. ACM, Melbourne, VIC, Australia (2019)

Spatial-Temporal Attention Network with Multi-similarity Loss for Fine-Grained Skeleton-Based Action Recognition

Xiang Li[1], Shenglan Liu[2]($\boxtimes$), Yunheng Li[1], Hao Liu[1], Jinjing Zhao[1],
Lin Feng[2]($\boxtimes$), Guihong Lao[1], and Guangzhe Li[1]

[1] Faculty of Electronic Information and Electrical Engineering,
Dalian University of Technology, Dalian, Liaoning 116024, China
[2] School of Innovation and Entrepreneurship, Dalian University of Technology,
Dalian, Liaoning 116024, China
{liusl,fenglin}@dlut.edu.cn

Abstract. In skeleton-based action recognition, the Graph Convolutional Networks (GCNs) have achieved remarkable results. However, in fine-grained skeleton-based action recognition, existing methods can't distinguish highly similar sample pairs well. This is because that the current methods do not pay attention to highly similar sample pairs and can't model complex actions through the spatial-temporal separation framework. In order to solve the above problems, we combine the multi-similarity loss function to weight the difficult sample pairs, and propose a novel ST-Attention module to construct the nodes connection between spatial and temporal. Finally, we used the Spatial-Temporal Attention Network with Multi-Similarity Loss (STATT-MS) to conduct experiments on NTU-RGBD-60, FSD-10 and UAV-human datasets and achieved state-of-the-art performance.

Keywords: Skeleton-based action recognition · Metric learning · Attention

1 Introduction

As action recognition is widely used in many fields, such as video surveillance, human-computer interaction, sports science, etc. Skeleton-based action recognition methods have received more and more attention because of their strong adaptability to the dynamic circumstance and complicated background. Compared with the RGB-based action recognition methods [25], the skeleton-based methods are

This study was funded by National Natural Science Foundation of Peoples Republic of China (61672130, 61972064). The Fundamental Research Fund for Dalian Youth Star of Science and Technology (No. 2019RQ035) and LiaoNing Revitalization Talents Program (XLYC1806006) and CCF-Baidu Open Fund (No. 2021PP15002000).

T. Mantoro et al. (Eds.): ICONIP 2021, LNCS 13109, pp. 620–631, 2021.
https://doi.org/10.1007/978-3-030-92270-2_53

more efficient and accurate for the complex action recognition [4,11,13,17,24,27]. Although the GCN-based action recognition methods [16,18,19,21,27] achieved remarkable results in skeleton-based action recognition, there are still some problems for GCN-based methods in fine-grained action recognition [15].

The major problem of fine-grained action recognition is the smaller inter-class variance and the larger intro-class variance, which may lead to misjudgment when faced with similar sample pairs of different classes. Due to the lack of adaptive blocks, most classification methods fail to recognize fine-grained actions accurately. Although some recent works push the positive sample pairs together, while negative sample pairs are pushed apart from each other in a metric embedding space through the pair-based loss function, they do not mine the hard sample pair and pay more attention to it. The second problem is that the node between frames cannot be connected dynamically under the GCN framework. Due to the complexity of fine-grained action, we generally cannot judge the action category by a single frame, so it is necessary to establish a node relationship between frames. Since the beginning of ST-GCN [27], many GCN-based methods [16,18,19,21] have followed its spatial-temporal modeling framework. Because the connection between the nodes is limited by the physical structure of the human body for ST-GCN, so it is difficult to model the action flexibly. In further works, [18,19,21] broke this restriction, and any pair of nodes can directly establish a connection by making the parameters of the Laplacian matrix learnable. We noticed that the nodes connection between the frames still cannot be established. And MS-G3D [16] establishes the relationship between frames through unfold operations. The G3D module in MS-G3D is only used as a branch and the connection between nodes are static. Many recent works have proved the effectiveness of Attention as a dynamic modeling module. So we need a module that can dynamically establish the node relationship between frames.

To solve the problem above. 1) This paper introduces the multi-similarity loss function [26] to mine hard sample pairs, and uses the General Pair Weighting (GPW) framework [26] to weight difficult samples. By combining with the softmax cross entropy loss function, it can increase the discrimination ability for similar actions. 2) We propose a novel ST-Attention module to connect the relationship between temporal and spatial dynamically and achieved state-of-the-art performance on FSD-10 [15], NTU-RGBD-60 [17], UAV-human datasets [12].

2 Related Work

GCN-Based Methods for Skeleton-Based Action Recognition. In the beginning, IndRNN [11] and TCN [22] both regarded skeleton data as time series, and they adopted a temporal modeling framework [17], but this did not model the spatial relationship compared to the GCN-based methods. As a pioneer, ST-GCN [27] used GCN for skeleton-based action recognition for the first time, and achieved remarkable results. The convolution kernel used by ST-GCN is $A \odot B$, where A is normalized Laplacian matrix and B is a learnable matrix. However, both 2S-AGCN [19] and DGNN [18] have found that the connection between the nodes is limited by the physical structure of the human body, so it is difficult to model the action flexibly, so they use a convolution kernel in the form of $A + B$, which make any pair of nodes can directly establish a connection. And MS-G3D [16] which establishes the connection of nodes between frames through the G3D module, using multi-scale time and space convolution kernel to achieve the state-of-the-art performance. In addition, there is a lot of work trying to construct more lightweight models, such as SGN [2], shift-GCN [1], which have achieved higher accuracy and less calculation.

Attention-Based Methods for Skeleton-Based Action Recognition. After No-Local [25] firstly applied self-attention [23] to the field of action recognition, many people discovered the effectiveness of self-attention and applied it in various fields, such as image classification [3], image segmentation [6]. As for skeleton-based action recognition, 2S-AGCN [19] applied self-attention for the first time and achieved remarkable performance. In details, 2S-AGCN uses an attention-based convolution kernel C, which dynamically constructs the relationship between nodes by calculating the attention map. But the convolution kernel of 2S-AGCN also combines the learnable Laplacian matrix, which contains the physical structure of the human body. Later, MS-AAGCN [21], DSTA-Net [20] and SGN [2] abandoned the Laplacian matrix and completely used the attention-based convolution kernel to achieve good performance. This implies that the attention-based convolution kernel may be more effective for modeling action.

3   Method

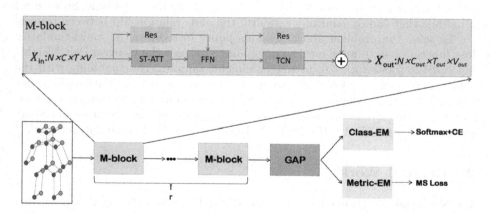

Fig. 1. The network consists of r M-blocks. For each M-block it has three core components: ST-ATT, FFN, and TCN. GAP stands for global average pooling layer, Class-EM and Metric-EM are two MLPs with different parameters.

As shown in Fig. 1, our architecture takes the skeleton data after extracting frames from raw data as the input X whose shape is $N \times C \times T \times V$. The four dimensions represent number of the input batch, input channels, input frames and input joints respectively. After input X through multiple M-blocks and global average pooling layers, we get a high-level semantic feature F. Then we get two representations F^C and F^M from F by using two different Multi Layer Perceptron (MLP). We update our network by minimize the following formula.

$$\mathcal{L} = \mathcal{L}_{CE}(F^C, Y) + \mathcal{L}_{MS}(F^M, Y) \tag{1}$$

Where Y is the category label of X, $\mathcal{L}_{CE}$ and $\mathcal{L}_{MS}$ are the cross-entropy loss function and the multi-similarity loss function respectively. λ is the weight of the MS loss. The specific form of the multi-similarity loss function will be given in the Sect. 3.1.

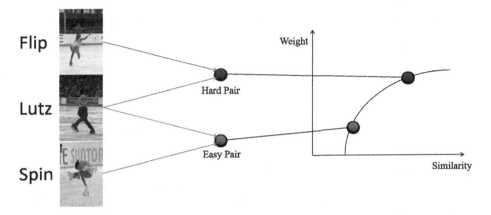

Fig. 2. Examples of hard pair weighting. The actions of "Flip" and "Lutz" are too similar, it is difficult for the network to distinguish. We call this sample pair hard pair and give it a higher weight, while "Lutz" and "Spin" are easy to distinguish. We call this sample pair easy pair and give it a lower weight.

3.1 Multi-similarity Loss

Multi-similarity Loss. In fine-grained action recognition, there are actions that are too similar, as shown in Fig. 2. The actions of "Flip" and "Lutz" are very similar, while the actions of "Lutz" and "Spin" are quite different. We call the former hard pair and the latter easy pair. Because the main challenge of fine-grained action recognition is to distinguish hard pairs, we use the multi-similarity loss function to make the network focus on this hard pair, so that the network can learn more detailed features to improve generalization ability. The formula for $\mathcal{L}_{MS}$ is as follows.

$$\mathcal{L}_{MS} = \frac{1}{m} \sum_{i=1}^{m} \left\{ \frac{1}{\alpha} log[1 + \sum_{k \in \mathcal{P}_i} e^{-\alpha(S_{ik} - \delta)}] + \frac{1}{\beta} log[1 + \sum_{k \in \mathcal{N}_i} e^{\beta(S_{ik} - \delta)}] \right\} \tag{2}$$

For a batch of m samples, each sample i need to calculate the cosine similarity with other samples k to get the S_{ik}. When the label of sample k is the same as the label of sample i, k is placed in the positive index set of sample i, namely $\mathcal{P}_i$ otherwise it is placed in the negative index set of sample i, namely $\mathcal{N}_i$. α and β are class balance parameters, and δ is a margin hyper parameter. Using the multi-similarity loss function can not only bring the features of the same category closer to each other, and move features of different categories away from each other, but also give extra weight to hard samples.

General Pair Weighting. We use the GPW framework [26] to analyze its ability of weighting hard pair. If we regard the form of the loss function based on the pair as the $\mathcal{L}(S,Y)$, where S is the similarity of each pair of samples, and Y represents whether the labels of this pair are equal. Then when we backpropagate to update the parameters in the network, there are the following formulas.

$$
\begin{aligned}
\frac{\partial \mathcal{L}(S,Y)}{\partial \theta} &= \frac{\partial \mathcal{L}(S,Y)}{\partial S} \cdot \frac{\partial S}{\partial \theta} \\
&= \sum_{i=1}^{m} \sum_{j=1}^{m} \frac{\partial \mathcal{L}(S,Y)}{\partial S_{ij}} \cdot \frac{\partial S_{ij}}{\partial \theta}
\end{aligned}
\tag{3}
$$

We notice that $\frac{\partial S}{\partial \theta}$ is determined by the structure of the network, and $\frac{\partial \mathcal{L}(S,Y)}{\partial S}$ is determined by the structure of the loss function. And the sign of $\frac{\partial \mathcal{L}(S,Y)}{\partial S}$ represents the direction of network optimization, and the absolute value of $\frac{\partial \mathcal{L}(S,Y)}{\partial S}$ represents the step size of this update, which is the weight of the sample pair. By letting $\mathcal{L}_{MS}$ to derive S_{ij}, we can prove that $\mathcal{L}_{MS}$ has such a property that when two samples belong to different categories, the greater their similarity S, the greater their weight. So this process mine the hard sample pair and pay more attention to it.

$$
\begin{aligned}
w_{ij}^{-} &= \left. \frac{\partial \mathcal{L}_{MS}(S,Y)}{S_{ij}} \right|_{Y_i \neq Y_j} \\
&= \frac{1}{e^{\beta(\lambda - S_{ij}) + \sum_{k \in \mathcal{N}_i} e^{\beta(S_{ik} - S_{ij})}}} \\
&= \frac{e^{\beta(S_{ij} - \lambda)}}{1 + \sum_{k \in \mathcal{N}_i} e^{\beta(S_{ik} - \lambda)}}
\end{aligned}
\tag{4}
$$

3.2   M-Block

In this section, we will introduce M-block, which is mainly composed of 3 parts, namely ST-ATT, FFN, TCN as shown in Fig. 3.

ST-ATT. Inspired by self-attention and MS-G3D, we propose a ST-Attenion operation, which can adaptively build the connection between frames and nodes. We firstly use three 1×1 convolutions to map X to Q, K and V as implemented

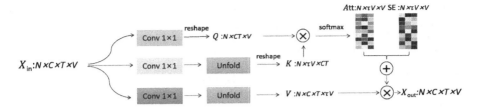

Fig. 3. Illustration of spatial-temporal sttention operation. Conv 1×1 represents a convolution operation with a convolution kernel size of $(1, 1)$, where different colors represent different parameters. Unfold is an up-sampling function. See Algorithm 1 for its specific process. Att represents the calculated attention map, and SE represents the structural code of the human body, which is a learnable matrix. $\oplus$ and $\otimes$ represents matrix addition and matrix multiplication respectively.

by No-Local [25]. The number of output channels of convolution is D. There are two main difference between self-attention [23] and ST-Attention. The first is that we use the unfold operation in MS-G3D for K and V to sample the features from adjacent frames. We denote this unfold operation as $\Gamma(\cdot)$ and the pseudo-code of it is in Algorithm 1. The second is that we define a learnable matrix SE to learn the structural relationship of the human body, as implemented in 2S-AGCN [19] and DGNN [18]. So the formula of ST-ATT is as follows.

$$ STATT(Q, K, V) = (Softmax(\frac{Q\Gamma(K)^T}{\sqrt{D}}) + SE)\Gamma(V) \qquad (5) $$

FFN & TCN. Following the popular transformer architecture [3,23], we also added the FFN layer to M-Block. Ours FFN operation consists of three fully connected layers and two residual layers. Its input consists of the input X of M-Block and the output Y of ST-ATT.

$$ FFN(X, Y) = FC_3(FC_2(FC_1(Y) + X) + X) \qquad (6) $$

FC is implemented by 1×1 convolution, and a Batchnorm layer and Relu activation function are appended after each FC layer. In order to better model the temporal context, we also added TCN operations, in detail, we perform a 9×1 convolution on the output feature map calculated by FFN.

4 Experiment

4.1 Dataset

FSD-10. FSD-10 [15] is a challenging fine-grained action recognition dataset, which provides three modalities: RGB, flow, and skeleton. This dataset contains 1,484 videos clips in 10 classes, which clips are sourced from World Figure Skating Championships. Since the duration of clips is up to 2000 frames, this is too long

---

Algorithm 1: Unfold Pseudocode, PyTorch-like

---

1 **Input:** x:the input data; τ:the temporal scale
2 **Initialization:** import torch.nn.functional.Unfold as F
3 N, C, T, V = $x.shape$
4 $x = F(x,kernel\_size{=}\tau,padding{=}((\tau - 1)//2,0))$
5 $x = x.view(N,\ C,\ \tau,\ T,\ V).permute(0,1,3,2,4).contiguous()$
6 $x = x.view(N,\ C,\ T,\ \tau \times V)$
7 **Return** x

---

for general network input. So we set a window size $S = 350$. If the number of frames of the action is less than S, we repeat the action until the length is equal to S. For actions with a length greater than S, we extract S frames by sampling at intervals to form our input data.

UAV-Human. UAV-human [12] is an action recognition dataset from the perspective of unmanned aerial vehicle, which contains 67,428 annotated video clips of 119 class for action recognition. Each action is captured by Azure Kinect DK. We will repeat the action that is less than 300 in length until the length reaches 300, and centralize and standardize the data, just like [21] did.

NTU-RGBD-60. NTU-RGBD-60 [17] is currently the most widely used indoor-captured action recognition dataset, which contains 56,000 action clips in 60 action classes. The clips are performed by 40 volunteers in different age groups ranging from 10 to 35. Each action is captured by 3 cameras at the same height but from different horizontal angles. The original paper [17] of the dataset recommends two benchmarks: 1) Cross-subject (X-Sub): the dataset in this benchmark is divided into a training set (40,320 videos) and a validation set (16,560 videos), where the actors in the two subsets are different. 2) Cross-view (X-View): the training set in this benchmark contains 37,920 videos that are captured by cameras 2 and 3, and the validation set contains 18,960 videos that are captured by camera 1. We use the same preprocessing method used in UAV-human to process the data.

4.2 Basline Setting

If there is no special parameter description, the following experiments are all performed under the settings of this section.

Optimizer. All our experiments use Stochastic Gradient Descent (SGD) as the optimizer. Our initial learning rate is set to 0.05, 0.1, 0.1 for FSD-10, UAV-human, and NTU-RGBD-60, respectively. The models are trained with bachsize 64 for 120 epochs with step LR decay with a factor of 0.1 at epochs {60, 90}. When training the network with MS loss, we randomly choose 8 categories, and then randomly sample 8 instances from each categories. The weight decay is 0.0001 and the SGD momentum is 0.9 for all datasets.

Model. There are 10 M-blocks in ST-ATT net, and the output dimension of TCN in the 3rd and 7th layers are increased from {64, 128} to {128, 256} respectively, and stride is set to 2. The unfold scale τ are 5. The Class-EM and the Metric-EM are set to a fully-connected layer with a 256-dimensional output channel. The δ, α, β in Eq. 2 are set to 0.1, 8, 56 respectively.

4.3   Ablation Study

Weight of Multi-similarity Loss Function. In Table 1, we study the influence of the weight of the multi-similarity loss function on FSD-10. We use ST-GCN as the baseline model. The first row represents the weight of MS loss, when λ is equal to 0, it represents the performance of the original ST-GCN. We noticed that the performance of the model on FSD-10 is improved when the λ range from 0.1 to 20. And when λ is equal to 10, the model achieves the best performance (86.86%). This shows that when the weight of the multi-similarity loss function is appropriately greater than the cross-entropy loss, its performance on the fine-grained action recognition dataset will be significantly improved.

Table 1. Ablation studies for weight of multi-similarity loss function on the FSD-10 dataset. The λ is the weight of multi-similarity loss function and the acc. is the recognition accuracy on the FSD-10 dataset.

λ	0.5	1	5	10	20	30
Acc. (%)	85.32	84.94	86.26	**86.86**	85.43	84.24

Class-EM & Metric-EM. In Table 2, we study the influence of the structurechanges of Class-EM and Metric-EM on NTU-RGBD-60 X-sub. Depth represents the number of FC layers in EM, and output dim represents the output dimension of the last FC layer. When depth is equal to 0, it is equivalent to directly putting the features output by GAP into the softmax cross entropy loss and MS loss. As shown in Table 2, when the depth changes from 0 to 1, the performance of the model has improved significantly. But when the depth increased from 1 to 3, the performance of the model continued to decline. This shows that it is necessary to set Class-EM and Metric-EM separately, but as the depth increases, over-fitting will occur.

Table 2. Ablation studies for Metric-EM structure on the NTU-RGBD-60 X-sub dataset. The depth and output dim represent the depth of the MLP and the output dimension of the feature vector respectively. the acc. is the recognition accuracy on the NTU-RGBD-60 X-sub dataset

Depth	0	1			2			3		
Output dim	256	64	128	256	64	128	256	64	128	256
Acc. (%)	83.16	84.39	84.57	**84.79**	83.24	83.34	84.51	82.35	82.56	82.81

M-Block. In Table 3 and Table 4, we study the influence of each part of M-block and the temporal scale τ in ST-ATT. We can see that the performance of the model reaches its best when the temporal scale τ is equal to 5, and then as the τ increases, the performance decreases. This may be because as the receptive field of ST-attention increases, the local features on the temporal dimension will be weakened, which is important for action recognition. We also study the effect of FFN and SE in Table 4. It can be seen that both FFN and SE have improved the generalization ability. And when we use the MS loss function, the model reached the best performance (89.37%).

Table 3. Ablation studies for temporal scale of ST-ATT operation on the FSD- 10 dataset. The τ is the temporal scale of ST-ATT operation and the acc. is the recognition accuracy on the FSD-10

τ	Acc. (%)
1	85.14
3	87.25
5	**87.89**
7	86.31
9	86.24
11	85.39

Table 4. Ablation studies on FSD-10 for different blocks. +X means adding the X module. STA denote using spatial-temporal attention module, SE denote using structure encoding, FFN denote using FFN module, MS denote training with multi-similarity loss

Method	Acc. (%)
TCN (base)	83.78
TCN+STA	85.24
TCN+STA+SE	86.97
TCN+STA+SE+FFN	87.89
TCN+STA+SE+FFN+MS	**89.37**

4.4   Comparison with Other Methods

We use ST-ATT network with multi-similarity loss and other state-of-the-art models to conduct comparative experiments on NTU-RGBD-60, FSD-10 and UAV-human in Table 5, 6, and 7. The methods used for comparison include the handcraft-feature-based methods Lie Group [24], RNN-based methods [11,17], CNN-based methods [7–9,14,22] and GCN-based methods [1,10,16,18–21,27]. Our model achieves comparable performance on NTU, and achieves state-of-the-art performance on FSD-10 and UAV-human. It should be noted that since the UAV-human dataset is not fully open source, the dataset split we use is not the same as that in the original paper, so we do not compare with the results in the original paper.

Table 5. Recognition accuracy comparison between our method and state-of-the-art methods on NTU-RGBD-60 dataset. CS and CV denote the cross-subject and cross-view benchmarks respectively.

Method	CS (%)	CV (%)
Lie Group [24]	50.1	52.8
Deep LSTM [17]	60.7	67.3
Ind-RNN [11]	81.8	88.0
TCN [22]	74.3	83.1
Clips+CNN [7]	79.6	84.8
Synthesized CNN [14]	80.0	87.2
Motion+CNN [9]	83.2	89.3
3scale ResNet152 [8]	85.0	92.3
ST-GCN [27]	81.5	88.3
2S-AGCN [19]	86.6	93.7
Shift-GCN [1]	87.8	95.1
MS-AAGCN [21]	88.0	95.1
DSTA-Net [20]	88.7	94.8
MS-G3D [16]	89.4	95.0
Ours	88.8	95.1

Table 6. Recognition accuracy comparison between our method and state-of-the-art methods on FSD-10 dataset

Method	Acc.(%)
TCN [22]	83.78
ST-GCN [27]	84.24
2S-AGCN [19]	86.94
DSTG-Net [5]	86.82
AS-GCN [10]	86.54
DGNN [18]	85.88
DSTA-Net [20]	87.22
MS-G3D [16]	88.72
Ours	89.37

Table 7. Recognition accuracy comparison of our method and state-of-the-art methods on UAV-human dataset

Method	Acc. (%)
TCN [22]	60.37
ST-GCN [27]	61.82
2S-AGCN [19]	63.29
DSTA-Net [20]	68.20
Ours	68.67

5 Conclusion

In this paper, we propose a novel ST-ATT network with multi-similarity loss functions and achieve significant performance on multiple datasets. The multi-similarity loss function allows the network to pay more attention to the hard sample pairs in fine-grained actions recognition, and the ST-ATT module allows network to simultaneously model temporal and spatial dynamically. Finally, experiments on three datasets prove the effectiveness of our proposed method.

References

1. Cheng, K., Zhang, Y., He, X., Chen, W., Cheng, J., Lu, H.: Skeleton-based action recognition with shift graph convolutional network. In: Proceedings of the IEEE/CVF Conference on Computer Vision and Pattern Recognition, pp. 183–192 (2020)
2. Ding, X., Yang, K., Chen, W.: A semantics-guided graph convolutional network for skeleton-based action recognition. In: Proceedings of the 2020 the 4th International Conference on Innovation in Artificial Intelligence, pp. 130–136 (2020)

3. Dosovitskiy, A., et al.: An image is worth 16×16 words: transformers for image recognition at scale. arXiv preprint arXiv:2010.11929 (2020)
4. Du, Y., Wang, W., Wang, L.: Hierarchical recurrent neural network for skeleton based action recognition. In: Proceedings of the IEEE Conference on Computer Vision and Pattern Recognition, pp. 1110–1118 (2015)
5. Feng, L., Lu, Z., Liu, S., Jiang, D., Liu, Y., Hu, L.: Skeleton-Based action recognition with dense spatial temporal graph network. In: Yang, H., Pasupa, K., Leung, A.C.-S., Kwok, J.T., Chan, J.H., King, I. (eds.) ICONIP 2020. CCIS, vol. 1333, pp. 188–194. Springer, Cham (2020). https://doi.org/10.1007/978-3-030-63823-8_23
6. Fu, J., et al.: Dual attention network for scene segmentation. In: Proceedings of the IEEE/CVF Conference on Computer Vision and Pattern Recognition, pp. 3146–3154 (2019)
7. Ke, Q., Bennamoun, M., An, S., Sohel, F., Boussaid, F.: A new representation of skeleton sequences for 3d action recognition. In: Proceedings of the IEEE Conference on Computer Vision and Pattern Recognition, pp. 3288–3297 (2017)
8. Li, B., Dai, Y., Cheng, X., Chen, H., Lin, Y., He, M.: Skeleton based action recognition using translation-scale invariant image mapping and multi-scale deep CNN. In: 2017 IEEE International Conference on Multimedia & Expo Workshops (ICMEW), pp. 601–604. IEEE (2017)
9. Li, C., Zhong, Q., Xie, D., Pu, S.: Skeleton-based action recognition with convolutional neural networks. In: 2017 IEEE International Conference on Multimedia & Expo Workshops (ICMEW), pp. 597–600. IEEE (2017)
10. Li, M., Chen, S., Chen, X., Zhang, Y., Wang, Y., Tian, Q.: Actional-structural graph convolutional networks for skeleton-based action recognition. In: Proceedings of the IEEE/CVF Conference on Computer Vision and Pattern Recognition, pp. 3595–3603 (2019)
11. Li, S., Li, W., Cook, C., Zhu, C., Gao, Y.: Independently recurrent neural network (INDRNN): building a longer and deeper RNN. In: Proceedings of the IEEE Conference on Computer Vision and Pattern Recognition, pp. 5457–5466 (2018)
12. Li, T., Liu, J., Zhang, W., Ni, Y., Wang, W., Li, Z.: UAV-human: a large benchmark for human behavior understanding with unmanned aerial vehicles. In: Proceedings of the IEEE/CVF Conference on Computer Vision and Pattern Recognition, pp. 16266–16275 (2021)
13. Liu, H., Tu, J., Liu, M.: Two-stream 3D convolutional neural network for skeleton-based action recognition. arXiv preprint arXiv:1705.08106 (2017)
14. Liu, M., Liu, H., Chen, C.: Enhanced skeleton visualization for view invariant human action recognition. Pattern Recogn. **68**, 346–362 (2017)
15. Liu, S., et al.: Fsd-10: a fine-grained classification dataset for figure skating. Neurocomputing **413**, 360–367 (2020)
16. Liu, Z., Zhang, H., Chen, Z., Wang, Z., Ouyang, W.: Disentangling and unifying graph convolutions for skeleton-based action recognition. In: Proceedings of the IEEE/CVF Conference on Computer Vision and Pattern Recognition, pp. 143–152 (2020)
17. Shahroudy, A., Liu, J., Ng, T.T., Wang, G.: NTU RGB+D: a large scale dataset for 3D human activity analysis. In: Proceedings of the IEEE Conference on Computer Vision and Pattern Recognition, pp. 1010–1019 (2016)
18. Shi, L., Zhang, Y., Cheng, J., Lu, H.: Skeleton-based action recognition with directed graph neural networks. In: Proceedings of the IEEE/CVF Conference on Computer Vision and Pattern Recognition, pp. 7912–7921 (2019)

19. Shi, L., Zhang, Y., Cheng, J., Lu, H.: Two-stream adaptive graph convolutional networks for skeleton-based action recognition. In: Proceedings of the IEEE/CVF Conference on Computer Vision and Pattern Recognition, pp. 12026–12035 (2019)
20. Shi, L., Zhang, Y., Cheng, J., Lu, H.: Decoupled spatial-temporal attention network for skeleton-based action-gesture recognition. In: Proceedings of the Asian Conference on Computer Vision (2020)
21. Shi, L., Zhang, Y., Cheng, J., Lu, H.: Skeleton-based action recognition with multi-stream adaptive graph convolutional networks. IEEE Trans. Image Process. **29**, 9532–9545 (2020)
22. Soo Kim, T., Reiter, A.: Interpretable 3D human action analysis with temporal convolutional networks. In: Proceedings of the IEEE Conference on Computer Vision and Pattern Recognition Workshops, pp. 20–28 (2017)
23. Vaswani, A., et al.: Attention is all you need. In: Advances in Neural Information Processing Systems, pp. 5998–6008 (2017)
24. Vemulapalli, R., Arrate, F., Chellappa, R.: Human action recognition by representing 3D skeletons as points in a lie group. In: Proceedings of the IEEE Conference on Computer Vision and Pattern Recognition, pp. 588–595 (2014)
25. Wang, X., Girshick, R., Gupta, A., He, K.: Non-local neural networks. In: Proceedings of the IEEE Conference on Computer Vision and Pattern Recognition, pp. 7794–7803 (2018)
26. Wang, X., Han, X., Huang, W., Dong, D., Scott, M.R.: Multi-similarity loss with general pair weighting for deep metric learning. In: Proceedings of the IEEE/CVF Conference on Computer Vision and Pattern Recognition, pp. 5022–5030 (2019)
27. Yan, S., Xiong, Y., Lin, D.: Spatial temporal graph convolutional networks for skeleton-based action recognition. In: Thirty-Second AAAI Conference on Artificial Intelligence (2018)

SRGAT: Social Relational Graph Attention Network for Human Trajectory Prediction

Yusheng Peng[1], Gaofeng Zhang[2,3], Xiangyu Li[1], and Liping Zheng[1,2,3]([✉])

[1] School of Computer Science and Information Engineering, Hefei University
of Technology, Hefei 230601, China
wisionpeng@mail.hfut.edu.cn, zhenglp@hfut.edu.cn
[2] School of Software, Hefei University of Technology, Hefei 230601, China
[3] Anhui Province Key Laboratory of Industry Safety and Emergency Technology,
Hefei 230601, China

Abstract. Human trajectory prediction is a popular research of computer vision and widely used in robot navigation systems and automatic driving systems. The existing work is more about modeling the interactions among pedestrians from the perspective of spatial relations. The social relation between pedestrians is another important factor that affects interactions but has been neglected. Motivated by this idea, we propose a Social Relational Graph Attention Network (SRGAT) via seq2seq architecture for human trajectory prediction. Specifically, relational graph attention network is utilized to model social interactions among pedestrians with different social relations and we use a LSTM model to capture the temporal feature among these interactions. Experimental results on two public datasets (ETH and UCY) prove that SRGAT achieves superior performance compared with recent methods and the predicted trajectories are more socially plausible.

Keywords: Social relations · Social Relational Graph Attention Networks (SRGAT) · Social interactions · Trajectory prediction

1 Introduction

As a key technology in robot navigation system and autonomous driving system, the human trajectory prediction has attracted considerable interests from both academia and industry over the past few years. The human trajectory prediction is full of challenges due to the subtle and intricate interactions among pedestrians.

Many scholars have worked to model these subtle and intricate interactions. The earlier works [8,12] attempt to use handcrafted energy functions to model these social interactions in crowded spaces. However, it is still full of challenges to overall consider various social behaviors. With the rapid development of artificial intelligence technology, the deep learning based human trajectory prediction

© Springer Nature Switzerland AG 2021
T. Mantoro et al. (Eds.): ICONIP 2021, LNCS 13109, pp. 632–644, 2021.
https://doi.org/10.1007/978-3-030-92270-2_54

approaches [1,6,10] has achieved great success. In related works, pooling mechanism [1,2,6], attention mechanism [5,16] and graph neural network [9,11] are widely used to model social interactions among pedestrians. Most of these methods model social interaction from the perspective of spatial relations, while the social relations between pedestrians have been neglected.

In view of the limitations of the above methods, we introduce the interpersonal distance to represent social relation. American anthropologist Edward Hall divides interpersonal distance into four kinds: intimate, personal, social and public [7]. We construct an social relational graph among pedestrians according to these four kinds of interpersonal distance. Besides, we introduce relational graph attention network (RGAT) [4] to model the social interactions among pedestrians within different interpersonal distances respectively.

Contributions: We propose a novel Social Relational Graph Attention Network (called SRGAT) with encoder-decoder architecture for trajectory prediction which respectively considers the social influence of neighboring pedestrians within different interpersonal distances. Firstly, we utilize RGAT to model social interactions among pedestrians with different social relations, and then, we adopt a gated attention mechanism to aggregate these social features to acquire social features. Secondly, we use a LSTM to explicitly capture the temporal correlations of these social features. This paper is the first attempt to use RGAT for social interaction modeling in human trajectory prediction. Experimental results demonstrate that the proposed SRGAT model successfully predicts future trajectories of pedestrians.

2   Related Works

2.1   Social Interaction Modeling

Handcrafted rules and energy parameters [8,12] have been used to capture social interactions but fail to generalize properly. In some recent approaches [1,2,6], pooling mechanisms have been used to model social interactions among pedestrians in local or global neighborhoods. In the view that pedestrians have different impacts on social interactions, Fernando et al. [5] introduced an attention mechanism in social interaction modeling. After that, existing approaches [15,16] adopt diverse attention mechanisms to improve performance of trajectory prediction. With the development of graph neural network, graph-based social interaction modeling has been utilized in various pedestrian trajectory models [9,11,18]. In our model, we utilize RGAT to capture spatial interaction features on the social graphs, and the spatial interaction features are fed to an LSTM to model the temporal correlation to capture spatio-temporal interaction features.

2.2   Social Relation Modeling

Social relations are the general term of mutual relations formed by people in the process of common material and spiritual activities, that is, all the relations between people. Recognizing the social relations between people can enable agents to better understand human behavior or emotions [17]. Sun et al. [14] directly annotate social relations as 0/1 which represents whether pedestrians are in the same group or not. The SRA-LSTM model [13] learned the social relation representations from the relative positions among pedestrians through social relationship encoder. However, the learned representation of social relation is lack of interpretability. In our work, we take the interpersonal distance theory [7] of sociological psychology as a standard and divide social relations according to four kinds of interpersonal distance (intimate, personal, social and public). We model the social interactions of pedestrians of each type of social relationship separately, and integrate the social interactions of four social relationships through gated attention mechanisms.

2.3 Graph Neural Networks

Graph Neural Networks (GNNs) are powerful neural network architecture for machine learning on graphs. Graph neural networks (GNNs) are effective neural networks for processing graph structure data. Recently, the variants of GNN including Graph Convolutional Network (GCN) and Graph Attention Networks (GAT) demonstrate breakthroughs on various tasks like social network prediction, traffic prediction, recommender systems and molecular fingerprints prediction [19]. In the pedestrian trajectory prediction task, GCN is used as message passing to aggregate motion information from nearby pedestrians to model social interactions [3,11,18]. While neighbor pedestrians have different impacts on the target pedestrian, GAT is more suitable for modeling such social interactions and achieved success in predicting future trajectory [9]. Inspired by Relational Graph Attention Network (RGAT) [4], we introduce a novel RGAT to model the social interactions among pedestrians with different social relations, and aggregate the social interaction features of different social relations through a gated attention mechanism.

3 Approach

The goal of human trajectory prediction is to predict the future trajectories from the given past trajectories of pedestrians. The goal of human trajectory prediction is to predict the future trajectories from the given past trajectories of pedestrians. The novel Social Relational Graph Attention Network (SRGAT) via seq2seq structure is proposed in this section (as shown in Fig. 1). For each pedestrian, an LSTM is employed to encode trajectory from relative positions to capture motion feature. Meanwhile, we create social relationship graph among neighbor pedestrians from absolute positions, and the RGAT was utilized to acquire the social interaction feature. Then an extra LSTM is used to encode social interactions of all time steps to capture the spatio-temporal social interaction features. Finally, we employ an LSTM as decoder to predict future positions from encoder feature.

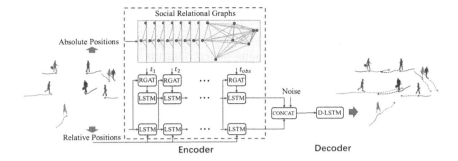

Fig. 1. Illustration of the overall approach. At each time-step, the pedestrians positions are used to calculate relative positions of each other, and the relative positions are processed through embed layer and LSTM to encode the social relationships of each pair of pedestrians. The social relationship attention module models the social interactions by attentively integrating the hidden states of neighbors. Then the social interaction tensor and the embedding vector of each pedestrian's position are treated as inputs of LSTM to output the current hidden states and infer the positions of next time-step.

3.1   Problem Formulation

This paper focuses on addressing the human trajectory prediction in surveillance video crowd scenarios. For better modeling the social interactions among pedestrians, we focus on two-dimensional coordinations of pedestrians in the world coordinate system at specific key frames. For each sample, we assumed that the surveillance video scene involved N pedestrians. Given certain observed positions $\{p_i^t | (x_i^t, y_i^t), t = 1, 2, ..., T_{obs}\}$ of pedestrians i of T_{obs} key frames, our goal is predicting the positions $\{p_i^{t'} | (\widehat{x}_i^{t'}, \widehat{y}_i^{t'}), t' = T_{obs} + 1, T_{obs} + 2, ..., T_{pred}\}$ of future T_{pred} key frames.

3.2   Trajectory Encoding

LSTM is often used to capture latent motion states in pedestrian trajectory prediction models [1,6,16]. By following these works, to capture the unique motion pattern, we also employ an LSTM denoted as Motion Encoder (ME-LSTM) to capture the latent motion pattern for each pedestrian. For each time-step, we embed the relative position into a fixed-length vector e_i^t, and the embedding vector is fed to the LSTM cell as follows:

$$e_i^t = \phi(\Delta x_i^t, \Delta y_i^t; W_e) \tag{1}$$
$$m_i^t = \text{ME-LSTM}(h_i^{t-1}, e_i^t; W_m) \tag{2}$$

where $(\Delta x_i^t, \Delta y_i^t)$ is the relative coordinate of pedestrian i at time-step t to the previous time-step, $\phi(\cdot)$ is an embedding function with ReLU nonlinearity, W_e is the weights of embedding function. The ME-LSTM weight is denoted by W_m. All these parameters are shared by all pedestrians involved in the current scene.

3.3   Social Interaction Modeling

To model the social interactions among pedestrians, pooling mechanism [1,2,6] is used to aggregate hidden states among pedestrians on occupancy map. Besides, GNNs are used to capture social interaction features in recent approaches [3,9] and achieve great successful performance. Inspired by the existed works [3,9], all pedestrians in the scene are treated as nodes on the graph. As illustrated in Fig. 2, the edge between each nodes represents latent social interaction between pedestrians.

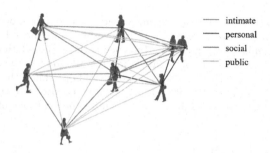

Fig. 2. Pedestrians in the scene are represented by nodes in the graph, and the social interaction between pedestrians is represented by edge between nodes. The different colored lines represent the interaction of different social relations between pedestrians. The intimate, personal, social and public relation are represented by red, blue, green and orange lines, respectively. (Color figure online)

Instead of the above works, we define a social relational graph to represent the social interactions among pedestrians in the scene. As shown in Fig. 2, We divided the social relations between pedestrians into four types: intimate, personal, social, and public [7]. For each social relational graph, all nodes under social relation r *(intimate, personal, social, and public)* represented by an adjacency matrix. The adjacency matrixs are calculated by 4 interpersonal distance ranges: $ranges = [0, 0.45], (0.45, 1.2], (1.2, 3.6], (3.6, 7.5]$:

$$A_{i,j}^{(r)} = \begin{cases} 1, \, dist(i,j) \in ranges^{(r)} \\ 0, \qquad otherwise \end{cases} \tag{3}$$

And then, we utilize an relational GAT [4] to model social interactions through the defined social relational graph. The input of RGAT is a graph with 4 relation types and N nodes. For each time-step t, the ith node is represented by the motion feature vector $m_i \in R^F$, and the features of all nodes are summarised in the feature matrix $M^t = [m_1^t m_2^t ... m_N^t] \in R^{N \times F}$. Through the operation of RGAT, we obtain the social interaction features under 4 kinds of relations:

$$\widehat{M^t} = \text{RGAT}(M^t, A^t) \tag{4}$$

where A^t is a summarised Adjacency matrix $A^t = [A^{r_1,t}A^{r_2,t}A^{r_3,t}A^{r_4,t}] \in R^{4 \times N \times N}$. The output $\widehat{M}^t$ is a summarised feature matrix $\widehat{M}^t = [\widehat{m}^{r_1,t}\widehat{m}^{r_2,t}\widehat{m}^{r_3,t}\widehat{m}^{r_4,t}] \in R^{4 \times N \times F}$.

For each relation r, we get a gate value $g^r = 0/1$ to represent whether there have neighbors of pedestrian i under this relation. And then we aggregate 4 social relational interaction features by the gate mechanism:

$$\mathcal{M}_i^t = \sum_{r \in \Re} g^r \cdot \widehat{m}_i^{r,t} \tag{5}$$

We only model the spatial interactions among pedestrians at the time-steps of the observation stage. To learn the temporal correlations between spatial interactions of pedestrians, we propose to employ an extra LSTM to encode these spatial interactions. We term this Social Encoder as SE-LSTM:

$$s_i^t = \text{SE-LSTM}(s_i^{t-1}, \mathcal{M}_i^t; W_s) \tag{6}$$

where W_s is the weight of SE-LSTM which is shared by all pedestrians in this scene.

3.4   Fusion and Prediction

In Encoder component of our proposed model, the ME-LSTM encoder is design to learn motion feature from observed trajectory, and the SE-LSTM encoder is designed to learn spatial-temporal interaction features. These two parts are combined together later to fusion of motion and social interaction features. At the last observed time-step T_{obs}, the encoder features of each pedestrian are represented by two hidden variables $(m_i^{T_{obs}}, s_i^{T_{obs}})$ from two LSTMs. The two variables will be fused to served as input of decoder to predict future trajectory. However, because of the uncertainty of pedestrian movement, it is necessary to predict multiple reasonable and socially acceptable trajectories. Thus, the latent code z from $\mathcal{N}(0,1)$ (the standard normal distribution) is added to encoding features. We concatenate three variables in our implementation:

$$d_i^{T_{obs}} = m_i^{T_{obs}} \parallel s_i^{T_{obs}} \parallel z \tag{7}$$

the concatenated state vector $d_i^{T_{obs}}$ then employed as the initial of the decoder LSTM hidden state (termed as D-LSTM). The inferred relative coordinate is given by:

$$d_i^{T_{obs}+1} = \text{D-LSTM}(d_i^{T_{obs}}, e_i^{T_{obs}}; W_d) \tag{8}$$

$$(\Delta x_i^{T_{obs}+1}, \Delta y_i^{T_{obs}+1}) = \delta(d_i^{T_{obs}+1}) \tag{9}$$

where W_d is D-LSTM weight, $\delta(\cdot)$ is a linear layer, $e_i^{T_{obs}}$ is from Eq. 1. The predicted relative coordinate from Eq. 9 at each time-step will be calculated by Eq. 1 to served as input to D-LSTM to infer the relative coordinate of next time-step.

The predicted relative positions are converted to absolute positions for calculating losses. As most previous works [2,6,9], we use a variety loss function that encourages the network to produce diverse samples. For each pedestrian, we generate k possible output predictions by randomly sampling z from $\mathcal{N}(0,1)$ and choosing the "best" prediction in L2 sense as our prediction to compute the loss:

$$L_{variety} = \min_{k} \parallel Y_i - \widehat{Y}_i^k \parallel_2 \tag{10}$$

where Y_i is the ground-truth of future trajectory, $\widehat{Y}_i^k$ is the future trajectory generated by SRGAT, and k is a hyperparameter. To train the network better, only the best trajectory is used to compute the loss to encourage the network to hedge its bets and generate multiple possible future trajectories which are consistent with past trajectory.

3.5   Implementation Details

The dimensions of the hidden state for encoder is 32 and decoder is 80. The input coordinates are embeded as 16 dimensional vectors. The dimension of noise z is set to 16. We use two graph attention layers in RGAT model and the dimensions of intermediate representations is set to 16 and 32 respectively. Adam optimizer with a learning rate of 0.001 is applied to train the model and the batch size is set to 64 in train and test stage.

4   Experiments

We evaluate our method on two public available human walking video datasets: ETH and UCY. These two datasets contain 5 crowd scenes, including ETH, HOTEL, ZARA1, ZARA2, and UNIV. All the trajectories are converted to the world coordinate system and then interpolated to obtain values at every 0.4 s.

Evaluation Metrics. Similar to prior works [6,18], the proposed method is evaluated with two types of metrics as follow:

1. *Average Displacement error (ADE)*: the Mean Square Error (MSE) between the ground-truth trajectory and predicted trajectory over all predicted time steps.
2. *Final Displacement error (FDE)*: the Mean Square Error (MSE) between the ground-truth trajectory and predicted trajectory at the last predicted time steps.

Baseline. As traditional approaches based on hand-crafted features perform not as well as social LSTM model [1], the traditional models are not listed as baseline. And we only compare SRGAT with the following deep learning based works:

1. *S-LSTM* [1]: An trajectory prediction method that combines LSTM with a social pooling layer, which can aggregate hidden states of the neighbor pedestrians.
2. *SGAN* [6]: An improved version of S-LSTM that the social pooling is displaced with a new pooling mechanism which can learn a "global" pooling vector. A variety loss function is proposed to encourage the GAN to spread its distribution and generate multiple socially acceptable trajectories.
3. *SR-LSTM* [18]: An improved version of S-LSTM by proposing a data-driven state refinement module. The refinement module can jointly and iteratively refines the current states of all participants in the crowd on the basis of their neighbors' intentions through a message passing mechanism.
4. *IA-GAN* [10]: A novel approach to pedestrian prediction that combines generative adversarial networks with a probabilistic model of intent.
5. *TAGCN* [3]: A three stream topology-aware graph convolutional network for interaction message passing between the agents. Temporal encoding of local- and global-level topological features are fused to better characterize dynamic interactions between participants over time.
6. *RSBG* [14]: A novel structure called Recursive Social Behavior Graph, which is supervised by group-based annotations. The social interactions are modeled by GCNs that adequately integrate information from nodes and edges in RSBG.
7. *SRA-LSTM* [13]: A novel social relationship encoder is utilized to learn the social relationships among pedestrians. The social relationship features are added to help modeling social interactions among pedestrians.

Evaluation Methodology. We use the leave-one-out approach similar to that from S-LSTM [1,6]. We train and validate our model on 4 sets and test on the remaining set. We take the coordinates of 8 key frames (3.2 s) of the pedestrian as the observed trajectory, and predict the trajectory of the next 12 key frames (4.8 s).

4.1 Quantitative Evaluations

Table 1 demonstrates the quantitative results between the proposed method and the above mentioned methods across five datasets. The SR-LSTM [18], TAGCN [3], IA-GAN [10], RSBG [14], and our proposed model adopted GNN models (like GCN and GAT) to model social interactions among pedestrians. To our knowledge, only the RSBG [14], SRA-LSTM [13] and our proposed model take social relations into account to model social interactions. On each column, the top three performing methods are highlighted in red, green, and blue. The last column of the table shows the average performance over the five crowd scenes. The SRA-LSTM achieves the minimum ADE and FDE on five dataset. Besides, the SR-LSTM and our proposed model achieve the minimum ADE and FDE respectively.

Due to the difference in motion patterns of pedestrians on each dataset, the performance of a model on the 5 datasets is also different. Thus, we evaluate these models with a point system on 5 datasets. For each dataset, the top-1, top-2, and top-3 models on ADE or FDE can score 3, 2, and 1 points, respectively. Based on this point system, our proposed model can win 17 points. The SRA-LSTM, RSBG, IA-GAN, TAGCN, SR-LSTM, SGAN, and S-LSTM models can score 14, 9, 8, 2, 12, 6, and 0 points, respectively. According to the scores, our model achieve the best performance on the five datasets.

Table 1. Quantitative results of all the baselines and the proposed method on ETH/UCY datasets. Top-1, top-2, top-3 results are shown in red, green, and blue. (GNN: Graph Neural Networks, SRS: Social Relations)

Method	Notes		Performance (ADE/FDE)					
	GNN	SRS	ETH	HOTEL	ZARA1	ZARA2	UNIV	AVG
S-LSTM	✗	✗	1.09/2.35	0.79/1.73	0.47/1.00	0.56/1.17	0.67/1.40	0.72/1.54
SGAN	✗	✗	0.87/1.62	0.67/1.37	0.35/0.68	0.42/0.78	0.76/1.52	0.61/1.21
SR-LSTM	✓	✗	0.63/1.25	0.37/0.74	0.41/0.90	0.32/0.70	0.51/1.10	0.45/0.94
TAGCN	✓	✗	0.86/1.50	0.59/1.15	0.42/0.90	0.32/0.71	0.54/1.25	0.55/1.10
IA-GAN	✓	✗	0.69/1.42	0.39/0.79	0.35/0.74	0.31/0.66	0.56/1.17	0.46/0.96
RSBG	✓	✓	0.80/1.53	0.33/0.64	0.40/0.86	0.30/0.65	0.59/1.25	0.48/0.99
SRA-LSTM	✗	✓	0.59/1.16	0.29/0.56	0.37/0.82	0.43/0.93	0.55/1.19	0.45/0.93
Ours	✓	✓	0.78/1.59	0.30/0.53	0.35/0.73	0.32/0.66	0.53/1.13	0.46/0.93

Table 2. Parameters size and inference time of different models compared to ours. The lower the better. Models were bench-marked using Nvidia GTX2080Ti GPU. The inference time is the average of several single inference steps. We notice that SRGAT has the least inference time compared to others. The text in blue show how many times our model is faster than others.

	Parameters count	Inference time
SGAN [6]	**46.4K** (0.83x)	0.0057 (1.84x)
SR-LSTM [18]	64.9K (1.17x)	0.0049 (1.58x)
SRA-LSTM [13]	67.1K (1.21x)	0.0045 (1.45x)
SRGAT	55.6K	**0.0031**

Table 2 lists out the speed comparisons between our model and publicly available models which we could bench-mark against. The size of SRGAT is 55.6K parameters. SGAN has the smallest model size with 46.4k parameters, which is about eight tenth of the number of parameters in SRGAT. The sizes of SR-LSTM and SRA-LSTM are 64.9K and 67.1K parameters respectively, which are very close. In terms of inference speed, SRGAT was previously the fastest method with an inference time of 0.0045 s per inference step. However, the inference time of our model is 0.0031 s which is about 1.45x faster than SRA-LSTM.

Observed Trajectory   Ground Truth   SR-LSTM   SRA-LSTM   SRGAT

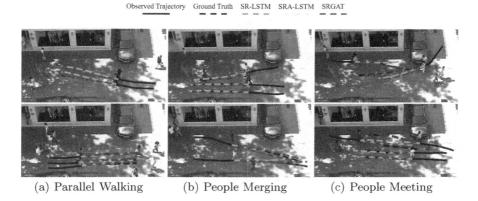

(a) Parallel Walking   (b) People Merging   (c) People Meeting

Fig. 3. Comparisons of our method with SRA-LSTM and SR-LSTM in 3 common social scenarios, which containing parallel walking, people merging, and people meeting. The blue solid line represents the observed trajectory, the dashed line represent the future trajectory (blue: ground truth, yellow: SR-LSTM, green: SRA-LSTM, red: our model). (Color figure online)

4.2   Qualitative Evaluations

Benefiting from the proposed social relational graph, SRGAT can learn the latent social relations and better model the social interactions among pedestrians. Thus, SRGAT can perform accurate trajectory prediction. Figure 3 illustrates the trajectory prediction results by using SR-LSTM, SRA-LSTM, and SRGAT in 3 common social scenarios. For the parallel walking cases, the trajectories predicted by SRA-LSTM and SRGAT model are more similar to ground truth. That benefits from the consideration of social relations in models. Furthermore, for more complex social scenarios such as people merging and people meeting, the SRGAT model can still predict the future trajectories which are more similar to the ground truth.

To verify the effect of the proposed model in multimodal trajectory prediction, we compare with the multimodal model SGAN. The multimodal trajectory predictions are shown in Fig. 4. In people meeting scenario, the multimodal trajectories generated by SRGAT are more agminated and tend to avoid the motions of each other. On the contrary, the trajectories predicted by the SGAN model are more dispersed. Similarly, the trajectories predicted by the SRGAT model in people merging scenario are also more agminated. The 3rd column shows the failure case, where neither SGAN nor SRGAT successfully predicted the future trajectories. Since the final destination is unknown, it is difficult to successfully predict the future trajectory of the pedestrian in this case only relying on the 8 time-steps' observed trajectory. That will be the focus of our future work.

People Meeting   People Merging   Failure

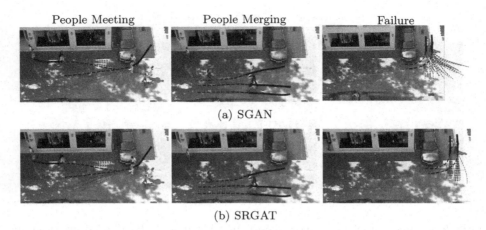

(a) SGAN

(b) SRGAT

Fig. 4. Comparisons of our method with SGAN in mutimodal trajectory predictions. The first two columns show the results of people merging and people meeting scenarios, and the last column shows a failure case.

5   Conclusions and Discussion

In the work, we designed a social relational graph and modeled the social interactions among pedestrians by relational graph attention network. Two LSTMs were employed to encode the movement of pedestrians and the social interactions among pedestrians to capture the latent motion features and the spatiotemporal interaction features among pedestrians. The encoding features and the random Gaussian noise are fused and then feed to the decoder to generate the multimodal future trajectories. Evaluations are performed in two commonly used metrics, namely, ADE and FDE, across five benchmarking datasets. Comparisons with baseline methods and state-of-the-art approaches indicate the effectiveness of the proposed SRGAT model. The qualitative results of some common social scenarios indicate the success of the use of social relation modeling in trajectory prediction research. To solve the failure cases, we will study the goal guidance of pedestrians and scene information guidance in the future work. In addition, we will study the SRGAT on dynamic interpersonal distance.

Acknowledgments. This work is supported by the National Natural Science Foundation of China (61972128), the Fundamental Research Funds for the Central Universities of China (Grant No. PA2019GDPK0071).

References

1. Alahi, A., Goel, K., Ramanathan, V., et al.: Social LSTM: human trajectory prediction in crowded spaces. In: 2016 IEEE Conference on Computer Vision and Pattern Recognition (CVPR), pp. 961–971. IEEE, Las Vegas (2016)
2. Amirian, J., Hayet, J.B., Pettre, J.: Social Ways: learning multi-modal distributions of pedestrian trajectories with GANs. In: 2019 IEEE/CVF Conference on Computer Vision and Pattern Recognition Workshops (CVPRW), pp. 2964–2972. IEEE, Long Beach (2019)
3. Biswas, A., Morris, B.T.: TAGCN: topology-aware graph convolutional network for trajectory prediction. In: Bebis, G., et al. (eds.) ISVC 2020. LNCS, vol. 12509, pp. 542–553. Springer, Cham (2020). https://doi.org/10.1007/978-3-030-64556-4_42
4. Dan, B., Dane, S., Pietro, C., et al.: Relational graph attention networks. In: International Conference on Learning Representations (ICLR) (2019)
5. Fernando, T., Denman, S., Sridharan, S., et al.: Soft + Hardwired Attention: an LSTM framework for human trajectory prediction and abnormal event detection. Neural Netw. **108**, 466–478 (2018)
6. Gupta, A., Johnson, J., Fei-Fei, L., et al.: Social GAN: socially acceptable trajectories with generative adversarial networks. In: 2018 IEEE/CVF Conference on Computer Vision and Pattern Recognition (CVPR), pp. 2255–2264. IEEE, Salt Lake City (2018)
7. Hall, E.T.: The Hidden Dimension. Doubleday, Garden City (1966)
8. Helbing, D., Molnar, P.: Social force model for pedestrian dynamics. Phys. Rev. E **51**, 4282 (1998)
9. Huang, Y., Bi, H., Li, Z., et al.: STGAT: modeling spatial-temporal interactions for human trajectory prediction. In: 2019 IEEE/CVF International Conference on Computer Vision (ICCV), pp. 6272–6281. IEEE, Seoul (2019)
10. Katyal, K.D., Hager, G.D., Huang, C.M.: Intent-aware pedestrian prediction for adaptive crowd navigation. In: 2020 IEEE International Conference on Robotics and Automation (ICRA), pp. 3277–3283. IEEE, Paris (2020)
11. Mohamed, A., Qian, K., Elhoseiny, M., et al.: Social-STGCNN: a social spatio-temporal graph convolutional neural network for human trajectory prediction. In: 2020 IEEE/CVF Conference on Computer Vision and Pattern Recognition (CVPR), pp. 14412–14420. IEEE, Seattle (2020)
12. Pellegrini, S., Ess, A., Schindler, K., et al.: You'll never walk alone: modeling social behavior for multi-target tracking. In: 2009 IEEE International Conference on Computer Vision (ICCV), pp. 261–268. IEEE, Miami (2009)
13. Peng, Y., Zhang, G., Shi, J., et al.: SRA-LSTM: social relationship attention LSTM for human trajectory. arXiv preprint arXiv:2103.17045 (2021)
14. Sun, J., Jiang, Q., Lu, C.: Recursive social behavior graph for trajectory prediction. In: 2020 IEEE/CVF Conference on Computer Vision and Pattern Recognition (CVPR), pp. 657–666. IEEE, Seattle (2020)
15. Vemula, A., Muelling, K., Oh, J.: Social Attention: modeling attention in human crowds. In: 2018 IEEE International Conference on Robotics and Automation (ICRA), pp. 4601–4607. IEEE, Brisbane (2018)
16. Xu, Y., Piao, Z., Gao, S.: Encoding crowd interaction with deep neural network for pedestrian trajectory prediction. In: 2018 IEEE/CVF Conference on Computer Vision and Pattern Recognition (CVPR), pp. 5275–5284. IEEE, Salt Lake City (2018)

17. Zhang, M., Liu, X., Liu, W., et al.: Multi-granularity reasoning for social relation recognition from images. In: 2019 IEEE International Conference on Multimedia and Expo (ICME), pp. 1618–1623. IEEE, Shanghai (2019)
18. Zhang, P., Ouyang, W., Zhang, P., et al.: SR-LSTM: state refinement for LSTM towards pedestrian trajectory prediction. In: 2019 IEEE/CVF Conference on Computer Vision and Pattern Recognition (CVPR), pp. 12077–12086. IEEE, Long Beach (2019)
19. Zhang, S., Tong, H., Xu, J., Maciejewski, R.: Graph convolutional networks: a comprehensive review. Comput. Soc. Netw. 6(1), 1–23 (2019). https://doi.org/10.1186/s40649-019-0069-y

FSE: A Powerful Feature Augmentation Technique for Classification Task

Yaozhong Liu[1(✉)], Yan Yang[1], and Md Zakir Hossain[1,2]

[1] The Australian National University, Canberra, ACT 2601, Australia
{u6686404,yan.yang,zakir.hossain}@anu.edu.au
[2] CSIRO Agriculture and Food, Black Mountain, Canberra, ACT 2601, Australia

Abstract. Neural networks are powerful at discovering the hidden relation, such as classifying facial expressions to emotions. The performance of the neural network is typically limited by the number of informative features. In this paper, a novel feature augmentation is proposed for generating new informative features in an unsupervised manner. Current data augmentation focuses on synthesizing new samples according to data distribution. Instead, our approach, *Feature Space Expansion (FSE)*, enriches data feature by providing their distribution information, which brings benefit based on model performance and convergence speed. To the best of our knowledge, FSE is the first feature augmentation method, which is developed based on feature distribution. We evaluate FSE performance on face emotion dataset and music effect dataset. We provide diverse comparisons with different alternative baselines. The experimental results indicate FSE provides significant improvement in model's prediction accuracy when the number of features in original dataset is relatively small, and less remarkable improvement when the number of features in original dataset is large. In addition, training on FSE augmented training set can have at least ten times faster convergence speed than training on original training set.

Keywords: Feature space expansion · Clustering · Fuzzy clustering · GMM

1 Introduction

Data augmentation is conventionally used during model training for increasing data diversity by synthesizing new data from existing ones [16]. It has been proven to be effective in multiple problems including image classification [14], object detection, and speech recognition. In this paper, we propose an alternative data augmentation definition, where focusing on automatically generating discriminate feature that provides lightweight distribution information. We name our approach as feature space expansion (FSE) which uses clustering to realize the goal.

Common data augmentation techniques aim to increase the model performance and generalization by training with extra synthesized samples enhancing

© Springer Nature Switzerland AG 2021
T. Mantoro et al. (Eds.): ICONIP 2021, LNCS 13109, pp. 645–653, 2021.
https://doi.org/10.1007/978-3-030-92270-2_55

the size of training set. In contrast, we strengthen the feature information by providing distribution characteristics. Our approach expands the feature space by using the membership allocation from the clustering algorithm. Clustering algorithm groups the data with similar distribution behaviour. And, the membership allocation assigned by the clustering algorithm summarises how the current data distributed with respect to the overall dataset. Neural Network acts as universal encoding/decoding functions, which maps input distribution (feature) to output distribution (target). Therefore, we can provide lightweight distribution information, clustering algorithm membership allocations, to the model directly, which can speed up convergence and potentially benefit performance.

- We propose a novel feature augmentation method which aims to augment feature space by introducing data distribution through clustering algorithm.
- We evaluate the proposed method based on prediction accuracy and convergence speed by using different algorithm setup including k-means, GMM and fuzzy clustering.
- We compare the proposed method with extensive baseline methods. It confirms that FSE is a powerful technique which can increase the prediction accuracy especially when the feature number is small in original dataset, the improvement is up to 64.62% and there are still 7.7% accuracy improvement achieved in the training set with large number of features.

2   Related Work

2.1   Artificial Neural Networks

Artificial neural networks (ANNs) allow modeling non-linear processes and they have turned into useful tools to solve the problems of decision making, classification and so on. Feed-forward neural network (FFNN) and Bidirectional neural network (BDNN) are two typical neural networks which are widely used. In FFNN, the information moves in only one direction which forwards from the input through hidden nodes to the output nodes and never goes backwards [20,21]. However in BDNN, information can be accepted from both directions, which means the input neurons in the forward direction can also be the output neurons in the backward direction [2]. Both of these neural network have self-learning capabilities that enable them to produce better results as more data or more valuable data features become available. Therefore, besides enlarge the size of the training samples, increasing the size of feature space in training data also strengthen the network's ability to achieve good classification accuracy.

2.2   Clustering

Clustering, as a powerful machine learning tool for detecting overall structures in datasets has been proven to be very suitable to discover intrinsic patterns of the labelled or unlabelled data [18]. It is the task of dividing data points into homogeneous classes or clusters making the elements in the same class are as similar as possible and the elements belong to different classes are as dissimilar as possible. The k-means is the centroid-based clustering which aims to partition 'n' observations into 'k' clusters in which each observation belongs to the nearest cluster center (hard clustering). The Gaussian mixture model (GMM) is the distribution-based clustering which assumes all the data points are generated from a mixture of a finite number of Gaussian distributions with unknown parameters. These randomly initialized Gaussian distribution will iterative optimized to fit the data better and every data point will has its memberiship to each Gaussian distribution eventually [19]. According to the demand of different target, the result of GMM can be converted into hard clustering or soft clustering. Nevertheless, the main information between these two methods are different. The k-means can be regarded as a partition which aims to chop data into relatively homogeneous pieces emphasizing the similarity and dissimilarity among whole data, but GMM is able to capture correlations and dependency between data attributes and the positional information of every data. Even though k-means and GMM have their focal points when they are used as clustering method, both of them describe the high level overall features that can not be observed directly from the original data, and thus may improve model's classification accuracy and are good feature augmentation candidates in our research.

Fuzzy clustering (also refers to soft k-means) allows the data points can belong to more than one cluster. For each of the points, associated membership values represent the degree of fitting into different clusters. Fuzzy clustering possesses the traits of the dividing clusters by similarity in k-means and the membership allocation in GMM at the same time. Fuzzy clustering can be used as a new feature augmentation technique to compare with k-means and GMM, which is the basic feature augmentation technique in this paper.

3   Method

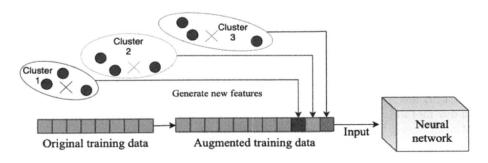

Fig. 1. Feature space expansion.

As the Fig. 1 shows, the whole process of our work which consists of two main parts: Augmentation and Training. We firstly implement the FSE procedure to the L-2 Normalized training set to produce extra features including overall information like data distribution and similarity. The new generated features will be concatenated with the original features for each training instance and form the new training set to feed into neural network afterwards. By using k-means, GMM, fuzzy clustering, we augment the original features to obtain three types of training set. The feature process of the k-means feature augmentation is simple since every instance will finally be divided into only one cluster. The number of extra features in the augmented training set is equivalent to the number of the labels we want to classify and we use one-hot encoding to indicate which cluster it belongs to for every data point. Using GMM will randomly initialized the Gaussian models with the number of the models is identical to the number of labels and used as extra features. Unlike k-means, a single data point can be generated by multiple Gaussian models and hence there are possibilities to represent the degree that this data point belongs to different Gaussian models. Fuzzy clustering includes the peculiarities from both of these two methods, not only focus on the similarity among the dataset but also not have restriction that one data point can belong to exactly one cluster while having membership value for each clusters to show the degree of the data points belongs to. This includes the distribution information of the dataset and allows data points near the cluster boundary which are usually too intangible to be divided into either cluster in k-means to have numerical measurement to indicate its belonging degree rather than be a binary case. We use three feature augmentation methods to enrich the training set and feed three kinds of new training sets into FFNN and BDNN training and then evaluate the corresponding models' performance including convergence speed and prediction accuracy respectively.

4   Experiment Setup

4.1   Dataset

The dataset we use is the same as the one in this paper [1] which is about distinguishing different face emotions. For each training instance, it has ten features and the type of these features are continuous value, including five features measured by local phase quantisation descriptor (`LPQ features`), five features measured by pyramid of histogram of oriented gradients descriptor (`PHOG features`), and a `label` which indicates its type. In order to exclude the possibility that our FSE methods is specialized in a certain dataset and validate the robustness of our method, we test FSE in another dataset which is about the music effect of brain waves [17] with twenty-four features and one label.

4.2   Network Training

At first, we set the 35 hidden neurons, 1200 epochs, and 0.005 learning rate to train the BDNN using original training set and three other augmented training

Table 1. The best accuracy of FDNN model in different types of datasets.

	Origianl	k-means	GMM	Fuzzy clustering
Face emotion dataset	24.58%	79.85%	89.2%	42.6%
Music effect dataset	38.8%	44.25%	46.5%	44%

sets. We continuously tune these hyper-parameters in BDNN aiming to achieve higher prediction accuracy in the test set. The BDNN consists of one hidden layer using the ReLu activation function and using the Softmax activation function in the output layer for classification. We choose to use cross entropy loss to instruct the training process and use Adam optimizer to update parameters of network.

4.3 Experiments w.r.t. the Network Backbone

We did not only use BDNN to train the model and make classification but also train simple feedforward neural network with one hidden layer using the same activation function, optimizer and loss measurement. Thus, the BDNN can make comparison within the aspects of training loss and test accuracy. The hyper-parameters of FFNN training are adjusted constantly so that a better prediction result can be obtained by the FFNN model.

4.4 Experiments w.r.t. the Clustering Algorithm

To investigate the effect on the performance of model when applying different clustering methods in FSE to augment the features in training data, we use GMM, k-means and fuzzy clustering to add extra features for training data respectively and test the prediction accuracy of the model and training convergence speed separately. The extra features generated by k-means is recorded using one-hot encoding while the extra features generated by GMM and fuzzy clustering are recorded directly using the membership result obtained from clustering. The models trained by training set using three clustering methods are compared with each other and also compared with the model trained by original training set without feature augmentation.

5   Results and Analysis

We take the test accuracy of the model trained by the original dataset as a baseline. We compare the baseline with alternative FSE strategies in Table 1. Figure 2 presents the FFNN models' average prediction accuracy of eight training results from two datasets respectively on the datasets which processed by FSE and original dataset. According to Fig. 2, three FSE methods have positive effect on promoting the prediction accuracy in two different datasets especially in the

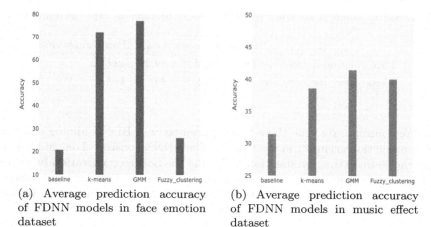

(a) Average prediction accuracy of FDNN models in face emotion dataset

(b) Average prediction accuracy of FDNN models in music effect dataset

Fig. 2. Average prediction accuracy in two different datasets

face emotion dataset where the biggest promotion has been reached nearly 60%, but the promotion in music effect dataset is less impressive. Table 1 shows the best accuracy obtained by FFNN models trained by different augmented training sets and the original training sets without being augmented. From the obtained results, we can find that the FSE method improves the model's performance in varying degrees in terms of two different types of training datasets. This is accord with our previous assumption that expanding the feature space in training set benefits the model's performance due to the new introduced features containing distribution or similarity information of the whole dataset. As a consequence, the model does not spend time on learning these information of overall distribution but can directly utilize them at the beginning of the training process.

For various clustering algorithms based FSE, GMM significantly strengthens the model's prediction ability by improving 64.62% accuracy in the face emotion dataset and 7.7% accuracy in the music effect dataset, respectively. By assuming the Gaussian modality of data samples, it provides the relative position of each sample at probabilistic aspects. Conventionally, the Gaussian distribution is widely taken as a dataset prior because of its generality. The GMM based FSE gains from the strong prior knowledge by supplying this information to the model directly.

Compared to GMM, the model improvement from k-means based FSE is smaller. K-means is a hard partition algorithm and potentially providing inaccurate information for data points near the cluster boundary. It assigns the data point to the most likely cluster and ignores other potential clusters. Therefore, the cluster-wise information of data point is lost, where the improvement is not as good as GMM based strategy. For fuzzy clustering, it has similar characteristics with k-means and GMM. However, the improvement to the model's accuracy is the most trifling. Though it captures more information than k-means and GMM, this information can be misleading. Within Fuzzy clustering, memberships allocation are almost uniformly assigned to each cluster, which can not provide refined location information to the model. This provides less useful information in model training and thus the accuracy improvement is not strong as the other two augmentation strategies.

From Table 1, We also notice the number of features in the original dataset correlates with the performance increase by using FSE. The face emotion dataset has 10 features. The prediction accuracy increase can range from 18.02% to three times the accuracy of the baseline model. In the music effect dataset, There are twenty-four features and the accuracy improvement caused by FSE is not remarkable. When the number of features becomes larger in the training set, the model is potentially learned from abundant features. Extra features from FSE contain high level information, which may not be capable of summarising informative distribution information. Large feature space with the small number of samples always reflects the sparsity. These extra features would be less effective and influential in strengthening model prediction ability than in the training set with the small number of original features.

Figure 3 displays the training loss of using the original feature and GMM augmented feature from the face emotion training set by using FFNN and BDNN architecture. Applying GMM based feature augmentation accelerates the converging speed to a large extent. From the FFNN training loss change plot in Fig. 3, we find the training loss of GMM augmented training drops quickly in the first hundred epochs. GMM augmented feature has similar training loss at 100 epoch with training with original feature at 1200 epoch, which is 12 times faster in terms of convergence speed. Additionally, after the first hundred training epochs, the training loss can decrease further but in a relatively steady tendency. BDNN model includes the loss in the forward direction and the loss in the backward direction, thus it doubles FFNN loss. Noticeably, the GMM accelerates BDNN convergence speed more conspicuously. The training loss of the GMM augmented training set achieves the same final training loss of the baseline training set by using 60 epochs in both experiments which are approximately 20 times faster than baseline. The final training loss of the GMM augmented dataset remains similar across two backbones. GMM augmented training set contains high level information about the overall distribution of the data which sufficiently helps the model to learn the implicit pattern in the small number of epochs.

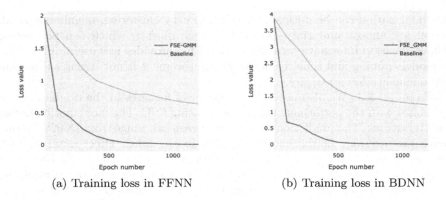

(a) Training loss in FFNN   (b) Training loss in BDNN

Fig. 3. The comparison of training loss between GMM augmented dataset and original dataset

6   Discussion

We provide a detailed analysis between the number of features and model prediction accuracy by applying FSE. Given limited data samples, the feature space becomes sparsity as the number of feature in increase. The membership allocation inferred by the clustering algorithm becomes less informative as each allocation corresponds with a large space. Model is less likely to benefit from FSE in this case. Future researches could focus on developing efficient feature reduction prior to applying FSE.

7   Conclusion

In this paper, we propose a novel data augmentation method called feature space expansion (FSE) which focuses on adding lightweight distribution information through introducing extra features to improve the prediction accuracy of the neural network models. FSE utilizes clustering algorithm to benefit model training by providing overall characteristics of the training set. It accelerates the convergence speed and helps model learning intrinsic pattern of training data. We experiment with k-means, GMM, and fuzzy clustering based FSE, and compare with possible benchmarks on several dataset. The FSE can potentially reduce the training cost and improve the prediction accuracy in different tasks which makes it as a powerful technique.

References

1. Dhall, A., Goecke, R., Lucey, S., Gedeon, T.: Static facial expressions in tough conditions: data, evaluation protocol and benchmark. In: 1st IEEE International Workshop on Benchmarking Facial Image Analysis Technologies BeFIT, ICCV2011, November 2011

2. Nejad, A.F., Gedeon, T.D.: Bidirectional neural networks and class prototypes. In: IEEE International Conference on Neural Networks. Proceedings, vol. 3, pp. 1322–1327. IEEE, November 1995
3. Zhang, T., Zheng, W., Cui, Z., Zong, Y., Li, Y.: Spatial-Temporal recurrent neural network for emotion recognition. IEEE Trans. Cybern. **49**(3), 839–847 (2018)
4. Liu, Y., Sun, C., Lin, L., Wang, X.: Learning Natural Language Inference using Bidirectional LSTM model and Inner-Attention. arXiv preprint arXiv:1605.09090 (2016)
5. Hansen, L., Salamon, P.: Neural network ensembles. IEEE Trans. Pattern Anal. Mach. Intell. **12**(10), 993–1001 (1990)
6. Shanker, M., Hu, M.Y., Hung, M.S.: Effect of data standardization on neural network training. Omega **24**(4), 385–397 (1996)
7. Wöllmer, M., Metallinou, A., Eyben, F., Schuller, B., Narayanan, S.: Context-sensitive multimodal emotion recognition from speech and facial expression using bidirectional LSTM modeling. In: Proceedings INTERSPEECH 2010, Makuhari, Japan, pp. 2362–2365 (2010)
8. Lin, C.T., Lee, C.S.G.: Neural-network-based fuzzy logic control and decision system. IEEE Trans. Comput. **40**(12), 1320–1336 (1991)
9. Ng, H.W., Nguyen, V.D., Vonikakis, V., Winkler, S.: Deep learning for emotion recognition on small datasets using transfer learning. In: Proceedings of the 2015 ACM on International Conference on Multimodal Interaction, pp. 443–449, November 2015
10. Dai, Z., et al.: CNN descriptor improvement based on L2-normalization and feature pooling for patch classification. In: 2018 IEEE International Conference on Robotics and Biomimetics (ROBIO), pp. 144–149. IEEE, December 2018
11. Maulik, U., Saha, I.: Automatic fuzzy clustering using modified differential evolution for image classification. IEEE Trans. Geosci. Remote Sens. **48**(9), 3503–3510 (2010)
12. Cai, W., Chen, S., Zhang, D.: Fast and robust fuzzy c-means clustering algorithms incorporating local information for image segmentation. Pattern Recogn. **40**(3), 825–838 (2007)
13. Shorten, C., Khoshgoftaar, T.M.: A survey on image data augmentation for deep learning. J. Big Data **6**(1), 1–48 (2019)
14. Perez, L., Wang, J.: The effectiveness of data augmentation in image classification using deep learning. arXiv preprint arXiv:1712.04621 (2017)
15. Schmidhuber, J.: Deep learning in neural networks: an overview. Neural Networks **61**, 85–117 (2015)
16. Xu, Y., Noy, A., Lin, M., Qian, Q., Li, H., Jin, R.: WeMix: How to Better Utilize Data Augmentation. arXiv preprint arXiv:2010.01267 (2020)
17. Rahman, J.S., Gedeon, T., Caldwell, S., Jones, R.: Brain melody informatics: analysing effects of music on brainwave patterns. In: 2020 International Joint Conference on Neural Networks (IJCNN), pp. 1–8. IEEE, July 2020
18. Alashwal, H., El Halaby, M., Crouse, J.J., Abdalla, A., Moustafa, A.A.: The application of unsupervised clustering methods to Alzheimer's disease. Front. Comput. Neurosci. **13**, 31 (2019). https://doi.org/10.3389/fncom.2019.00031
19. Reynolds, D.A.: Gaussian mixture models. Encycl. Biometrics **741**, 659–663 (2009)
20. Auer, P., Burgsteiner, H., Maass, W.: A learning rule for very simple universal approximators consisting of a single layer of perceptrons. Neural Networks **21**(5), 786–795 (2008)
21. Zell, A.: Simulation neuronaler netze, vol. 1, no. 5.3. Addison-Wesley, Bonn (1994)

AI and Cybersecurity

FHTC: Few-Shot Hierarchical Text Classification in Financial Domain

Anqi Wang, Qingcai Chen[✉], and Dongfang Li

Harbin Institute of Technology Shenzhen, Shenzhen, China
19s051040@stu.hit.edu.cn, qingcai.chen@hit.edu.cn

Abstract. As an extensively applied task in the domain of natural language processing, text classification has moved a long way since deep learning technology develop rapidly. Especially after the pre-trained models arrived, the classification performance has been tremendous improved. However, complicated financial text often has multiple structured labels, and there are also many difficulties to have large amounts of labeled samples to ensure high-quality predictions. The existing competitive classification models can only solve one of the problems. To address these issues, we propose a hierarchical classification structure with two level. In the first level, the basic classifier is enhanced by label confusion algorithm to mine the dependency between labels and samples. In the second level, a few-shot classification model under meta-learning framework can complete the classification task based on the predictions from the previous level and a few labeled training samples. We explain our model on two large Chinese financial datasets, and find that it has superiority in both performance and computational expenditure compared to existing competitive classification model, few-sample classification model and hierarchical classification model.

Keywords: Natural language processing · Text classification · Label confusion · Few-shot learning

1 Introduction

Text classification, which is classic in natural language processing tasks, has been widely applied to sentiment analysis [9–11] to determine or calculate the sentiment contained, information filtering [12] such as spam filtering, question answering [13,14] to answer concise answers according to the questions input by the user in natural language. In these scenarios, text can be automatically classified into specific categories to assist human or machine decision-making. As a typical supervised learning task, it needs a large number of labeled training samples to train. However, it usually takes a lot of financial and human consumption to produce labeled training samples, which is the so-called labeling bottleneck. Besides, the distribution of sample size for all the categories in the dataset is often unbalanced [15], which is also the main difficulty for classifier to achieve a better fitting.

© Springer Nature Switzerland AG 2021
T. Mantoro et al. (Eds.): ICONIP 2021, LNCS 13109, pp. 657–668, 2021.
https://doi.org/10.1007/978-3-030-92270-2_56

Text classification is also widely used in financial domain in recent years [16]. At present, many products based on this have been applied, such as automatic analysis of financial reports, public opinion analysis of financial markets, etc. However, the performance of the competitive methods such as BERT [2]and CNN [25] is not satisfactory [17,18] owing to the higher semantic complexity and professionalism. There are at least three challenges:

- The error of annotation. We have conducted a manual sampling check on the existing Chinese financial datasets, and the result is that the manual labeling error rate is above 10%.
- The extreme imbalance of data. As shown in **Fig.** 1, the distribution of financial samples is obviously unbalanced.
- The complicacy of labeling system. After a lot of research, we found that the labeling system of financial texts is often complex and hierarchical, which leads to a variety of categories in **Fig.** 1.

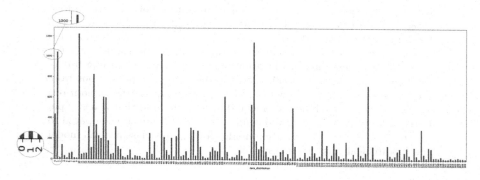

Fig. 1. Unbalanced categories distribution: some of the hundreds of categories contain thousands of samples, and some only have less than 10.

Under such a dilemma, we find that it is even more difficult to solve the classification problem of these few sample categories text. First of all, the main thing is that it is difficult to accurately capture the characteristics due to the small number of samples and the presence of noise. Secondly, there are many categories of small samples to be classified, and there are a semantic dependencies between them, which makes the direct application of the few-shot classification methods will not get good learning effect. Therefore, we propose a hierarchical text classification model, which captures basic semantic features in the first level, and strengthens the classification effect in the second level.

In this work, our contributions are summarized as follows:

- We propose a innovative hierarchical text classification model to efficiently solve the classification problem of financial few-sample categories.

- The label confusion algorithm is applied to alleviate the dependencies between labels and the interference of noise. This provides a reference idea for solving similar tasks.
- We put forward the idea of meta-learning as a low-level learning task, and proved its feasibility and high value.
- Experimental results shows that the model proposed outperforms other comparison methods, and the computational cost has not increased significantly.

2   Related Work

2.1   Label Distribution Learning

LDL is a general learning framework [20], and it has been demonstrated through experiments that it is important to design specifically for the characteristics of LDL problems. Recently, some methods to improve LDL have been proposed. The method called Logistic Boosting Regression (LogitBoost) can be considered as an additive weighted function regression from the statistical viewpoint is provided by [21]. Apart from image datasets, it was also a noteworthy idea in the field of text datasets.

2.2   Meta-learning

Meta learning has been proved to be highly effective framework in the application of computer vision. In machine translation tasks, [22] constructs the low-resource translation problem as a meta-learning problem. In addition, [23] proposes an adaptive metric learning method in a text classification task, which automatically determines the best weighted combination from a set of metrics obtained from meta-training tasks for a newly seen few-shot task. Different from the conventional mapping from training set to test set, meta learning captured the overall information across tasks, which made it perform better in few sample tasks.

3   Method

Because we are facing multiple challenges at the same time such as the complex relationship between labels and few training samples, we try to use a two-level model to overcome these difficulties one by one. The 1st-level model is a classification model for non-few sample categories, which takes the dependence between financial labels and the interference of noise caused by human labeling into account. The 2nd-level model is a few-shot learning model on the basis of meta-learning framework and the prediction from the last level.

3.1 The 1st-Level

The model will complete the classification of the first label firstly. We all know that many few-sample categories have the same first label. Therefore, the impact of complex financial labels and labeling noise need attention in the first level. In fact, it is a structure which cannot only capture the relationship between instances and labels, but also learn the dependency between different labels.

It is well-known that one-hot vector is usually used to uniquely represent the label distribution. In fact, it completely ignores the meaning of labels and will lead to the serious consequences in the case of labeling errors.

Consequently, we design a model structure to learn the simulated label distribution by calculating the semantic relationship between samples and labels. Then it is regarded as the real label distribution and is compared to the predicted label distribution to get the loss. The structure is shown in Fig. 2.

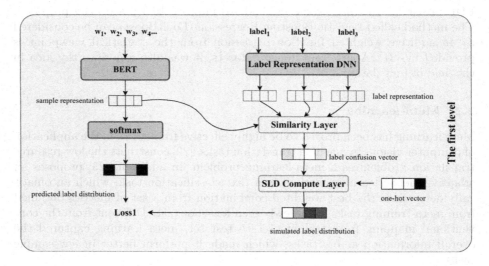

Fig. 2. The 1st-level model includes two parts: the basic BERT model part and the label confusion part. The former completes tasks of model training and model prediction, and the latter fully considers the potential information of the labels during the training process.

On the left is the basic BERT model, and the classification process can be expressed as

$$v^{(i)} = f_{BERT}(w) = f_{BERT}(w_1, w_2 \ldots w_n) \tag{1}$$

$$y^{(p)} = softmax\left(v^{(i)}\right) \tag{2}$$

where w_i is the ith input word and f_{BERT} is the BERT input encoder function.

On the right side of the Fig. 2, firstly, a deep neural network (DNN) is used to yield the label representation, which is used as the input of similarity layer together with the current instance representation to calculate the similarity, and the label confusion distribution is obtained by softmax classifier. Then, the final distribution is acquired by normalizing the result that is from the label confusion distribution and one-hot vector with a control parameter γ. Subsequently, it replaces one-hot vector as a new training target.

$$\mathbf{V}^{(s)} = f_{DNN}(l) = f_{DNN}(l_1, l_2 \ldots l_m) \tag{3}$$

$$y^{(b)} = softmax\left(v^{(i)^T} \mathbf{V}^{(s)} W + b\right) \tag{4}$$

$$y^{(s)} = softmax\left(\gamma y^{(o)} + y^{(b)}\right) \tag{5}$$

In the formula above, $y^{(b)}$ is the representation of label confusion distribution, and $y^{(s)}$ can reflect the simulated label distribution. Finally, we calculate loss by KL-divergence [24] of $y^{(p)}$ and $y^{(s)}$.

3.2 The 2nd-Level

The reasons for the proposed method are as follows:

- The number of samples in each 2nd-level label is much little. If we use the ordinary text classification model, it will greatly affect the effect of the model training and easily cause over-fitting.
- Generally speaking, the classifier needs to be retrained in different scenarios. However, as the 2nd-level classification model, its computational and storage costs are unbearable. Therefore, the meta-learning framework perfectly solves this problem.
- For the classification task with less labeled-sample categories, we consider using other corpus with more labeled-sample categories to enhance the training process.

Therefore, meta learning framework is chosen. The reason is that, first of all, the training model does not need to be adjusted for the various task, which ensures lower cost. Secondly, the existence of other multi-sample categories reduces the probability of overfitting the training set, which pave the way for generalization on new tasks. In addition, we don't make use of the semantics of the words directly, but select its distribution characteristics that are easier to generalize. In order to adapt to this adjustment, we also add source pool which contains all the samples under categories not be chosen and label pool composed of the samples whose 1st-level label predicted are same as the current instance's. Specifically, the model is divided into two parts (Fig. 3):

- Attention generator. It transforms the distributional characteristics of words into attention weight. The module generates class specific attention by combining the distribution statistics of source pool, label pool and support set.

- Ridge regressor. For each episode, this module uses the attention from the attention generator to construct the lexical representation. The goal is to predict on the query set after learning from the support set.

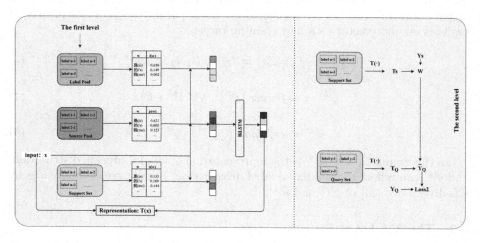

Fig. 3. The structure of 2nd-level model: The left side shows the process of constructing a sample representation based on distributional characteristics, and the right side explains the process of completing the prediction on query set after the learning on the support set.

In the first stage, we use the following method to generate the representation:

$$s\left(x_i\right) = \frac{\sigma}{\sigma + P\left(x_i\right)} \tag{6}$$

$$p\left(x_i\right) = \mathcal{H}\left(P\left(y^{(1)} \mid x_i\right)\right)^{-1} \tag{7}$$

$$l\left(x_i\right) = \mathcal{H}\left(P\left(y^{(2)} \mid x_i\right)\right)^{-1} \tag{8}$$

$P\left(x_i\right)$ is the unigram likelihood of the ith input. $\varepsilon = 10^{-3}$). $s\left(x_i\right)$ reflects the word importance from frequency. $y^{(1)}$ and $y^{(2)}$ are the 1st-level label and 2nd-level label respectively. $\mathcal{H}(\cdot)$ is the entropy operator. $P\left(x_i\right)$ and $l\left(x_i\right)$ measures the ability of the ith input to make the first label and the second label distinguishable. In order to better integrate this information, we concatenate them by BiLSTM [7]:

$$h = BiLSTM([s(x); l(x)]) \tag{9}$$

$$h = BiLSTM([h; p(x)]) \tag{10}$$

Finally, attention score of each word x_i can be calculated. We get the representation $T(x_i)$. In the second stage, we first construct the representation of the support set based on this, and then begin fast training and prediction, and calculate the learned weight W by minimizing the regularized squared loss which is used to predict on the query set.

$$\hat{Y}_Q = \alpha \Phi_Q W + \beta \tag{11}$$

where α and β is the meta-parameters learned through meta-training.

4   Experiments Setup

4.1   Datasets and Data Preprocessing

The research of financial few-shot hierarchical text classification needs the support of large-scale corpus. To simulate the real scene, we verify the applicability of our method on two large Chinese financial text corpora.

The first is to build datasets. In order to ensure that samples from the same document do not appear respectively in the training set, validation set, or test set, we directly divide all the original documents into these three sets. Second, the data preprocessing module needs to complete the construction of the sample. In addition to the extraction, deduplication and noise removal of the text to be classified, it is also necessary to distinguish between multi-sample categories and few-sample categories. Among them, the samples of the multi-sample categories are used for meta training in the second level, and the samples of the small-sample categories are the source of each set of the first level model and the support set and query set of the second level model. The two datasets will be introduced in more detail from Table 1.

Financial Public Opinion. Screened by the preprocessing module, it contains 12765 public opinion news from various financial websites. There are 17 few-sample categories whose training samples is less than 15 for experimentation in total. In order to reduce the cost of model training, the 2nd-level task also can be regard as four subcategories (N = 2) and two tricategories (N = 3) tasks.

Zbd Data. Compared to the classification of financial events, this task for financial entities is obviously more challenging. Similarly, the 2nd-level classification for 12 few-sample categories can be divided as four tricategories tasks (N = 3). Other relevant parameters are shown in Table 1.

4.2   Baselines

Considering the particularity of scenarios and tasks, we compare the approach (denoted as OUR) to three different task domain models.

Table 1. Parameters of datasets.

Datasets	Financial public opinion	Zbd data
Subject of classification	Financial events	Financial elements
1st-level labels	Debt, Administrative Penalties, Corporate Personnel, $\cdots$	Staging, Financing Entity, Credit Enhancement, $\cdots$
2nd-level labels	Suspension, New Bond Issuance, Shareholder, $\cdots$	Exercise Period, Investment Amount, Internal Rating, $\cdots$
Maximum text length	173	519
Number of 1st-level categories	34	21
Number of 2nd-level categories	170	95
Few-sample categories in 1st-level	7	8
Few-sample categories in 2nd-level	17	12
1st-level training samples	98	92
1st-level validation samples	82	83
Test samples	242	620

Competitive Classification Model. BERT [2] and CNN [25] represent the competitive classification models in this scenario task. That is to say, their experimental results reflect the performance if not considering the impact of few-labeled samples.

Few-Shot Classification Model. If only consider the case of a small number of samples and not the multi-category classification requirements, we will get the approximate results of this model (denoted as FS). In other words, in the N-way K-shot problem, we directly set n as all the categories that need to be classified [3].

Hierarchical Classification Model. Considering the hierarchical structure, we will compare the approach with the excellent hierarchical classification model, which is used to classify documents into the most relevant categories by integrating text and hierarchical category structure [4] (denoted as HARNN).

4.3 Implementation Details

We use word embeddings pre-trained on Financial News [1] using skip-gram [5] algorithm for our 2nd-level model and baselines except BERT model. In order to reduce the loss efficiently, we choose Adam Optimizer [6] with a learning rate of 0.001 [8]. All the models are implemented using Keras and are trained on GPU Tesla V100.

For the 1st-level, we experiment with pre-trained BERT embeddings. The relevant parameters of the training process are shown in Table 2. Besides, the embedding size and hidden size are same as the basic BERT model.

For the 2nd-level, we present the relevant parameters in Table 3. During training, we extract 100 training episodes per epoch. We apply for early termination if the verification loss is not improved within 20 epochs. At the same time, we try to use the data of the same first label to form support set.

Table 2. Hyper parameters of first level.

Hyper parameters	Values
Dense units	64
Initial learning rate	0.001
Batch size	16
Epoch	30

Table 3. Hyper parameters of second level.

Hyper parameters	Values
N	2,3
K	5
Query	25

5   Results and Analysis

5.1   Results

We have divided the dataset to ensure that in the training set, validation set and test set there don't exist two samples being from the same document, and made experimental comparison of multiple models from multiple dimensions on the two datasets. Therefore, Table 4 clearly shows the advantages of the approach in classification task and also the parts that need to be strengthened. In order to prove the validity of the experimental results, the values in the Table 4 are the average of 5 times runs.

In the 2nd-level model of FS and OUR, value of K is 5. In the traditional N-way K-shot problem, K is usually 1 (one-shot problem) or 3, or even 0 (zero-shot problem). However, there are great differences between the field of natural language processing and image processing. First of all, we are faced with the environment of complex semantics and unsophisticated vocabulary. If K value is too small, the learning efficiency will be greatly affected, Secondly, the datasets is satisfied with the condition of $K = 5$, which is also the value of better learning outcomes tested.

Table 4. Experiments results

	Finance				Zbd			
	Acc	Precision	Recall	F1-score	Acc	Precision	Recall	F1-score
BERT [2]	0.760	0.527	0.558	0.517	0.566	**0.609**	0.575	0.577
FS [3]	0.769	**0.600**	0.619	0.587	0.575	0.556	0.581	0.552
HARNN [4]	0.488	0.495	0.504	0.500	0.484	0.492	0.524	0.507
CNN [25]	0.190	0.109	0.156	0.094	0.245	0.283	0.252	0.249
OUR	**0.802**	0.592	**0.623**	**0.594**	**0.609**	0.603	**0.615**	**0.600**

The experimental results have objectively confirmed the following points:

- It can be inferred from the performance of CNN that it is difficult to classify these obscure financial texts.
- In terms of general experimental results, there is no doubt that the approach has its superiority not only in accuracy, but also in overall performance of each category.
- If paying more attention to the characteristic of few samples than the feature of hierarchical structure, we will get a slightly better accuracy.
- The traditional hierarchical classification model has apparent weakness in the task of few-sample classification.

5.2 Analysis

The respective results from the two levels provide the basis for more analysis.

The 1st-Level. We strengthen the general classification model in the first level in order to predict more accurate label and decrease the overall predictive error. **Table** 5 shows us the 1st-level classification performance of several models.

Table 5. Performance of 1st-level models

	Finance				Zbd			
	Acc	Precision	Recall	F1-score	Acc	Precision	Recall	F1-score
BERT [2]	**0.893**	0.791	**0.863**	0.813	0.818	0.816	0.824	0.817
HARNN [4]	0.818	0.821	0.835	0.828	0.806	0.809	0.818	0.813
OUR	0.835	**0.838**	0.840	**0.835**	**0.955**	903	**0.893**	**0.894**

It can be clearly seen the assistance of the approach in general classification task especially complex and unbalanced financial task from Table 5. As we all know, financial data has a strong skilled requirement. That is to say, the annotation is very likely to contain mistake, which is almost fatal for categories classification with few samples. However, the approach can slow down the impact of these problems, and thus open up some new thinking for hierarchical classification.

The 2nd-Level. First, the approach has better performance in the accuracy of classification and the stability of each category. Second, the new structure proposed has a high value of practical application because we only need to train the model once when n takes the same value. Third, through meta-learning framework, the difficulty of classification caused by few training samples is well alleviated, not only in the financial field, but also in other few-sample datasets with complicated context.

6 Conclusion

In this paper, we propose a new hierarchical model for complex financial text processing scenarios with few samples. The first level weighing the degree of dependence between labels and content, accomplish the prediction for the first label. While the second level employ the distributional characteristic about the dataset and the predicted first label, adopting the few-shot learning to generate the 2nd-level label. We have verified the importance in dependence between the same level or different level labels for classification on two few-sample Chinese financial datasets with complicated con-text. Besides, the breakthrough experimental results demonstrate that our hierarchical classification structure has extremely high application value in reality.

References

1. Li, S., Zhao, Z., Hu, R., Li, W., Liu, T., Du, X.: Analogical reasoning on chinese morphological and semantic relations. arXiv preprint arXiv:1805.06504 (2018)
2. Devlin, J., Chang, M.W., Lee, K., Toutanova, K.: Bert: pre-training of deep bidirectional transformers for language understanding. arXiv preprint arXiv:1810.04805 (2018)
3. Bao, Y., Wu, M., Chang, S., Barzilay, R.: Few-shot text classification with distributional signatures. arXiv preprint arXiv:1908.06039 (2019)
4. Huang, W., et al.: Hierarchical multi-label text classification: an attention-based recurrent network approach. In: Proceedings of the 28th ACM International Conference on Information and Knowledge Management, pp. 1051–1060 (2019)
5. Mikolov, T., Sutskever, I., Chen, K., Corrado, G.S., Dean, J.: Distributed representations of words and phrases and their compositionality. In: Advances in Neural Information Processing Systems, pp. 3111–3119 (2013)
6. Kingma, D.P., Ba, J.: Adam: a method for stochastic optimization. arXiv preprint arXiv:1412.6980 (2014)
7. Hochreiter, S., Schmidhuber, J.: Long short-term memory. Neural Comput. **9**(8), 1735–1780 (1997)
8. Srivastava, N., Hinton, G., Krizhevsky, A., Sutskever, I., Salakhutdinov, R.: Dropout: a simple way to prevent neural networks from overfitting. J. Mach. Learn. Res. **15**(1), 1929–1958 (2014)

9. Maas, A., Daly, R.E., Pham, P.T., Huang, D., Ng, A.Y., Potts, C.: Learning word vectors for sentiment analysis. In: Proceedings of the 49th Annual Meeting of the Association for Computational Linguistics: Human Language Technologies, pp. 142–150 (2011)
10. Tai, K.S., Socher, R., Manning, C.D.: Improved semantic representations from tree-structured long short-term memory networks. arXiv preprint arXiv:1503.00075 (2015)
11. Zhu, X., Sobihani, P., Guo, H.: Long short-term memory over recursive structures. In: International Conference on Machine Learning, pp. 1604–1612. PMLR (2015)
12. Sriram, B., Fuhry, D., Demir, E., Ferhatosmanoglu, H., Demirbas, M.: Short text classification in twitter to improve information filtering. In: Proceedings of the 33rd International ACM SIGIR Conference on Research and Development in Information Retrieval, pp. 841–842 (2010)
13. Kalchbrenner, N., Grefenstette, E., Blunsom, P.: A convolutional neural network for modelling sentences. arXiv preprint arXiv:1404.2188 (2014)
14. Liu, P., Qiu, X., Chen, X., Wu, S., Huang, X.J.: Multi-timescale long short-term memory neural network for modelling sentences and documents. In: Proceedings of the 2015 Conference on Empirical Methods in Natural Language Processing, pp. 2326–2335 (2015)
15. Li, Y., Sun, G., Zhu, Y.: Data imbalance problem in text classification. In: 2010 Third International Symposium on Information Processing, pp. 301–305. IEEE (2010)
16. Ciravegna, F., et al.: Flexible text classification for financial applications: the FACILE system. In: ECAI, pp. 696–700 (2000)
17. Stamatatos, E.: Author identification: using text sampling to handle the class imbalance problem. Inf. Process. Manag. **44**(2), 790–799 (2008)
18. Thabtah, F., Hammoud, S., Kamalov, F., Gonsalves, A.: Data imbalance in classification: experimental evaluation. Inf. Sci. **513**, 429–441 (2020)
19. Guo, B., Han, S., Han, X., Huang, H., Lu, T.: Label confusion learning to enhance text classification models. arXiv preprint arXiv:2012.04987 (2020)
20. Geng, X.: Label distribution learning. IEEE Trans. Knowl. Data Eng. **28**(7), 1734–1748 (2016)
21. Xing, C., Geng, X., Xue, H.: Logistic boosting regression for label distribution learning. In: Proceedings of the IEEE Conference on Computer Vision and Pattern Recognition, pp. 4489–4497 (2016)
22. Gu, J., Wang, Y., Chen, Y., Cho, K., Li, V.O.: Meta-learning for low-resource neural machine translation. arXiv preprint arXiv:1808.08437 (2018)
23. Yu, M., et al.: Diverse few-shot text classification with multiple metrics. arXiv preprint arXiv:1805.07513 (2018)
24. Kullback, S., Leibler, R.A.: On information and sufficiency. The Ann. Math. Stat. **22**(1), 79–86 (1951)
25. Rakhlin, A.: Convolutional neural networks for sentence classification. GitHub (2016)

JStrack: Enriching Malicious JavaScript Detection Based on AST Graph Analysis and Attention Mechanism

Muhammad Fakhrur Rozi[1,2]([⊠]), Tao Ban[1], Seiichi Ozawa[2], Sangwook Kim[2], Takeshi Takahashi[1], and Daisuke Inoue[1]

[1] National Institute of Information and Communications Technology, Koganei, Tokyo, Japan
{fakhrurrozi95,bantao,takeshi_takahashi,dai}@nict.go.jp
[2] Kobe University, Kobe, Hyogo, Japan
ozawasei@kobe-u.ac.jp, kim@eedept.kobe-u.ac.jp

Abstract. Malicious JavaScript is one of the most common tools for attackers to exploit the vulnerability of web applications. It can carry potential risks such as spreading malware, phishing, or collecting sensitive information. Though there are numerous types of malicious JavaScript that are difficult to detect, generalizing the malicious script's signature can help catch more complex JavaScripts that use obfuscation techniques. This paper aims at detecting malicious JavaScripts based on structure and attribute analysis of abstract syntax trees (ASTs) that capture the generalized semantic meaning of the source code. We apply a graph convolutional neural network (GCN) to process the AST features and get a graph representation via neural message passing with neighborhood aggregation. The attention layer enriches our method to track pertinent parts of scripts that may contain the signature of malicious intent. We comprehensively evaluate the performance of our proposed approach on a real-world dataset to detect malicious websites. The proposed method demonstrates promising performance in terms of detection accuracy and robustness against obfuscated samples.

Keywords: Cyber security · Malicious JavaScript · Abstract syntax tree · Graph neural network

1 Introduction

Javascript payload injection into legitimate or fake websites has been one of the largest attack on the web. The malicious script can exploit the vulnerability of the web applications to perform a drive-by download attack [2] or cross-site scripting (XSS) [19]. When the attack is succesful, attackers distribute malware to clients, which can cause damage such as sensitive data leakage, wire transfer, or integrating into distributed denial-of-service (DDoS) attacks [3]. For instance, one of the most famous examples of XSS vulnerability is the Myspace Samy

© Springer Nature Switzerland AG 2021
T. Mantoro et al. (Eds.): ICONIP 2021, LNCS 13109, pp. 669–680, 2021.
https://doi.org/10.1007/978-3-030-92270-2_57

worm by Samy Kamkar in 2005 [9]. He exploited a vulnerability on the target that could give him priviledge to store a JavaScript payload on his Myspace profile. Moreover, web technology improvement helps attackers use the latest method to avoid detection, such as the obfuscation techniques.

Researchers have identified the malicious JavaScript payload, which is typically used by attackers as part of a web security attack. A variety of detection systems has been proposed that use JavaScript features to detect malicious intent. We can take many approaches to create a detection system for malicious JavaScript, such as strings, function calls, bytecode sequences, abstract syntax tree (ASTs), outputs of dynamic analysis tools. Among these features, AST gives the most notably excellent performance. Fass et al. [6] use this feature for their static analysis and use the N-gram model to detect malicious obfuscated JavaScripts. However, their work focused on the frequency analysis of the specific patterns with the connection between syntactic units of AST feature ignored. We have to analyze it at the tree level instead of the sequence level when we want to capture the semantic meaning of the code.

We propose JStrack, a malicious JavaScript detection system using a graph-based approach on the AST features to capture the whole semantic meaning which has not been considered in previous works. We hypothesize that the style of malicious code tends to be better structured due to the decryption or deobfuscation process that should exist inside the code instead of having an abstract structure. Analyzing the whole AST as a graph structure also gives us more information about the actual intent of the source code. To capture that information thoroughly, we use a supervised graph neural network (GNN), known as a graph convolutional neural network (GCN) model. This model can capture the connections between nodes in the graph structures and formulate them as vectorial features to be used in a neural network model. Moreover, we try to combine it with the attention layer to know which parts of AST carry a significant information to detect malicious JavaScript code.

To summarize, our contributions are as follow:

- We introduce JStrack, a static analysis method, to detect malicious Java-Script using the AST features as a graph. We applied GCN to capture the typical structure and attribute of the AST representation from malicious JavaScript samples. The GCN model is built by stacking multiple convolutional layers to be used as a layer-wise linear model in our detection system.
- We track the suspicious part of the AST graph, which corresponds to the actual JavaScript code, by using the attention layer in our proposed model. The attention scores give us significant code segments that can lead us to the signature of a malicious script.
- We evaluate our proposed approach using real-world malicious samples and collected JavaScript files from the top domain list as benign. We show that our graph-based approach can accurately detect malicious JavaScript even with the presence of the obfuscation techniques to evade the detection system. Moreover, our approach detects the obfuscation pattern of AST-graph by

observing the similarity of graph structures and attributes among malicious or benign samples.

The rest of the paper is organized as follows. Section 2 provides the background of JavaScript-based attack and related works. Then, we will explain how we parse JavaScript code to get the AST representation and how we construct the graph based on that. Section 3 explains our proposed approach, which uses a graph-based model to extract the AST feature. Section 4 presents our experiment and evaluation result of our JStrack in Sect. 5. Finally, we provide our concluding remarks.

2   Background and Related Works

In this section, we explain the background of JavaScript-based attacks and how attackers use obfuscation technique to hide their malicious intent. We also give an overview of the AST feature as an abstract representation of JavaScript and the derivation of the graph from the characteristics of the AST features.

2.1   JavaScript-Based Attack

According to Web Technology surveys [16], JavaScript is the most used client-side programming language on websites, reaching about 97.4%. Because of that, malicious JavaScript code is one of the most common web security vulnerabilities that are frequently found in buttons, text, images, or pop-up pages. For instance, if a website does not sanitize angle brackets (< >), attackers can insert `<script></script>` to inject payload, which this tag instructs the browser to execute the JavaScript between them [21]. The injected script can be triggered when a single HTTP request runs the malicious payload and attackers did not store it anywhere on the website or when a site saves and renders it unsanitized [21].

The malicious JavaScript code generally contains some function calls that attackers usually use to execute their intended action. Examples of function calls include `document.write()`, `eval()`, `unenscape()`, `SetCookie()`, `GetCookie()`, or `newActiveXObject()` [7]. Attackers will activate the malicious payload by altering the document object model (DOM) to drop the malware or steal users' sensitive data. Due to many malicious samples have these functions, we can assume that this part of the code gives more important information about the maliciousness of code. However, in practice, attackers hide the malicious code by particular means to take advantage of the security flaw. It won't be easy to detect such kinds of payload that it can bypass the system. In addition, they utilize obfuscation techniques to hide their malicious code, making it harder to find the signature.

2.2   Related Works

Previous researches have thoroughly explored the machine learning-based method for detecting malicious JavaScript. They used various features of

JavaScript and applied a different approach to increase the performance. Ndichu et al. [12] applied the FastText model to detect the malicious JavaScript based on AST features. They tried to deobfuscate the source code to catch the identical actual malicious payload before modeling. However, their approach handles the short relationship between syntactic units in AST that they forgot to consider the edge connection. Besides that, Fass et al. [6] did a similar work that they proposed a syntactical analysis approach using a low-overhead solution that mixes AST feature extraction sequences and a random forest classifier model.

Differently, Rozi et al. [14] used bytecode sequences as the main features of JavaScript code, which is the middle language between machine and high-level code. Due to the super long problem in the bytecode sequence, they used a deep pyramid convolutional neural network (DPCNN) that contains a pyramid shape network to get a more straightforward representation. The limitation is that they have to declare all possible DOM objects in every sample to generate the sequences.

Moreover, Song et al. [15] and Fang et al. [5] used recurrent neural networks (RNNs) architectures to capture the semantic meaning of JavaScript. Song et al. [15] tried to use the Program Dependency Graph (PDG), AST, and control flow diagram (CFG), which preserve the semantic information of JavaScript. However, Fang et al. [5] only relied on AST features to capture the sequence patterns of syntactic unit sequences. Both of them applied Bidirectional Long-Short Term Memory (BiLSTM) and Long-Short Term Memory (LSTM) to learn the long-term dependencies.

3   Proposed Approach

To overcome such challenges from malicious JavaScript, we propose a detection system that can predict the label of a given source code, whether it is malicious or benign. Our proposed approach uses AST as the feature of JavaScript that can define the style and semantic meaning of the source code. By analyzing this feature, we can capture the malicious intent based on the typical structure and attribute of the AST graph. We use GCN to learn the graph to have the generalization of malicious and benign samples.

3.1   Overview

We can see the entire detection system framework in Fig. 1. It begins with a JavaScript file that we want to predict the malicious intent. After that, we parse it using a parser to get the AST representation, describing how programmers write the code. The output is a JSON format file where each record is a syntactic unit object based on ESTree standardization [4]. We can construct graph objects from a JSON file as a simplification of its data structure. The graph generator creates syntactic unit types as finite nodes, and the hierarchical connection among nodes is an edge of the AST graph. Next, we create two matrices, feature matrix $\mathbf{X}$ and adjacency $\mathbf{A}$, representing the feature value of each node and

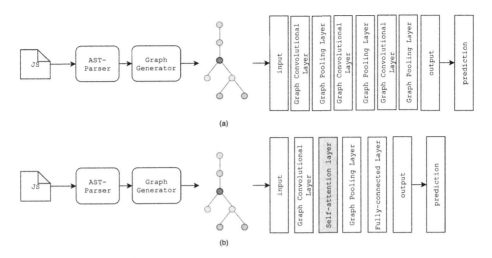

Fig. 1. The overview of proposed approach. (a) The original architecture consists of three layers of convolutional and pooling layers. (b) The combination of GCN and attention mechanism to locate the suspicious codes of JavaScript. To get the whole information of nodes, we put the pooling layer after attention layer before going to fully-connected layer.

all connections of edges, respectively. The GCN is similar to the convolutional neural network (CNN) in that it consists of two main layers, the convolutional and pooling layers. The difference is that GCN applies these layers on a graph structure to get a suitable vector representation for the graph. The output is the prediction score to determine the JavaScript label.

3.2   AST Graph Construction

We often find many systems around us that use graph representation to solve many problems. Graph representation can render a complex system become more structured so that the problem will be easier to solve. A graph is a ubiquitous data structure and universal language consisting of a collection of objects, including a set of interactions between pairs of objects [8].

Formally, we can define graph $\mathcal{G}(\mathcal{V}, \mathcal{E})$ as a set of nodes $v \in \mathcal{V}$ and edges $e \in \mathcal{E}$. (u, v) denotes an edge going from node $u \in V$ to node $v \in V$ [8]. We can represent a finite graph $\mathcal{G}$ in a squared matrix called adjacency matrix $\mathbf{A} \in \mathbb{R}^{|\mathcal{V}| \times |\mathcal{V}|}$. Each row and column indicates all nodes that a finite graph $\mathcal{G}$ has. Furthermore, edges represent entries in $\mathbf{A}$ where $\mathbf{A}[u, v] = 1$ if $(u, v) \in \mathcal{E}$ and otherwise $\mathbf{A}[u, v] = 0$. Matrix $\mathbf{A}$ will not necessarily be symmetric if graph $\mathcal{G}$ has directed edges. Some graphs also have weighted edges, where the entries in the adjacency matrix are real-values. Besides that, a graph may have an attribute or feature information for each node that using a real-valued matrix $\mathbf{X}^{|\mathcal{V}| \times m}$ where m is the feature size of nodes, and the ordering of the nodes is consistent with the adjacency matrix $\mathbf{A}$. In some cases, edges also have real-valued features in addition to discrete edge types.

We can use a graph-based approach to represent the AST feature with a tree graph structure. AST is a top-down parsing structure in which each syntactic unit has at least one hierarchical connection where the root is always a 'program' type. Based on that, we consider each syntactic unit as a node and hierarchical link as an edge. Using graph representation simplifies the AST feature in a fixed form to help the feature extraction process. This representation also allows us to capture the big picture of the source code, which shows the complexity yet the programmer's obfuscation style.

3.3 Learning AST Graph Feature

Suppose we have $G = \{\mathcal{G}_1, \mathcal{G}_2, \mathcal{G}_3, ..., \mathcal{G}_N\}$, a set of all graphs in our dataset. We can define a graph $\mathcal{G}_i(\mathcal{V}_i, \mathcal{E}_i)$ consisting of nodes $\mathcal{V}$ and edges $\mathcal{E}$. In our problem, we assume our target for the model is $t \in \{0, 1\}$ which 0 as benign and 1 as the malicious.

Graph Convolutional Neural Networks. The basic idea of GCN is actually from convolutional neural networks (CNNs), where it also uses the convolution and pooling function for getting feature information of each node in the graph. Originally, Kipf et al. proposed GCN to solve semi-supervised classification tasks such as graph Laplacian regularization include label propagation [22], manifold regularization [1], and deep semi-supervised embedding [20]. The basic idea is to generate embedding information of nodes via neural message passing to aggregate information from all neighborhoods. GCN consists of a stack of graph convolution layers, where a point-wise non-linearity follows each layer. The number of layers is the farthest distance that node features can travel. The number of layers also influence the performance. More layers are not guaranteed to get a good result because it makes the aggregation less meaningful if it goes further.

The multi-layer network in GCN follows layer-wise propagation rule:

$$\mathbf{H}^{(l+1)} = \sigma\left(\tilde{\mathbf{D}}^{-\frac{1}{2}} \tilde{\mathbf{A}} \tilde{\mathbf{D}}^{-\frac{1}{2}} \mathbf{H}^{(l)} \mathbf{W}^{(l)}\right). \tag{1}$$

Where $\tilde{\mathbf{A}} = \mathbf{A} + \mathbf{I}_N$ is the adjacency matrix of the undirected graph $\mathcal{G}$ with added self-connections. $\tilde{\mathbf{D}}_{ii} = \sum_j \tilde{\mathbf{A}}_{ij}$ and $\mathbf{W}^{(l)}$ is trainable weight matrix in specific layer. $\mathbf{H}(l) \in \mathbb{R}^{N \times m}$ is the matrix of activations in the lth layer with m is the feature size of nodes; $\mathbf{H}^{(0)} = \mathbf{X}$. $\sigma(\cdot)$ stands for an activation function, such as the $\text{ReLU}(\cdot) = \max(0, \cdot)$.

Attention Mechanism. This mechanism is basically about paying more focus on some component that significantly influences the system. Precisely, the attention function map a query and a set of key-value pairs to an output, where the query, keys, values, and output are all vectors [17]. The computation of attention function as follows:

$$Attention(\mathbf{Q}, \mathbf{K}, \mathbf{V}) = softmax(\frac{\mathbf{Q}\mathbf{K}^T}{\sqrt{d_k}})\mathbf{V} \qquad (2)$$

where $\mathbf{Q}, \mathbf{K}, \mathbf{V}$ are query, key, and value matrices, respectively. d_k is the key of dimensions.

In this work, the attention mechanism can leverage the learning process of GCN by giving attention weight to concentrate selectively on a discrete aspect of the graph convolutional layer. We use a self-attention layer to handle long-range dependencies and have lower complexity than other layer types (e.g., convolutional or recurrent).

4   Experiments

In this section, we present our experiments to evaluate our proposed approach for detecting malicious JavaScript samples. We evaluated our framework's performance by adjusting the maximum number of nodes in each graph. Then, we compared our results with some related works that have a similar task. Finally, we give some analysis discussion to find out our limitations.

4.1   Setup

Dataset. We collect malicious and benign JavaScript datasets, where the malicious samples are from two different sources due to the difficulties of getting the real-world dataset. For our malicious samples, we mixed the dataset from Rozi et al. [14] and Ndichu et al. [12] that use some different time stamps of files from 2015 until 2017. We also confirmed that all those datasets are dangerous scripts based on the VirusTotal scanner [18]. Meanwhile, we collected JavaScript codes for benign samples by scrapping from the top domain list on the Majestic website [10], and we combined it with the benign dataset from SRILAB [13]. We consider all JavaScript codes inside popular websites as safe code without any attacking intent.

We split our dataset into two parts: training and testing. We used the training dataset for the learning purpose of our graph learning model. Otherwise, we evaluated our model with the testing dataset. We conducted 10-folds cross-validation to see our model's average performance that generalizes to an independent dataset. Because of that, the proportion between training and testing is 80% and 20%, respectively. Table 1 summarizes the number of JavaScript files that we use in our experiments.

Hyper-parameters and Setup. We set optimal hyper-parameters to conduct our experiments to control the learning process. We used the Adam algorithm optimization with a 0.01 learning rate and 32 for the batch size. In addition, the feature size of the convolutional layer in GCN is 32 and using rectified linear unit (ReLU) as the activation function. For the pooling layer, we used a 50% ratio to downsample the matrix node.

Table 1. The description of our dataset that is used for training and testing process.

Label	Dataset		
	Training	Testing	Total
Benign	97,361	24,341	121,702
Malicious	31,560	7,890	39,450
Total	128,921	32,231	**161,152**

Unlike the usual deep learning model, adding more layers does not correlate with the performance. When we work with the GNNs, this model will significantly lose the ability to learn if we have too deep layers, where we call this problem over-smoothing [23]. The main idea of over-smoothing is that all node representations look identical and uninformative after too many message passing rounds due to too many layers. Zhou et al. [22] recommended using between 2 and 4 layers to achieve an optimal solution. Therefore, we used the middle range number, three layers, in our experiments.

Moreover, we applied a data loader with disjoint mode for creating mini-batches of data in graph learning. It represents a batch of graphs with a disjoint union that gives us one big graph [11]. Figure 2 illustrates how the disjoint loader works.

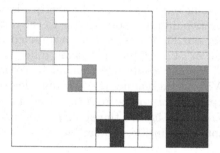

Fig. 2. Disjoint loader is a method to load dataset in graph learning process that represents batch of graphs via disjoint union. It uses zero-based indices to keep track of the different graphs.

5   Evaluation and Discussion

Due to the memory capacity reason, we could not include all nodes in the learning process. Because of that, we evaluated six different maximum nodes of the AST graph: 50, 100, 200, 500, 1000, and 2000. This experiment aims to find the sufficient nodes that we need to detect the maliciousness of JavaScript. Table 2 shows the performances (precision, recall, F1 score) for each maximum nodes setting. We can see that the performance of our method will increase in line with the number of nodes in the AST graph that we can capture. This result is in accordance with our hypothesis that AST nodes give an abstraction of the source code where all nodes give essential information. However, using 2000 nodes still give high performance even though we did not include all information. It is because AST uses the hierarchical structure that each node has summarized its successor.

Table 2. Overall performances of our detection system using graph-based approach on accuracy, precision, recall, F1 score, and AUC.

Max nodes	Accuracy	Precision	Recall	F1 score	AUC
50	0.9864	0.9872	0.9878	0.9875	0.9878
100	0.9877	0.9881	0.9901	0.9891	0.9901
200	0.9906	0.9929	0.9937	0.9933	0.9937
500	0.9933	0.9940	0.9956	0.9948	0.9956
1000	**0.9941**	0.9953	0.9965	0.9959	0.9965
2000	0.9940	**0.9956**	**0.9971**	**0.9963**	**0.9971**

Table 3 shows the comparison between previous works and our proposed method. GCN has around 98% in terms of F1 score for our dataset with the maximum 50 nodes of the AST graph. Meanwhile, adding attention layers before fully connected layers can improve the performance by 99%. Our approaches outperform the previous works that use the FastText model based on frequency analysis of syntactic AST units. Even though the difference is relatively small, our proposed method can predict the part of the source code which gives more attention to detect malicious intent. This information will be valuable for further analysis of malicious code. Figure 4 is one of the malicious samples in our dataset that shows the attention score for each node in a graph. Moreover, the bytecode sequences feature cannot be implemented on every JavaScript samples because we have to declare all possible DOM objects.

Moreover, we found in our experiments that the malicious JavaScript has its obfuscation technique to hide the actual source code. Figure 3(a) shows the graph visualization of malicious JavaScript code. The structure of the AST graph for malicious JavaScript has many repetitions of the subgraph that we rarely find in benign samples. Some similar styles appear many times within the same time range, indicating that attackers consistently use their obfuscation function that normal programmers will not use. On the other hand, most benign samples in Fig. 3(b) have an arbitrary structure of AST and inconsistent subgraph patterns. This result is in line with our hypothesis that benign JavaScript mostly does not use obfuscation techniques, or if it has obfuscated parts, it uses more complicated methods to protect from reverse engineering.

Table 3. Performance comparison with closely related works.

Model	Feature	F1
DPCNN [14]	Bytecode sequence	0.9684
DPCNN+LSTM [14]	Bytecode sequence	0.9657
DPCNN+BiLSTM [14]	Bytecode sequence	0.9683
LSTM [12]	AST	0.9234
FastText [12]	AST	0.9873
GCN (3-layers;max 50 nodes)	AST	0.9875
GCN (w/attention; max 50 nodes)	**AST**	**0.9935**

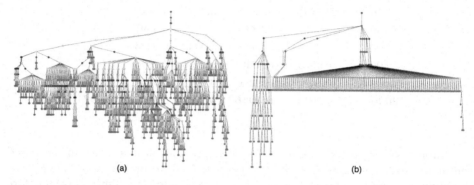

Fig. 3. A sample of AST graph that is constructed from a benign (a) and malicious (b) JavaScript file.

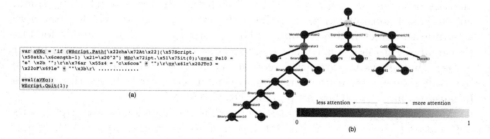

Fig. 4. (a) A malicious sample where the highlight parts are the vital parts to execute the code. (b) The AST representation of the malicious code that each node has a color represents the attention score. Some nodes have high scores that correlate to the vital part of malicious code.

However, there are two limitations to our proposed method that we are considering. First, we lose detailed information about malicious code due to using the AST feature to represent JavaScript. In the AST graph, we merely use the syntactic units and omit component details for each unit, which may contain the essential information for our detection system. Then, the use of deep/machine learning does not always consider uncertainty in the prediction task. It relies on statistical assumptions about the distribution of the dataset to train the model. Consequently, adversaries-based attacks can exploit the machine learning model to disrupt the analysis process and make false detection.

6 Conclusions and Future Works

In this paper, we proposed an alternative approach to detect malicious JavaScript based on the analysis of AST representation. The syntactical structure of Java-Script can give more comprehensive information about the source code's semantic meaning to capture the generalization of malicious signatures to overcome future attacks. GCN successfully encodes the whole AST graph via a neural message from its local neighborhood that leads to high detection performance. Additionally, the attention layers also help us locate suspicious parts of the malicious samples, significantly contributing to the detection system. As future plan, we will extend our research for future work to detect malicious websites based on encoded JavaScript information. We will explore more about other JavaScript features that probably increase the performance.

Acknowledgements. This research was partially supported by the Ministry of Education, Science, Sports, and Culture, Grant-in-Aid for Scientific Research (B) 21H03444.

References

1. Belkin, M., Niyogi, P., Sindhwani, V.: Manifold regularization: a geometric framework for learning from labeled and unlabeled examples. J. Mach. Learn. Res. **7**, 2399–2434 (2006)
2. Cova, M., Kruegel, C., Vigna, G.: Detection and analysis of drive-by-download attacks and malicious JavaScript code. In: Proceedings of the 19th International Conference on World Wide Web, WWW 2010, pp. 281–290. Association for Computing Machinery, New York (2010). https://doi.org/10.1145/1772690.1772720
3. Douligeris, C., Mitrokotsa, A.: DDoS attacks and defense mechanisms: classification and state-of-the-art. Comput. Netw. **44**(5), 643–666 (2004)
4. The estree spec. https://github.com/estree/estree. Accessed 20 Jan 2021
5. Fang, Y., Huang, C., Liu, L., Xue, M.: Research on malicious JavaScript detection technology based on LSTM. IEEE Access **6**, 59118–59125 (2018)
6. Fass, A., Krawczyk, R.P., Backes, M., Stock, B.: JAST: fully syntactic detection of malicious (obfuscated) JavaScript. In: Giuffrida, C., Bardin, S., Blanc, G. (eds.) DIMVA 2018. LNCS, vol. 10885, pp. 303–325. Springer, Cham (2018). https://doi.org/10.1007/978-3-319-93411-2_14
7. Gupta, S., Gupta, B.: Enhanced XSS defensive framework for web applications deployed in the virtual machines of cloud computing environment. Procedia Technol. **24**, 1595–1602 (2016). https://doi.org/10.1016/j.protcy.2016.05.152. https://www.sciencedirect.com/science/article/pii/S2212017316302419. International Conference on Emerging Trends in Engineering, Science and Technology (ICETEST - 2015)

8. Hamilton, W.L.: Graph representation learning. In: Synthesis Lectures on Artificial Intelligence and Machine Learning, vol. 14, no. 3, pp. 1–159 (2020)
9. Kamkar, S.: phpwn: attacking sessions and pseudo-random numbers in PHP. In: Blackhat (2010)
10. Majestic. https://majestic.com/. Accessed 26 Jan 2021
11. Data modes. https://graphneural.network/data-modes/. Accessed 17 Apr 2021
12. Ndichu, S., Kim, S., Ozawa, S.: Deobfuscation, unpacking, and decoding of obfuscated malicious JavaScript for machine learning models detection performance improvement. CAAI Trans. Intell. Technol. **5**, 184–192 (2020)
13. Raychev, V., Bielik, P., Vechev, M., Krause, A.: Learning programs from noisy data. SIGPLAN Not. **51**(1), 761–774 (2016)
14. Rozi, M.F., Kim, S., Ozawa, S.: Deep neural networks for malicious JavaScript detection using bytecode sequences. In: 2020 International Joint Conference on Neural Networks (IJCNN), pp. 1–8 (2020)
15. Song, X., Chen, C., Cui, B., Fu, J.: Malicious JavaScript detection based on bidirectional LSTM model. Appl. Sci. **10**(10), 3440 (2020). https://doi.org/10.3390/app10103440. https://www.mdpi.com/2076-3417/10/10/3440
16. Usage statistics of JavaScript as client-side programming language on websites. https://w3techs.com/technologies/details/cp-javascript. Accessed 08 May 2021
17. Vaswani, A., et al.: Attention is all you need. In: Proceedings of the 31st International Conference on Neural Information Processing Systems, NIPS 2017, pp. 6000–6010. Curran Associates Inc., Red Hook (2017)
18. Virustotal. https://www.virustotal.com/gui/. Accessed 15 Jan 2021
19. Wassermann, G., Su, Z.: Static detection of cross-site scripting vulnerabilities. In: 2008 ACM/IEEE 30th International Conference on Software Engineering, pp. 171–180 (2008). https://doi.org/10.1145/1368088.1368112
20. Weston, J., Ratle, F., Collobert, R.: Deep learning via semi-supervised embedding. In: Proceedings of the 25th International Conference on Machine Learning, ICML 2008, pp. 1168–1175. Association for Computing Machinery, New York (2008). https://doi.org/10.1145/1390156.1390303
21. Yaworski, P.: Real-world bug hunting: a field guide to web hacking **14**(3) (2019)
22. Zhou, K., et al.: Understanding and resolving performance degradation in graph convolutional networks. arXiv e-prints arXiv:2006.07107, June 2020
23. Zhu, X., Ghahramani, Z., Lafferty, J.D.: Semi-supervised learning using gaussian fields and harmonic functions. In: Fawcett, T., Mishra, N. (eds.) Proceedings of the Twentieth International Conference on Machine Learning (ICML 2003), Washington, DC, USA, 21–24 August 2003, pp. 912–919. AAAI Press (2003). http://www.aaai.org/Library/ICML/2003/icml03-118.php

Author Index

Adhikary, Abhijit   491, 502
Agarap, Abien Fred   250
Agarwal, Sonali   393
Aguilera, Miguel   228
Anh Tu, Nguyen   513
Anoop, C. S.   536
Azam, Basim   285
Azcarraga, Arnulfo P.   250

Ban, Tao   669
Ben Said, Aymen   145
Bulanda, Daniel   238

Cao, Xin   405
Chen, Lin   348
Chen, Qingcai   657
Chen, Weizheng   121
Chen, Ya   133
Chen, Yu   359
Cheng, Xin   309
Cheng, Xinmin   3
Cheng, Yongli   99
Cui, Li   560

Dai, Genan   608
Dai, Hao   381
Das, Anik   526
Deng, Lingfei   321
Deng, Yiran   428
Do, Tu   335
Dong, Jiwen   585
Dong, Zhicheng   348
Dong, Zhuben   597
Du, Wenhui   3

Feng, Lin   573, 620
Feng, Rui   439
Fu, Hao   133
Fu, Heyu   573
Fusauchi, Tomohiro   170

Gao, Xizhan   585
Gaschi, Félix   216
Gedeon, Tom   193, 463

Gong, Jun   40
Gong, Mingming   439
Gong, Saijun   348

Han, Chuanqi   560
He, Changxiang   15
He, Dan   99
He, Haoran   548
He, Yifan   52
He, Zheng   87
Higashita, Risa   479
Horzyk, Adrian   238
Hossain, Md Zakir   645
Hu, Chenhao   608
Hu, Lianyu   573, 597
Huang, Dong   181, 417
Huang, Ruoran   560

Ikeda, Yuki   170
Imai, Michita   158
Inoue, Daisuke   669

Jiang, Ning   309
Jiang, Wanrong   133
Jose, Babita Roslind   297
Ju, Shenggen   359

Kabir, Muhammad Ashad   526
Kanazawa, Soma   158
Kim, Sangwook   669
Komorowski, Jacek   451
Kong, Weiyang   608
Krishna, Aneesh   193

Lao, Guihong   597, 620
Lekshmi, R.   297
Li, Chun-Hong   417
Li, Dongfang   657
Li, Guangzhe   620
Li, Hao   369
Li, Jiajun   111
Li, Rong   479
Li, Shuaicheng   439
Li, Xiang   573, 620

Li, Xiangyu 121, 632
Li, Xiaohui 585
Li, Xiu 405
Li, Xu 111
Li, Xueming 111
Li, Yuan 64
Li, Yun 206
Li, Yunheng 620
Li, Zhiyuan 52
Li, Ziqiang 28
Liao, Yihao 52
Linda, R. J. 297
Ling, Hua-Bao 181
Liu, Gaoshuo 359
Liu, Guiquan 133
Liu, Hao 573, 620
Liu, Jiang 479
Liu, Jingjing 439
Liu, Lin 548
Liu, Shenglan 573, 597, 620
Liu, Shuting 15
Liu, Tingting 585
Liu, Yaozhong 645
Liu, Yubao 608
Long, Guodong 479
Lu, Jianwei 273
Luo, Ye 273
Luong, Ngoc Hoang 335
Lv, Hairong 75

Ma, Junhua 111
Mandal, Ranju 285
Mathew, Jimson 297
Matsumori, Shoya 158
Meng, Ming 52
Miao, Shuyu 439
Minh Phuong, Tu 513
Mohammed, Emad A. 145
Mouhoub, Malek 145

Nakajima, Kohei 262
Niu, Sijie 585

Ou, Jinxiang 75
Ozawa, Seiichi 669

Pan, Yongping 262
Peng, Yusheng 632
Pham, Duc-Son 193
Plested, Jo 463

Poc-López, Ángel 228
Pondit, Ashish 526
Punn, Narinder Singh 393

Qiang, Jipeng 206
Qin, Xiaofei 15
Qin, Zengchang 87
Quan, Xueliang 321

Ramakrishnan, A. G. 536
Rastin, Parisa 216
Rozi, Muhammad Fakhrur 669
Rukon, Muhammad Eshaque Ali 526

Samura, Toshikazu 170
Sanodiya, Rakesh Kumar 297
Santo, Daisuke 479
Shen, Xiaobo 206
Shen, Xuyang 463
Singha, Tanmay 193
Sonbhadra, Sanjay Kumar 393
Sudhanshu 393
Sun, Chao 206

Takahashi, Takeshi 669
Tanaka, Gouhei 28
Tang, Jialiang 309
Thi Thu Uyen, Hoang 513
Toussaint, Yannick 216
Trzcinski, Tomasz 451

Verma, Brijesh 285

Wan, Kejia 64
Wan, Tao 87
Wang, Anqi 657
Wang, Fang 99
Wang, Feng 75
Wang, Hao 502
Wang, Yu 548
Wen, Junhao 548
Wen, Wujun 597
Wójcik, Maciej 238
Wu, Dongrui 321
Wu, Ruihong 262
Wu, Wenxiong 99
Wysoczanska, Monika 451

Xia, Kun 321
Xiao, Chen 15

Xu, Huangzhen   3
Xu, Xianghao   99
Xu, Xinhai   64
Xu, Yifang   273
Xuan Bach, Ngo   513

Yan, Daisong   40
Yan, Siyuan   491
Yang, Xiaoyan   309
Yang, Yan   645
Yin, Hongwei   3
Yu, Dan   273
Yu, Jing   99
Yu, Wei   439
Yu, Wenxin   40, 309
Yuan, Yun-Hao   206

Zeng, Jun   548
Zhang, Fukang   3

Zhang, Gaofeng   632
Zhang, Peng   309
Zhang, Xuedian   15
Zhang, Yang   479
Zhang, Yansen   608
Zhang, Zhiqiang   40
Zhang, Zhiyong   428
Zhao, Chenchen   369
Zhao, Jinjing   620
Zhao, Ying   15
Zheng, Lin   439
Zheng, Liping   632
Zhou, Wei   548
Zhou, Yingjie   428
Zhou, Yuguang   87
Zhu, Enbei   273
Zhu, Mengliang   321
Zhu, Yi   206

Printed in the United States
by Baker & Taylor Publisher Services